CLEANER FISH BIOLOGY AND AQUACULTURE APPLICATIONS

CLEANER FISH BIOLOGY AND AQUACULTURE APPLICATIONS

Edited by Jim Treasurer

5m Publishing

First published 2018

Published by
5M Publishing Ltd,
Benchmark House,
8 Smithy Wood Drive,
Sheffield, S35 1QN, UK
Tel: +44 (0) 1234 81 81 80
www.5mpublishing.com

A Catalogue record for this book is available from the British Library

ISBN 9781912158218

Book layout by Servis Filmsetting Ltd, Stockport, Cheshire
Printed by CPI Anthony Rowe Ltd, UK
Photos by the contributors of the book unless otherwise indicated

Cover photos (front) Main photo Ballan wrasse, credit Jim Greenfield; smaller images from the left lumpfish, Claire Treasurer; goldsinny wrasse, Mark Woombs, WA Marine & Environment; lumpfish male and female, Claudia Wittwer, Arctic Cleanerfish AS; (back) Atle N. Kristoffersen, OK Marine

Contents

Contributors

Adam Powell, University of Gothenburg – Kristineberg, Department of Biological and Environmental Sciences, Kristineberg, Sweden

Agnar Steinarsson, The Aquaculture Research Station of the Marine Research Institute of Iceland, Grindavík, Iceland

Albert K. D. Imsland, Akvaplan-niva Iceland Office, Akralind 4, 201 Kópavogur, Iceland; Department of Biology, University of Bergen, High Technology Centre, 5020 Bergen, Norway

Amber Irwin Moore, Carna Research Station, National University of Ireland, Galway, Ireland

Andrew Davie, Institute of Aquaculture, University of Stirling, Stirling, FK9 4LA, UK

Ane Vigdisdatter Nytrø, Department of Biology, Norwegian University of Science and Technology, Trondheim, Norway

Anette Lekva, Marine Harvest Norway, Cleaner fish Øygarden, Ljøsøyvegen 74, 5337 Rong, Norway

Anita Rønneseth, Department of Biology, University of Bergen, Norway

Anne Marie Power, School of Natural Sciences, Ryan Institute, National University of Ireland, Galway, Ireland

Antonios Chalaris, Otter Ferry Seafish Ltd, Tighnabruaich, Argyll PA21 2DH, UK

Ása Johannesen, Fiskaaling, við Áir, Hvalvík, Faroe Islands

Bridie Grant, Institute of Aquaculture, University of Stirling, Stirling, FK9 4LA, UK

Carlos Garcia de Leaniz, Centre for Sustainable Aquatic Research CSAR, Department of Biosciences, Swansea University, Singleton Park, Swansea, UK

Carolina Gutierrez Rabadan, Marine Harvest Scotland, Farms Office, Claggan, Fort William

Chris Hempleman, Scottish Sea Farms, South Shian, Oban, Argyll

Chris Noble, Nofima, Tromsø, Norway

Claudia Wittwer, Arctic Cleanerfish A/S, Steineveien 87, 8340 Stamsund, Lofoten Islands, Norway

Craig Pooley, Centre for Sustainable Aquatic Research, Department of Biosciences, Swansea University, Singleton Park, SA2 8PP, UK

Danny Boyce, Department of Ocean Sciences, Memorial University of Newfoundland, St. John's, NL, Canada, A1C 5S7

Dave Jackson, Marine Institute Oranmore, Co. Galway, Ireland

Espen Grøtan, Marine Harvest Norway , Cleaner fish Øygarden, Ljøsøyvegen 74, 5337 Rong, Norway

Felix Scholz, FishVet Group Ireland, Oranmore Business Park, Oranmore, Co. Galway, Ireland;

Marine and Freshwater Research Centre, Galway–Mayo Institute of Technology, Dublin Rd, Galway, Ireland

Felix Sproll, Marine Harvest Ireland, Letterkenny, Co. Donegal, Ireland

Gyri Teien Haugland, Department of Biology, University of Bergen, Norway

Harald Sveier, Lerøy Seafood Group ASA, Bontelabo 2, 5020 Bergen, Norway

Heidrun I. Wergeland, Department of Biology, University of Bergen, Norway

Henriette Glosvik, Marine Harvest Norway AS, Pb. 4102 Sandviken, N-5835 Bergen, Norway

Herve Migaud, Institute of Aquaculture, University of Stirling, Stirling FK9 4LA UK

Ingrid Lein, Nofima, Sjølsengvegen 22, NO-6600 Sunndalsøra, Norway

Juan Carlos Sánchez, Fundación Chile, Estación Experimental Quillaipe, Ruta 7 Km 23, 8 Carretera Austral, Quillaipe – Puerto Montt, Chile

Jorge Mancilla, Marine Harvest Chile, Camino El Tepual 8 – Puerto Montt, Chile

Jack D'Arcy, Marine Institute, Oranmore, Co. Galway, Ireland

James Kennedy, Marine Research Institute, Iceland

Jim Watt Treasurer, FAI Aquaculture, Ardtoe Marine Research Facility, Acharacle, Argyll PH36 4LD UK

Joana Robalo, MARE – Marine and Environmental Sciences Centre, ISPA – Instituto Universitário de Ciências Psicológicas, Sociais e da Vida, Rua Jardim do Tabaco 34, 1149-041 Lisbon, Portugal

Jon Sherwood, FAI Aquaculture, Ardtoe Marine Research Facility, Acharacle, Argyll PH36 4LD UK

Katerina Kousoulaki, Nofima AS, Norway

Keng Pee Ang, Cooke Aquaculture Inc., 1 Fundy Bay Drive, Saint George, NB, Canada, E5C 3E2

Kirstin Eliasen, Fiskaaling, við Áir, Hvalvík, Faroe Islands

Luca Mirimin, Marine and Freshwater Research Centre, Galway–Mayo Institute of Technology, Dublin Road, Galway, Republic of Ireland

Luke Wilson, Marine Harvest Ireland, Letterkenny, Co. Donegal, Ireland

Martin Hevia, Fundación Chile, Estación Experimental Quillaipe, Ruta 7 Km 23, 8 Carretera Austral, Quillaipe – Puerto Montt, Chile

Majbritt Bolton-Warberg, Carna Research Station, National University of Ireland, Galway, Ireland

Mar Marcos-López, FishVet Group Ireland, Oranmore Business Park, Oranmore, Co. Galway, Ireland; Marine and Freshwater Research Centre, Galway–Mayo Institute of Technology, Dublin Rd, Galway, Ireland

Maria Scolamacchia, Centre for Sustainable Aquatic Research, Department of Biosciences, Swansea University, Singleton Park, SA2 8PP, UK

Martin Haugmo Iversen, Faculty of Biosciences and Aquaculture, Nord University, Bodo, Norway

Matthew Zietz, Wester Ross Salmon, Ullapool, UK

Mette Remen, Akvaplan-niva, High North Research Centre, 9296 Tromsø, Norway

Mohd Faizzi Bin Mohd Yusof, Centre for Foundation Studies in Science, University of Malaya, Kuala Lumpur, Malaysia

Olav Breck, Marine Harvest Norway, Sandviksboder 77AB, Pb. 4102 Sandviken, 5835 Bergen, Norway
Ólöf Dóra Bartels Jónsdóttir, Akvaplan-niva Iceland Office, Akralind 4, 201 Kópavogur, Iceland
Patricio J. Saez, Fundación Chile, Estación Experimental Quillaipe, Ruta 7 Km 23, 8 Carretera Austral, Quillaipe – Puerto Montt, Chile
Patrick Reynolds, GIFAS AS, Gildeskål, 8140 Inndyr, Norway
Regin Arge, Marine Harvest, Eysturoy Faroe Islands
Richard Fitzgerald†, Carna Research Station, University of Galway, Republic of Ireland
Richard Prickett, Native Marine Centre; 3RSP Services Limited, 120 Netherton Road, Appleton, Abingdon, Oxon, OX13 5LA, UK
Sandra Schlittenhardt, Marine Harvest Ireland, Letterkenny, Co. Donegal, Ireland
Sonia Rey Planellas, Institute of Aquaculture, University of Stirling, UK
Stephen Murphy O'Sullivan, Carna Research Station, National University of Ireland, Galway, Ireland
Thor Magne Jonassen, Akvaplan-niva, High North Research Centre, 9296 Tromsø, Norway
Tómas Árnason, The Aquaculture Research Station of the Marine Research Institute of Iceland, Grindavík, Iceland
Velmurugu Puvanendran, Nofima Marine, Tromsø
William Clark, Institute of Aquaculture, University of Stirling, Stirling, FK9 4LA, UK

Preface

When did we first encounter a fish and when did we first engage in a lifelong relationship with these denizens of the deep? I am sure that this fascination sets fish people aside from many of the other biologists, researchers, and farmers the world over. Innately we have a common bond with farmers and ichythologists in any corner of the planet. My first experience with a fish was when I was aged four and helping haul my grandfather's creel set off the rocks in the small Aberdeenshire heritage fishing village of Crovie. The catch was not our much-anticipated lobster but a monster Ballan wrasse glistening, colourful, spikey, eye-balling, full of menace and locally called a 'sea sue'. This specimen was so large that we wondered how it managed to get through the small entrance. The next visit saw us hitch a lift on a salmon fisher's 'cobble' (flat-bottomed and wide-beamed boat) to the stake nets set in deep water on the Banffshire headlands of Rattray and Troup Head. Large lumpfish were caught along with the salmon and had the mystifying local name of 'paddlecock'. These fish were never eaten (salmon, haddock and herring being then plentiful). Looking back, I often wondered what the role of these fish was in the marine foodchain. Little did I know that these relatively obscure fish would find worldwide attention and be bred in millions to become a biological control for sea lice on farmed salmon. The more recent journey from low-scale use of cleaner fish in the 1990s to the integration of cleaner fish in IPM sea lice control strategies has been remarkable. Adaptations to fitting the cleaner fish lifestyle and habitats to in-cage, man-made habitats with environmental enrichment has been rapid. There may be hundreds of researchers and research students working on cleaner fish in northern Europe and now globally, and it is hard to keep pace with the increasing wealth of grey literature, technical advances, local adaptations, and rising academic output. Recent conferences and summits have given great opportunities to exchange views and move the technology on. Developments have indicated major concerns such as in fish health, vaccination, welfare, nutrition and in better deployment of cleaner fish.

This volume highlights much that is known about current knowledge. However, this knowledge is expanding rapidly. I am grateful to the team of enthusiastic workers and writers for giving of their time and data to contribute to this unique, and I hope, stimulating volume. It has been more than 20 years since the first dedicated publication on the biology of wrasse (Sayer *et al.*, 1996) and the time is right for a contemporary overview of the biology and aquaculture of commercial cleaner fish species. All those who work with cleaner fish are fish enthusiasts in the truest sense and passionate about their subject, and in their interest in this fascinating and colourful group of fishes. Speaking collectively for all contributors, we hope that this book will be a discussion point in the improvement of the cleaner fish technique for sea lice control and will prompt continuing questioning, assessment, feedback, development and collaboration. Also it raises questions for fish health and welfare, both in hatcheries and in deployment, and in the ethics of fishing, sometimes shipping wrasse some distance from fishing areas to salmon farms, and on the utilisation and reuse of cleaner fish in commercial settings.

I am grateful to so many contributors to chapters who have generously given of their time. The chapters were peer-reviewed by two and sometimes three commentators in the field. These included Adam Powell, John Davenport, Lesley McEvoy, Matt Longshaw, Hamish Rodger, Carlos Garcia de Leaniz, Albert Imsland, Sonia Planellas, Chris Secombes, Antonios Chalaris, Danny Boyce, Agnar Steinarsson, Tómas Árnason, Ása Johannesen, Majbritt Bolton-Warberg, Anette Lekva, Craig Pooley, James Kennedy, and Mette Remen. Although it is difficult to identify specific individuals, I would like to acknowledge the encouragement and sound advice provided by the late Richard FitzGerald, and Katerina Kousoulaki, and also the energetic Adam Powell and Albert Imsland for their insight and helpful comments. In a specific sense thanks to the staff of FAI Ardtoe in Ardnamurchan for their comradeship.

I would like to thank Alessandro Pasini, Jeremy Toynbee, Rachel Shepherd, Paul Middleton, staff of 5M Publishing, and the printing and production teams for their contributions. I am also grateful to my wife Wilma for her patience and support. My thanks to Marie-Clare Russell and the staff of FAI Ardtoe in Ardnamurchan for their comradeship and support in this cleaner fish review, and also to Tim Atack for our former collaboration in cleaner fish hatchery production.

Jim Treasurer, Ardtoe

Reference

Sayer, M.D.J., Treasurer, J.W. and Costello, M.J. (eds.) *Wrasse biology and use in aquaculture.* Fishing News Books, Oxford, 283 pp.

This publication is dedicated to the late Richard FitzGerald, whose wit and sound advice are sorely missed by the community, and whose inspiration and passion live on in the students, staff, colleagues and friends who knew him.

PART I

Biology and rearing of wrasse and lumpfish

CHAPTER 1

An introduction to sea lice and the rise of cleaner fish

Jim Treasurer

FAI Aquaculture, Ardtoe, Acharacle UK

1.1 Introduction

The salmon farming industry is an international success story for sustainable and reliable food production, and a source of essential fatty acids during a period of huge growth in the human population and decline/stagnation globally in world fisheries (FAO, 2016). Salmon (e.g. *Salmo salar* and *Oncorhynchus* spp.) production was 1.7 million tonnes in 2006, worth US$8.4 billion (FAO), and this has increased to 2.5 million tonnes valued at $15.8 billion in 2014 (FAO, 2016). Sea lice have been a problem in most fish farming areas with production of more than 15,000 tonnes (Costello, 2009) and remain the main health issue affecting the industry (Pike and Wadsworth, 1999; Johnson *et al.*, 2004). Sea lice consist of several species of parasitic copepod, for example *Lepeophtheirus salmonis* (Fig. 1.1), a species affecting mainly salmonids, and the generalist *Caligus elongatus* from the northern hemisphere, and *C. rogercresseyi* in Chile (Boxshall and Bravo, 2000). Farmed salmon are initially infected from migratory wild fish and lice numbers increase towards the end of the first summer. In the case of *C. elongatus,* infections can be almost immediate when wild fishes are shoaling nearby. The motile sea lice are attached with the cephalothorax and graze the epidermis of the salmon with a rasping strigil organ (Pike and Wadsworth, 1999). If sea lice numbers are not controlled by medicinal and other means eventually epidermal damage occurs, and morbidity ensues from the loss of body fluids. Close health and veterinary scrutiny is required to monitor sea lice numbers and to administer treatments. Sea lice infection can predispose fish to other health conditions such as Infectious Pancreatic Necrosis (IPN), pancreas disease (PD) and gill diseases. The development of reduced sensitivity of sea lice to medicines (Aaen *et al.*, 2015) due to gene mutations may encourage detoxifying enzymes. Treatments become less effective and, with a limited number of medicinal options, further resistance may occur. The overall cost of sea lice to the fish farming industry in 2006 was estimated as US$480 million (Costello, 2009) and this included the medicine costs, staff time, downgrading of fish after harvest and reduced growth rates. The farming situation can therefore be difficult and

Fig. 1.1 Sea lice, (a) adult female and male *Lepeophtheirus salmonis*, (b) gravid female *L. salmonis*, (c) chalimus stages of *L. salmonis* on salmon gills. Credit: J. Treasurer

regulatory authorities can place restrictions on the expansion of the industry and in the size of pens used until adequate means of sea lice control are in place. This may lead to regulatory change and the fish health inspectorate may have to be contacted if lice numbers exceed a certain limit (e.g. Anon, 2017).

However, sea lice can be an issue for wild stocks as well as farmed fish (Butler, 2002). This places sea lice as a universal parasite problem for fish, hampering the development of salmon farming, and impacting one of the world's most successful farming operations and, in salmon in particular, one of the world's most iconic fishes, animal and sporting species. Potential environmental impacts of both the parasite and associated medicinal treatments have engendered systematic and intensive forms of sea lice control, and the search for greater efficacy, sustainability and non-medicinal alternatives. The adaptive responses of the parasite and the development of reduced sensitivity to medicines (Aaen *et al.*, 2015) have necessitated good farming practice (Grant and Treasurer, 1993) to reduce the risk of infection and development of a multiple approach including fallowing and single year classes, and increasingly mechanical removal of sea lice in the fight to control sea lice numbers. This is commonly termed Integrated Pest Management (IPM). Fish that graze or 'clean' sea lice from farmed fish are now an integral and widely adopted part of this. In locations such as Scotland in 2016, cleaner fish use and mechanical methods led to a large decline in medicinal treatments, with the costs of chemotherapeutants dropping from 71% to 28% of fish health management costs (Hjul, 2017). The use of cleaner fish with all species of farmed salmonids is being examined in this volume, although cleaner fish have mainly been stocked with Atlantic salmon to date.

1.2 What is a sea louse?

Sea lice, persistently the greatest health challenge for salmon farmers over many years (Pike and Wadsworth, 1999), are ectoparasitic copepods. *L. salmonis* and *C. elongatus* are classified in the copepod suborder Siphonostomatoida and family Caligidae (Kabata, 1979). Sea lice have flattened bodies and appendages that enable them to attach to the surface of hosts. Many caligid copepods have a direct life cycle of up to ten stages with moults between each stage. Johnson and Albright (1991) described these stages in *L. salmonis* (Fig. 1.2). These successively comprise two planktonic nauplius stages in caligids, a free-swimming infective copepodid stage, followed by two to six chalimus stages fixed to the host with a filament, one or two pre-adult and finally an adult stage. Exceptions to this include *Caligus punctatus* and *C. elongatus* that are reported to have no pre-adult stage (Kim, 1993; Piasecki, 1996). The duration of each stage is dependent on water temperature, from a minimum of 93 days at 5°C to 19 days at 15°C for the entire life cycle in

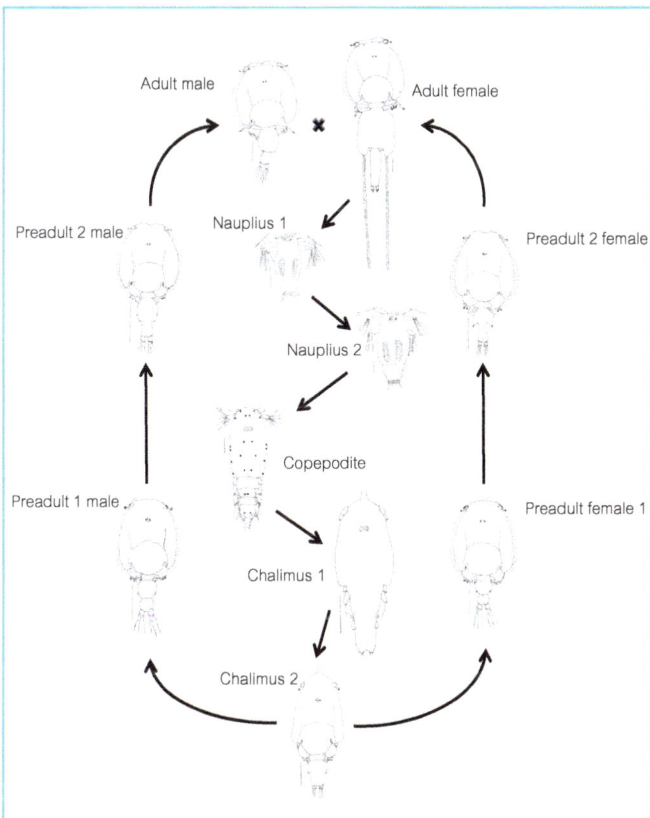

Fig. 1.2 Lifecycle of *Lepeophtheirus salmonis.* Nauplius 1 and 2 stages are free swimming and the copepodite stage is infective to their salmonid host. Following attachment, the copepodid stages undergo metamorphosis and moulting to become non-motile, attached chalimus stages. Separate sexes can be discerned at late chalimus 2 stage but are not shown here for the sake of brevity. Following a further two moults as pre-adult 1 and 2, adult stages develop and continue the cycle. Different stages are not shown to scale. Nauplius 1 and 2 after Johannessen (1978), copepodite, chalimus 1 and 2, preadults 1 and 2 are after Johnson and Albright (1991), adult male and female after Kabata (1973). Scale bars: Nauplius 1 and 2 and copepodite: 100 µm; chalimus 1 and 2: 500 µm; preadult 1 and 2, adult: 1 mm. Redrawn and credit: Matt Longshaw, Benchmark Animal Health

L. salmonis. Each adult female may produce up to 300 eggs per egg strand. Sea lice feed on host mucus, blood and tissues. Fish mortalities are caused directly, due to secondary infections and there are economic costs of lost growth, downgrading of product, and treatment costs (Pike and Wadsworth, 1999; Costello, 2009). Heavy infection may give rise to extensive erosion of skin and haemorrhaging (Johnson *et al.*, 2004).

1.3 Sea lice control measures

There have been several approaches to sea lice control (Table 1.1, Fig. 1.3). Initial methods were based on medicines, firstly dichlorvos, then hydrogen peroxide and in the late 1990s both cypermethrin (Excis) and emamectin benzoate (SLICE) became available. Increasing resistance to medicines reinforced the concept of integrated pest management (IPM). This included preventive measures, single year classes, 'all in, all out' stocking, and fallow periods were introduced from 1993 (Grant and Treasurer, 1993). This was effective in breaking the cycle of infection and generally lice numbers were lower in the first year and numbers did not increase to a treatable level until eight to ten months into the production cycle (Revie *et al.*, 2002). Part of this focus also included national sea lice control strategies such as introduced in Scotland under the auspices of the Scottish Salmon Growers' Association, now the Scottish Salmon Producers' Organisation SSPO (Rae, 2002). The fish farming areas in Scotland were divided into health management zones largely based on hydrography. The intention was to treat fish when a threshold of sea lice numbers was breached, normally one gravid female per salmon in the spring. Two coordinated sea lice treatments were carried out in Scotland (Wadsworth, 1998), normally in week ten before increased copepodid release in spring, and also at week 17, which preceded a second peak in copepodid release.

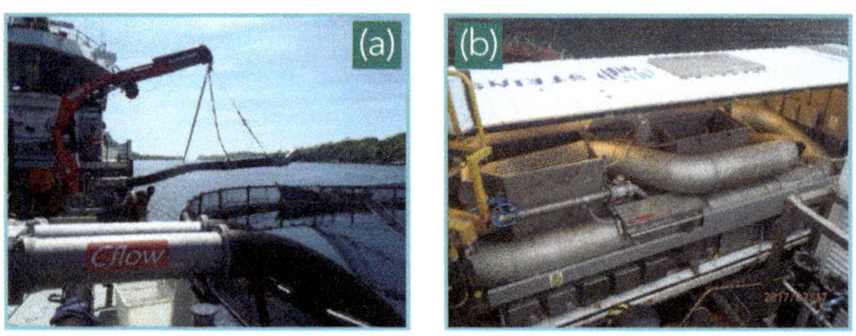

Fig. 1.3 Sea lice therapeutic treatments: a) preparation for pumping fish into wellboat for hydrogen peroxide treatment, b) treatment using heated water in a thermolicer, c) d) and e) Skamic wash brusher module.

Table 1.1 Medicines used to treat salmonids with sea lice burden as of 2017. Blue shading indicates where medicine authorised.

				COUNTRY:							
Medicine/ Techniques	Brand name	Dosage	Treatment time	Norway	Iceland	Faroes	Scotland	Ireland	Canada BC	Canada NB	Chile
Emamectin benzoate	SLICE	25 mg Slice kg^{-1} day^{-1}	7 days oral		none						
Cypermethrin	BMX (Excis)	0.3 ppm (0.3 ml m^{-3})	30-60 mins								
Deltamethrin	AMX Alphamax	0.2 ml AMX m^{-3} for 30 min	30-60 mins							used previously	
Teflubenzuron	Calicide, Ectobann	10 mg kg^{-1} fish (tot 70 mg)	7 days daily oral							used previously	
Diflubenzuron	Releeze (Cargill)		14 days								
Azamethiphos	Salmosan	0.2 ppm	30-60 mins					AR16 licence		1 year	
Hydrogen peroxide	Paramove 49.5%	1500 mg^{-1}, 20 min	20 mins max, less if higher temp						local agreement		

Sources: Patti Jones, Canada; Mar Lopez, Ireland; Sarah Pflaum, UK; Ása Johannesen, the Faroes; all Personal communications April 2017. For the UK, more info on some of these can be found on the NOAH Compendium website: www.noahcompendium.co.uk Chile: Manual de Buenas Practicas. (2015). 3rd edition. Sernapesca. pp. 34; and Sandra Bravo, personal communication, 2018).

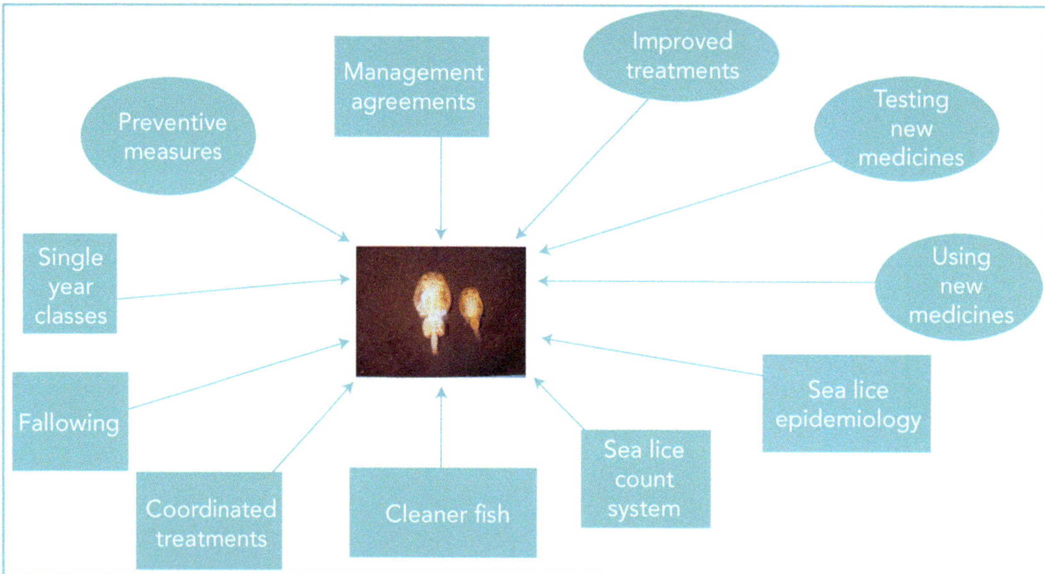

Fig. 1.4 Control of sea lice by an Integrated Pest Management strategy.

There have been many different approaches to sea lice control. The industry has been affected by the development of reduced sensitivity of sea lice to medicines and a limited arsenal of treatment medicines in much of the past 30 years. The industry started treatments with dichlorvos (Nuvan), a now banned organophosphate, and then a derivative Aquagard. Early testing of azamethiphos (Salmosan) followed but it was several years before a licence was granted. The industry was aided by the introduction of hydrogen peroxide as a bath treatment in 1993, which was the only effective method for several years, but required large volumes of the medicine to provide a therapeutic dose of >1500 ppm. Issues also arose with lice not being killed, i.e. causing only a 'knock down' effect, although studies showed that reattachment to hosts was not significant (Treasurer and Grant, 1997). Subsequently cypermethrin (Excis) and then in feed emamectin benzoate (SLICE) were used, followed by avermectin derivatives such as Alphamax. A wide range of novel approaches was assessed including mechanical grading, onions masking semiochemical signals from the host, and, more promisingly in recent years, mechanical treatments, thermal warming termed thermolicing, and laser technology (Powell and Pooley, 2015). Selective breeding (Raynard *et al.*, 2002) was examined and also pheromonal control (Invgarsdottir *et al.*, 2002) but these approaches require further development to prove their efficacy at commercial scale. The key to infestation management has rested with better management of farms, synchronised sea lice treatments, national sea lice treatment strategies, improved staff training, and regular lice counts and monitoring. The model of IPM is therefore being applied (Fig. 1.4), which has provenance in other areas of agriculture.

1.4 What is biological control?

Biological control is defined as the use of living organisms to control pests and weeds (Anon, 2000). Another definition of biological control is the utilisation of natural 'enemies' to control a pest species. From an ecological standpoint, it has been considered as one of the most important forces in nature to reduce the damage caused by noxious organisms to tolerable levels (DeBach and Rosen, 1991). That is, the regulation of plant and animal numbers by natural enemies, and practically all living species are attacked by natural enemies. These enemies (such as parasites, predators or pathogens) feed on them and frequently regulate their population densities. The term biological control should perhaps be restricted to the control of pest populations by natural enemies. The first modern example of biological control was the spectacular control of the cottony cushion scale on citrus plants in California in 1888 by introduced natural enemies (DeBach and Rosen, 1991). The method has formed an important part of pest control since the 1930s (Pimentel, 1991) with many examples from terrestrial agriculture.

Examples of biological control in aquatic environments have been reviewed (Treasurer, 2002), and include, for example, cyclopoid copepods controlling mosquito larvae (Marten *et al.*, 2000), ciliates affecting seastar *Asterias amurenis* (Goggion and Bouland, 1997), and freshwater prawn *Macobrachium rosenbergii* controlling mollusc population in fish ponds (Mahapatra and Datta, 1996). Many pathogens and epibionts of Crustacea have been recorded (Treasurer, 2002), including algae such as Ulva on *L. salmonis* (Minchin and Jackson, 1993), fungus such as *Aphanomyces* on the copepod *Boeckella dilitata* (Burns, 1985), Dinoflagellates such as *Gyrodinium undulans* on copepod and rotifer eggs (Drebes and Schnepf, 1998), *Microspora* sp. on copepods (Vidtmann, 1990), ciliates such as *Ephelota* sp. (Stone and Bruno, 1989) (Fig. 1.5a), *Epistylis* spp., *Trichiliodes*, and *Licnophora* (Xu and Burns, 1991). The monogenean *Udonella caligorum* (Fig. 1.5b) is reported from sea lice in Ireland (Minchin and Jackson, 1993) and Scotland. The worms are found on the dorsal shield of the louse but move to the genital segment before oviposition (Kabata, 1973). The genital segment of the louse is raised above the surface of the fish host and the eggs receive a good waterflow, which aids in development. As *Udonella* is distributed on the dorsal surface of lice they have direct contact with the epidermis of the fish. However, *Udonella* is only a commensal of sea lice as it grazes on epidermal tissue that has been removed from the host fish (Kabata, 1973). The impact on lice is thought to be minimal, although it may have a marginal effect on lice egg production. None of these biological agents therefore has a deleterious effect on sea lice.

The only reported successful biological control of sea lice is therefore cleaner fish (Treasurer, 2002). Examples of sea lice control and how wrasse selectively consumed the larger, maturing lice stages were given by Bjordal (1991) and Darwall *et al.* (1992a). There have been several quantitative studies demonstrating the cleaning activity of wrasse (Deady *et al.*, 1995), on salmon trial group pens (Treasurer, 1994), and from analyses of wrasse alimentary tracts. Cleaning activity of lumpfish has also been shown in the laboratory and has been filmed (Imsland *et al.*, Chapter 9, this volume). The presence of large numbers of sea lice in the stomachs of lumpfish has been shown on many occasions (Prickett, 2016).

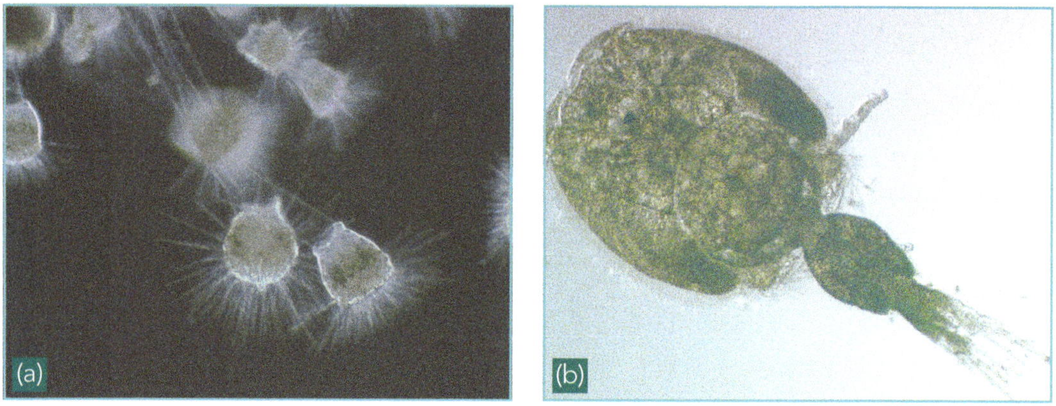

Fig. 1.5 Sea lice symbionts, (a) Ephelota sp. (length = 470 μm) (credit: David Bruno) and (b) *Udonella caligorum* (280 μm) from *Lepeophtheirus salmonis.*

1.5 What is a cleaner fish?

'Cleaners' can be fish (Fig. 1.6) or shrimps that remove ectoparasites and dead tissue (Feder, 1966; Losey *et al.*, 1990). The highest degree of cleaner fish specialisation is on coral reefs where Labridae and Gobiidae families are cleaners that feed almost exclusively on material that they remove from the host fish (Losey *et al.*, 1999). 'Client' fish can visit cleaners several times each day (Grutter, 1995). The family Labridae includes 453 species in 68 genera (Parenti and Randall, 2015). Of these fish species 208 have been assessed as being cleaner fish (Vaughan *et al.*, 2017). The diverse Labrid family has structural changes of the skull and especially in the jaws and dentition that enable wrasse to consume a wide variety of food organisms (Westneat *et al.*, 2005; Parenti and Randall, 2015), including ectoparasites, zooplankton and shellfish. The diet of wild wrasse includes amphipods, isopods, and other small fish, and wrasse therefore utilise a range of invertebrates of rocky shores (Hillden, 1978). Wrasse are a browsing and grazing species with an extendable lower jaw that enables them to clean invertebrates from rocks and weed.

Many studies of wild fish initially held the view that cleaner and client was not a mutual relationship and that host fish did not receive benefits from the cleaner fish. This led to the hypothesis that cleaner fish are 'behavioural parasites' that exploit the sensory system of the clients to obtain food, rather than to increase the client's fitness. However, cleaner fish studies disproved this (Grutter, 1999). The cleaner fish *Labroides dimidiatus* eats parasitic gnathiid isopods, which decline in number on the client fish *Hemigymnus melapterus* daily between dawn and sunset. Cleaner fish reduced parasite abundance, resulting in a 4.5-fold difference within 12 hours, and this supported the hypothesis that cleaning behaviour is mutualistic (Bshary et al., 2006). While Grutter disproved the behavioural parasite theory, wrasse are opportunists that eat any available prey, live on a range of prey, and sea lice are edible crustaceans. It was proposed that the definition of cleaning symbiosis should be amended to distinguish cleaning symbiosis from incidental cleaning by identifying communication between client and cleaner fish (Vaughan *et al.*, 2017).

Fig. 1.6 Some examples of cleaner fish from several geographical areas. Gallery compiled by Katerina Kousoulaki.

1.6 Cleaner fish innovation

Cleaner fish (Labrid wrasse) were used at a remarkably early stage in salmon farming, from 1988 in Norway (Bjordal, 1988) and in Scotland from 1990 (Rae, 2002). Initially the technique was hampered by a preliminary finding that Ballan wrasse *Labrus bergylta* could on occasion pick scales and eyes of salmon, and so smaller less robust species were often used. All wrasse were captured in the wild and subjected to the stress of capture, storage and transfer. When alternative medicines were introduced in the late 1990s and infectious salmon anaemia (ISA) was detected on several farms, cleaner fish use fell out of favour. However, wrasse continued to be stocked on a small scale on farms, especially with organic accreditation, throughout the next few years.

The second phase of cleaner fish use commenced around 2008, initiated in Norway and Scotland. The Scottish Salmon Company Ltd (SSC) asked for wrasse to be reared at Ardtoe in 2008. The farming of wrasse had been examined in 1991 with early rearing trials in Norway, at Golden Sea Produce Ltd (Hunterston, Scotland) and at the Sea Fish Industry Authority (SFIA) Marine Farming Unit. Although the idea of farming wrasse was considered, none of the companies took this on, and it was not until 2008 that farming of wrasse commenced on a commercial scale. In 2013 lumpfish (*Cyclopterus lumpus*) were assessed as potential cleaner fish, and farming quickly accelerated when it became clear that it was possible to rear larger numbers of these fish, that there were high survival rates in the hatchery to stocking, and reduced cold torpor during winter and spring. Now the use of cleaner fish appears as one of the mainstays in sea lice control in an integrated pest management approach with preventive measures and strategic use of medicines when required (Fig. 1.4).

Of course, the use of cleaner fish does not come without technical problems, ethical considerations, and occasional criticism. In the early days, many farms were sceptical and insufficient attention was paid to detail to ensure the best conditions for wrasse and lumpfish introduced to pens. Consequently, efficacy of the technique was reduced. Questions have been raised about cleaner fish health and welfare in the pens (Chapter 15, this volume) and the appropriateness of fishing large numbers of wild fish and translocating them between geographic zones (Chapter 17, this volume). These issues have to be addressed and improved, the fishery has to be managed (Darwall *et al.*, 1992b), and defined handling practices used by fishermen are now being encouraged through the industry and accreditation schemes, for example RSPCA 'Freedom Foods' guidelines in the UK. The availability of vaccines for cleaner fish (Chapter 14, this volume) and the maintenance of good welfare (Chapter 15, this volume) are therefore priorities to ensure that biological control of sea lice remains an attractive and successful technique. Some means of reuse and use of cleaner fish at the end of the production cycle for economic benefit also needs to be addressed.

The use of cleaner fish brings with it many economic benefits, and a lifeline and opportunity for the marine finfish farming sector, and especially in the wake of the downturn in cod *Gadus morhua* farming. The numbers of hatcheries involved in wrasse and lumpfish farming is increasing so rapidly that it is difficult keeping track of the numbers of hatcheries in Europe, now at least 41 in Norway, two in Iceland, one in Ireland, one in the Faroes and nine in the UK. This re-energises expertise in finfish culture, in hatchery capacity and in the farming of new species. Spin-off effects have included the development of new areas in fisheries science, in fish biology and in genetics, fish nutrition, fish transport, pathology, health management, hatchery practice and cage technology (or rather, the use of furniture, hides, and enriched environments).

The next developments are likely to include the spread of the technique to other continents and especially in the Americas in Canada (Chapter 23) and in Chile (Chapter 24). Creative approaches may one day lead to the application of cleaner fish in south-east Asia and in the Mediterranean, should the need arise.

1.7 Biological interest in the cleaner fish species and aquaculture innovations

For fish biologists the use of cleaner fish has led to many opportunities and development of their science. The field of animal behaviour is particularly interesting: For instance, how much do we really know about fish behaviour in pens and how different fish species interact? With differences in the wild habitat and the pen environment, habitats have to be introduced to pens and environmental enrichment. Detailed study of cleaner fish behaviour in pens is required to establish good living conditions and environments. There was little literature or knowledge of cleaner fish in Europe in the early 1990s. Of course, wrasse species were a favourite with divers and anglers, but there was no commercial fishery in most areas and there was a limited amount of publications. The works of Dipper and Pullin (1979) on the Manx Labridae captured the imagination because of the cleaning behaviour of the fish, their longevity, colourful appearance, interesting behaviour, and ability for sex reversal. Overnight these wrasse species have become more visible and attracted much interest from biologists, making these and lumpfish now among the most studied and researched fishes, certainly in the north Atlantic. They are also the main finfish species, other than salmonids, reared in northern Europe. Interestingly, because wrasse are not regarded as a commercial species these inshore fishes were not of great interest to fisheries biologists involved in fishery management and there is no Total Allowable Catch (TAC) for them in the UK. Therefore, the effects of fishing effort on wrasse species were initially not evaluated thoroughly.

Recent management measures have, of course, been applied in the Norwegian fishery (Chapter 17, this volume) (and this has spilled over to Swedish waters) because of the large numbers of fish being captured and it will be interesting to learn from recent studies whether the fishery is resilient and whether further management measures are required. This appears to be outside of EU fishery agreements as wrasse is not used for human consumption on a large scale. Increasing pressure has therefore been placed on hatcheries to establish their own broodfish, both for wrasse and lumpfish rearing. It can be envisaged that this will lead to efforts to selectively breed wrasse and lumpfish for traits such as growth, disease resistance, and cleaning ability (Chapters 11 and 12, this volume).

There is the phenomenon of 'reduced sensitivity to medicines' when medicines become ineffective. The mechanism is thought to be for selection for lice with higher levels of detoxifying enzymes. Generally farms will rotate the use of medicines to ensure that resistance does not develop so quickly.

From the 1990s the concept of using wrasse as a form of biological control gained acceptance. The commitment by companies to use cleaner fish is now seen as part of a long-term strategy: so much so, that major players have invested heavily in the opening of cleaner fish hatcheries and funding long-term research. For example, Marine Harvest Norway Ltd (MHN) operates six hatcheries in Norway dedicated to wrasse production and nine lumpfish producers. Both Scottish Sea Farms Ltd (SSF) and MH (Scotland) Ltd leased the marine fish hatchery at Machrihanish, Kintyre, from 2010 with considerable support from the Technology Strategy Board and joint funding from both companies. Latterly there has been support from the Scottish Aquaculture Innovation Centre (SAIC) to examine some of the biological and fish health issues of wrasse use in pens. The health of farmed wrasse in pens is now improved due to understanding of the immunology of the fish and the development of effective vaccines (Chapter 14). Also, welfare of wrasse in pens has been enhanced by the use of hides and supplementary feeding when lice numbers are low. Operational welfare indicators have been developed for wrasse and there is now a greater understanding of basic welfare needs of cleaner fish. Cleaner fish in a sense have come full circle and they are treated in the same fashion as salmon and any other farmed fish. However, beyond the company support, there is a confidence, belief and commitment by farm personnel in the use of cleaner fish and the many small points of detail that enhance cleaner fish survival and efficacy have helped greatly in the success of the technique. So, it appears that cleaner fish are now an established and mainstream technique with salmon farming companies in Europe. The technique has also been adopted beyond the initial natal location for cleaner fish use, with active cleaner fish developments taking place in the Americas (Canada) and also for the first time in the southern hemisphere (Chile).

1.8 History of cleaner fish use in aquaculture

The first recognised proponent of the cleaner fish technique in aquaculture, to control sea lice, was Åsmund Bjordal with his first experiments with wrasse of a range of species (Fig. 1.7) in 1989 (Bjordal, 1991). This followed from observations years previously in aquaria that species such as corkwing *Symphodus melops* were cleaner fish of other marine fishes (Potts, 1973, 1974). If Bjordal had not thought of this idea would cleaner fish have been used over the last 25 years?

Fig. 1.7 Cleaner fish species stocked with Atlantic salmon. (a) Ballan wrasse (Credit: Felix Sproll, Marine Harvest Ireland), (b) goldsinny, (c) rock cook and (d) cuckoo wrasse (all, Mark Woombs), (e) corkwing wrasse (Felix Sproll) (f) and lumpfish (Åsa Johannesen).

There were obstacles in this first use of wrasse as suitable handling and use of the species were required. The uptake in the 1990s of cleaner fish was popular (Bjordal, 1988, 1991; Bjordal and Costello, 1990; Lysaght *et al.*, 1995) but there were setbacks because of health issues and unidentified losses of wrasse from pens. As Ballan wrasse were not stocked in the 1990s (Bjordal, 1991), potentially this hampered the efficacy of the cleaner fish technique as evidence suggests that Ballan wrasse is the most effective cleaner species and it is also robust with good survival rates (Prickett, 2016). It can be questioned why lumpfish were not stocked with salmon until recently, from 2013 in Norway. Why was this species not considered when it has been shown to give higher levels of cleaning activity and survival at lower temperatures and more northerly latitudes than wrasse? It could be that it has not been a species of aquaria and divers noticed cleaning activity only relatively recently (A. Powell, Sweden, March 2017, personal communication). Would lumpfish have been considered as a suitable cleaner species if the rearing of large numbers of wrasse had been an easier task? Certainly, the economics and certainty of lumpfish production are attractive and they have been able to fill the void in the large numbers of cleaner fish that are required, sometimes as high as one to every ten salmon. While the use of cleaner fish declined towards the end of the 1990s, this was in large part due to the licensing of new medicines, principally Cypermethrin (Excis) as bath treatments and also the in feed treatment Emamectin benzoate (SLICE). However, the recent success of cleaner fish due to stocking them at higher density (as much as 10% by number) and improved survival of wrasse in pens due to the use of vaccination has led to a resurgence of biological control and almost universal adoption of the technique in salmon farming in northern Europe. Lumpfish faced the first initial difficulties shared by wrasse with mortalities due to diseases, poor acclimation to pens, losses due to lack of adaptation to pens, and lack of suitable substrates.

1.9 Economic benefits

There are several costs related to sea lice control (Costello, 2009), including the purchase of medicines and equipment; reduced fish growth, which was estimated from 5 to 15% (Sinnott, 1999); reduced food conversion efficiency with 5% more feed required, downgrades due to physical damage from 1% to 15% (Mustafa *et al.*, 2001); adverse publicity and losses due to secondary infections, and increased fish mortalities. The global cost of sea lice control in 2006 was estimated as \$480 million. Published estimates of the cost have been expressed as production costs per kg, ranging from £0.153–£0.230 kg^{-1} in Scotland to \$0.3 kg^{-1} in Chile (Carvajal *et al.*, 1998). The economic benefits of using cleaner fish have been estimated on several occasions, with figures offsetting the cost of purchasing medicines equivalent to 1. 5 treatments (depending on the medicine being used).

1.10 Public and retailer attitudes to cleaner fish use

In general, perception has been positively enhanced and brought to the public knowledge through television programmes. The use of biological methods to control sea lice has been received with interest by the public, and has also received press coverage. Any reduction in the use of medicines and advent of reliable methods of controlling sea lice are seen as positive by the public, NGOs and environmentalists. Concerns continue regarding the long-term use of sea lice medicines, although there have been some intensive studies indicating no residual effect of medicines (SAMS, Post Authorisation Monitoring Programme, 2005). Retailers see the use of cleaner fish as a positive marketing image.

1.11 Quality schemes

The RSPCA 'Freedom Foods' (animal welfare assurance accreditation scheme) did not initially give permission for the use of cleaner fish in pens, but subsequently participating farms could apply for and receive a derogation to enable cleaner fish to be used and included in the certification scheme. Freedom Foods has issued guidance and recommendations on the use of cleaner fish on farms and these cover fish capture, giving guidance on holding and handling of fish, environmental conditions such as temperature, oxygen levels and water exchange, netting of fish, transport of cleaner fish, stocking, and in management of fish in pens. These were drafted following discussion and feedback from farmers and give a good framework for the maintenance of welfare of cleaner fish in pens. Extended guidance conditions on cleaner fish were implemented for Freedom Food accreditation from November 2017.

1.12 The context and regulation of cleaner fish use

Given that fish farms and fisheries are subject to a range of regulations, codes of good practice, and certification schemes, what regulates how cleaner fish are reared and deployed and in context of the interaction with the food chain? Generally the attitude of authorities has been supportive and encouraging in the use of cleaner fish. The environmental authority in Scotland (SEPA) probably sees their use as an alternative to medicines, and as a way to help reduce the potential development of resistance. Marine Scotland has shown increasing support for cleaner fish technology, with the provision of advice and specialist health diagnostic support and information resource. In Scotland permission has to be obtained from the Fish Health Inspectorate to stock cleaner fish on salmon farms and a risk assessment of introducing these species to salmon farms has to be compiled and considered. The health screening of wild-caught fish prior to stocking is not mandatory. There are recommendations on good practice regarding the movement of fish between management areas and there is a requirement to remove or cull cleaner fish after two salmon production cycles to reduce the carry-over of disease issues between salmon year classes (Murray, 2014). This was modelled but the industry may still query the logic and arguments for culling these experienced cleaner fish. In other regards, the health requirements required of cleaner fish are equal to those applied to salmon by Marine Scotland. Good health practices, as described in the Code of Good Practice (SSPO) must be maintained, vaccination encouraged, and removal of mortalities and reporting of any notifiable health issues ensured. Wrasse were initially regarded as a low-risk species, but the outbreak of viral haemorrhagic septicaemia (VHS) in wild-caught wrasse, stocked in Shetland, was a setback. After extensive investigation, the testing of farmed and wild fish, and epidemiological modelling it was concluded that VHS had been transmitted from wild fish in the vicinity of the affected farms (Wallace *et al.*, 2015).

The fishery for wrasse and lumpfish now involves dedicated inshore fishermen with shrimp pots or fyke nets, and fish farming companies, in some cases, also have their own licensed boats. The minimum length of fish and condition, and conditions under which fish are captured, handled and transported, are instructed by the salmon farmers, and in most cases suppliers adhere to these requests. This complements the stipulations on capture and transport of cleaner fish also made by the RSPCA Freedom Foods scheme. Fishing for cleaner fish in Scotland requires that the fishing boat has to be registered. The regulatory considerations that apply in various countries are described elsewhere in more detail in this book, but eight specific questions were asked of participating countries (Table 1.2).

Marine Scotland indicates that wrasse are currently excluded from the list of Total Allowable Catch (TAC) species; i.e. wrasse do not have a TAC quota as applied under The Common European Fisheries Policy (Marine Scotland, personal communication). As such, the numbers that can be caught are unrestricted and there also does not appear to be a regulatory minimum legal landing size, although an agreement to implement catch size restrictions was reached between Marine Scotland and salmon farmers. RSPCA Freedom Foods also specifies lower and upper size limits, while farmers apply their own as they do not accept fish that will pass through the net mesh. The majority of commercial wrasse fishing is on the west coast of Scotland, by vessels under 10 m, and these have to be licensed. These vessels, given their size, have limited gear restrictions, although they do need to be registered sellers and those who purchase the fish from them need to be registered buyers. There is no requirement for vessels under 10 m to carry an electronic log book. Currently they complete a Marine Scotland 'FISH 1 form', which details the catch. This is then returned to the fishery office for the area in which the vessel is registered and the information is passed to the relevant department for statistical recording. Fishermen must also complete sales notes for fish sold, and buyers must supply details of purchases to the regulator. As this information is commercially sensitive, it is not a matter of public record.

England and Wales do not currently farm salmon and regulations only concern cleaner fish production within general farm regulations (not deployment), and capturing wild broodstock within general fishing operations (Table 1.2). In this regard, they largely follow Scottish regulations under the 2009 Aquatic Animals Act. The reader is recommended to check with the national regulatory authority on latest guidance, as regulations in relation to developments in fishing and farming cleaner fish are changing rapidly in light of increased demand for cleaner fish.

The Marine Institute governs the regulations relating to the movement of fish in Ireland (Bolton-Warberg, Galway, 2017, personal communication). Only facilities with approved Fish Health Authorisation (FHA) can apply for movement orders. Applications for movement orders are submitted to the Fish Health Unit of the Marine Institute in advance of moving fish offsite (five days in advance for exports, three days for internal movements) and importing fish (five days in advance). A registered veterinarian is required to provide a pre-movement report on eggs or fish being moved into or around the country. For movements into Ireland from another country, health screening is also required to ensure the stock is disease free. To facilitate this, the Trade Control and Expert System (TRACES) is used. This web-based system is run by the European Commission's Directorate-General for Health and Consumer Protection. It processes all necessary documentation and sends copies to the appropriate authorities in both countries.

There are general laws for aquaculture and harvesting of fish in Norway, and these are also valid for wrasse. Farming of wrasse is regulated by the aquaculture regulation for operation of aquaculture facilities (https://lovdata.no/dokument/SF/forskrift/2008-06-17-822) Chapter 1 and 7 (see Chapter 17 this volume for details of aquaculture and fisheries regulations as they affect cleaner fish in Norway). Fishermen keeping cleaner fish alive before hatchery supply is regulated by the general regulations on the exercise of fishing in the sea (https://lovdata.no/dokument/SF/forskrift/2004-12-22-1878). Various regulations relate to fishing cleaner fish in Norway, including the general regulations on the exercise of fishing in the sea, and these are quoted in (https://lovdata.no/dokument/SF/forskrift/2004-12-22-1878). It states that the operations involved in fishing, keeping fish alive during recovery and transferring of cleaner fish are to be conducted responsibly in terms of fish welfare. All possible measures that will reduce the risk and extent of suffering, injuries and stress must be taken in relation to fishing gear, including periods of exposure to the air, crowding, handling, fishing depth, feeding, and health controls.

Iceland has a long history of catching lumpfish for roe, and the gillnet fishery is accredited by the MSC (Anon., 2015). General farming licences are required from the government and the local health authorities in Iceland. Since the VHS diagnosis in 2015, all wild lumpfish broodstock must be reared in a specialised quarantine area and tested for a range of diseases (A. Steinarsson, Iceland, 2017, personal communication). The capture of lumpfish as the main catch using lumpfish nets must be done within the specified local lumpfish season, and it requires a special licence from the government. However, special licences from the government can also be

Table 1.2 Regulatory and national guidelines regarding the use of cleaner fish screening and fishing at November 2017. Producers should check with national authorities. For more information see text and country chapters, this volume.

Location	Notification of rearing of cleaner fish	Health regulations	Certification schemes and cleaner fish
Norway	General Aquaculture regulations of aquaculture facilities	General Aquaculture regulations	General Aquaculture regulations
Scotland	Notification to Marine Scotland	Health screening not mandatory Rick assessment prior to stocking; CoGP applies; Remove cleaner fish after two production cycles	RSPCA Freedom Foods guidelines on capture, transport and stocking in pens
England and Wales	Species specific licence application to Fish Health Inspectorate for commercial rearing	None currently enforced (deployment not applicable). Fish Health Authority certificate required for imported eggs and fish. (Currently VHS free only)	n/a legally but usually follow Scottish regulations
Guernsey	n/a	n/a	n/a
Ireland	Notification of rearing of cleaner fish: aquaculture sites must have aquaculture licence (Department of Agriculture, Food and the Marine – DAFM) and a Fish Health Authorisation (Marine Institute)	Fish health authorisation required for facilities; health screening of eggs and fish from other countries; veterinarian screens and approves egg and fish movements Fish Health Handbook and Risk Assessments as guidelines for producers	None
Faroes	Aquaculture licence required from government	Salmon farming regulations apply; broodstock have to be tested for ISAV; mortality and veterinary reports to health authority	None but it is possible that salmon farmer may ask lumpfish farmers to be Global Gap certified
Iceland	General aquaculture licence from government	Broodstock reared in specialised quarantine unit; broodstock tested for range of diseases	Codes for salmon and other farmed species but not yet for lumpfish
Canada	A hatchery licence is required to rear or hold cleaner fish in Canada, similar to a salmon hatchery.	Full health screening is carried out during rearing at the facility and prior to transport from the hatchery facility to the farms Pre- transfer involves bacteriology, histology and virology samples. No use of wild-caught fish.	Provincial and federal authority guidelines. Biosecurity audits annually. All facilities have to follow guidelines prescribed by institutional, provincial and federal authorities.
Chile	None and rearing of cleanerfish still in early stages	None	None

Source: K. Kousoulaki, R. Prickett, M. Bolton-Warberg, Á. Johannesen, A. Steinarsson, D. Boyce, J. Treasurer, all personal communications.

Table 1.2 (continued)

Fishing licence for wrasse or broodfish	Quota and size limits on fisheries/ fisheries management	Fish welfare guidelines for fishermen
Must have a supply agreement with a farmer in place before applying for a licence to fish for cleaner fish.	Quotas for size limits per species and numbers per area in Norway	General guidelines from fisheries regulations on capture of fish which will be kept alive
Boat licence required from Marine Scotland; catches reported to MS; fish farming companies report numbers purchased to MS	Non TAC species Fisheries statistics only collected from 2014; weights only required; fish farm companies required to keep record of cleaner fish purchased	Salmon farming companies require adherence to RSPCA welfare standards, Nov 2017
Ambiguous. As fishery is commercial, fishermen should have boat licence (<10m or >10m) and Cat C permit	Non TAC species	Nothing formal
Import documents required upon entry to mainland UK		
	Local quota enforced	
No specific fishing licence to use pots for wrasse or broodfish but fishermen would need a general inshore fishing These licences would be issued by DAFM	No quotas for wrasse or lumpsucker	Fish welfare guidelines in Fish Health Handbook produced by Marine Institute and Marine Harvest Ireland have a manual translated from their Norwegian counterparts that they give to the fishermen that they work with
Licence required for commercial fishing; also transport regulations for transfer of fish from one area to another apply	No quotas for lumpfish but there is a fishing day limit per annum for boats	None
Only within the specified lumpfish fishing season; requires a special license from the government.	No special quotas for lumpfish	No formal arrangements
Require a collection permit. Collectors licence and designated fishermen. Commercial fishing licence can be used and fish caught comes off quota.	Larger Cunner kept for broodstock. Caught by trap. Although there is no size limit the lumpfish fishery has a minimum 9 inch mesh size (227 mm)	Welfare is considered in fishing but is not closely monitored
None	None	Not at present

obtained to set lumpfish nets out of season. Lumpfish are also caught out of season as bycatch in cod nets, and all such bycatch must be reported to the government. Quotas are just being considered in Iceland and there are no official catch records for aquaculture quota. The government issues a quota (6200 tons in 2015) based on stock size. The catch is managed by restricting the length of the season (32 days in 2015). The maximum size of boat is 15 tons, with a maximum of 300 nets per boat, and a mesh size of 265–290 mm (females) and 175–200 mm (males). The maximum duration for the net in the water is six days. There are quality codes and schemes for the major species such as salmon and char but not yet for lumpfish. There are also no national welfare schemes, and just the standard rules for animal welfare.

Lumpfish are stocked on Faroese salmon farms according to the same regulations that are in place for salmon (A. Johannesen, Faroes, 2017, personal communication). This means that the salmon farmer needs to have permission to farm lumpfish and follow the same procedures as with salmon in terms of monitoring, quarantine procedures and fallowing. However, when keeping or breeding fish for research purposes, a custom research licence is usually granted. There are no specific health regulations covering lumpfish. At the moment salmon farming rules are followed, for example lumpfish have to be tested for ISAV. A licence is required for any commercial fishing in the Faroes. In addition to that, a permit to transport fish from one area to another is required (salmon farming ISAV precautionary rules). There is no fishery quota for lumpfish. Much of the fishing industry works on the 'fishing days' system, where boats are not assigned quotas, but they are given a number of days they are permitted to fish. The days depend on the size of boat and type of gear they use. As lumpfish are usually caught using gillnets, and this gear is not included in the usual coastal fishing fleet regulations, there are no general rules that can be applied to lumpfish. Boat owners may apply for a licence to fish using gillnets, specifying which species of fish they aim to catch and, if given permission, may be alloted fishing days or a quota.

All the current farming areas in the Faroes are already taken by salmon farmers. There are no rules in place to permit lumpfish farming within the same area as a salmon farm. A temporary permission to farm lumpfish can be allocated if the incoming and outgoing water is treated with UV light if the farm is land based. However, a commercial lumpfish farmer would struggle to obtain permission to farm lumpfish anywhere in the Faroes. There are new regulations being written, which it is understood will make land-based farming possible (A. Johannesen, Faroes, 2017, personal communication), if the water is treated. This would work on the basis that these land-based farms are isolated from the surrounding farm area, allowing for movement from the site to other farming sites.

A hatchery licence to rear or hold cleaner fish is required in Canada, similar to a salmon hatchery. Full health screening is carried out prior to transport from the hatchery facility to the farms. Pre-transfer involves histology, RT-PCR, and virology samples. The collection of wild-caught juvenile fish and transport to cage sites is not permitted. A collection permit/licence is required to fish. It is possible to obtain an authorised collector's licence and the fishermen are designates. If fishermen also have a commercial fishing licence any fish caught comes off their quota. There is no catch quota for the fishery at present. Cunner *Tautogolabrus adsperus* are captured by trap and the larger fish are kept for broodstock. Although there is no size limit, the lumpfish fishery has a minimum 227 mm mesh. Welfare is considered in fishing but is not closely monitored. All facilities have to follow guidelines prescribed by provincial and federal authorities. Biosecurity audits take place each year. There is no welfare scheme covering cleaner fish use. However, the use of cleaner fish has just started in Canada with two facilities growing cleaner fish from eggs and another one being used as a grow-out site.

The use of cleaner fish in Chile is still at too early a stage to have entailed any regulatory measures and guidance to farmers.

1.13 Advances in rearing technology

The rearing of wrasse and lumpfish utilises marine finfish technology developed over many years, with early life stages reared on rotifers (wrasse only) and *Artemia* spp. as live feed. However, the technology of cleaner fish production is continuously tested and improved to increase survival, growth and quality of cleaner fish. For lumpfish, aquaculture and deployment research is at a relatively early stage. While many of the details remain commercially sensitive, a recent review outlines the current technical status and likely strategies (Powell *et al.*, 2017; Chapter 6).

Wrasse production has not accelerated as fast as expected although reasonable numbers of fish have been grown by Marine Harvest Norway, now around 1 million per annum, and also by MHS and SSF at Machrihanish, with an estimated 200,000 wrasse produced in 2016. The rearing of wrasse is therefore technically feasible, although achieving high survival rates from egg to stocking in pens has been elusive. The establishment of broodstocks and collection of eggs has been possible, and rearing systems and equipment have been developed. However, the larvae are notoriously time-consuming to wean and growth is slow in hatcheries, therefore up to two years is required to stock wrasse in most situations. This is to rear fish to a minimum size around 10 to 12 cm to prevent escape through typical mesh sizes in pens.

1.14 Aim of this volume

There are many unanswered questions about the use of cleaner fish, how to improve hatchery rearing techniques, the best strategy for deployment, making cleaner fish more effective in pens, and in maintaining good health and welfare. There is much new information on the biology of cleaner fish and much of the information is in the grey literature and disparate sources. This volume attempts to bring all the main knowledge and practical experiences together. It is not intended to be exhaustive but to summarise current knowledge and to discuss present challenges. There is a hatchery manual already published for Ballan wrasse (Helland *et al.*, 2014) but this current volume is not intended to be in a manual format; rather, it is intended to provide a manual format where necessary but mainly to review technology in some chapters and also to provide new and unpublished information in others. The volume is in three sections: the first concentrates on basic biology and rearing of two of the main cleaner fish species; the second section is thematic and covers specialisms such as cleaner fish nutrition, genetics, immunology, health, welfare, transport, and the fisheries; and the final section deals with and encapsulates the status of cleaner fish rearing and use and strategy in different geographical areas. The intention is to give an overview of cleaner fish biology advances, in rearing, and in strategic use, and we collectively hope that you enjoy the read. Feedback, partnerships and further goal setting are always welcome.

References

Aaen, S.M., Helgesen, K.O., Bakke, M.J., Kaur, K. and Horsberg, T.E. (2015) Drug resistance in sea lice: a threat to salmonid aquaculture. *Trends in Parasitology* 31, 72–81.

Anonymous (2000) *Henderson's Dictionary of Biological Terms*. Prentice Hall, London.

Anonymous (2015) Traditional small boat Icelandic lumpfish fishery is MSC certified. www.msc.org/newsroom/news/traditional-small-boat-icelandic-lumpfish-fishery-is-msc-certified. Consulted 16/2/17.

Anonymous (2017) Imminent action on £300 million sea lice problem. http://files.heraldscotland.com/news/environment/15009983.Imminent-action-on-300m-Sea-lice-problem

Bjordal, Å. (1988) Cleaning symbiosis between wrasse (Labridae) and lice infested salmon (*Salmo salar*) in mariculture. ICES Mariculture Committee 1988/F17, 8 pp.

Bjordal, Å. (1991) Wrasse as cleaner fish for farmed salmon. *Progress in Underwater Science* 16, 17–28.

Bjordal, Å. and Costello, M. (1990) How good is this natural control of sea-lice? *Fish Farmer* 13 (3), 44–46.

Black, K. (2005) Ecological effects of sea lice treatments www.sams.ac.uk/kenny-black/PAMP%20 Laymans.pdf/view

Boxshall, G.A. and Bravo, S. (2000) On the identity of the common Caligus (Copepoda: Siphonostomatoida: Caligidae) from salmonid netpen systems in southern Chile. *Contributions to Zoology* 69, 137–146.

Bshary, R., Redouan, and Grutter, A.S. (2006) Image scoring and cooperation in a cleaner fish mutualism. *Nature* 441, 975–978.

Burns, C.W. (1985) Fungal parasitism in a copepod population: the effects of Aphanomyces on the population dynamics of *Boeckella dilitata* Sars. *Journal Plankton Research* 7, 201–205.

Butler, J. (2002) Wild salmonids and sea louse infestations on the west coast of Scotland: sources of infection and implications for the management of marine salmon farms. *Pest Management Science* 58, 595–608.

Carvajal, J., Gonzalez, L. and George-Nascimento, M. (1998) Native sealice (Copepoda: Caligidae) infestation of salmonids reared in netpen system in the southern Chile. *Aquaculture* 66, 241–246.

Code of Good Practice (2015) Scottish Finfish Aquaculture. www.thecodeofgoodpractice.co.uk

Costello, M.J. (2009) The global economic cost of sea lice to the salmonid farming industry. *Journal of Fish Diseases* 32, 115–118.

Darwall, W., Costello, M.J. and Lysaght, S. (1992a) Wrasse: how well do they work? *Aquaculture Ireland* 50, 26–29.

Darwall, W.R.T., Costello, M.J., Donnelly, R. and Lysaght, S. (1992b) Implications of life-history strategies for a new wrasse fishery. *Journal of Fish Biology 41 (Supplement B)*, 111–123.

Deady, S., Varian, S.J.A. and Fives, J.M. (1995) The use of cleaner fish to control sea lice on two Irish salmon (*Salmo salar*) farms with particular reference to wrasse behaviour in salmon cages. *Aquaculture* 131, 73-90.

DeBach, P. and Rosen, B. (1991) *Biological Control by Natural Enemies*. Cambridge University Press, New York. 440 pp.

Dipper, F.A. and Pullin, R.S.V. (1979) Gonochorism and sex-inversion in British Labridae (Pisces). *Journal of Zoology* 187, 97–112.

Drebes, G. and Schnepf, E. (1998) *Gyrodinium undulans* Hulbert, a marine dinoflagellate feeding on the bloom-forming diatom *Odontella aurita*, and on copepod and rotifer eggs. *Helgoländer Meeresuntersuchungen* 52, 1–14.

FAO (2016) The state of world fisheries and aquaculture 2016. FAO, Rome, pp 200. www.fao.org/3/a-i5555e.pdf

Feder, H.M. (1966) Cleaning symbiosis in the marine environment. In: Henry, S.M. (ed.) *Symbiosis* Vol. 1, Academic Press, New York, pp. 327–380.

Federation of European Aquaculture Producers (FEAP) (2014) Production Report 2005–2014.

Goggin, C.L. and Bouland, C. (1997) The ciliate *Orchitophyra cf stellarum* and other parasites and commensals of the northern Pacific seastar *Asterias amurensis* from Japan. *International Journal of Parasitology* 27, 1415–1418.

Grant, A. and Treasurer, J.W. (1993) The effects of fallowing on caligid infestation on farmed Atlantic salmon (*Salmo salar* L.) in Scotland. In: Boxshall, G.A. and Defaye, D. (eds.) *Pathogens of wild and farmed fish: sea lice.* Ellis Horwood, Chichester, pp. 255–260.

Grutter, A.S. (1995) Relationship between cleaning rates and ectoparasite loads in coral reef fishes. *Marine Ecology Progress Series* 118, 51–58.

Grutter, A.S. (1999) Cleaner fish really do clean. *Nature* 398, 672–673.

Helland, S., Wiborg, D.S., Hamre, K., Skiftesvik, A.B., Lein, I., Kousoulaki, K., Øie, G., Browman, H., Evjemo, J.O., Kjørsvik, E. and Grøtan, E. (2014) Best practices for farming of ballan wrasse. *Production of ballan wrasse, science and practice.* The Norwegian Seafood Research Fund, Norway, pp. 123–128.

Hillden, N-O. (1978) On the feeding of the Goldsinny *Ctenolarbrus rupestris* L. (Pisces: Labridae) *Ophelia* 17, 195–198.

Hjul, J. (2017) Medicine costs drop in sea lice fight. FISH update 21 June. fishupdate.com/medicine-costs-drop-sea-lice-fight

Imsland, A. (2014) The use of lumpfish (*Cyclopterus lumpus* L.) to control sea lice (*Lepeophtheirus salmonis* Kroyer) infestations in intensively farmed salmon (*Salmo salar* L.). *Aquaculture* 424, 18–23.

Ingvarsdottir, A., Birkett, M.A., Duce, I., Genna, R.L., Mordue, W., Pickett, J.A., Wadhams, L.J. and Mordue (Luntz), A.J. (2002) Semiochemical strategies for sea louse control: host location cues. In: Mordue, A.J. and Pike, A.W. (eds.) *Pest Management Science* 58, 537–545.

Johannessen, A. (1978) Early stages of *Lepeophtheirus salmonis* (Copepoda, Caligidae). *Sarsia* 63, 169–176.

Johnson, S.C. and Albright, L.J. (1991) The developmental stages of *Lepeophtheirus salmonis* (Kroyer, 1837) (Copepoda: Caligidae). *Canadian Journal of Zoology* 69, 929–950.

Johnson, S., Treasurer, J.W., Bravo, S., Nagasawa, K. and Kabata, K. (2004) A review of the impact of parasitic copepods in marine aquaculture. *Zoological Studies* 42, 229–243.

Kabata, Z. (1973) The species of *Lepeophtheirus* (Copepoda: Caligidae) from fishes of British Columbia. *Journal of the Fisheries Research Board of Canada* 30, 729–759.

Kabata, Z. (1979) *Parasitic copepoda of British Fishes.* Ray Society, London. 468 pp.

Kim, I-H. (1993) Developmental stages of *Caligus punctatus* Shiino, 1955 (Copepoda: Caligidae). In: Boxshall, G.A. and Defaye, D. (eds.) *Pathogens of wild and farmed fish Sea lice.* Ellis Horwood, Chichester, pp. 5–15.

Losey, G.S. (1987) Cleaning symbiosis. *Symposium on Marine Symbioses* 4, 229–256.

Losey, G. S., Grutter, A., Rosenquist, G., Mahon, J. L. and Zamzow, J. (1999) Cleaning symbiosis: A review. In: V.C. Almada, R.F. Oliveira and E.J. Goncalves (eds.) *Behaviour and Conservation of Littoral Fishes*, pp. 379–395.

Lysaght, S., Costello, M.J. and Deady, S. (1995) *A technical guide to the use of wrasse (Labridae) in the biological control of sealice on salmon.* Dublin, Bord Iascaigh Mhara.

Mahapatra, B.K. and Datta, N.C. (1996) *Macrobrachium rosenbergii* (de Man) as biological agent to control molluscan population in fish culture ponds of the Sundarbans. *Environment and Ecology* 14, 968–972.

Marten, G.G., Nguyen, M., Giai, N. and Ng, G. (2000) Copepod predation on *Anopheles quadrimaculatus* larvae in rice fields. *Journal of Vector Ecology* 25, 1–6.

Minchin, D. and Jackson, D. (1993) *Udonella caligorum* Johnston, 1835 (Platyhelminthes: Udonellidae) associated with caligid copepods on farmed salmon. In: Boxshall, G.A. and Defaye, D. (eds.) *Pathogens of wild and farmed fish: sea lice.* Ellis Horwood, Chichester, pp. 346–355.

Murray, A.G. (2014) A modelling framework for assessing the risk of emerging diseases associated with the use of cleaner fish to control parasitic sea lice on salmon farms. *Transboundary and Emerging Diseases* 63, 270–277.

Mustafa, A., Rankaduwa, W. and Campbell, P. (2001) Estimating the cost of sea lice to salmon aquaculture in eastern Canada. *Canadian Veterinary Journal* 42, 54–56.

Parenti, P. and Randall, J.E. (2011) Checklist of the species of the families Labridae and Scaridae: an update. *Smithiana Bulletin* 13, 29-44.

Piasecki, W. (1996) The developmental stages of *Caligus elongatus* Nordmann van Nordmann, 1832 (Copepoda, Siphonostomatoida, Caligidae). *Canadian Journal of Zoology* 74, 1459–1478.

Pike, A.W. and Wadsworth, S.L. (1999) Sea lice on salmonids: their biology and control. *Advances in Parasitology* 44, 233–337.

Pimentel, D. (1991) Diversification of biological control strategies in agriculture. *Crop Protection* 10, 243–253.

Potts, G.W. (1973) Cleaning symbiosis among British fish with special reference to *Crenilabrus melops* (Labridae). *Journal of the Marine Biological Association of the UK* 53, 1–10.

Potts, G.W. (1974) The colouration and its behavioural significance in the corkwing wrasse, *Crenilabrus melops*. *Journal of the Marine Biological Association of the UK* 54, 925–938.

Powell, A. and Pooley, C. (2015) *Innovative technologies for control of Salmon Lice.* Conference Poster: Biopesticides – innovative technologies and strategies for pest control; Swansea, UK.

Powell, A., Howes, P.N., Stringwell, R. and Garcia de Leaniz, C. (2016) Aquatic research at Swansea University. *Hatchery International*, May/June, pp. 24–25.

Powell, A., Treasurer, J.W., Pooley, C.L., Keay, A.J., Lloyd, R., Imsland, A.K. and Garcia de Leaniz, C. (2017) Use of lumpfish for sea-lice control in salmon farming: challenges and opportunities. *Reviews in Aquaculture* DOI:10.1111/raq.12194

Prickett, R. (2016) A review of cleaner-fish production in the UK. *Bulletin Aquaculture Association of Canada.* 2016–2, 53–58.

Rae, G.H. (1991) *The use of wrasse as cleaner fish in salmon farming.* SSGA report to members.

Rae, G.H. (2002) Sea louse control in Scotland, past and present. *Pest Management Science* 58, 515–520.

Raynard, R.S., Bricknell, I.R., Billingsley, P.F., Nisbet, A.J., Vigneau, A. and Sommerville, C. (2002) Development of vaccines against lice. In: Mordue, A.J. and Pike, A.W. (eds.), *Pest Management Science*, 58, 569–575.

Revie, C.W., Gettinby, G., Treasurer, J.W., Rae, G.H. and Clark, N. (2002) Temporal, management and environmental factors influencing the epidemiological patterns of sea lice (*Lepeophtheirus salmonis*) infestations on farmed Atlantic salmon (*Salmo salar*) in Scotland. In: Mordue, A.J. and Pike, A.W. (eds.), *Pest Management Science* 58, 576–584.

SAMS (2005) Ecological effects of sea lice medicines in Scottish sea lochs. Scottish Association of Marine Science, Oban. 268 pp.

Sherwood, J.P.B. (1990) Wrasse breeding trial. Sea Fish Industry Authority Marine Farming Unit. 11 pp.

Sherwood, J. (1991) Wrasse breeding trial. Sea Fish Industry Authority Marine Farming Unit. 10 pp.

Sinnott, R. (1999) Cost of sea lice to Scottish salmon farmers. *Outlook*, Trouw publication 11, 8–10.

Stone, J. (1996) Preliminary trials on the culture of goldsinny and corkwing wrasse. In: Sayer, M.D.J.,

Treasurer, J.W. and Costello, M.J. (eds), *Wrasse biology and use in aquaculture*. Fishing News Books, Oxford, pp 142–170.

Stone, J. and Bruno, D.W. (1989) A report of *Ephelota* sp., a suctorian found on the sea lice *Lepeophtheirus salmonis* and *Caligus elongatus*. *Bulletin of the European Association of Fish Pathologists* 9, 113–115.

Treasurer, J.W. (1994) Prey selection and daily food consumption by a cleaner-fish, *Ctenolabrus rupestris* (L.), on farmed Atlantic salmon, *Salmo salar* L. *Aquaculture* 122, 269–277.

Treasurer, J.W. (1996) Capture techniques for wrasse in inshore waters of west Scotland. In: Sayer, M.D.J., Treasurer, J.W. and Costello, M.J. (eds) *Wrasse biology and use in aquaculture*. Fishing News Books, Oxford, pp. 74–90.

Treasurer, J.W. (1997) Parasites of wrasse (Labridae) in inshore waters of west Scotland stocked as cleaner fish of sea lice (Caligidae) on farmed Atlantic salmon. *Journal of Fish Biology* 50, 895–899.

Treasurer, J.W. (2002) A review of potential pathogens of sea lice and the application of cleaner fish in biological control. In: Mordue, A.J. and Pike, A.W (eds), *Integrated pest management in salmon farming*, 58, 546–558.

Treasurer, J.W. (2013) *Use of wrasse in sea lice control*. SARF Report, Birnam.

Treasurer, J.W. and Grant, A. (1997) The efficacy of hydrogen peroxide for the treatment of farmed Atlantic salmon, *Salmo salar* L., infested with sea lice (Copepoda: Caligidae). *Aquaculture* 148, 265–275.

Treasurer, J.W. and Laidler, L.A. (1994) *Aeromonas salmonicida* infection in wrasse (Labridae) used as cleaner fish, on an Atlantic salmon, *Salmo salar* L., farm. *Journal of Fish Diseases* 17, 155–161.

Vaughan, D.B., Grutter, A.S., Costello, M.J. and Hutson, K.S. (2017) Cleaner fishes and shrimp diversity and a re-evaluation of cleaning symbioses. *Fish and Fisheries* 18, 698–716.

Vidtmann, S. (1990) Microsporodina fauna (Protozoa, Microsporidia) of crustaceans (Cladocera and Copepoda) in waters of the Lithuanina SSR, with description of three new species. *Ekologija* 1990, 104–111.

Wadsworth, S.L. (1998) *The control of Lepeophtheirus salmonis (Copepoda: Caligidae) (Kroyer, 1837) on Atlantic salmon Salmo salar L. production sites*. PhD thesis, Department of Zoology, University of Aberdeen.

Wallace, I.S., Donald, K., Munro, L.A., Murray, W., Pert, C.C., Stagg, H. and Hall, M. (2015) A survey of wild marine fish identifies a potential origin of an outbreak of viral haemorrhagic septicaemia in wrasse, Labridae, used as cleaner fish on marine Atlantic salmon, *Salmo salar* L., farms. *Journal of Fish Diseases* 38, 515–521.

Westneat, M.W. and Alfaro, M.E. (2015) Phylogenetic relationships and evolutionary history of the reef fish family Labridae. *Molecular Phylogenetics and Evolution* 36, 370–390.

Xu, Z. and Burns, C.W. (1991) Effects of the epizoic ciliate, *Epistylis daphnia*, on growth, reproduction and mortality of *Boeckella triarticulata* (Thomson) (Copepoda: Calanoida). *Hydrobiology* 209, 183–189.

CHAPTER 2

The Ballan wrasse (*Labrus bergylta*) reproductive physiology, broodstock management and spawning behaviour

Andrew Davie, Bridie Grant, William Clark and Herve Migaud

Institute of Aquaculture, University of Stirling, Stirling, UK

2.1 Introduction

As a predominantly inshore species adult Ballan wrasse are found closely associated with rock faces, boulder slopes, off-shore reefs and macroalga beds in sublittoral zones to 30 m (Treasurer, 1994). They are strongly diurnal and so are only active during the day, then hide at night in rock crevices or seaweed beds and show a high degree of home range fidelity (Lipej *et al.*, 2009; Morel *et al.*, 2013; Villegas-Ríos *et al.*, 2013c). Ballan is a slow-growing (Artüz, 2005), and long-lived species that can achieve a total length of 65.9 cm and 4.35 kg over a 29-year lifespan (Ottesen *et al.*, 2012). Their natural prey items are diverse but in the adult stages include crab *sp.*, prawns, and bi-valve molluscs (Deady and Fives, 2009). The variability in colour and pattern morphisms is notable in the species (Dipper, 1981), but this does not clearly relate to gender, origin or seasonality (Dipper and Pullin, 1979; Leclercq *et al.*, 2014; Porteiro *et al.*, 1996; Villegas-Ríos *et al.*, 2013b).

In accordance with other temperate high-latitude teleosts; the reproductive cycle of Ballan wrasse in Northern Europe follows clear annual cycles (Muncaster *et al.*, 2010). The sex ratio is highly skewed (circa 10% male) (Quignard and Pras, 1986; Dipper *et al.*, 1977; Leclercq *et al.*, 2014) as Ballan wrasse are monandric protogynous hermaphrodites and as such are born female and then a subset of animals undergo sex change to become male. Field evidence suggests that individuals can undergo sex change from five years of age and a length greater than 28 cm (Dipper *et al.*, 1977; Leclercq *et al.*, 2014, Muncaster *et al.*, 2013), though the environmental and social drivers of this change remain undetermined in the species. Dipper and Pullin (1979) suggested that social cues could be the main driver, which is similar to related species including Bluehead wrasse (*T. bifasciatum*) (Lamm *et al.*, 2015), and other sequentially hermaphroditic species (Choat and Robertson, 1975). In such a situation physical size and social dominance are most likely the main primers governing transition. In accordance with the Size Advantage Hypothesis (SAH) (Munday *et al.*, 2006), where a hierarchy becomes unstable (i.e. in male absence), the largest and most dominant female will undergo transition to redress the balance

(Semsar and Godwin, 2003), and thereby increase reproductive success relative to the primary gender (Erisman *et al.*, 2013). Thus, in Ballan wrasse, sex change is normally observed between the ages of five and 14 years (Dipper *et al.*, 1977) with age at 50% sex change evaluated at 7.4–11.8 years (Villegas-Ríos *et al.*, 2013a; Leclercq *et al.*, 2014).

Information on spawning behaviour in the natural environment is somewhat limited for Ballan wrasse, though a more detailed picture can be inferred when known behaviour in other comparable wrasse species is considered. During the spawning season, male Ballan wrasse become highly territorial. They are reported to defend an area of up to 300 m^2 (0.0003 km^2) (Sjolander *et al.*, 1972), which is much smaller than their proposed home range (0.091 km^2) (Villegas-Ríos *et al.*, 2013c). It is suggested that this reduced territory during the spawning season can be explained through the constant demand on the male to guard active nest sites, which in turn restricts range in the spawning season (Villegas-Ríos *et al.*, 2013c). In other comparable wrasse species including the Bluestreak wrasse (*Labroides dimidiatus*) such 'spawning territories' form the outer spatial boundary of a nested hierarchy of territories with dominant males being central within these overlapping boundaries (Robertson and Choat, 1974). It is common in these situations that there is continual testing and re-affirmation of relationships within the group through high levels of social interaction (Godwin *et al.*, 2003). Though ultimately, once the spawning season is complete in Ballan wrasse, as with other species, this strong territoriality is relaxed when normal foraging activities are resumed (Thangstad, 1999) over a wider home range (Villegas-Ríos *et al.*, 2013c).

The spawning season of Ballan wrasse under natural environmental conditions is reported to range between April and June (Muncaster *et al.*, 2010; Grant *et al.*, 2016a) with there being a suggestion of a latitudinal cline in the onset of spawning, but this remains unsubstantiated as yet. Spawning is believed to be complete by the summer solstice with ovarian atresia observed from August (Muncaster *et al.*, 2010). In preparation for the subsequent spawning season, gonad maturation and ovarian recrudescence begins in November with the appearance of oocytes with cortical alveoli and a concomitant rise in sex steroids, which is followed by the oocytes progressing to vitellogenesis by late winter (Muncaster *et al.*, 2010). While early research classed Ballan wrasse as a group-synchronous, single spawning species (Dipper and Pullin, 1979), more recent research has redefined the species as group-synchronous, multiple batch spawners, based on both histological evidence (Muncaster *et al.*, 2010; Villegas-Ríos *et al.*, 2014) as well as direct observation of spawning productivity over a spawning season and subsequent assignment of parentage using microsatellite markers (Grant *et al.*, 2016a).

2.2 Establishment of captive broodstock populations

To date, all commercial hatcheries are reliant on wild captured mature adults. Animals are generally sourced through the use of baited traps or trammel net systems, targeting mature animals >25 cm total length. There are two significant concerns relating to the sourcing of wild broodstock, being the sourcing of an adequate number of males and considerations for genetic background of the animals being used. With respect to the first point, sexually mature males are a key limiting factor that will determine the scale of a broodstock facility. In the wild, the sex ratio is highly skewed (*circa* 10% male) (Quignard, 1986; Dipper *et al.*, 1977; Leclercq *et al.*, 2014),

Fig. 2.1 Typical examples of the diversity in colour and pattern variation of wild-caught Ballan wrasse (*Labrus bergylta*). Top to bottom left-hand images are females, right-hand images are males. Scale bar boxes in image are 20 mm. Credit: Andrew Davie, Institute of Aquaculture, University of Stirling, Stirling

thus sourcing and assuring the capture of functional males can be challenging. Externally, Ballan wrasse have highly variable colour and pattern phenotypes (Porteiro *et al.*, 1996), but ultimately this appears to have no relation to gender (Villegas-Ríos *et al.*, 2013b), thus identifying males is challenging (Fig. 2.1). A number of methods have been tested to aid gender identification, methods such as endocrine profiling (Muncaster *et al.*, 2010), ultrasound scanning (Talbot *et al.*, 2012) or gonadal biopsy can work, but have a limited functional window and can have high technical demands. Clark *et al.* (2016) further confirmed that the plasma bile pigment bilvierdin is not a reliable indicator of gender in Ballan wrasse, whereas plasma pigment type and/or concentration has been shown to have such gender resolving power in other cleaner fish species including lumpsuckers *Cyclopterus lumpus* (L. 1758) (Mudge and Davenport, 1986), peacock wrasse *Crenilabrus tinca* (L. 1758) (Abolins, 1961), cuckoo wrasse *Labrus mixtus* (L. 1758), axillary wrasse *Crenilabrus mediterraneus* (L. 1758) (Abolins, 1961), as well as the blue-throated wrasse *Notolabrus tetricus* (Richardson 1840, cited in Gagnon, 2006). In response, the most reliable gender identification method, proven to work in the field and throughout the spawning season, is the use of discriminatory morphometric selection. Leclercq *et al.* (2014) clearly demonstrated that in addition to the male size bias, which is to be expected with protogynous hermaphrodites (Fig. 2.2), there is an overlaying size dimorphism where males, compared to females of a similar age, are generally heavier and longer (*circa* 30% and 10 % respectively). As such, at the point of capture it is possible to discriminate the genders with >90% accuracy through a simple combined analysis of individual weight, total length and Fulton's condition factor, which should be perfectly possible in most field situations (Leclercq *et al.*, 2014).

The second point to be considered in sourcing wild broodstock is that of the genetic profile of the founder animals. The natural range of Ballan wrasse reportedly extends along the Eastern

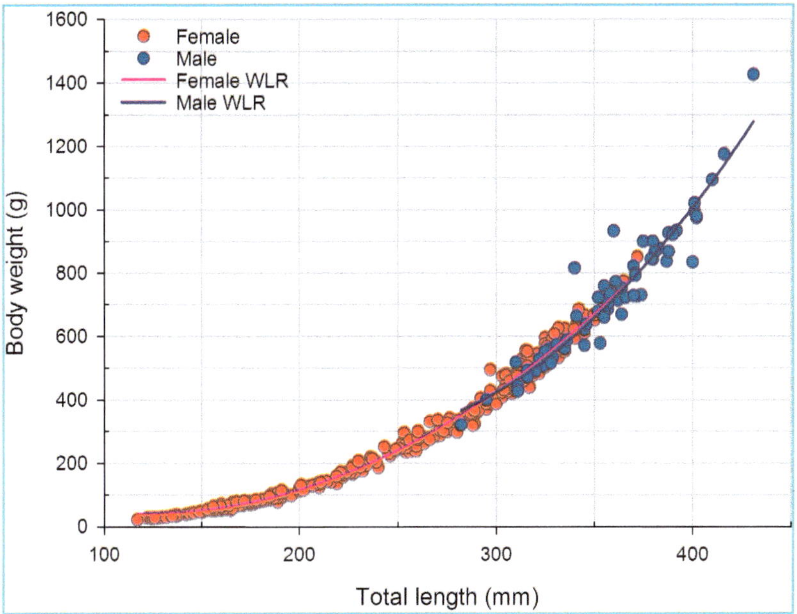

Fig. 2.2
Demonstration of a typical weight length relationship by gender of wild captured Ballan wrasse (*Labrus bergylta*). For full details of dataset and analysis of the gender growth relationship see Leclercq *et al.*, (2014).

Atlantic seaboard from Morocco to Norway (Quignard and Pras, 1986). Given what is known about the behavioural phenotype of the species, the potential for phylogenetic structuring over such a range is real and any such structuring could have implications for broodstock choice and subsequent fish deployment (Johnson, 2000; Doupé and Lymbery, 2000). Research to date on the phylogeographic structuring of Ballan wrasse is limited. D'Arcy *et al.* (2013) presented evidence of structuring across Ireland, Scotland and southern Norway, reporting the presence of two divergent clades (an Atlantic and a Scandinavian clade) based on the analysis of mitochondrial DNA haplotypes. However, the structuring was not clear cut as both clades had representation from individuals from all sampling locations, which, the authors suggested, may be a result of these two main lineages already existing before wrasse colonised the sampled areas. The application of more powerful genotyping methods, e.g. application of microsatellite or single nucleotide polymorphism (SNP) markers, have yet to be published, though tools to enable such studies are in development, e.g. Quintela *et al.*, (2014). Comparable studies conducted on corkwing wrasse (*Symphodus melops*) (Robalo *et al.*, 2011), a species with similar reproductive characteristics (Potts, 1984), revealed comparable grouping of Atlantic and Scandinavian populations. While, Sundt and Jørstad (1998), were able to find local genetic differentiation in goldsinny wrasse (*Ctenolabrus rupestris*) populations between geographically separated fjord systems within the same regions of Norway. Based on the limited available literature, the pragmatic management decision would be to source broodstock local to the production site and area of deployment to minimise the perceived risk of subsequent possible genetic introgression. In the long term it is not unreasonable to consider working within a closed life cycle management plan where F1(+) animals are retained for future broodstock. To date there are no F1 captive broodstock that have spawned in captivity, the slow growth and late age/size at first maturation (Villegas-Ríos *et al.*, 2013a; Leclercq *et al.*, 2014) requires long-term investment in the holding of such animals. It must also be recognised that any such animals will be all female at the point of first maturation (Leclercq *et al.*, 2014), thus a reliable methodology to induce sex change to functional male needs to be proven in the species. It is for these reasons that selective breeding has not been actively pursued to date.

2.3 Broodstock holding and spawning management

Commercial hatcheries currently rely on the natural spawning of broodstock maintained under controlled photo-thermal conditions. The exact environmental parameters that entrain reproduction and determine quality of spawning are yet to be defined in the species (i.e. photoperiod and thermal cues). It is fair to assume that there will be thermal optima and critical seasonal windows as have been determined in other temperate marine teleosts (Migaud *et al.*, 2013), and the suggested latitude cline in spawning season is suggestive of a temperature determinant to the timing of final maturation (Dipper and Pullin, 1979; Muncaster *et al.*, 2010; Villegas-Ríos et al., 2013a; Grant *et al.*, 2016a). However, despite the lack of such data, Ballan wrasse appear to respond well to photo-thermal manipulation, with commercial hatcheries in both the UK and Norway reporting the successful adjustment of spawning seasons by up to three to four months (both advanced and delayed) in controlled environmental conditions (P. Featherstone, Marine Harvest Scotland, and E. Grotan, Marine Harvest Norway, January 2017, personal communications).

Routine husbandry practice is to hold broodstock in large communal holding tanks for the majority of the year where animals are held in low stocking densities (5–10 kg m^{-3}) and the tanks are populated with a diverse range of 'furniture', e.g. plastic pipes, artificial kelp, etc. in which the wrasse will actively seek refuge. In anticipation of a pending spawning season broodstock will be individually inspected and sorted into spawning tanks (generally about 7–10 m^3 volume tanks) in groups of approximately 20–30 individuals. Within this spawning population males are generally stocked at the 10–20% ratio (Grant *et al.*, 2016a) and rapidly a dominance hierarchy becomes evident with the dominant male(s) being seen to dominate subordinate males and regularly inspect female maturation status. These spawning tanks have a large amount of 'furniture' to allow fish to rest and hide from dominant males, however an area is left open (generally 30–50% of the tank surface area) to act as the spawning arena within the tank (Fig. 2.3).

The courtship behaviour reported in the wild by Sjolander *et al.* (1972) appears to be recreated in captive stocks where the dominant male will visit females regularly that show an interest in

Fig. 2.3 Typical Ballan wrasse (*Labrus bergylta*) broodstock tank set-up. Refuge is provided at the centre of the tank using a combination of 'artificial kelp' as well as pipe furniture. On the periphery of the tank are four spawning arenas each comprising of four short pile carpet tiles (0.28 m^2 area per tile). Credit: Dr Bridie Grant

spawning over the spawning substrate. Just prior to successful copulation the female will lay on her side while quivering caudal and pectoral fins and the male will then align and the eggs and milt are released on to the chosen spawning site. The female may remain briefly after spawning over the chosen site. However, there is no reported active defence of the spawning sites in comparison to reports in the wild (Sjolander *et al.*, 1972). Ballan wrasse spawn adhesive, spherical, negatively buoyant eggs of approximately 1 mm in diameter (D'Arcy *et al.*, 2012). Hatcheries use artificial spawning substrates (short-pile carpet tiles or artificial turf) laid on the tank floor in the open arena area to focus spawning effort and aid the subsequent recovery of the adhesive eggs. During the spawning season mats are collected daily from broodstock tanks, inspected for egg burden and, if deemed viable, the egg-bearing mats are surface disinfected and passed on to the incubation facility. New clean mats are then placed into the spawning tanks. All hatcheries incubate egg mats in aerated incubation tanks and collect the larvae, which hatch directly from the substrate after a period of six days or approximately 72 degree days (dd) post-fertilisation (Ottesen *et al.*, 2012). Such an incubation method is labour intensive and has biosecurity concerns, and commercial experience proves that effective disinfection of the spawning substrates is challenging. In response Grant *et al.* (2016b) have demonstrated that enzyme alcalase® can be used to effectively remove the adhesive gum layer from the eggs without compromising hatch rates. Such a methodology could allow a hatchery to extract embryos from spawning substrates and thus return to pelagic egg incubation as is used in other temperate species, but scaling up of the methodology to a commercial scale remains to be optimised (Grant *et al.*, 2016b).

The spawning season typically lasts two to three months under ambient photoperiod and thermal conditions, and this would mean spawning starting in April and ending in June with peak spawning activity being in May (Muncaster *et al.*, 2010; Grant *et al.*, 2016a). There are limited data on the spawning productivity of captive stocks. However, Grant *et al.* (2016a) provide evidence that there may be an underlying rhythmicity in spawning productivity where a broodstock population will exhibit a structured spawning activity where there is a spawning window of two to six days during which there is daily spawning activity to be followed by an interspawning interval of eight to 12 days where no activity is evident. This spawning period (which lasts on average 13–15 days) will be repeated four to six times over the spawning season. Similar structured spawning has been reported in commercial hatcheries but is not always clearly evident. Grant *et al.* (2016a) proposed this could be indicative of a natural semi-lunar spawning cycle in the species as has been reported for other labrid species including the saddle wrasse (*Thalassoma dupery*) and paddlefin wrasse (*Thalasoma lucasanum*) (Warner, 1982; Ross, 1983; Taylor, 1984), though the social and environmental drivers of this rhythmicity and their possible implication for captive stock management remain to be clearly determined.

There is currently no reliable method to quantify spawning productivity directly from broodstock tanks. Based on a gravimetric estimation of oocyte abundance, Villegas-Ríos *et al.* (2014) suggest that batch fecundity of Ballan wrasse could range between 11,000–135,000 eggs per female (not corrected for female size) which is comparable to the estimated egg number recorded per spawning event (25,000–74,000 eggs) reported by Grant *et al.* (2016a). In practice, most commercial hatcheries gather the mats from a given day that are deemed to contain a viable number of embryos to be incubated together. It is at the point of hatch that the first accurate numeric estimation can be made using a routine volumetric estimation method. With respect to egg quality indicators, fertilisation rate can be estimated if eggs are physically removed from the spawning mats and examined under a microscope, and normal embryo development can also be checked using this method (D'Arcy *et al.*, 2012). Experience has proven that more accurate estimates of fertilisation success are realised if fertilisation rate is estimated at least 24 hours post-fertilisation (Grant *et al.*, 2016a) when the fertilised embryos have reached the gastrula stage (D'Arcy *et al.*, 2012). Egg size appears to remain relatively constant over the spawning season. However, the thickness of the gum-layer does apparently reduce by up to 30%, though the functional significance of this change remains to be determined (Grant *et al.*, 2016a).

The omnivorous grazing nature of wild Ballan wrasse is well documented, with Deady and Fives (2009) describing adult diets typically to include crab spp., prawns, and bi-valve molluscs. In terms of commercial broodstock feeds there are no dedicated wrasse specific feeds being marketed to date. To reflect the natural feed preferences, current industry practice is to feed a standard cold water broodstock formulated feed in either pellet or moist sausage form and then supplement this with natural prey types including crustacea, e.g. *Pandalus borealis* and *Nephrops norvegicus,* or molluscs, e.g. Mytilus spp. To aid biosecurity these natural prey types should be frozen for an appropriate length of time before being offered to the broodstock. Hamre *et al.* (2013) reported that the gross proximate composition of wild and captive reared broodstock were generally comparable. However, there remained some observable differences in specific components such as taurine, water and lipid soluble vitamins and minerals, etc. However, due to the fundamental lack of understanding of nutritional requirements in the species and the subsequent impact this could have on broodstock spawning and performance it is hard to conclude, as yet, if feeds are appropriate. Surprisingly, Grant *et al.* (2016a) observed no direct correlation between egg fatty acid content and fertilisation or hatch rates. This is unusual as strong links (especially in DHA and EPA) to fertilisation and hatching success have been reported in many other marine teleost species, including cod (Pickova *et al.*, 1997), sea bass (*Dicentrarchus labrax*) (Bruce *et al.*, 1999) and Atlantic halibut (*Hippoglossus hippoglossus*) (Mazorra *et al.*, 2003). Therefore it is fair to assume that this could be suggestive that these components are not limiting in the current broodstock feed practices. Lipid content and FA composition of fish eggs are known to vary considerably between species (Sargent *et al.*, 2002). With the exception of high levels of ARA (~3.8 ± 0.5 % of total fatty acids) compared to ~2.5 % of the total fatty acids in other marine species, the Ballan wrasse egg fatty acids profile observed by Grant *et al.* (2016a) fits the general profile for marine fish (Tocher *et al.*, 1985; Fraser *et al.*, 1988; Sargent *et al.*, 2002). The relative levels of EPA observed in captive Ballan eggs are similar to those reported for wild Ballan wrasse, while the levels of ARA, DHA, and DHA:EPA ratio were lower than those previously reported for wild Ballan wrasse (6 ± 2; 30 ± 4; and 2.5 ± 0.5 %, respectively) (Hamre *et al.*, 2013). Clearly a more comprehensive investigation is required to ensure there are not some critical nutritional factors that compromise broodstock spawning performance that are being overlooked in current feed formulations.

2.4 Conclusion

Ballan wrasse culture and captive rearing is still in its infancy, with commercial broodstock operations having been in operation since approximately 2010. Ballan wrasse appear to acclimatise well to captivity and can be considered a reliable natural spawning species. A captive production cycle is the preferred, and more 'sustainable', production model for cleaner fish supply and as such it can be anticipated that there will continue to be demand for Ballan wrasse production and with it a concomitant expansion of broodstock facilities. In terms of future research priorities, focus should be directed towards aspects of nutritional requirements, feed formulation and its impacts on spawning performance, optimisation of environmental rearing conditions and understanding the social/behavioural aspects of spawning and gender change. All of this could contribute to improving broodstock facilities productivity and reliability. In the long term, it should be possible to close the life cycle and have broodstock facilities utilising F1(+) broodstocks but, given the species slow growth rate and late age/size to first maturation, aspiring to move towards selective breeding would be challenging.

References

Abolins, L. (1961) The existence of sexual dimorphism of blood plasma in fishes of the family *Labridae*. *Ark Zoology* 13, 541–544.

Artüz, M.L. (2005) Embryonic and larval development of the ballan wrasse *Labrus bergylta* Ascanius 1767. *Hidrobiologica* 98–101.

Bruce, M., Oyen, F., Bell, G., Asturiano, J.F., Farndale, B., Carrillo, M., Zanuy, S., Ramos, J. and Bromage, N. (1999) Development of broodstock diets for the European sea bass (*Dicentrarchus labrax*) with special emphasis on the importance of n – 3 and n – 6 highly unsaturated fatty acid to reproductive performance. *Aquaculture* 177, 85–97.

Choat, J.H. and Robertson, D.R. (1975) Protogynous hermaphroditism in fishes of the family *Scaridae*. In: Reinboth, R. (ed.) Intersexuality in the animal kingdom. Springer Berlin Heidelberg, pp. 263–283.

Clark, W., Leclercq, E., Migaud, H., Nairn, J. and Davie, A. (2016). Isolation, identification and characterisation of ballan wrasse *Labrus bergylta* plasma pigment. *Journal of Fish Biology* 89(4), 2070–2084.

D'Arcy, J., Mirimin, L. and FitzGerald, R., (2013) Phylogeographic structure of a protogynous hermaphrodite species, the ballan wrasse *Labrus bergylta*, in Ireland, Scotland, and Norway, using mitochondrial DNA sequence data. *ICES Journal of Marine Science: Journal du Conseil,* 70(3), 685–693.

D'Arcy, J., Dunaevskaya, E., Treasurer, J.W., Ottesen, O., Maguire, J., Zhuravleva, N., Karlsen, A., Rebours, C. and Fitzgerald, R.D. (2012) Embryonic development in ballan wrasse *Labrus bergylta*. *Journal of Fish Biology* 81, 1101–1110.

Deady, S. and Fives, J. M. (2009) Diet of ballan wrasse, *Labrus bergylta*, and some comparisons with the diet of corkwing wrasse, *Crenilabrus melops*. *Journal of the Marine Biological Association of the United Kingdom* 75(03), 651.

Dipper, F.A., Bridges, C.R. and Menz, A. (1977) Age, growth and feeding in the ballan wrasse *Labrus bergylta* Ascanius 1767. *Journal of Fish Biology* 11(2), 105–120.

Dipper, F. A. (1981) The strange sex lives of British wrasse. *New Scientist*, 90(1253), 444–445.

Dipper, F.A. and Pullin, R.S.V. (1979) Gonochorism and sex-inversion in British *Labridae* (Pisces). *Journal of Zoology 187*(1), 97–112.

Doupé, R.G. and Lymbery, A.J. (2000) Managing translocations of aquatic species. *Aquaculture Research* 31, 151–156.

Erisman, B. E., Petersen, C. W., Hastings, P. A. and Warner, R. R. (2013) Phylogenetic perspectives on the evolution of functional hermaphroditism in teleost fishes. *Integrative and Comparative Biology* 53(4), 736–54.

Fraser, A.J., Gamble, J.C., and Sargent, J.R. (1988) Changes in lipid content, lipid class composition and fatty acid composition of developing eggs and unfed larvae of cod (*Gadus morhua*). *Marine Biology* 99, 307–313.

Gagnon, M.M., (2006) Serum biliverdin as source of colouration upon sexual maturation in male blue-throated wrasse *Notolabrus tetricus*. *Journal of Fish Biology*, 68(6), 1879–1882.

Godwin, J., Luckenbach, J. A., and Borski, R. J. (2003) Ecology meets endocrinology: Environmental sex determination in fishes. *Evolution and Development* 5(1), 40–49.

Grant, B., Davie, A., Taggart, J.B., Selly, S.L., Picchi, N., Bradley, C., Prodohl, P., Leclercq, E. and Migaud, H., (2016a) Seasonal changes in broodstock spawning performance and egg quality in ballan wrasse (*Labrus bergylta*). *Aquaculture* 464, 505–514.

Grant, B., Picchi, N., Davie, A., Leclercq, E. and Migaud, H., (2016b). Removal of the adhesive gum layer surrounding naturally fertilised ballan wrasse (*Labrus bergylta*) eggs. *Aquaculture* 456, 44–49.

Hamre, K., Nordgreen, A., Grøtan, E. and Breck, O. (2013) A holistic approach to development of diets for Ballan wrasse (*Labrus bergylta*) – a new species in aquaculture. *PeerJ, 1*, p.e99.

Johnson, M.S., (2000) Measuring and interpreting genetic structure to minimize the genetic risks of translocations. *Aquaculture Research* 31, 133–143.

Lamm, M. S., Liu, H., Gemmell, N. J., and Godwin, J. R. (2015) The Need for Speed: Neuroendocrine Regulation of Socially-controlled Sex Change. *Integrative and Comparative Biology* 55, 307–322.

Leclercq, E., Grant, B., Davie, A. and Migaud, H. (2014) Gender distribution, sexual size dimorphism and morphometric sexing in ballan wrasse *Labrus bergylta*. *Journal of Fish Biology* 84(6), 1842–1862.

Lipej, L., Orlando-Bonaca, M., Ozebek, B. and Dulčić, J. (2009) Nest characteristics of three *labrid* species in the Gulf of Trieste (northern Adriatic Sea). *Acta Adriatica* 50(2), 139–150.

Mazorra, C., Bruce, M., Bell, J.G., Davie, A., Alorend, E., Jordan, N., Rees, J., Papanikos, N., Porter, M. and Bromage, N. (2003) Dietary lipid enhancement of broodstock reproductive performance and egg and larval quality in Atlantic halibut (*Hippoglossus hippoglossus*). *Aquaculture* 227(1), 21–33.

Migaud, H., Bell, G., Cabrita, E., McAndrew, B., Davie, A., Bobe, J., Herraez, M.P. and Carrillo, M. (2013) Gamete quality and broodstock management in temperate fish. *Reviews in Aquaculture*, *5*(s1), S194–S223.

Morel, G.M., Shrives, J., Bossy, S.F., and Meyer, C.G. (2013) Residency and behaviour rhythmicity of ballan wrasse (*Labrus bergylta*) and rays (*Raja spp.*) captured in Portlet Bay, Jersey: implications for Marine Protected Area design. *Journal of the Marine Biological Association of the United Kingdom* 93, 1407–1414.

Mudge, S.M., and Davenport, J. (1986). Serum Pigmentation in *Cyclopterus lumpus* L. *Journal of Fish Biology* 29(6), 737–745.

Muncaster, S., Andersson, E., Kjesbu, O.S., Taranger, G.L., Skiftesvik, A.B., and Norberg, B. (2010) The reproductive cycle of female Ballan wrasse *Labrus bergylta* in high latitude, temperate waters. *Journal of Fish Biology* 77(3), 494–511.

Muncaster, S., Norberg, B. and Andersson, E. (2013) Natural sex change in the temperate protogynous Ballan wrasse *Labrus bergylta*. *Journal of Fish Biology* 82(6), 1858–1870.

Munday, P.L., Buston, P.M. and Warner, R.R. (2006) Diversity and flexibility of sex-change strategies in animals. *Trends in Ecology and Evolution* 21(2), 89–95.

Ottesen, O., Dunaevskaya, E. and D'Arcy, J. (2012) Development of *Labrus Bergylta* (Ascanius 1767) Larvae from Hatching to Metamorphosis. *Journal of Aquaculture Research and Development* 3(03), 3–6.

Pickova, J., Dutta, P.C., Larsson, P.O. and Kiessling, A. (1997). Early embryonic cleavage pattern, hatching success and egg lipid fatty acid composition: comparison between two cod stocks. *Canadian Journal of Fisheries and Aquatic Science* 54, 2410–2416.

Porteiro, F.M., Barreiros, J.P. and Santos, R.S. (1996) Wrasses (Teleostei: Labridae) of the Azores. ARQUIPÉLAGO. *Ciências Biológicas e Marinhas Life and Marine Sciences* 14, 23–40.

Potts, G.W. (1984) Parental behaviour in temperate marine teleosts with special reference to the development of nest structures. In: Potts, G.W. and Wootton, R.J., (eds.) *Fish reproduction: strategies and tactics.* Academic Press, London, pp. 223–224.

Quignard, J.-P. and Pras, A. (1986) Wrasses: Family *Labridae.* In: P. J. P. Whitehead, M.-L. Bauchot, J.-C. Hureau, J. Nielsen, and E. Tortonese (eds.), *Fishes of the north-eastern Atlantic and the Mediterranean.* Paris: UNESCO. pp. 919–942.

Quintela, M., Danielsen, E.A., Svåsand, T., Knutsen, H., Skiftesvik, A. B., and Glover, K.., (2014). Isolation and characterisation of twenty microsatellite loci for the ballan wrasse, *Labrus bergylta. Conservation Genetic Resources* 6, 425–428.

Robalo, J.I., Castilho, R., Francisco, S.M., Almada, F., Knutsen, H., Jorde, P.E., Pereira, A.M., and Almada, V.C. (2011) Northern refugia and recent expansion in the North Sea: the case of the wrasse *Symphodus melops* (Linnaeus, 1758). *Ecology and Evolution* 2, 153–164.

Robertson, D.R. and Choat, J.H. (1974) Protogynous hermaphroditism and social systems in *labrid* fish. In: Proceedings of the second international symposium on coral reefs Vol. 1, pp. 217–225.

Ross, R.M. (1983) Annual, semilunar and diel reproductive rhythms in the Hawaiian labrid *Thalassoma duperrey. Marine Biology* 72, 311–318.

Sargent, J.R., Tocher, D.R., and Bell, J.D. (2002) The lipids, In: Halver, J.R. and Hardy R.W. (eds.) *Fish Nutrition*, 3rd edition, Academic Press, San Diego, CA, pp. 181–257.

Semsar, K., and Godwin, J. (2003) Social influences on the arginine vasotocin system are independent of gonads in a sex-changing fish. *The Journal of Neuroscience*: The Official Journal of the Society for Neuroscience 23(10) 4386–4393.

Sjolander, S., Larson, H. and Engstrom, J. (1972) On the reproductive behaviour of two labrid fishes, the ballan wrasse (*Labrus bergylta*) and Jago's goldsinny (*Ctenolabrus rupestris*). *Revue du comportement animal* 6, 43–51.

Sundt, R.C., and Jørstad, K.E. (1998) Genetic population structure of goldsinny wrasse, *Ctenolabrus rupestris* (L.), in Norway: implications for future management of parasite cleaners in the salmon farming industry. *Fisheries Management and Ecology* 5, 291–302.

Talbot, C., Medeiros, M.V. and Davie, A. (2012) *In vivo gender determination in captive ballan wrasse (Labrus bergylta Ascanius 1767).* Final Report to the Scottish Salmon Producers Organisation, 54.

Taylor, M.H. (1984) Lunar synchronization of fish reproduction. *Transactions of the American Fisheries Society* 113, 484–493.

Thangstad, T. (1999) *Spatial and temporal distribution of three wrasse species (Pisces: Labridae) in Masfjord, western Norway: habitat association and effects of environmental variables.* MSc Thesis, University of Bergen.

Tocher, D.R., Fraser, A.J., Sargent, J.R., and Gamble, J.C. (1985) Fatty acid composition of phospholipids and neutral lipids during embryonic and early larval development in Atlantic herring (*Clupea harengus* L.). *Lipids* 20, 69–74.

Treasurer, J. W. (1994) The distribution, age and growth of wrasse (*Labridae*) in inshore waters of west Scotland. *Journal of Fish Biology* 44(5), 905–918.

Villegas-Ríos, D., Alonso-Fernández, A., Dominguez-Petit, R., and Saborido-Rey, F. (2013a) Intraspecific variability in reproductive patterns in the temperate hermaphrodite fish, *Labrus bergylta. Marine and Freshwater Research* 64, 1156–1168.

Villegas-Ríos, D., Alonso-Fernández, A., Fabeiro, M., Banon, R., and Saborido-Reu, F. (2013b) Demographic variation between colour patterns in a temperate protogynous hermaphrodite, the Ballan wrasse, *Labrus bergylta. PLOS ONE* 8, e71591.

Villegas-Ríos, D., Alos, J., March, D., Palmer, M., Mucientes, G., and Saborido-Rey, F. (2013c) Home range and diel behaviour of the ballan wrasse, *Labrus bergylta*, determined by acoustic telemetry. *Journal of Sea Research* 80, 61–71.

Villegas-Ríos, D., Alonso-Fernández, A., Domínguez-Petit, R. and Saborido-Rey, F. (2014) Energy allocation and reproductive investment in a temperate protogynous hermaphrodite, the ballan wrasse *Labrus bergylta. Journal of Sea Research* 86, 76–85.

Warner, R.R. (1982) Mating systems, Sex Change and Sexual Demography in the Rainbow Wrasse, *Thalassoma lucasanum. Copeia 1982*, 653–661.

CHAPTER 3

Rearing of Ballan wrasse

Anette Lekva and Espen Grøtan

Marine Harvest Norway, Øygarden, Norway

3.1 Hatcheries and research groups

Marine Harvest Norway, located outside Bergen, began producing Ballan wrasse in 2009, and the goal is to produce 5 million juveniles each year. The facilities utilised were built in 2001, and were formerly used as a cod hatchery from 2002 to 2009. While producing wrasse the company has participated in different R&D projects, with the Norwegian Seafood Research Fund (FHF), Nofima AS, Institute of Marine Research (IMR) and National Institute of Nutrition and Seafood Research (Nifes).

Other hatcheries that rear wrasse, or have done until recently, are located at:

- Marine Harvest and Scottish Sea Farms, Machrihanish, Scotland (2010–present)
- Otter Ferry Seafish, Scotland (2010–present)
- Havlandet Marin Yngel, Norway (2014–present)
- Profunda, Norway (2010–2013)
- Nordland leppefisk, Norway (2010–2012)
- Cleanfish, Norway (2011–2012)
- FAI Aquaculture, Ardtoe, Scotland (2008–present)
- Nord University, Norway (research)
- Institute of Marine Research, Norway (research)

Various wrasse research and rearing manuals have been produced such as:

- The 'Production of Ballan wrasse (*Labrus bergylta*) – Science and practice' (Helland *et al.*, 2014) is a wrasse manual developed by scientists and partners from the industry. The three-year project was supported by the Norwegian Seafood Research Fund (FHF). The manual has prioritized the most important challenges in the industry, and includes summaries of several trials, protocols and recommendations for the farming of Ballan wrasse.

- EcoFish. This was part of an EU Northern Periphery Programme project involving Bodø University, University of Galway and Ardtoe Marine Laboratory (chapter 20, this volume).

3.2 Hatchery design

Marine Harvest Norway, Øygarden: The sea water intake is at 150 m depth, with an annual temperature variation of 7 to 9°C, and salinity is 34 ppt. The water is first filtered (40 µm), and then UV treated. In addition, there is an extra UV treatment in every section of the facility, as well as vacuum degassing. Heating of water is mostly done by heat exchanging, both with used water from the facility and cooling water from a nearby power station. If necessary, additional heating is facilitated by heatpumps or boilers. In the live feed section the water is further filtered (5 µm) and heated to 24–25°C. The different production areas of the facility are: live feed, broodfish, spawning, incubation, hatchery and nursery. The nursery and hatchery capacity is 25 m^3 ×16 tanks and 9 m^3 × 8, although the nursery period is mostly at two other Marine Harvest facilities at Stord and Rissa. There are eight spawning tanks (9 m^3 capacity), and six broodfish tanks (40 m^3).

3.3 Egg incubation and hatching

The Ballan wrasse spawns benthic sticky eggs, in portions over several weeks. Therefore felt mats are placed on the bottom of the spawning tanks for collecting eggs (Fig. 3.1). Each spawning tank contains up to 50 wrasse each where approximately 15 to 20% are males, quite similar to Scotland. The mats are collected daily during the spawning period. Some eggs are probably lost in the drain and between mats. The stripping of eggs and sperm, and elimination of egg stickiness, have been tested (Lein *et al.*, 2014); however, more trials are needed with timing of treatment and effects on egg and larval development before these methods are implemented in commercial wrasse production. The mats with eggs are hung vertically in incubators (Fig. 3.2), with flow-through water, at 11–12°C. Continuous light has a negative effect on hatching rates

Fig. 3.1 Spawning tanks with mats. Credit: Marine Harvest Norway, Øygarden

Fig. 3.2 Incubation tanks with mats. Credit: Marine Harvest Norway, Øygarden

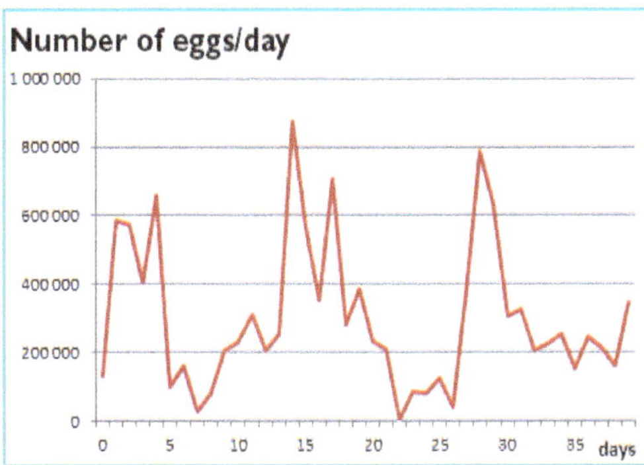

Fig. 3.3 Example of spawning rates, estimated number of eggs for 40 days; total number of eggs from eight tanks with 50 wrasse in each. Credit: Marine Harvest Norway, Øygarden

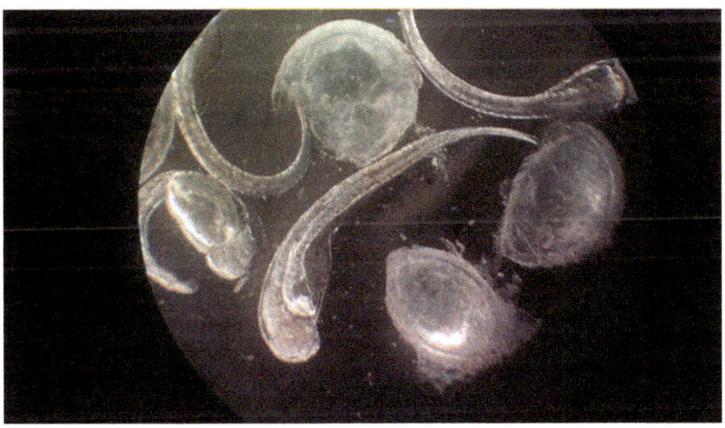

Fig. 3.4 Egg and newly hatched larvae with yolk sac. Credit: Marine Harvest Norway, Øygarden

(Rembelski, 2014), and therefore incubation light is timed to 12 hours of light and 12 hours of darkness. Hatching rates probably also vary with egg quality (Fig. 3.3). The eggs are almost 1 mm in diameter, and hatch after eight to nine days in the incubator; the larvae are 3.7 mm total length at hatching (Fig. 3.4).

Disinfection of eggs

The mats with eggs were hung directly in the larval tanks in an early production trial, which resulted in high hatching rates but poor survival of larvae through the rotifer phase (E. Grøtan, Marine Harvest, Norway, March 2017, personal communication). Survival of larvae improved considerably after introducing incubation tanks and better disinfection routines. Due to sticky eggs, disinfection is more difficult compared with non-sticky pelagic eggs such as cod, therefore daily cleaning of spawning tanks and disinfection of mats are especially important. The incubated mats with eggs are disinfected daily with bronopol as pyceze (Novartis), for the seven first days of incubation; 130 ml pyceze for 45 minutes with no flow. On the first day of incubation the mats are also treated with formalin (38%) for 45 minutes with no flow. The water turnover in 1 m^3 tanks is every one hour. The disinfection protocol at Otter Ferry is similar, i.e. formalin on day 0:150 ppm for one hour, and daily 100 ppm pyceze for one hour with no flow.

Transfer to larval tanks

The incubator is connected to a smaller collection tank when the eggs start to hatch. During the night the larvae flow slowly over to the collection tank. Egg hatching can also be controlled by physical stimulation at about 75–80 dd, scraping with a spatula or light stimulation (A. Chalaris, Scotland, March 2017, personal communication). The next day larvae are disinfected with pyceze to eliminate bacteria in the water or on the surface of the larvae (0.1 ml l^{-1} for 30 minutes). Afterwards the larvae are carefully scooped over to a rearing tank. In the collector, during the disinfection, the larvae are evenly distributed with an airstone, and then the number of larvae are calculated from at least five counted samples of 10 ml. The egg numbers are also estimated on the mats by eye. These estimates are not exact, and vary with experience, egg quality and hatching rates.

3.4 Hatchery period

Larval rearing, tanks and equipment

Tanks of different sizes (9 m^3 and 30 m^3) are used at the start of the feeding stage (Fig. 3.5). Starting feeding in larger volume tanks seems to reduce stress, although this needs to be investigated further. The tanks are fitted with a flat bottom sieve and a surface strainer. The bottom sieve is not used the first week. Water is slightly aerated using five airstones to ensure that the larvae are evenly distributed in the tank. However, only sufficient air or flow should be used, since the wrasse might be especially sensitive to high current and aeration (compared with cod). The water flow currently used in the beginning is 40 l min^{-1} (9 m^3 tank) (Table 3.1). Temperature, oxygen and water flow are recorded daily, and total gas is measured every week. The tanks are cleaned daily, utilising an automated cleaning arm on the bottom to collect faeces, dead larvae and sedimented clay. The surface skimmer is especially important to clean when the larvae inflate their swim bladder, starting at day seven to ten (see 'development and growth', this chapter). The

Fig. 3.5 Hatchery tank and equipment. Credit: Marine Harvest Norway, Øygarden

surface strainer is also cleaned daily. The bottom sieve is not in use the first two weeks (except for short flushing periods).

About 1,000,000–1,500,000 larvae are stocked in each tank (9 m^3), and it takes one to four days of egg collection to fill one tank, depending on spawning rates (Fig. 3.3). Ideally all the larvae should be the same age in one tank. Sudden changes in the larval environment must be avoided, since this may induce stress, and as it appears that the wrasse is especially sensitive to changes (see 'hiding and stress behaviour', this chapter).

Temperature and light in hatchery period

The yolk sac larvae are kept in the dark until feeding starts, four days after hatching. Continuous light is used as the Ballan wrasse is a visual feeder and the natural spawning period is in early summer. Different light intensities are used at different hatcheries (Marine Harvest: lux: 500-1000 and testing 75–150 in another room). Reduced light might reduce stress, therefore light intensity and different light regimes (day and night) will be investigated further. There have been some more light and temperature trials in the weaning and growing phase (see 'temperature and light regimes', this chapter).

Table 3.1 Rearing protocol for Ballan wrasse. Credit: Marine Harvest Norway, Øygarden

<table>
<tr><td>Days after hatching</td><td>0</td><td>4</td><td>10</td><td>20</td><td>30</td><td>40</td><td>50</td><td>60</td><td>70</td><td>80</td><td>90</td><td>100</td></tr>
<tr><td>Rotifers</td><td></td><td></td><td></td><td></td><td></td><td></td><td></td><td></td><td></td><td></td><td></td><td></td></tr>
<tr><td>Artemia</td><td></td><td></td><td></td><td></td><td></td><td></td><td></td><td></td><td></td><td></td><td></td><td></td></tr>
<tr><td>Dry feed</td><td></td><td></td><td></td><td></td><td></td><td></td><td></td><td></td><td></td><td></td><td></td><td></td></tr>
<tr><td>Flow (1/min – in 9 m^3) tank</td><td>40 l</td><td></td><td></td><td>60 l</td><td></td><td>70 l</td><td></td><td>90 l</td><td></td><td></td><td></td><td></td></tr>
<tr><td>Clay</td><td></td><td colspan="7">some hours before first feeding</td><td></td><td></td><td></td><td></td></tr>
<tr><td>Light</td><td></td><td colspan="11">some hours before first feeding</td></tr>
<tr><td>Bottom sieve</td><td colspan="6">0.5 mm</td><td colspan="5">1.5 mm</td><td>3 mm</td></tr>
<tr><td>Surface strainer</td><td colspan="6">450 µm</td><td colspan="6">1200 µm</td></tr>
</table>

Ballan wrasse larvae do not thrive at low temperatures. Feeding is usually started with temperatures from 12 to 14°C, that is a maximum of two degrees above incubator temperature, and increasing gradually to 16°C. Temperatures for the larvae and juveniles have been tested at 12 and 16 degrees, in combination with feeding trials (Breck *et al.* 2013). The results showed that both the larvae and juveniles grow much faster at 16°C than 12°C.

Turbidity

Clay is added continuously by pumps to increase the contrast between the background and live feed (approximately 1 kg clay per day per 9 m^3 tank). This improves feeding and improves the distribution of larvae in the water column. Algal paste was first used as green water, but trials early in the production showed that clay had similar or better results (E. Grøtan, Marine Harvest, Norway, March 2017, personal communication), and is cheaper and easier to use. The clay also has less biological load compared to fresh or frozen algae. In addition, clay may to some extent reduce the bacterial load (Attramadal *et al.*, 2012). Live algae would probably be the best alternative, but it is expensive and difficult to cultivate in large volumes. The duration and amount of clay have been tested in different regimes (0, 30 and 60 days). No clay affected the fish negatively (high mortalities), but it is difficult to conclude which is better between 30 and 60 days. In addition to improving feeding, clay might also give some concealment for the wrasse, and help reduce stress, although this should be examined further. However, some other hatcheries continue to use greenwater techniques with microalgae, such as FAI Ardtoe.

Disinfection and cleaning

The daily cleaning of tanks and equipment is important to ensure good water quality (see 'tanks and equipment', this chapter). Pyceze (Bronopol, Novartis Animal Health) is used to disinfect rotifers and artemia, and is mixed in the water in the larval tanks. Wrasse seem to have lower tolerance to pyceze than cod (E. Grøtan, Marine Harvest, pers. comm.). However, pyceze has been shown to be an effective treatment with some developmental stages at lower concentrations and with careful distribution in the tanks (around 30–40 ppm). That is, mixing the pyceze with water before it is slowly added to the tanks, and with no changes in the flow. There are different disinfection strategies of water in the larval tanks; added continuously to the water to prevent bacterial overload, or added as a treatment when necessary, for example if the larvae stop eating or if there seems to be another problem with the larvae. Regular pyceze treatments also seem to keep the tanks cleaner than tanks without them, as it appears to deter clay or other substances from attaching to the tank walls.

Live feeding and feeding regime

In the present feeding regime Ballan wrasse larvae are first fed enriched rotifers, *Brachionus sp.*, then *Artemia sp.* and then the juveniles are weaned to dry feed (Table 3.1). The optimal feeding regime and the length of co-feeding have still to be refined. However, higher temperatures gave faster larval growth and then also reduced the live feed period (Breck *et al.*, 2013). Cultivated copepods (*Acartia tonsa*) gave faster growth (Dahle *et al.*, 2014) and stress tolerance (Øie *et al.*,

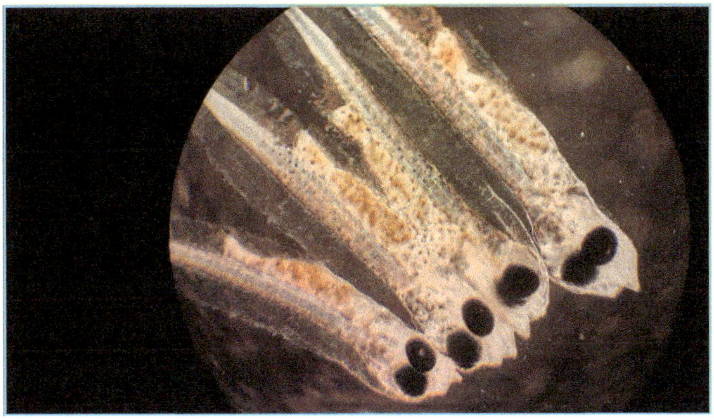

Fig. 3.6 Larvae with rotifer in gut. Credit: Marine Harvest Norway, Øygarden

2014), and wrasse larvae preferred to prey on copepods compared with rotifers (Dahle *et al.*, 2014). Although intensive production of copepod eggs is successful, the production is still insufficient to replace rotifers in a commercial wrasse hatchery. Live feed is stored in a cooling tank (4°C), to increase the shelf life of the live feed, and is pumped to the larval tanks in four meals in one day. In the current feeding regime one tank is fed 16,000 rotifers per litre, per meal, overfeeding in the beginning, to ensure that the larvae can encounter the feed. Gut filling is inspected regularly, and should be almost 100% shortly after start feeding (Fig. 3.6). Artemia is co-fed for two to five days, and the amount increases with increasing larval size.

Production of live feed

The rotifer species Brachionus and the crustacean Artemia sp. (Fig. 3.7) are commonly used as live feed in aquaculture since they can be cultivated in high densities and are available in large quantities in a fairly short period of time (Lavens and Sorgeloos, 1996). Brachionus species such as Nevada, Austria and Cayman are commonly used in hatcheries. Mixed cultures should be avoided since they have different sizes, growth rates and different preferences for salinity and temperatures. Rotifers can be cultivated using different methods; batch, continuous or a

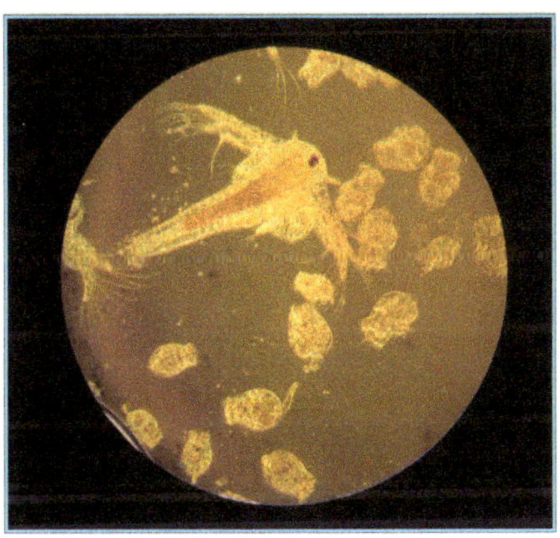

Fig. 3.7 Artemia sp. and Brachionus sp. magnified. Credit: Marine Harvest Norway, Øygarden

combination of these. Feeding and cultivation methods will influence the nutritional value and the bacterial load, and small changes may cause crashes in the production, therefore good washing and production procedures are important. Rotifers are often fed freshwater chlorella or yeast and then enriched with a nutritious diet before being fed to the larvae; several hatcheries are using Multigain (Biomar).

Artemia are hatched from resting cysts, so there is no need to maintain a culture as with the rotifers, only hatching and then enriching before the Artemia are fed to the larvae. Artemia cysts and other products needed for Artemia production are commercially available (INVE).

Weaning on dry feed and feeding

Feeding Ballan wrasse with dry feed is challenging as this species is very particular when it comes to taste and texture of the feed. It seems important to have a distinct taste of shrimp or krill and as little taste of fish as possible. The feed also needs to be relatively soft and easy to digest as the wrasse has a short intestine and no stomach (Chapter 10, this volume). Problems with digesting the feed have led to increased mortalities and deformities in production (Chapter 13, this volume).

As the Ballan wrasse is especially difficult to wean, there have been several weaning trials at Marine Harvest Norway (2010–2014), in collaboration with NIFES, Nofima and IMR. Shrimp meal is necessary for a high initial feeding response (Kousoulaki *et al.*, 2015), and replacing regular fish meal with cod muscle meal, which has a more neutral taste, makes the wrasse wean to artificial diets with high survival, but also increases the risk of developing severe head deformities (Kousoulaki *et al.*, 2014). The deformities are a serious welfare issue and might also reduce the potential for cleaning salmon lice in the cages. Therefore it is important that the wrasse is given a diet containing fishmeal, which has a better nutritional quality than cod muscle meal, shortly after weaning is finished. OFS and FAI Ardtoe have been weaning wrasse with the Nofima diet, but no severe head deformities were noticed. However, rearing at these locations was at lower temperatures and with lower growth rates than at MHN Øygarden.

In the current hatchery feeding regime the wrasse is weaned to a shrimp and cod muscle meal-based feed made by Nofima. These meals are of very high quality and delivered by Seagarden AS. This weaning feed is used for ten days; OFS and FAI Ardtoe use a longer weaning period. While weaning on Nofima, artemia feeding is gradually reduced, before switching to Otohime diet from Marubeni Nisshin. Otohime contains a high proportion of krill as well as fish meal and squid meal. Ballan wrasse get more tolerant to fish meal after weaning is completed and krill meal works well as a good substitute for shrimp meal.

Otohime is fed from 60 days after hatching, until the fish goes out to the salmon cages. In trials this feed gave the best survival, growth and fewer deformities (Kousoulaki *et al.*, 2014). This is mainly because the Ballan wrasse prefers the taste, but probably also partly because it is cold extruded and granulated, which could give higher digestibility of nutrients (Kousoulaki, 2017, unpublished). On the other hand, this feed seems to affect the water quality more than other feeds tested, and contaminated water seems to increase problems with fin damage, especially when slightly overfeeding. Other on-growing diets are used by other hatcheries, such as Biomar and Skretting.

Several post-weaning diets have been tested. Otohime from Marubeni Nisshin had less head deformities compared to other feeds and with the same survival (Fig. 3.8).

Fig. 3.8 Wrasse with and without head deformities. Credit: Karen Kvalheim, Marine Harvest Norway, Øygarden, 2011

The composition of diets for Ballan wrasse has been developed and improved by Hamre *et al.*, 2013 and Kousoulaki *et al.*, 2015. However, on-growing performance is still not satisfactory, and further development of a nutritionally complete diet that suits the short stomach and basic pH digestive system of wrasse is needed.

High quality shrimp and krill meals are costly and experiments have been carried out to try to replace them as main attractants in the feed for Ballan wrasse. Squid meal, blue mussel meal and cheaper shrimp meals (made from head and shells instead of the whole shrimp) did not give such a favourable feeding response (Breck *et al.*, 2013).

Different feeding regimes have been tested (that is continuous, portions and self-feeding), and continuous feeding gave slightly higher survival, although the wrasse seemed to be adaptable to different regimes without affecting growth (Helland *et al.*, 2014). Moist feeds may induce higher feeding response, but not necessarily higher growth (Kousoulaki *et al.*, 2014). The present feeding regime at commercial hatcheries is continuous feeding, with some hand feeding to ensure good appetite. Appetite control is slightly difficult due to large variation of behaviour and fish size in the same tank, some fish clump in the surface and might seem hungry all day, and at the same time some fish swim deeper and seem satiated.

3.5 Nursery period

Ballan wrasse juveniles are not strong swimmers like cod or salmon and therefore have a low tolerance for high water current during most of the nursery period. This makes self-cleaning of tanks difficult, and daily manual cleaning of tanks is needed, preferably with automated cleaning arms at the bottom. Feeding is mostly performed with automatic feeders; see feeding strategies in 'weaning on dry feed and feeding', this chapter.

Temperature and light regimes

Temperature is one of the most important factors for growth, and in production the optimum growth is found at around 16°C, both in the larval and nursery period. Higher temperatures may increase growth further, but this has yet to be tested and is dependent on heating capacity.

However, a weaning trial in Ballan wrasse did demonstrate best growth and survival at 16°C where temperatures ranged from 12 to 18°C (Opstad and Skiftesvik, unpublished, mentioned in Helland *et al.*, 2014). Good growth is achieved by maintaining the temperature above 14°C and temperatures lower than 12°C should be avoided. Larger juveniles, above 20 g, can be held at 8°C and are still healthy, but the growth rate is low. Fish exposed to ambient winter water temperature (<7°C) lost weight and elevated mortalities became apparent (Antonios Chalaris, Otter Ferry, February 2017, personal communication).

Too intense a light can be stressful or even lethal, but too low a light can give slower growth in fish species that are visual feeders (Boeuf and Le Bail, 1999). For Ballan wrasse very low light (below 20 lux) in the artemia and weaning phases reduced growth and survival (E. Grøtan, Marine Harvest, personal communication). After weaning, low light intensity (below 60 lux) seemed to improve the fin condition, and this could therefore be beneficial. Transfer from continuous light (24:0) to discontinuous light (16:8) reduced growth (Helland *et al.*, 2014). There were no clear effects on skin and fish health when continuous light was used, except for some increased erosion on the dorsal fin. Abrupt changes in the light regime are most likely quite stressful and should be avoided. It should also be noted that Ballan wrasse 'sleep' on the bottom of the tank if the lights are turned off completely, something that might lead to loss of fish in tanks with high numbers due to suffocation. The continuous light regime is therefore still utilised because of this and the above mentioned trials. However, different light regimes should be tested further.

Hiding and stress behaviour

Clumping behaviour starts around 40 to 50 days after hatching and continues through the nursery phase (Fig. 3.9). Clumping is believed to be natural stress behaviour for this species when there are no other places to hide. Several types of hides have been tested such as artificial kelp, shreds of tarpaulin, branched pines, PVC tubes and plates. In small-scale trials these seemed to calm the fish and gave improved production performance. However, when applied in production tanks with high numbers of fish no positive effect on production was seen (E. Grøtan, Marine Harvest, personal communication). All obstacles in the tanks impair the water flow and tank cleaning and increase the bacterial load. In addition, the cleaning of shelters gives a greater tank workload, and is most likely stressful for the wrasse. Designing optimal shelters with anti-fouling could be a way to reduce stress and anti-fouling surfaces have been tested (Dahle *et al.*, 2014).

Fig. 3.9 Clumping wrasse in fish tank. Credit: Marine Harvest Norway, Øygarden

Transfer, grading and stocking of juvenile wrasse

The transfer and grading is a challenge due to the low tolerance of Ballan wrasses to change and disturbance. Aqui-S® (Isoeugenol 540 mg ml^{-1}) has been tested at several stages and seems to lower stress, at least on broodfish and with wrasse transportation, where the stress reduction is notable (H. Glosvik, Marine Harvest, Norway, 2017, personal communication; see also transport of wrasse, Chapter 16, this volume).

Grading machines (Milanese Alpha, www.milaneseitalia.com) seem to work well with wrasse from around 1 g but earlier grading must be done by grid grading with a manual box. The optimum frequency of grading has still to be determined. Regular grading with other fish species from small sizes often reduces aggression and fin damage, for example in cod *Gadus morhua* (Treasurer *et al.*, 2011). This effect is not so clear for smaller juveniles of Ballan wrasse, at least not with the grading strategies tested at Marine Harvest. The negative effects of stress related to the grading process seem to be greater than the positive effects of getting a more uniform size of fish. Larger fish are less affected by handling stress and fish exceeding 10 g can be graded with no great problems.

There have been no significant differences in survival and fin damage between fish in tanks that have been graded and those in non-graded tanks in grading trials. The smallest fish in the tank is normally the fish that succumbs, regardless of whether the fish are graded or not, and this may indicate aggression. However, more frequent grading to further reduce the size difference might be a better strategy.

Stocking densities in nursery tanks probably need to be at a certain level. Too low densities in the tanks seem to be stressful for the fish, since the wrasse seem to need to be able to clump when there are no places to hide, and 15–20 kg per m^3 is currently recommended. However, this can potentially be increased.

3.6 Production results and discussion

Development and growth

The developmental stages of Ballan wrasse from embryo to hatching and to metamorphosis have been studied by D'Arcy *et al.* (2012) (Ireland Chapter 20, this volume) and Ottesen *et al.* (2012). The larvae were divided in four age-related stages; yolk sac (0–9 dph), preflexion (10–25 dph), flexion (26–33 dph) and postflexion (34–49 dph). In commercial hatcheries growth of Ballan wrasse larvae is slightly difficult to monitor as there can be different ages in the same tank (see 'larval rearing' and 'egg incubation'), and slightly different temperatures between tanks. However, monitoring growth and development is important to compare and improve results.

The time and duration of the window for swimbladder inflation has not been confirmed in Ballan wrasse, although some observations are available. Swimbladders were first observed dilated at 7 dph, and inflated at 13 dph, by Ottesen *et al.* (2012). At Marine Harvest inflated swim bladders are first observed from around 7 dph. Not all larvae have inflated swimbladders, and the inflation success is variable at 10–40 dph, but after 30–40 dph swimbladder filling is good, indicating some mortality. In many finfish cultures there are several biotic and abiotic factors

Fig. 3.10 Red and blue/green wrasse. Credit: Marine Harvest Norway, Øygarden

that can be manipulated to achieve higher swimbladder inflation and reduce mortality during the early larval stages (Woolley and Qin, 2010). Surface skimmers to ensure a clean water surface are especially important for the larvae to fill their swimbladders (Woolley and Qin, 2010). Light period, light intensity, temperature, water turbulence and salinity also affect swimbladder inflation in other species, and should be further investigated with Ballan wrasse.

Growth rates are very different between different individuals. From the early larval stage the large difference in individual size becomes very clear.

Photoperiod, temperature, initial length, initial swim bladder inflation, and survival from different hatcheries and research facilities are presented (Table 3.2). It should be noted that direct comparisons in performance and benchmarking are difficult because the hatcheries have different environmental and functional systems.

During the period of feeding artemia some of the juveniles become blue in colour (Fig. 3.10). These fish are weak and have a slower growth rate and higher mortality than the typical red-coloured fish. Most of lower quality fish are lost in weaning or are graded out at a later stage because of low performance. However, most of the blue fish change to a red colour when weaned to dry food (Hovland, 2016). The reasons behind these colour changes are not known, and it may be caused by nutrition, environmental factors or genetics. Strongly blue-coloured wrasse are probably undergoing some sort of challenges. Clark *et al.* (2016) identified the Ballan wrasse plasma pigment biliverdin, which causes the blue–green coloration in plasma fractions of Ballan wrasse. Biliverdin has some effect on external coloration. Clark *et al.* also found a potential association with gender inversion processes.

Health issues

Mortalities are highest during the first weeks of startfeeding (Fig. 3.11), and feed uptake varies considerably during this period. These high mortalities are probably connected to live feed, or water quality. Rod-shaped bacteria (probably Vibrio species) are often seen in large numbers in the larval gut in the first few weeks. Bacterial problems might be reduced by introducing probiotics, by improving the disinfection routines of the water (see 'disinfection and cleaning', this chapter), or improving live feeding strategies, and this should be investigated further. There might also be some mortalities due to failed swimbladder filling.

Table 3.2 Photoperiod, temperature, initial length, initial swim bladder inflation and survival in ballan wrasse rearing (calculated from weighing at grading and larvae stocked in tanks). Note: for the Rembelski data, 5 incubation temperatures are quoted, and the incubation period, hatch rate and larval length at each temperature are quoted in the respective columns.

Reference	Incubation temp.	Incubation photoperiod: hours light (L) and dark (D)	Incubation duration, days from fertilization until hatching and total degree days	Hatching success/ rate %	Larvae length at hatching, total length mm ± SD	Photo-period in larval tanks; hours light (L) and dark (D)	Rearing temp. °C	First observed swim-bladder SB	SB inflation success	Survival after weaning at 100 dph (0.5 g)	Water system
Marine Harvest Norway, Øygarden	11.3°C	12L:12D	about 9 days, 100°C	unknown	3.7	24L:0D from 4 dph	start at 12, increase to 16°C after 4–20 days	7 dph (first observation)	50–90%	10–20%	Flow through
Otterferry, Scotland		24L:0D	about 80 degree days	95%	3.94 ± 0.4 (SL)		10–12°C	7–8 dph	>90%	5–23%	Flow through
Havlandet Marin Yngel AS	11.5°C	12L:12D	8–9 days, 88–100 degree days	unknown	Average of 3.9	24L:0D	start at 12, increase to 16 over 14 days	8 dph	60%+	unknown	
FAI Ardtoe	8.5°C started to hatch 8/5/2013 and hatching over 5 days to 11/5/2013	0L:24D	85 dd +	30–50%	4.52 ± 0.06	24L:0D from 3 dph	10–12°C			c.10%	
Ottesen et al. (2012)	11°C	0L:24D	7 days, 72°C		3.9	18h light (1100 lux) and 6h total darkness	increasing from 11–14°C over 4 days	7 dph first observed dilated swim bladder			
Rembelski, L. (2014)	8 – 10 – **12** – 14 – 16	12L:12D	x – 10 – **8** – 5 – 4 days	0 – 16 – **55** –35 – 15	x – 3.57 – **3.28** – 3.13 – 3.15	–	–	–	–	–	
	8 – 10 – **12** – 14 – 16	24L:0D	x – 10 – 8 – 5 – 4 days	0 – 2 – 6 – 20 – 16	x – 3.42 – 3.39 – 3.54 – 2.85	–	–	–	–	–	
Ireland					3.63 ± 0.12	16L:8D	12.9 ± 1.0				
Norway					3.64 ± 0.05	18L:6D	13.9 ± 0.9				
D'Arcy (2012) Scotland, batch 1					4.83 ± 1.20	18L:6D	14.4 ± 1.3	unknown	unknown	unknown	
Scotland, batch 2					4.10 ± n/a	18L:6D	14.7 ± 1.2	unknown	unknown	unknown	
Scotland, batch 3					4.07 ± 0.05	18L:6D	14.5 ± 0.3	unknown	unknown	unknown	

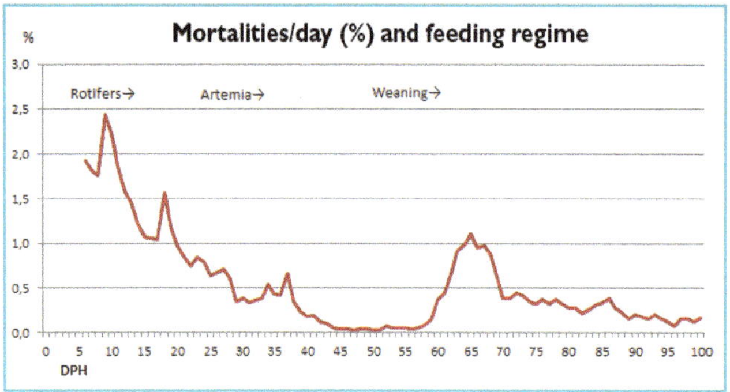

Fig. 3.11 Average daily recorded mortalities on days post hatch (DPH) and feeding regime from one cycle in Marine Harvest Norway (this is consistent with almost all cycles). Credit: Marine Harvest Norway, Øygarden

One cycle had especially high mortalities early in the larval stage at Marine Harvest Norway in 2015. Almost all larvae had a loss of appetite, energy and intestine changes, and then high mortality around 20 dph (Glosvik, 2016). Although the cause of this was not determined, a change in the bacterial load in the live feed or the larval rearing water were the most likely causes of the problem as a complete deep cleansing of the facilities seemed to normalise production in the next cycle. This high mortality incidence at 20 dph and the typical high mortality in the first weeks demonstrate that the larvae are very sensitive at this stage, and this stage should be examined further.

Mortalities are somehow lowered after introducing artemia. There is again a mortality peak when weaning, indicating the scope for further improving of weaning strategies (see weaning on dry feed and feeding, this chapter).

Deformities have been examined by radiography in juveniles of 4–8 g in size, and all the early production cycles at MH had a high occurrence of malformations in the spine, as for example lordosis (Bæverfjord, 2012). Lordosis may be caused by high water current in the tank, as with other species such as sea bass. Malformations may also be caused by nutritional factors or incorrect temperature regime. Temperature is an important factor in the early stages with salmon (Wargelius *et al.*, 2005) and should also be considered with wrasse. There have also been incidents with head and jaw deformities caused by the lack of bespoke on-growing feeds during these first production cycles. There have been fewer of these problems in the later production cycles and deformities are no longer considered a big issue.

Feed composition and production methods are probably the main challenges at several stages, and more research on optimal feed composition would probably improve fish performance in several areas (see weaning on dry feed and feeding).

Erosion of pectoral and tail fins is a problem for some Ballan wrasse juveniles and this starts just before weaning. Fin problems are likely to be caused by bacterial problems, resulting from poor water quality or nutritional deficiencies. Stress might also affect fin status as there are indications that low light improves the condition of pectoral fins (Breck *et al.*, 2013).

Stress is one of the major problems in wrasse rearing, from early larvae when clumping starts, and throughout the nursery phase. This stress results in a lower tolerance for other production operations, and the lower tolerance for pyceze might also indicate a high stress level. There are several ways to reduce stress, with environmental techniques such as hides, low flow rates, the use of automatic cleaners, less handling and lower light intensity, and also an effective dimming system that might reduce stress from abrupt light changes. Better nutritional quality of the feed can also give the wrasse higher stress tolerance (Kjørsvik *et al.,* 2014).

Breeding is probably the best solution for stress reduction, survival and growth in the long term. Marine Harvest Norway and Havlandet Marin Yngel started selecting individuals for breeding in 2015. Progress is thought to be noticable after only one or two generations of breeding. Growth is most likely to be the most relevant factor to achieve domestication as the larger fish appear less stressed, and the largest fish (30 g) in a group are ready for transfer to sea cages at around nine months after hatching (Hovland, 2016). Breeding will result in higher survival and shorter production time.

Technical issues

Incubation of eggs on mats produces a high workload for several months in every production cycle. Stripping or induced spawning would probably reduce the workload considerably (see egg incubation and hatching, this chapter). Breeding cycles could be better planned, and there could be more egg production cycles, so this could be an important priority for trial work.

Individual temperature control of larval rearing tanks would make the growth difference between tanks less as there is a large difference between incubation temperature and optimal larval rearing temperature, and since it currently takes some weeks to fill several tanks with larvae, hence the oldest tank will grow slower in the beginning than the youngest (see temperature and light in hatchery period, this chapter).

Light regimes should be investigated further as lower light intensity might reduce stress and improve fin status (see light and temperature in nursery period, this chapter).

Hides reduce stress, however, anti-fouling would also be needed to reduce the bacterial load and the stress caused by cleaning the hides (see hiding and stress behaviour this chapter).

Counting and grading of larvae should be performed early to be able to feed them more accurately, as feeding to appetite is difficult, due to clumping behaviour, and over-feeding might result in increased fin erosion. The optimal time to grade and how to grade without stressing the wrasse should be tested further.

Production results and future activity

Ballan wrasse for use in salmon cages are produced in two sizes in Marine Harvest Norway; 20 g Ballan wrasse are stocked with smolts and 50 g fish are used with larger salmon later in the production cycle. The rearing time from egg hatch to stocking is nine to 12 months for 20 g Ballan wrasse, and 12–18 months for fish of 50 g.

The goal is to produce enough cleaner fish to be independent of wild-caught wrasse, including both Ballan wrasse and lumpfish production. Marine Harvest Norway has increased the Ballan wrasse production in Øygarden from a few thousand 0.5 g fry in 2009 to about 4 million in 2016. How many of these fry go to the sea depends on size needed and at what time. The increase in rearing production has been achieved through both increased survival and production capacity. The number of facilites increased from one in 2009 in Øygarden to three in 2014, with on-growing capacity at Stord and Rissa. Stord also established a Ballan wrasse hatchery in 2015. Work started on construction of new on-growing facilities in 2016 to increase both the hatchery and on-growing capacity in Øygarden. Marine Harvest also rented on-growing capacity in Tjeldbergodden in 2016.

Currently there are two production cycles each year, with two broodfish groups with two different light regimes and natural spawning. Individuals are already selected for breeding and a breeding programme will be established. MHN and MHS remain commited to Ballan wrasse production and, with good farming techniques, the species can be used over repeated production cycles in the future.

References

Attramadal, K.J.K., Tøndel, B., Salvesen, I., Øie, G., Vadstein, O. and Olsen, Y. (2012) Ceramic clay reduces the load of organic matter and bacteria in marine fish larval culture tanks. *Aquacultural Engineering* 49, 23–34.

Bæverfjord, G. (2012) *X-ray examination of ballan wrasse, at Marine Harvest, Norway.* Nofima AS.

Boeuf, G. and Le Bail, P.Y. (1999) 'Does light have an influence on fish growth?' *Aquaculture* 177, 129–152.

Breck, O., Grøtan, E., Nordgreen, A., Kousoulaki, K., Opstad, I., Browman, H., Skiftesvik, A.B., Hamre, K. and Kvenseth, P.G. (2013) *Optimized production, nutrition and use of the cleaner fish ballan wrasse (Labrus bergylta).* Summary of NFR project 200523/S40. Marine Harvest Norway AS.

Clark, W.M., Leclercq, E., Migaud, H., Nairn, J. and Davie, A. (2016) Isolation, identification and characterisation of ballan wrasse *Labrus bergylta* plasma pigment. *Journal of Fish Biology* 89 (4), 2070–2084.

Dahle, S.W. and Hagemann, A. (2014) Feeding preference of ballan wrasse larvae at first feeding. *Production of ballan wrasse, science and practice.* The Norwegian Seafood Research Fund, Norway, pp. 49–50.

Dahle, S.W., Hagemann, A., Attramadal, Y., Stavrakaki, M.G., Evjemo, J.O. and Kjørsvik, E. (2014) Long term feeding with copepods. *Production of ballan wrasse, science and practice.* The Norwegian Seafood Research Fund, Norway, pp. 46–47.

Dahle, S.W., Hagemann, A. and Stenstad, P. (2014) Anti-fouling surface materials for ballan wrasse shelters. *Production of ballan wrasse, science and practice.* The Norwegian Seafood Research Fund, Norway, pp. 104–106.

D'Arcy, J. (2013) Studies on the biology of the ballan wrasse, *Labrus bergylta*. PhD thesis. The National University of Ireland, Galway.

D'Arcy, J., Dunaevskaya, E., Treasurer, J.W., Ottesen, O., Maguire, J., Zhuravleva, N., Karlsen, A., Rebours, C. and FitzGerald, R.D. (2012) Embryonic development in ballan wrasse *Labrus bergylta*. *Journal of Fish Biology* 81, 1101–1110.

Glosvik, H. (2016) Health and quality with farmed cleaner fish in Marine Harvest. Cleaner fish conference (www.fhf.no), Norway.

Hamre, K., Nordgreen, A., Grøtan, E. and Breck, O. (2013) A holistic approach to development of diets for ballan wrasse (*Labrus bergylta*) – a new species in aquaculture. *PeerJ* 1:e99 https://doi.org/10.7717/peerj.99

Helland, S., Wiborg, D.S., Hamre, K., Skiftesvik, A.B., Lein, I., Kousoulaki, K., Øie, G., Browman, H., Evjemo, J.O., Kjørsvik, E., and Grøtan, E. (2014) Best practices for farming of ballan wrasse. *Production of ballan wrasse, science and practice.* The Norwegian Seafood Research Fund, Norway, pp. 123–128.

Helland, S., Grøtan, E. and Noble, C. (2014) Effects of transfer from continuous light to different light regimes on growth, survival and fin and skin health. *Production of ballan wrasse, science and practice.* The Norwegian Seafood Research Fund, Norway, pp. 94–95.

Helland, S., Lein, I. Sæle, Ø., Lie, K.K., Kousoulaki, K., van Dalen, S.C.M, Klaren, P.H.M., Bakke, A.M. and Krogdahl, Å. (2014) Effects of feeding frequency on growth and gut health of Ballan wrasse juveniles. *Production of ballan wrasse, science and practice.* The Norwegian Seafood Research Fund, Norway, pp. 83–89.

Hovland, H. (2016) Challenges in ballan wrasse farming. Havlandet Marin Yngel, cleaner fish conference (www.fhf.no), Norway.

Kjørsvik, E., Øie, G., Hansen, M.H. and Sørøy, M.O. (2014) Handling stress tolerance test – sampling and analysis. *Production of ballan wrasse, science and practice.* The Norwegian Seafood Research Fund, Norway, pp. 52–53.

Kousoulaki, K., Grøtan, E., Kvalheim, K., Høstmark, Ø., Klinge, M., Bogevik, A. (2014a) Protein quality, commodity options and ballan wrasse weaning performance. *Production of ballan wrasse, science and practice.* The Norwegian Seafood Research Fund, Norway, pp. 96–100.

Kousoulaki, K., Grøtan, E., Bjelland, R., Kvalheim, M., Bogevik, A. and Skiftesvik, A.B. (2014b) Protein quality and feed technical quality effects on ballan wrasse on-growing performance. *Protein quality and feed technical quality effects on ballan wrasse on-growing performance.* The Norwegian Seafood Research Fund, Norway, pp. 101–102.

Kousoulaki, K., Bogevik, A., Skiftesvik, A.B., Jensen, P.A. and Opstad, I. (2015) Marine raw material choice, quality and weaning performance of ballan wrasse (*Labrus bergylta*) larvae. *Aquaculture Nutrition* 21, 644–654.

Lavens, P. and Sorgeloos, P. (1996) Manual on the production and use of live food for aquaculture. *FAO Fisheries Technical Paper.* No. 361. Rome, FAO. 295 pp.

Lein, I. and Barr, Y. (2014) Elimination of egg stickiness. *Production of ballan wrasse, science and practice.* The Norwegian Seafood Research Fund, Norway, pp. 26–33.

Lein, I., Helland, S. and Tveiten, H. (2014) Initial tests of procedures for stripping of eggs and sperm and induction of final maturation in ballan wrasse. *Production of ballan wrasse, science and practice.* The Norwegian Seafood Research Fund, Norway, pp. 16–19.

Ottesen, O.H., Dunaevskaya, E. and Arcy, J.D. (2012). Development of *Labrus bergylta* (Ascanius 1767) Larvae from Hatching to Metamorphosis. *Aquaculture Research and Development* 3:3.

Rembelski, L. (2014). *Temperature dependent egg development and impact of light regimes on incubation of ballan wrasse (Labrus bergylta) egg.* M.Sc. thesis. The University of Bergen, Norway.

Treasurer, J., Atack, T., Rolton, A., Walton, J. and Bickerdike, R. (2011) Social, stocking density and dietary effects on the failure of farmed cod *Gadus morhua. Aquaculture* 322, 241–248.

Øie, G., Dahle, S.W., Hagemann, A., Gagnat, M.R., Søry, M.O., Berg, M., Attramadal, K.J.K., Evjemo, J.O., Bardal, T. and Kjørsvik, E. (2014). Short term feeding with cultivated copepods and rotifers of different quality. *Production of ballan wrasse, science and practice.* The Norwegian Seafood Research Fund, Norway, pp. 44–45.

Wargelius, A., Fjelldal, P.G. and Hansen, T. (2005) Heat shock during early somitogenesis induces caudal vertebral column defects in Atlantic salmon (*Salmo salar*). *Development Genes and Evolution* 215, 350 357.

Woolley, L.D. and Qin, J.G. (2010) Swimbladder inflation and its implication to the culture of marine finfish larvae. *Reviews in Aquaculture* 2, 81–190.

CHAPTER 4

Improving the use of wrasse (Labridae) stocked in commercial sea cages with Atlantic salmon (*Salmo salar*) for sea lice control: Key features for a successful deployment

Carolina Gutierrez Rabadan

Marine Harvest Scotland, Rosyth, UK

Deployment of wrasse – key features for a successful deployment

This chapter focuses on all the significant aspects to take into consideration at the time the wrasse arrive on a commercial salmon sea site (normally by lorry or boat) to be used as a delousing control. It is important that there is an agreed delivery date between the transporter and the responsible person on site (normally the farm manager) before the arrival of the fish on the farm. The farm must have all the equipment ready on site such as hatchery hand nets, suitable and clean bins and lids, oxygen and sufficient aeration systems, oxygen, temperature, salinity and pH meters, ammonia analysis kit, seawater supply, hoses and water pumps.

In the case of farmed wrasse, fish should be acclimatised in the hatcheries to ambient water temperature, and also hides and feed around five to six weeks before they are deployed to commercial sea cages. For wild wrasse, the fisherman should give details of the site where the wrasse have been caught and their health should be checked on arrival on site before deployment. Any wrasse with lesions, unhealthy or moribund fish should be humanely euthanized on arrival. If the percentage of compromised fish is high, the fisherman should be asked to review the methods of catching and maintaining wrasse to ensure the health and welfare of the fish.

4.1 Unloading wrasse

The transporter should give the farm manager all the required documentation including the Animal Transport Certificate (ATC) before unloading the fish. The manager should also present a Cleaner fish Stock Record that includes fish numbers with average weights, stocking densities in each lorry tank, health status of the fish, source, stock, vaccination, treatments, starvation period, and measurements of the basic water quality parameters such as temperature, dissolved oxygen, salinity and pH. Once all documentation has been checked and found to be correct, the site manager should visually check the fish and ensure they are in good condition and ready for unloading on arrival. The site manager must also check the size of the fish and ensure that the minimum size of the net used on site matches with the size of the fish, so no escapees are possible.

A delivery stocking plan should be in place before the arrival, hence fish can be stocked at the percentages recommended by the cleaner fish manager. This is dependent on the sea lice challenge, the history of the farm, the location of the site and the environmental conditions such as tides, currents, and other factors.

All tanks on the lorry must be drained to a suitable level to reduce the volume of water and thus to facilitate the catch of the fish. There are two alternatives to transfer the fish to the cages: the lorry goes into a landing craft and the fish are transferred from the lorry to the pen directly through pipes; or, using fish nets, the fish are transferred in a first stage to bins, and then the bins are transferred to the pens by boat. The fish will be netted again or transferred directly into the pens with the help of a crane.

At the moment of the unloading, oxygen should be left during the draining of the tanks and until the fish are completely transferred out of the tank. After that, staff should prepare everything for the discharge. Ideally, the discharge should be done through pipes and with water pumps to minimise handling, which significantly stresses the fish. If not possible, fish must be hand-netted slowly and carefully into the bins using hatchery nets; each netting should not be more than a third full to avoid increase of pressure out of the water and they should not be out of the water for more than eight seconds. It is recommended that for these operations at least four people are available.

The stocking densities in each bin for the transfer should be also considered. A bin of 1000 l should never hold more than 1,500 farmed or wild wrasse. Once most of the fish are already into the bins, the hand net should be placed into the mucon valve and the remaining fish should be released to avoid mixing transfer water with fresh seawater. It is important that wrasse do not suffer from a thermic or a pH shock and that both waters – from the transfer and at the sea – are similar to each other. Wrasse can usually tolerate fluctuations of temperature and pH if they are progressive. Once the bin is filled with fish this should be placed immediately on the boat and transported to the cages with oxygen and aeration systems. Fish must be visually checked regularly for behaviour and activity until they are deployed. Oxygen levels must be monitored during

Fig. 4.1 Wrasse delivery into a salmon net cage. Wrasse are being held in a harvest bin, which is being used to transfer wrasse from the lorry tanks to the cages with the help of a crane.

the complete operation at all times, and never fall below 80% (80–130%) but also avoiding short-term hyper-oxygenation (>130%). Depending on the weather conditions, such as strong sunlight or windy conditions, bins should be covered with a lid during the transport from shore to the cages. Lids are important as they create shade and darkness areas in the bins where the wrasse can hide.

When discharging wrasse into seawater cages it is important that salmon are fed before wrasse are deployed, and so some possible aggression can be avoided. Ideally, wrasse should be deployed directly into the hides or as close as possible to them. If a crane is available on site, the bin containing the fish could be carefully lifted into the pen so wrasse can be tipped out gently by slowly lifting the base of the bin with the crane. If a crane is not available, fish should be transferred using hand nets and placed close to the hides, although such nets could be more stressful for them.

4.2 Deployment and acclimatisation of the wrasse

The conditioning or acclimatisation of farmed Ballan wrasse *Labrus bergylta* when transferred into seawater cages is one of the key features in achieving success. Several trials conducted at Marine Harvest Scotland Ltd by researchers from the University of Stirling have shown that Ballan wrasse are more active and have a better coverage of the area around the perimeter of a pen if they have had enough time to acclimatise, compared with non-conditioned farmed wrasse. As these findings were positive MHS uses conditioning pens when stocking farmed wrasse into seawater cages. Conditioning pens are normally between 6-12 m^3 and contain four hides made of artificial kelp in each corner along with four feed bags where agar blocks are supplied. Hides in the conditioning pen are crucial for welfare reasons and also to avoid aggression, as the wrasse is a very territorial species.

The main purpose of the conditioning pens (Fig. 4.2) is to ensure that fish acclimatise to the environment of the cage before they are released. Also, fish can be easily monitored for the first two weeks if any sampling is needed and also to ensure that they get accustomed to the feed. They will also be offered the same food in the hatchery before deployment. Wrasse are released into the pen after one to two weeks.

4.2.1 Fish and net size

The mesh size of the cage nets must be adapted to the size of the wrasse. This will vary depending on the size and the body shape of the different species; i.e., cuckoo wrasse *Labrus mixtus* is longer and slimmer whereas goldsinny wrasse *Ctenolabrus rupestris* is smaller and shorter. The minimum mesh size on site should match the size of the fish to maintain the population, avoiding escapees and minimising unregistered losses. For instance, wrasse should be a minimum of 12 cm in length for an 18 mm mesh net for all species, unless cuckoo wrasse, and these should be taken at a minimum of 15 cm length. Table 4.1 suggests different mesh sizes to be used for different wrasse species.

Fig. 4.2 Different sizes (6 m^3, 12 m^3 and 15 m^3; respectively) and models of conditioning pens used for the acclimatisation of the farmed Ballan wrasse in commercial sea cages stocked with Atlantic salmon.

Table 4.1 Recommended mesh sizes (stretched and half, mm) for different wrasse (Labridae) species. Mesh size is measured in ohms.

Mesh size (ohms)	Half mesh (mm)	Whole mesh (mm)	Ballan wrasse (*Labrus bergylta*)	Corkwing wrasse (*Symphodus melops*)	Goldsinny wrasse (*Ctenolabrus rupestris*)
40	15.5	27.5		≥11 cm	≥11 cm
38	16.5	33		≥11.5 cm	≥12 cm
35	18	36	≥14 cm	≥12 cm	≥13 cm
32	20	40	≥15 cm	≥13 cm	≥13.5 cm
30	21	42	≥16 cm	≥14 cm	
28	22.5	45	≥17 cm	≥15 cm	
25	25.5	50	≥19 cm	≥17 cm	
22	29	58	≥20 cm	≥18 cm	

4.2.2 Stocking period, numbers and ratios

Wrasse are essentially acclimated to warm summer temperatures (Sayer *et al.*, 1996) so it is recommended that they are stocked in sea cages with Atlantic salmon smolts from early May to late September/early November, when the seawater temperatures are above 10 degrees. The industry policy is to stock wrasse in seven months of the year. The two important factors that significantly affect the delousing performance of the wrasse are water temperature and the availability of wrasse. A long cold spring may be a predictor of poor efficacy during May, for example, compared with a long hot summer. Even if the availability of fish is lower, delousing can be more successful when water temperatures are higher because wrasse are more active.

Wrasse stocking proportions commonly range between two and ten cleaner fish per salmon but this will depend on the species used, water temperature and geographical region. The welfare of the wrasse stocked in remote and oceanic sites can be significantly compromised by the weather, the currents and the tide. Also, the historical challenges of the farm should be considered. Stocking densities currently used for wrasse in Scotland are between 2 and 8% of the salmon stock depending on which species are used. A supplementary stocking (between 1 and 2%) should be considered on a cage-by-cage basis, based on the time of the year, the lice challenge, salmon numbers (biomass), proximity to the final harvest and availability of different cleaner fish, either wrasse or lumpfish.

4.2.3 Use of hides

Wrasse are widely distributed in inshore waters of west Scotland in rocky and algal habitats (Treasurer, 1994). This is the main reason why wrasse should have a wide access to shelters or hides in the cages when stocked with salmon. Hides provide protection to wrasse from environmental changes both in water quality parameters and also from currents and tides. Also, they reproduce the natural habitat and offer them a place to rest during the night as they have a common behavioural (circadian) rhythmicity. Also, wrasse used as a cleaner fish with farmed Atlantic salmon can be subjected to large and rapid temperature and salinity fluctuations in late autumn and early winter when the surface of the water is affected by snowmelt episodes (Sayer *et al.*, 1996), so it is very important that they can rest in the hides at different depths to find an area with stable water parameters.

Different formats of hides (Fig. 4.3) have been used for wrasse, from car tyres and lanterns to different combinations of pipes, tyres and artificial kelp. Pipes and artificial kelp appear to be the most effective hides for wrasse compared with others, although lanterns are still used successfully in some cases such as small cages. It is crucial that hides must be in place before the wrasse arrive and are kept clean at all times. To achieve this, hides must be designed not only to be stable in the sea but also to be easy to clean. Enough hides should be available to enable replacement when they are being cleaned and also to reflect the total population of wrasse on a pen-by-pen basis. Hides must be provided at a rate that allows all wrasse to rest or hide at all times. As a guide, at least 1 m of artificial kelp should be considered for 30 small or 15 large wrasse, dependent on their size.

Fig. 4.3 Examples of hides used for cleaner fish, especially wrasse, in commercial sea cages when stocked along with Atlantic salmon.

Wrasse should be transferred directly to the hides or as close as possible to them as it has been proven that a higher survival is achieved compared with transferring wrasse along the edge of the net cage. Hides should also be adapted to the site, the format and the size of the cage (square or circular cages) and the current conditions, especially in sites with strong tides, although cleaner fish should not be stocked on sites with extreme energy conditions. Artificial kelp of 10 m length can be set up in circular cages from one support to the next alongside the side of the net. Hides in square cages should be located in the corners, but at least 1 m from the cage net.

4.3 Nutrition

Nutrition is one of the most important points to maintain healthy and active wrasse, but feed management has to be strategic. The Ballan wrasse lacks a functional stomach (Hamre *et al.*, 2013), but has a very long intestine and this is one of the reasons they are termed 'grazers' as they need to eat constantly to maintain their nutritional requirements and their natural behaviour. If sea lice levels are low, supplementary diets for Ballan wrasse are necessary to maintain a good condition factor, which contributes to better health and welfare, as sea lice may not be nutritionally sufficient. However, the feeding regime is very controversial when sea lice numbers increase and sea lice strategic treatment regimes should be applied depending on the sea lice challenge history of the farm, availability of other treatments, cleaner fish health and environmental parameters.

4.3.1 *Supplementary feeding for wrasse in sea cages*

Supplementary feeding is considered when sea lice numbers are low or not sufficient to satisfy the entire wrasse population in sea cages. This is presently quite controversial, as it has been reported recently that lice consumption by wrasse is not affected by supplementary feeding on, for instance, fresh crushed blue mussels (Leclercq *et al.*, 2015). In addition, supplementary feeding did not satiate based on the functional predatory response of wrasse to sea lice density, which could lead to the possibility that wrasse should receive supplementary feed at all times, independently of the sea lice counts.

Seafood such as fresh or frozen green crab, crushed blue mussels, and prawns has been used as a supplementary feed for wrasse in the past but, because of the limited availability as well as all the biosecurity concerns, the use of manufactured dry pellets for marine species has increased. However, they were found to be impractical as they disintegrate within hours. To reduce a significant wastage of these dry pellets for wrasse in net pens, a water-stable wrasse feed based on commercially available dry feed was proposed and used successfully. Agar blocks made with marine premix have been found to be physically water-stable for one week with no nutrient leaching and are well accepted by the wrasse (Leclercq *et al.*, 2015). Such water-stable, agar-based diet is presented to wrasse within fine mesh feed bags at different points in the net pen and close to the hides. Each feed bag contains 1 kg of agar block (Figs 4.4, 4.5). Depending on the water temperatures, which will be directly related to wrasse activity, feed stations should be allocated at different depths. Feed bags are usually placed between 5 and 7 m depth in the summer, based on experience, when the water temperature is above 9°C and bags are lowered from 10 to 12 m depth when the temperature drops below 9°C.

Fig. 4.4 An example of a feed bag containing a marine agar block (Biomar Marine Symbio®) used as a supplementary feed for wrasse within a pen.

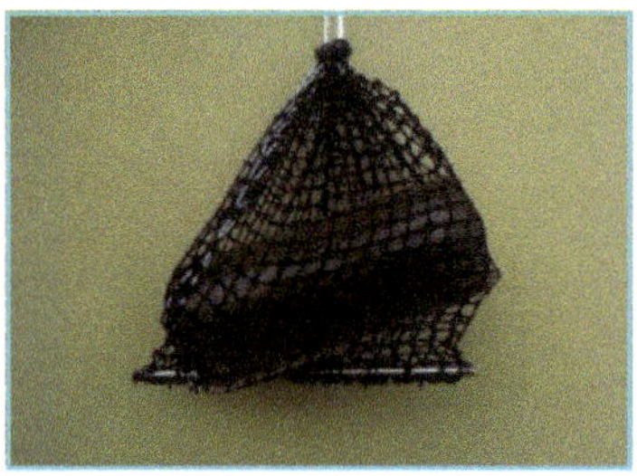

Fig. 4.5 A marine agar block in a feed bag before deployment.

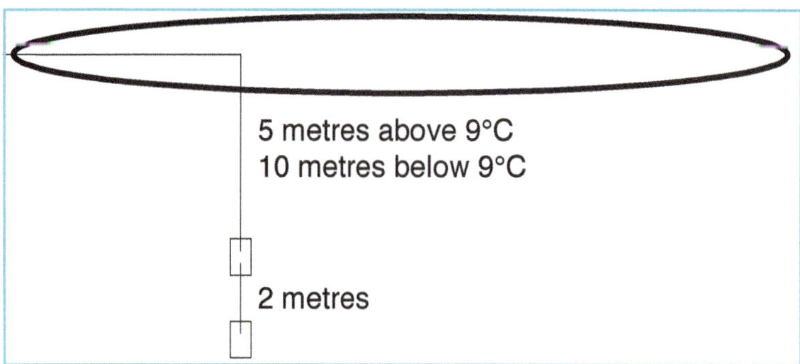

Fig. 4.6 An example of the set-up for a wrasse feed station in a sea cage.

Fig. 4.7 Farmed wrasse eating water-stable, agar-based diet in early trials.

Feed stations (bags) should be monitored on a daily basis, replenished if they are found empty, and the amount of feed eaten recorded daily. This data can be used as an additional tool to better understand not only the wrasse requirements in each cage, but also to give information on the remaining wrasse population and evidence of any significant change in appetite. This also indicates the level of wrasse activity in a net pen and the depth at which wrasse are located in the cage.

Agar blocks should be fed at a minimum feed rate of 2% body weight (total biomass) daily and a minimum of four feeding points should be placed per cage. All uneaten agar blocks should be removed within a week to avoid rancidity.

4.3.2 Feeding preferences of wrasse in sea cages

The biofouling that grows in the nets on the sea pens acts as an alternative feed source for the cleaner fish stock (Deady *et al.*, 1995) and this is the main reason why nets and hides should be kept clean to maintain a good delousing performance on the cages (Figs 4.8, 4.9).

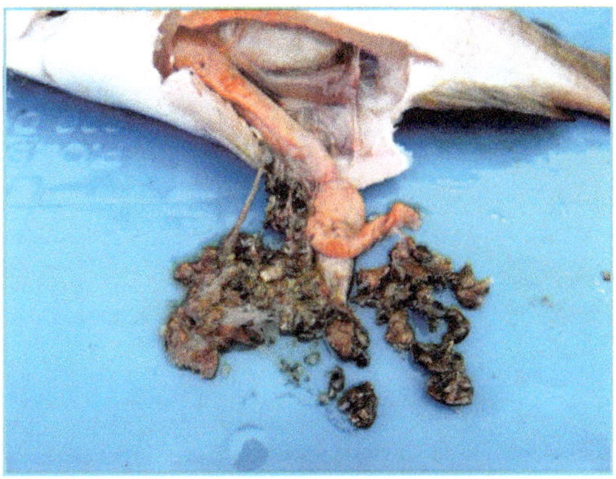

Fig. 4.8 The intestinal contents of a Ballan wrasse, showing mostly crab and mussel shells, when nets and hides are not strictly clean.

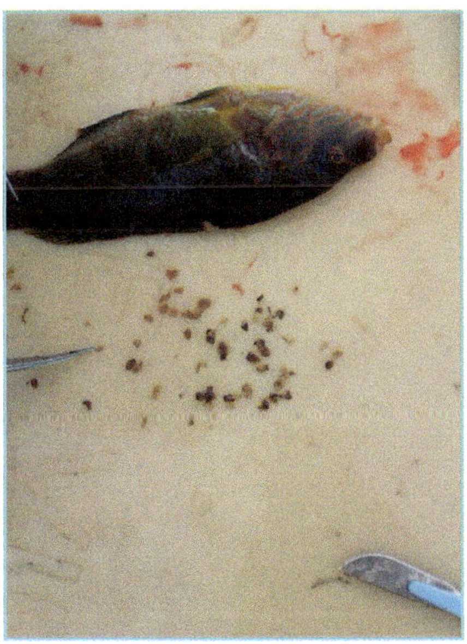

Fig. 4.9 Different stages of sea lice found in the intestine of a corkwing wrasse *Symphodus melops* where nets and hides were clean.

4.4 Basic husbandry

It is important to follow good husbandry measures to ensure a good delousing efficacy by wrasse eating sea lice. Staff attitude is a key feature of success, and a sense of ownership gives improved effectiveness of wrasse on farms. Farmers must insist that fishermen adhere to the aim of good fish welfare to ensure the cleaner fish work efficiently in the salmon cages. A strict schedule is necessary in maintaining clean nets and collecting fish mortalities on a daily basis is crucial for the wrasse to be effective and to avoid possible distractions.

After nets and pen furniture were properly cleaned (no other actions taken), a remarkable reduction in lice counts was noticed, starting with numbers of gravid females declining to zero within a month and other stages (females and males) dropping afterwards (Fig. 4.10). This is an example that shows wrasse delousing efficacy when there is no alternative feed source available.

Also, when hides are removed to be washed, they should be lifted very slowly (for instance, at most 1 m every 30 seconds). Fish with closed swim bladders such as wrasse (physoclist), in contrast to salmon (physostome), face different challenges in sea cages as swim bladder inflation and deflation rates are slower (Korsøen *et al.*, 2010). Hence, it is critical that hides and pen nets are lowered/lifted slowly during farming operations to avoid any buoyancy issues such as swim bladder over-inflation and/or prolapse, which could lead to other health issues that may lead to mortalities. Cage nets and hides should be lifted at a speed that does not compromise the swim bladder (even more care required in lower temperatures), and any compromised fish should be placed in a creel or open fish trap and lowered slowly to 5 m.

It is very important that all the site staff are aware of the management techniques and welfare needs of the cleaner fish at all times but especially during farming operations. A dedicated member of the farm staff, who is competent and has the responsibility for the management and welfare of the wrasse, should be named on each site (RSPCA Standards Appendix 3).

During crowding operations, site staff must be made available to transfer any wrasse from the crowded to the non-crowded section of the pen. When transferring wrasse from pen to pen (in case of swim through, grading events, treatments or harvest), a plan should first be in place and

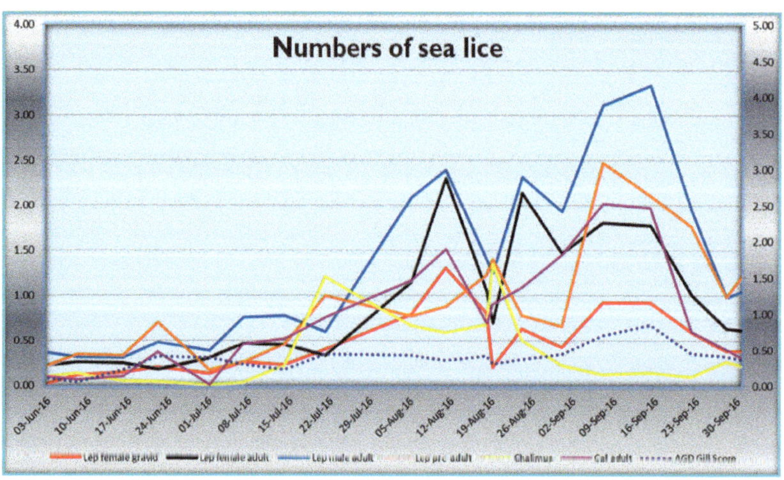

Fig. 4.10 Weekly sea lice counts by stage in a Scottish salmon farm stocked with 6% wrasse.

all movements should be recorded on site. Also, supplementary oxygen must be available at all times during transfers to maintain a minimum of 7 mg·l^{-1} (RSPCA Standards Appendix 3).

For dead haul harvests, any wrasse that are transferred to a ship must be returned to the pen as soon as possible. Wrasse should be fished out before complete pen transfer for a minimum of three days using at least 20 creels or fish traps per cage and lifting them twice each day. Alternative and efficient methods of removal of cleaner fish from pens will be formulated. An example of a recovery of fish during a husbandry procedure is to collect wrasse when salmon are pumped aboard a wellboat for a freshwater bath treatment. Wrasse should be captured before they are pumped into the wellboat as they are intolerant of pure freshwater (~0ppt). Ballan wrasse transferred directly into freshwater are reported to become disorientated within three to 21 minutes and mortalities of 50% occur after 24 hours. In comparison, there were no mortalities when the treatment involved a gradual decrease of salinity (Skiftesvik *et al.*, 2015).

It would be ideal to capture all wrasse at least three days before mechanical treatments such as flushing and thermolicing to ensure they are not pumped into the treatment boats. Some of the boats have cleaner fish returning pipes that transfer fish straight back to the pen, and de-watering bars can be checked for the presence of wrasse being stuck on secondary bars. Ideally wrasse should be removed first from the pen to be treated. A plan must be in place prior to freshwater treatments in wellboats but it will depend if the boat has the capacity for grading fish. If so, wrasse can be redirected through cleaner fish pipes into bins and have an alternative bath treatment, hydrogen peroxide for instance.

Wrasse should not be removed from the pens for bath treatments such as Salmosan or Alphamax used for sea lice, or hydrogen peroxide used to treat AGD, unless they have compromised swim bladders. In this case, the use of closed creels or barrels located outside the pen could be beneficial until the treatment finishes and they recover, and then they can be transferred slowly back to the pen.

4.5 Post-transfer mortality records

Mortalities should be recorded directly after the transfers to assess the overall transfer performance and to ensure that fish are not suffering from any health issue after the stress of the travel. If possible, water samples should be taken from the transport tanks before departure, midway and on arrival. Water quality might be checked if there is an incident and the results can be compared with the measurements (readings) on the lorry to ensure everything was well monitored.

4.6 Delousing efficacy after deployment

Wild and farmed Ballan wrasse are confirmed as highly effective therapeutic and preventive biological controls against sea lice. Leclercq *et al.* (2014) obtained a reduction in mean sea lice numbers from 12 to 0.4 within a maximum of 84 hours, using farmed wrasse naïve of previous louse exposure at a commercially relevant stocking rate of 5%. Sea lice numbers for three consecutive generations (salmon S1 (Quarter 1 and Q2)) are represented in Fig. 4.11. Wrasse were

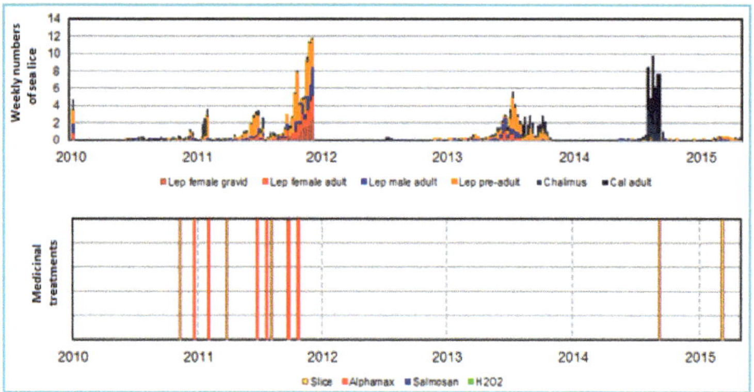

Fig. 4.11 Sea lice numbers by stage and timing of medicinal treatments in a Scottish commercial salmon farm before and after stocking with wrasse.

stocked at a ratio of 4.6% to salmon numbers in May 2012 and an additional top-up of 1% was added in June 2013. This graph shows how wrasse can control sea lice numbers and eliminate the use of medicinal treatments, and provides a suitable and environmentally friendly alternative.

When wrasse are stocked in salmon cages, they should start their cleaning activity within the first two to three weeks, depending on the water temperatures, fish numbers and the sea lice challenge. A small trial was carried out on a commercial farm (Fig. 4.12) to assess the interval for farmed and wild wrasse to commence consuming sea lice after deployment. All wrasse were stocked at the same time, between 9 and 14 August. Pens 1, 2, 3 and 15 (in blue) were stocked with trained wild wrasse that had been harvested and collected from another farm in the same farm management area. Pens 5 and 7 (in yellow) were stocked with farmed Ballan wrasse. Pens 4 and 14 (in red) were not stocked with any wrasse and were used as a control.

There was a clear reduction in *Lepeophtheirus salmonis* adult numbers (gravid females, females and males) in three to four weeks after deployment (Fig. 4.12) in those pens stocked with trained wild wrasse. Pens stocked with farmed Ballan wrasse (in yellow) took a little longer to reduce adult lice numbers significantly, but were still remarkable at four to five weeks after deployment. This is evidence that both wild and farmed wrasse are efficient delousing tools and it highlights the need for acclimatisation before deployment of farmed wrasse, which will be beneficial to initiate earlier cleaning behaviour.

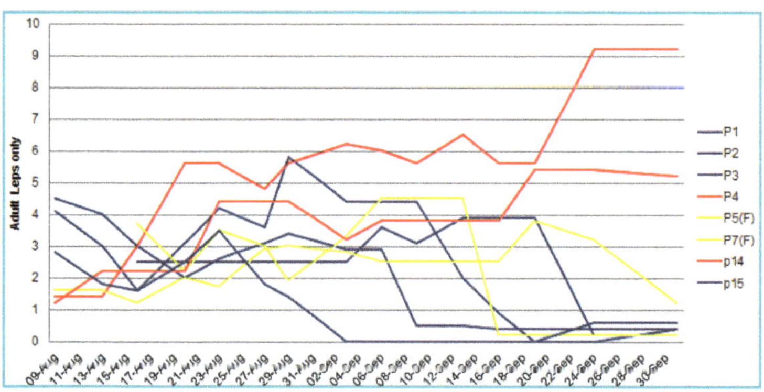

Fig. 4.12 Sea lice numbers in different pens stocked with trained wild farmed wrasse (blue), farmed Ballan wrasse (yellow) and pens with no cleaner fish (control, red) during a two-month period.

4.7 Health management of wrasse populations

It is crucial that a veterinarian or a health manager carries out a pre-transfer health check, especially to ensure that fish are initially fit to travel and secondly, to minimise the risk of pathogen transmission to salmon. Cleaner fish, as salmon, have also to be monitored closely on a regular basis. Currently, systemic bacterial infections (sometimes, secondary to physical damage caused by aggression or during routine farming operations) and Amoebic Gill Disease (AGD) are the two main causes of mortality in wrasse populations. On one hand, all farmed wrasse at MHS are vaccinated with autogenous vaccines for significant pathogens that are considered to be pathogenic to wrasse. The wrasse receive two dip/bath immersions with an aqueous-based emergency vaccine and, when they are minimum 15 g in weight, they receive an intraperitoneal injection with an oil-based vaccine. Finally, they are deployed in the sea cages at a minimum of 400 degree days C after the intraperitoneal vaccination. Due to the lack of commercial and authorised vaccines for wrasse on the market, these autogenous vaccines are continually evolving and being formulated with the current pathogens that are isolated from the different farms. Efficient vaccines are fundamental to mitigate losses and improve the health status but isolates have to be selected carefully as candidates for the vaccine after they have proven to be pathogenic, as many of the infections in wrasse are caused by opportunistic bacteria.

On the other hand, AGD is controlled in the hatcheries with brackish bath treatments and in the commercial cages with hydrogen peroxide baths, normally for a maximum of 20 minutes at 1200–1400 ppm.

4.7.1 On-site health surveillance

Survival of wrasse in sea cages can be low due to different factors, but mainly to the known susceptibility of wrasse to bacterial, viral and parasitic diseases (Menezes, 1992; Costello *et al.*, 1996; Treasurer, 2012). This is the main reason why a cleaner fish health screening surveillance must be in place as part of the Veterinary Health and Welfare Plan, designated for salmon, for each commercial site.

Mortality episodes in cleaner fish are often related to bacterial infections, and atypical furunculosis, caused by atypical strains of *Aeromonas salmonicida*, is perhaps the most commonly diagnosed infectious disease (Gulla, 2015). In addition, gill lesions such as patches and histological changes characteristic of AGD have been found in both clinically healthy and moribund wrasse, which confirms that wrasse can suffer and also act as carriers of *Paramoeba perurans* (Karlsbakk *et al.*, 2013).

Weekly sea lice counts on a pen-by-pen basis is considered an indicator of cleaner fish health and activity, although it is also true that there is high variability between sites and cleaner fish can be healthy and not efficient in different areas. Regular gill sampling for parasites (especially *Paramoeba perurans* causing AGD), tissue swabs for bacteriology screening (mostly for atypical furunculosis) and tissue samples for histopathology analysis should be taken to assess any health issue that could lead to a lack of efficacy. Moreover, this screening is used to diagnose (or determine mortality causes) to control disease outbreaks through alternative treatments and all together to improve survival of wrasse on sea cages and avoid transmission of diseases to salmon.

Intestine content, physical condition (presence of fin/tail erosions or skin lesions) and condition factor (including weights and lengths) are also important parameters in assessing wrasse welfare (Treasurer and Feledi, 2014).

Physical damage is sometimes observed in sea cages and it is related most of the time with farm operations. Thus it is very important that care is provided to wrasse during grading, fish movements, net cleaning (including swim through) and treatments.

4.8 Interaction between wrasse and lumpfish

The lumpfish, *Cyclopterus lumpus L.*, has been suggested as an alternative cold water cleaner fish, and its effectiveness in biological delousing of Atlantic salmon has recently been documented (Imsland *et al.,* 2014a, b, c). Lumpfish are usually stocked in commercial sea cages during the winter season, between November and April, when the wrasse are not sufficiently active to maintain sea lice numbers. Thus, lumpfish may have the potential to survive winters in cages with salmon (Imsland *et al.,* 2016) and contribute to the cleaner fish delousing ability. Lumpfish and different species of wrasse, which include Ballan, goldsinny, rock cook *Centrolabrus exoletus*, corkwing and cuckoo, cohabitate in the seawater cages with Atlantic salmon during some months. It has been reported from some farms (MHS) that a few juvenile lumpfish are seen attached to wrasse and sometimes lumpfish are slightly more aggressive than wrasse in grazing the sea lice from salmon.

Conclusions

The use of wrasse in aquaculture as a green tool for sea lice control started in 1989. However, there is still a great variability in delousing performance between different sites and areas and significant knowledge gaps. Good husbandry in sea cages is necessary to succeed with wrasse, but staff training and preparation are almost more important. Delousing efficiency is highly variable and must be assessed in each commercial sea site at different periods of time, and all the data collected must be reviewed and analysed to improve performance in future years. Keeping the nets and the hides clean, collecting mortalities every day and feeding the wrasse with supplementary feeding are key features for a successful performance. To improve wrasse survival in sea cages, strict health monitoring should be in place to be able to control any possible infectious outbreak, and further research carried out into effective vaccines that can mitigate and minimise losses due to bacterial infections.

Acknowledgements

I thank Marine Harvest Scotland Ltd, especially Ronnie Hawkins for his time and support and Dave Cockerill for his thoughts, and to the staff at the cleaner fish hatcheries such as Machrihanish and Otter Ferry. Also all the people that work on a daily basis in the sea sites taking care of the cleaner fish and surprising us with new suggestions to improve the husbandry. I am grateful to

Herve Migaud, Eric Leclerq, Andrew Davie and Sandra Adams of the University of Stirling for their association with MHS.

References

Costello, M.J., Deady, S., Pike, A. and Fives, J.M. (1996) Parasites and diseases of wrasse being used as cleaner-fish on salmon farms in Ireland and Scotland. In: Sayer, M., Treasurer, J. and Costello, M. (eds) *Wrasse Biology and Use in Aquaculture*, Fishing News Books, Blackwell. pp. 211–227.

Deady, S., Varian, S.J.A. and Fives, J.M. (1995) The use of cleaner-fish to control sea lice on two Irish salmon (*Salmo salar*) farms with particular reference to wrasse behaviour in salmon cages. *Aquaculture* 131, 73–90.

Gulla, S., Lund, V., Kristoffersen, A.B., Sørum, H. and Colquhoun, D.J. (2016) vapA (A-layer) typing differentiates *Aeromonas salmonicida* subspecies and identifies a number of previously undescribed subtypes. *Journal of Fish Diseases* 39, 329–342.

Hamre, K. Nordgreen, A., Grøtan, E. and Breck, O. (2013) A holistic approach to development of diets for ballan wrasse (*Labrus bergylta*) a new species in aquaculture. *PeerJ b*, 99.

Imsland, A.K., Reynolds, P., Eliassen, G., Hangstad, T.A., Foss, A., Vikingstad, E. and Elvegård, T.A. (2014a) The use of lumpfish (*Cyclopterus lumpus* L.) to control sea lice (*Lepeophtheirus salmonis Krøyer*) infestations in intensively farmed Atlantic salmon (*Salmo salar* L.). *Aquaculture* 424–425, 18–23.

Imsland, A.K., Reynolds, P, Eliassen, G., Hangstad, T.A., Nytrø, A.V., Foss, A., Vikingstad, E. and Elvegård, T.A. (2014b) Assessment of growth and sea lice infection levels in Atlantic salmon stocked in small-scale cages with lumpfish. *Aquaculture* 433, 137–142.

Imsland, A.K., Reynolds, P., Eliassen, G., Hangstad., TA, Nytrø, A.V., Foss, A., Vikingstad, E. and Elvegård, T.A. (2014c). Notes on behaviour of lumpfish in sea pens with and without Atlantic salmon. *Journal of Ethology* 32, 117–122.

Imsland, A.K., Reynolds, P., Eliassen, G., Hangstad, T.A., Jonsdottir, O.D.B., Elvegård, T.A., Lemmen, S.C.A., Rydland, R. and Nytrø, A.V. (2016) Investigation of behavioural interactions between lumpfish (*Cyclopterus lumpus*) and Goldsinny wrasse (*Ctenolabrus rupestris*) under controlled conditions. *Aquaculture International* 24, 1509–1521.

Karlsbakk, E., Olsen, A.B., Einen, A.B., Mo, T.A., Fiksdal, I.U., Aase, H., Kalgraff, C., Skår, S. and Hansen, H. (2013) Amoebic gill disease due to *Paramoeba perurans* in ballan wrasse (*Labrus bergylta*). *Aquaculture* 412–413, 41–44.

Korsøen, Ø.J., Dempster, T., Fosseidengen, J.E., Fernö, A., Heegaard, E. and Kristiansen, T.S. (2010) Behavioural responses to pressure changes in cultured Atlantic cod (*Gadus morhua*): Defining practical limits for submerging and lifting sea-cages. *Aquaculture* 308, 106–115.

Leclercq, E., Davie, A. and Migaud, H. (2014) Delousing efficiency of farmed ballan wrasse (*Labrus bergylta*) against *Lepeophtheirus salmonis* infecting Atlantic salmon (*Salmo salar*) post-smolts. *Pest Management Science* 70 (8), 1274–1282.

Leclercq, E., Graham, P. and Migaud, H. (2015) Development of a water-stable agar-based diet for the supplementary feeding of cleaner fish ballan wrasse (*Labrus bergylta*) deployed within commercial Atlantic salmon (*Salmo salar*) net-pens. *Animal Feed Science and Technology* 208, 98–106.

Menezes, J. (1992) Hazards from pathogens carried by wild fish particularly wrasse used as lice cleaners. *Bulletin of the European Association of Fish Pathologists* 12, 194–195.

Sayer, M.D.J. and Reader, J.P. (1996) Exposure of goldsinny, rock cook and corkwing wrasse to low temperature and low salinity: survival, blood physiology and seasonal variation. *Journal of Fish Biology* 49, 41–63.

Skiftesvik, A.B., Bjelland, R., Durif, C., Hjellum, R.B. and Moltumyr, L. (2015) Cleaner fish: behaviour, mixing species, AGD treatments. Havforskningsinstituttet, Institute of Marine Research.

Treasurer, J.W. (1994) The distribution, age and growth of wrasse (Labridae) in inshore waters of west Scotland. *Journal of Fish Biology* 44, 905–918.

Treasurer, J.W. (2012) Diseases of north European wrasse (Labridae) and possible interactions with cohabited farmed salmon, *Salmo salar* L. *Journal of Fish Diseases* 35, 555–562.

Treasurer, J. and Feledi, T. (2014) The Physical Condition and Welfare of Five Species of Wild-caught wrasse stocked under Aquaculture Conditions and when stocked in Atlantic salmon, *Salmo salar*, production cages. *Journal of World Aquaculture Society* 45 (2), 213–219.

CHAPTER 5

Annual reproductive cycle and egg production in wild-caught wrasse goldsinny *Ctenolabrus rupestris* and rock cook *Centrolabrus exoletus*, and realised egg collection and rearing of goldsinny wrasse

Jim Treasurer, Jon Sherwood, Antonios Chalaris

FAI Aquaculture, Ardtoe, Acharacle, UK

5.1 Introduction

There has been considerable interest in northern Europe, Chile and Canada in the use of cleaner fish to control sea lice numbers on salmon farms, both *Lepeophtheirus salmonis* and Caligus species (Bjordal, 1990; Bjordal and Costello, 1990; Treasurer, 1993; Chapter 23, Canada). Sea lice continue to be the main health threat to farmed salmon and medicinal treatments may have limitations because of the universal development of resistance by the sea lice to medicines (Bravo *et al.*, 2008; Sevatdal *et al.*, 2015). Although wrasse are seen as having a valuable part to play in controlling sea lice, cleaner fish are viewed as part of an integrated strategy IPM to protect the health of salmon along with strategic use of medicines and preventive measures such as fallowing of farms (Grant and Treasurer, 1993). Many of the cleaner fish that have been supplied to salmon farms have been harvested from the sea with 18 million wild-caught wrasse stocked on farms in Norway in 2016, and a higher than expected catch of 26 million fish in 2017. This fishery compares with 1 million reared wrasse in 2016 and a forecast of 2 million in 2017 (Sveier and Breck, chapter 18, Norwegian Government Department of Fisheries statistics, 2017), and >500,000 wild caught wrasse in Scotland in 2015 and more than 1 million in the English Channel (Riley *et al.*, 2017). The industry is resourcing the farming of wrasse to mitigate potential impacts on wild fish but also to ensure a supply of disease-free fish and at a time that suits the salmon production cycle. While five species of wild wrasse have been stocked, salmon farmers have focused on the rearing of Ballan wrasse *Labrus bergylta* (eco-fish.org) and, to a lesser extent goldsinny wrasse *Ctenolabrus rupestris* (Treasurer, 2011). Most scientific and technical attention has been given to the rearing of Ballan wrasse with several studies on reproduction in Ballan (Muncaster, 2010; D'Arcy *et al.*, 2012; www.ecofish.org; Grant *et al.*, 2016).

Less attention has been given to the biology of other wrasse species. This chapter examines the reproductive biology and potential egg production of goldsinny wrasse and compares this with that of rock cook wrasse *Centrolabrus exoletus*, a similar-size fish with the same geographical distribution on the west coast of Scotland (Sayer and Treasurer, 1996). Basic statistics on the two species and on their life history are given in Darwall *et al.* (1992). Little has been published about reproduction in goldsinny wrasse. Stone (1996) carried out preliminary rearing of goldsinny at Golden Sea Produce, Hunterston, North Ayrshire, and was able to strip eggs and rear goldsinny and corkwing wrasse *Symphodus melops*. Goldsinny was the species of interest reared by the Sea Fish Industry Authority (Sherwood 1990, 1991) and Skiftesvik *et al.* (1996) also reared goldsinny and corkwing wrasse in Norway.

Goldsinny (Fig. 5.1) is a small wrasse species of 100 to 140 mm length and with maximum ages of 14 years for males and 20 years for females (Sayer *et al.*, 1995), although another study over a wider geographical range recorded a maximum of 16 years for males and 12 years for females (Treasurer, 1994). Females can be distinguished readily from males by the drab brown appearance, larger ventral plates and black cloaca at any time of the year. The male has a gold lateral band in the breeding season but, even without this, can be distinguished by the closer scale pattern on the ventral surface. Goldsinny are found in the Mediterranean and in the Atlantic from Morocco to Norway (Wheeler, 1969). The habitat is mainly rocky shores with a requirement for shelter holes (Sayer *et al.*, 1996) and they are found in the intertidal zone to depths of 50 m. Postlarvae settle in shallow rocky subtidal areas (Sayer and Treasurer, 1996). Goldsinny eat molluscs and crustacea (Quignard, 1966) and have been reported to eat amphipods and small crabs (Hillden, 1978). Goldsinny have maximum growth in the first two years in the wild with 3 cm increment and around 2 cm thereafter (Quignard, 1966). Males mature at two to three years' old with females a year later (Sayer *et al.*, 1996). Goldsinny spawn from April to September (Dipper and Pullin, 1979), dependent on latitude, but in Scotland the spawning period is from mid-June to the end of July (this study). The male and female swim vertically during spawning and shed the planktonic eggs in mid-water (Hillden, 1981), but fecundity has not been estimated. In years with high densities they may shoal but, when numbers are less and after spawning, some females and sub-adult males may become territorial. They appear less active at lower temperatures and in winter it is thought that they take refuge in rocks. Goldsinny are territorial and a male may defend an area of 0.5 to 2 m^2 (Hillden, 1981). Each territory has a central shelter hole. Cleaning behaviour in goldsinny has been documented by Potts (1973), Hillden (1983), and Bjordal (1988).

Fig. 5.1 Wild-caught (a) goldsinny broodstock (Ardtoe) and (b) rock cook. Credit: Mark Woombs.

Early larval rearing of goldsinny was carried out at the Sea Fish Industry Authority at Ardtoe on Scotland's west coast from 1990 and this was largely unpublished and reported internally. Janet Stone working at Golden Sea Produce reared more than 5000 juvenile goldsinny wrasse (Stone, 1996). The broodstock were manually stripped. The first feed was the marine algae Isochrysis, followed after several days with the rotifer *Brachionus plicatilis* and finally with the copepod *Eurytemora velox*. Artemia nauplii and on-grown Artemia were provided from 30 days post-hatch and then juveniles were gradually weaned to dry feed. Survival from hatching to the juvenile phase, a period of 12 to 18 months, was up to 7%. Goldsinny eggs were also hand-stripped by Skiftesvik *et al.* (1996) and milt collected by syringe. Eggs were incubated in 10 l containers. *Tetraselmis* and *Brachionus plicatilis* were supplied after filtering to 180 µm to select the smaller prey items. Rotifers were replaced with Artemia nauplii.

There have been fewer publications on rock cook wrasse *Centrolabrus exoletus* (Fig. 5.1) concerned with sex ratio, maturation age or fecundity, although the biology of both species has been studied on the west coast of Scotland (Treasurer, 1994; Sayer *et al.*, 1996) including distribution, age, growth and diet. This is considered to be one of the less well investigated wrasse species with limited biological information (Costello, 1991). Rock cook are commonly found in the size range 100–140 mm, with a maximum length of 165 mm (Sayer and Treasurer, 1996) and a maximum age in Scotland of nine years (Treasurer, 1994). Rock cook are normally red–brown on the dorsal surface and lighter on the ventral. The sexes cannot be distinguished by external features. The distribution is from the Bay of Biscay (Lythgoe and Lythgoe, 1991), largely on the west coasts of the British Isles, and on the southern coasts of Norway and Sweden. Rock cook are found mainly on rocky coasts from shallow water to a depth of 40 m (Sayer and Treasurer, 1996).

This chapter examines reproductive development in females of the two wrasse species and compares egg production and egg development cycles, and duration of the spawning period in wild-caught fish on the west coast of Scotland. A comparison is also drawn with spontaneous spawning of hatchery-held goldsinny wrasse over a four-year period. The timing and duration of spawning and potential egg deposition are calculated from a known number of hatchery females. This study of wild-caught fish enabled examination of the natural reproductive cycle and actual annual fecundity.

5.2 Methods

5.2.1 Fish capture

Wrasse were captured from April to October in fyke nets on rocky ground close inshore in water of 3 to 20 m depth by commercial fishermen. The nets were set unbaited for 24 hours before retrieval. The wrasse were captured in Brodick Bay, Isle of Arran (55°45'N, 5°3'W). A random sample of 30 captured fish (>90 mm) was retained monthly over a one-year cycle as far as wrasse activity at lower water temperatures would allow and twice monthly from June to August when the fish were spawning. The sex ratio varied but generally at least ten fish of each gender were sampled on each occasion. Fish were captured when water temperatures exceeded 10°C and few fish were captured at temperatures below that limit. Therefore selected samples were taken of wild-caught wrasse held in 500 l tanks at FAI Ardtoe during this period from November to March to augment the sampling of various developmental stages.

5.2.2 Fish sampling

Fish were euthanised by overdose with the anaesthetic MS_{222} (Sandoz). The fork length in mm and weight to 0.01 g of the fish were measured. The fish were then dissected by incision in the abdomen. Viscera were removed and weighed, and the liver and gonad were weighed separately to 0.01 g. The gender of the fish was identified and a maturation stage assigned macroscopically after Rinchard and Kestemont (1996). A sample of the gonad was fixed in 10% neutral buffered formalin NBF, and later embedded in paraffin wax and sectioned at 5 µm, stained in haematoxylin and eosin H&E, and examined at 100× and 400× magnification to describe the histological appearance and development of the gonad. The hepatosomatic index HSI was calculated as (liver wt/eviscerated wt × 100), and the gonadosomatic index GSI as (gonad wt/eviscerated wt × 100). The ovary was removed and fixed in Gilson's fluid to assist breakdown of connective tissue and separation of oocytes to enable estimation of egg number. A sample of the gonad was taken at the anterior, mid and posterior part of the ovary, the total value weighed and fixed. After three months the eggs were gently washed with distilled water and filtered with a 30 µm mesh to retain only developing oocytes. The eggs were transferred to a 2 l beaker, mixed thoroughly, and a 0.5 ml sample was taken immediately using a Stempel pipette. The eggs were placed in a marked grid counting chamber and the number of eggs in excess of 60 µm diameter were counted. The total egg number was calculated from N counted × 2000/0.5 ml. Three replicate samples were counted for each fish. The diameter of eggs was measured using a binocular microscope at 40× magnification. The histological sections were assigned to a stage of maturation after Rinchard and Kestemont (1996) after first scanning all slides to determine the range of developmental stages that were present. A random sample of 100 oocytes sectioned through the nucleus was measured and oocyte size distribution was plotted, with the proportion of oocytes present of each maturation developmental stage calculated.

5.2.3 Statistics

Mean values were calculated for HSI, GSI, oocycte diameter and SD calculated. Data were analysed for homogeneity of variances using Bartlett's test and, where there was no homogeneity of variance, data were transformed to $\log_{10}$ to enable statistical comparison using ANOVA and also t-test. The distributions of oocyte diameters were compared using the Kolmogorov–Smirnov test. Potential egg production was plotted against fish length and weight and the relationship described by a regression equation, F = *a*X*b*, where F= absolute fecundity, X is an independent variable and *a* and *b* are constants.

5.3 Spawning season and ovarian development

The maximum GSI is considered the point of maximum ovary weight just prior to spawning (egg release) and following this the gradual decline over several weeks to values less than 3% indicate atresia of some eggs and this can be combined with histological examination of the ovaries and also the oocyte diameters and the distribution of the maturation stages to give clear indication of the spawning period.

The GSI of female goldsinny was 2.4 in January and 2.7 in March and increased rapidly to 10.1 in June (Fig. 5.2). Maximum GSI was 17.6 on 28 June and it is likely that there would have been no spawning prior to that. GSI declined to 7.8 by 17 July and intense spawning activity therefore occurred in that three-week period with a decline in GSI through July and a low of 3.6 on 9 August. It continued to drop to 1.1 on 11 October, perhaps due to resorption of oocytes, and remained low until January. The annual cycle of GSI of rock cook females was very similar with an annual maximum of 18.3, also on 28 June, and declining to 9.1 on 17 July and 3.2 by 9 August. GSI was at a minimum over winter from 0.6 to 1.6. The GSI of males

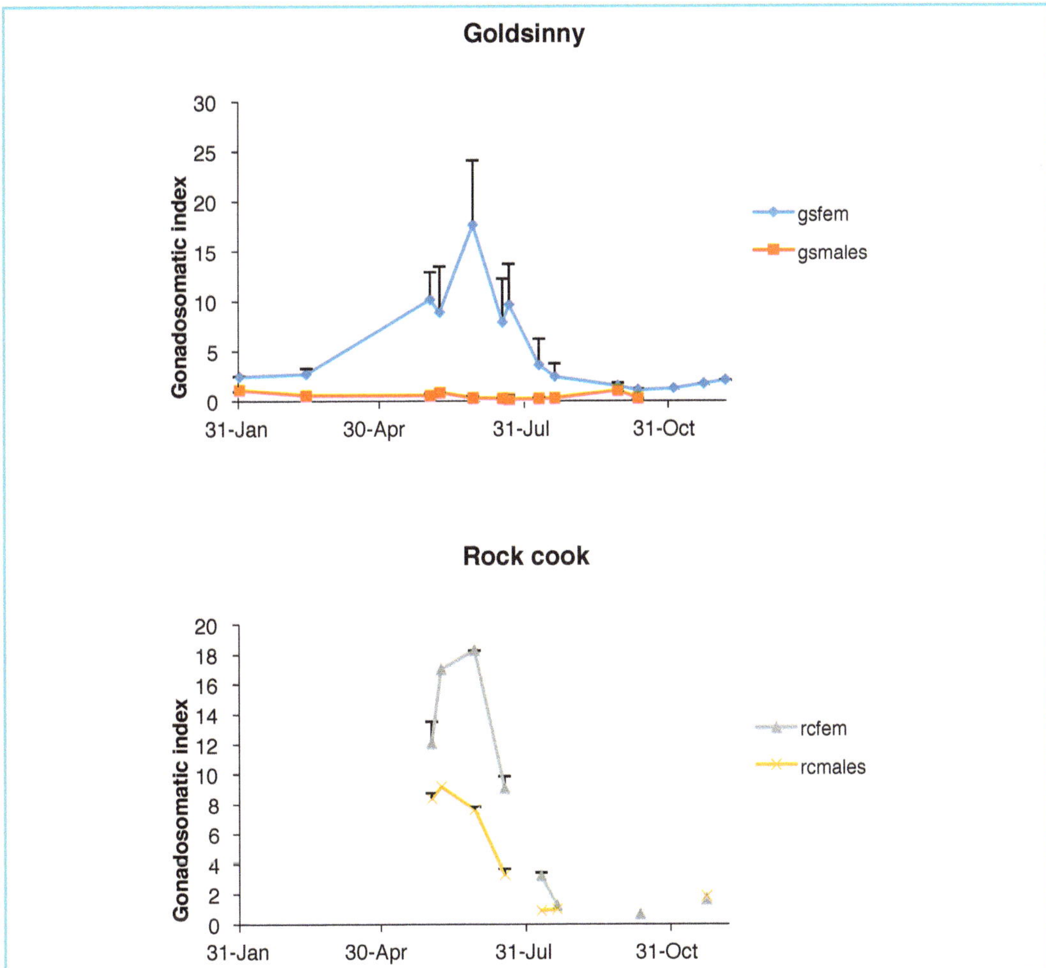

Fig. 5.2 The cycle of gonad development in male and female goldsinny and rock cook wrasse as the gonadosomatic index (GSI). Error bars indicate SD. These wrasse were captured in Brodick Bay on the Isle of Arran (55°45′N, 5°3′W).

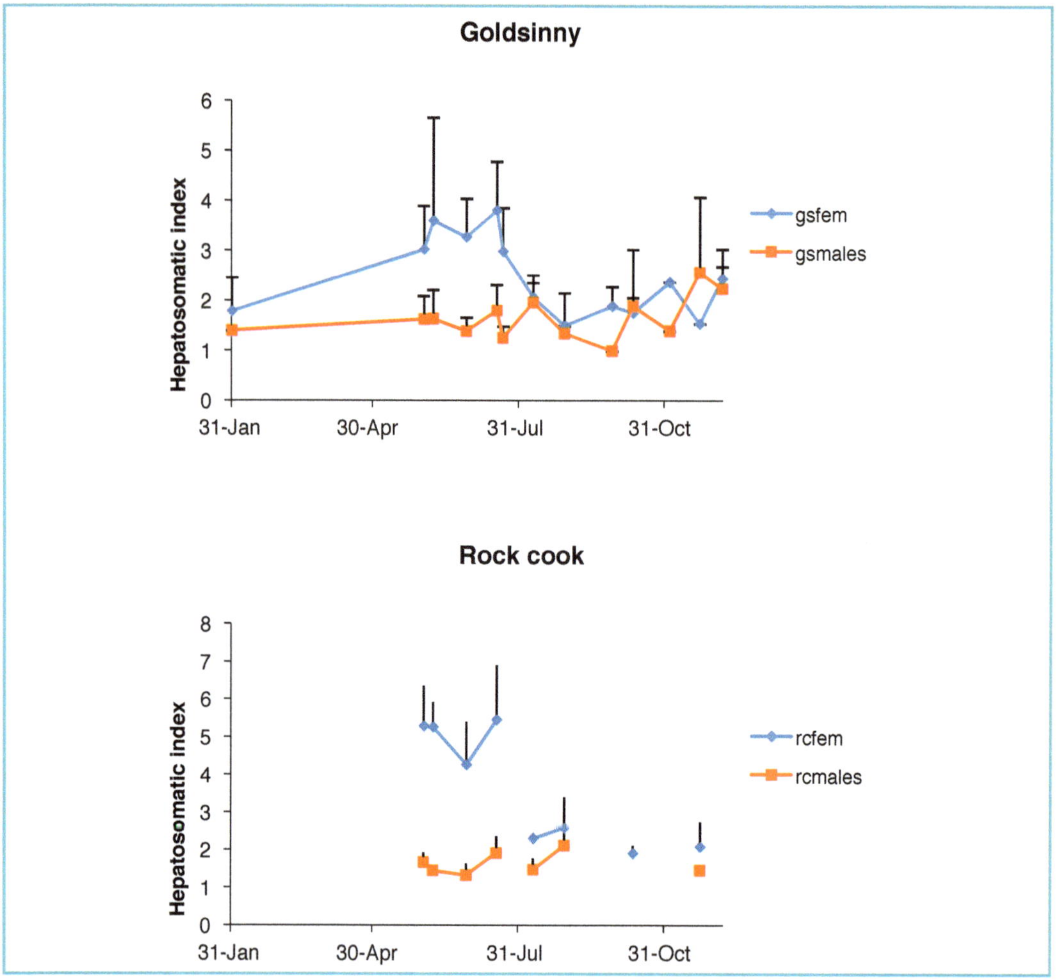

Fig. 5.3 The annual cycle in hepatosomatic indexes (HSI) of goldsinny and rock cook wrasse. Error bars indicate SD.

was much lower in goldsinny and this is in agreement with low volumes of milt obtained when hand-stripping of males. GSI was 0.15 in January with a seasonal maximum of 0.97 on 1 June and 28 June. The GSI had declined to 0.1 by 19 August. In contrast, the testes of rock cook were much larger with maximum GSI values of 9 on 7 June declining to 3.2 by 17 July and 0.8 by 9 August. These data indicate that spawning activity in goldsinny wrasse and rock wrasse in Scotland peaked in the last weeks in June and through July, and was contemporaneous in both species.

Hepatosomatic index

The HSI of female goldsinny peaked at 3.8 on 17 July, as did that of rock cook at 5.4 (Fig. 5.3), whereas the annual pattern of HSI varied little in males with 2 in August and maximum values of 2.5 in November and December in goldsinny, and an HSI maximum of 2 in July and August in rock cook.

5.4 Ovary developmental stages

Histological samples of ovaries of goldsinny and rock cook taken through the year formed a representative range of ovary development and visual stages as defined by Rinchard and Kestemont (1996) (Table 5.1). These samples were examined at ×100 and ×400 magnification to provide a histological description of ovarian development. The main pre-spawning, spawning and post-spawning stages are shown in (Fig. 5.4). Ovaries in fish that had not spawned are compared with goldsinny and rock cook from the same capture location, Lamlash Bay on the Isle of Arran, on 1 June. These show comparable developmental stages (Table 5.1), with oocytes in a range of sizes and development from stages 3= the oocytes full of yolk vesicles

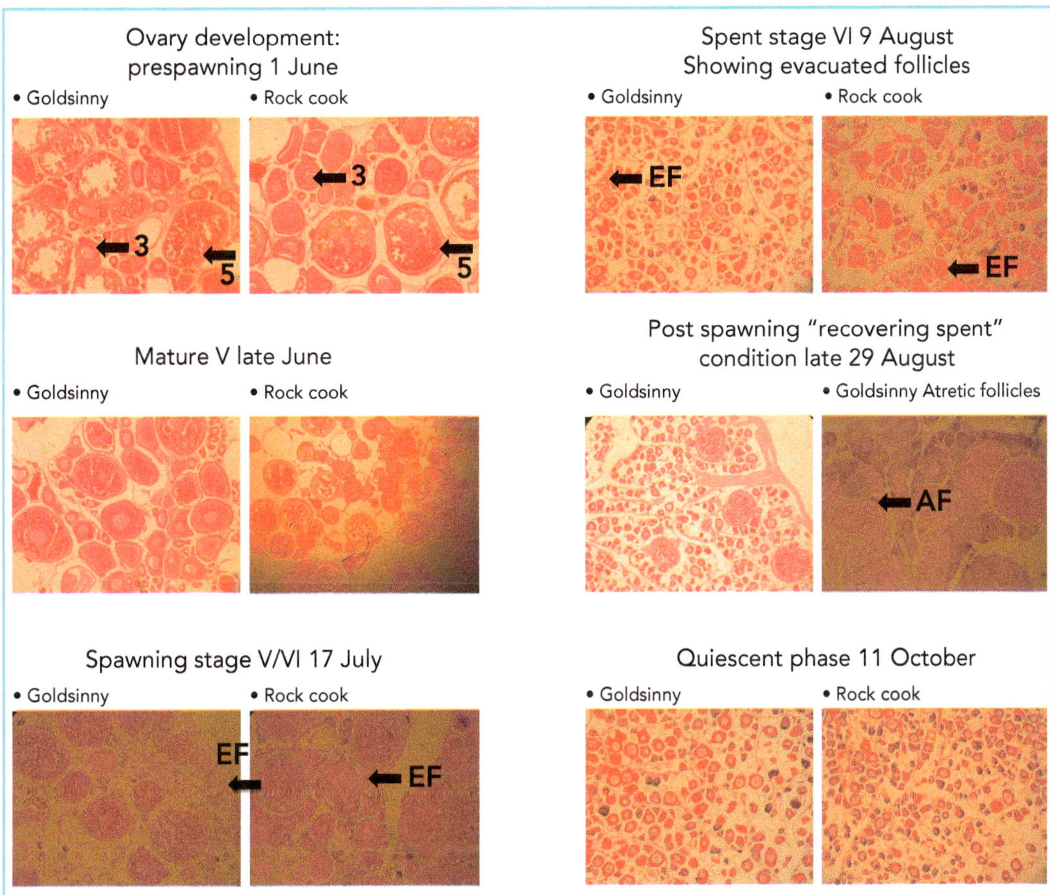

Fig. 5.4 The stages of ovary development in goldsinny and rock wrasse. See scale of development in Table 5.1 and full explanation of stages in text. Arrows indicate oocytes of stages 3 and 5 in development, EF = evacuated follicle, AF = atretic follicle.

Table 5.1 Ovarian developmental stages in goldsinny and rock wrasse, after scale by Rinchard and Kestemont (1996).

Development stage	Description: Macroscopic	Histological
1, previtellogenic	Small pink-coloured organ, c. 10 mm in length, occupying about one quarter of the body cavity Lying under the swimbladder and above the gut.	Previtellogenic oocytes Oocytes with vacuole free cytoplasm
2, Beginning of endogenous vitellogenesis	Ovary opaque occupies about 1/3 length and 1/5 volume of body cavity no eggs seen by eye	Previtellogenic oocytes and oocytes in endogenous vitellogenesis Yolk vesicles apparent in the periphery of the cytoplasm
3, Completion of endogenous vitellogenesis	Occupies about 1/2 to 3/4 length body cavity, 1/4 volume, eggs granular. Oocytes just visible as granular	Previtellogenic oocytes and oocytes having complete endogenous vitellogenesis Oocytes are full of vesicles Follicular and cellular layers are differentiated
4, Exogenous vitellogenesis	Ovary fills 2/3 – 3/4 body cavity golden eggs seen by eye	Previtellogenic oocytes and oocytes at different stages of exogenous vitellogenesis. Oocytes accumulate yolk globules and yolk vesicles are at the periphery of the cytoplasm In late stage 4 the oocytes are almost hyaline
5, Final maturation	Ovary fills body cavity Granular eggs seen	Previtellogenic oocytes and oocytes in final maturation Appearance of the micropyle and migration of the germinal vesicle to the micropyle. Oocytes are difficult to section and appear hyaline.
6, Intermediate spawning stage	Ovary lies above gut c. 1/2-3/4 length of body cavity, large oocytes visible. Most eggs opaque but some transparent eggs Atretic eggs can be seen with low power microscope Much thinner than stage 5, about 1/6 volume of body cavity	Oocytes at different stages of endogenous and exogenous vitellogenesis and postovulatory follicles. The postovulatory follicles hypertrophy and multiply showing phagocytosis
7, Post-spawning	Occupies 2/3 body cavity length Only about 1/5 bc volume. Thin, translucent	Previtellogenic oocytes and pre- and post-ovulatory follicles The pre- and post-ovulatory follicles hypertrophy, the yolk substance degenerates

and differentiation of the follicle cell layers, stage 4= the oocytes are accumulating yolk globules and yolk vesicles are at the periphery of the cytoplasm, and stage 5 the hyaline oocytes in a very mature stage of development and showing many yolk vesicles that coalesce and the oocytes appear hyaline. These sections demonstrated a range of oocyte sizes and evidence that both goldsinny and rock cook were multiple spawners producing several egg batches. Few pre-vitellogenic oocytes were present.

The ovarian samples in late June showed fish that were in spawning condition but no post-ovulatory follicles were evident, indicating that these goldsinny and rock cook were in a very advanced maturation condition but had not spawned. The ovaries of several wrasse were sampled for histology on each sampling date and all were in the same developmental stage. The ovaries of both goldsinny and rock cook were in the spawning phase on 17 July. On this date there was lamellar disorganisation and post-ovulatory follicles were very noticeable and in various stages of hypertrophy and phagocytosis. At this point there were two distinct size groups of oocytes visible that would be spawned in the current season.

All the ovaries were in the post-spawning stage by 9 August (Fig. 5.4). The ovaries were considerably disorganised with post-ovulatory follicles that were degenerating. The ovaries were characterised by pre-vitellogenic oocytes and pre-ovulatory follicles. The ovaries were in the recovering spent condition on 29 August with mainly pre-vitellogenic oocytes and occasional resorbing eggs. In the example shown the ovary was egg-bound, the fish had failed to spawn and the eggs were undergoing autolysis.

The ovaries of goldsinny in September and October were in a recovering and quiescent phase respectively. Both were classified as stage 1 as the ovary was clear and translucent and occupied about a quarter of the body cavity. The oocytes were all previtellogenic. A comparison of goldsinny and rock cook ovaries in October showed the same maturation stage 1 and a similar appearance. Therefore there was little change and development of ovaries from the end of August to January with only oocytes present and vacuole-free cytoplasm. Throughout the one-year study the GSI and maturation stages and appearance of all ovaries of both species on any date were very similar and synchronised.

5.5 Distribution of maturation stages in the two species

The percentage of oocytes in each maturation stage was calculated. In all fish examined through the year stage 1 or pre-vitellogenic oocytes dominated the frequency of maturation stages with a similar distribution in both goldsinny and rock cook wrasse (Fig. 5.5). The developing oocytes and oocytes in final maturation, from stages 3 to 5, were prominent from June to July. This indicates that several egg batches were spawned in the same season. From 9 August few hyaline oocytes were present and by 29 August only resorbing oocytes were present with almost all oocytes being stage 1 in development. All the oocytes from August to December were pre-vitellogenic.

Distribution of oocyte diameters in goldsinny and rock cook wrasse

Distinct size distributions of oocytes were present from 1 June corresponding to oocytes in stage 3 and 5 of development in both goldsinny and rock cook (Fig. 5.6). Oocytes of goldsinny in stage 3 on 1 June were of mean diameter range from 140–224 µm. By 28 June stage 3 was 102–183 µm, while stage 5 mature oocytes were 306–367 µm. There was a similar pattern in oocyte diameter on 21 July from 122–183 µm for stage 3 and 275–428 µm for stage 5, and by 9 August this was 132–194 µm for stage 3 and 265–326 µm in stage 5. From September to December the pre-vitellogenic oocytes were 22 to 65 µm. The oocytes of rock cook in different developmental

Fig. 5.5 The distribution of developmental maturation stages of wild-caught goldsinny and rock cook wrasse through an annual cycle. Each bar represents examination of one ovary.

stages were also in distinct diameter groups. On 1 June three developmental stages were present, stage 1 from 30–70 µm range, stage 3 from 122 to 193 µm, and stage 5 from 285 to 367 µm. On 28 June oocytes in stages 1, 2, 3 and 5 were 30–91, 102–112, 142–183, and 285–367 µ m respectively. By 17 July oocytes in stages 1, 2, 3 and 5 were 20–91, 91–122, 183–244, and 265–346 µm.

5.6 Potential egg production: fecundity from egg counts

The number of eggs in each ovary is shown for goldsinny (n=27 fish) and rock cook (n=21 fish) (Table 5.2) in samples taken from the Isle of Arran on 1 and 28 June. The mean fecundity of goldsinny was 25,547, with a range from 6312 to 51,138 per fish in a length range 93 to 139 mm, 12 to 48 g weight. Mean egg number in rock cook was 26,286 with a similar range of 8419 to 51,860 eggs per fish, for fish of 90 to 140 mm length and 11 to 49 g. This can be compared with realised fecundity in goldsinny of 3000 to 80,000 reported by Stone (1996). The relative fecundity (N eggs g^{-1} fish) was 1067 and 1256 for goldsinny and rock cook respectively. Egg number for fish weight was described by the equations $y = 951x + 6095$ $r^2 = 0.44$ where y= and

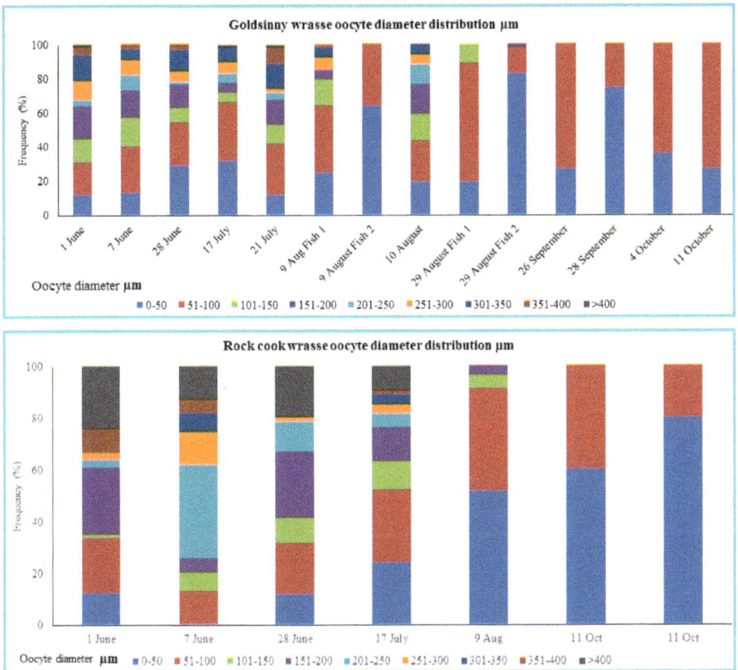

Fig. 5.6 Oocyte diameters of goldsinny and rock wrasse through the spawning cycle. Each bar represents one fish.

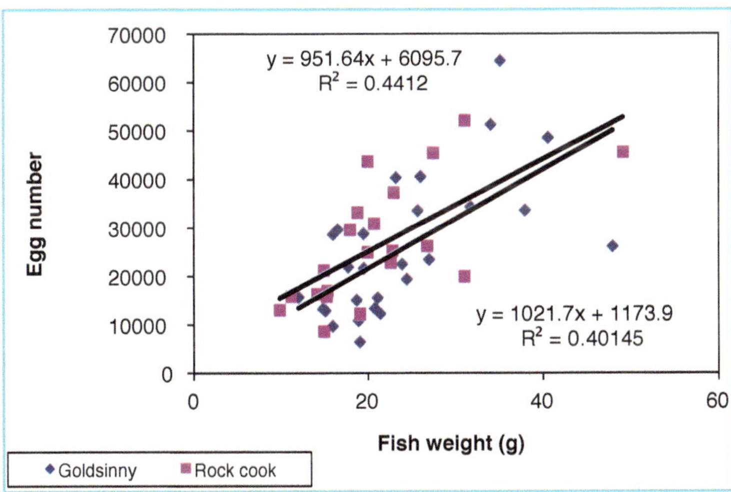

Fig. 5.7 Fecundity of goldsinny and rock cook wrasse plotted against fish weight. The upper line is goldsinny.

x= intercept on the y axis for goldsinny and y= 1021x + 1173 r^2 =0.40 for rock cook (Fig. 5.7), although there was no significant correlation of egg number with weight. Egg diameter of the ovaries fixed in Gilson's fluid was similar to that found in histological sections and ranged from 0.26 to 0.37 mm in goldsinny wrasse.

Table 5.2 Annual fecundity (F) of goldsinny and rock cook estimated by gravimetric counts of ovary samples (see text).

Goldsinny				Rock cook			
Length mm	Weight g	Fecundity	Relative F	Length mm	Weight g	Fecundity	Relative F
102	19.0	6312	332	125	31.0	51860	1671
93	12.0	15553	1295	90	11.2	15620	1397
116	24.4	19206	786	114	22.9	37037	1615
126	38.0	33432	881	121	31.0	19709	636
102	17.7	21770	1228	95	15.3	16732	1092
110	21.4	12120	568	102	17.9	29363	1637
112	23.9	22320	934	110	20.0	24756	1241
104	18.7	14920	800	97	14.9	20966	1406
139	47.9	26000	543	108	19.1	12037	632
112	27.0	23240	861	104	15.3	15604	1023
105	18.9	10720	568	103	14.9	8419	564
100	20.7	13240	639	108	20.7	30640	1482
96	15.1	12760	847	112	22.6	22560	999
97	15.9	9600	603	98	14.2	16040	1133
126	35.1	64344	1832	90	9.8	12800	1301
128	31.7	34156	1077	115	22.8	25032	1098
103	14.8	13073	883	119	27.4	45189	1647
106	16.5	29475	1789	140	49.0	45333	924
112	21.1	15377	730	122	26.8	25958	970
115	26.0	40419	1553	107	20.0	43425	2177
118	25.7	33387	1300	111	18.8	32931	1752
108	19.5	21477	1101				
108	19.5	28640	1471				
128	40.6	48401	1193				
103	16.0	28470	1784				
125	34.0	51138	1502				
115	23.2	40226	1736				

5.7 Realised egg collection in hatchery maintained goldsinny by hand-stripping eggs and milt compared with natural spawning

5.7.1 Egg collection set-up

Adult goldsinny were captured by fishermen in baited prawn creels in the Sound of Mull. The breeding stock comprised 40 males and 47 females of 100–140 mm length and 15–50 g in weight. The gender of the fish could be readily determined at any time of the year, with females displaying a dark brown colour while the males had a distinct gold band on the flanks during the breeding season and had a tighter, less distended abdominal area. These fish were stocked in a 1440 l volume tank supplied with ambient temperature seawater that had been passed through a sand filter and treated with UV light. Water was filtered locally to 5 µm and supplied at a rate of 10 l min^{-1}. Oxygen, maintained between 7 and 10 mg l^{-1} using aeration only, and temperature were measured daily. This was in the range 12.5 to 17.3°C and temperatures increased in the recirculation system as the water was heated in a header tank. The tank was placed adjacent to a window and therefore natural light and day length were provided. Plastic pipes and artificial seaweed fronds were placed in the tanks to provide shelter and environmental enrichment. The fish were fed *ad lib* three times daily to satiation with a Skretting marine broodstock diet. Fish were given a supplemental diet of crushed mussel or defrosted shore crab for eight weeks prior to spawning.

5.7.2 Egg collection and disinfection

Eggs were collected by three methods. In method one eggs and milt were stripped from the fish manually by gently massaging the belly and a 1 ml syringe was used to collect the milt while the eggs were extruded into a petri dish (Fig. 5.8a). In all cases the volume of milt obtained was minimal, around 0.1 to 0.2 ml (Fig. 5.8b). All developed eggs in a female were stripped in one event in fish anaesthetized with MS_{222}, and no further egg batches were obtained from the fish. The milt, which was a light yellow colour, was added to the 'dry' eggs that had been transferred to a 2 l jug. Some seawater, ca. 200 ml, was added to activate the milt and to cover the eggs. More seawater was added after two minutes to fill the jug to 2 l. Fertilisation rates were in the range from 60 to 70%. The fertilised eggs were rinsed with clean seawater after 30 minutes, disinfected with 50 ppm Pyceze (50% active bronopol, Novartis). The viable eggs were allowed to float and dead material was removed by siphoning. The eggs were then transferred to 10 l volume Perspex incubation units with a (recirculating) flow through seawater supply at 1 l min^{-1} and gentle aeration.

In the second method broodstock goldsinny were allowed to spawn naturally in a tank with an overflow to an egg chamber for the eggs to avoid predation in the main tanks. Eggs were collected by natural spawning to increase egg supply and to reduce stress in the broodstock. Broodstock were stocked in five aquaria of 140 l volume. Water was supplied at the bottom of the tank and flow was used to carry eggs over a partition in the tank to a separate egg collection area where they could be collected with a fine hand net. Four females and one male were stocked in each tank. Light was provided by white strip light to simulate ambient photoperiod conditions and slow transition of a dawn and dusk period was provided by a Sunmatch control unit.

The third method was to allow fish to spawn naturally within a 500 l tank of 120 × 60 × 50 cm dimensions. The eggs were pelagic and were carried from the surface water of the tank to an

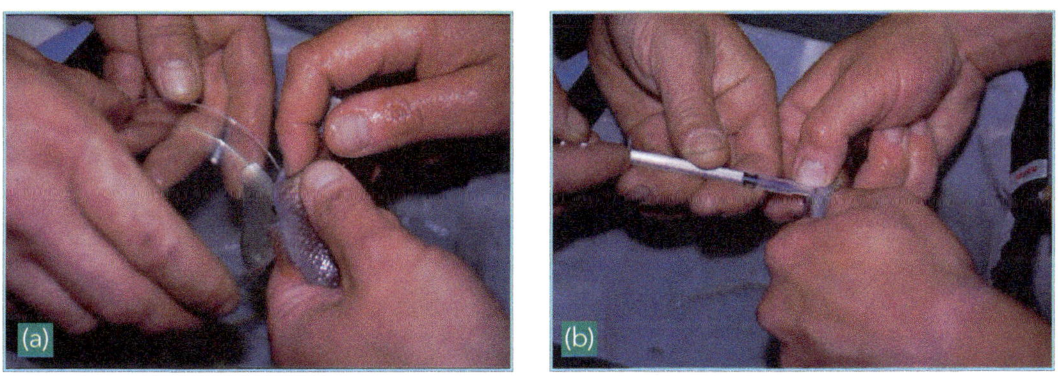

Fig. 5.8 Eggs (a) and milt collection (b) from goldsinny broodstock. Credit: Ardtoe

egg collection unit fitted with a plankton mesh of 68 µm diameter net held within a smaller tank that received the overflow from the broodstock tank. The eggs were collected every day at 1600 hours and transferred directly to an experimental egg incubation room, where the 10 l incubation tanks were held in racks and provided gentle aeration to ensure adequate egg oxygenation. The system was part of a recirculation unit with a bio-filter and heater. The temperature over the spawning period was in the range 12.5–15°C.

5.7.3 Egg disinfection

Eggs were disinfected in a hand net held in a basin with a 1:250 solution of Kick Start™ (CD Lines), a mixture of periacetic acid and hydrogen peroxide, for 30 seconds and then rinsed in clean seawater. After disinfection the eggs were moved to 10 l aquaria. Eggs were disinfected daily during incubation with Pyceze™ (Bronopol BP 50% w/v) at 50 ppm for 30 minutes in the egg incubator. During the treatment the waterflow was stopped and aeration provided. The temperature was in the range from 12.5°C to 19.1°C and the salinity from 32.3 to 34.2 ppt throughout. Continuous lighting was used at a level of 400 lux. This can be compared with Marine Harvest Norway where there was low hatch with continuous lighting (A. Lekva, MHNorway, personal communication, 2017). Dead eggs ('drop out') were removed daily by siphon. The proportion of floating viable eggs as a percentage of the total egg collection varied from 40 to 100% with an average of 59%. Egg diameter was in the range 0.74 mm to 0.82 mm and this also declined significantly ($P<0.05$) as the spawning season progressed. The fertilisation rate of the net collected eggs was variable and in the range 20–98%, and it also declined with time. The stages in egg development were assessed by removing a sample of ca. 50 eggs at one hour intervals for 12 hours and six hourly thereafter (Fig. 5.9 developmental stages). The incubation period to hatch was three days at 15°C (45 degree days C) with the hatch rate of batches varying from 10 to 90%.

5.7.4 Manual stripping of eggs

Fifteen batches of eggs were successfully stripped and fertilised. Egg diameter was 0.7 to 0.8 mm. There was no pigment or oil globule and eggs were positively buoyant. The larvae hatched at an early stage in development and eyes were not pigmented, the mouth was not fully formed, and there was no pigmentation or developed gut. At four days post-hatch (dph) the eyes became pigmented and by 5 dph the mouth had developed and the yolk sac was diminishing. By six days the larvae showed some interest in the feed and green pigment in the gut suggested that the larvae were ingesting algae.

5.8 Larval rearing trials

5.8.1 Larvae from hand-stripping

Ten egg batches were incubated. The egg production was in the range 640–3400 per egg batch, and was similar to realised fecundity of 2712 per female obtained from natural spawning in 2008.

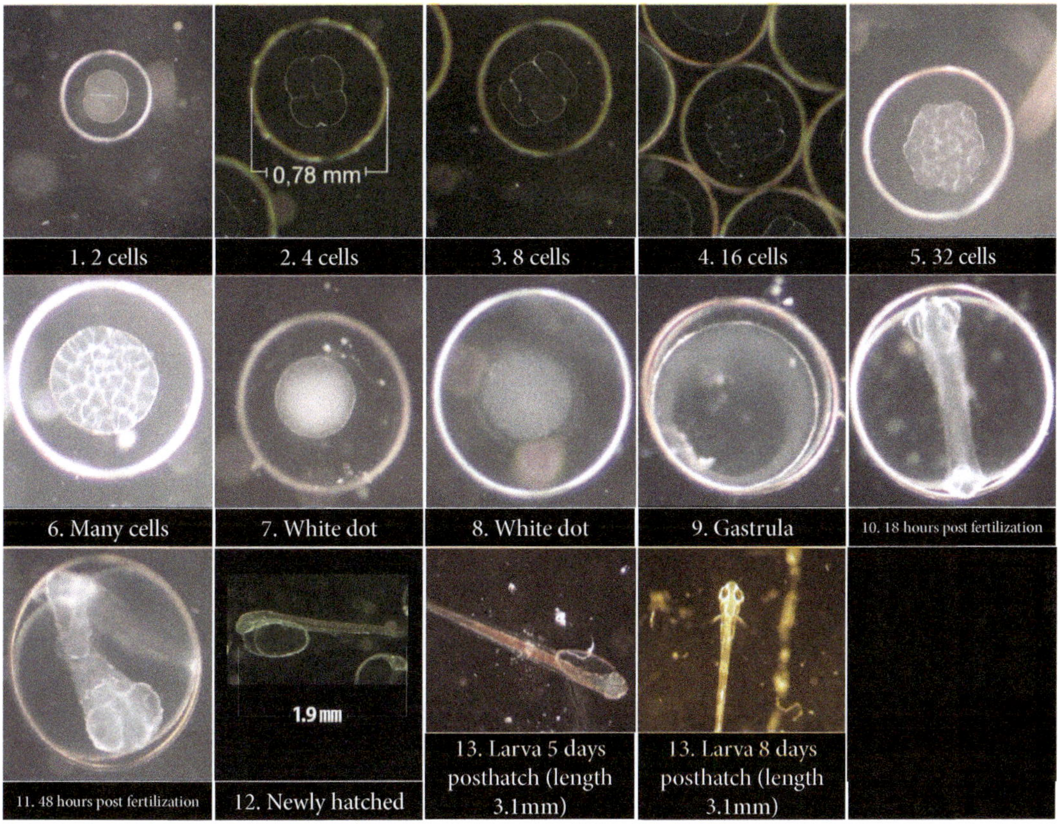

Fig. 5.9 Egg developmental stages of goldsinny wrasse (adapted from Chalaris and Treasurer, 2010).

The length of larvae on hatching three days after spawning was 1.9 mm. There was no eye pigmentation and the mouth and gut were poorly developed. Algae was added to the 2 m^3 tanks and larvae were observed to be feeding within five to six days of hatch. The yolk sac was comparatively large with a mean depth of 0.8 mm. A range of larval feeds was offered five days post-hatch, with the algae *Nannochloropsis oculata* and *Dunaliella tertiolecta* supplied to the tanks in the morning and the rotifer *Brachionus plicatilis* in the afternoon. It appeared that the jaw gape of the larvae was too small to allow ingestion of rotifers so mussel larvae (*Mytilus edulis*) and protozoa were therefore fed. Wild mussels were induced to spawn by injection of potassium chloride and trochophores of 0.063 mm length were successfully obtained. These were fed together with protozoans, comprising mainly vorticellids. The protozoans hatched from dry samples were not abundant and did not meet the feed requirements of the larvae.

Larvae were fed with Isochrysis and Nannochloropsis, followed by copepod nauplii. Best survival was obtained from tanks containing adult copepods and Artemia nauplii until larvae attained a length of 30 to 35 mm and then small live adult Artemia and frozen adult Artemia were offered. By three to four weeks the larvae developed vertical bands of dark pigment giving them a striped appearance. The larvae were swimming actively, ingested rotifers and copepods and were 8 to 9 mm total length. After six weeks the larvae abandoned the pelagic swimming behaviour throughout the entire tank area and took refuge in cover on the sides of the tanks, and behind airlines and air stones. Artificial hides made by attaching biological filter bobbins to thin string were added to one tank and these were utilized by the larvae. Coloration of the larvae changed with increased pigmentation and they changed to a brick red colour with a distinctive white eye stripe. The larvae accepted Artemia nauplii readily at this point. This was followed by offering dry feed. The use of newly hatched mussel (*Mytilus edulis*) larvae was also tested successfully. Locally sourced mussels were injected with KCL (potassium chloride) allowing artificial fertilisation of their eggs. The newly hatched mussel larvae (79 µm total length) were introduced (60 larvae l^{-1}) to half of the small goldsinny held in 60 l volume egg incubation tanks and later larvae rearing tanks were used. However, the use of mussel larvae as live feed was deemed sub-optimal for goldsinny larval rearing due to limitations to this live feed supply. After ten months the mean length of juvenile goldsinny in two batches was 68 mm (n=80) and 57 mm respectively, and mean weights were 4.9 g and 2.9 g. The respective ages were 316 and 282 dph.

5.8.2 Larvae from natural spawning

The duration of the spawning season was from 7 July until 14 August 2008 at a water temperature range of 12.5 to 14.7°C. The daily egg production ranged from 1 g to 12 g (Fig. 5.10), and the volume of eggs spawned and egg diameter (Fig. 5.11) declined as the season progressed. A total of 127 g of eggs was collected by natural spawning. There were 850 eggs g^{-1} and this represented 108,545 eggs from 47 fish, which is equivalent to a realised fecundity of 2310 eggs per fish. This is less than the numbers estimated by egg counts from ovaries. It would suggest a loss of eggs in the broodfish tank from predation by the broodstock, sinking eggs being lost in the system and not being carried from the broodstock tank, or fecundity of hatchery-reared fish falling behind wild stocks because of nutritional deficiencies.

Survival to the juvenile stage was low, ranging from zero in some egg batches to 2.9 to 6% in the remainder (n=9). In 2007 and 2008 survival was variable between egg batches from zero survival to 3%. However, these losses were compensated for by restocking tanks and by stocking

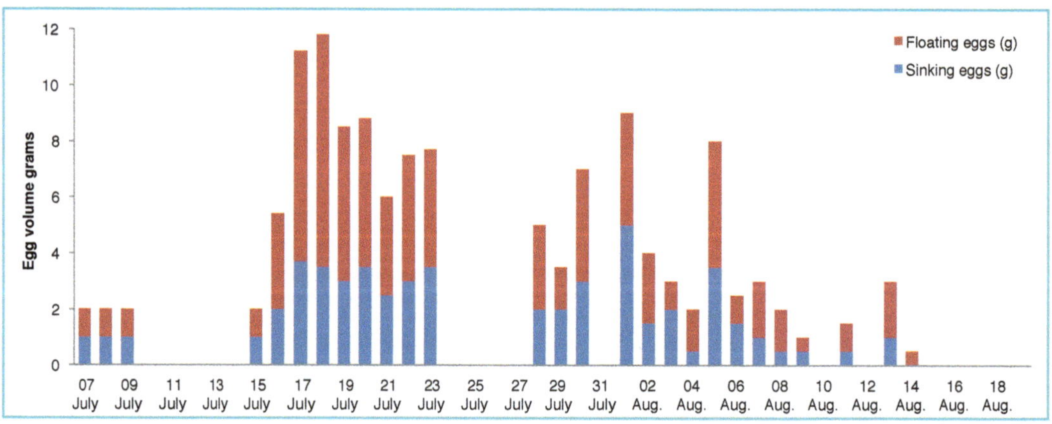

Fig. 5.10 Pattern of natural spawning in goldsinny wrasse through the season.

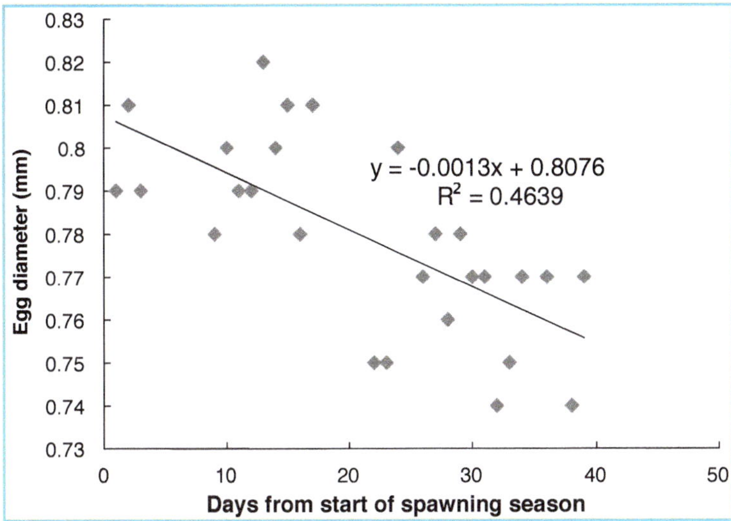

Fig. 5.11 Egg diameter of goldsinny wrasse through the spawning season.

of large numbers of eggs. Clearly further work has to be carried out in first feeding to find small rotifers or very small copepod nauplii that would enhance survival. Once at the post-larval period goldsinny were weaned to dry feed more readily than Ballan wrasse and over a shorter co-feeding period, with three to four weeks of Artemia co-feeding sufficient.

5.9 Diseases and quality issues

Few mortalities were recorded in the goldsinny broodstock that were maintained for nine years. There was no incidence of amoebic gill disease AGD or bacterial infection during these rearing trials. Likewise, although survival of larvae was low to weaning, this was related to the provision of suitable items for first feeding rather than health issues. A sample of 150 fish was examined for vertebral anomalies in 2008 and none was found. Similarly, the mouth and the jaws were normal in larval rearing over five years. However, many fish had jaw deformations in 2013, in 76% and 72% of fish in two separate rearing tanks (Fig. 5.12). Of 46 fish examined, nine had

Fig. 5.12 Goldsinny wrasse with head and jaw abnormalities, from the top: normal, mild abnormality, fish with pug head, fish with pug head and lower jaw deformation.

a shortened upper jaw, 23 with a moderate level of lower jaw downturn, and nine fish had pug heads. Some of the fish had a combination of these deformations. The abnormalities in five of the 46 fish could be classified as severe.

5.10 Discussion

The annual gonad development and maturation stages in goldsinny and rock cook wrasse in wild populations are described in this chapter. The development of the ovary was described well by the maturation scheme after Rinchard and Kestemont (1996). The timing of the seasonal cycle in ovarian development was reflected in developmental stage distribution and the oocyte size distribution. The spread of developmental stages and oocyte size distribution indicated that there were several batches of oocytes developing in goldsinny over the spawning season. A feature was the similarity in timing of each ovarian developmental stage in both goldsinny and rock cook. The onset and completion of spawning was also synchronous between the species. Although GSI and fecundity were similar between the two species, reproductive investment in rock cook males was much higher at 6% compared with less than 1% GSI in goldsinny. This made the hand-stripping of milt from goldsinny difficult. This cycle in GSI and timing of peak values in June were similar to that found by Sayer *et al.* (1996) for goldsinny and rock cook.

5.10.1 Comparison of egg collection methods

The manual stripping of gametes from broodstock goldsinny was successful and has been attempted by several hatcheries (Stone, 1996; Skiftesvik, 1996; Sherwood, 1990, 1991, current study) but the technique is time-consuming and frequent handling may affect the fish. Repeat batches of eggs could have been expected with stripping but this did not appear to happen in practice. However, there were no mortalities after manual stripping and eggs were obtained from these broodfish over a nine-year period. Sherwood (1991) found variable results with manual stripping. The egg stripping had to be carried out at the optimum point in egg development in the ovarian developmental cycle. It was not possible to calculate the spawning cycle of individual females, and many of the stripped egg batches were either over- or under-ripe and therefore not viable.

Natural spawning in aquaria with an overflow egg collecting section had limited success as it was only possible to stock four female fish and one male in the restricted volume (Sherwood, 1991), thus giving limited egg supply. Aggression was seen on several occasions between and within sexes in this setting. However, the size of the aquaria used at Ardtoe (0.33 m^2) should not have inhibited natural spawning (Sherwood, 1991) as goldsinny have been reported to spawn in glass tanks with a base of as little as 0.4 m^2 (Hillden, 1981). A complicating factor could have been the late stocking of broodstock into the aquaria. Males may not have established territories and this may have caused enough disruption to prevent natural spawning. Natural spawning was highly successful when goldsinny were maintained in larger broodstock tanks of 500 l, with dimensions of 120 × 80 × 60 cm. The overflow water from the tank was collected in a large plankton net of 1 m diameter and 68 µm mesh held in an adjacent overflow tank. Spawning commenced on 7 July at a water temperature of 12.5°C and continued for 43 days with the last egg batch collected on 19 August at 17°C. All egg batches were fertilised with a range of 20–83%. Only half the eggs were buoyant and viable, and this may have been due to nutritional deficiencies in broodstock diets. However, a proportion of sinking eggs has also been reported for goldsinny in the wild (Quignard, 1966).

5.10.2 Larval rearing

There was a wide variation in larval survival between tanks over several years. This was related to deficiencies in the initial feed offered as rotifers were too large. Further work is required to look at alternative small feed. Microalgae appeared an important initial feed item and the quality and species and nutrient properties may have been important. However, in contrast to Ballan wrasse (www.eco-fish.org, 2013) goldsinny adapted rapidly to dry feeds. Growth rates were slow, around 30 mm per year in the first two years and slower thereafter. Growth to around 100 mm minimum is required to ensure no escapement from 15 mm mesh salmon net and in Norway a minimum length of 110 mm is stocked (Halvorsen *et al.*, 2017), so focus on the use of heated water and cultivation in recirculation systems is required to improve growth rate. A technical data sheet for goldsinny culture is given in Appendix 1.

5.10.3 Future of goldsinny wrasse rearing, fisheries and deployment

Goldsinny wrasse are regarded as effective cleaner fish and they comprise the dominant species in the wrasse fisheries in Scotland (Treasurer, 1994; 1996) and also in Norway at Skaggerak (Halvorsen *et al.*, 2017). Therefore, goldsinny was the main wrasse species stocked in initial cleaner fish trials in Scotland in the 1990s, when Ballan were wrongly considered undesirable due to the fear of eye picking. Slow growth in the hatchery and low early survival have to be tackled. Nevertheless, goldsinny wrasse have been reared in recent years in Scotland at Otter Ferry Seafish, Machrihanish and at FAI Aquaculture Ardtoe (Chalaris and Treasurer, 2010), and in Norway in various hatcheries (e.g. Marine Harvest Labrus, A. Lekva, MHN, March 2017, personal communication). However, efforts are currently focused on lumpfish production and also on Ballan wrasse rearing, and it remains to be seen whether goldsinny wrasse will be reared in commercial numbers in the future. Despite this, goldsinny will remain an important species in wild fish collection and stocking and more than 40% of wild-caught wrasse stocked in some farms in Scotland were goldsinny (Chapter 17). Fishery management measures have been introduced in Norway, where more than 26 million wrasse were captured in 2017 (Sveier and Breck, Chapter 18). The recommendations for wrasse fisheries (Norwegian Department of Fisheries) were to have a lower catch limit of 110 mm for goldsinny and a TAC of 4 million wrasse in north Norway, 10 million in mid and 4 million in south in 2016 (Halvorsen *et al.*, 2017). A cleaner fish-stocking strategy followed by several salmon farming companies is to stock goldsinny and other small wrasse species with smaller salmon in the first year of production and to introduce larger Ballan wrasse in the second year. Although this policy will have to be evaluated and refined, there is no doubt that the 'smaller' wrasse species will continue to be stocked in large numbers and will have an important part to play in the biological control of sea lice as farming of Ballan wrasse is currently insufficient to meet demand.

Acknowledgements

We thank the staff at Ardtoe and Tim Atack for support over the various years of goldsinny wrasse farming development. We are grateful to Craig Pooley and Anette Lekva for reviewing the chapter and suggesting improvements. Antonios Chalaris was a student of the University of Thessalonica while carrying out his portion of the work at Ardtoe.

References

Bjordal, Å. (1988) *Cleaning symbioses between wrasses Labridae and lice infested salmon Salmo salar in mariculture.* ICES Mariculture Committee 1988/F: 17, 8 pp.

Bjordal, Å. (1990) Sea lice infestation on farmed salmon: possible use of cleaner-fish as an alternative method for delousing. *Canadian Technical Report of Fisheries and Aquatic Sciences* 14 pp.

Bjordal, A. and Costello, M. (1990). How good is this natural control of sea-lice? *Fish Farmer* 13 (3), 44–46.

Bravo, S., Sevatdal, S and Horsberg, T.J. (2008) Sensitivity assessment of *Caligus rogercresseyi* to emamectin benzoate in Chile. *Aquaculture* 282, 7–12.

Chalaris, A. and Treasurer, J.W. (2010) Developing the commercial rearing of goldsinny wrasse for the biological control of sea lice. *Fish Farmer* 33, 59–61.

Costello, M. (1991). Review of the biology of wrasse (Labridae: Pisces) in northern Europe. *Progress in Underwater Science* 16, 29–51.

D'Arcy, J., Dunaevskaya, E., Treasurer, J., Ottesen, O., Maguire, J., Zuraleva, N., Karlsen, A., Rebours, C. and FitzGerald, R.D. (2012) Embryonic development in ballan wrasse *Labrus bergylta. Journal of Fish Biology* 81, 1101–1110.

Darwall, W.R.T., Costello, M.J., Donnelly R. and Lysaght, S. (1992) Implications of life-history strategies for a new wrasse fishery. *Journal of Fish Biology* 41 (Supplement B), 111–123.

Dipper, F.A. and Pullin, R.S.V. (1979) Gonochorism and sex inversion in British Labridae (Pisces). *Journal of Zoology, London* 187, 97–112.

Grant, A. and Treasurer, J.W. (1993) The effects of fallowing on caligid infestation on farmed Atlantic salmon (*Salmo salar* L.) in Scotland. In: Boxshall, G.A. and Defaye, D. (eds.). *Pathogens of wild and farmed fish: sea lice.* Ellis Horwood, Chichester, pp. 255–260.

Grant, B., Davie, A., Taggart, J.B., Selly, S.-L.C., Picchi, N., Bradley, C., Prodohl, P., Leclercq, E. and Migaud, H. (2016) Seasonal changes in broodstock spawning performance and egg quality in ballan wrasse. *Aquaculture* 464, 505–514.

Halvorsen K.T., Larsen T., Sørdalen T.K., Vøllestad, L.A., Knutsen, H. and Olsen, E.M. (2017) Impact of harvesting cleaner fish for salmonid aquaculture assessed from replicated coastal marine protected areas. *Marine Biology Research* www.tandfonline.com/doi/abs/10.1080/17451000.2016.1262042

Hillden, N-O. (1978) On the feeding of the goldsinny *Ctenolabrus rupestris L.* (Pisces, Labridae). *Ophelia* 17, 195–198.

Hillden, N-O. (1981) Territoriality and reproductive behaviour in the goldsinny *Ctenolabrus rupestris* L. *Behavioural Processes* 6, 207–221.

Hillden, N-O. (1983) Cleaning behaviour of the goldsinny (Pisces: Labridae) in Swedish waters. *Behavioural Processes* 8, 87–90.

Lythgoe, J.N. and Lythgoe, G.I. (1991) *Fishes of the Sea: The North Atlantic and Mediterranean.* Blandford Press, London.

Muncaster, S. (2010) The reproductive cycle of female ballan wrasse *Labrus bergylta* in high latitude, temperate waters. *Journal of Fish Biology* 77, 494–511.

Norwegian Government Department of Fisheries Statistics (2016) http://www.fiskeridir.no/English for data from 1998.

Potts, G. (1973) Cleaning symbiosis among British fish with special reference to *Crenilabus melops* L. *Journal of the Marine Biological Association of the UK* 53, 1–10.

Quignard, J.-P. (1966) Recherches sur les Labridae (Poissons, Teleostéens, Perciformes) des côtes européennes: systématique et biologie. *Naturalia Monspeliensia* (Zoologie) 5, 7-248.

Riley, A., Jeffery, K., Cochrane-Dyet, T., White, P. and Ellis, J. (2017) *Northern European Wrasse-Summary of commercial use, fisheries and implications for management.* CEFAS, Lowestoft. 18 pp.

Rinchard, J. and Kestemont, P. (1996) Comparative study of reproductive biology in single- and multiple-spawner cyprinid fish. I. Morphological and physiological features. *Journal of Fish Biology* 49, 883–894.

Sayer, M.D.J, Gibson, R.N. and Atkinson, R.J.A. (1993) Distribution and density of populations of goldsinny wrasse (*Ctenolabrus rupestris*) on the west coast of Scotland. *Journal of Fish Biology* 43 (Supplement A), 157–167.

Sayer, M.D.J, Gibson, R.N. and Atkinson, R.J.A. (1995) Growth, diet and condition of goldsinny on the west coast of Scotland. *Journal of Fish Biology* 46, 317–340.

Sayer, M.D.J. and Treasurer, J.W. (1996) North European wrasse: identification, distribution and habitat. In: Sayer, M.D.J., Treasurer, J.W. and Costello, M.J. (eds.) *Wrasse biology and use in aquaculture.* Fishing News Books, Oxford, pp. 3–12.

Sayer, M.D.J., Gibson, R.N. and Atkinson, R.J.A. (1996) Seasonal, sexual and geographical variation in the biology of goldsinny, corkwing and rock cook on the west coast of Scotland. In: Sayer, M.D.J., Treasurer, J.W. and Costello, M.J. (eds.) *Wrasse biology and use in aquaculture.* Fishing News Books, Oxford, pp. 13–46.

Sevatdal, S., Copley, L., Wallace, C., Jackson, D., and Horsberg, T.E. (2015) Monitoring of the sensitivity of sea lice (*Lepeophtheirus salmonis*) to pyrethroids in Norway, Ireland and Scotland using bioassays and probit monitoring. *Aquaculture* 244, 19–27.

Sherwood, J.P.B. (1990) *Wrasse breeding trial.* Sea Fish Industry Authority Marine Farming Unit. 11 pp.

Sherwood, J.P.B. (1991) *Wrasse breeding trial.* Sea Fish Industry Authority Marine Farming Unit. 10 pp.

Skiftesvik, A.B., Boxaspen, K. and Parsons, A. (1996) Preliminary breeding trials of wrasse in an intensive system. In: Sayer, M.D.J., Treasurer, J.W. and Costello, M.J. (eds.) *Wrasse biology and use in aquaculture.* Fishing News Books. Oxford. 283 pp. 136–141.

Stone, J. (1996) Preliminary trials on the culture of goldsinny and corkwing wrasse. In: Sayer, M.D.J., Treasurer, J.W. and Costello, M.J.(eds.) *Wrasse biology and use in aquaculture.* Fishing News Books, Oxford, pp. 142–167.

Treasurer, J.W. (1993) Management of sea lice (Caligidae) with wrasse (Labridae) on Atlantic salmon (*Salmo salar* L.) farms. In: Boxshall, G.A. and Defaye, D. (eds.) *Pathogens of wild and farmed fish: sea lice.* Ellis Horwood, London, pp. 335–345.

Treasurer, J.W. (1994) The distribution, age and growth of wrasse (Labridae) in inshore waters of west Scotland. *Journal of Fish Biology* 44, 905–918.

Treasurer, J.W. (1996) Capture techniques for wrasse in inshore waters of west Scotland. In: Sayer, M.D.J., Treasurer, J.W. and Costello, M.J., (eds) *Wrasse: Biology and Use in Aquaculture*, Fishing News Books, Oxford, pp. 74-90.

Treasurer, J.W. (2011). *Advances in cleaner fish rearing technology.* Presentation to the BMFA Annual Conference, SAMS Oban.

Wheeler, A. (1969) *Fishes of the British Isles and North-East Atlantic.* Macmillan, London.

CHAPTER 6

Review of lumpfish biology

Adam Powell,[1] Craig Pooley,[2] Maria Scolamacchia[2] and Carlos Garcia de Leaniz[2]

[1]University of Gothenburg – Kristineberg, Department of Biological and Environmental Sciences, Sweden

[2]Centre for Sustainable Aquatic Research, Department of Biosciences, Swansea University, UK

6.1 Distribution

The lumpfish, *Cyclopterus lumpus,* occupies different habitats depending on life stage. Adults are generally solitary and spend much of the year far from land, and at up to several hundred metres depth (Kudryavtseva and Karamushko, 2002; Parin *et al.*, 2002). The following distribution describes hatchlings, juveniles and adults during the breeding season, which takes place generally closer to shore, in shallow coastal water (after Davenport, 1985; Stein, 1986; Aquamaps, 2017).

Lumpfish may be regarded as abundant, potentially inhabiting *c.* 32,000 km of coast across both sides of the Atlantic Ocean (see Davenport, 1985 for a summary of archived studies). Lumpfish are distributed in the boreal region of the east and west North Atlantic coasts (Fig. 6.1). For the western Atlantic, the most northerly occurrence has been found on the island of Disko off north-western Greenland; lumpfish are distributed from there southwards to Chesapeake Bay (range: 70° – 37°N). This distribution incorporates most of eastern Canada, including Nunavut, Hudson Bay, James Bay, Labrador, Newfoundland (Stevenson and Baird, 1988), Gulf of St Lawrence, New Brunswick and Nova Scotia.

On the western side of the Atlantic Ocean, the species has been recorded over a wider range of latitudes than the European coast (see Davenport, 1985). However, in Europe, recent records of the species have been recently reported from lower latitudes off Spain, southern Portugal (36°N) and a probable vagrant in the Mediterranean Sea (Bañón *et al.*, 2008; Vasconcelos *et al.*, 2004; Dulčić and Golani, 2006). The most northerly occurrences in Europe include Jan Mayen (north of Iceland), Svalbard, and the White and Barents Seas including Novaya Zemlya (Kudryavtseva and Karamushko, 2002), with populations found easterly into the Baltic Sea. The species is commonly found off Iceland, the southerly tip of Greenland and the Faroes, Norway (Holst, 1993) and countries bordering the North Sea, particularly France, the UK and Ireland.

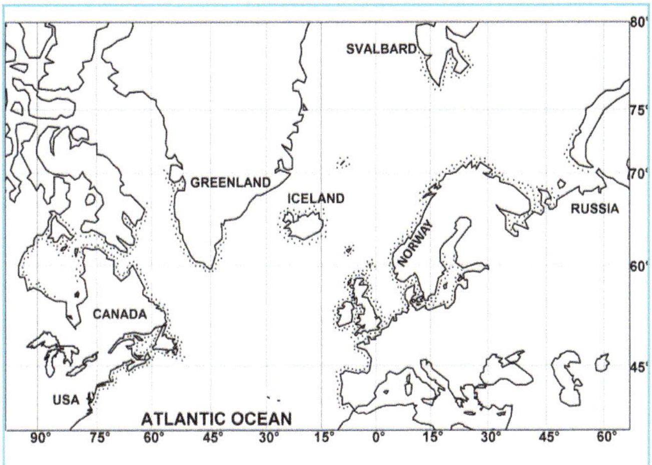

Fig. 6.1 Map showing coastal distribution of *Cyclopterus lumpus* (shaded areas showing probable extent of adult spawning grounds and habitat of hatchlings and substrate associated juveniles). Credit: Redrawn by the authors (after Davenport, 1985)

6.2 Description of the appearance of fish and taxonomy

The lumpfish or lumpsucker was formally named and initially characterised by Linnaeus (1758) as *Cyclopterus lumpus* ('round fin'). Other common names in current usage across Europe (in direct translation to English) include stone biter, sea chicken and fatfish, with names occasionally differing for each sex due to pronounced sexual dimorphism (Davenport, 1985). The species is a bony fish (class: Osteichthyes, infraclass: Teleostei) belonging to the order Scorpaeniformes, family Cyclopteridae. *C. lumpus* is unique and morphologically distinct, and is the only species of the genus *Cyclopterus*. Related taxa include the less common *Eumicrotremus spinosus* (Atlantic spiny lumpfish, found in the western Atlantic Ocean) and *Aptocyclus ventricosus* (Smooth lumpfish, northern Pacific Ocean; Froese and Pauly, 2017). These other species have no recorded commercial value and have not been investigated for use as cleaner fish to our knowledge.

Davenport (1985) and Bigelow and Schroeder (2002) provide a detailed generic and specific description of adult *C. lumpus*. Briefly, in profile the body is about twice as long as it is deep, and it is compressed anteriorly and posteriorly. The first dorsal fin forms a high crest with large compressed tubercles, which increase in size with age, although these may be reduced in specimens inhabiting particularly cold or low salinity habitats, such as the Baltic Sea. There are three longitudinal ridges along the length of the body marked by a line of pointed tubercles, with the most obvious as a dorsal crest. The head is short with lateral moderate-sized eyes, while the opercula have slit-like openings. The snout is blunt with a terminal, slightly upturned mouth containing small teeth. In cross section, the body is vaguely triangular with a flattened ventral surface (with the exception of gravid females, when distended with roe), with a round, broad, muscular sucking disc (Davenport and Thorsteinsson, 1990) that gives the species its generic name. The disc constitutes approximately 20% of the body length and is a specialised modification of the pelvic fins. The vivid skin colouring in adults and associated sexually dimorphic characteristics (see section 6.12) and skin texture (rubbery, tough and scaleless) are also diagnostic features (Fig. 6.2).

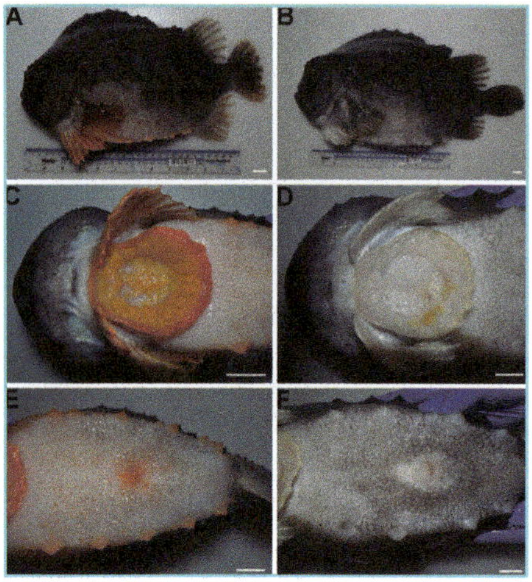

Fig. 6.2 Adult *Cyclopterus lumpus*. Top, A. Male, B. Female. Centre, ventral detail of sucking disc. C. Male, D. Female. Bottom, ventral detail of vent. E. Male, F. Female. Scale bars = 2cm. Credit: Photographs taken by Craig Pooley and Maria Scolamacchia, CSAR, Swansea University

6.3 The lumpfish fishery: vulnerability to fishing and sources of broodstock

Historically, the lumpfish fishery has entered three distinct phases. Prior to the 1940s, there was a small-scale artisanal fishery for males that were sold smoked, salted or dried. This was then largely replaced by a commercial fishery emerging in the late 1940s, focusing on roe from females (as a cheap alternative to caviar from sturgeon; see Davenport, 1985). This has continued to the present day at scales of many thousand tonnes per year (Fig. 6.3; Johannesson, 2006). More recently, the fishery has diversified somewhat to include male and female broodstock for the emerging cleaner fish industry. Aquaculture production of the species currently depends on the capture of wild broodstock, which together with current fisheries may impact on natural populations.

The species has a trophic level of 3.9, typical of secondary consumers, and a low resilience to fishing pressure, with an estimated time for doubling of population size of 4.5 to 14 years. It also has a moderate to high vulnerability to fishing (47 of 100; Cheung *et al.*, 2005; Froese and Pauly, 2017). Although the conservation status of the species has not been assessed by the IUCN (International Union for Conservation of Nature), the characteristics outlined above suggest a species vulnerable to overexploitation. A significant decrease in some spawning stocks has been recorded in recent decades, suggesting that some stocks may already be overexploited (Pampoulie *et al.*, 2014). Global lumpfish production from fisheries is seasonably variable (Johannesson, 2006; FAO, 2017), potentially influenced by the combined impacts of exploitation, with emerging problems such as invasive species that feed on lumpfish eggs (Mikkelsen and Pedersen, 2012), climate change (e.g. Perry *et al.*, 2005) and emerging diseases (e.g. Freeman *et al.*, 2013).

The reported catch of lumpfish has steadily increased from a few hundreds of tonnes per year in the early 1950s to 20,365 tonnes in 2013 (FAO 2017, Fig. 6.3). Mean catch since the year 2000 has been 15,997 tonnes (SE = 1,002) with unreported catches probably varying between 1,600 and 4,600 tonnes (i.e. 9–22% of the total). However, historical catch records are probably misleading because most of the former catch was unreported, mostly due to discards on the

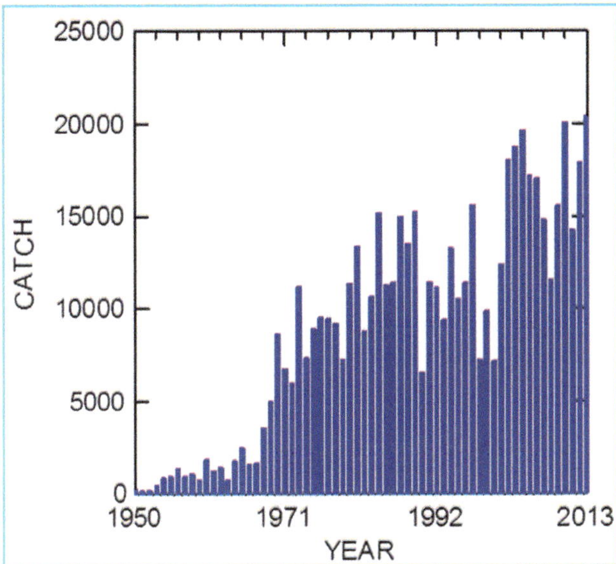

Fig. 6.3 Reported global catch for lumpfish (tonnes), 1950–2013. Credit: FAO Fishstat

east coast of Canada. Data reconstruction indicates that unreported catches may have reached *c.* 124,000 tonnes (98% of total) by 1965, when the declared catch was only 2,400 tonnes (Pauly and Zeller, 2015).

Nowadays, the most important lumpfish fisheries operate in the Canadian eastern Atlantic–west Greenland (*c.* 70% of catch) and Iceland (*c.* 23% of catch), which are exploited by the Greenlandic and Icelandic fleets. Data from 2010 indicate that these are mostly mixed artisanal (57% of catch) and commercial fisheries (43% of catch), with only 30 tonnes being classified as subsistence (0.1% of catch; Pauly and Zeller, 2015).

The artisanal fisheries consist of small boats, fishing close to shore in up to ~30 m depth, and which are typically manned by one to three people. Gears vary among locations (Johannesson, 2006). In Norway, nets with 252 mm mesh size are used, these typically being *c.* 47 m wide with a 3.4 m drop (Bertelsen, 1994). Mesh size in Iceland varies between 267 and 286 mm (Thorsteinsson, 1996). Monofilament nets with a 256–281 mm mesh size are used in Canada to catch lumpfish, nets being 100 m long (Benson *et al.*, 1998). Their round shape and short fins means that lumpfish do not entangle in the nets to the same extent that other species do, and this is the reason lumpfish nets have as much 'bag' as possible, as this is thought to increase catch efficiency. Typically, nets are set up on 45 m float and foot lines, with 90–125 cm verticals to maintain the bags and prevent lumpfish escape (Johannesson, 2006).

Traditionally, lumpfish have been harvested for their roe, which can be processed into an alternative to caviar (Johannesson, 2006). No reliable data are available for the number of females removed from the wild for the caviar industry, but this probably amounts to several million fish annually; efforts to harvest lumpfish eggs non-destructively have met with limited success (Grant, 2001). The number of adult lumpfish taken by the incipient cleaner fish industry (*c.* 300 fish in the UK in 2014, personal observation) is currently small compared with the commercial and artisanal fisheries (20,365 tonnes in 2013, FAO 2017), but there is a growing concern from some conservation bodies regarding the sustainability of the catch (Anon, 2013a). Maximum lumpfish production is attained when males reach 20 cm and females reach 32 cm, equivalent to two and three years of age, respectively (Hedeholm *et al.*, 2014), and that means that removing broodstock older than two to three years of age may have a disproportionately high impact on wild populations. Both Iceland and Greenland have adopted strict fishing regulations (Stevenson and Baird, 1988) and their lumpfish fisheries have recently been awarded MSC certification (Anon, 2013b; Lassen *et al.*, 2015).

While it is possible to use means other than gillnets to capture lumpfish, for example by scuba diving (Killen *et al.*, 2007a), these tend to be more labour intensive and much less efficient. Lumpfish lack a swim bladder, and adults hauled quickly to the surface may experience barotrauma, even using static gear. This can be rectified by returning animals to depth in a cage, followed by gradual decompression.

6.4 Commercial uses of lumpfish

Up until the 20th century lumpfish had little economic value. Small fisheries existed on both sides of the North Atlantic for local consumption, but fish caught as bycatch were often used as animal feed or bait (Stevenson and Baird, 1988). A dedicated fishery targeting lumpfish roe started in the late 20th century. Ripe females yield 15–36% of roe by weight (Davenport, 1985; Stevenson and Baird, 1988; Mitamura *et al.*, 2007; Hedeholm *et al.*, 2014) and were rapidly targeted by the fishery. Thus, in Newfoundland, roe production grew from 21 tonnes in 1970 to 3,000 tonnes by 1989 (Stevenson and Baird, 1988). Similarly in Norway, 100 tonnes of roe were taken annually in the 1950s, compared with 500 tonnes of roe by the middle of the 1980s (*c.* 2,500 tonnes of fish).

Lumpfish eggs are marketed in two ways: either as whole roe, which is then dried, salted, or smoked, or as processed eggs, which are separated from the ovaries and then further elaborated into lumpfish caviar. Annual production of lumpfish caviar is about 4 million kg and has a market value of about $ 60 million, with Canada (35%), Iceland (31%) and Norway (15%) as the main producers (Johannesson, 2006). The price of lumpfish caviar fluctuates depending on supply, but is currently sold in supermarkets at $36–72 per kg.

Excluding the sale of caviar, the fishery has a low value (always below $5 million since 2000), and the species is classed as a 'low value' fish in the prize category of FishBase (Sumaila *et al.*, 2007; Froese and Pauly, 2017). This is in stark contrast to the situation during 1950–1970, when the value of the fishery peaked at *c.* $200 million in 1961, mostly generated by the European fleet (Spain, Portugal, Germany) operating in eastern Canada. However, in some places such as Iceland, the UK and Norway, this situation has changed in recent years, with the development of the cleaner fish industry. Fishermen are currently paid US $87 (GBP £60) for ripe females in some parts of the UK, and a kilogram of fertilised eggs for export to the cleaner fish industry may cost several hundred US dollars in Iceland (personal observation), which may put new pressures on wild stocks.

Currently, all lumpfish used as cleaner fish are culled at the end of the salmon production cycle, when they reach about 500 g. Although this is done on biocontainment grounds, to prevent transmission of diseases between salmon and cleaner fish, some have considered this practice as wasteful and have urged the industry to find alternative uses for farmed lumpfish (Anon

2013a; Powell *et al.*, 2017). This has prompted research into new markets and a consideration of possibilities for nutrient recycling, either as animal feed or into more valuable products, particularly in the lucrative Asian markets, where the species could perhaps find a market niche as a delicacy (Nøstvold, 2017). Another alternative to culling would be to recondition large juveniles in captivity and use them as broodstock for the production of eggs, either for caviar or for the cleaner fish industry (Powell *et al.*, 2017), although such animals would need to be certified free from disease.

6.5 Abundance

Lumpfish occupy a wide distribution, and have been recorded in 24 countries in Europe and North America. Animal distribution also varies according to life history stage and season, with many historic reports unaware of the semi-pelagic adult phase outside the spawning season. Davenport (1985) summarised this as a fundamental problem with assessments of the total area occupied by the species. A recent study of stocks in the Barents Sea (Eriksen *et al.*, 2014) suggested a mean annual biomass (48,000–143,000 t) and mean annual abundance (53–132 million individuals) since 1980. Some data on stock assessment are available for the Icelandic and Greenland fisheries that were recently awarded MSC certification, but not for other areas. On the other hand, reliance on catch data (Fig. 6.3) to infer trends in abundance is fraught with difficulties, as the majority of lumpfish caught during 1950–1970 likely went unreported.

6.6 Food

The diet of lumpfish was summarised by Davenport (1985), who concluded that the overall impression was of a species that subsisted mainly on large planktonic organisms living in surface/mid waters, but which sometimes browsed upon benthic organisms, particularly those dwelling upon weed. For adults, historic studies found a high proportion (*c.* 70–80%) of adults with empty stomachs in sampled populations. A similarly high proportion of empty stomachs has also been reported in juveniles (Ingólfsson and Kristjánsson, 2002). The intestine is long, being more than twice the length of the body in adult fish, with many bends and numerous pyloric caecae (Davenport, 1985) suggesting efficient digestion and absorption of food. Gut contents of adults are variable across all studies, with small crustaceans (mysids, amphipods, euphausids, isopods, decapod zoeae), ctenophores, polychaetes, seagrass, insects, small fish and fish eggs recorded (Davenport, 1985; Davenport and Rees, 1993).

Juvenile lumpfish (*c.* 5–55 mm length) are apparently year-round seaweed specialists, inhabiting and feeding on surface plankton after hatching, and weed-associated invertebrate fauna when larger (Daborn and Gregory, 1983; Vandendriessche *et al.*, 2007). Motile crustaceans larger than 0.5 mm dominate diet, with the natural abundance of these prey items on seaweed correlating with their abundance in the gut of juvenile lumpsuckers (Ingólfsson and Kristjánsson, 2002). Furthermore, as hatchlings increased in size and yolk reserves became exhausted, the proportion of larvae with full guts increased. Hatchlings tend to consume harpacticoid copepods or halacrid mites, with older juveniles switching to physically bigger and more active crustacean prey such as amphipods or decapod larvae (Daborn and Gregory, 1983; Tully and Ó Ceidigh, 1989; Ingólfsson and Kristjánsson, 2002). Cannibalism has been observed between individuals of similar age and size. Rather, a change in behaviour corresponding to age and feed density may allow them to enter a different trophic group (See section 6.15; Brown, 1986; Tully and Ó Ceidigh, 1989; Killen, 2007a, b).

More recently, interest has switched to the gut contents of lumpsuckers deployed in salmon pens. Large lumpfish juveniles 54 g in weight deployed in sea pens are seemingly highly opportunistic, not limiting themselves to one food item (Imsland *et al.*, 2015a). Over a 77-day trial, the most common food items observed were salmon pellets, although crustaceans, hydrozoans, mussels and, as the study progressed, sea lice, became an increasing part of the diet. Further research suggests that food preference in pens may also depend on genetic provenance, and differ between juveniles originating from distinct families (Imsland *et al.*, 2016a), size and co-existence with other species during deployment (Imsland *et al.*, 2016b, 2016c).

6.7 Physiology

Lumpfish anatomy is unusual, with most studies focusing on the skeleton, skin and ventral sucker, tissue composition and buoyancy, and reproduction.

Although the species has no swim bladder and contains large quantities of dense eggs, gravid females have a body density very similar to seawater (Davenport and Kjörsvik, 1986). The low density is achieved by extensive subcutaneous jelly, low osmolarity ovarian fluid, and a dorsal musculature, which is loose-fibred and has high water content. Males do not present these characteristics to such an extent, but have a higher lipid composition. This enables adult lumpfish to reside in the open ocean for long durations. However, during the breeding season, they return to shallow, rocky waters that experience strong wave surge. The low body density, large surface area and limited swimming power demands an extremely powerful and efficient muscular, cartilaginous sucker (or disc), which enables lumpsuckers to attach themselves to substrate, and to rest without being swept away (Davenport and Kjorsvik, 1986; Davenport and Thorsteinsson, 1990). Lumpsuckers can also colour-match their skin to local substrates within minutes (Davenport and Bradshaw, 1995). Suitable resting places and substrates have also been recommended during deployment in salmon pens, to promote fish welfare (Imsland *et al.*, 2015b).

Unfertilised eggs in the large ovaries are bathed and protected in copious ovarian fluid, which has a low divalent ion concentration. As portions of ripened eggs are released during spawning events, they harden and clump together in seawater due to the presence of divalent ions, particularly calcium and magnesium. Copious fluid and a strong sphincter muscle reduces the risk of seawater reflux into the oviduct, hardening remaining eggs inside the fish and blocking further egg release. The kidney likely removes divalent ions, while female fish have large urinary bladders that may assist with urine storage and water reabsorption (Davenport and Lönning, 1983; Lönning *et al.*, 1984).

Sperm extracted from males was found to remain viable for several days after removal of the fish, although Davenport (1983) suggested that stripping eggs or milt from adults was not possible and euthanasia was required to extract gametes for artificial fertilisation. Powell *et al.* (2017)

found that it was possible to strip eggs from very ripe females, while recently caught males often release milt upon handling (although this was never achieved a few days later, in aquaria) requiring culling and manipulation of male gonadal tissue. However, with well-considered logistics, it may be possible to store sperm from wild 'running' males and, using enhancers or cryopreservation techniques (Norðberg *et al.*, 2015), to reduce the necessity for maintaining or euthanising adults.

6.8 Environmental requirements and ecotoxicology

Overall, *C. lumpus* is regarded as 'hardy' (i.e. robust); with the southern and northern limits of distributional range approximating to the 20°C and 0°C August surface water isotherms, indicating that the lumpsucker is eurythermal but capable of inhabiting very cold water (overview by Davenport, 1985). Adults are generally not found in low salinities within their latitudinal range, although populations exist in the Baltic Sea and Hudson Bay that have permanently low salinities; these populations also show some differences in skin, body shape and smaller comparative size (Davenport, 1985), suggesting a discrete race or even subspecies inside a particular oceanographic area that experiences low salinity.

Hatchlings are precocious with a functional sucker, allowing immediate attachment to rock or weed substrate, or even momentarily to the tending male shortly after hatching (Davenport, 1985). Hatchlings and juveniles appear to be almost obligate inhabitants of seaweed, living and feeding on and around weed beds (Vandendriessche *et al.*, 2007), although young juveniles are also commonly found in parts of the intertidal zone such as rockpools (Moring, 2001).

Few ecotoxicological studies have been made on the species, although very low concentrations of mercury have been found in the flesh of wild adult lumpsuckers (Freeman *et al.*, 1974). Juveniles proved relatively resilient to experimental oil exposures designed to simulate oil spills (Frantzen *et al.*, 2015). An antifoulant candidate (medetomidine) has been tested on lumpfish larvae; sub-lethal effects (a reduction in respiration rate and skin colour adaptation rate) were found when exposed at nanomolar concentrations for 72–98 hours. However, larvae returned to baseline levels after 48 hours in untreated seawater (Bellas *et al.*, 2005). This finding may have implications for deployment near antifoulant-treated structures.

6.9 Movements and migrations

The species is benthopelagic and can be found at depths <868 m (Parin *et al.*, 2002), although they usually live in waters between 50 and 150 m deep (Stein, 1986). The semi-pelagic status of adult females has been confirmed with a recent tagging study (Kennedy *et al.*, 2016). Lumpfish may undertake extensive annual migrations between their feeding grounds found in deeper waters in the winter and the shallower waters preferred for spawning in spring and summer (Blacker, 1983; Davenport, 1985; Kasper *et al.*, 2014). Tagging studies indicate that the species displays homing behaviour and may return to breed in the same areas more than once (Davenport, 1985; Stevenson and Baird, 1988; Kennedy *et al.*, 2014). Tagged females were found to reside in a fjord for up to a week and then disappeared, possibly returning offshore after spawning (Mitamura *et al.*, 2012). Following spawning, females can travel up to 49 km per day (Kennedy *et al.*, 2014), unlike males, which remain in the same location for several weeks to tend the eggs (Davenport, 1983).

6.10 Genetic diversity and genetic stock differentiation

Lumpfish have 25 haploid chromosomes (n) and 50:50 diploid chromosomes (2n) (Li and Clyburne, 1977; Klinkhardt *et al.*, 1995), but little is known about its genetic diversity or extent of population differentiation. Twenty-two novel microsatellite DNA loci were characterised recently for the species (Skirnisdottir *et al.*, 2013) and these have revealed three distinct genetic groups in the North Atlantic: Maine–Canada–Greenland; Iceland–Norway, and the Baltic Sea (Pampoulie *et al.*, 2014; Garcia-Mayoral *et al.*, 2016) with little evidence of gene flow among these zones. However, data for other parts of the range are currently lacking. Preliminary results from populations in the English Channel suggest that lumpfish there have low to moderate levels of genetic diversity and low genetic differentiation (Expected Heterozygosity, He = 0.53–0.61; Pooley *et al.*, 2015). These initial data suggests indistinct genetic populations in this particular geographic area. This demands further investigation to inform a sustainable fishing strategy, and potential end uses after deployment as cleaner fish.

6.11 Behaviour

Adult lumpfish are typically solitary in the wild, though hatchlings tend to aggregate in clumps in tanks during captivity (personal observation). Adult spawning behaviour has been described by Goulet *et al.* (1986), Goulet and Green (1988) and is summarised in section 6.13. In brief, during spawning, the male exhibits parental care and tends the eggs, which are kept oxygenated through fanning and air puffing.

Unlike most other Actinopterygians (ray-finned fish), lumpfish lack Mauthner neurons that are involved in the startle response and escape behaviour (Hale, 2000). They have a longer startle response latency than other fishes, but are nevertheless capable of performing a fast anti-predatory response, albeit not as effective. They appear to have a relatively small brain for their size (encephalisation coefficient = 0.07–0.21; Albert *et al.*, 1999), though this may be due to the unusual shape and large body mass of the species. The species is apparently an intelligent, inquisitive fish that can be trained to accept feed (personal observation) and can change colour rapidly for camouflage (Davenport and Thorsteinsson, 1989; Davenport and Bradshaw, 1995).

The ventral sucking disc enables lumpfish to forage differently from most other fishes. They can cling and feed passively, or forage actively in pursuit of prey. The larvae become more active a few weeks post-hatch (Brown, 1986) and their foraging mode appears to depend on prey abundance. Thus, they adopt a 'passive cling' foraging mode when food is abundant, and switch to a more 'active swim' mode when food is more scarce (Killen *et al.*, 2007a).

Little is known about the welfare requirements for the species. Some information is available on the substrate and colour preferences following deployment in salmon pens (Imsland *et al.*, 2014, 2015b), but studies are also needed during the juvenile phase. In the wild, lumpfish match the colour of seaweed, suggesting that light intensity, photoperiod and tank colour may also affect juvenile growth, since these factors have been observed to alter melanin concentrations under experimental conditions (Davenport and Bradshaw, 1995). In salmon pens at night, they prefer to aggregate together on smooth plastic and concrete substrates (thought to be similar to seaweed), rather than on stones or car tyres (Imsland *et al.*, 2015b). In comparison with Ballan wrasse, *Labrus bergylta* (Helland *et al.*, 2014), availability of suitable substrates appears to be important for health and welfare.

The behaviour of lumpfish in cages has been studied via underwater cameras (Imsland *et al.*, 2015b), and an ethogram has been constructed that identified a range of different behavioural elements, which varied with the presence of Atlantic salmon (Imsland *et al.*, 2014a). When salmon were present, lumpfish were more active and spent less time resting, but little antagonistic behaviour was observed between the two species. However, lumpfish did show some aggression to co-deployed goldsinny wrasse (*Ctenolabrus rupestris*), particularly across disparate size grades between the two species (Imsland *et al.*, 2016b).

Lumpfish are opportunistic omnivores in salmon cages, eating not only sea lice, but also salmon pellets and organisms attached to the cages (Imsland *et al.*, 2015a, 2016b). Significant variation appears to exist among families in the efficacy of sea lice grazing (Imsland *et al.*, 2016a), and this may provide the basis for the selective breeding of lumpfish with improved delousing efficiency.

6.12 Sexual dimorphism and sex ratio

Lumpfish display pronounced sexual dimorphism, allowing straightforward differentiation of sexually mature specimens. Males are typically smaller (30 ± 10 cm length) and during the spawning season attain startling red, orange or purple coloration of the fins, eyes and ventral surfaces (e.g. Davenport and Thorsteinsson, 1989). Females are larger (42 ± 10 cm length), less colourful and usually grey or blue-green, and are markedly rotund, likely due to increased subcutaneous gelatinous tissue, urinary bladder and copious ovarian tissue, eggs and fluid; they also have a larger vent (Davenport and Lönning 1983; Davenport, 1985; Goulet *et al.*, 1986; Fig. 6.2).

Although the available data on sex ratio is scant, it is clear that the species deviates from the Fisherian 1:1 ratio, the extent of which is likely to differ in space and time. The apparent itinerant, oceanic and solitary nature of sexually mature lumpfish outside spawning seasons, and contrasting habitat favoured by juveniles (and difficulty to assign gender) makes an accurate and meaningful estimation of the sex ratio somewhat challenging. It is also likely that different arrival and residence times between sexes during the spawning period (females arrive later and for only a few days; Davenport, 1985; Mitamura *et al.*, 2007) would further complicate the description since sex ratio would change during the season. For instance, the reported sex ratio of 1.57:1 (M:F; n=54; Gregory and Daborn, 1982) is likely to be specific for the sample taken in shallow water, in the Bay of Fundy, in early summer.

Furthermore, some have speculated that males outnumber females since they are apparently capable of guarding more than one egg mass at a time; however, females may spawn their eggs in a number of batches at different locations (Mochek, 1973; Davenport, 1985; Davenport and Kjørsvik, 1986). In addition, it has yet to be confirmed if sexually mature females spawn annually, at longer intervals or if they are effectively semelparous due to potentially low recovery rates after spawning (which becomes progressively more expensive with increasing size; Thorsteinsson, 1983; Kasper *et al.*, 2014). An added complication is that males mature one to two years earlier than females, however females appear to experience greater longevity (Davenport, 1985; Albert *et al.*, 2002; Hedeholm *et al.*, 2014). Therefore, it seems the operational sex ratio (i.e., the local ratio of fertilisable females to sexually active males at any given time) for lumpfish is apparently quite complex.

In areas of their range where a roe fishery is significant, there may be further bias toward recorded numbers due to the difference in size and behaviour between genders, and stronger swimming ability of males. Males are likely to be underestimated in the gillnet fishery since they may be able to evade capture or are unlikely to move significant distances while guarding eggs (Davenport, 1985). Furthermore, the very presence of a fishery that targets gravid female fish will bias the sex ratio. Landing data are often expressed by biomass or tonnage rather than by number (e.g. Marine Research Institute, 2015) although analysis of long-term survey data from Newfoundland found that the sex ratio has decreased progressively from approximately 2.24:1 to 1.09:1 (F:M) between 1985–1994, despite an increase in catch per unit effort (CPUE; Stansbury *et al.*, 1995; Hoenig and Hewitt, 2005). However, human activity may not be the sole reason for female mortality. Emerging diseases such as *Nucleospora cyclopteri* were more prevalent in females in a recent survey of the species off Iceland (Freeman *et al.*, 2013).

6.13 Reproduction, spawning and geographical patterns

Early in the breeding season, males establish territories in shallow water on rocky and/or seaweed-covered substrate, prior to the arrival of females; females arrive asynchronously, allowing more than one pairing and breeding event to occur per animal. This process occurs in spring and early summer (Davenport, 1985) with a likely latitudinal variation. In Iceland, northern Norway and Newfoundland, females start to move inshore during March and April, with evidence for spawning occurring in early July and hatching until late August (Brown *et al.*, 1992; Kennedy *et al.*, 2014; Mitamura *et al.*, 2012) whereas in the English Channel gravid females have been caught from early January until early May only (personal observation).

Davenport (1985) and Goulet *et al.* (1986) provide accounts of courtship, which appears to be extended and of several hours' duration. This includes showing flanks to one another, pectoral fin brushing, quivering, and long periods of sucker attachment in close proximity; there was also evidence for olfactory (perhaps pheromone) communication. In the wild, this included cleaning a nest site (which may be a simple crevice or depression in bedrock, boulders and/or vegetation). Mating appears to correlate with night (or darkness in aquaria) and potentially with high tide.

Fertilisation is external, with females releasing eggs freely into the water on to the surface of a nest (or in the aquaria, near to any shallow depression or tank structure, personal observation) with attendant males immediately fertilising the eggs, with the overall act lasting only five to ten seconds (Goulet *et al.*, 1986). Females are thought to lay two to three batches of eggs over *c.* one to two weeks (Davenport *et al.*, 1985), and in one study actively moved in and out of a fjord within days of spawning (Mitamura *et al.*, 2012). On contact with sea water, the eggs adhere to one another to form large ovoid masses, which are fertilised by the male. Before the mass hardens, the male may mould the eggs into the nest. Males also create funnel-like depressions that likely assist gaseous exchange and removal of nitrogenous waste from the centre of the egg mass.

The eggs undergo paternal care throughout the incubation, with the females expending no time in egg care after fertilisation. The most frequent behaviour includes water circulation by fanning with the pectoral, but also dorsal and caudal fins (for a few seconds to hours in duration), and puffing (expelling water from the mouth). This appears to occur more frequently for a few hours after fertilisation, and also near the end of the incubation period when the embryos are more developed, and to disseminate emergent larvae (Davenport, 1985; Goulet *et al.*, 1986). If eggs are exposed by low tides, males have been observed to spout water from the mouth to maintain them (Davenport, 1985). The male also becomes aggressive and removes or defends the egg mass from conspecifics and predators, including sea urchins, periwinkles and even large predatory fish, although this is more challenging against schools of ocean pout *Zoarces americanus*, cunners *Tautogolabrus adspersus* (Davenport, 1985; Goulet *et al.*, 1986) or large crabs such as the invasive red king crab *Paralithodes camtschaticus* (Mikkelsen and Pedersen, 2012). Males expend much effort in guarding egg masses, apparently not eating during this period, potentially guarding more than one mass at a time or guarding successive masses over the spawning season. The amount of time spent in parental care, or number of eggs guarded, is independent of male size, while nest characteristics such as depth, distance from shore, and topography do not correlate with hatching success (Davenport, 1985; Goulet *et al.*, 1986; Goulet and Green, 1988).

6.14 Egg stages and environmental preferences

Lumpfish eggs are relatively uniform in size (2.0–2.6 mm diameter) across their range (see Davenport, 1985; Benfey and Methven, 1986; Brown *et al.*, 1992) and after extrusion undergo a colour change from pink to a variety of colours that are homogenous within a particular egg mass. Colour is lost with time as the pigments move from the yolk into the chromatophores of the developing embryo. The hardening process occurs via a transparent adhesive material, secreted by the ovary to coat the eggs. This becomes viscous and elastic on exposure to seawater and condenses within an hour at 5°C to form a dense layer around the eggs. Within 48 hours the eggs themselves harden and become dense (Zhitenev, 1970; Lönning *et al.*, 1984; Davenport, 1985).

Embryonic development has been pictorially described for the related smooth lumpsucker *Aptocyclus ventricosus* (Kyûshin, 1975) and for *C. lumpus* in this chapter (Fig. 6.4). With the naked eye, the initial egg colour fades three to four days after fertilisation, and eggs 'eye-up' after 10–12 days, with full embryo pigmentation and a darker egg mass existing from about day 16 onwards. Infertile or undeveloped eggs remain unpigmented and opaque. Initial development can be slow and variable in the egg mass (Davenport, 1983). In our studies, the following stages were observed post-fertilisation at 10°C: morula (one to two days); blastula and blastodisc (two to four days); gastrulation; initial somites visible (five to six days); otic capsule visible, continued segmentation (seven days); eye pigmentation and heart beat (eight to ten days); yolk vascularisation and head pigmentation (11–12 days); development of eyes, head and extensive vascularisation (15 days); uniform pigmentation, regular heart beats, mouth opening, movement (16–19 days).

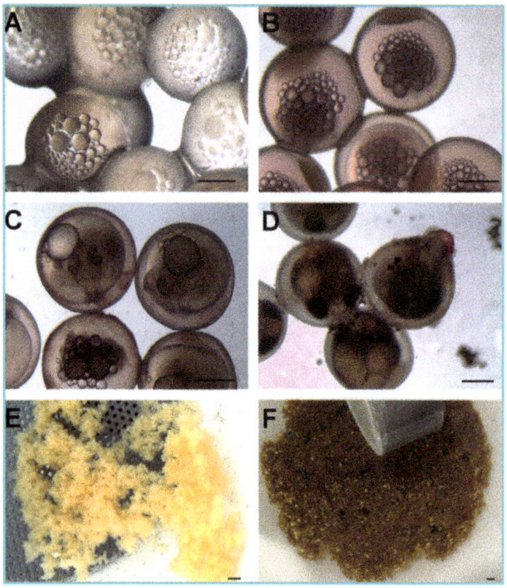

Fig. 6.4 Overview of embryogenesis of C. *lumpus* at 10°C. A. Morula stage, *c.* two days post-fertilisation. B. Blastula stage, *c.* four days post-fertilisation. C. Eye pigmentation, *c.* ten days post-fertilisation. D. Full pigmentation and development at hatch. Scale bar = 1mm. E. Part of egg mass, corresponding to Blastula stage; eye pigmentation has not developed but mass has lost initial colour. F. Part of egg mass, just before hatch. Note dark colour of eggs and occasional light eggs denoting undeveloped embryos. Scale bar = 10 mm. Credit: Description and images by Maria Scolomacchia

Early observations showed that eggs develop more rapidly as temperature rises (Davenport, 1985), with studies showing a temperature of greater than 3.8°C is required for development, with hatching after 40, 31 and 25 days at 5°C, 6.4°C and 9.8°C respectively (Collins, 1978; Davenport, 1983) with hatching occurring mainly at night (Benfey and Methven, 1986). Studies at Swansea University have found a time to hatch of 20–30 days at 10°C (i.e. 200–300 degree days C) with hatching occurring over a seven- to ten-day period. Egg masses manipulated into flattened layers 5–10 mm thick immediately after fertilisation (but prior to hardening) have generally led to better hatching success and more even development rate, perhaps due to improved water circulation. This is in agreement with Brown *et al.* (1992), who had better success with 'monolayers' rather than fragmented egg masses, and which developed at 28.5 days at 10°C. Lumpfish embryos at early developmental stages are capable of extracting oxygen at low oxygen tensions (*c.* 40% of air saturation), or even withstand 30 minutes of anoxia, although sensitivity is increased in more developed embryos nearing hatching (Davenport, 1983). Correct egg development also demands salinities between *c.* 20–34 ppt (Kjörsvik *et al.*, 1984). Little appears to be known regarding optimal photoperiod or pH. Further details on lumpfish broodstock, hatchery and deployment are reviewed by Powell *et al.* (2017) and are considered in Chapters 7, 8 and 9.

6.15 Juveniles

Juvenile lumpfish hatchlings are *c.* 5–6 mm length and 2–3 mg weight (Benfey and Methven, 1986; Brown *et al.*, 1992). The sucker is present at hatching and allows immediate attachment to surfaces, although in other respects newly hatched lumpfish do not resemble adults. A posterior, protocercal fin is present dorsally and ventrally that terminates at the tail. The median fin breaks up into separate fins by the length of about 8–9 mm. The first dorsal fin is gradually overgrown

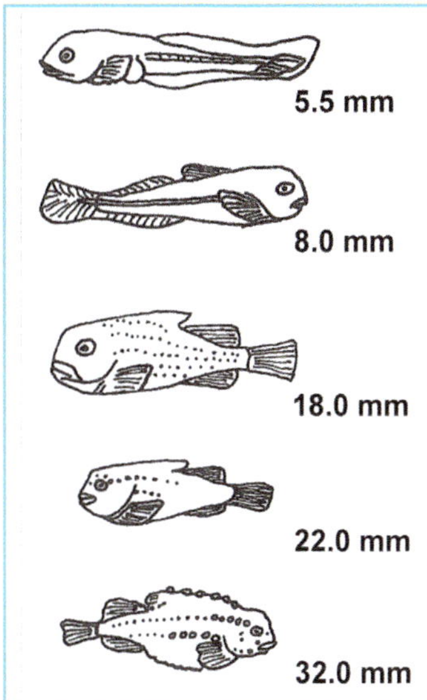

Fig. 6.5 C. *lumpus*. Larval hatchlings and juveniles (not to scale). Credit: Davenport, 1985 after Cox, 1920

by the characteristic dorsal 'hump', and a 32 mm specimen is essentially a miniature of the adult fish (Fig. 6.5; Davenport, 1985). Feminisation of lumpsucker hatchlings has been demonstrated, either by immersion in oestradiol or via enrichment in live feeds, which could promote production of monosex populations (Martin-Robichaud *et al.*, 1994).

Wild juveniles feed on plankton, and substrate-associated small invertebrates, and have been offered *Artemia* and/or small dry feed pellets in hatcheries (e.g. Benfey and Methven, 1986; Brown, 1986; Nytrø *et al.*, 2014; Powell *et al.*, 2017). Juvenile lumpfish show limited aerobic scope compared with other fish species (Killen, 2007a, b) and thus feed and behave differently from most other cultured fish due to the ventral sucker. This enables hatchlings to choose a 'passive cling' mode or a more costly 'active swim', with the latter change in behaviour induced at low prey densities. Higher prey densities allow scope for other physiologically demanding processes such as growth and digestion (Killen, 2007a, b). However, lumpfish larvae grew faster when food was administered in short pulses than when it was administered continuously (Brown *et al.*, 1997).

A few weeks post-hatch, hatchlings become more active (Brown, 1986), and also have reduced yolk stores (Ingólfsson and Kristjánsson, 2002) allowing the start of weaning. Immobile dry feed pellets <800 μm have been offered after 21 days, and hatchlings are maintained in shallow raceways (Nytrø *et al.*, 2014), or circular tanks (Powell *et al.*, 2017). The use of live feeds early in the ontogeny of lumpfish larvae seems crucial, although the use of preserved or dry feeds in older hatchlings switched on genes responsible for lipid metabolism and in some cases yielded improved growth and survival (Belova, 2015). A strategy for species-specific improvement of live feed enrichments or dry feeds is to match their composition with that of early life stages of the species. Indeed, vitellogenin amino acid composition has been described for lumpfish eggs (Yao and Crim, 1996) and proximate carcass analysis for juveniles (Sayer *et al.*, 2000). Further research effort could yield a range of specific feeds for lumpfish hatcheries, particularly as the sector increases in scale.

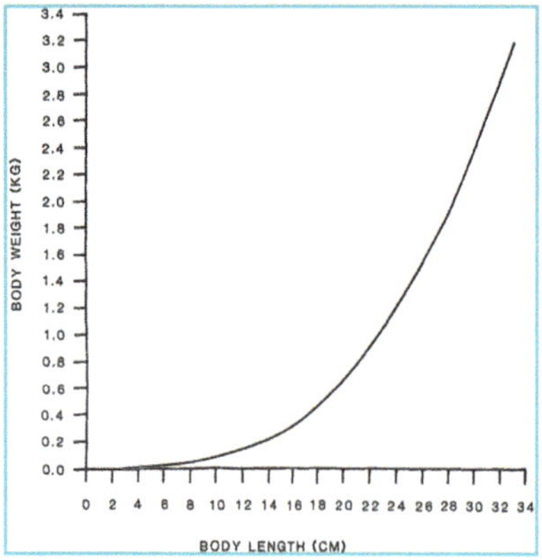

Fig. 6.6 C. *lumpus*. Length–weight relationship of adult lumpsuckers from Newfoundland waters. Credit: Davenport, 1985 after Cox, 1920

6.16 Adults

Davenport (1985) provides an overview on the length–weight relationship of lumpfish (Fig. 6.6) and age of sexual maturity. Historic studies suggest that spawning lumpfish are at least four years of age but most common at five to seven years up to nine to ten years, while males in the North Sea may attain sexual maturity one or two years earlier. More recent studies (using updated otolith readings, age estimates and length frequencies) suggest that males may actually spawn for the first time at age two to three, and females at age three to four in the North Sea, Norway, Iceland and Greenland (Albert *et al.*, 2002; Hedeholm *et al.*, 2014).

Off Greenland, the maximum total production rate (somatic and gonadal tissue) for males and females was 0.47 and 0.92 kg wet weight per year, attained at 20 and 32 cm total length respectively. For both sexes, somatic production declined steeply after the onset of maturation (Hedeholm *et al.*, 2014; Kasper *et al.*, 2014). The anatomy of the gonads is described by Davenport and Lönning (1983). In mature, ripe females, up to two-thirds of the visceral cavity contains pink roe. The ovaries and oviducts are fused to form a single sac that is strongly bifurcated anteriorly, although the left horn is reduced to allow space for other organs. The dorsal and dorso–lateral portions of the ovary-oviduct are lined by a thick, viscous layer that contains whiteish opaque, unripe eggs in most females. Ripe eggs, which are clear and rose pink, fall from this layer into the lumen of the structure, which is filled with 200–500 ml protective ovarian fluid. Egg numbers (100,000 to 400,000 per female) vary with the overall size of the fish, and geographic population. The gonads of the male lumpsucker are simple and unremarkable white structures.

Mortality appears to be size dependent. In Icelandic populations, Thorsteinsson (1983) showed that mean length increased with age until a certain length interval (*c.* 42–44 cm in Icelandic females). Thereafter, mean length decreased with age, indicating death of larger fish. Other than fishing pressure, adults are predated upon by different animals depending on their location. During pelagic stages there is evidence of predation by sharks, seals and sperm whales. In shallow water during spawning season, males (which have a longer residency) are taken by gulls, fish eagles and otters (see Davenport, 1985 for an overview). Thorsteinsson (1983) also suggested that reproduction becomes progressively more expensive with increasing size, and that eventually large fish do not recover from spawning (i.e. gonad production, spawning migration and the associated period of prolonged starvation). Davenport and Kjörsvik (1986) suggested that spent female fish might be positively buoyant (because the eggs are much denser than seawater) and more vulnerable to predation. This is probably more likely for larger females following spawning.

6.17 Growth rate in the wild and hatcheries

The itinerant nature of adult lumpfish outside the breeding season makes growth assessments challenging, with tagging experiments highlighting some limitations of this approach in the species (e.g. Kaspar *et al.*, 2014). Archive data on tagging, summarised by Davenport (1985) suggests that large adults show a very small linear growth increment of a few centimetres after about one year between tagging and recapture. Juveniles, however, are thought to possess rapid allometric growth prior to leaving surface waters and follow a LW relationship of $W = 8.7L^{3.36} \times 10^{-6}$, (Newfoundland) and $W = 2.309L^{3.053}$ (Norway; L in mm) with 55 mm length juveniles thought to be about one year old (Myrseth, 1971; Daborn and Gregory, 1983). A long-term study of wild juveniles found in summer rockpools by Moring (2001) reported that average lengths increased by 23–43% and wet weight by 280–342% per month. Studies of both wild and cultured larval and juvenile lumpfish show a rapid increase in growth rate from mid-July to August, before decreasing in August–September (Benfey and Methven, 1986; Moring, 2001; Ingólfsson and Kristjánsson, 2002).

For adults, the largest reported lumpfish was an individual of body mass 9.5 kg, (maximum total length 70 cm, maximum age of 14 years). However, the bulk of spawning females caught in commercial gillnets are between 35 and 50 cm in length (roughly 2–5 kg) with slow growth after maturity (Bagge, 1964; Thorsteinsson, 1983; Davenport, 1985).

Rearing studies started in the mid-1980s and attempted to culture larvae from the egg, offering dry (Benfey and Methven, 1986) and live (Brown, 1986) *Artemia* to hatchlings. After experimental periods of approximately one month, larvae reached a standard length of *c.* 12 mm on live feed, compared to 7 mm on dry feed. Brown *et al.* (1992) grew lumpfish from eggs and reported a growth of 29 g after one year, and 510 g after two years at ambient temperatures (between -1.8 and 14°C), although survival had declined to 0.5% after the first year. Lumpfish cultured in the laboratory became sexually mature at the end of their second year.

Sayer *et al.* (2000) attempted on-growing of wild caught juveniles, most success being achieved by use of a low lipid diet formulated specifically for the species (higher lipid diets allowed rapid growth, but fat accumulated in brain and liver, causing early mortality). For fish under 350 g in weight, a SGR of up to 3.8% per day was recorded, although the temperature varied between 6.1°C and 15.4°C during the on-growing period. The most recent studies of juvenile growth have been derived from juveniles originating from eggs reared in-house and weaned on to appropriate dry feeds (Nytrø, 2013; Nytrø *et al.*, 2014). The highest growth rates were comparable to those reported by Sayer *et al.* (2000), and were observed for juveniles under 120 g in warmer 13 and 16°C treatments, which resulted in an SGR of 3.65 and 3.60% per day respectively. However, the effect of rearing temperature on juvenile growth was reduced in size grades over 120 g. For lumpfish deployed in fish pens at sea, a recent study has shown differences in growth rate between juveniles originating from different families (*c.* 1.2–2.1%/day; Imsland *et al.*, 2016a).

Although there is likely genetic variation between families (Imsland *et al.*, 2016a), given optimal temperature, there is a high growth potential for juveniles (i.e. potentially increasing their weight by an order of magnitude in under 80 days), thus allowing animals to reach deployable size (10–15 g) within months, and achieving some annual synchrony with commercial salmon farms.

6.18 Diseases and health implications

Maintaining high health status of all life stages of lumpfish is key during hatchery operations to promote welfare and to attain numbers of deployable juveniles, and is particularly important given the goal of closing the life cycle in captivity. Furthermore, understanding the susceptibility of lumpfish to notifiable diseases (or the potential to act as 'carriers') and thus onward transfer to salmonid species during deployment, is also of clear importance to understand and manage potential risks (Murray, 2016).

Lumpfish are a relatively novel candidate for aquaculture, with diseases and parasite records found in academic literature (summarised in Powell *et al.*, 2017), but also increasingly from 'grey' literature (such as veterinary reports) as cultured stock present with symptoms. The knowledge of lumpfish diseases has increased in recent years, alongside research into the species immune system that may foster vaccine development (e.g. Rønneseth *et al.*, 2015). The topic will be considered in further detail in Chapters 13 and 14.

Briefly, general (and treatable or preventable) pathogenic and non-pathogenic diseases in hatcheries include ciliates (*Trichodinia* sp.), flukes (*Gyrodactylus* sp.), fungi (*Exophiala* sp), cataracts and post-capture stress (physical handling, transport, barotrauma, fatty deposits around organs with high lipid diets) have been recorded in adults and reared juveniles (Scyborska, 1948; Dawit, 2015; Sayer *et al.*, 2000; personal observation).

Whilst parasitic worms (Rolbiecki and Rokicki, 2008) and myxosporeans (Cavin *et al.*, 2012; Kristmundsson and Freeman, 2014) have been recorded in wild lumpfish, endemic diseases are likely more damaging. The fungal microsporidian *Nucleospora cyclopteri* (Mullins *et al.*, 1994) has been found at high (*c.* 25%) prevalence in wild adults (Freeman *et al.*, 2013) although there is currently no confirmed route of transmission or effective treatment. Lumpfish also appear to suffer from a high prevalence (61–100%) of sea lice such as *Caligus elongatus* (Boxshall, 1974; Heuch *et al.*, 2007). More recently, amoebic gill disease (Perry and Treasurer, 2015), pasteurellosis and other bacterial infections (e.g. Alarcón *et al.*, 2015; Smage *et al.*, 2016), and viral diseases (Towers, 2015) have been detected.

6.19 Summing up: gaps in knowledge and research needs

Knowledge of the biology of lumpfish has advanced greatly over the last few years in response to the needs of the new cleaner fish industry, but information on many critical areas is still missing. A detailed gap analysis is provided in Powell *et al.* (2017) but, in brief, the ultimate objective is to produce disease-free juveniles that adapt well in captivity, do not pose a risk to salmon or other fishes, and are efficient at delousing.

To work toward this goal, the production of lumpfish needs to be closed in captivity, without dependence on wild broodstock. This will require better knowledge of artificial reproduction, particularly with respect to control of maturation, gamete collection and storage. Commercial production will need to be derived entirely from farmed strains, and this will require the development of a genetic breeding programme, one that can produce elite lines with superior performance and desirable traits, including disease resistance and delousing efficiency. Targeting slow growing lumpfish strains may be advantageous in order to prolong the time lumpfish can graze on sea lice. Larval production also needs to be optimised, along with improved diets to help reduce high post-weaning mortality. Vaccines, as well as more effective therapeutants, are also required to combat emerging infectious diseases. Finally, reuse of deployed lumpfish needs to be considered, perhaps through their use as broodstock, in animal feeds, or for human consumption.

Acknowledgements

The authors would like to thank an anonymous reviewer for constructive comments during the preparation of this chapter.

References

Alarcón, M., Gulla, S., Røsæg, M.V., Rønneseth, A., Wergeland, H., Poppe, T.T., Nilsen, H. and Colquhoun, D.J. (2015) Pasteurellosis in lumpsucker *Cyclopterus lumpus*, farmed in Norway. *Journal of Fish Diseases* 39, 489–495.

Albert, O.T., Torstensen, E., Bertelsen, B., Jonsson, S.T., Pettersen, I.H. and Holst, J.C. (2002) Agereading of lumpsucker (*Cyclopterus lumpus*) otoliths: dissection, interpretation and comparison with length frequencies. *Fisheries Research* 55, 239–252.

Albert, J.S., Froese, R., Bauchot, R. and Ito, H. (1999) Diversity of brain size in fishes: preliminary analysis of a database including 1174 species in 45 orders. In: Séret, B. and Sire, J-Y. (eds.) Proceedings of the 5th Indo-Pacific Fisheries Conference, Noumea, New Caledonia, 3–8 November 1997. La Société française d'ichtyologie, Paris, France. pp. 647–656.

Anon (2013a) Cleaner Fish Position Paper. Available at: www.mcsuk.org/downloads/fisheries/Cleaner_Fish_Position_Paper.pdf (Accessed 30 April 2016)

Anon (2013b) MSC Assessment for Icelandic Gillnet lumpfish Fishery. 5M Publishing.

Aquamaps (2017) Computer generated distribution maps for *Cyclopterus lumpus* (Lumpfish), with modelled year 2100 native range map based on IPCC A2 emissions scenario. www.aquamaps.org, version of August 2013. Accessed 13 February 2017.

Bagge, O. (1964) Some observations on the biology of the lumpsucker (*Cyclopterus lumpus*). *International Council for the Exploration of the Sea Council Meeting document 1964.Baltic Belt Seas Committee.* No. 150: p. 7.

Bañón, R., Garazo, A. and Fernández, A. (2008) Note about the presence of the lumpsucker *Cyclopterus lumpus* (Teleostei, Cyclopteridae) in Galician waters (NW Spain). *Journal of Applied Ichthyology* 24, 108–109.

Blacker, R.W. (1983) Pelagic records of the lumpsucker, *Cyclopterus lumpus* L. *Journal of Fish Biology* 23, 405–417.

Bellas, J., Hilvarsson, A. and Granmo, A. (2005) Sublethal effects of a new antifouling candidate on lumpfish (*Cyclopterus lumpus* L.) and Atlantic cod (*Gadus morhua* L.) larvae. *Biofouling* 21, 207–216.

Belova, I. (2015) Nutritional programming in lumpsucker, *Cyclopterus lumpus* – a transcriptomic approach. MSc Thesis. Faculty of Biosciences and Aquaculture, University of Nordland.

Benfey, T.J. and Methven, D.A. (1986) Pilot-scale rearing of larval and juvenile lumpfish (*Cyclopterus lumpus* L.), with some notes on early development. *Aquaculture* 56, 301–306.

Benson, R., Bailey, D., Mercer, J., Murphy, T., Mills, R., Churchill, G., Callahan, R., Allen, N., Whitaker, R. and Way, E. (1998) Lumpfish caviar production technology for the Newfoundland seafood processing industry. Research and Development Report. Newfoundland and Labrador Institute of Fisheries and Marine Technology, St John's, Newfoundland, Canada.

Bertelsen, B. (1994) Bestandsutvikling og bestandsstruktur hos rognkjeks og rognkall (*Cyclopterus lumpus L.*) i Sifjorden i Troms, og maskeseleksjon og fangsteffektivitet ved garn-ogrusefiske etter arten. BS thesis (Fishing Technology). Norwegian Fishery University, Tromsö, Norway. (cited in Johannesson, 2006).

Bigelow, H.B. and Schroeder, W.C. (2002) Fishes of the Gulf of Maine. Revision 1.1. *Fishery Bulletin of the Fish and Wildlife Service* 53 (74), 459–462.

Boxshall, G. (1974) Infections with parasitic copepods in North Sea marine fishes. *Journal of the Marine Biological Association of the United Kingdom* 54, 355–372.

Brown, J.A. (1986) The development of feeding behaviour in the lumpfish, *Cyclopterus lumpus. Journal of Fish Biology* 29, 171–178.

Brown, J.A., Somerton, D.C., Methven, D.A. and Watkins, J.R. (1992) Recent advances in lumpfish *Cyclopterus lumpus* and ocean pout *Macrozoarces americanus* larviculture. *Journal of the World Aquaculture Society* 23, 271–276.

Brown, J.A., Wiseman, D. and Kean, P. (1997) The use of behavioural observations in the larviculture of cold-water marine fish. *Aquaculture* 155, 297–306.

Cavin, J.M., Donahoe, S.L., Frasca, S., Innis, C.J., Kinsel, M.J., Kurobe, T., Naples L.M., Nyaoke A., Poll, C.P. and Weber E.P. 3rd. (2012) *Myxobolus albi* infection in cartilage of captive lumpfish (*Cyclopterus lumpus*). *Journal of Veterinary Diagnostic Investigation* 24, 516–524.

Cheung, W.W.L., Pitcher, T.J. and Pauly, D. (2005) A fuzzy logic expert system to estimate intrinsic extinction vulnerabilities of marine fishes to fishing. *Biological Conservation* 124, 97–111.

Collins, M.A.J. (1978). Experiments on the hatching period of the eggs of the lumpsucker *Cyclopterus lumpus L.* in Newfoundland waters. *Le Naturalist Canadien* 105, 169–71.

Cox, P. (1920) Histories of new food fishes. 2. The lumpfish. *Bulletin of the Biological Board of Canada* 2, 1–28.

Daborn, G.R. and Gregory, R.S. (1983) Occurrence, distribution, and feeding habits of juvenile lumpfish, *Cyclopterus lumpus* L. in the Bay of Fundy. *Canadian Journal of Zoology* 61, 797–801.

Davenport, J. (1983) Oxygen and the developing eggs and larvae of the lumpfish *Cyclopterus lumpus. Journal of the Marine Biological Association of the United Kingdom* 63, 633–640.

Davenport, J. (1985) Synopsis of biological data on the lumpsucker, *Cyclopterus lumpus* (Linnaeus, 1758). *FAO Fishery Synopsis* Vol. 147, p. 31.

Davenport, J, and Bradshaw, C. (1995) Observations on skin colour changes in juvenile lumpsuckers. *Journal of Fish Biology* 47, 143–154.

Davenport, J. and Kjørsvik, E. (1986) Buoyancy in the lumpsucker *Cyclopterus lumpus. Journal of the Marine Biological Association of the United Kingdom* 66, 159–174.

Davenport, J. and Lönning, S. (1983) On the structure and function of the urogenital system of the female lumpsucker *Cyclopterus lumpus* L. (Teleostei: Scorpaeniformes). *Proceedings of the Royal Society of London B: Biological Sciences* 218, 201–210.

Davenport, J. and Rees, E.I.S. (1993) Observations on neuston and floating weed patches in the Irish Sea. *Estuarine and Coastal Shelf Science* 36, 395–411.

Davenport, J. and Thorsteinsson, V. (1989) Observations on the colours of lumpsuckers, *Cyclopterus lumpus* L. *Journal of Fish Biology* 35, 829–838.

Davenport, J. and Thorsteinsson, V. (1990) Sucker action in the lumpsucker, *Cyclopterus lumpus* L. *Sarsia* 75, 33–42.

Dawit, B.G. (2015) Adaptive humoral immune response in lumpfish, *Cyclopterus lumpus* L. MSc Thesis, University of Bergen.

Dulčić, J. and Golani, D. (2006) First record of *Cyclopterus lumpus* L., 1758 (Osteichthyes: Cyclopteridae) in the Mediterranean Sea. *Journal of Fish Biology* 69, 300–303.

Eriksen, E., Durif, C.M.F. and Prozorkevich, D. (2014) Lumpfish (*Cyclopterus lumpus*) in the Barents Sea: development of biomass and abundance indices, and spatial distribution.*Oxford Academic PubMed Google Scholar ICES Journal of Marine Science* 71, 2398–2402.

FAO, 2017. Species Factsheets: *Cyclopterus lumpus.* Available at www.fao.org/fishery/species/2537/en (downloaded February 2017).

Frantzen, M., Hansen, B.H., Geraudie, P., Palerud, J., Falk-Petersen, I.-B., Olsen, G.H. and Camus, L. (2015) Acute and long-term biological effects of mechanically and chemically dispersed oil on lumpsucker (*Cyclopterus lumpus*). *Marine Environmental Research* 105, 8–19.

Freeman, H.C., Horne, D.A., McTague, B. and McMenemy, M. (1974) Mercury in some Canadian Atlantic coast fish and shellfish. *Journal of the Fisheries Research Board of Canada* 31, 369–72.

Freeman, M.A., Kasper, J.M. and Kristmundsson, Á. (2013) *Nucleospora cyclopteri* n. sp., an intranuclear microsporidian infecting wild lumpfish, *Cyclopterus lumpus* L., in Icelandic waters. *Parasites & Vectors* 6, 1–13.

Froese, R. and Pauly, D. (2017) *FishBase.* Available at: www.fishbase.org

Garcia-Mayoral, E., Olsen, M., Hedeholm, R., Post, S., Nielse, E.E. and Bekkevold, D. (2016) Genetic structure of West Greenland populations of lumpfish *Cyclopterus lumpus. Journal of Fish Biology* 89, 2625–2642.

Goulet, D. and Green, J.M. (1988) Reproductive success of the male lumpfish (*Cyclopterus lumpus* L.) (Pisces: Cyclopteridae): evidence against female mate choice. *Canadian Journal of Zoology* 66, 2513–2519.

Goulet, D., Green, J.M. and Shears, T.H. (1986) Courtship, spawning, and parental care behavior of the lumpfish, *Cyclopterus lumpus* L., in Newfoundland. *Canadian Journal of Zoology* 64, 1320–1325.

Grant, S.M. (2001) Alternative harvesting strategies for lumpfish: the feasibility of collecting lumpfish roe using non-lethal techniques. Report prepared for Fisheries Diversification Program, Department of Fisheries and Aquaculture, Newfoundland and Labrador. vi + p. 63.

Gregory, R.S. and Daborn, G.R. (1982) Notes on adult lumpfish *Cyclopterus lumpus* from the Bay of Fundy. *Proceedings of the Nova Scotia Institute of Science* 32, 321–326.

Hale, M.E. (2000) Startle responses of fish without Mauthner neurons: escape behavior of the lumpfish (*Cyclopterus lumpus*). *Biological Bulletin* 199, 180–182.

Hedeholm, R., Blicher, M.E. and Gronkjaer, P. (2014) First estimates of age and production of lumpsucker (*Cyclopterus lumpus*) in Greenland. *Fisheries Research* 149, 1–4.

Helland, S., Dahle, S.W., Hough, C. and Borthen, J. (2014) Production of ballan wrasse (*Labrus bergylta*). Science and Practice. The Norwegian Seafood Research Fund (FHF), p. 136.

Heuch, P.A., Øines, Ø., Knutsen, J.A. and Schram, T.A. (2007) Infection of wild fishes by the parasitic copepod *Caligus elongatus* on the south east coast of Norway. *Diseases of Aquatic Organisms* 77, 149–158.

Hoenig, J.M. and Hewitt, D.A. (2005) What can we learn about mortality from sex ratio data? A look at lumpfish in Newfoundland. *Transactions of the American Fisheries Society* 134, 754–761.

Holst, J.C. (1993) Observations on the distribution of lumpsucker (*Cyclopterus lumpus*, L.) in the Norwegian Sea. *Fisheries Research* 17, 369–372.

Imsland, A.K., Reynolds, P., Eliassen, G., Hangstad, T.A., Nytrø, A.V., Foss, A., Vikingstad E. and Elvegård, T.A. (2014) Notes on the behaviour of lumpfish in sea pens with and without Atlantic salmon present. *Journal of Ethology* 32, 117–122.

Imsland, A.K., Reynolds, P., Eliassen, G., Hangstad, T.A., Nytrø, A.V., Foss, A., Vikingstad E. and Elvegård, T. A. (2015a) Feeding preferences of lumpfish (*Cyclopterus lumpus* L.) maintained in open net-pens with Atlantic salmon (*Salmo salar* L.). *Aquaculture* 436, 47–51.

Imsland, A.K., Reynolds, P., Eliassen, G., Hangstad, T.A., Nytrø, A.V., Foss, A., Vikingstad E. and Elvegård, T. A. (2015b) Assessment of suitable substrates for lumpfish in sea pens. *Aquaculture International* 23, 639–645.

Imsland, A.K., Reynolds, P., Eliassen, G., Mortensen, A., Hansen, Ø.J., Puvanendran, V., Hangstad, T.A., Jónsdóttir, Ó.D.B., Emaus, P.-A., Elvegård, T.A., Lemmens, S.C.A., Rydland, R., Nytrø, A.V. and Jonassen, T.M. (2016a) Is cleaning behaviour in lumpfish (*Cyclopterus lumpus*) parentally controlled? *Aquaculture* 459, 156–165.

Imsland, A.K., Reynolds, P., Eliassen, G., Hangstad, T.A., Jónsdóttir, Ó.D.B., Elvegård, T.A., Lemmens, S.C.A., Rydland, R. and Nytrø, A.V. (2016b) Investigation of behavioural interactions between lumpfish (*Cyclopterus lumpus*) and goldsinny wrasse (*Ctenolabrus rupestris*) under controlled conditions. *Aquaculture International* 24, 1509–1521.

Imsland, A.K., Reynolds, P., Nytrø, A.V., Eliassen, G., Hangstad, T.A., Jónsdóttir, Ó.D.B., Emaus, P.-A., Elvegård, T.A., Lemmens, S.C.A., Rydland, R. and Jonassen, T.M. (2016c) Effects of lumpfish size on foraging behaviour and co-existence with sea lice-infected Atlantic salmon in sea cages. *Aquaculture* 465, 19–27.

Ingólfsson, A. and Kristjánsson, B.K. (2002) Diet of juvenile lumpsucker *Cyclopterus lumpus* (Cyclopteridae) in floating seaweed: effects of ontogeny and prey availability. *Copeia* 2, 472–476.

Johannesson, J. (2006) Lumpfish caviar – from *vessel* to *consumer*. FAO Fisheries Technical Paper. No. 485. Rome, FAO. p. 60.

Kasper, J.M., Bertelsen, B., Ólafsson, H.G., Holst, J.C., Sturlaugsson, J. and Jónsson, S.P. (2014) Observations of growth and postspawning survival of lumpfish *Cyclopterus lumpus* from mark-recapture studies. *Journal of Fish Biology* 84, 1958–1963.

Kennedy, J., Jónsson, S., Ólafsson, H.G. and Kasper, J.M. (2016) Observations of vertical movements and depth distribution of migrating female lumpfish (*Cyclopterus lumpus*) in Iceland from data storage tags and trawl surveys. *ICES Journal of Marine Science* 73, 1160–1169.

Kennedy, J., Jónsson, S., Kasper, J.M. and Ólafsson, H.G. (2014) Movements of female lumpfish (*Cyclopterus lumpus*) around Iceland. *ICES Journal of Marine Science* 72, 880–889.

Killen, S.S., Brown, J.A. and Gamperl, A. (2007a) The effect of prey density on foraging mode selection in juvenile lumpfish: balancing food intake with the metabolic cost of foraging. *Journal of Animal Ecology* 76, 814–825.

Killen, S.S., Costa, I., Brown, J.A. and Gamperl, A.K. (2007b) Little left in the tank: metabolic scaling in marine teleosts and its implications for aerobic scope. *Proceedings of the Royal Society of London B: Biological Sciences* 274, 431–438.

Kjörsvik, E., Davenport J. and Lönning, S. (1984) Osmotic changes during the development of eggs and larvae of the lumpsucker, *Cyclopterus lumpus* L. *Journal of Fish Biology* 24, 311–321.

Klinkhardt, M., Tesche, M. and Greven, H. (1995) Database of fish chromosomes. Magdeburg (Westarp-Wissenschaften), Germany, 237 pp.

Kristmundsson, A. and Freeman, M.A. (2014) Negative effects of *Kudoa islandica* n. sp. (Myxosporea: Kudoidae) on aquaculture and wild fisheries in Iceland. *International Journal for Parasitology: Parasites and Wildlife* 3, 135–146.

Kudryavtseva, O.Y. and Karamushko, O.V. (2002) Biology of the lumpfish *Cyclopterus lumpus* during reproduction in the Murman near shore zone. *Journal of Ichthyology* 42, 309–313.

Kyûshin, K. (1975) The embryonic and larval development, growth, survival and changes in body form, and the effect of temperature on these characteristics of the smooth lumpsucker, *Aptocyclus ventricosus* (Pallas). *Bulletin of the faculty of fisheries Hokkaido University* 26, 49–72.

Lassen, H., Lockwood, S., Cappell, R. and Pedersen, G.M. (2015) Full Assessment Report. Public Certification Report – Initial Assessment of the Greenland Lumpfish Fishery. Sustainable Fisheries Greenland. Report no.: 2014–018, rev. 5. Certificate no.: f-dnv-165369, 35 pp. https://www.msc.org/track-a-fishery/fisheries-in-the-program/fisheries-by-species/fisheries-by-species#lumpfish

Li, M.F. and Clyburne, S. (1977) New cell line from the marine lumpfish, *Cyclopterus lumpus*. *Journal of the Fisheries Research Board of Canada* 34, 134–139.

Lönning, S., Kjörsvik, E. and Davenport, J. (1984) The hardening process of the egg chorion of the cod, *Gadhus morhua* L., and lumpsucker, *Cyclopterus lumpus* L., *Journal of Fish Biology* 24, 505–522.

Marine Research Institute. (2015) State of Marine Stocks in Icelandic Waters 2014/2015 and Prospects for the Quota Year 2015/2016. Marine Research in Iceland 182. 217 pp.

Martin-Robichaud, D.J., Peterson, R.H., Benfey, T.J. and Crim, L.W. (1994) Direct feminization of lumpfish (*Cyclopterus lumpus* L.) using 17ß-estradiol enriched Artemia. *Aquaculture* 123, 137–151.

Mikkelsen, N. and Pedersen, T. (2012) Invasive red king crab affects lumpsucker recruitment by egg consumption. *Marine Ecology Progress Series* 469, 87–99.

Mitamura, H., Thorstad, E.B., Uglem, I., Bjørn P.A., Økland, F., Næsje T.F., Dempster, T. and Arai, N. (2007) Movements of female lumpsucker *Cyclopterus lumpus* in a Norwegian fjord during the spawning period. NINA Report 288: 20 pp.

Mitamura, H., Thorstad, E.B., Uglem, I., Bjørn P.A., Økland, F., Næsje T.F., Dempster, T. and Arai, N. (2012) Movements of lumpsucker females in northern Norwegian fjord during the spawning season. *Environmental Biology of Fishes* 93, 475–481.

Mochek, A.D. (1973) Spawning behaviour of the lumpsucker *Cyclopterus lumpus* L.. *Journal of Ichthyology* 13, 615–619.

Moring, J.R. (2001) Intertidal growth of larval and juvenile lumpfish in Maine: A 20-year assessment. *Northeastern Naturalist* 8, 347–354.

Mullins, J.E., Powell, M., Speare, D.J. and Cawthorn, R.J. (1994) An intranuclear microsporidian in lumpfish *Cyclopterus lumpus*. *Diseases of Aquatic Organisms* 20, 7–13.

Murray, A.G. (2016) A Modelling Framework for Assessing the Risk of Emerging Diseases Associated with the Use of Cleaner Fish to Control Parasitic Sea Lice on Salmon Farms. *Transboundary and Emerging Diseases* 63, e270–e277.

Myrseth, B. (1971) Fekunditet, vekst, levevis og ernaering hos *Cyclopterus lumpus* L. Thesis. University of Bergen, Norway.

Norðberg, G., Johannesen, A. and Arge, R. (2015) Cryopreservation of lumpfish *Cyclopterus lumpus* (Linnaeus, 1758) milt. *PeerJ* 3:e1003.

Nøstvold, B.H. (2017) Fra lusespiser til middagsmat. Cleaning Fish Conference and drug lice control in 2017; Trondheim. https://nofima.no/en/nyhet/2016/02/lumpfish-from-louse-eater-to-the-dinner-plate

Nytrø, A.V. (2013) The effect of temperature and fish size on growth of juvenile lumpfish (*Cyclopterus lumpus* L.). Master Thesis. University of Tromsø.

Nytrø, A.V., Vikingstad, E., Foss, A., Hangståd, T.A., Reynolds, P., Eliassen, G., Elvegård, T.A.,

Falk-Petersen, I.B. and Imsland, A.K. (2014) The effect of temperature and fish size on growth of juvenile lumpfish (*Cyclopterus lumpus* L.). *Aquaculture* 434, 296–302.

Pampoulie, C., Skirnisdottir, S., Olafsdottir, G., Helyar, S.J., Thorsteinsson, V., Jónsson, S.P., Fréchet, A., Durif, C.M.F., Sherman, S., Lampart-Kaluzniacka. M., Hedeholm, R., Ólafsson, H., Danielsdottir, A.K. and Kasper, J.M. (2014) Genetic structure of the lumpfish *Cyclopterus lumpus* across the North Atlantic. *ICES Journal of Marine Science* 71, 2390–2397.

Parin, N.V., Fedorov., V.V. and Sheiko, B.A. (2002) An annotated catalogue of fish-like vertebrates and fishes of the seas of Russia and adjacent countries: Part 2. Order Scorpaeniformes. *Journal of Ichthyology* 42 (Suppl.1), S60–S135.

Pauly, D. and Zeller, D. (2015) So long, and thanks for all the fish: The Sea Around Us, 1999–2014, A fifteen year retrospective. 172 pp. Available at www.seaaroundus.org/wp-content/uploads/2015/01/small_SO_LONG_-_report_to_Pew-Dec-03-w-cover.pdf (accessed 30 April 2016).

Perry, B. and Treasurer, J.W. (2015) An overview of cleanerfish production at Ardtoe. Proceedings of the Cleanerfish Workshop, Inverness, 1 December 2015. Scottish Salmon Producers Organisation.

Perry, A.L., Low, P.J., Ellis, J.R. and Reynolds, J.D. (2005) Climate change and distribution shifts in marine fishes. *Science* 308, 1912–1915.

Pooley, C., Consuegra, S., Gray, C., Tew, I. and Garcia de Leaniz, C. (2015) Genetic screening of lumpfish (*Cyclopterus lumpus*) for the control of sea lice in salmon farming. Proceedings of the Aquaculture Europe 2015 (Poster) Rotterdam, 2015.

Powell, A., Treasurer J.W., Pooley C.L., Keay A.J., Lloyd, R. and Garcia de Leaniz, C. (2017) Use of lumpfish for sea-lice control in salmon farming: challenges and opportunities. *Reviews in Aquaculture* (doi: 10.1111/raq.12194).

Rolbiecki, L. and Rokicki, J. (2008) Helminths of the lumpsucker (*Cyclopterus lumpus*) from the Gulf of Gdańsk and Vistula Lagoon (Poland). *Oceanological and Hydrobiological Studies* 37, 53–59.

Rønneseth, A., Ghebretnsae, D.B., Wergemand, H.I. and Haugland, G.T. (2015) Functional characterization of IgM+ B cells and adaptive immunity in lumpfish (*Cyclopterus lumpus* L.). *Developmental and Comparative Immunology* 52, 132–143.

Sayer, M.D.J., Gibson, S., Thomas, G., Joyce, J.A., Reader, J.P., dePomerai, D.I., Green, A.C., Davenport, J., McKinney, R., Bradbury, N., Morris, P., Twelves, E., Twelves, J., Goward, E., Campbell-Preston, R., Whitmarsh, D. and Pickering, H. (2000) The background and economic potential of lumpsucker culture: Diversification using existing infrastructure. Final report to the MAFF/NERC LINK Aquaculture Programme and to the PESCA FIFG. *Scottish Association for Marine Science/Dunstaffnage Marine Laboratory Internal Report* 225. 59 pp.

Scyborska, T.A. (1948) The parasitofauna of *Cyclopterus lumpus* in the White Sea. *Raboty Morsk Biologii Stat Karelo-Finskogo Universiteta* 1, 145–151.

Skirnisdottir, S., Olafsdottir, G., Olafsson, K., Jendrossek, T., Dafydd Lloyd, Ha., Helyar, S., Pampoulie, C., Danielsdottir, A.K. and Kasper, J.M. (2013) Twenty-two novel microsatellite loci for lumpfish (*Cyclopterus lumpus*). *Conservation Genetics Resources* 5, 177–179.

Smage, S.B., Frisch, K., Brevik O.J., Watanabe, K. and Nylund, A. (2016) First isolation, identification and characterisation of *Tenacibaculum maritim* in Norway, isolated from disease farmed sea lice cleaner fish *Cyclopterus lumpus L. Aquaculture* 464, 178–184.

Stansbury, D.E., Murphy, E.F. and Bishop, C.A. (1995) *An update of stock status of 3KLP lumpfish*. Department of Fisheries and Oceans Canada, Atlantic Fisheries Research Document 95/65 Ottawa.

Stein, D.L. (1986) Cyclopteridae. In: Whitehead, P.J.P., Bauchot, M.-L., Hureau, J.-C., Nielsen, J. and Tortonese, E. (eds.) *Fishes of the North-eastern Atlantic and the Mediterranean*. UNESCO, Paris. Vol. III. 1269–1274.

Stevenson, S.C. and Baird, J.W. (1988) The fishery for lumpfish (*Cyclopterus lumpus*) in Newfoundland waters. *Canada Department of Fisheries and Oceans Canadian Technical Report of Fisheries and Aquatic Sciences Report* 1595. p. 26.

Sumaila, U.R., Marsden, A.D., Watson R. and Pauly, D. (2007) A global ex-vessel fish price database: construction and applications. *Journal of Bioeconomics* 9, 39–51.

Thorsteinsson, V. (1996) Hrognkelsi. pp. 6–11, in: Lifriki sjavar [Marine life]. (In Icelandic). Reykjavík, Iceland: Namsgagnastofnun and Hafrannsoknastofnun. (cited in Johannesson, 2006).

Thorsteinsson, V. (1983) Some aspects of the biology and the fisheries of the lumpfish (*Cyclopterus lumpus*). Thesis. State University of New York at Stony Brook.

Towers, L. (2015) Viral hemorrhagic septicemia reported for first time in Icelandic fish. Available at: The Fish Site www.thefishsite.com/fishnews/26621/viral-hemorrhagic-septicemia-reported-for-first-time-in-icelandic-fish (accessed 30 April 2016).

Tully, O. and Ó Ceidigh, P. (1989) The ichthyoneuston of Galway Bay (west of Ireland). II. Food of post-larval and juvenile neustonic and pseudo-neustonic fish. *Marine Ecology Progress Series* 51, 301–310.

Vandendriessche, S., Messiaen, M., O'Flynn, S., Vincx, M. and Degraer, S. (2007) Hiding and feeding in floating seaweed: Floating seaweed clumps as possible refuges or feeding grounds for fishes. *Estuarine, Coastal and Shelf Science* 71, 691–703.

Vasconcelos, P., Monteiro, C.C., Santos, M.N. and Gaspar, M.B. (2004) First record of the lumpfish (*Cyclopterus lumpus* Linnaeus, 1758) off the Algarve coast (southern Portugal): southward extension of the species distributional range. *Journal of Applied Ichthyology* 20, 159–160.

Yao, Z. and Crim, L.W. (1996) A biochemical characterization of vitellogenins isolated from the marine fish ocean pout (*Macrozoarces americanus* L.), lumpfish (*Cyclopterus lumpus*) and Atlantic cod (*Gadus morhua*). *Comparative Biochemistry and Physiology Part B: Biochemistry and Molecular Biology* 113, 247–253.

Zhitenev, A.N. (1970) Ecological and morphological affinities of the lumpsucker. *Voprosy Ikhtiologii* 10, 77–84.

CHAPTER 7

Hatchery management of lumpfish

Thor Magne Jonassen,[1] Ingrid Lein[2] and Ane Vigdisdatter Nytrø[3]

[1]Akvaplan-niva, Tromsø, Norway
[2]NOFIMA, Sunndalsøra, Norway
[3]Norwegian University of Science and Technology, Trondheim, Norway

7.1 Environmental regulation of sexual maturation and spawning

The control of sexual maturation can be useful in broodstock management for improved synchronisation of spawning to deliver eggs at a desired time for juvenile production independent of season. For salmonids and many marine species the photoperiod is a common and efficient way of controlling this. In addition, photoperiod (extended daylight) has in many species been efficient in enhancing growth that, for broodstock recruits, reduces the time to reach sexual maturity. Increased growth also increases size at spawning and therefore fecundity.

Similar light regimes as those previously used in commercial breeding programmes for cod (J.-E. Eliassen, Marine Breed, Norway, personal communication) have been tested for lumpfish, where groups of fish were exposed to continuous light for a minimum of three months prior to the planned spawning time. This was then switched to a three-month compressed natural photoperiod running a cycle from the midsummer solstice to the next midsummer in three months. Spawning was expected to start at the end of the three months compressed light cue ('autumn–spring') at day length of approximately 17 hours. This light regime was used in an experiment with lumpfish for testing the effect of growth and timing of spawning (Fig. 7.1), with one group regulated for three months delayed spawning (PP3), one for six months delayed spawning (PP6) and one group under simulated natural photoperiod (SLDN). There was a natural variation in temperature during the experiment, ranging from 3.0 to 8.6°C.

During the period with continuous light (LD24:0) prior to the light cue in the PP3 and PP6 groups there was a growth-enhancing effect of continuous light compared to fish at SLDN. However, when males and females were analysed separately this effect was only evident for females and not for males (Fig. 7.2). There seems to be a clear sex-dependent growth response to photoperiod in lumpfish above 200 g. Irrespective of photoperiod, females grew faster than males and the growth rates for males slowed down after they reached approx. 600 g (Fig. 7.2). At that size (in September) the first change in coloration (red) started to appear, indicating

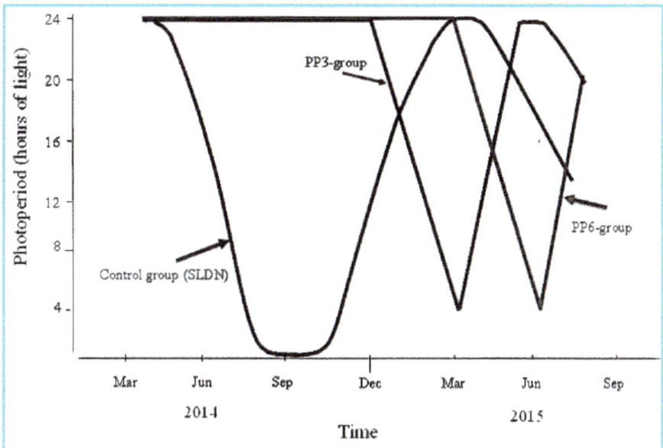

Fig. 7.1 Photoperiod regimes for natural spawning (SLDN), three (PP3) and six (PP6) months delayed spawning. Arrows indicate expected start of spawning season.

maturation in males. Sex determination of females based on coloration was five months later than in males (in February).

The effect of photoperiod on regulation of spawning time was not as clear in lumpfish as seen for other marine species such as the Atlantic cod *Gadus morhua*. Most of the lumpfish males showed external signs of maturation very early, already at around 600 g, irrespective of light regimes.

In the PP3 and PP6 groups, which were expected to start spawning in April and July respectively, some female fish spawned in February, mostly from the PP6 group, prior to the onset of the light cue that was expected to initiate the maturation process (Fig. 7.3). The total volume of eggs spawned in PP6 was 3,240 ml compared to 10,360 ml in PP3.

For PP3 spawning frequency and the sizes of the egg batches had a clear peak in July, approximately six months after the onset of the 'autumn–spring' light cue (Fig. 7.3). An explanation for this may be that the 'autumn–spring' light cue that under natural conditions takes six months was compressed to only three. This suggests that female lumpfish need six months to mature and get ready to spawn. According to this hypothesis the absence of any peak spawning for PP6 could be that this should be expected in October, three months after the experiment was terminated.

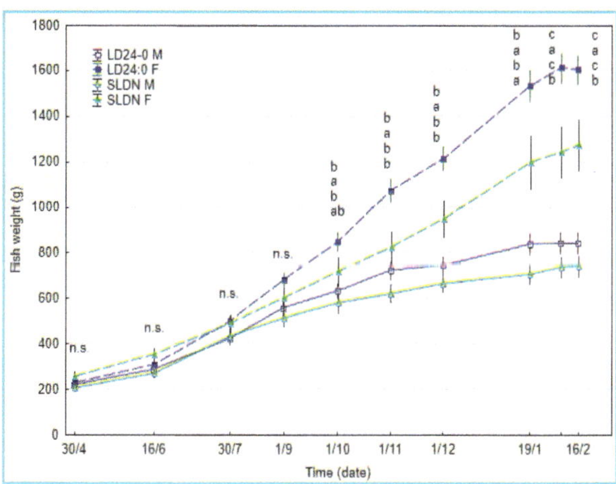

Fig. 7.2 Sex-dependent growth in lumpfish at two different photoperiod regimes; LD24:0 and SLDN. F=females, M=males. Different letters indicate that weights for trial groups are statistically significant.

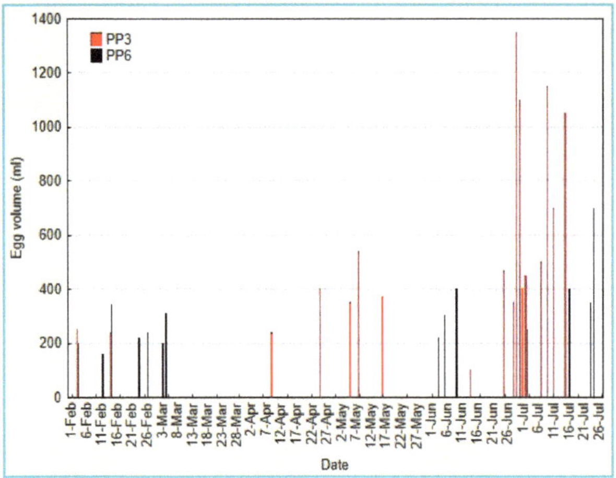

Fig. 7.3 Timing for individual spawnings and egg volume (ml) in lumpfish at two different photoperiod regimes. (PP3=3 months delayed and PP6=6 months delayed compared to natural photoperiod cycle)

No female fish spawned in the SLDN group. Since this light regime gave less growth stimulation than in PP3 and PP6, the fish size was significantly smaller when females in PP3 and PP6 started spawning. Most importantly, the 'autumn–spring' cue that is expected to stimulate the start of final maturation came earlier for SLDN when the fish was small. At winter solstice the fish in SLDN were around 400 g, while PP3 and PP6 were 1,600 g and 1,900 g respectively at that stage. It is common theory that fish need to reach a minimum threshold size or energy status before entering the subsequent maturation process, and it may be speculated that the SLDN group had not reached the threshold size for female lumpfish.

Temperature may also have played a role in the regulation of final maturation and spawning, both as a rate-controlling factor, but also as a directional cue for season and timing of spawning since there was a seasonal variation in temperature during the experiment. The temperature regime may have acted as a confounding factor since it was out of phase with the simulated light regimes. Such interactions of size and temperature on photoperiod manipulation for regulating sexual maturation and spawning in lumpfish need further research.

Lumpfish seem to have a wide temperature tolerance, and it is common practice to keep lumpfish broodstock under varying temperatures, commonly from 10°C and down to approx. 4°C, which is the normal spawning temperature in the wild. The effect of temperature on ovulation and synchronisation of spawning was investigated on farmed broodstock acclimated to approx. 5 and 10°C for three months prior to spawning. Spawning for both temperature groups lasted for approx. 1.5 months (Fig. 7.4). All eggs were hand stripped and the size of the egg batches was generally larger at 5°C (363 ml) compared to 10°C (247 ml).

The egg size (diameter) was also larger at 5°C compared to 10°C (Fig. 7.5), which may result in larger larva at hatching. Although the eggs were smaller at 10°C the higher temperature did not have a negative impact on egg survival (Fig. 7.6). The higher egg survival at 10°C compared to 5°C may be due to a smoother temperature acclimation at incubation since all eggs were incubated at approx. 10°C (Fig. 7.6).

The number of ripe and stripped fish was higher at 10°C giving a higher total egg production (2,425 ml) compared to the 5°C group (1,500 ml). This higher spawning activity at 10°C indicates that temperature may be an important stimulator for triggering spawning and higher

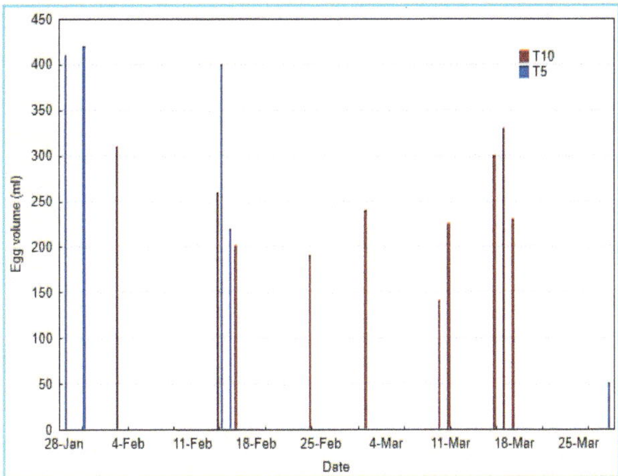

Fig. 7.4 Spawning time and sizes of egg batches spawned (ml) in lumpfish at two different temperatures (T5= 5°C, T10= 10°C).

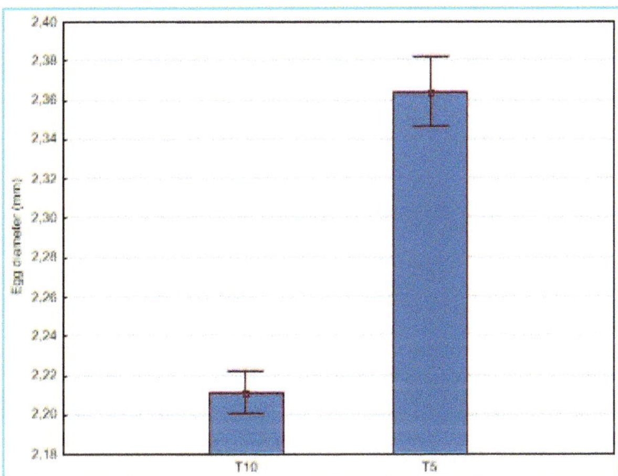

Fig. 7.5 Mean egg diameter (± SEM) for eggs stripped at 10°C (T10) and 5°C (T5). Data are based on ten eggs from six fish in group T10 and ten eggs from three fish in group T5.

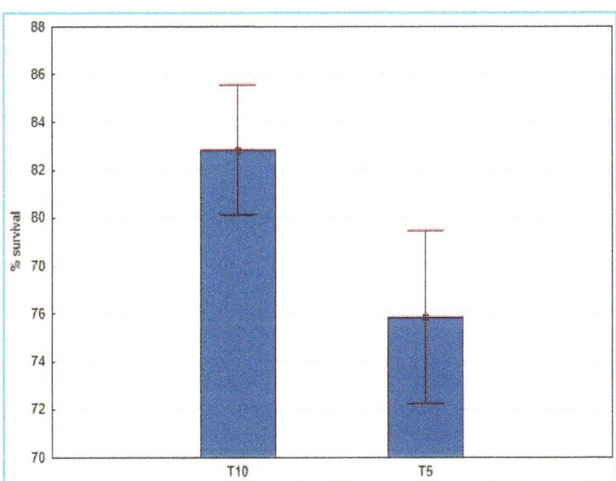

Fig. 7.6 Mean survival (± SEM) for eggs stripped at 10°C (T10) and 5°C (T5). Data are based on ten eggs from six fish in group T10 and ten eggs from three fish in group T5.

temperature may increase the number of females spawning, hence increasing the total egg production from a spawning population. A possible trade-off from the higher temperature is the impact on larval quality, which requires further investigations.

7.2 Harvest and utilisation of lumpfish eggs and sperm

Lumpfish (*Cyclopterus lumpus L.)* is a new species in aquaculture. Initially, for any new species, it is common to use eggs and sperm from wild caught broodstock to produce larvae. When the production of lumpfish started a few years ago, broodfish were transported live from the fishing sites to hatcheries along the coast. Over the last couple of years, it became evident that unfertilised eggs and male gonads could be transported during the first hours after harvesting, rather than transporting live broodfish over long distances. However, wild fish are often carriers of pathogens, and thus represent a risk of bringing pathogens into the hatchery. Therefore, it is important to establish broodstocks of lumpfish of farmed origin as soon as possible. In addition to reducing the risk of disease, domestication increases the robustness of the offspring, and reduces the stress of the fish under farm conditions. Until farmed broodstocks can be established, good hygiene routines and disinfection protocols are of special importance to avoid introducing diseases from the wild into the hatcheries. Lumpfish has proven to be susceptible to various diseases such as atypical furunculosis, *Pasteurella sp.*, *Vibrio anguillarum*, VHSv (viral hemorrhagic septicemia virus) and the blood parasite *Nucleospora cyclopteri.*

The following describes methods for harvesting eggs and sperm, and for storing sperm. Preliminary results from egg disinfection and egg transport trials are also included.

7.2.1 Harvest of eggs

Lumpfish is a batch spawner, and preliminary results indicate that each female spawns twice during the spawning season with an interval of six to ten days. Final egg maturation, which is vital for successful fertilisation and hatching of the eggs, takes place during the last few days before spawning. Therefore, to ensure successful stripping of eggs and reduce stress on the female, it is important to make sure that she is fully mature at the time of stripping. Traditionally, hand-stripping involves the eggs being expressed from the female simply by applying gentle pressure along the flanks of the fish towards the gonadal pore (Fig 7.7).

Pathogens on the skin of the spawning fish can potentially contaminate eggs and milt. Therefore, the procedure of harvesting eggs has been refined over the last couple of years. Before stripping of eggs, the female is placed on a disposable cloth, which covers the whole body. Thereafter, an incision is made in the cloth to display the gonadal pore. Spraying with alcohol (70%) is used to disinfect the gonadal pore before stripping the eggs (Fig 7.8).

Several farmers now routinely take tissue samples from both females and males for PCR-analysis to make sure that the broodfish are pathogen-free (Fig 7.9). Fertilised batches of eggs must be incubated separately and samples of milt stored in the fridge until this analysis is complete. Eggs from females proven to be carriers of pathogens or that have been fertilised from pathogen-carrying males can then be discarded, to prevent transfer of diseases to the offspring.

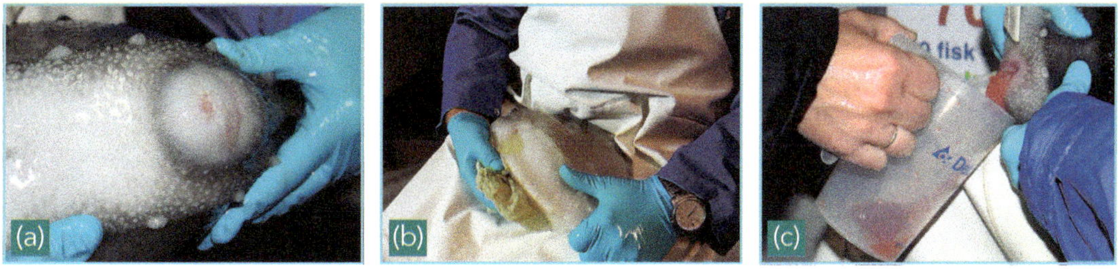

Fig. 7.7 Traditional hand-stripping of lumpfish eggs. Fully mature gonadal pore (a), drying off the gonadal pore (b) and stripping of eggs (c). Images: Ingrid Lein.

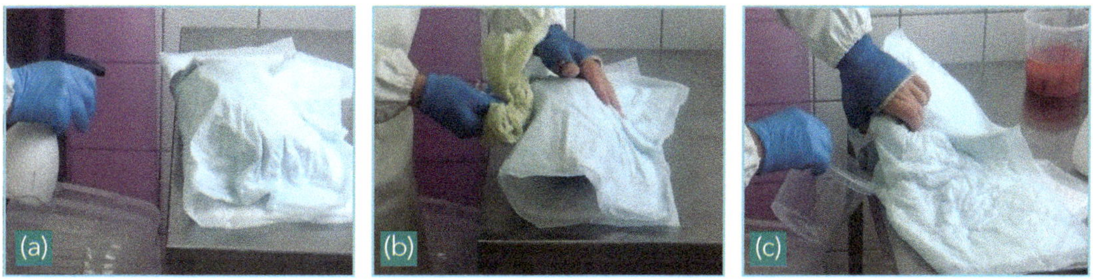

Fig. 7.8 Harvesting of eggs. Disinfection with 70% alcohol (a), drying of the gonadal pore (b) and stripping of eggs (c).

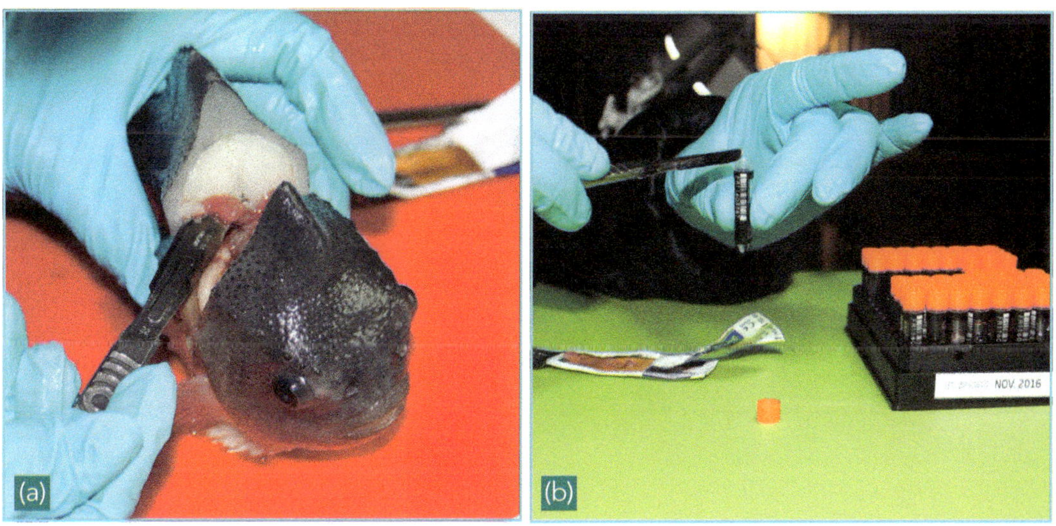

Fig. 7.9 Sampling of kidney tissue for PCR-analysis for known pathogens.

7.2.2 Harvest of sperm

Many producers of lumpfish eggs use milt extracted from gonads rather than hand stripped from the males, as is the practice in many other species. The reason for this is that few males release milt spontaneously. The use of milt extracted from gonads involves sacrificing wild males. In the near future, males from farmed lumpfish broodstock will be available. For economic reasons, it will become important to restrict the number of broodfish. Therefore, methods for an efficient utilisation of the milt from each male are crucial. Today, there are two methods for storage of milt available, short-term storage at 3–4°C or, long-term storage by cryopreservation (-190°C). Salmon producers use both methods frequently. Experiments performed at Nofima show very high fertilisation (>90%) for both methods.

Preparation of milt before the stripping of the eggs makes it possible to ensure good milt quality and saves labour and time. The males are killed by an overdose of anaesthetic before opening the abdomen. The gonads are cut out, ground using a parsley grinder and sieved through a tea rinse to remove tissue debris before dilution (1:1) in a physiological solution (AquaBoost® SpermCoat, Cryogenetics as) (Fig. 7.10). Use of this diluent not only facilitates oxygenation of the sperm cells during storage, it also prevents sperm activation.

Sperm motility can be checked under the microscope by adding a small amount of seawater to activate the sperm cells. The number of sperm cells per volume of milt varies considerably between males (3–30 billions ml^{-1}) and can be determined either using a haemocytometer (counting chamber) or, a photometer calibrated for lumpfish milt, which is less time-consuming.

7.2.2.1 Short-term storage of milt

After dilution in the physiological solution, milt can be stored in the fridge at 3–4°C for at least two weeks. Results from experiments at Nofima using milt stored this way showed fertilisation rates >90% up to 14 days after preparation of the milt. The milt should be stored in a thin layer in a flat-bottomed container such as a cell culture flask to ensure that all sperm cells get access to oxygen. The containers must be rocked gently by hand two to three times a day, or stored on a moving device such as a shaking table, to periodically re-suspend the sperm cells (Fig. 7.10d).

The largest supplier of lumpfish eggs and gonads in Norway now uses this method routinely. Samples of male kidneys and milt can then be analysed for the presence of pathogens, while

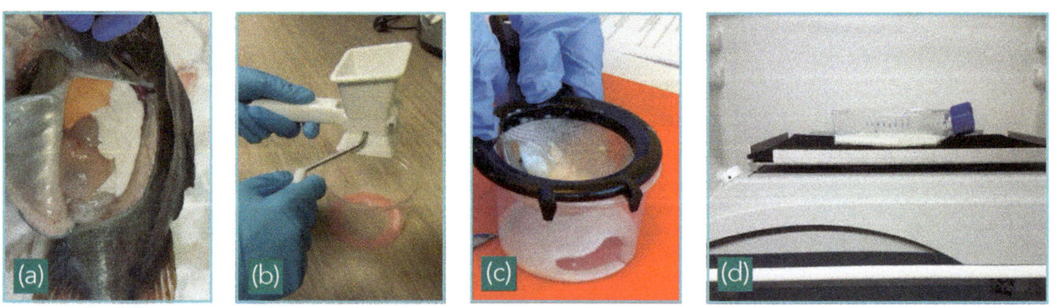

Fig. 7.10 Preparation of milt. Surgical harvesting of male gonad (a), grinding of gonad (b), rinsing of milt (c) and cold storing in cell culture flasks in the fridge (d).

the milt is stored in the fridge. Milt proven to be positive for pathogens is discarded to avoid bringing pathogens into the hatcheries.

Recent results indicate a faster reduction in the fertility of milt harvested at higher seawater temperatures (>15°C) during the autumn, compared to milt harvested during the winter. This may suggest that eggs fertilised at higher ambient temperatures should use milt that has been harvested at lower temperatures (i.e. during wintertime) and cryopreserved, rather than using milt freshly prepared in the autumn and stored in the fridge.

7.2.2.2 Cryopreservation of lumpfish milt

Cryopreservation is a method for the long-term storage of milt. This method includes dilution of the milt and addition of an anti-freeze medium before freezing in liquid nitrogen at -190°C. A series of trials on cryopreservation of lumpfish milt at Nofima showed that lumpfish milt is suitable for cryopreservation, as both short-term cold-storage (3–4°C) and cryopreservation of lumpfish milt resulted in >90% fertilisation. This is in contrast to findings by Nordberg *et al.* (2015), who found a 26.8% reduction in fertilisation rate for cryopreserved sperm compared to fresh sperm. The difference may be caused by the use of different anti-freeze media and/or freezing curves, as well as differences in sperm quality.

Cryopreservation allows analysis of the milt for known pathogens before it is used to fertilise eggs. Milt proven positive for any known pathogen can be discarded, i.e. only pathogen-free milt is used to fertilise eggs. Before cryopreservation, the sperm density is standardised to 2 billion ml^{-1} and the motility of the sperm is checked using a microscope. Only milt with good motility and high sperm density is used for cryopreservation. In Norway, a commercial company (Cryogenetics AS) undertakes the cryopreservation. The use of this storage method is now increasing as it saves the farmer time and effort and means disease-free milt is available at any time. This method is also well suited to use in genetic selection programmes because milt can be harvested from males from different geographic areas and at different times of the year.

Cryopreserved milt is relatively easy to use. It is stored at the farm in a thermos flask filled with liquid nitrogen. A water bath with an accurate thermostat is necessary to ensure correct thawing of the sperm cells. Thawing happens after immersion for 30 secs at 25°C. Then, the milt is ready to use. Because the sperm density and the volume of the packages are standardised, it is easy to calculate the volume needed to fertilise a given number of eggs. Lumpfish egg numbers can be estimated from their volume using the concentration factor of 80,000 $litre^{-1}$. However, since the diameter of the eggs varies between females, this factor also varies and is only approximate. The suggested number of sperm used per egg is 200,000 to 300,000 sperm egg^{-1}.

7.3 Egg disinfection and transport

7.3.1 Egg disinfection

Disinfection of unfertilised or newly fertilised eggs combined with good hygiene procedures is the best method to prevent bringing pathogens into the hatchery. The development of disinfection procedures for lumpfish eggs is still being examined, but some preliminary results are presented here.

In the production of salmonid eggs, disinfection with iodine both before and during egg hardening is commonly used. This method has also been tested for disinfection of marine eggs such as halibut *H. hippoglossus* and cod eggs, but resulted in very low survival. Disinfection with glutaraldehyde, on the other hand, has been used successfully for both halibut and cod eggs. It should be emphasized that disinfection of marine eggs is most often done at the time of transport, i.e. from the eyed egg stage onwards.

Disinfection of lumpfish eggs using iodine (50 ppm for 30 minutes or 100 ppm for ten minutes) before egg hardening resulted in almost no fertilisation. However, when the same disinfectant, doses and durations were applied after egg hardening, >90% of the eggs were fertilised. In addition, disinfection with glutaraldehyde (200 ppm for eight minutes or 400 ppm for four minutes) after egg hardening resulted in more than 90% fertilisation. However, only disinfection with 200 ppm glutaraldehyde for eight minutes gave successful hatching in the same range as the untreated control. Because these experiments were done during the autumn season when the temperature was high (>15°C), the overall survival and hatching of eggs was poor. Therefore, the experiment is being repeated during the winter to see if the preliminary results can be confirmed.

7.3.2 Egg transport

Until now, unfertilised lumpfish eggs have been transported over long distances with good results. However, the need to transport fertilised lumpfish eggs increases as some hatcheries start to incubate eggs for sale. It is well known that fish eggs are more susceptible to handling at different stages during egg incubation. The susceptibility of lumpfish eggs to handling was tested in a transport experiment at Nofima. Newly fertilised eggs were incubated in separate incubators. Every three days eggs were moved from their incubators (three incubators per time), and packed 'dry' (i.e. without seawater) in Styrofoam transport boxes. These boxes are normally used for transport of salmon eggs. The top shelf was filled with wet seawater ice to ensure stable low temperatures. The box was agitated on a moving table for six hours to imitate transport (Fig. 7.11).

After six hours of agitation, the eggs were returned to their incubators and followed until hatching. The results showed that lumpfish eggs can be transported as eyed eggs with no dramatic reduction in hatching success between 100 degree days post-fertilisation (ddpf) and 160 ddpf. After 160 ddpf the hatching rate decreased. In conclusion, the results showed that lumpfish eggs can successfully be transported 'dry' between 100–160 ddpf as long as the temperature during transport is kept low.

7.4 Incubation and hatchery set-up

7.4.1 Hatchery set-up

Norway's largest lumpfish hatcheries today deliver several sizes of small juveniles for on-growing in third-party facilities alongside lumpfish ready for the sea phase to maximise production capacity. Full production capacity ranges to up to 6 million lumpfish juveniles per year with one to four generations of lumpfish maintained within the same production plant. Batches of eggs

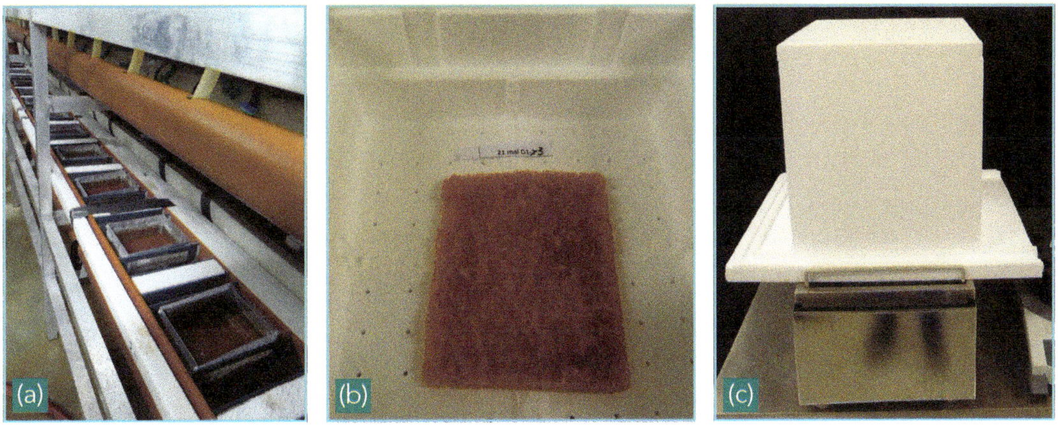

Fig. 7.11 Lumpfish eggs in hatching units (a), eggs in transport box (b) and transport box on moving table (c).

are usually obtained from wild-caught broodstock but some hatcheries keep their own farmed broodstock, though the latter is less common due to several unsolved bottlenecks in terms of recruitment and rearing.

Eggs are usually fertilised directly in the hatchers. Incubator set-up varies but common approaches include standard salmon hatching trays submerged in shallow raceways or family hatchers. In the latter, water flows through the hatcher from bottom to top. The outlet is located on the rim of the hatcher, and allows the farmer to utilise two different strategies for transferring larvae into the first feeding tanks. One solution is that the hatchers are delivered with perforated screens or cages with a mesh size of 1 mm or smaller. This will retain the newly hatched larvae, allowing transfer to first feeding tanks by either a dip net or siphoning. The other and most commonly used system for collecting hatched larvae is to use the hatchers without the 'cage', and to incubate the eggs directly on a perforated plate in the bottom of the hatchers (Fig. 7.12). This will allow the newly hatched larvae to flow out through the hatcher outlet and directly into a first feeding tank or a tank for accumulating larvae before transfer to first feeding. The family hatcher also allows for a better separation of families, reduces the risk of contamination between groups of eggs and enables easier monitoring of water quality as all incubators are facilitated with their own inlet and outlet. If a traditional system of raceways and trays is used the hatched larvae need to be transferred to a first feeding tank by siphoning or dip nets. If the latter is used, the raceways should be thoroughly checked for escapees from hatching trays, as the optimal mesh size for hatchers in terms of flow-through may to some degree coincide with the size of the larvae.

Irrespective of the set-up of incubators it should be noted that larvae are relatively well developed upon hatch, displaying the characteristic suction disc. This causes the larvae to attach to all available surfaces from day one, including the water surface in incubators and start feeding tanks. Lumpfish larvae seemingly floating upside down on the surface should therefore not be disregarded as deformed or ill. Furthermore, it is important to remember that the clinging behaviour is an important part of the species' larval life strategy, and must be taken into consideration, both in terms of available surface in the incubator upon hatch and also when it comes to handling and transfer operations.

Fig. 7.12
Incubation takes place in family hatchers or traditional raceways after dry, artificial fertilisation. Eggs are spread in a thin layer and treated with buffodine (2.5 ml l^{-1} water for ten minutes). Hatching commences at approximately 280 d°.

7.4.2 Incubation and cleaning

The early stages of embryo development are relatively slow during incubation compared with other marine species, hypothesising a very low metabolic rate compared to other marine species during the early developmental stages. The incubation period typically varies from 280–300 degree days (d°) and the hatching rate is usually above 80%. The hatching success of a batch of eggs incubated under adequate water conditions is usually up to 100% (Goulet *et al.,* 1986) but, in aquaculture, hatching success varies and in some cases hatching rates down to 30% have been observed. The causes for such low hatching rates are complex but premature stripping of roe and acute changes in temperature between stripping and incubation are believed to influence egg quality and hatching success to some degree. Low temperatures during incubation have also been shown to reduce hatching success (Brown *et al.,* 1992).

Lumpfish eggs are adhesive, and are under natural conditions observed as compact 'lumps' of eggs. In the wild the male lumpsucker will aerate and clean the batch of eggs throughout the incubation period, fanning the eggs using its pectoral, caudal or dorsal fins and pushing the egg ball around with his snout to ensure good water conditions for eggs throughout the whole mass. During this cleaning process males will also rhythmically expand and contract the branchial chamber, 'puffing' water into the egg mass. This behaviour is essential for normal hatching and development of the eggs (Goulet *et al.,* 1986). This, together with the turbulence in the lumpfish's natural spawning habitat, is sufficient to ensure satisfactory ventilation and cleaning of the egg mass under natural conditions (Davenport, 1983; Davenport *et al.,* 1983).

In aquaculture fertilised lumpfish eggs are commonly moulded into a pancake-like structure after stripping and dry fertilisation, and before the eggs harden. This is usually done by pouring fertilised eggs and ovary fluid directly into the incubator, where eggs are spread out prior to the saltwater-activated hardening of the chorion (Brown *et al.,* 1992). This facilitates uniform ventilation and environmental conditions for all eggs in the batch, contrary to naturally spawned batches of eggs where the oxygen tension in water-filled spaces within the egg mass declines rapidly compared to better ventilated eggs on the surface of the egg mass. This may cause slow development, mortality and bacterial growth in the centre of the egg mass (Davenport *et al.,* 1983), and may reduce hatching success (Brown *et al.,* 1992).

For similar reasons, dead eggs must be removed from the egg batch as bacterial growth in unfertilised eggs will spread rapidly to eggs surrounding the dead patch, especially at high water temperatures (Davenport *et al.*, 1983). The surface of the chorion of the lumpfish egg is very hard and similar to that of demersal species such as salmon, and the adherence between eggs is strong (Lønning *et al.,* 1984). In the wild, a male lumpfish will clasp egg debris with his mouth and spit it out and away from the nesting site (Goulet *et al.,* 1986). The spreading out of newly fertilised eggs in the incubator will facilitate easier removal of debris from non-viable eggs, but removing the debris is challenging without stressing nearby eggs. Therefore, methods for dispersing eggs prior to salt water activation has been a point of interest, but currently no viable methods are available.

Hatching usually happens within the same day or in the subsequent days, dependent on temperature (Goulet *et al.,* 1986). There is therefore no need to remove dead egg shells in the hatching period as the whole batch of eggs can usually be discarded as one.

7.4.3 Water quality and environmental factors during incubation

Water inlets for lumpfish hatcheries are usually situated at depths below 40 m to reduce impacts of local environmental factors such as temperature, turbidity and potential pathogens. In Norway, most production facilities utilise treated, ambient temperature seawater with normal temperatures ranging from 2–17°C throughout the year. Salinities vary between 30 and 35‰.

Most hatcheries use column or vacuum degassers for their main water supply but variations with degassers directly in connection to single tanks in the on-growing phase also exist. Additionally, some producers prefer the option of adjusting the oxygen levels through a separate oxygen diffusor system, especially in the fast-growing juvenile stages after metamorphosis when the biomass in tanks may increase rapidly.

Water filtration methods include sand, drum and Bernoulli filters or combinations of the three. The screen size of the main water filtration systems (drum or Bernoulli) is usually 90–100 microns while that of the secondary filtration systems (mostly drum) is usually 40–60 microns. In addition, most lumpfish producers use UV treatment for their water, but it has been questioned whether such intensive treatment may cause reduced stimulation of the larval and juvenile lumpfish immune system. There are usually separate water treatment systems for the hatcheries and on-growing facilities for larger fish as eggs and larvae are less robust due to poorly developed immune systems.

Immediately after spawning the ammonia levels within a batch of eggs are at the highest and decrease throughout the incubation period. During the early stages of development lumpfish eggs are tolerant of hypoxic conditions and show a high oxygen uptake, despite low oxygen tensions. This ability decreases throughout the later developmental stages. The oxygen requirement for lumpfish eggs increases during the incubation period with the initial oxygen uptake being very low at 0.0109 µl O_2 dry wt^{-1} h^{-1}, increasing to around twentyfold the initial value close to hatching, and premature hatching may occur when eggs are exposed to abrupt changes in temperature or oxygen close to hatching (Davenport, 1983).

Upon hatching the oxygen requirement of the larvae increases to 0.625µl wt^{-1} h^{-1} per individual. It has been observed that 275 degree days (d°) old larvae weighing 0.863 mg are found to use 3.24 µl wt^{-1} h^{-1} per individual (Davenport, 1983). Temperature-related effects on the development of lumpfish embryos and hatching have to a very little extent been explored, though some data do exist. In most hatcheries lumpfish eggs are mostly produced from wild-caught broodstock reared at ambient temperature until spawning. Temperatures during incubation are usually ambient, with temperatures ranging from 2 to 12°C throughout the year or 8–13°C (heated). The latter causes an immediate increase in egg temperature when transferred to incubators and, although the effect of such strategies are little documented, results from an unpublished master thesis give indications that such an abrupt rise is unfavourable, and that a gradient or cold water regime may be favourable in terms of deformities and larval size at hatch (Danielsen, 2016, unpublished).

7.4.4 Photoperiod and light management

A variety of different approaches exist when it comes to light regimes through the incubation period though a common approach is avoiding direct or strong lighting. Light is usually dimmed and set to a 24-hour regime. Some producers prefer to use lids over the incubators to completely shut out all light except for during observation and cleaning. Red lights are also reportedly used in some incubation rooms. Strong light has been observed to affect egg coloration, and in exposed eggs colour fades more rapidly than eggs kept incubated in dim lighting or darkness. This has, in some farms, been observed to produce larvae of poorer quality than larvae from egg batches reared in the dark, though the impact of this in commercial production has not yet been explored.

7.4.5 Start feeding and feed selection

The yolk sac period of the lumpfish larvae is relatively short and larvae are relatively large and well developed upon hatching, displaying functional mouth and eyes (Brown, 1986; Davenport and Kjorsvik, 1986). It is therefore assumed that the species also displays a well-developed gut morphology from hatching (Brown *et al.*, 1997). The larvae are approximately 5.7 mm and 2.4 mg (Benfey and Methven, 1986; Brown, 1986). Complete absorption of the yolk sac occurs within 140 degree days C (dd) from hatching (Benfey and Methven, 1986) and lumpfish actively start feeding three to seven days post-hatch, depending on water temperature (Benfey and Methven, 1986; Brown *et al.*, 1992).

Start feeding takes place in silo tanks, where larvae are fed formulated feed. Feed granules and water are usually added together to ensure sufficient distribution of feed in the whole water column. Juvenile lumpfish come in all colours (Fig. 7.13), but usually turn blue-turquoise in different sizes with increasing size, though some pink exceptions can be observed in the on-growth period.

Due to the active use of the suction disc for adherence to surfaces from hatch, very little swimming activity is observed until start feeding takes place. When feeding does commence with live or formulated feed, swimming activity increases greatly (Brown *et al.*, 1992). At high prey densities larvae will also feed while adhered to surfaces (Brown *et al.*, 1992), but this behaviour is reduced with ageing of the larvae (Brown *et al.*, 1997). Three different strategies have mainly been utilised for start feeding lumpfish although no optimised larval feed currently exists. Natural plankton has to a great degree been observed to be associated with outbreaks of disease from several marine bacteria and other live feeds such as enriched *Artemia* nauplii have been known to accumulate organic matter in the tanks, causing bacterial growth and reduced water quality. The most common approach in start feeding today is therefore introducing a formulated feed diet three to five days post-hatch, sometimes in combination with *Artemia* for the first four to eight weeks post-hatching. Hand feeding is usually utilised as a supplement to automatic feeding.

Lumpfish larvae initially fed *Artemia* have been shown to display better growth than larvae

Fig. 7.13 The lumpfish larvae will readily adhere to all surfaces in the tank from hatching.

fed only a commercial formulated dry feed (Belova, 2015; Benfey and Methven, 1986) and start feeding with *Artemia* for the first seven to 21 days before weaning onto formulated feed has been shown to decrease mortality during the larval and juvenile stages (Belova, 2015). Depending on the temperature, *Artemia* are therefore offered for four to eight weeks before weaning on to formulated feed. Formulated feed is introduced gradually over two to three days before the *Artemia* are excluded from the diet.

Belova (2015) found that larvae fed natural plankton had lower survival and developmental rates in comparison to larvae fed formulated feed or preserved copepods. Feeding larvae a diet comprising of preserved copepods has been hypothesised to be associated with increased activity of genes associated with cell differentiation and immune response (Belova, 2015). Despite this, copepods are seldom part of the farmed lumpfish diet as they are less available and more time-consuming to produce than *Artemia* and dry feeds.

The most commonly used commercial feed brands for lumpfish as of today were originally developed for cod or other marine species. There are formulated dry feeds marketed specially for lumpfish, but due to the lack of knowledge on the nutritional requirement, there is currently no feed specifically adapted for lumpfish. It is well known that the risk of developing cataracts in salmonids may be related to malnutrition (Breck *et al.,* 2005). A recent study on the relationship between cataracts and nutrition in lumpfish (Jonassen *et al.*, 2017) suggests that cataracts in lumpfish may be related to primary or secondary disturbed nutrient metabolism or malnutrition, shown by high levels of specific amino acids in different tissues, which may cause osmotic imbalance and cataract development. Hence, a less nutrient-dense feed could be a better alternative for lumpfish, e.g. a moist feed. Cataracts as a nutritional- or environmental-related welfare problem deserves further research.

Pellet sizes for lumpfish larvae are usually 150 µm when first introduced before gradually adjusting to sizes around 300 µm at metamorphosis. After metamorphosis pellet sizes up to 500 µm are gradually introduced and gradually increased to approximately 1.5 mm when the lumpfish juvenile reaches 10 g.

7.4.6 Feeding strategy

Survival through first feeding is generally high but mortality increases as metamorphosis commences at approximately 30 mm TL (Brown *et al.,* 1992). The lumpfish changes its feeding strategy with increasing size indicating an ontogenetic change in terms of prey size in juvenile lumpfish that coincides with the species migration from shallow rock pools to the open sea (Daborn and Gregory, 1983; Moring, 1989). This change from a benthic to pelagic life strategy at 40–50 mm greatly affects the species feeding preferences (Daborn and Gregory, 1983). Yolk sac larvae will feed on a selection of harpacticoid copepods and increase their interest in amphipods and other food sources of increasing size (Daborn and Gregory, 1983; Ingolfsson and Kristjansson, 2002) as their swimming behaviour increases with increasing body size (Brown, 1986; Jobling *et al.,* 2012). Feeding is usually achieved through pulse feeding every one to two hours through the day and night. For small juvenile lumpfish of <0.5 g a feeding rate of approximately 10% of the total biomass is utilised. For larger fish of between 0.5–10.0 g, a feeding rate of 3–4% per day is common.

Lumpfish larvae are often observed in association with macroalgae (Moring, 1990), in particular eelgrass (Moring, 1989) and *Laminaria* (Moring and Moring, 1991), providing shelter, camouflage and easy access to prey in the pelagic zone (Williams and Brown, 1991). In terms of metabolic cost the juvenile lumpfish's foraging strategy is 6–12% more efficient in comparison to the ambush strategy displayed by most other marine larvae (Jobling *et al.,* 2012) underlining the importance of the species' adhesive disc in the larval stage. Jobling (1997) found that the limited aerobic scope of the juvenile lumpfish limits the species' ability to hunt for prey, favouring a foraging strategy based on minimising prey encounter rate through a sit-and-wait foraging strategy where the larvae cling to the substrate and wait for prey to drift by rather than actively searching for food (Brown, 1986; Ingolfsson and Kristjansson, 2002; Jobling *et al.,* 2012). Pulse feeding may therefore increase feeding behaviour by mimicking patterns of food abundance under natural conditions (Brown *et al.,* 1997). Brown et al. (1997) found that pulse feeding with *Artemia* rather than constant exposure to prey produces faster growing larvae hypothesising that the ingestion efficiency is reduced when prey is constantly being ingested.

Lumpfish show little interest in slow-moving particles or those that settle on the tank bottom (Ingolfsson and Kristjansson, 2002; Nytrø *et al.,* 2014). This may cause fish in shallow tanks to be unable to feed to satiation despite feed being offered in abundance. A slow-sinking pellet or live feed is therefore preferred. Smaller lumpfish deploy the sit-and-wait foraging strategy to a greater extent than their larger counterparts and larger specimens display a more active hunting strategy by utilising the whole water column (Killen *et al.,* 2007). Nytrø, et al. (2014) found that strong size-related hierarchies do exist between size rankings of fish causing larger, more dominant individuals to dominate feeding stations in the tank and therefore increasing the size differences between large and smaller individuals. Frequent grading of the fish within a tank is therefore important to maximise growth within a batch of lumpfish.

7.5 On-growing phase

7.5.1 Tank set-up and cleaning

After hatching larvae are transferred to start feeding units. These are usually conically shaped tanks holding 200–1000 l, with a surface skimmer and a centre outlet covered in 500 μm mesh to prevent larvae from escaping. For larger units aerators are sometimes used to facilitate water circulation. Lumpfish larvae are retained in these tanks for the first two to three weeks before being transferred to larger conical or flat-bottomed tanks. The latter enables the use of automatic methods of cleaning the tank bottom such as an automatic cleaning arm.

The main challenge with small juvenile lumpfish is their limited aerobic scope and almost constant need to cling to a smooth, vertical surface in order to maximise growth. When designing a start feeding facility for lumpfish the density in tanks always comes second, while the available surface area and sufficient water quality comes first. Insufficient surface area frequently causes increased aggression, tail-biting and cannibalism. A short-term solution when such problems arise and grading or splitting of the tank is not possible is to increase the surface area by adding plastic panes to the open water area of the tank. This may reduce the self-cleaning effect in the tank but instantly increases the available surface for attachment in the tanks. Another solution to this is the use of 'doughnut' tanks. Though not commercially available, many lumpfish farmers use variations of the concept to increase the available area in start feeding tanks. Doughnut tanks are made by inserting a second tank wall in the middle of the tank, reducing the water mass in the tank but increasing the available horizontal wall space greatly and promoting self-cleaning. Larger, conical tanks can also be used during this period of on-growth and are usually custom-made to order with a central outlet covered with a fitted sieve.

Juvenile lumpfish become increasingly pelagic with increasing size and begin to utilise the whole water column, swimming freely or schooling. Larger juveniles are also more frequently observed swimming freely in the water and hence dominating feeding stations. A secondary consequence of this is the increase of hierarchies in the tank as smaller, subordinate lumpfish more frequently are observed to be at the tank bottom, feeding on slower-moving and sessile food sources. This causes an automatic cleaning arm to be less efficient but, with increasing pellet size, brushing excess pellets and faeces towards the tank outlet with a squeegee or broom becomes less of a challenge. In the on-growth phase tanks are usually flushed every day.

Tank walls must be cleaned frequently to prevent bacterial growth. This is done by using a squeegee or a windscreen wiper attached to a stick, preferably after flushing the tank to prevent lumpfish from sticking to the tank walls. During the start feeding phase tanks are usually cleaned every two to three days. Another strategy to reduce negative effects of bacterial films in tanks is to move the entire tank-population to a clean, dried out tank every three to four weeks.

7.5.2 Water quality and environmental factors

When planning production in lumpfish hatcheries, the estimation of growth rates throughout the production period is crucial. At present, it is assumed that the cleaning efficacy of the lumpfish is greatly correlated to size and possibly the onset of sexual maturation (Imsland *et al.*, 2014) causing small lumpfish to be high in demand.

The commercially available vaccines for lumpfish today require a minimum fish size of 7 g and a retention time of 500 d° until the vaccine is active (AMARINE micro 4-2, PHARMAQ), causing the lumpfish to reach sizes of 40–60 g before the lumpfish is ready to cohabit with the salmon. This results in the demand for on-growth capacity to increase greatly as lumpfish have previously been sent to sea at 13–25 g at four to five weeks after vaccination.

7.5.2.1 Temperature

Juvenile lumpfish reared at suboptimal temperatures show growth rates 81% lower than fish reared at optimum temperatures, though juveniles persist to actively feed and grow at temperatures below 4°C (Nytrø *et al.*, 2014). In the wild, lumpfish spend the first few years in the intertidal zone before becoming pelagic and migrating to the open sea. It is assumed that this migration is size-dependent and larger specimens leave the intertidal zone before their smaller conspecifics (Myrseth, 1971). Juvenile lumpfish show very high growth potential and large ontogenetic variation in optimum temperatures for growth. Growth rates of juvenile lumpfish under 120 g are greatly influenced by temperature and fish size, showing a stepwise increase in growth rate with increasing temperature between 4 and 16°C (Nytrø *et al.*, 2014), similar to that of juvenile halibut in the same size range (Hallaraker *et al.*, 1995). Larger individuals (120–200 g) display a modest response in terms of temperature-related growth and optimum temperatures for growth. Findings indicate an observed decrease in optimum temperatures for growth of 7°C

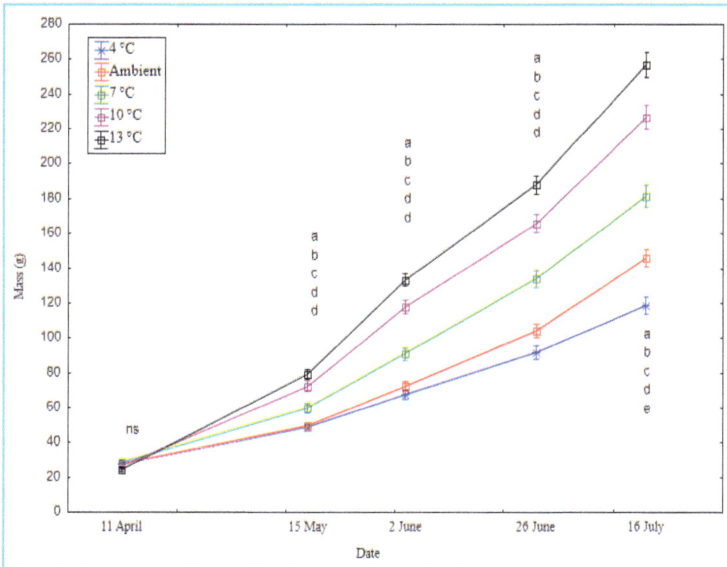

Fig. 7.14 Weight development of juvenile lumpfish reared at five different temperatures (n = 68–72 in each experimental group). Vertical whiskers indicate standard error of mean (± S.E.) of individually tagged fish. The results for two replicates are combined. Different letters indicate statistical differences between groups, with 'a' as the highest value. Vertical lines indicating S.E. may be obscured by symbols. Adapted from (Nytrø, 2013).

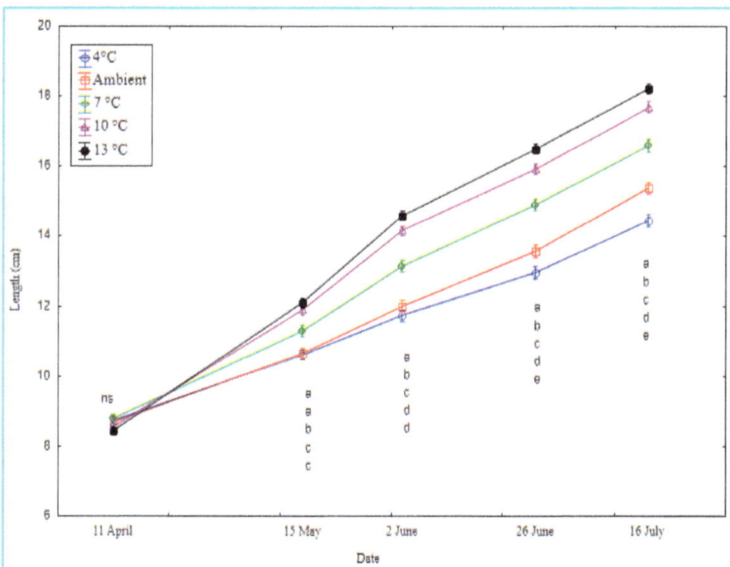

Fig. 7.15 Length development of juvenile lumpfish reared at five different temperatures (n = 348).Vertical whiskers indicate standard error of mean (± S.E.) of individually tagged fish. The results of two replicates are combined. Different letters indicate statistical differences between groups, with 'a' as the highest value. Vertical lines indicating S.E may be obscured by symbols. Adapted from (Nytrø, 2013).

in the size interval 11–200 g but no effect of temperature on growth above 160 g. This reduction in temperature response with increasing body size may therefore be a secondary effect of the migration to open sea where smaller lumpfish exploit the higher temperatures associated with the intertidal zone (Nytrø, 2013). When reared at optimum temperatures juveniles are potentially large enough for vaccination and co-culture with salmon within six months from hatching (Nytrø *et al.*, 2014).

Studies of potential negative long-term effects of rearing at optimum temperatures for growth have not been undertaken, but indications of occurring cataracts and increased mortality have been observed at high temperatures (Nytrø *et al.*, 2014). A non-comprehensive mapping of the occurrence of cataracts in farmed and wild-caught lumpfish conducted in 2015 (Jonassen *et al.*, 2016) showed that the extent of the challenges related to cataracts are mostly associated with farmed broodstock and that the cataract score amplified with size. Cataracts were also observed during the juvenile stages but these scored lower in terms of severity. It may be hypothesised that a part of the underlying cause of this problem may be related to the rapid growth in juvenile lumpfish as rapidly growing species of fish such as the lumpfish are more prone to developing cataracts (Bjerkås *et al.*, 2001; Waagbø *et al.*, 1996; Waagbø *et al.*, 1998).

7.5.2.2 Salinity and flow through

Flow-through is usually set to 100–400% per hour in the on-growth phase, dependent on biomass and density. Flow-through usually does not exceed 150% per hour until the lumpfish reaches 3 g. Most commercial hatcheries utilise full flow-through systems with good quality seawater, though some variations occur. In hatcheries utilising freshwater periodically to combat the gill parasite *Gyrodactylus cyclopteri* in juvenile lumpfish and localities with salinities below 32‰, increased occurrences of cataracts have been observed (Jonassen *et al.*, 2016). As a consequence,

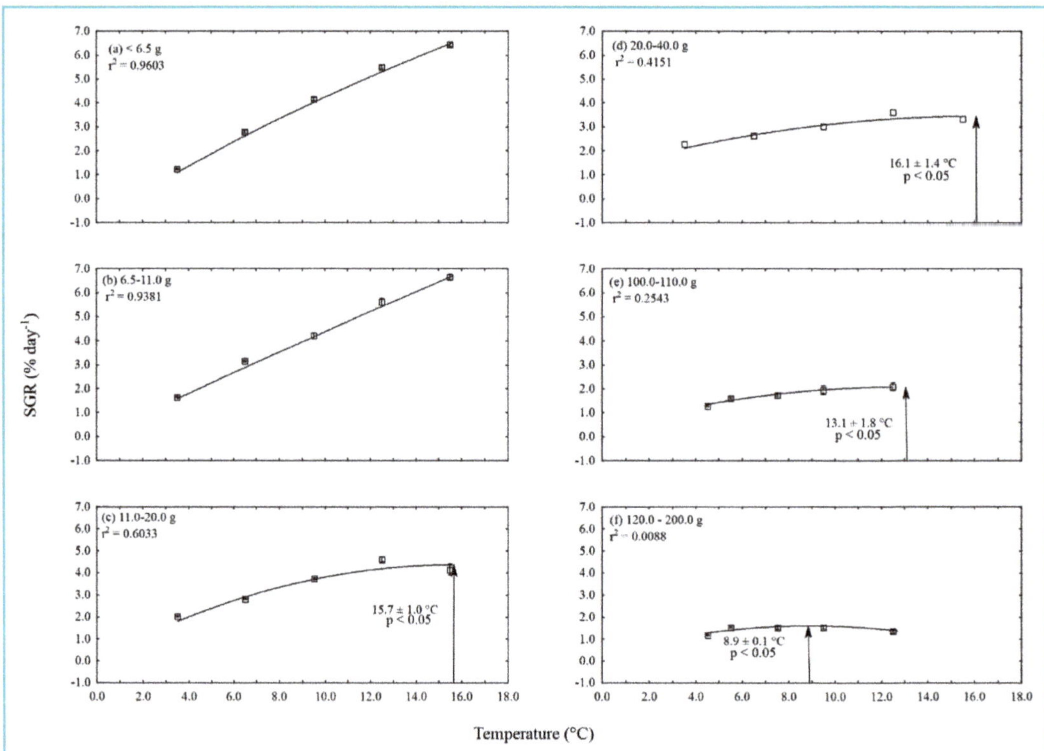

Fig. 7.16 Changes in specific growth rate with temperature for six different size classes of juvenile lumpfish. When possible the optimum temperature for specific growth rate (ToptSGR, ±S.E) was calculated from the first order derivate of the parabolic regressions. All calculated ToptSGR are significant (P <0.05). Temperature optimums are indicated by arrows in the figure (Nytrø *et al.*, 2014).

exposure to low salinities and brackish water over longer periods of time are not recommended until further conclusions can be drawn on the long-term effects.

7.5.2.3 Oxygen saturation

Juvenile lumpfish are highly sensitive to suboptimal levels of oxygen saturation. Most commercial hatcheries have good control in terms of biomass, density, water supply, CO_2 and O_2 saturation but, in the period between vaccination and relocating to net pens with salmon, fish grow rapidly and therefore push the capacity of the on-growth facility to the limit. Suboptimal values of environmental factors such as CO_2 and O_2 saturation may cause chronic stress as density increases, which may cause cataracts in lumpfish (Jonassen *et al.*, 2016). At 69% saturation Haatuft (2015) observed an increase in ventilation activity, indicating oxy-regulation as a means to increase the gradient for oxygen uptake.

Oxygenated water up to 110% is frequently added in on-growing tanks during time periods when temperatures or densities are high. At 69–81% oxygen saturation Haatuft (2015) found significantly reduced growth rates and final weight in juvenile lumpfish when compared to a group of fish reared at 96% oxygen saturation after 57 days. It is unknown to what extent lumpfish are exposed for oxidative stress as a result of increased oxygen levels (Jonassen *et al.*, 2016).

Haatuft (2015) found that long-term exposure to oxygen saturations below 96% reduced feeding rate, growth rates and body mass in juvenile lumpfish. Suboptimal oxygen levels also caused increased plasma cortisol levels, especially in fish exposed to acute hypoxia and handling (Barnes *et al.*, 2011).

7.5.2.4 Tank colour

In terms of tank colour, lumpfish prefer the shade. Therefore, dark colours such as green, dark grey or black are preferred. If part of the tank is shaded the lumpfish will almost immediately choose to rest in this area. To prevent clumping the tank should therefore be of uniform colour and with uniform lighting conditions. These colour preferences can also be utilised in terms of facilitating cleaning efficiency. Bright colours such as white are seemingly less attractive. Some farmers therefore paint the tank bottom white to stop lumpfish adhering to the tank bottom, so that an automatic cleaning arm can operate freely without harming bottom-dwelling lumpfish.

7.5.2.5 Photoperiod and light management

Direct exposure to UV light (sunlight) has been hypothesised to induce cataracts in lumpfish over time (Jonassen *et al.*, 2016) and a continuous photoperiod has been shown to have great influence on growth in female broodstock. Light sources should be placed so that shadows or reflections on the tank wall are minimised and light sources should not be placed directly over fish tanks. Most commercial producers of lumpfish use continuous light during all stages, though some variations do occur. A regime of 18:6 regimes for start feeding are known to be utilised, and during the on-growing phase some hatcheries prefer a 12:12 day and night variation.

7.5.3 Grading and handling

Findings by Nytrø (2013) indicated a strong size rank correlation within groups of juvenile lumpfish, indicating that lumpfish keep their relative size position through the on-growing phase in the hatchery. Size rank correlation was higher in low temperature groups than high (r_{sp} = 0.95 and r_{sp} = 0.67, 4°C and 13°C respectively), indicating that some individuals may have a larger scope for temperature-related growth than others. Juvenile lumpfish are observed to be highly territorial and the impact of hierarchies on growth rates has an increasing impact on growth rates over time. Smaller, subordinate juveniles become forced to the bottom of the tank while larger dominant individuals spend more time swimming and hunting for feed pellets, therefore increasing their energy expenditure. This in turn causes neither the dominant nor the subordinate lumpfish to fully maximise their growth potential (Haaland, 1996). This is shown through significant size correlation throughout the on-growth period (Nytrø *et al.,* 2014).

Erosion, bruising and wounds are observed to occur frequently in lumpfish hatcheries, causing potential conditions for opportunistic bacteria such as *Tenacibacculum,* which is commonly observed when the temperature is low. The cause of these wounds may have several underlying explanations but social interaction and aggression between conspecifics have been reported to occur frequently when size hierarchies are established in the on-growth facilities, especially in periods with rapid growth. The most important strategy to combat this problem is frequent grading of the lumpfish to reduce aggression and competition for feed pellets. It is common to reduce light intensity in tanks to reduce such behaviour.

Grading begins at 0.5–1.0 g and is repeated at one- to three-week intervals throughout the production cycle. Automated grading machines developed for lumpfish do exist but, more commonly, traditional grading boxes and grids are used, especially for fish under 3 g. It is thought that increased focus towards breeding programmes and more uniform sizing between individuals from the same batch will decrease the need for grading greatly. Handling is usually reduced to a minimum during periods of high or variable temperatures.

Although frequent grading helps to reduce the observed aggression and feeding competition in on-growing tanks, it is also a contributing factor to increased stress and has been noted to increase ocular disease in farmed species (Williams and Brancker, 2006). A stress test conducted by Haatuft (2015) indicated that short-term exposure to handling and low oxygen levels (30%) caused an increase from 0.97 ng ml^{-1} to 33 ng ml^{-1} in plasma cortisol levels after 30 minutes with maximum levels of stress observed at 120 minutes where peak plasma cortisol measured to ~47 ng ml^{-1} ($p > 0.05$). This differs greatly to a similar study where fish were crowded (280–300 kg m^{-3}) for 20 minutes before monitoring at optimal oxygen conditions, as plasma cortisol was measured peaking at 72 ng mL^{-1} 60 minutes post-stressor and it reduced to approximately 65 ng ml^{-1} after 120 minutes (Iversen *et al.,* 2015). This suggests that short-term exposure to crowding may produce a larger stress response than short-term exposure to low oxygen levels.

References

Barnes, R.K., King, H. and Carter, C.G. (2011) Hypoxia tolerance and oxygen regulation in Atlantic salmon, *Salmo salar* from a Tasmanian population. *Aquaculture* 318, 397–401.

Belova, I. (2015) Nutritional programming in lumpsucker, *Cyclopterus lumpus* – a transcriptomic approach, Faculty of Biosciences and Aquaculture. University of Nordland, Bodø, 71 pp.

Benfey, T.J. and Methven, D.A. (1986) Pilot-scale rearing of larval and juvenile lumpfish (*Cyclopterus lumpus* L.), with some notes on early development. *Aquaculture* 56, 301–306.

Bjerkås, E., Bjørnestad, E., Breck, O. and Waagbø, R. (2001) Water temperature regimes affect cataract development in smolting Atlantic salmon, *Salmo salar* L. *Journal of Fish Diseases* 24, 281–291.

Breck, O., Bjerkås, E., Campbell, P., Rhodes, J.D., Sanderson, J. and Waagbø, R. (2005) Histidine nutrition and genotype affect cataract development in Atlantic salmo, *Salmo salar* L. *Journal of Fish Diseases* 28, 357–371.

Brown, J.A. (1986) The development of feeding behaviour in the lumpfish, *Cyclopterus lumpus* L. *Journal of Fish Biology* 29, 171–178.

Brown, J.A., Wiseman, D. and Kean, P., (1997) The use of behavioural observations in the larviculture of cold-water marine fish. *Aquaculture* 155, 297–306.

Brown, J.A., Somerton, D.C., Methven, D.A. and Watkins, J.R. (1992) Recent advances in lumpfish *Cyclopterus lumpus* and ocean pout *Macrozoarces americanus* larviculture. *Journal of the World Aquaculture Society* 23, 271–276.

Daborn, G.R. and Gregory, R.S. (1983) Occurrence, Distribution, and Feeding-Habits of Juvenile Lumpfish, *Cyclopterus-Lumpus* L in the Bay of Fundy. *Canadian Journal of Zoology* 61, 797–801.

Danielsen, M. (2016) *The effect of incubation on eggs and larvae of lumpsucker,* (*Cyclopterus lumpus* L.). Department of Arctic and Marine Biology. University of Tromsø, Tromsø, 83 pp.

Davenport, J. (1983) Oxygen and the developing eggs and larvae of the lumpfish, *Cyclopterus lumpus. Journal of the Marine Biological Association of the United Kingdom* 63, 633–640.

Davenport, J. and Kjorsvik, E. (1986) Buoyancy in the Lumpsucker *Cyclopterus-Lumpus. Journal of the Marine Biological Association of the United Kingdom.* 66, 159–174.

Davenport, J., Lønning, S. and Kjørsvik, E. (1983) Ammonia output by eggs and larvae of the lumpsucker, *Cyclopterus lumpus,* the cod, *Gadus morhua* and the plaice, *Pleuronectes platessa. Journal of the Marine Biological Association of the United Kingdom* 63, 713–723.

Goulet, D., Green, J.M. and Shears, T.H. (1986) Courtship, Spawning, and Parental Care Behavior of the Lumpfish, *Cyclopterus-Lumpus* L, in Newfoundland. *Canadian Journal of Zoology* 64, 1320–1325.

Haaland, T. (1996) Oppforing av rognkjeks. Universitetet i Nordland, Bodø.

Haatuft, A.C. (2015) Effects of reduced water oxygen saturation on growth and plasma cortisol levels in juvenile lumpfish (*Cyclopterus lumpus* L.) in aquaculture, Norwegian college of fisheries science. University of Tromsø, Tromsø, 55 pp.

Hallaraker, H., Folkvord, A. and Stefansson, S.O. (1995) Growth of juvenile halibut (*Hippoglossus hippoglossus*) related to temperature, day length and feeding regime. *Netherlands Journal of Sea Research* 34, 139–147.

Imsland, A.K., Reynolds, P., Eliassen, G., Hangstad, T.A., Foss, A., Vikingstad, E. and Elvegård, T.A. (2014) The use of lumpfish (*Cyclopterus lumpus* L.) to control sea lice (*Lepeophtheirus salmonis* Krøyer) infestations in intensively farmed Atlantic salmon (*Salmo salar* L.). *Aquaculture* 424–425, 18–23.

Ingolfsson, A. and Kristjansson, B.K. (2002) Diet of juvenile lumpsucker *Cyclopterus lumpus* (Cyclopteridae) in floating seaweed: Effects of ontogeny and prey availability. *Copeia* 2002, 472–476.

Iversen, M.H., Jakobsen, R., Eliassen, R. and Ottesen, O. (2015) Sedasjon av berggylt og rognkjeks for å redusere stress og dødelighet, Norsk Fiskeoppdrett, pp. 42–46.

Jobling, M. (1997) Temperature and growth: Modulation of growth rate via temperature change. Cambridge University Press, Cambridge (UK).

Jobling, M., Alanärä, A., Kadri, S. and Huntingford, F. (2012) Feeding Biology and Foraging. Wiley-Blackwell.

Jonassen, T.M., Hamadi, M. and Imsland, A.K. (2016) *Kartlegging av katarakt hos rognkjeks i oppdrett.* Akvaplan-niva, Aqua Kompetanse, Fiskeri- og havbruksnæringens Forskningsfond.

Jonassen, T.M., Hamadi, M., Remø, S.C. and Waagbø, R. (2017) An epidemiological study of cataract in wild and farmed lumpfish (*Cyclopterus lumpus* L.) and the relation to nutrition. *Journal of Fish Diseases* 40, 1903–1914.

Killen, S.S., Brown, J.A. and Gamperl, A.K. (2007) The effect of prey density on foraging mode selection in juvenile lumpfish: balancing food intake with the metabolic cost of foraging. *Journal of Animal Ecology* 76, 814–825.

Lønning, S., Kjørsvik, E. and Davenport, J. (1984) The hardening process of the egg chorion of the cod, *Gadus morhua* L., and lumpsucker, *Cyclopterus lumpus* L. *Journal of fish Biology* 24, 505–522.

Moring, J.R.(1989) *Food-Habits and Algal Associations of Juvenile Lumpfish, Cyclopterus-Lumpus L, in Intertidal Waters. Fish B-Noaa* 87, 233–237.

Moring, J.R. (1990) Seasonal Absence of Fishes in Tidepools of a Boreal Environment (Maine, USA). *Hydrobiologia* 194, 163–168.

Moring, J.R. and Moring, S.W. (1991) Short-term movements of larval and juvenile lumpfish, *Cyclopterus lumpus* L., in tidepools. *Journal of Fish Biology* 38, 845–850.

Myrseth, B. (1971) *Fekunditet, vekst, levevis og ernæring hos Cyclopterus lumpus* L.. University of Bergen, Norway, pp. 113 bl.

Norðberg, G., Johannesen, A. and Arge, R. (2015) Cryopreservation of lumpfish *Cyclopterus lumpus* (Linnaeus, 1758) milt., *PeerJ*, 3, pp. e1003. DOI:10.7717/peerj.1003.

Nytrø, A.V. (2013) The effect of temperature and fish size on growth of juvenile lumpfish (*Cyclopterus lumpus* L.). University of Tromsø, Tromsø, 80 pp.

Nytrø, A.V., Vikingstad, E., Foss, A., Hangstad, T.A., Reynolds, P., Eliassen, G., Elvegård, T.A., Falk-Petersen, I.B. and Imsland, A.K. (2014) The effect of temperature and fish size on growth of juvenile lumpfish (*Cyclopterus lumpus* L.). *Aquaculture* 434, 296–302.

Waagbø, R., Bjerkås, E., Sveier, H., Breck, O., Bjørnestad, E. and Maage, A. (1996) Nutritional status assessed in groups off smolting Atlantic salmon, *Salmo salar* L., developing cataracts. *Journal of Fish Diseases* 19, 365–373.

Waagbø, R., Sveier, H., Breck, O., Bjørnestad, E., Maage, A. and Bjerkås, E. (1998) Cataract formation in smolting Atlantic salmon, *Salmo salar*, fed low and high energy diets. *Bulletin of the European Association of Fish Pathologists* 18, 201.

Williams, D.L. and Brancker, W.M. (2006) Aggravating factors in the development of ocular abnormalities in farmed Atlantic halibut (*Hippoglossus hippoglossus*). *The Veterinary Journal* 172, 501–505.

Williams, P.J. and Brown, J.A. (1991) Developmental changes in foraging-predator avoidance trade-offs in larval lumpfish *Cyclopterus lumpus. Marine Ecology Progress Series* 76, 53–60.

CHAPTER 8

Egg collection by hand-stripping and natural spawning in lumpfish *Cyclopterus lumpus* broodstock

Claudia Wittwer[1] and Jim Treasurer[2]

[1]Arctic Cleanerfish AS, Stamsund, Norway
[2]FAI Aquaculture, Ardtoe, UK

8.1 Introduction

The supply of lumpfish eggs would be more predictable and there would be less disease risk in establishing hatchery-reared lumpfish broodstock. There may also be potential impacts of taking wild caught broodstock each year and sacrificing these at the time of egg collection. These fish are normally stripped manually of eggs and it is probable that capture and handling stress combined might have an impact on survival of hand-stripped broodfish. However, wild-caught broodfish have been difficult to domesticate with many not surviving for more than a few weeks after capture. This could be due to the trauma of capture when fish caught by passive gear such as gill and trammel nets may be damaged. In addition, health authorities, as in Iceland where broodstock were affected by a VHS outbreak in 2015 (Jónasson, 2017), may call on all fish to be sacrificed after egg stripping to check the health status of the fish and guarantee disease-free classification. Other issues in maintaining broodstock appear to have been low-level mortalities due to disease and females becoming egg bound (Powell *et al.*, 2017). The long-term goal of the cleaner fish hatcheries is therefore to domesticate wild-caught broodstock or to rear F1 generation fish and, of course, to avoid in-breeding by following a selection programme. Ultimately a selective breeding programme would assist in breeding lumpfish resistant to diseases such as amoebic gill disease (AGD) (Jónasson, 2017; Mommens, 2017) and also with a high proportion of fish that clean lice and are more effective as cleaner fish. The size of a viable breeding population has to be determined. Hatchery experiences in egg collection by both hand-stripping of broodfish and permitting natural spawning are described and compared here, and the advantages and disadvantages of the contrasting egg management approaches are discussed.

8.2 Hand collection of lumpfish eggs

8.2.1 Egg collection

Lumpfish eggs were collected in the Lofoten Islands by hand-stripping from local fished wild stocks (January until July) or from broodstock reared in the Arctic Cleanerfish hatchery (November until April). Since 2015 it has been possible to collect eggs from hatchery-reared broodstock and future objectives are all-year controlled spawning of good quality eggs with no reliance on wild lumpfish. An advantage of the hand-stripping method compared with natural spawning is that eggs are not contaminated with bacteria or debris from the spawning tank. However, it is very important in the hand-stripping method that females are absolutely ready for spawning to get good quality eggs and a high fertilisation rate. Females show a greatly distended abdomen and a swollen, red-coloured urogenital region close to spawning, while males show a red coloration of the body (Figs. 8.1, 8.2). Hatchery-reared and wild-caught broodfish show both those external characteristics.

The fish are anaesthetized (with MS_{222} or Eugenol) and the urogenital region is dried with paper towels before hand-stripping to avoid any contact between eggs or milt and seawater. First the urogenital opening of the female is pushed in with the finger to open it. Afterwards the abdomen can be pressed gently to collect the eggs in a beaker. One female gives from 0.4 to up to 1 l of eggs, which represents approximately 50,000 to 120,000 eggs. Good quality eggs should run freely and smoothly from the opening without great effort. Eggs that are still attached to each other or are bloody have shown bad quality during incubation. The colour of the eggs varies widely between females, from light rose, to dark pink to yellow or orange. The colour is due to the presence of carotenoid pigments in the yolk and this is slowly lost during the embryonic development (Davenport, 1985). No correlation has been noticed between egg quality and colour.

Fig. 8.1 Male (left) and female (right) hatchery-reared lumpfish.

Fig. 8.2 Mature female lumpfish ready for hand-stripping the eggs.

Continuous pressing of the abdomen laterally and ventrally showed great success in collecting the milt of the male (Fig. 8.3). With this method the male does not need to be sacrificed and can be used to fertilise eggs throughout a whole spawning season. A few drops of the milt from two different males are collected directly into the beaker with the eggs. After hand-stripping the fish, the eggs and milt are mixed well to fertilise all eggs. The eggs are transferred to the incubator tray after 15 minutes. Another way used to collect the milt on other farms is to euthanise the male and collect the whole testes, which can be stored in the fridge for several days.

8.2.2 Egg incubation

Lumpfish eggs are incubated in salmon egg incubator troughs each with four incubator trays (Fig. 8.4). Trays are equipped with an extra sieve with a smaller mesh size than for salmon

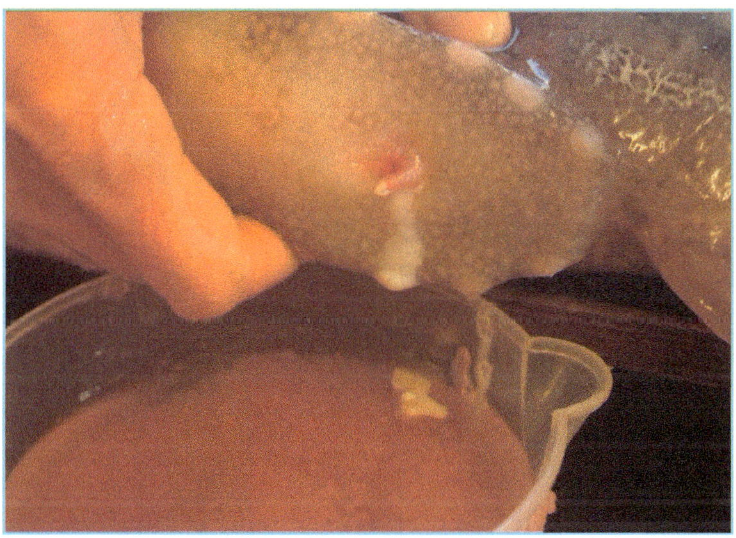

Fig. 8.3 Collecting the milt of a male lumpfish.

Fig. 8.4 The hatchery at Arctic Cleanerfish AS.

(1000 µm) to prevent the escape of hatched larvae. One beaker of fertilised eggs from one female is moved into one incubator tray and then transferred into the incubator trough filled with seawater. After the tray is fixed the water flow is increased to 6 l min^{-1}. Eggs form a hard compact egg mass from a few hours to 48 hours later.

Eggs are incubated under continuous red light conditions. Temperature during the year can range from 3 to 16°C depending on the ambient seawater temperature. Incoming temperature can be increased or decreased to accelerate or decelerate hatching and control the production cycle. Temperatures over 10 degrees are avoided to reduce the potential for bacterial development. The water flow is slowly increased during incubation period starting with 6 l min^{-1} and increased up to 10 l min^{-1} depending on the oxygen level. Oxygen is maintained over 90% during the whole incubation period and incoming water can be also oxygenated if levels are too low.

The condition of the eggs is controlled daily and includes a check on the appearance of dead eggs, bacteria or fungi. The dead eggs are collected by siphoning, normally only once a few days after fertilisation and then once or twice in the incubation period. Small amounts of bacteria or fungi coverage can also be removed by siphoning. Disinfection of the eggs is carried out with Pyceze and is useful if bacteria or fungi occur on a larger scale. A concentration of 1:10,000 for 30 minutes is applied and was successful without harming the eggs. However, eggs should not be treated within three to four days of hatching.

The egg diameter is between 2.2 and 2.5 mm (Fig. 8.5) and can vary in one spawning batch. Also, the average diameter can vary between different female batches. The fertilisation rate is over 95% if the quality of the eggs during hand-stripping is good. After 15 to 30 minutes the zygote period is visible under the microscope and it is possible to determine if eggs are fertilised (Fig. 8.5). The cleavage, blastula and gastrula period can be seen in the next hours and days, dependent on temperature (Wittwer, 2017). After around 50 degree days C (dd) the embryo is visible (Fig. 8.5). The incubation time is between 260 to 300 dd. Hatching can take place over several days. To counteract this a sudden increase of the light intensity when the first 100 hatched larvae are visible can trigger and synchronise the hatching of the larvae and consequently the main hatch

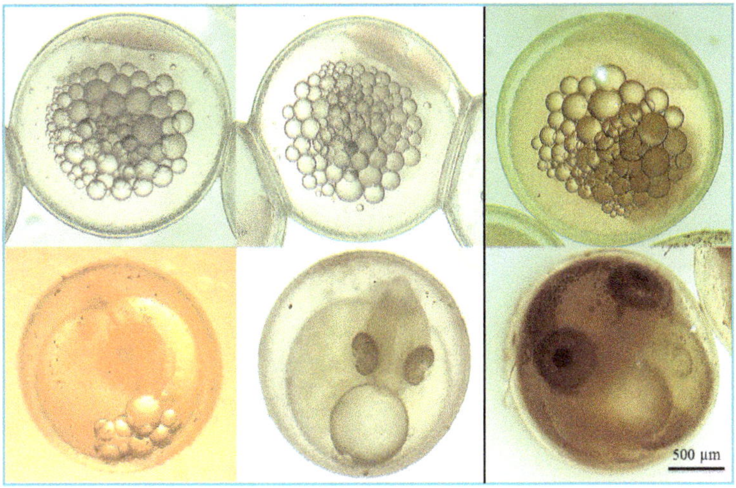

Fig. 8.5 Embryonic development of lumpfish, from top left clockwise: zygote, cleavage, gastrula, segmentation, pigmentation of the eyes, close to hatch. Egg diameter = 2.2 mm.

can occur over just a few hours. The average hatch rate is 80%. Larvae are collected with a small aquarium net from the incubation tray into a bucket filled with seawater. The number of larvae is calculated and then transferred into the start feeding tanks.

8.3 First feeding

The start feeding department consists of thirteen 600 l start feeding tanks designed internally by Arctic Cleanerfish AS (Fig. 8.6). The tanks are cylindrical and flat-bottomed, with a sieve of a mesh of 1000 µm on the bottom. Approximately 80,000 newly hatched larvae are stocked in the first feeding tank. Larvae are reared under continuous daylight conditions. The temperature during the year can range from 4 to 16°C depending on the ambient seawater temperature. Usually temperature is increased or decreased to control the production cycle and bacteria development. The water flow is kept at 8 l min^{-1} at the start and increased to 20 l min^{-1} throughout the rearing cycle. The special design of the tanks helps to keep the tank environment clean. A cleaning disc is underneath the bottom sieve and can be rotated manually. This removes feed waste quickly and easily from the bottom of the tank. Additionally, the tanks are cleaned daily by siphoning to remove larval mortalities and remaining feed waste.

Newly hatched larvae are 6.3 ± 0.4 mm in total length (Fig. 8.7). Although the yolk sac is totally depleted after ten to 15 days (Ingólfsson and Kristjánsson, 2002), larvae are capable of feeding exogenously earlier. After three days the micro feed is introduced into the tanks via a Sterner automatic feeding system and also hand feeding. Larvae were fed with live feed before 2015, which was wild harvested copepods from the adjacent pond. There was then a switch to micro feed due to the high risk of transferring diseases into the first feeding tanks and also because of good results using micro feed as the first starter diet.

First feeding of marine fish larvae is a big challenge and this is one of the main issues in marine hatcheries (Holt, 2011). Many marine fish larvae are cultivated using live feed and this is costly and labour intensive. In contrast, lumpfish larvae can be fed with micro diets from the start.

Nevertheless, some farms still use live feed such as *Artemia*. Bacterial growth in the first feeding tank can be a problem while using micro diets due to the high leakage of proteins (Langdon, 2003; Hamre, 2006). This could be one of the reasons why some farms choose to use *Artemia* instead. It has been noticed that there is more bacterial development while using micro diets than using copepods at Arctic Cleanerfish AS. However, this has been counteracted by tank design and good cleaning and feeding routines that have limited bacterial growth.

The diet influences the survival and growth of lumpfish during first feeding. Four different feeds were tested and this showed that growth was significantly increased during 14 days of feeding *Artemia* compared with the micro diets (Wittwer, 2015). There was no significant difference in survival in larvae fed with *Artemia* or solely with micro diets. In the latest development a new non-commercial micro diet did improve both growth and survival of lumpfish, compared with previous diets, indicating that there is potential to improve the first feeding diet of lumpfish (Wittwer, 2015).

Lumpfish are able to attach themselves to the surface area of the tank with their ventral adhesive disc which is functional at hatch. Therefore, next to the water quality and the diet, sufficient resting or sitting places are important during the early stages (Wittwer, 2016). Larvae are transferred after approximately 400 dd to the nursery unit for on-growing. Survival of larvae in first feeding is over 90%.

Fig. 8.6 Start-feeding tank at Arctic Cleanerfish AS.

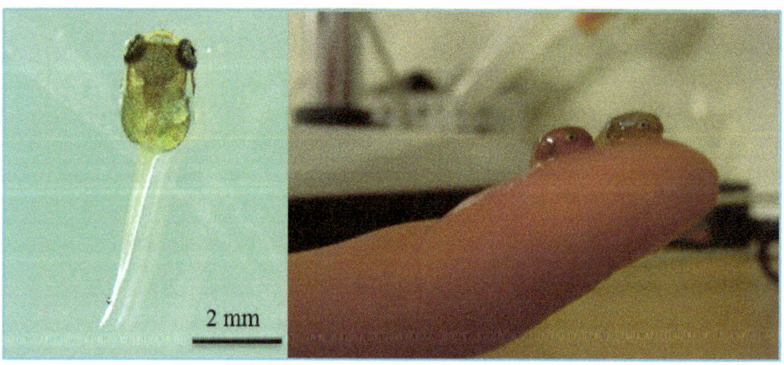

Fig. 8.7 Left: lumpfish larvae at hatch. Length = 6.4 mm. Right: lumpfish larvae moved from first feeding tanks after 40 days.

8.4 Natural spawning of lumpfish

8.4.1 Environment and tanks

Lumpfish broodstock were reared from wild-caught broodfish captured by trammel net a few days before spawning in the English Channel. The fish were held in small holding containers of 2 m^3, and then stripped by hand a few days later. The eggs were 88 dd old when received on 4 March 2014 at a temperature of 8°C. The weight of the egg masses was 500, 500 and 650 g and, from a mean egg number of 136 eggs g^{-1}, the estimated number of eggs stocked was 224,400. The fertilisation rate was 100%. Each egg batch was stocked in a separate incubator of 60 l volume. These were conical tanks suspended alongside the larval rearing tank and with an overflow pipe to convey hatched larvae into the rearing tank. The eggs were incubated from 8°C on receipt at the hatchery on a rising water temperature to 10°C at hatch after 290 dd days. The larvae were reared in 1300 l volume circular black polypropylene tanks, 1.5 m diameter and 1 m deep. Water that had been sand filtered and treated with UV light was initially supplied at 5 l min^{-1} on a flow-through system. Light was provided at 300 Lux and oxygen maintained at 8 to 9.5 mg l^{-1}. The larvae were fed for 28 days on Artemia nauplii enriched with Multigain (Reed Mariculture) and then weaned over several days to 150 µm LARVIVA Prowean diet. Fifty of the fastest growing fish (sex not identified at this stage) from each of three egg batches were selected after six months on the basis of good appearance, no external damage, and large size (100 to 125 g). These fish were transferred to three 1600 l black polypropylene tanks in a nursery building. The sex ratios had not been determined but the gender of the fish could be identified from January 2015 from the size differentiation between males and females, with the latter up to three times heavier than the males. The males could also be identified from January by a vivid gold to red coloration on the ventral surface compared with the overall grey appearance of females.

The broodfish were held in an ambient temperature regime (from 6 to 15°C) over the year. The water was flow-through (one use) and was provided by a seawater pump passed through a sand filter to 60 µm, degassed, treated with 360 KJl ultraviolet light and filtered locally to 10 µm. Water was supplied to tanks at 10 l per minute. After one year the fish were transferred to 28 m^3 volume glassfibre circular green tanks. A screen cover and shading was provided over the tank, aeration, and oxygenation to give 8 mg l^{-1} oxygen. The fish were checked regularly from January 2015 throughout the day for condition and the appearance of egg strands. Rocks of 30 to 40 cm length were added to the tanks on 27 April after spawning commenced at regular spacing to create horseshoe-shape spawning nests (Fig. 8.8).

8.4.2 Nutrition

The fish were fed first on Larviva Prowean diet and this was followed with Biomar Inicio Plus through a range of pellet sizes from 1.2 to 3 mm. Then fish were offered Biomar Symbio wrasse feed, which contains shrimp meal and other crustacean material, but no marine finfish oil was fed for two months. Some of the fish were developing cataracts after four months post-hatch and so the feed was changed to a standard marine oil finfish diet 50:50 shrimp-based feed. Only Biomar EFICO diet was offered from January 2015.

Fig. 8.8 Photo of spawning nests comprising natural stone blocks. Credit: Institute of Marine Research, Iceland

8.4.3 Gonadosomatic index of broodstocks

Twelve broodstock died during a low oxygen episode just at the start of the spawning season in February 2015 and it was therefore possible to assess the developmental stage of the ovary and the GSI. The GSI was calculated as (Wt gonad/Eviscerated wt) ×100. Ovary weight was in the range 79 to 505 g with a mean of 308 g. This gave an eviscerated GSI from 23 to 51% with an average of 33.4. This compared with GSIs of wild capture broodstock at spawning of 32.3. The average testes weight was 19.5 g, which represented a GSI of 4.3 compared with 18.2 g (8.9%) in wild-caught males from the east coast of Scotland.

8.4.4 Comparison of realised and potential fecundity

Realised fecundity is the number of collected eggs that were spawned naturally by the lumpfish. From histology it was assessed that these small first-time spawners produced only one egg mass in the year as oocytes were of a uniform diameter and there was no indication of later developing batches of smaller oocytes. However, literature indicates that females produce two to three egg batches at intervals of eight to 14 days (Andriyashev, 1954). These naturally spawned eggs were collected from the broodstock tanks by hand net, disinfected in 4000 ppm Kickstart for 45 secs (CD Lines), and transferred to small collection baskets with a water exchange of 5 l min^{-1}. A small sample of eggs was taken from each batch, *c.* 150 eggs. This was examined under a microscope at ×40 magnification and the fertilisation rate was estimated after 50 dd. The diameter of 30 eggs from each egg batch was measured. The eggs, in a size range from 1.9 to 2.2 mm, were incubated and embryonic development was observed until hatch in 250 to 300 dd. Three subsamples of *c.* 50 eggs were blot-dried with paper towel and weighed on a balance to 0.00001 g. The average wet wt (mg) of each egg was calculated. The wet weight of the total egg mass was divided by this

Table 8.1 Numbers of eggs per batch of naturally spawned eggs. NF = not fertilised.

Date	Egg batch	Batch egg weight g	Average egg wt g	Egg number	Eggs g^{-1} of ovary	Fertilised
5/3/2015	1	216				NF
22/3/2015	2	150				NF
25/3/2015	3	215	0.003	63578	296	NF
26/3/2015	4	207	0.005	44797	216	NF
28/3/2015	5	169	0.005	32607	193	NF
3/4/2015	6	35	0.004	9815	280	NF
3/4/2015	7	245	0.005	50819	207	NF
13/4/2015	8	242	0.007	36476	151	NF
14/4/2015	9	54	0.003	15704	291	NF
15/4/2015	10	47	0.005	8844	188	NF
16/4/2015	11	113	0.006	18254	162	NF
16/4/2015	12	247	0.005	53434	216	NF
19/4/2015	13	247	0.006	39409	160	Fertilised
25/4/2015	14	171	0.005	32605	191	Fertilised
26/4/2015	15	370	0.005	72613	196	Fertilised
26/4/2015	16	93	0.005	17985	193	Fertilised
28/4/2015	17	151	0.006	25046	166	Fertilised
28/4/2015	18	100	0.006	16637	166	Fertilised
30/4/2015	19	54	0.006	9747	181	Fertilised
2/5/2015	20	117	0.006	19719	169	Fertilised
2/5/2015	21	300	0.006	49910	166	Fertilised
8/5/2015	22	178				

to give total egg number in the egg mass and the SD was calculated from three sub-samples. For comparative purposes, potential total annual egg production was estimated from the weight of ovaries of reared lumpfish that had died or were sampled prior to spawning and fecundity was estimated by dividing ovary weight by mean egg wet weight of *c.* 50 eggs taken from three replicates.

Realised egg production for these first-time spawners was in the range 8,844 to 72,613 with a mean of 32,526 per egg mass (Table 8.1). This can be compared with potential fecundity from measurement of eggs in three sampled fish in 2015. The fish mean weight was 1281, 789 and 760 g, with respective ovary weights 427, 135, 139 g, and the total number of eggs from the three fish was 90,912, 27,060 and 30,569, with a mean potential fecundity of 49,154 eggs, and fecundities of 211, 200, and 219 eggs g^{-1} ovary respectively.

8.4.5 Spawning period

The juvenile potential broodfish grew rapidly and attained a weight of 6.5 g by August 2014, 397 g (range 253–682 g, n = 150) in December (gender not identified) and 462 g average for males and 1234 g for females at the end of March 2015 (n = 60). Fish in three separate families

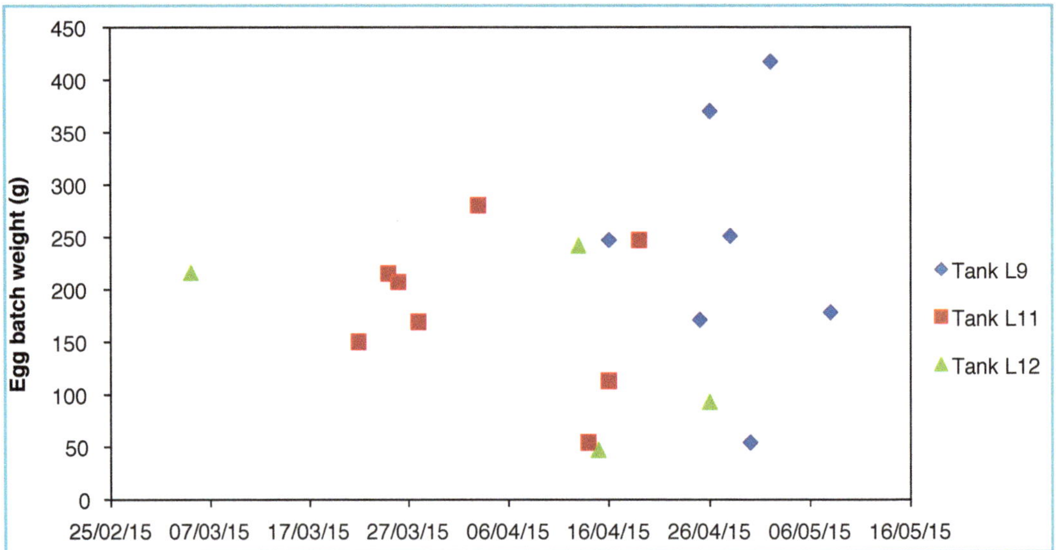

Fig. 8.9 The spawning pattern of captive lumpfish broodstock in three 3 m diameter tanks in 2015.

(tanks) spawned naturally from 5 March to May (Fig. 8.9) at water temperatures from 8 to 9.5°C. Naturally spawned egg batches of these first-time spawners varied from 35 g to 370 g weight, and egg number from 8,844 to 72,613 per batch. Further work remains to be carried out on tank habitats, optimum gender ratios, genetic studies, crossing of different lumpfish families, and assessing performances of stocks of different origins.

The fish did not spawn synchronously in all three tanks. Spawning commenced in tank L12 on 5 March 2015 at a water temperature of 10.3°C and a day length of 11 hours (Fig. 8.9). There was no further spawning for the next 22 days. Only three other egg batches were spawned, with the last on 26 April. The fish in tank L11 commenced spawning on 22 March at 11.2°C degrees and finished on 19 April with nine egg batches in total. The fish in tank L9 started spawning on 16 April at 10.5°C degrees and finished on 8 May with a daylength of 16 hours with a total of seven egg batches. The total length of the spawning seasons was 23, 29 and 53 days respectively, and for all three tanks was 64 days (Fig. 8.9).

Twenty-two egg batches were spawned in total with eight from tank L9, nine from tank L11 and five from tank L12. Most of the males were mature evidenced by the vivid coloration and spawning condition following examination of mortalities. The proportion of female first-time spawners at 1 year old was 32%, 36% and 25% in tanks L9, L11 and L12 respectively, assuming fish produced only one egg batch. Of these spawners, realised egg production was measured in 19 egg batches. Egg batches varied in weight from 35 to 370 g with a mean weight of 169 g.

Mean egg wet weight was 5.2 mg with a range from 3.3 to 6.6 mg. There was no significant difference in diameter between egg batches and with date of spawning (Fig. 8.10). Egg weight increased slightly during the spawning season (Fig. 8.11) but the positive correlation with date was not significant and was described by the relationship y = 4E-05x – 1.848, R^2 = 0.3392 where y = wt of egg and x = intercept on the y axis.

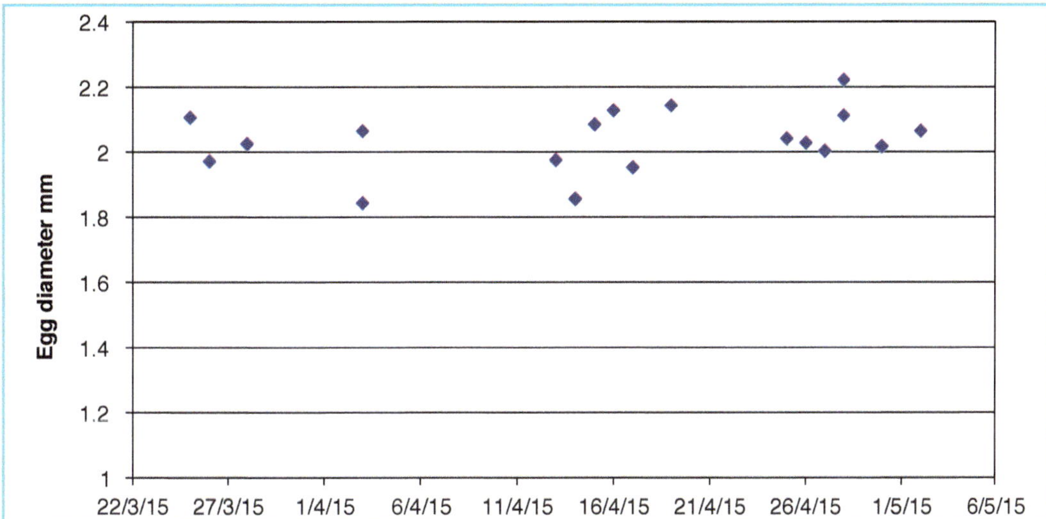

Fig. 8.10 Egg diameter of lumpfish eggs through the spawning season.

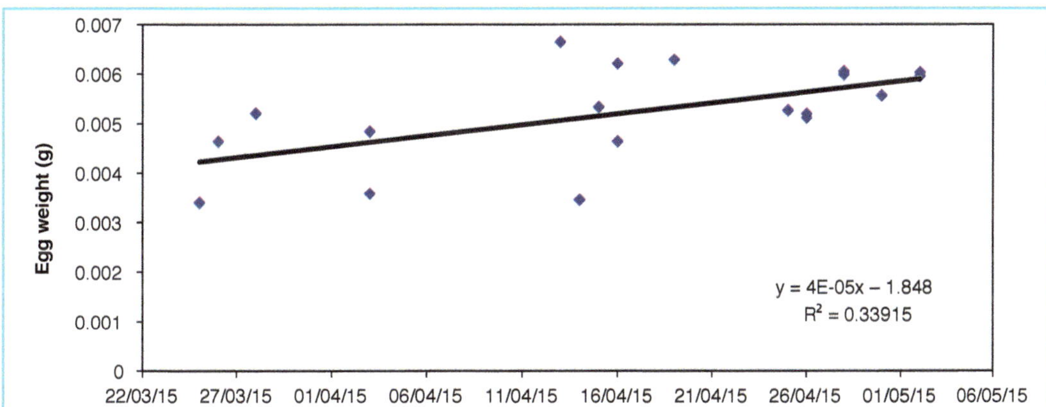

Fig. 8.11 Mean egg wet weight of naturally spawning lumpfish through the spawning season.

8.4.6 Survival of broodstock

There were 19 lumpfish mortalities from 146 broodstock during the spawning period from January to May 2015 (13%). This was attributed mainly to fungus infection and egg-bound fish. There was no specific pattern to mortalities and they were evenly spaced through the year.

8.5 Early maturation in hatchery bred lumpfish

Most wild fish of all species mature at later ages than hatchery-reared fish because growth rates in farmed lumpfish are higher, up to 5% per day in the first year. Faster growth has been reported to encourage earlier maturation in fishes. Lumpfish are reported to mature in the wild at four to five years old (Davenport, 1985). Cox (1920) reported spawning lumpfish from five to seven years old. Other reports were in agreement with five to six years for first breeding (Saemundsson, 1926), with spawning fish from four to nine years in the North Sea and Baltic (Bagge, 1964, 1967). Bagge (1967) found only one male to be mature at four years old at a length of 31 cm. Thorsteinsson (1981) found all females to be recruited to the spawning stock at five years old with only one female mature at four years old. In the current study a large proportion of males and 31% of female lumpfish were mature and spawning at 1 year old, and this is very unusual for a north temperate teleost. Earlier maturation in male fish is a feature of many fish species, for example halibut *H. hippoglossus*, cod *Gadus morhua* and salmon *Salmo salar*, and reflects the smaller resource investment in males compared with females. However, it is unusual for females of any commercial temperate fish species in northern Europe to mature in their first year.

Lumpfish in the current study were 462 g weight in males and 1234 g in females after one year compared with literature reports of lower growth of wild male lumpfish and females (Davenport, 1985). Early maturation may have wider life history implications for a commercial finfish species in cold temperate waters. Sexual maturation in fishes is controlled by a combination of environmental (Yoneda and Wright, 2004) and genetic factors (Thorpe *et al.*, 1980; Campton and Gall, 1988). However, in exploited stocks such as cod, there have been long-term consequent changes on age and size at maturity (Trippel, 1995).

8.6 Egg production

The lumpfish in the present study were spawning for the first time and were small, and therefore egg production from 8,844 to 72,613 per female was less than reported for older fish. For example, Zhitenev (1970) estimated 38,600 eggs in one lumpfish egg mass and 52,700 in another. Goulet (1986) estimated an average of 69,000 eggs per female of 1 kg in Canada in 1982, and fecundity was 160,000 per spawner in 1983 and 107,000 in 1984 when the sample size was larger, n = 99. It was reported that lumpfish in Canada lay from two to three egg batches at intervals of eight to 14 days (Cox and Anderson, 1922; Anon, 2011). Egg masses were reported to contain up to 140,000 eggs. Fecundity of lumpfish from the Bay of Fundy ranged from 54,000 to 159,000 and GSIs as a percentage of body weight were 3% in males and 26% in females.

8.7 Comparison of spawning methods and benefits and disadvantages of each method

Hand-stripping, even though carried out gently, means considerable handling of the fish and is consequently stressful. However, survival and second spawning after hand-stripping has still not been observed. This has to be studied further to see if lumpfish broodstock females can be hand-stripped over several years with continuous spawning. In contrast, there were only a few mortalities in fish that spawned naturally, and these were in egg-bound fish, see above. Another benefit is that females come in to spawning condition naturally, and eggs are usually developed and ready for fertilisation. Natural spawning of lumpfish in this study determined the pattern of egg deposition in three populations that were allowed to spawn without handling and intervention. Egg batches were present in the tanks but none of the eggs were fertilised until rocks were added to act as spawning nests for males to use and attract females. From that point egg batches were fertilised. In hand-stripped fish an assessment has to be made of the external characteristics of the female to ensure that the eggs are mature. Stripping of eggs gave the benefits of synchronising egg collection from a range of females, higher and more predictable fertilisation rates of >95%, and eggs were cleaner and in less contact with waterborne bacteria and detritus contamination. During natural spawning eggs were obviously covered in detritus in the water and therefore egg batches incubated in the hatchery were disinfected daily with 20 ppm Pyceze. Stripping of eggs therefore gives clean eggs that do not have to be handled so frequently during incubation time.

Maturation of reared lumpfish at one year old here was extremely early compared with the reported age of maturation of wild-caught fish, and this compares with reported ages at maturation of three to five years (Davenport, 1985; also Chapter 6). Maturation at one year old in females is unusual in north temperate commercial teleosts. The maturation of lumpfish at one year old as in the present study is therefore notable. The cause of this is most likely due to regular feeding to satiation. This encouraged rapid growth of lumpfish and in this study lumpfish grew rapidly in farm conditions with weights of up to 462 g in males and 1240 g in females after one year. A proportion of the fish, all of the males and up to 31% of females were maturing after seven months and were mature at the end of one year post-hatch. Further improvements can be made by examining the numbers of fish in each tank and the sex ratio. It appeared that a high number of males may have led to competition between them.

The present study indicates that lumpfish broodstock reared in the hatchery can provide successful broodfish with no recourse to wild fish collection with rapid growth rates and maturation of fish from age one year. Initial findings suggested a natural spawning season over two months from February to April. Future work on broodstock may involve photoperiod control of spawning, degumming of eggs, improvement in disinfection protocols, development of a more suitable broodstock diet, advances in disease treatments, and improvement of stock through selective breeding. This will enable more regular and reliable egg production that will not be dependent on the vagaries of wild fish supply, will permit disease screening of fish, and will relieve fishing pressure on wild-sourced broodstocks.

Acknowledgements

We thank the staff of Arctic Cleanerfish AS and FAI Aquaculture Ardtoe for their support and care of the broodfish during this study. We are especially grateful for the technical inputs and encouragement provided by Daniel Carcajona, Tim Atack and Dag Hansen.

References

Andriyashev, A.P. (1954) Ryby severykh morey S.S.S.R. (Fishes of the northern seas of the U.S.S.R.) Moscow An S.S.S.R.

Anon. (2011) Assessment of lumpfish in the Gulf of St Lawrence (in 2010). DFO Canadian Science Advisory Secretariat, Sci. 20111/005.

Bagge, O. (1964) Some observations on the biology of the lumpsucker *Cyclopterus lumpus*, ICES CM. 1964. Baltic Belt Seas Committee No 150, 7 pp. (mimeo).

Bagge, O. (1967) Some preliminary results from tagging of the lumpsucker *Cyclopterus lumpus* 1966. ICES CM 1967/F:23 Demersal (N) Committee, 3 pp. (mimeo).

Campton, D.E. and Gall, G.A.E. (1988) Effect of individual and group rearing on age and size at maturity of male mosquitofish *Gambusia affinis*. *Journal of Fish Biology* 33, 203–212.

Cox, P. (1920) Histories of new food Fishes. 2. The lumpfish. *Bulletin Biological Board Canada* 92, 1–28.

Cox, P. and Anderson, M. (1922) A study of the lumpfish (*Cyclopterus lumpus* L.). *Contributions Canadian Biology* 1 (1), 3–20.

Davenport, J. (1985) *Synopsis of biological data on the lumpsucker Cyclopterus lumpus Linnaeus 1758.* FAO Fisheries Synopsis 147.

Goulet, D.C. (1986) *Reproductive behaviour, spawning success and mate choice of lumpfish, Cyclopterus lumpus L., in Newfoundland.* MSc thesis. Department of Biology, University of Newfoundland.

Hamre, K. (2006) Nutrition in cod (*Gadus morhua*) larvae and juveniles. *ICES Journal of Marine Science* 63(2), 267–274.

Holt, G.J. (2011) *Larval fish nutrition.* John Wiley & Sons.

Ingólfsson, A. and Kristjánsson, B.K. (2002) Diet of juvenile lumpsucker *Cyclopterus lumpus* (Cyclopteridae) in floating seaweed: effects of ontogeny and prey availability. *Journal Information* 2002 (2).

Jónasson, J. (2017) Lumpfish production in Iceland. Presentation at International Cleaner Fish Summit, FHF/SAIC, Glasgow.

Langdon, C. (2003) Microparticle types for delivering nutrients to marine fish larvae. *Aquaculture* 227(1), 259–275.

Mommens, M. (2017) Lumpfish breeding: new species, new traits. Presentation at International Cleaner Fish Summit, FHF/SAIC, Glasgow.

Powell, A., Treasurer, J., Pooley, C.L., Keay, A.J., Lloyd, R., Imsland, A.K. and Garcia de Leaniz, C. (2017) Use of lumpfish for sea-lice control in salmon farming: challenges and opportunities. *Reviews in Aquaculture* DOI: 10.1111/raq.12194

Saemundsson, B. (1926) *Islensk Dyr I. Fiskarnir (Pisces Isladiae).* Reykjavik, Bokaverslum Sigfusar Eymundssonar, 528 pp.

Thorpe, J.E., Morgan, R.I.G., Ottaway, E.M. and Miles, M.S. (1980) Time of divergence of growth groups between potential 1+ and 2+ smolts among sibling Atlantic salmon. *Journal of Fish Biology* 17, 13–21.

Thorsteinsson, V. (1981) *The ageing validation of the lumpsucker Cyclopterus lumpus and the age composition of the lumpsucker in Icelandic lumpsucker fisheries.* ICES CM. 1981/G58. Demersal Fish Committee, 26 pp. (mimeo).

Trippel, E. (1995) Age at maturity as a stress indicator in fisheries. *Bioscience* 45, 759–771.

Wittwer, C. (2015) *A new micro diet with big potential to improve first feeding of lumpsucker Cyclopterus lumpus.* University of Nordland, 64 pp.

Wittwer, C. (2016) Improving survival in early stage lumpfish *Cyclopterus lumpus* with additional substrate. Fish Farmer 39 (12), 48-51.

Wittwer, C. (2017) Embryonic development of lumpfish *Cyclopterus lumpus*. Poster and abstract, European Aquaculture Society, EAS 17.

Yoneda, M. and Wright, P.J. (2004) Temporal and spatial variation in reproductive investment of Atlantic cod *Gadus morhua* in the northern North Sea and Scottish west coast. *Marine Ecology Progress Series* 276, 237–248.

Zhitenev, A.N. (1970) Ecological and morphological affinities of the lumpsucker. *Journal of Ichthyology* 10, 77–84.

CHAPTER 9

Lumpfish deployment on salmon farms

Albert K. D. Imsland,[1,2] and Patrick Reynolds[3]

[1]Akvaplan-niva Iceland Office, Kópavogur, Iceland
[2]Department of Biology, University of Bergen, High Technology Centre, Bergen, Norway
[3]GIFAS AS, Gildeskål, Norway

9.1 Introduction

9.1.1 Biological control of sea lice with cleaner fish

The biological control of sea lice through the use of 'cleaner fish' has recently become a feasible alternative due to the increased occurrence of resistant lice, the reduced public acceptance of chemotherapeutic use in food production, and the urgent need for an effective and sustainable method of parasite control in Atlantic salmon aquaculture (Denholm *et al.*, 2002; Treasurer, 2002). Currently, some farms are using wrasse species to aid in the control of sea lice (Boxaspen, 2006). Of those, the Ballan wrasse (*Labrus bergylta*) being the biggest and most robust of the available wrasse species has the greatest potential for large-scale biological delousing, particularly for use on larger production grade salmon. However, the wrasse species currently in use for biological delousing are temperature sensitive, making them unfit for use under low temperatures, less than 6°C, (Sayer and Reader, 1996). As a cold water alternative, the common lumpfish (*Cyclopterus lumpus*) has been suggested. However, until recently, no published work existed describing the use of lumpfish as biological delousers, despite the fact that such a practice would most likely lead to major economic gains through improved growth and quality of slaughtered fish.

9.1.2 Has cleaning behaviour evolved between lumpfish and Atlantic salmon?

Cleaning behaviour is considered to be a classical example of mutualism. Very limited information exists on the stomach content of wild lumpfish, but *L. salmonis* have been reported in the stomachs of wild juvenile lumpfish and other copepods are a substantial part of the diet of juvenile lumpfish. Moreover, wild Atlantic salmon in the sea can have a large number of adult *L. salmonis* attached on their surface during winter. Given the opportunity lumpfish will effectively graze on *L. salmonis* on Atlantic salmon (Imsland *et al.*, 2014a–c) and it may be speculated that this parasite grazing behaviour has developed between the two species in a similar way as seen for parasite cleaning fish in the tropics. This is an exciting research avenue that we will try to investigate further. Lumpfish are distributed throughout the North Atlantic Ocean, and adults migrate considerable distances between offshore feeding areas and shallow inshore spawning areas (Thorsteinsson, 1983; Davenport, 1985). Very limited information has been published concerning behaviour of lumpfish until now. A limited number of studies has been performed with wild lumpfish, including larval lumpfish (Brown, 1986; Williams and Brown, 1991), juvenile lumpfish (Killen *et al.*, 2007) and adult lumpfish (Mitamura *et al.*, 2012), whereas little is known about the behavioural interaction between lumpfish and Atlantic salmon under controlled conditions in sea cages.

9.1.3 Behavioural interaction between lumpfish and wrasse

One of the most frequently used options as cleaner fish until recently has been goldsinny wrasse, *Ctenolabrus rupestris* L., which has been the most common cleaner species used so far for controlling lice numbers in young salmon (Deady *et al.*, 1995). Goldsinny are more temperature sensitive than other wrasse species and are not suitable for delousing at low temperatures (< 6°C, Sayer and Davenport, 1996; Sayer and Reader, 1996). Juvenile lumpfish persist to actively feed and grow at temperatures close to 4°C (Nytrø *et al.*, 2014), providing a head start in comparison with the slower growing goldsinny wrasse at similar temperatures (Sayer and Reader, 1996). They may therefore have the potential to survive winters in net pens with salmon, even in the northernmost localities, due to the broad geographical distribution of the species in the wild (Blacker, 1983). Behavioural interactions between lumpfish and Atlantic salmon under controlled conditions have been described (Imsland *et al.*, 2014c). However, it is possible that goldsinny wrasse and lumpfish could be used simultaneously for delousing as this would allow a larger size interval of Atlantic salmon to be treated at a broader temperature range. Wrasses can be kept in mixed species groups (Treasurer and Feledi, 2014), but at present no published studies exist describing the behavioural interactions between goldsinny and lumpfish.

9.1.4 Salmon growth with and without lumpfish

Duoculture of Atlantic salmon and other fish species has mainly been investigated along two lines. Firstly, there have been trials with Atlantic salmon in duoculture with other salmonids in an attempt to decrease aggression and enhance growth (Holm, 1989; Nortvedt and Holm, 1991). More recently Atlantic salmon have been reared with other candidate cleaner fish species to delouse the salmon (Imsland *et al.*, 2014a, b, c; 2015a, b). However, an important prerequisite for the use of duo-culture or the use of any cleaner fish species is that important performance traits, e.g. growth, feed intake and feed conversion of the target species (here Atlantic salmon), are not affected by the presence of the cleaner fish. At present no comparisons of the growth performance of Atlantic salmon with and without lumpfish exist.

9.1.5 Feeding preferences of lumpfish in sea cages

The natural feeding ecology of lumpfish juveniles during their first year has previously been studied (e.g. Daborn and Gregory, 1983; Davenport and Rees, 1993; Ingólfsson and Kristjánsson,

2002). Ingólfsson and Kristjánsson (2002) investigated the feeding ecology of juvenile lumpfish during their first year in natural floating seaweed clumps and experimental floating seaweed branches in south-west Iceland. The juveniles appeared to ignore sessile, slow-moving, and small animals (e.g. ostracods, bivalves, gastropods, oligochaetes, polychaetes, turbellarians, rotifers, and nematodes), while taking most other prey organisms in approximate proportion to availability. Crustaceans were the dominant prey of young lumpfish. However, feeding preferences of lumpfish when reared in sea pens along with other species have not been investigated. As it is intended to use lumpfish as a biological delousing agent it is of interest to investigate the incidence of sea lice separately from the rest of the crustaceans group.

9.1.6 Lice grazing of lumpfish: Possible parental effects

Large individual differences in feeding preferences between lumpfish and their sea lice grazing activity have been indicated (Imsland *et al.*, 2014a). In a semi-commercial trial it was found that only 10% of the lumpfish grazed on lice from salmon at the beginning of the trial, increasing to 36% toward the end. Data from the trial showed there were individuals grazing more actively on sea lice compared with their conspecifics. This indicates there is a large variation in foraging behaviour, and hence delousing efficiency of lumpfish. Currently any cause or influencing factors are unknown.

The process of domestication has only just begun in salmonids, and considerable effort is devoted to selecting for and estimating the heritability of economically important traits in these fish, such as growth rate, size at maturity, muscle composition, disease resistance and stress responsiveness (see e.g. Gjerde, 1993; Gjedrem, 2000; Midtlyng *et al.*, 2002; Øverli *et al.*, 2006). Research has indicated that a number of behavioural traits in fish are under genetic control and that behaviour can be modified by selection (e.g. Huntingford, 2004; Huntingford and Adams, 2005). In other organisms it is known that behavioural traits respond to both natural and sexual selection, show geographic adaptation, and that such traits may be heritable (reviewed in Roff and Mousseau, 1987; Meffert *et al.*, 2002; Stirling *et al.*, 2002). We propose to investigate the observed variation in sea lice foraging behaviour of lumpfish (Imsland *et al.*, 2014a, c) to reveal potential correlations between inclination to graze sea lice and genetic composition.

9.2 The use of lumpfish to control sea lice (*Lepeophtheirus salmonis* Krøyer) infestations in intensively farmed Atlantic salmon

9.2.1 Background

To assess the efficacy of lumpfish grazing on attached sea lice on Atlantic salmon, six sea cages (5 × 5 × 5 m) were each stocked with 120 Atlantic salmon with a mean (± SD) weight of 619 ± 49 g. Two of the cages were further stocked with 12 lumpfish (10% density), and two with 18 lumpfish (15% density) with a mean (± SD) weight of 54.0 g ± 7.2. Two cages without lumpfish acted as controls. Sea lice numbers were recorded every other week for 54 days. Gastric lavage was performed every two weeks to determine the diet preferences of lumpfish in the cages. No differences in salmon or lumpfish growth between test groups and controls were observed. There were clear signs of lumpfish grazing on sea lice, with significantly lower average numbers of pre-adult, mature males and females stages of *Lepeophtheirus salmonis* per salmon. Lumpfish reduced the mature female stage of *L. salmonis* to levels equal to or lower than the counts recorded prior to the start of the study. There were no significant differences between the treatments (10 and 15% densities) in grazing efficacy. There was clear evidence of grazing from the results of gastric lavage, with 28% of all lumpfish found to have ingested sea lice on the last sampling day. Overall, the results indicate that lumpfish is a suitable cold water option for biological delousing of Atlantic salmon.

9.2.2 Sea lice infestation levels

The stage composition of the *L. salmonis* population changed over time with a natural succession from predominately chalimus to pre-adult to adult lice. At day 11 of the trial period there were indications that lumpfish were grazing on some of the stages of *L. salmonis*. The evidence became more conclusive as the trial progressed beyond 11 days – but it was not until day 38 that there were clear differences between the treatments, with both the 10% and 15% stocked cages showing less attached chalimus stages on salmon from these cages compared to the control cages. The size of the lice at this stage of development of *L. salmonis* (1 to 2 mm) could also be too small to be attractive as a preferred food item for lumpfish, as none was found in the stomachs of the fish when gastric lavage was undertaken.

For pre-adult stages, and particularly the mature male stage of *L. salmonis*, there were significantly fewer lice attached to Atlantic salmon in cages with lumpfish as compared to the controls. There was clear evidence of grazing from day 11 onwards. The average number of mature female lice remained low throughout the study period for both the 10% and 15% stocked cages. At day 54, the average number of adult female lice per fish had increased to 1 per fish in the control group, whereas the 10% and 15% treatments had 93% and 97% fewer mature female lice present compared to the control (Fig. 9.1).

At present, five species of wrasse (family Labridae) have been used as cleaner fish, and stocking densities of approximately 5% have been suggested to provide good results in reducing the

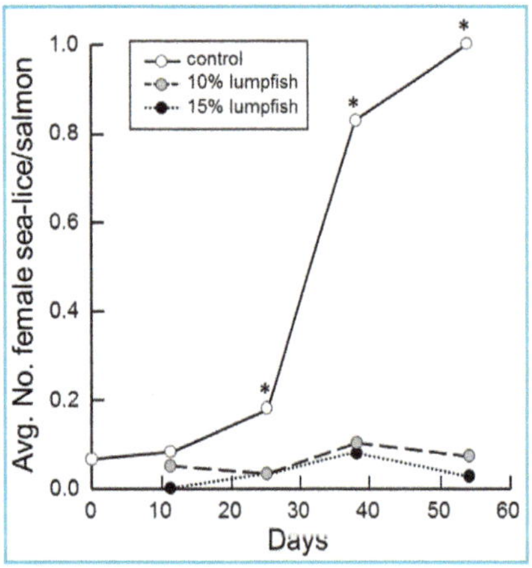

Fig. 9.1 Mean number of adult female sea lice (*L. salmonis*) found on Atlantic salmon reared without lumpfish (control) or with 10% and 15% of lumpfish. Asterisks denote significant differences (modified from Imsland *et al.*, 2014a).

Table 9.1 Mean weights (g), condition factor and specific growth rate (SGR, % day^{-1}) of lumpfish at two stocking densities (10 and 15%) together with Atlantic salmon in sea cages. Values are presented as mean (SD). No significant differences were found between the groups.

Weight g			
	Date	10% group	15% group
	12 June	53.0 (6.3)	55.0 (7.9)
	23 June	73.8 (13.2)	72.8 (13.9)
	7 July	115.3 (23.0)	108.0 (24.7)
	20 July	139.7 (29.8)	135.7 (35.8)
	5 August	181.9 (47.6)	172.3 (54.7)
Condition factor			
	Date	10% group	15% group
	12 June	4.2 (0.5)	4.2 (0.5)
	23 June	4.1 (0.6)	4.1 (0.5)
	7 July	4.5 (0.8)	4.5 (0.6)
	20 July	4.2 (0.7)	4.1 (0.4)
	5 August	4.1 (0.4)	4.2 (0.5)
SGR			
	Period	10% group	15% group
	12-23 June	2.9 (1.2)	2.5 (1.3)
	24 June – 7 July	3.1 (0.6)	2.7 (0.8)
	8-20 July	2.5 (0.5)	2.3 (0.7)
	21 July – 5 August	2.2 (0.5)	2.0 (0.7)

number of attached sea lice (Treasurer, 1993; Costello, 2006). In this trial, 10% and 15% stocking densities produced the same efficiency for adult male and female sea lice removal. We suggest that forthcoming trials with lumpfish should focus on testing effect of sea lice infections levels on salmon using stocking density of lumpfish lower than 10%.

9.2.3 Growth and performance of lumpfish

Lumpfish from both density groups attained similar levels of growth and both groups more than tripled their weight in the course of 54 days (Table 9.1).

The level of growth and cleaning performance of the lumpfish used in this study were in contrast to results obtained using larger lumpfish (> 200 g) in a previous pilot study (P. Reynolds, GIFAS, unpublished data). Most of the fish lost weight at certain points during that study and only attained slight increases in weight by the end of the study period. Lumpfish showed signs of sexual maturation at 450 g and exhibited weak lice grazing efficacy after reaching sexual maturation. Wild lumpfish become sexually mature between two and six years of age (Thorsteinsson, 1983; Albert *et al.*, 2002) and previous studies have shown that males can mature up to two years prior to the females (Davenport, 1985; Albert *et al.*, 2002). If maturation is triggered at a certain size threshold as seen in many teleosts (Imsland *et al.*, 1997), the onset of maturation may occur much earlier in cultured lumpfish. If the upper limit for the size of lumpfish that can be stocked is assessed at around 400 to 500 g, then only fish below this potential upper limit will be of use commercially. Based on initial findings (P. Reynolds, Gifas, A. Imsland, Akvaplan-niva, unpublished data), a 10 g juvenile would take approximately between 14 to 16 months to attain a body weight close to the 400 to 500 g potential upper size limit. Given this, the total marine production phase for Atlantic salmon is close to the time period that lumpfish may be useful as delousing agents. Lumpfish populations may have to be introduced into commercial cages at the time that smolts are transferred to seawater, and be maintained with the salmon throughout their marine culture.

9.2.4 Gastric lavage of lumpfish

The gastric lavage method used to assess food choices in lumpfish displayed the presence of both *L. salmonis* and *C. elongatus* in the stomachs of several fish throughout the study period. The proportion of lumpfish with sea lice increased from 10% at day 11 to 28% at day 54 (Fig. 9.2).

The number of lumpfish eating lice may in fact be much higher as these values were only determined from lavaging fish every 14 days throughout the trial period. The number of days between sample points allowed for lumpfish to consume sea lice and fully digest them. The relatively large increase in numbers of lumpfish found with ingested sea lice in their stomachs suggests that the level of grazing intensified throughout the study period. Comparable information for wrasse consumption of sea lice is unavailable; however, if 30% or more of the lumpfish introduced into commercial cages consumed sea lice on a regular basis then infestation levels would be greatly supressed. The low average numbers of mature female lice found in both treatment groups throughout the study period suggest that this stage is actively selected by lumpfish as a preferred prey item. If lumpfish are preferentially selecting the larger mature females, then the potential for re-infestation is greatly diminished. It has been shown that mature females can live for up to seven months and produce approximately six to ten new egg strings during this time (Mustafa *et al.*, 2000). Removing or reducing mature female lice has the greatest effect in reducing the size of the sea lice population (Boxaspen, 2006). The most common food item

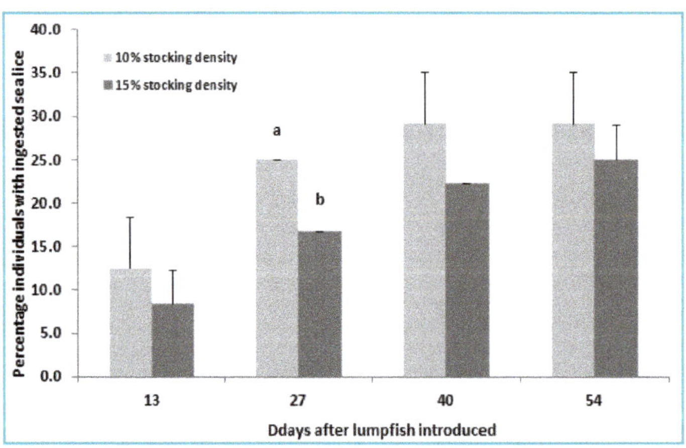

Fig. 9.2 Results from gastric lavage of lumpfish showing the percentage of fish from each group with presence of sea lice in their stomachs. Values are presented as means ± S.D. Mean values that do not share a letter were found to be significantly different (Tukey's multiple range test ($P = 0.03$)) (modified from Imsland *et al.* 2014a).

identified in the stomachs of lumpfish throughout the study period was fragments of salmon pellets (data not shown). The amount of salmon pellets eaten by the lumpfish, however, did not affect the growth of salmon in the sea pens as growth was almost identical in the sea cages with and without lumpfish.

9.3 Assessment of suitable substrates for lumpfish in sea pens

9.3.1 *Introduction*

The pelvic fins of lumpfish constitute a ventral suction disc, allowing fish to rest on rocky substrate, kelp and vegetation. The lumpfish is notable among fishes for hatching with a functional mouth and a fairly well-developed digestive system (Timeyko, 1986). Juveniles are typically found among kelp during their first year of life, both attached and free-floating (Ingólfsson and Kristjánsson, 2002). About one year after hatching, they are found in pelagic habitats, but frequently found in floating seaweed (Ingólfsson and Kristjánsson 2002; Vandendriessche *et al.*, 2007) using their suction disc to adhere to the seaweed surface when resting. Mimicking their natural requirements for surface adhesion, it is reasonable that when placing juvenile lumpfish in sea pens alongside Atlantic salmon, the lumpfish will need some kind of shelter or substrate to attach to when resting. Substrates may also provide shelter for lumpfish during periods of inactivity and/or extreme environmental perturbations. We assessed the use of different types of substrates (concrete tube; two sizes of polyethylene plastic plates; car tyre and stone substrate) for lumpfish in sea pens. Observations were recorded in the early morning before first light and the number of lumpfish attached to each surface type was recorded.

9.3.2 *Results and discussion*

Lumpfish seemed to prefer flat and smooth vertical surfaces as a significantly higher proportion (Fig. 9.3) was attached to the concrete tube and the plastic plates as compared to stones and car tyres.

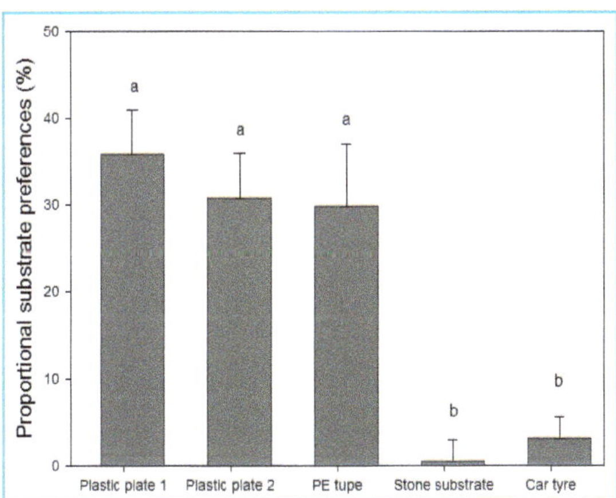

Fig. 9.3 Average proportional substrate preferences of juvenile lumpfish during the study period. Values (N = 42) are presented as means ± SE. Different letters indicate significant difference among treatment groups ($\chi^2 = 6.1$, $P < 0.05$). Modified from Imsland *et al.* (2015b).

Throughout the study period the lumpfish were evenly dispersed on all substrates. In the wild juvenile lumpfish are typically found among kelp, both attached and free floating during their first year of life (Ingólfsson 2000; Ingólfsson and Kristjánsson 2002), but are also found attached to flat and rigid substrates including blue mussel *Mytilus edulis* and stones (Moring 1989a, b). In general, members of the family Cyclopteridae use their ventral adhesive disc to adhere to rocks, vegetation and other available substrates (Brown 1986; Moring 1989a, b). Ingólfsson (2000) investigated the colonisation of lumpfish in coastal floating seaweed clumps, originating from the intertidal zone. A negative correlation between the density of lumpfish and clump weight was found and the author speculated if this might indicate some form of territorialism in lumpfish (i.e. fry excluding other individuals from their immediate vicinity) or cannibalism, common in floating seaweed or kelp. Based on the data, Ingólfsson (2000) deduced that floating seaweed and kelp are an important habitat for the species during the first year, while older fry, although pelagic, are not found associated with floating seaweed. In sea pens, however, lumpfish show a clear preference for smooth artificial surfaces such as plastic plates.

9.4 Examples of interactions between lumpfish and wrasse

9.4.1 Background

A series of studies (study 1, 2 and 3, below) were undertaken to determine the behavioural interactions between three different size classes (110, 70 and 32 g) of lumpfish, and one size class (30 g) of goldsinny wrasse. These studies attempted to determine whether goldsinny wrasse could coexist with juvenile lumpfish to enhance lice grazing potential of Atlantic salmon by using both species simultaneously.

9.4.2 Results and discussion

Indication of size-dependent dominance behaviour of lumpfish against goldsinny wrasse was found. Behavioural observations in study 1, which assessed behaviour with the largest lumpfish, recorded aggression towards goldsinny 13% of the time, while for study 2 and 3, aggressive behaviour accounted for only 6% of all observations (Fig. 9.4). Lumpfish were observed chasing goldsinny away from available food and were in some instances observed biting the caudal fins of goldsinny. When excess food was available, lumpfish would maintain position close to the excess food, thus preventing goldsinny access to it. The different levels of aggression observed between lumpfish and goldsinny would indicate that smaller lumpfish may be able to cohabit with goldsinny. However, if the behaviour is entirely size dependent then at some point if stocked together, lumpfish would become much larger than goldsinny due to their superior growth rates (Imsland *et al.*, 2014b; Nytrø *et al.*, 2014; Skiftesvik *et al.*, 2014) particularly at lower water temperatures, and the antagonistic feeding behaviour may increase as lumpfish weight increases.

The spatial distribution of both species observed during studies 2 and 3 showed that goldsinny were found to be positioned towards the bottom of the tank while lumpfish were recorded as being located more near the surface and more spread throughout the water column (Fig. 9.5). These observations may have been influenced by the aggressive behaviour exhibited by the largest

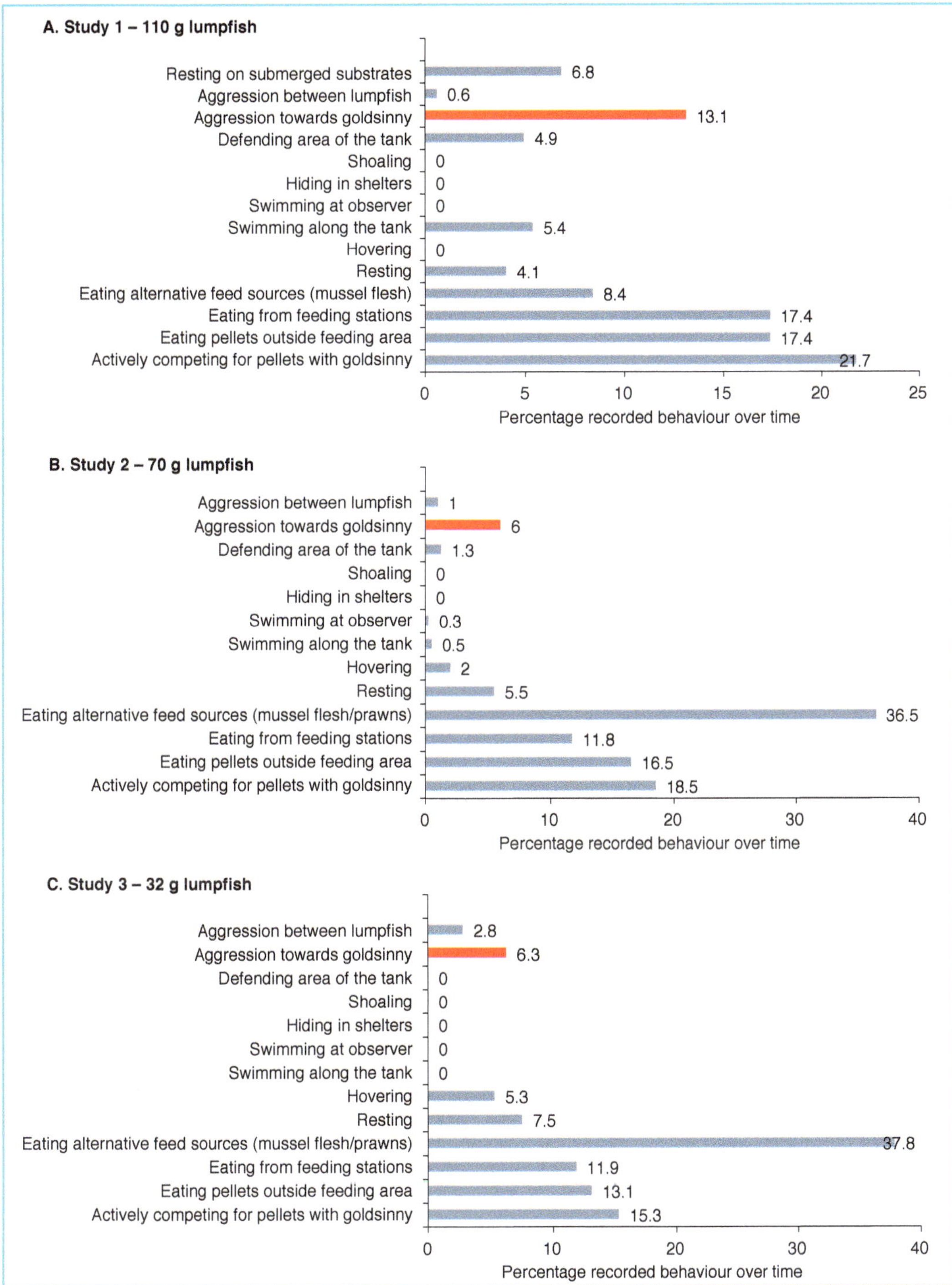

Fig. 9.4 Behaviour of different size classes of lumpfish when reared together with goldsinny wrasse in tanks. Size-dependent level of aggression was observed between lumpfish and goldsinny ($\chi^2 = 11.3$, $df = 1$, $P < 0.01$). Modified from Imsland *et al.* (2016a).

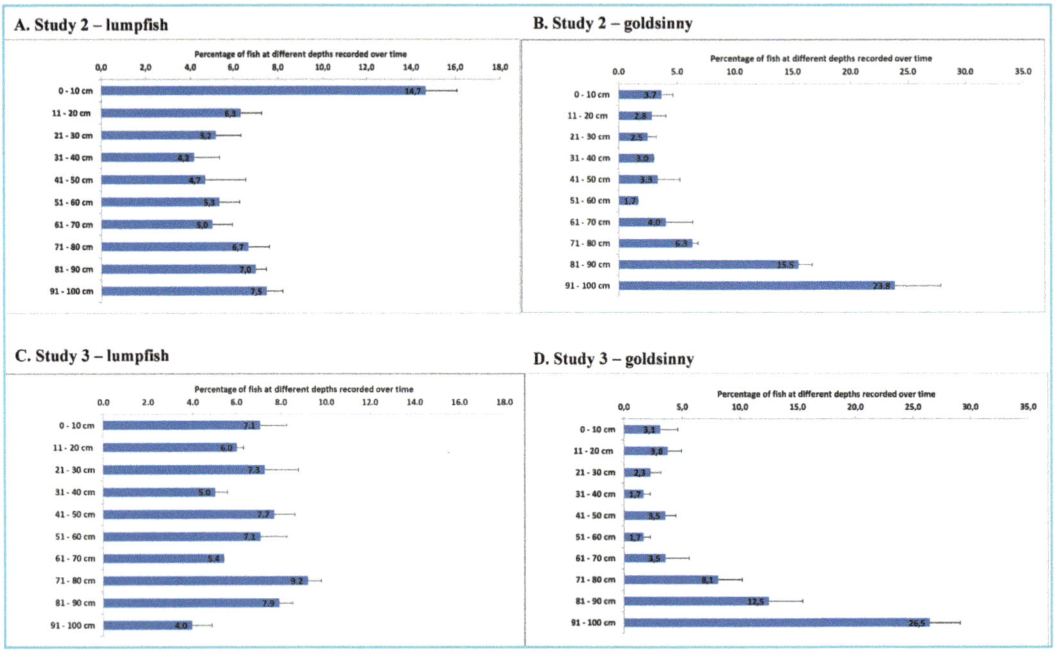

Fig. 9.5 Percentage of lumpfish (A and C) and goldsinny wrasse (B and D) recorded at different depth when reared together in tanks. Values represent means ± S.D. Modified from Imsland *et al.* (2016a).

lumpfish, but not for the fish assessed during studies 2 and 3 where aggression was much less evident. Given that if goldsinny were used to treat sea lice in commercial salmon cages, particularly in northern areas, they would be exposed to lower water temperatures than recorded during these studies and if exhibiting a torpid state, they would not be effective in suppressing sea lice infestation levels during winter periods. In addition, the aggressive behaviour observed by lumpfish towards goldsinny would infer that the use of these two species coexisting in salmon cages may be disadvantageous. Six species of wrasse are found in Norway, four of which have been used as cleaner fish for delousing salmonids, namely: goldsinny wrasse, corkwing wrasse, *Symphodus melops* L., rock cook, *Centrolabrus exoletus* L., and juvenile Ballan wrasse *Labrus bergylta* Ascanius. The maximum size of the different wrasse species has been reported to be up to 60 cm for Ballan wrasse (Quignard and Pras, 1986), 35 cm for cuckoo wrasse (Darwall *et al.*, 1992), 16.5 cm for rock cook wrasse (Sayer *et al.*, 1996), 18 cm for goldsinny and 28 cm for corkwing (Darwall *et al.*, 1992). The aggression exhibited by lumpfish towards goldsinny may have been as a direct result of size and larger wrasse species such as Ballan wrasse may be a more suitable candidate for use with lumpfish. However, the present trial was performed in 1.5 m^3 land-based tanks and behaviour may be different in large open net pens with Atlantic salmon where it is easier for the species to avoid behavioural interactions.

Based on the present data, the feeding behaviour of lumpfish observed can be classified as strongly opportunistic and the fish do not appear to restrict themselves or rely on a single food source if others are present. These findings are consistent with previous observations of lumpfish activity in sea pens with Atlantic salmon (Imsland *et al.*, 2014a, c). Lumpfish were also observed remaining stationary on the surfaces of the tanks using their ventral adhesive disk and taking food as it passed over them. Killen *et al.* (2007) studied the behaviour of wild juvenile lumpfish and found that the juvenile fish forage using one of two modes: they can actively search for prey while swimming or they can 'sit and wait' for prey while clinging to the substrate using a ventral adhesive disk. The study suggested that juvenile lumpfish forage in a manner that reduces activity and conserves space in their limited aerobic scope. The authors noted that this behavioural flexibility is of great benefit to this species, as it allows young individuals to divert energy towards growth as opposed to activity. The present data are in line with these observations. It has also been documented that goldsinny are also opportunistic in their feeding behaviour. It has been shown that amphipods, shrimps or prawns, crabs, gastropods, and bivalves formed a considerable part of the diet of goldsinny wrasse, although in contrasting proportions in different investigations (Fjøsne and Gjøsæter, 1996). During this study, most goldsinny observed eating were choosing prawns rather than pellets when available. Only a small number of goldsinny were observed eating pellets during all three studies even though, prior to the studies, all goldsinny assessed were fed marine profile pellets and were frequently observed eating them.

The amount of time spent resting by goldsinny increased in studies 2 (28.3%) and 3 (36.9%) compared to study 1 (1%), whereas lumpfish spent a similar amount of time resting in all three studies (4.1–7.5%). Less time spent resting by goldsinny in study 1 could be linked to the higher level of aggression by lumpfish towards goldsinny in that study. But the increased amount of resting in study 2 and 3 by the goldsinny could also be a temperature effect as the mean suggested that the hypometabolic torpor observed is an adaptive strategy for goldsinny survival at low environmental temperatures (Sayer and Davenport, 1996). However, to maintain low sea lice infestation levels in Atlantic salmon cages, cleaner fish stocked in cages should have the potential of grazing all year round to maintain good levels of suppression. Previous studies have shown that juvenile lumpfish are active at the lowest seasonal water temperatures recorded in salmon cages and continue to graze attached stages of *L. salmonis* from infested fish (Imsland *et al.*, 2014a, b) and the present study demonstrates that lumpfish are consistently active even during the lowest water temperatures recorded, making them suitable candidates for use in sea cages even during winter time.

9.5 Behavioural activity of lumpfish in salmon pens

9.5.1 Background

The behaviour of lumpfish in sea pens, with and without Atlantic salmon present, was assessed by underwater camera technology. Behaviour was classified (Table 9.2) by recording for 30 second intervals the principal activity of individual fish. Thirty fish were observed during each observation time period, giving a total of 30 activity records on each observational time point. All lumpfish were individually tagged with a floy tag. Each floy tag had a unique number on the shaft and this was used to identify each tagged fish. In addition eight different colours of tag were used to assist in individual identification (white, white with black strip, blue, green, yellow, red,

Table 9.2 Behavioural types and detailed description of lumpfish behaviour in sea pens. Based on Tully *et al.* (1996).

Behaviour type	Description
Feeding on net fouling	Observed feeding on side of nets
Feeding on free-swimming organisms	Observed feeding on organisms entering the sea pen
Actively competing for pellets	Observed competing with Atlantic salmon for pellets
Eating pellets outside salmon feeding zone	Observed eating pellets on the periphery of feeding Atlantic salmon
Eating from feeding stations	Observed eating from offered feed stations
Resting	Observed resting in folds of the sea pen net
Hovering	Near motionless in sheltered areas of the sea pen
Swimming along net side	Swim up or down at the sea pen net
Swimming at observer	Observing diver
Swimming in between Atlantic salmon	Observed swimming amongst Atlantic salmon
Shoaling	Groups of Lumpfish swimming together
Inspecting Atlantic salmon	Swimming along individual Atlantic salmon
Cleaning Atlantic salmon	Displaying cleaning of *L. salmonis* from Atlantic salmon

black and orange) (five of each for eight fish). There was no algae growth on the tags so they were always visible. The tags were sited on the highest point of the back just beside the dorsal array. As each fish was identified either with the camera or directly (if on or near the surface) the number was normally seen. If the number was not visible, then another fish would be observed until all 30 recordings were achieved by different fish. Behavioural observations were classified using behavioural indices as used by Tully *et al.* (1996).

9.5.2 Results and discussion

The majority of daylight time was spent actively foraging for food (Fig. 9.6). Based on present data, the feeding behaviour in lumpfish in sea pens can be classified as strongly opportunistic and the fish do not restrict themselves or rely on a single food source if others are present. This omnivorous feeding behaviour has been reported with wild juvenile lumpfish (Ingólfsson and Kristjánsson, 2002; Vandendriessche *et al.*, 2007). Our data indicate that the majority of daylight time is spent foraging for food and, if not feeding or foraging, the fish were found to be either resting within the floating seaweed within the cage or hovering just under the weed. In the absence of Atlantic salmon lumpfish spent more time resting.

No antagonistic behaviour between the two species was observed and they seemed to co-exist with each other in the sea pens. Wild lumpfish and Atlantic salmon share feeding grounds and Sheedan *et al.* (2012) found large numbers of juvenile and adult lumpfish together with Atlantic salmon when sampling Atlantic salmon with surface trawl in the Northwest Atlantic (Labrador Sea). In the Northeast Atlantic (Norwegian Sea) lumpfish and Atlantic salmon inhabit much the same areas (Hansen and Jacobsen, 2000; Bjelland and Holst, 2004) feeding on similar prey

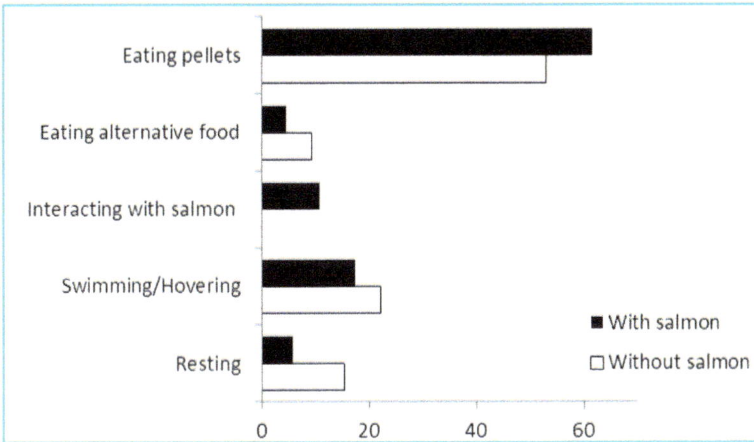

Fig. 9.6 Behaviour of lumpfish in sea pens with and without Atlantic salmon present. Significant differences were found in the amount of resting as the lumpfish spent more time resting when reared alone (modified from Imsland *et al.* (2014b).

items (Jacobsen and Hansen, 2001; Vandendriessche *et al.*, 2007). Holst (1993) investigated the geographic distribution of lumpfish in the Norwegian Sea and found that lumpfish is widely distributed throughout it, with the largest concentrations occurring in the areas close to the polar front, i.e. north of 72°N. In their study of the spatial and temporal distribution of post-smolts of Atlantic salmon in the Norwegian Sea, Holm *et al.* (2000) found that they appear to follow the main surface currents northwards into the Norwegian Sea. There they spread in a fan-like distribution over an area covering most of the international waters between the exclusive economic zone (EEZ) of Norway, the Faroes and Iceland up to about 73–75°N, i.e. similar to the geographic distribution of lumpfish (Holst, 1993). The fact that lumpfish and Atlantic salmon share feeding grounds in the wild may help to explain the non-antagonistic behaviour seen between the two species in this study.

Cleaning behaviour is considered to be a classical example of mutualism. Imsland *et al.* (2014a) investigated the efficacy of lumpfish grazing on attached *L. salmonis* on Atlantic salmon in sea pens during summer (June to August) when *L. salmonis* are more common on Atlantic salmon. There were clear signs of lumpfish grazing of *L. salmonis* with significantly lower average numbers of pre-adult, mature males and females stages of *L. salmonis* per Atlantic salmon. Gastric lavaging of lumpfish showed that 28% of all lumpfish were found to have ingested *L. salmonis* on the last sampling day (Imsland *et al.*, 2014). Very limited information exists on the stomach content of wild lumpfish, but *L. salmonis* have been reported in the stomachs of wild juvenile lumpfish (Ingólfsson and Kristjánsson, 2002) and other copepods are a substantial part of the diet of juvenile lumpfish (Ingólfsson and Kristjánsson, 2002; Vandendriessche *et al.*, 2007). Moreover, wild Atlantic salmon in the sea can have a large number of adult *L. salmonis* attached on their surface during winter (Jacobsen and Gaard, 1997). Given the opportunity lumpfish will effectively graze on *L. salmonis* on Atlantic salmon (Imsland *et al.*, 2014; present study) and it may be speculated that this parasite grazing behaviour has developed between the two species in a similar way as seen for parasite cleaning fish in the tropics (Grutter, 1995; Clague *et al.*, 2011).

9.6 Salmon growth with and without lumpfish

9.6.1 Introduction

Growth and performance of Atlantic salmon in the presence of lumpfish was assessed in two studies. In the first study, six sea cages (5 × 5 × 5 m) were each stocked with 120 Atlantic salmon with a mean (± SD) weight of 619 (± 49) g and reared for 159 days. Control cages were without lumpfish, while two of the cages were stocked with 12 lumpfish (10% density), and two with 18 lumpfish (15% density) with a mean (± SD) weight of 54.0 (± 7.2) g. The lumpfish were removed from the cages after 56 days, but feed consumption and growth of the salmon were monitored for another 12 weeks to investigate the possible long-term effect of the presence of lumpfish on growth performance of the salmon. In the second study, six sea cages (5 × 5 × 5 m) were each stocked with 80 Atlantic salmon with a mean (± SD) weight of 2400 (± 220) g and reared for 60 days. Two of the cages were further stocked with four lumpfish (5% density), and two with eight lumpfish (10% density) with a mean (± SD) weight of 360 (± 30) g. Two cages without lumpfish acted as controls. In both studies sea lice infestation levels were recorded every other week.

9.6.2 Results and discussion

There were no significant differences in growth, feed consumption or feed conversion ratio between the treatments over time in the trial with small salmon and lumpfish (Fig. 9.7).

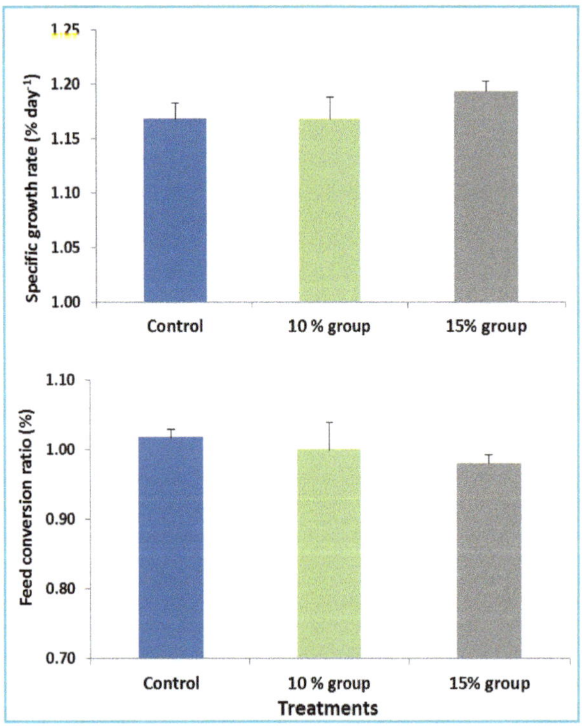

Fig. 9.7 Specific growth rate (SGR, mean ± SD) and feed conversion ratio (FCR, mean± SD) of small (mean initial weight ± SD, 619 ± 49 g) Atlantic salmon reared with 10% and 15% density of small (mean initial weight ± SD, 54.0 ± 7.2 g) lumpfish. No significant differences were seen (modified from Imsland *et al.*, 2014c).

Growth rates calculated for the salmon were comparable to what would be expected given the size of fish and water temperature (Austreng *et al.*, 1987; Oppedal *et al.*, 1999; Folkedal *et al.*, 2012). Observations during the trial showed that the salmon competed for pellets in the centre of the cage and did not appear discouraged by the presence of lumpfish during meals. Imsland *et al.* (2014b) investigated the behaviour of lumpfish in sea pens for two months, with and without Atlantic salmon present, using underwater cameras. Feeding behaviour in lumpfish was classified as strongly opportunistic and the fish do not restrict themselves or rely on a single food source if others are present. This omnivorous feeding behaviour has been reported in wild juvenile lumpfish (Ingólfsson and Kristjánsson, 2002; Vandendriessche *et al.*, 2007). Imsland *et al.* (2014b) further found that the majority of daylight time is spent foraging for food and, if not feeding or foraging, the fish were found to be either resting within the floating seaweed within the cage or hovering just under the weed. In the absence of Atlantic salmon, lumpfish spent more time resting. Although no antagonistic behaviour between the two species was observed, lumpfish spent 29% of the time actively competing for pellets (Imsland *et al.*, 2014b). In the present study lumpfish were regularly seen competing for pellets with salmon within the feeding zone and only a few fish appeared to select pellets outside this zone of feeding, but the data indicate that this had limited effect on the growth or feed intake in salmon.

In contrast to the findings for the smaller salmon, differences in growth and feed conversion were seen in the trial with large (> 2.4 kg) salmon (Fig. 9.8). The presence of lumpfish at the two different stocking densities did affect overall growth and performance of Atlantic salmon in the cages as a result of directly competing for pellets. The small-scale cages used in the present study, with approximately 123 m^3 of water volume, allowed for both species to be in close contact with

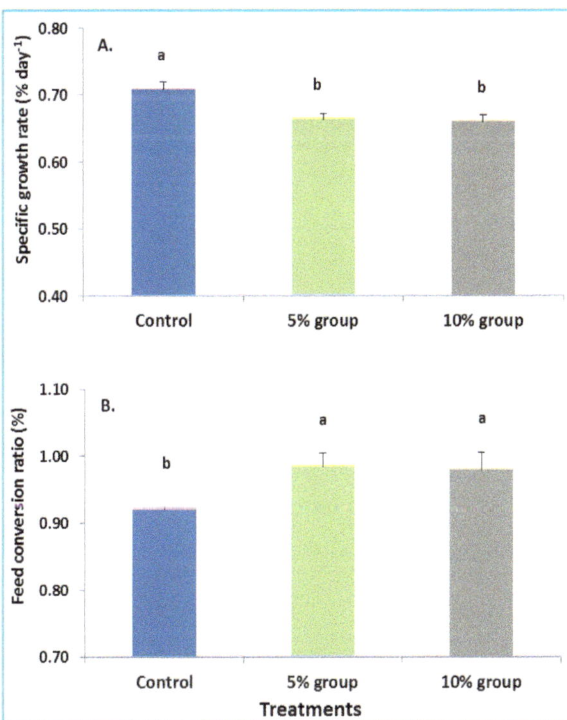

Fig. 9.8 Specific growth rate (SGR, mean ± SD) and feed conversion ratio (FCR, mean ± SD) of large (mean initial weight ± SD, 2400 ± 220 g) Atlantic salmon reared with 5% and 10% density of large juvenile (mean initial weight ± SD, 360.0 ± 30 g) lumpfish. Different letters indicate significant difference among treatment groups (Student-Newman-Keuls test, $P < 0.05$). Modified from Imsland *et al.* (2014c).

each other and the lumpfish were in close proximity to salmon feed on a daily basis. It is therefore not entirely surprising that the opportunistic behaviour of lumpfish would enable exploitation of salmon pellets as a food source. However, in larger commercial-scale sea cages with higher densities of salmon and a greater water volume (Dempster *et al.*, 2008; Fredheim and Langan, 2009) it would be unlikely that lumpfish at 5–15% density in the same cages could access the feed delivered to the same extent as seen in the present study. There is more competition for pellets in commercial Atlantic salmon farms (see Oppedal *et al.*, 2011 for a recent review) and lumpfish would probably be forced to feed on the periphery of the feeding zone more than observed in this study. This should be investigated in future studies.

Another factor that may explain the lower growth rates observed in salmon is that many of the lumpfish were also noted undergoing sexual maturation, with male fish showing characteristic morphological changes associated with spawning (Thorsteinsson, 1983). Sexual maturation may have triggered adaptations in feeding behaviour of the larger lumpfish as appetite may be suppressed (Davenport *et al.*, 1983; Davenport, 1985). Consequently, they may have stopped foraging for food sources that required an output of energy otherwise sequestered for spawning, e.g. sea lice. The sea lice grazing data from this study (Fig. 9.9) support this as sea lice grazing by lumpfish was found to be size dependent. Overall there were significantly fewer lice attached to Atlantic salmon in cages with lumpfish as compared to the controls for the small fish study. Earlier research has indicated that lumpfish is opportunistic when selecting food items (Ingólfsson and Kristjánsson, 2002; Vandendriessche *et al.*, 2007; Imsland *et al.*, 2014a, b). Data

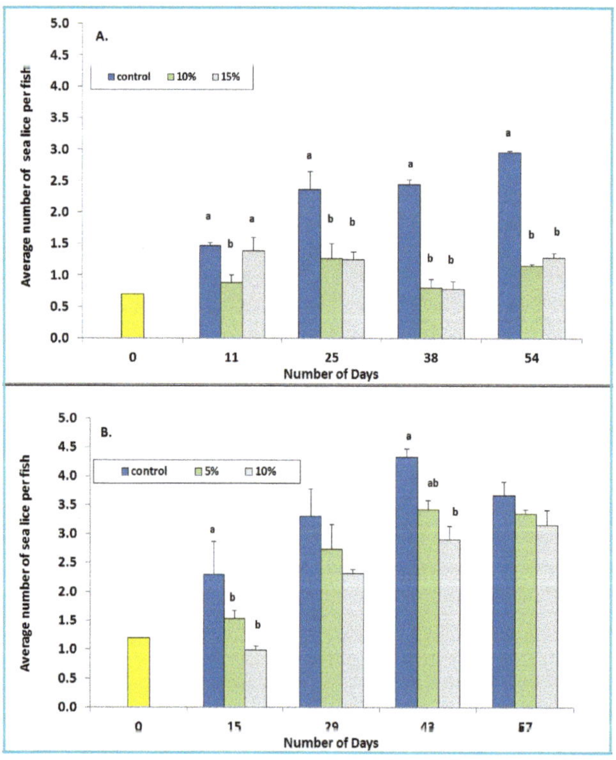

Fig. 9.9 Average number of *L. salmonis* per salmon in sea cages with small (A, 54.0 ± 7.2 g) and large (B, 360.0 ± 30 g) lumpfish recorded for each duplicate treatment during each of the sampling dates. Values are presented as means ± S.D. Different letters indicate significant difference among treatment groups (Student-Newman-Keuls test, $P < 0.05$). Modified from Imsland *et al.*, 2014c).

on stomach contents of the lumpfish in the both studies (P. Reynolds, GIFAS, unpublished data) indicate that salmon pellets were the dominant feed source for the lumpfish in the trial with large salmon, and may explain the reduction in growth seen in trial 2.

9.7 Feeding preferences of lumpfish in sea cages

9.7.1 Background

Biofouling is a major problem and expense factor in finfish aquaculture worldwide. In Norwegian waters and other temperate regions, the most common species are the blue mussel *Mytilus edulis*, the ascidian *Ciona intestinalis*, the hydroid *Ectopleura larynx* (syn. *Tubularia larynx*) and algae of the genus Ulva and Ectocarpus. In addition, caprellid amphipods can reach critical abundances (Bloecher *et al.*, 2013). Accordingly, the most common biofouling agents previously found on the experimental cages used (blue mussel and hydrozoans) were included as separate food groups when investigating the feeding preferences of lumpfish.

Feeding preferences of juvenile lumpfish reared together with Atlantic salmon in sea pens were assessed. Juvenile lumpfish, initial mean (± SE) weight of 54.0 ± 0.9 g, were reared in small sea cages (125 m^3, 5 × 5 × 5 m) together with Atlantic salmon with an initial mean (± SE) weight of 619 ± 2.5 g. Lumpfish were stocked at 10 and 15% density (12 and 18 lumpfish vs. 120 salmon, respectively) in replicate sea pens. Gastric lavage was performed every two weeks to determine the feeding preferences of lumpfish in the cages and the feeding items were identified and categorised as: a) consumed sea lice, b) formulated feed fragments, c) crustacean species, d) hydrozoan species, e) *Mytilus edulis*, f) unidentified material or no contents found.

9.7.2 Results and discussion

Based on the present data the feeding behaviour in lumpfish in sea pens can be classified as strongly opportunistic and the fish did not restrict themselves or rely on a single food source if others were present (Fig. 9.10). Lumpfish seem to switch their choice of food to whatever becomes available to them within their environment (see 9.6.2).

Gastric lavaging of lumpfish in the present study showed that 33–38% of all lumpfish were found to have ingested sea lice at day 77 compared to 13–17% at day 11. The relatively large increase in numbers of lumpfish found with ingested sea lice in their stomachs also suggests that the level of grazing intensified throughout the study period (see also 9.2.4). The number of lumpfish eating lice may in fact be much higher as these values were only determined from lavaging fish every 14 days throughout the trial period. The number of days between sampling points allowed for lumpfish to consume sea lice and fully digest them.

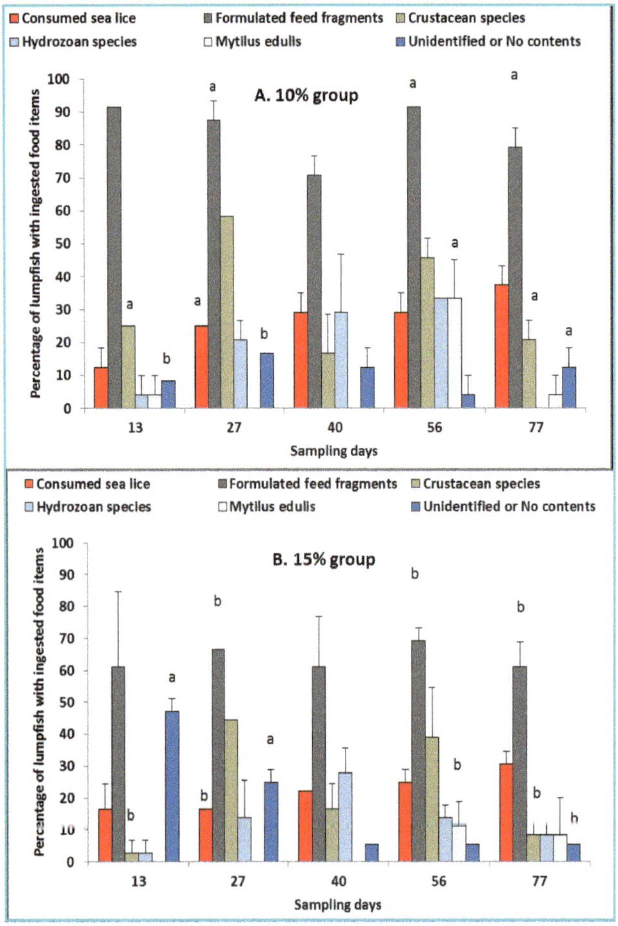

Fig. 9.10 Percentage food choices for lumpfish on days from stocking with salmon. Values are presented as means ± S.D. Different letters indicate significant differences between the two lumpfish groups (Tukey's multiple range test, $P < 0.05$) within the same food category at each sampling date. Modified from Imsland *et al.* (2015a).

9.8 Lice grazing of lumpfish. Possible parental effects

9.8.1 Background

The observed variation in sea lice foraging behaviour of lumpfish was investigated and revealed potential correlations between inclination to graze sea lice and genetic composition (Imsland *et al.*, 2014a, c). It may be possible with this information to predict inclination for sea lice grazing in individual fish, and families, with clear applications for the aquaculture industry. These include the ability to select fish for breeding purposes, which have not been in contact with salmon, and to introduce a breeding programme for continuous improvements of the cleaning efficiency of lumpfish.

The possible family influence on sea lice grazing of lumpfish on Atlantic salmon was investigated with nine families of lumpfish (N = 360 total) with a mean (± SD) weight of 169.4 ± 8.3 g distributed among nine sea cages (5 × 5 × 5 m) each stocked with 400 Atlantic salmon with a mean (± SD) weight of 123.3 ± 12.3 g. For each family, 20 lumpfish were stocked into one of nine sea cages and 20 into another cage, thus establishing duplicate treatments for each genetic family giving two families stocked per cage and a total of 18 cages.

9.8.2 Results and discussion

9.8.2.1 Sea lice grazing in different families

Although the sea lice infestation rate was low in the study (Fig. 9.11), the percentage of lumpfish found to have consumed pre-adult and mature female stages of *L. salmonis* and/or *C. elongatus* varied between the families. A Restricted Maximum Likelihood Estimate based on variance component analysis of sea lice grazing showed a significant paternal and maternal effect on sea lice grazing, indicating the effect is additive. Both a paternal and a maternal effect on lice grazing activity suggests that this family effect is likely to have a genetic effect that can be selected for in future selection programmes for lumpfish behaviour.

Despite low sea lice infestation rates there was a high delousing effect of lumpfish in two cages, 402 and 408, particularly, and both contained family 2 (Fig. 9.12). There was >70% reduction

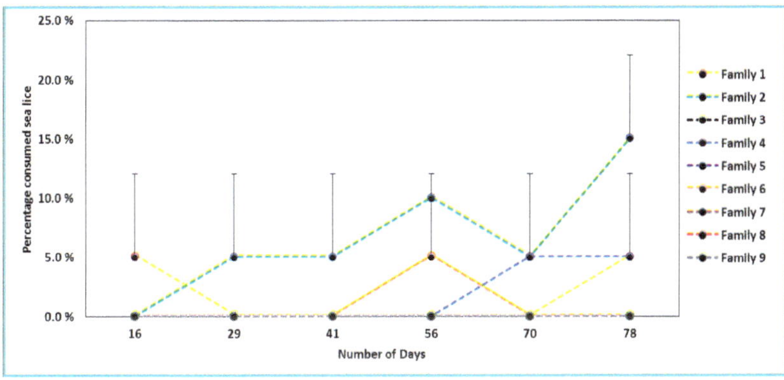

Fig. 9.11 Percentage of lumpfish with sea lice in stomach contents in nine different families of lumpfish. Sea lice in stomach was found in four families. Values are presented as mean percentage of two sea cages for each family ± S.D.

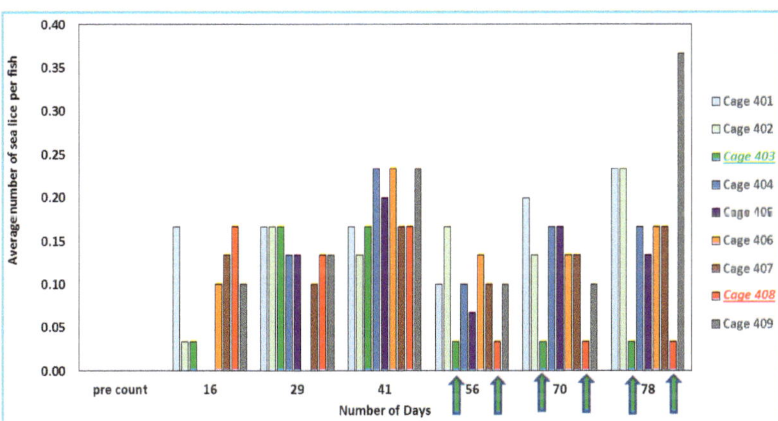

Fig. 9.12 Average number of sea lice (all stages *L. salmonis* and *Caligus elongatus*) in each experimental sea pen throughout the experimental period. Each histogram represents one sea cage. Arrows indicate the sea pens where family 2 was reared (sea cage 403 and 408) resulting in 65–90% lower sea lice infections rate on the salmon. Modified from Imsland *et al.* (2016b).

of infestation from the maximum lice numbers at day 42 to a stable minimum of less than 0.05 from day 56 onwards. Efficient delousing capability when sea lice abundance is low is an extremely important trait for ecological control with sea lice in salmon farming.

9.8.2.2 Difference in disease resistance of lumpfish

The mortality rates observed in lumpfish were attributed to *Pasteurella* spp. outbreak confirmed by histopathological examination of formalin-fixed tissues and by bacteriological investigation of kidney tissues. There were significant differences in mortality rates over the study period (78 days) between lumpfish families, with three of the families showing higher survival rates (92–98% survival) compared to the other six (65–78% survival). The results may potentially indicate a degree of genetic resistance to this bacterium and further research is required to fully elucidate this potential. The next step would be to carry out controlled challenge studies. Selection for disease resistance is commonly used for aquaculture species (Midtlyng *et al.*, 2002), but lumpfish is a new species for aquaculture so no selection work towards disease resistance has yet been performed, whereas present data are promising in relation to possible selection on a family and/or individual basis.

9.8.2.3 Difference in behaviour of lumpfish by family

There were significant differences in the frequency of behavioural types between the nine families throughout the study (Fig. 9.13) period as fish in family 5 had a higher degree of behavioural interaction with salmon, including competing for pellets, and spent less time resting. On the other hand, fish from family 2 were more predisposed to choosing to eat natural food sources as compared to feed pellets and this coincided with higher growth for this family. Such greater scope of growth linked with reduced activity has been indicated for turbot, *Scophthalmus maximus* (Imsland *et al.*, 1995) and Atlantic halibut, *Hippoglossus hippoglossus* (Kristiansen *et al.*, 2004).

9.9 Lumpfish deployment on salmon farms - concluding remarks

These current studies have clearly indicated that lumpfish at a 10–15% density (i.e. 1–1.5 lumpfish for every ten salmon) actively grazed upon sea lice, resulting in lower average numbers per fish of pre-adult, mature males and mature females stages of *L. salmonis* when lumpfish were present in the cages. Lumpfish suppressed the mature females stage of *L. salmonis* to levels equal to or lower than the pre-treatment count. The studies have shown very strong opportunistic feeding behaviour of lumpfish. The species appears to follow spatial and temporal variations in naturally occurring food sources. Lumpfish also did not feed on a single food source and targeted multiple sources. Size-dependent levels of aggression observed between lumpfish and goldsinny wrasse indicated that smaller lumpfish may be able to cohabit with goldsinny. Successful duoculture of Atlantic salmon and lumpfish is dependent on the size of the lumpfish. Large (>350 g) lumpfish can have a negative effect on overall growth and feed conversion of Atlantic salmon. Significant differences in feeding preferences, behaviour and mortality between different genetic families of lumpfish were seen. Both maternal and paternal effects were found for sea lice grazing activity and this suggests that this family effect is likely to have a genetic component that may be selected in future breeding programmes for lumpfish.

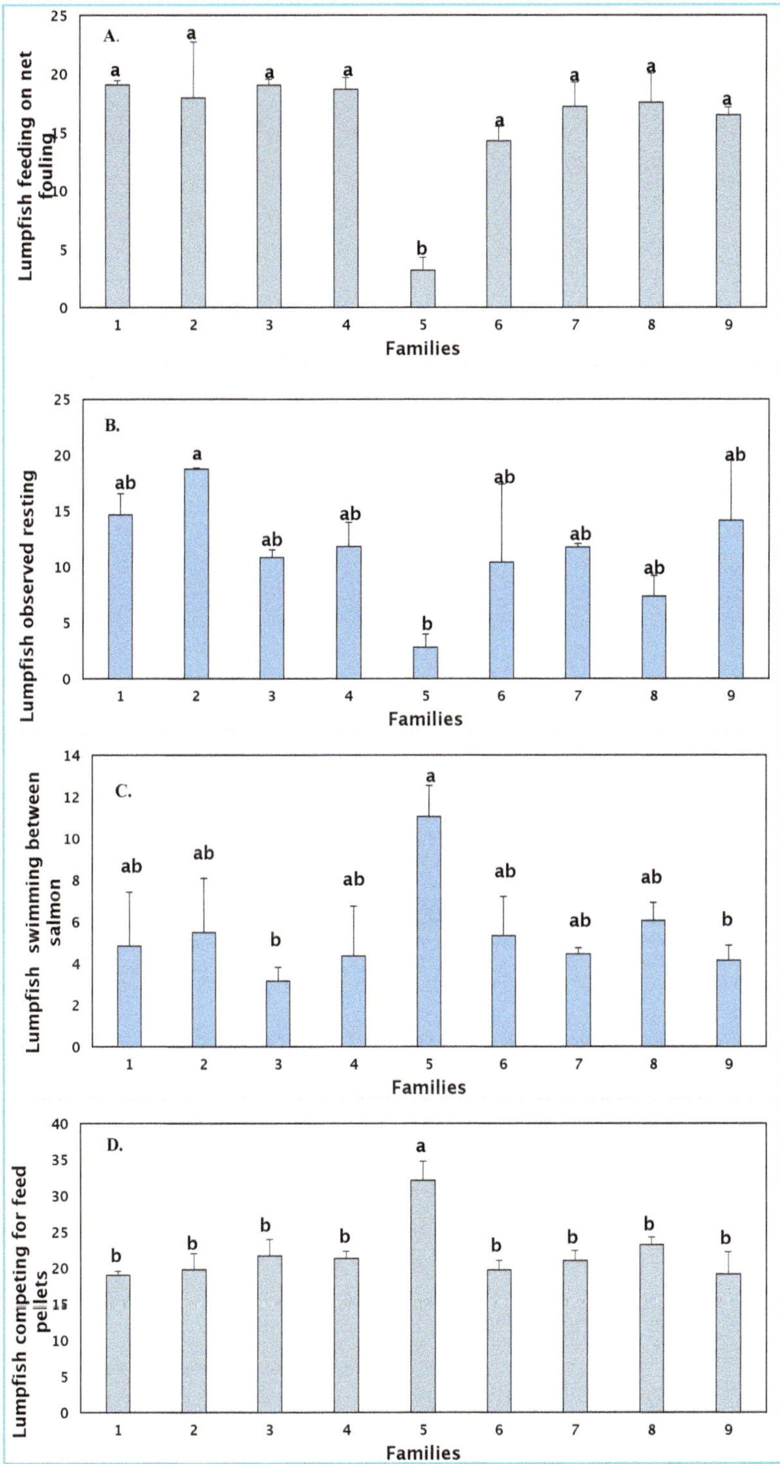

Fig. 9.13 Behaviour of lumpfish from nine families throughout the experimental period of 78 days. A: feeding on net fouling; B: resting; C: swimming between the salmon; D: competing for pellets with Atlantic salmon. Values are presented as means ± S.D. Mean values that do not share a letter were found to be significantly different (SNK test, $P < 0.05$). Modified from Imsland *et al.* (2016b).

References

Albert, O.T., Torstensen, E., Bertelsen, B., Jonsson, S.T., Pettersen, I.H. and Holst, J.C. (2002) Age-reading of lumpsucker (*Cyclopterus lumpus*) otoliths: dissection, interpretation and comparison with length frequencies. *Fisheries Research* 55, 239–252.

Austreng, E., Storebakken, T. and Åsgård, T. (1987) Growth rate estimates for cultured Atlantic salmon and rainbow trout. *Aquaculture* 60, 157–160.

Bjelland, O. and Holst, J.C. (2004) Other fish species and fish communities. In: Skjoldal, H. (Ed.), The Norwegian Sea ecosystem. Tapir Academic Press, Trondheim, Norway, pp. 357–369.

Blacker, R.W. (1983) Pelagic records of the lumpsucker, *Cyclopterus lumpus* L. *Journal of Fish Biology* 23, 405–417.

Bloecher, N., Olsen, Y. and Guenther, J. (2013) Variability of biofouling communities on fish cage nets: A 1-year field study at a Norwegian salmon farm. *Aquaculture* 416-417, 302–309.

Boxaspen, K. (2006) A review of the biology and genetics of sea lice. *ICES Journal of Sea Research* 63, 1304–1316.

Brown, J.A. (1986) The development of feeding behaviour in lumpfish, *Cyclopterus lumpus*. *Journal of Fish Biology* 29, 171–178.

Clague, G.E., Cheney, K.L., Goldizen, A.W., McCormick, M.I., Waldie, P.A. and Grutter, A.S. (2011) Long-term cleaner fish presence affects growth of a coral reef fish. *Biological Letters* 7, 863–865.

Costello, M.J. (2006) The ecology of sea lice parasitic on farmed and wild fish. *Trends in Parasitology* 22, 475-483.

Daborn, G.R. and Gregory, R.R. (1983) Occurrence, distribution, and feeding habits of juvenile lumpfish, *Cyclopterus lumpus* L. In the Bay of Fundy. *Canadian Journal of Zoology* 6, 97–801.

Darwall, W.R.T., Costello, M.J., Donnelly, R., Lysaght. S. (1992) Implications of life-history strategies for a new wrasse fishery. *Journal of Fish Biology* 41, 111–123.

Davenport J (1985) Synopsis of biological data on the lumpsucker *Cyclopterus lumpus* (Linnaeus, 1958). FAO Fisheries Synopsis no. 147, 31 pp.

Davenport, J, and Rees, E.I.S. (1993) Observations on neuston and floating weed patches in the Irish Sea. *Estuarine, Coastal and Shelf Science* 36, 395–411.

Davenport, J., Lønning, S. and Kjørsvik, E. (1983) Ammonia output by eggs and larvae of the lumpsucker, *Cyclopterus lumpus*, the cod, *Gadus morhua* and the plaice, *Pleuronectes platessa*. *Journal of the Marine Biological Association of the UK* 63, 713–723.

Deady, S., Varian, S.J.A. and Fives, J.M. (1995) The use of cleaner-fish to control sea lice on two Irish salmon (*Salmo salar*) farms with particular reference to wrasse behaviour in salmon cages. *Aquaculture* 13, 73–90.

Dempster, T., Juell, J.E., Fosseidengen, J.E., Fredheim, A. and Lader, P. (2008) Behaviour and growth of Atlantic salmon (*Salmo salar* L.) subjected to short-term submergence in commercial scale sea-cages. *Aquaculture* 276, 103–111.

Denholm, I., Devine, G.J., Horsberg, T.E., Sevatdal, S., Fallang, A., Nolan, D.V. and Powell, R. (2002) Analysis and management of resistance to chemotherapeutants in salmon lice *Lepeophtheirus salmonis* (Krøyer) (Copepoda: Caligidae). *Pest Management Science* 58, 528–536.

Fjøsne, K. and Gjøsæter, J. (1996) Dietary composition and the potential of food competition between 0-group cod (*Gadus morhua* L.) and some other fish species in the littoral zone. *ICES Journal of Marine Science* 53, 757–770.

Folkedal, O., Stien, L.H., Nilsson, J., Torgersen, T., Fosseidengen, J.E. and Oppdal, F. (2012) Sea caged Atlantic salmon display size-dependent swimming depth. *Aquatic and Living Resources* 25, 143–149.

Fredheim, A. and Langan, R. (2009) Advances in technology for offshore and open ocean aquaculture. In: Burnell, G., Allan, G. (Eds), New technologies in aquaculture: improving production efficiency, quality and environmental management. Woodhead Publications, Cambridge, pp. 914–944.

Gjedrem, T. (2000) Genetic improvement of cold-water fish species. *Aquaculture Research* 31, 25–33.

Gjerde, B. (1993) Breeding and selection. In: Heen, K., Monahan, R., Utter, F. (Eds.), Salmon Aquaculture, Blackwell, Oxford, pp. 187–208.

Grutter, A.S. (1995) Relationship between cleaning rates and ectoparasite loads in coral reef fishes. *Marine Ecology Progress Series* 118, 51–58.

Hansen, L.P. and Jacobsen, J.A. (2000) Distribution and migration of Atlantic salmon (*Salmo salar* L.) in the sea. In: Mills, D. (Ed.), The ocean life of Atlantic salmon—environmental and biological factors influencing survival. Blackwell Science, Oxford, UK, pp. 75–87.

Holm, J.C. (1989) Mono- and Duoculture of juvenile Atlantic Salmon (*Salmo salar*) and Arctic Char (*Salvelinus alpinus*). *Canadian Journal of Fisheries and Aquatic Science* 46, 697–704.

Holm, M., Holst, J.C. and Hansen, J.P. (2000) Spatial and temporal distribution of post-smolts of Atlantic salmon (*Salmo salar* L.) in the Norwegian Sea and adjacent areas. *ICES Journal of Marine Science* 57, 955–964.

Holst, J.C. (1993) Observations on the distribution of lumpsucker (*Cyclopterus lumpus*, L.) in the Norwegian Sea. *Fisheries Research* 17, 369–372.

Huntingford, F.A. (2004) Implications of domestication and rearing conditions for the behaviour of cultivated fishes. *Journal of Fish Biology* 65, 122–142.

Huntingford, F.A. and Adams, C. (2005) Behavioural syndromes in farmed fish: implications for production and welfare. *Behaviour* 142, 1207–1221.

Igboeli, O.O, Fast, M.D., Heumann, J. and Burka, J.F. (2012) Role of P-glycoprotein in emamectin benzoate (SLICE (R)) resistance in sea lice, *Lepeophtheirus salmonis*. *Aquaculture* 344, 40–47.

Imsland, A.K., Folkvord, A. and Stefansson, S.O. (1995) Growth, oxygen consumption and activity of juvenile turbot (*Scophthalmus maximus*) reared under different temperatures and photoperiods. *Netherlands Journal of Sea Research*, 149–159.

Imsland, A.K., Folkvord, A., Grung, G.L., Stefansson, S.O. and Taranger, G.L. (1997) Sexual dimorphism in growth and maturation of turbot, *Scophthalmus maximus* (Rafinesque 1810). *Aquaculture Research* 28, 101–114.

Imsland, A.K., Reynolds, P., Eliassen, G., Hangstad, T.A., Foss, A., Vikingstad, E. and Elvegård, T.A. (2014a) The use of lumpfish (*Cyclopterus lumpus* L.) to control sea lice (*Lepeophtheirus salmonis* Krøyer) infestations in intensively farmed Atlantic salmon (*Salmo sala*r L.). *Aquaculture* 425–426, 18–23.

Imsland, A.K., Reynolds, P., Eliassen, G., Hangstad, T.A., Nytrø, A.V., Foss, A., Vikingstad, E. and Elvegård, T.A. (2014b) Notes on behaviour of lumpfish in sea pens with and without Atlantic salmon. *Journal of Ethology* 32, 117–122.

Imsland, A.K., Reynolds, P., Eliassen, G., Hangstad, T.A., Nytrø, A.V., Foss, A., Vikingstad, E. and Elvegård, T.A. (2014c) Assessment of growth and sea lice infection levels in Atlantic salmon stocked in small-scale cages with lumpfish. *Aquaculture* 433, 137–142.

Imsland, A.K., Reynolds, P., Eliassen, G., Hangstad, T.A., Nytrø, A.V., Foss, A., Vikingstad, E. and Elvegård, T.A. (2015a) Feeding preferences of lumpfish (*Cyclopterus lumpus* L.) maintained in open net-pens with Atlantic salmon (*Salmo salar* L.). *Aquaculture* 436, 47–51.

Imsland, A.K., Reynolds, P., Eliassen, G., Hangstad, T.A., Nytrø, A.V., Foss, A., Vikingstad, E. and Elvegård, T.A. (2015b) Assessment of suitable substrates for lumpfish in sea pens. *Aquaculture International* 23, 639–645.

Imsland, A.K., Reynolds, P., Eliassen, G., Hangstad, T.A., Jónsdóttir, Ó.D.B., Elvegård, T.A., Lemmens, S.C.A., Rydland, R. and Nytrø, A.V. (2016a) Investigation and quantification of behavioural interactions between lumpfish (*Cyclopterus lumpus*) and Goldsinny (*Ctenolabrus rupestris*) under controlled conditions. *Aquaculture International* 24, 1509–1521.

Imsland, A.K., Reynolds, P., Eliassen, G., Mortensen, A., Hansen, Ø.J., Puvanendran, V., Hangstad, T.A., Jónsdóttir, Ó.D.B., Emaus, P.A., Elvegård, T.A., Lemmens, S.C.A., Rydland, R., Nytrø, A.V. and Jonassen, T.M. (2016b) Is cleaning behaviour in lumpfish (*Cyclopterus lumpus*) parentally controlled? *Aquaculture* 459, 156–165.

Ingólfsson, A. (2000) Colonization of floating seaweed by pelagic and subtidal benthic animals in southwestern Iceland. *Hydrobiologia* 440, 181–189.

Ingólfsson, A. and Kristjánsson, B.K. (2002) Diet of juvenile lumpsucker (*Cyclopterus lumpus*) in floating seaweed: effect of ontogeny and prey availability. *Copeia* 2, 472–476.

Jacobsen, J.A. and Gaard, E. (1997) Open ocean infestation by salmon lice (*Lepeophtheirus salmonis*): comparison of wild and escaped farmed Atlantic salmon (*Salmo salar* L.). *ICES Journal of Marine Science* 54, 1113–1119.

Jacobsen, J.A. and Hansen, L.P. (2001) Feeding habits of wild and escaped farmed Atlantic salmon *Salmo salar* L., in the Northeast Atlantic. *ICES Journal of Marine Science* 58, 916–933.

Johansen, L.H., Jensen, I., Mikkelsen, H., Bjorn, P.A., Jansen, P.A. and Bergh, Ø. (2011) Disease interaction and pathogens exchange between wild and farmed fish populations with special reference to Norway. *Aquaculture* 315, 229–243.

Killen, S.S., Brown, J.A. and Gamperl, A.K. (2007) The effect of prey density on foraging mode selection in juvenile lumpfish: balancing food intake with the metabolic cost of foraging. *Journal of Animal Ecology* 76, 814–825.

Kristiansen, T.S., Ferno, A., Holm, J.C., Privitera, L., Bakke, S. and Fosseidengen, J.E. (2004) Swimming behaviour as an indicator of low growth rate and impaired welfare in Atlantic halibut (*Hippoglossus hippoglossus* L.) reared at three stocking densities. *Aquaculture* 230, 137–151.

Meffert, L.M., Hicks, S.K. and Regan, J.L. (2002) Nonadditive genetic effects in animal behavior. *American Naturalist* 160, 198–213.

Midtlyng, P.J., Storset, A., Michel, C., Slierendrecht, W.J. and Okamoto, N. (2002) Breeding for disease resistance in fish. *Bulletin of the European Association of Fish Pathologists* 22, 166–172.

Mitamura, H., Thorstad, E.B., Uglem, I., Bjorn, P.A., Okland, F., Naesje, T.F., Dempster, T. and Arai, N. (2012) Movements of lumpsucker females in a northern Norwegian fjord during the spawning season. *Environmental Biology of Fishes* 93, 475–481.

Moring, J.R. (1989a) Seasonal Absence of Fishes in Tidepools of a Boreal Environment (Maine, USA). *Hydrobiologia* 194, 163–168.

Moring, J.R. (1989b) Food habits and algal associations of juvenile lumpfish, *Cyclopterus lumpus* L., in intertidal waters. *Fisheries Bulletin* 87, 233–237.

Mustafa, A., Conboy, G.A. and Burka, J.F. (2000) Lifespan and reproductive capacity of sea lice, *Lepeophtheirus salmonis*, under laboratory conditions. *Aquaculture Association of Canada, Special Publication* 4, 113–114.

Nortvedt, R. and Holm, J.C. (1991) Atlantic salmon in duoculture with Arctic charr – decreased aggression enhances growth and stocking density potential. *Aquaculture* 98, 355–361.

Nytrø, A.V., Vikingstad, E., Foss, A., Hangstad, T.A., Reynolds, P., Eliassen, G., Elvegård, T.A., Falk-Petersen, I.B. and Imsland, A.K. (2014) The effect of temperature and fish size on growth of juvenile lumpfish (*Cyclopterus lumpus* L.). *Aquaculture* 434, 296–302.

Øverli, Ø., Sørensen, C., Kiessling, A., Pottinger, T.G. and Gjøen, H.M. (2006) Selection for improved stress tolerance in rainbow trout (*Oncorhynchus mykiss*) leads to reduced feed waste. *Aquaculture* 261, 776–781.

Oppedal, F., Taranger, G.L., Juell, J.E. and Hansen, T. (1999) Growth, osmoregulation and sexual maturation of underyearling Atlantic salmon smolt *Salmo salar* L. exposed to different intensities of continuous light in sea cages. *Aquaculture Research* 30, 491–499.

Oppedal, F., Dempster, T. and Stien, L.H. (2011) Environmental drivers of Atlantic salmon behaviour in sea-cages: A review. *Aquaculture* 311, 1–18.

Quignard, J.P. and Pras, A. (1986) Labridae. In: Whitehead P.J.P., Bauchor M.-L., Hureau J.-C., Nielsen J., Tortonese E. (eds) Fishes of the North-eastern Atlantic and the Mediterranean. Volume II. Paris: UNESCO, pp. 919–942.

Roff, D.A. and Mousseau, T.A. (1987) Quantitative genetics and fitness: lessons from Drosophila. *Heredity* 58, 103–118.

Sayer, M.D.J. and Davenport, J. (1996) Hypometabolism in torpid goldsinny wrasse subjected to rapid reductions in seawater temperature. *Journal of Fish Biology* 49, 64–75.

Sayer, M.D.J. and Reader, J.P. (1996) Exposure of goldsinny, rock cook and corkwing wrasse to low temperature and low salinity: Survival, blood physiology and seasonal variation. *Journal of Fish Biology* 49, 41–63.

Sayer, M.D.J., Gibson, R.N. and Atkinson, R.J.A. (1996) Seasonal, sexual and geographical variation in the biology of goldsinny, corkwing and rock cook on the west coast of Scotland. In: Sayer, M.D.J., Treasurer, J.W., Costello, M.J. (eds) *Wrasse: Biology and Use in Aquaculture.* Oxford: Blackwell Science, pp. 13–46.

Sheedan, T.F., Reddin, D.G., Chaput, G. and Renkawitz, M.D. (2012) SALSEA North America: a pelagic ecosystem survey targeting Atlantic salmon in the Northwest Atlantic. *ICES Journal of Marine Research* 69, 1580–1588.

Skiftesvik, A.B., Blom, G., Agnalt, A.L., Durif, C.M.F., Browman, H., Bjelland, R.M., Harkestad, L.S., Farestveit, E., Paulsen, O.I., Fauske, M., Havelin. T., Johnsen, K. and Mortensen, S. (2014) Wrasse (Labridae) as cleaner fish in salmonid aquaculture – The Hardangerfjord as a case study. *Marine Biology Research* 10, 289–300.

Stirling, D.G., Reale, D. and Roff, D.A. (2002) Selection, structure and the heritability of behaviour. *Journal of Evolutionary Biology* 15, 277–289.

Thorsteinsson, V. (1983) Some aspects of the biology and the fisheries of the lumpfish (*Cyclopterus lumpus* L.). *Journal of Ichthyology* 10, 77–83.

Timeyko, V.N. (1986) The digestive system of White Sea cod, *Gadus morhua marisalbi*, and lumpfish *Cyclopterus lumpus*, at different stages of ontogeny. *Journal of Ichthyology* 26, 72–82.

Treasurer, J.W. (1993) Management of sea lice (Caligidae) with wrasse (Labridae) on Atlantic salmon (*Salmo salar* L.) farms. In: Boxshall, G.A., Defaye, D. (Eds.), *Pathogens of Wild and Farmed Fish: Sea Lice.* Ellis Horwood, Chichester, UK, pp. 335–345.

Treasurer, J.W. (2002) A review of potential pathogens of sea lice and the application of cleaner fish in biological control. *Pest Management Science* 58, 546–558.

Treasurer, J. and Feledi, T. (2014) The physical condition and welfare of five species of wild-caught wrasse stocked under aquaculture conditions and when stocked in Atlantic salmon, *Salmo salar*, production cages. *Journal of the World Aquaculture Society* 45, 213–219.

Tully, O., Daly, P., Lysaght, S., Deady, S., and Varian, S.J.A. (1996) Use of cleaner-wrasse (*Centrolabrus exoletus* (L.) and *Ctenolabrus rupestris* (L.)) to control infestations of *Caligus elongatus* Nordmann on farmed Atlantic salmon. *Aquaculture* 142, 11–24.

Vandendriessche, S., Messiaen, M., O'Flynn, S., Vincx, M., Degraer, S. (2007) Hiding and feeding in floating seaweed: Floating seaweed clumps as possible refuges or feeding grounds for fishes. *Estuarine, Coast and Shelf Science* 71, 691–703.

Williams, P.J. and Brown, J.A. (1991) Developmental-changes in foraging-predator avoidance trade-offs in larval lumpfish *Cyclopterus lumpus*. *Marine Ecology Progress Series* 76, 53–60.

PART II

Advances in cleaner fish science (thematic)

CHAPTER 10

Cleaner fish species nutrition and feeding practices

Katerina Kousoulaki,[1] Herve Migaud[2] and Andrew Davie[2]

[1]Nofima AS, Bergen, Norway
[2]Institute of Aquaculture, University of Stirling, UK

10.1 Introduction

Fish species currently used as cleaner fish in sea-based salmon aquaculture also exhibit the same behaviour in nature. Wrasse as well as lumpfish *Cyclopterus lumpus* have long been observed to pick parasites off the skin of other fish. Over the past few years, both wild and farmed cleaner fish have also proven effective in cleaning farmed salmon of sea lice. Farming of the most effective species is imperative for sustainable use of cleaner fish in salmon aquaculture. Thus, there is a need for appropriate diets for the different cleaner fish species. Feeding cleaner fish is also necessary in the cages to maintain their condition and welfare during periods of low lice occurrence. In practice, both compound feeds and wild bait are used. There are only a few published studies focusing on nutrition and feeding of cleaner fish species in wild and farmed populations, thus our knowledge in this area is still very limited. In this chapter, we review the currently available literature and provide an overview of farmers' experiences regarding nutritional aspects related to the most commonly used cleaner fish species.

10.2 Natural feeding habits of cleaner fish species

10.2.1 Wrasses

Wrasse are carnivorous generalist predators with specially designed jaws for very rapid movement while they capture their prey (van Hasselt, 1979), armed with a single row of teeth, except in goldsinny (*Ctenolabrus rupestris*) that have several. Wrasses are also armed with strong maxillary and pharyngeal teeth, but have a short oesophagus, no stomach or pyloric caeca (Quignard, 1966) and a short intestine, ⅔ of the fish's body length in size (Hamre and Sæle, 2011). They feed mainly on invertebrates (molluscs, crustaceans, annelids, hydrozoans, bryozoan, Echinodermata), and to a much less extent on fish, algae (*Antithamnion plumula*) and sea grass (*Posidonia oceanica*;

Cymodocea nodosa). The two latter are most likely ingested during consumption of prey species confined in the seaweeds and the rocky substrata (Quignard, 1966; Hilldén, 1978; Shepherd and Clarkson, 2001; Ben *et al.*, 2007). Wrasse larvae up to 2 cm long feed on copepods, cladocerans and nauplii and, as they grow, consume larger prey such as amphipods, isopods, decapod larvae and polychaetes (Quignard, 1966; Dipper *et al.*, 1977).

The blue-throated wrasse (*Notolabrus tetricus*) is a diurnal predator with a defined territory that increases with increasing size of the wrasse. Foraging activity in this species commences soon after dawn, reaches a peak during mid-morning, declines during the afternoon, and ceases at dusk. In this wrasse species the observed inter- and intraspecific agonistic interactions are few and size-related (Shepherd and Clarkson, 2001).

Spring-caught wild Ballan (*Labrus bergylta*) and corkwing wrasse (*Symphodus melops)* in Galway Bay (Ireland) and Dinard (Brittany, France), appear to feed mainly on decapods and bivalves and smaller quantities of algae and gastropods (Deady and Fives, 1995a) whereas on the west coast of Scotland, corkwing wrasse diet is dominated by gastropod molluscs (Sayer *et al.*, 1996). Larger fish, longer than 20 cm, consume more bivalves than the smaller ones, that prey mostly on small crustaceans (Deady and Fives, 1995b). Similarly, in the shallow waters around New Zealand and southern Australia, the larger fish (>18 cm) of the wild banded wrasse (*Notolabrus fucicola*), also feed mainly on bivalves, gastropods but also crabs, whereas the smaller fish (10–18 cm) prey mainly on smaller crustaceans such as isopods and amphipods, with no significant sex variation (Denny and Schiel, 2001). Dietary preference variation among male and female wrasse has been reported in goldsinny wrasse (*Ctenolabrus rupestris*) caught in Millport, Oban and Luing on the west coast of Scotland, which may be due to sexual dimorphism in size (Sayer *et al.*, 1995). During the breeding season, the dietary items of wild wrasses are more diverse (Deady and Fives, 1995), with high occurrence of digestive vacuity during winter and spring. This is, for instance, observed in wild-caught peacock wrasse (*Symphodus tinca)*, a cleaner fish species common along the coast of Algeria (Boughamou *et al.*, 2016), and brown wrasse (*Labrus merula)* from northern Tunisian coasts (Ben *et al.*, 2007). Peacock and brown wrasse feed mainly on crustaceans but also consume peanut worms (sipunculans), polychaetes (*Pontogenia chrysocoma*) and molluscs as secondary prey species, at similar relative amounts between season, sex and maturity stage, though smaller crustaceans (peracarida), as in other studied species, are more important dietary items for smaller fish (Ben *et al.*, 2007; Boughamou *et al.*, 2016).

Wild wrasse, used in net pens to clean farmed Atlantic salmon (*Salmo salar*) of sea lice (*Lepeophtheirus salmonis*), exhibit lice cleaning behaviour in nature towards other fish, contributing to the health of fish communities. However, some wrasse individuals set out in salmon cages appear not to consume lice. Accordingly, the North Atlantic (Henriques and Almada, 1997), as well as the temperate wrasse species (Navarez *et al.*, 2015) appear to be facultative cleaner fish, assumed to clean only during their juvenile phase, and have a broader diet than obligatory cleaner fish in the tropics. For instance, cleaning acts of rock cook (*Centrolabrus exoletus*), a facultative cleaner fish, represented only 7% of its observed feeding activity (Henriques and Almada, 1997). Moreover, it appears that facultative cleaner fish develop special preferences, becoming very species selective in their cleaning behaviour. For instance, the temperate water species *Coris julis,* consumed both gnathiid larvae and caligid copepods, whereas *Thalassoma pavo* was not found to consume any ectoparasites, though caligid copepods were abundant on client species of the area

(Navarez *et al.*, 2015). Wrasse species are also known to clean parasites from pelagic thresher sharks (*Alopias pelagicus*) (Cadwallader *et al.*, 2015), but also among each other and other wrasse species. For example, rock cook perform cleaning behaviour on 12 species, and among these are corkwing and Ballan wrasse (Henriques and Almada, 1997). Goldsinny in Swedish waters is also know to clean Ballan wrasse (Hilldén, 1983). Often, wild-caught wrasse introduced in the salmon cages may also be infected by sea lice. Nevertheless, observations from the British Isles showed that the risk of sea lice cross-infection among wild wrasse species and farmed salmon is apparently low, as they are hosts to different species of sea lice. *Caligus centrodonti* Baird and *Caligus elongatus* Nordmann were identified on Ballan, rock cook and goldsinny wrasse, while salmon is infected with *Lepeophtheirus salmonis* (Kroyer) (Bron and Treasurer, 1992) as well as *Caligus* sp. in northern Norway and Chile.

Wrasse are known to have a taste for the slime of client fish and the incentive to 'cheat' and exploit their partner. This can cause wounds to farmed salmon when using wrasse as cleaner fish of sea lice (*Lepeophtheirus salmonis*) in net pens. Nevertheless, in nature it is observed that cleaner wrasse *Labroides dimidiatus* can resist the temptation of 'cheating' the client fish by biting on its slime and, instead, cooperatively eat ectoparasites if the threat of early termination of an interaction is sufficient (Gingins *et al.*, 2013).

10.2.2 Lumpfish

Lumpfish is a cold water cleaner fish species, distributed throughout the North Atlantic Ocean. Lumpfish migrate considerable distances between offshore feeding areas and shallow inshore spawning areas, and do not feed during spawning migrations (Davenport, 1985). Wild lumpfish, unlike wrasse species, have no particular preference for slow-moving or sessile animals such as bivalves, gastropods and annelids. In the yolk larvae and juvenile stage, they feed on organisms associated with floating seaweed species such as *Ascophyllum nodosum* and *Himanthalia elongate* (Vandendriessche *et al.*, 2007), where they often stay during the first and second year of their life (Davenport and Rees, 1993; Ingólfsson and Kristjánsson, 2002). Such prey organisms are crustaceans (calanoid and harpacticoid copepods; gammarid amphipods, e.g. *Gammarus locusta* and *Gammarus crinicornis*; isopods, e.g. *Idotea balthica*; swimming crab larvae, e.g. *Liocarcinus holsatus*; shrimp post-larvae e.g. *Palaemon elegans*), mites, fish eggs and plankton. As they grow larger, lumpfish can also exhibit cannibalistic feeding behaviour on smaller lumpfish (Ingólfsson and Kristjánsson, 2002; Vandendriessche *et al.*, 2007). Adult lumpfish have also been observed to feed on jellyfish. These are composed mainly of water, which may explain the difficulty in identifying the gut content composition of this species (Davenport, 1985).

10.3 Feeding practices in farming of cleaner fish species

Of all wrasse species used as cleaner fish in salmon pens, the focus is mainly on Ballan wrasse produced in captivity. This protogynous wrasse (Dipper *et al.*, 1977) is the largest in the north-east Atlantic (Quignard, 1966; Froese and Pauly, 2012) and considered as the most efficient wrasse species at delousing sea lice (*Lepeophtheirus salmonis*) from Atlantic salmon. More recently, an even larger effort in farming lumpfish has been observed. The following sections describe the known nutritional requirements and feeding practices in farming Ballan wrasse and lumpfish, as well as feeding practices of both wild and farmed Ballan wrasse and lumpfish when deployed in salmon net pens.

10.4 Broodstock nutrition

The nutritional requirements of on-grown and broodstock cleaner fish are little known. Broodstock fish are often fed frozen crayfish or shrimp. As in Ballan wrasse farming, the commercial production of lumpfish is almost exclusively reliant on wild broodstock fish. Nevertheless, first generation farmed broodstock groups are now available in several production companies and research institutes (i.e. at Nofima AS). Given the lack of available knowledge of broodstock nutritional requirements and the limited availability of commercial diets, the nutritional composition of the whole body and gonad more specifically differ in Ballan wrasse kept in captivity compared with wild-caught broodstock (Hamre *et al.*, 2013; Grant *et al.*, 2016). This indicates that the current commercial diets used during captive on-growing and broodstock maintenance should be further improved in terms of their nutrient composition. In the study of Hamre *et al.* (2013), farmed Ballan wrasse fed commercial diets had higher whole body levels of most vitamins, but vitamin A, K and D were comparable to the levels analysed in wild individuals. The broodfish kept in captivity also had similar or higher levels of most minerals except iron (Fe). On the other hand, female gonads contained lower levels of taurine, higher levels of most vitamins except vitamin K, and higher levels of most minerals except iodine (I). Grant *et al.* (2016) assessed Ballan wrasse egg fatty acid profiles across an entire spawning season from broodstock maintained on a combination of extruded pellets (Symbio Wrasse Diet, 6.5 mm diameter; Biomar Ltd, Grangemouth, Scotland UK) supplemented with langoustine (*Nephrops norvegicus*) tails. The study demonstrated that egg fatty acid (FA) profiles remained quite consistent across the spawning season and did not clearly correlate with batch fertilisation or hatch rates. The most abundant saturated (SFA), monounsaturated (MUFA), and polyunsaturated fatty acids (PUFA) observed in Ballan wrasse eggs were palmitic acid (16:0), oleic acid (18:1n-9), and docosahexaenoic acid (DHA) (22:6n-3), respectively. SFA accounted for an average of 32.5 ± 4.0% of the total fatty acids in Ballan wrasse eggs. MUFA ranged from 18.4 to 29.4% of the total fatty acids. With the exception of the high levels of ARA (~3.8 ± 0.5% of total FA) compared with ~2.5% total FA in other marine species, Ballan wrasse egg FA profile observed by Grant *et al.* (2016) fits the general profile for marine fish (Tocher *et al.*, 1985; Fraser *et al.*, 1988; Sargent *et al.*, 2002).

A comparison of the chemical composition of roe obtained from wild and farmed lumpfish broodstock revealed that the roe of farmed fish was richer in all analysed components (including dry matter, total lipids, vitamins, minerals, essential amino acids, taurine and total free amino acids) compared with roe from wild fish, with the exception of iodine and n-3 fatty acids, which were found at higher levels in wild lumpfish roe (Hamre, 2016). Another difference observed was the higher levels of ethanolamine and lower levels of phosphatidylcholine in wild lumpfish roe compared to roe from farmed broodfish.

10.5 Larval rearing: Live feed and weaning cleaner fish

10.5.1 First feeding wrasse larvae with live feed

Four to five days post hatch (dph) Ballan wrasse larvae have fully absorbed their yolk sac (Skiftesvik *et al.*, 2011) and are commonly first fed DHA and EPA (20:5n-3) enriched rotifers, mainly *Brachionus plicatilis*, for approximately three weeks. They are then fed *Artemia* sp. nauplii, for approximately six weeks. Clay or algae are often added to the larval tanks during the first 90 days of rearing to stimulate start feeding and control the bacterial growth and water quality in the tanks (Skiftesvik *et al.*, 2011). During the Artemia feeding period water temperature is usually raised from 11.5°C to 16°C and approximately 30% of the fish become blue, as opposed to the skin coloration of the majority of the fish population, which is red (Hovland, 2016). Such fish belong to the lowest size grade and they are usually removed. A recent study in Ballan wrasse investigated the bile pigment biliverdin, which is the driver of this coloration change and has a range of bioactive properties (Clark *et al.*, 2016). Ballan wrasse appear to mobilise this pigment in association with tissue damage and repair mechanisms (Clark *et al.*, 2016), suggesting these bottom-grade fish are subject to an as yet undetermined challenge or stressor.

The nutritional quality of rotifers has been questioned as first feeding Ballan wrasse larvae given rotifers often develop skeletal anomalies, and have poor growth and low survival rates (Grøntvedt, 2010; Helland *et al.*, 2012). This is common in many other newly farmed marine fish species. Rotifers can be substituted by live copepods, with prey organism size ranging between 0.11 and 0.9 µm, offered to Ballan wrasse larvae of respective mouth opening size (Hagerman, 2014). This practice is more successful in rearing larvae of other marine fish species, such as Atlantic cod (*Gadus morhua*) (Imsland *et al.*, 2016). Research results showed better growth, stress resistance, predation capacity, earlier onset of ossification and lower skeletal anomalies in Ballan wrasse larvae fed copepods (e.g. *Acartia tonsa*) instead of rotifers (*Brachionus ibericus* Cayman). Nevertheless, survival can be higher when Ballan wrasse larvae are fed Chlorella-enriched rotifers compared with copepods (Sørøy, 2012). Copepods contain higher amounts of total protein, lipids, taurine, trimethylamine N-oxide (TMAO) and marine phospholipids compared with both Artemia and rotifers, and higher amounts of betaine compared with Artemia, but lower compared with rotifers (Almli, 2012). Thus, copepods appear to be richer than rotifers in the necessary building blocks for harmonic fast fish larval growth, as well as functional nitrogenous compounds both in terms of lipid metabolism, osmoregulation and feeding attractants. Nevertheless, the contribution of live prey to the digestive enzymes of Ballan wrasse appeared to be insignificant (Hansen *et al.*, 2014). Ballan wrasse larvae are producing similar levels of most digestive enzymes during the live prey and weaning phases, exhibiting, besides trypsin, phospholipase and lipase activities, also significant enzyme activity of both amylase and chitinase (Hansen *et al.*, 2014), which is consistent with their crustacean-rich natural diet. However, despite their high potential value to marine larvae rearing, to date, copepod production is not practised to the same extent as that of rotifers, which are easier to cultivate (Øie, 2014). Thus, there are still insufficient available volumes of copepods to meet the commercial demand of marine hatcheries.

10.5.2 Wrasse larvae weaning diets and protocols

Ballan wrasse larvae are weaned using specially developed crustacean meal-based artificial diets (Skiftesvik, 2009; Skiftesvik and Bjelland, 2003). These are high-protein and low-lipid shrimp, krill, cod fillet and squid meal-based feeds, devoid of fish meal and rich in plant and marine phospholipids (Diet CF-S in Table 10.1; Nofima AS, Bergen, Norway). Ballan wrasse weaning performance is positively correlated with increasing levels of dietary DHA/EPA, cholesterol, free amino acids, total water-soluble nitrogenous compounds and phosphorus, whereas it is negatively correlated to total dietary lipids, dietary ethoxyquin and lipid oxidation metabolites or fish meal freshness (Kousoulaki *et al.*, 2014a; Bogevik *et al.*, 2016).

In practice, weaning protocols are still not yet standardised across hatcheries operating in Europe. In some cases, the above mentioned whole fishmeal-free weaning diet is only given during a few days (eight to ten), followed by other commercial fish and krill meal-based diets (weaning protocol used by hatcheries in Norway, Kousoulaki and Grøtan, unpublished results). When fish meal or fish meal fractions were included from the start in the weaning diet of Ballan wrasse larvae reared in these hatcheries, feed acceptance, as well as weaning survival, was significantly reduced (Kousoulaki *et al.*, 2014b). A suggestion was that ethoxyquin, present in fishmeal and/or the secondary metabolites of fish oil oxidation, may act as feeding repellents for the particular Ballan wrasse larvae or mask the attractants derived from the crustacean feed component (Bogevik *et al.*, 2016). However, best results have been obtained in other hatcheries in Scotland when using the fishmeal-free weaning diet throughout weaning and during early on-growing also. Differences in the Ballan wrasse stocks and hatchery-specific husbandry regimes and feed specifications may explain these contrasting results. Further research is clearly required.

10.5.3 Deformities and fin damage development during weaning

The long-term feeding of Ballan wrasse larvae on a fishmeal-free extruded weaning diet, though achieving up to 95% weaning survival, has consistently lead to nearly 100% head–jaw deformity occurrence (Kousoulaki *et al.*, 2014b). As the feeding apparatus of marine fish larvae ossifies first (Koumoundouros *et al.*, 1999; Sæle *et al.*, 2004), it is thus expected that nutritional deficiencies early in development will be first evident in the mouthparts and related head structures of wrasse larvae. The above mentioned head–jaw deformity may not occur or be less prominent when agglomerated diets are used instead of extruded crumbled pellets, or when fishmeal or the water-soluble part of fish meal alone (stickwater) is included in the diet (approximately 12% inclusion level of a 30–40% dry matter raw material from herring or blue whiting) (Fig. 10.1; Kousoulaki *et al.*, 2014a, b).

The mechanisms behind these deformities are not yet studied and may be related to either the specific nutritional quality of whole fishmeal combined with the digestive capacity of the stomach-less Ballan wrasse larvae for artificial diets of different physical and nutritional qualities, or the cumulative effects of other husbandry-related factors during larvae rearing. The diets containing whole stickwater have high levels of water-soluble nitrogenous compounds (40.5 g/100 g protein). However, this property alone does not provide the solution to the problem as the observed head and jaw deformity was not prevented either by use of hydrolysed fish muscle

(CPSP) nor by krill hydrolysate (Fig.10.2; Nofima-Marine Harvest Labrus unpublished data). The water-soluble part of whole fishmeal contains soluble phosphorus and free amino acids, which are key micronutrients for the development of organisms as well as other metabolites stimulating feed intake and growth such as nucleotides, trimethylamine oxide (TMAO), creatinine, organic acids (Carr *et al.*, 1996; Wu and Bechtel, 2012). Ballan wrasse consume only marginal amounts of fish in nature (Figueiredo *et al.*, 2005), and it is therefore unlikely that they have a specific requirement for fish-derived components.

On the other hand, extruded diets (Table 10.1) may be difficult to digest and to cover the requirement of fast-growing larvae. The extrusion process involves relatively high temperature and pressure, which alters the physicochemical properties of the dietary nutrients. Ballan wrasse larvae are very sensitive to small freshness differences, as a result of the applied processing technology (drying, pelleting), of otherwise high-quality marine raw materials. Freeze drying results in higher quality raw materials compared to warm air drying and agglomeration results in better performing Ballan wrasse larvae weaning feed compared to cold pelleting and extrusion (Fig. 10.3; Kousoulaki and Opstad, 2012).

Marine fish larvae have a high quantitative essential fatty acids requirement (EPA/DHA/ARA) due to their high metabolic rate and the importance of these fatty acids for neural tissue formation, osmoregulation and organ development (Izquierdo and Koven, 2011). However, the requirement of Ballan wrasse larvae for EPA, DHA and ARA is not yet defined.

Table 10.1 Ballan wrasse weaning diets main raw material composition (MHL-Nofima unpublished trial data). Treatments were cod fillet–shrimp meal (CF-S), cod fillet–shrimp meal–stickwater (CF-S-SW), fishmeal–shrimp (FM-S) and OTOHIME (commercial).

	CF-S	CF-S-SW	FM-S	OTOHIME
Fish meal			x	x
Fishmeal stickwater		x		
Cod fillet	x	x	x	
Shrimp meal	x	x	x	
Squid meal	x	x		x
Krill meal				x
Krill hydrolysate	x	x		
Krill oil	x	x		
Fish oil				x
Soy lecithin	x	x	x	x
Binder (wheat), mineral–vitamin mix, amino acids, pigment, yeast or yeast extracts, etc.	x	x	x	x

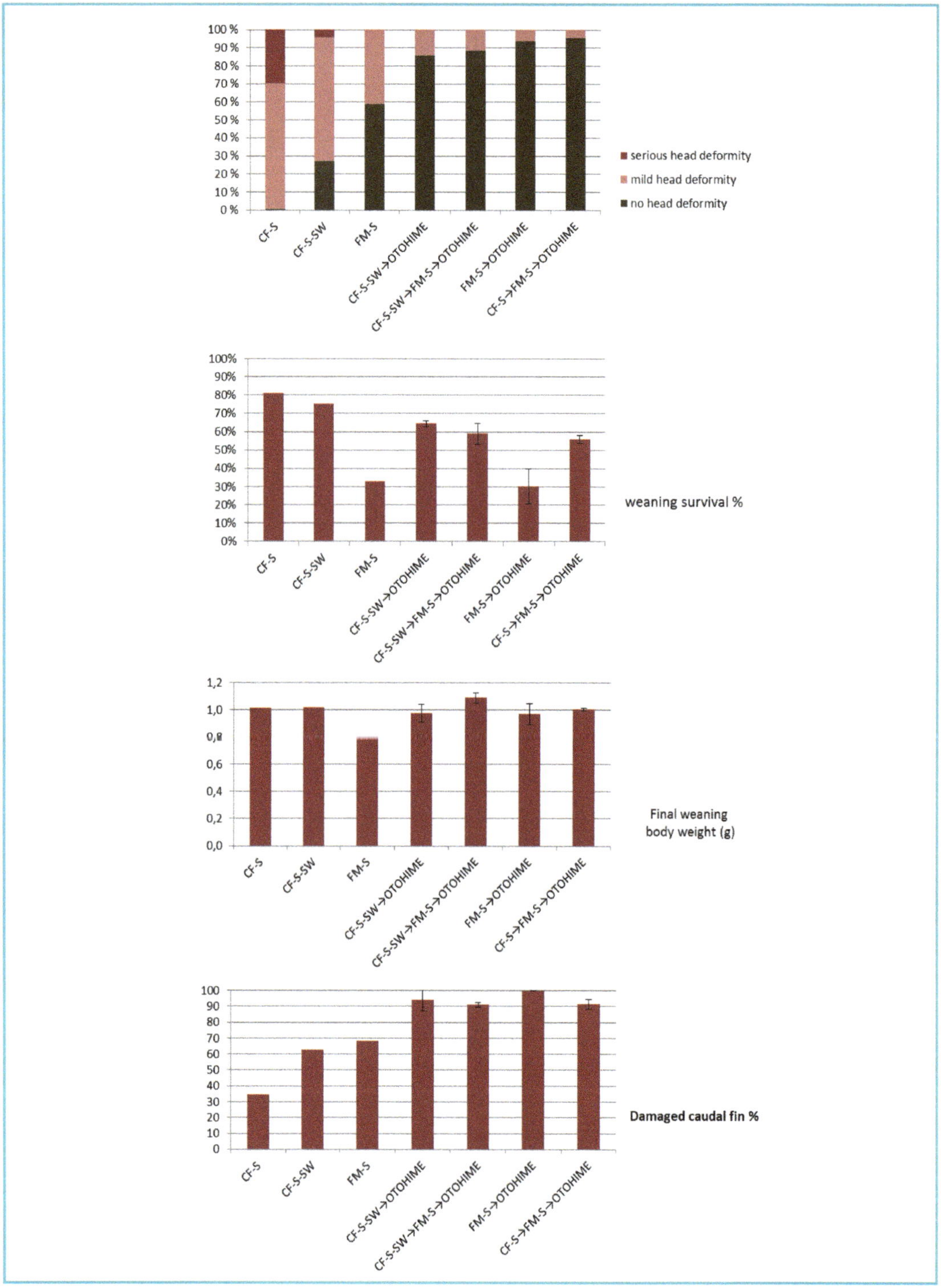

Fig. 10.1 Ballan wrasse weaning survival, growth, head and jaw deformity and caudal fin damage occurrence, using different weaning diets and protocols (MHL-Nofima unpublished trial data).

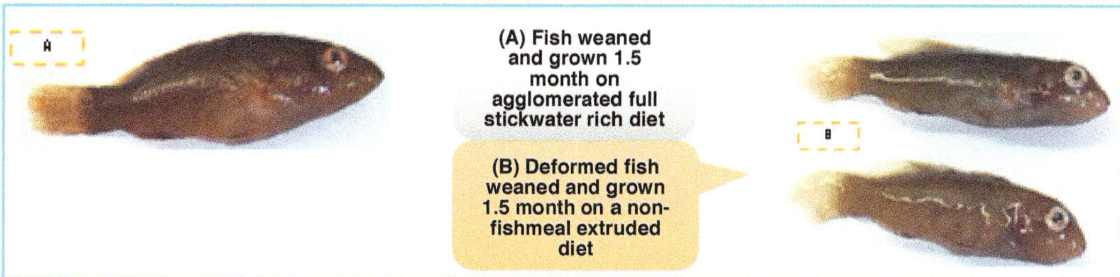

Fig. 10.2 Ballan wrasse larvae weaned on fishmeal whole stickwater (A) supplemented agglomerated or fishmeal-free extruded diet (B), the latter with typical obvious head deformity.

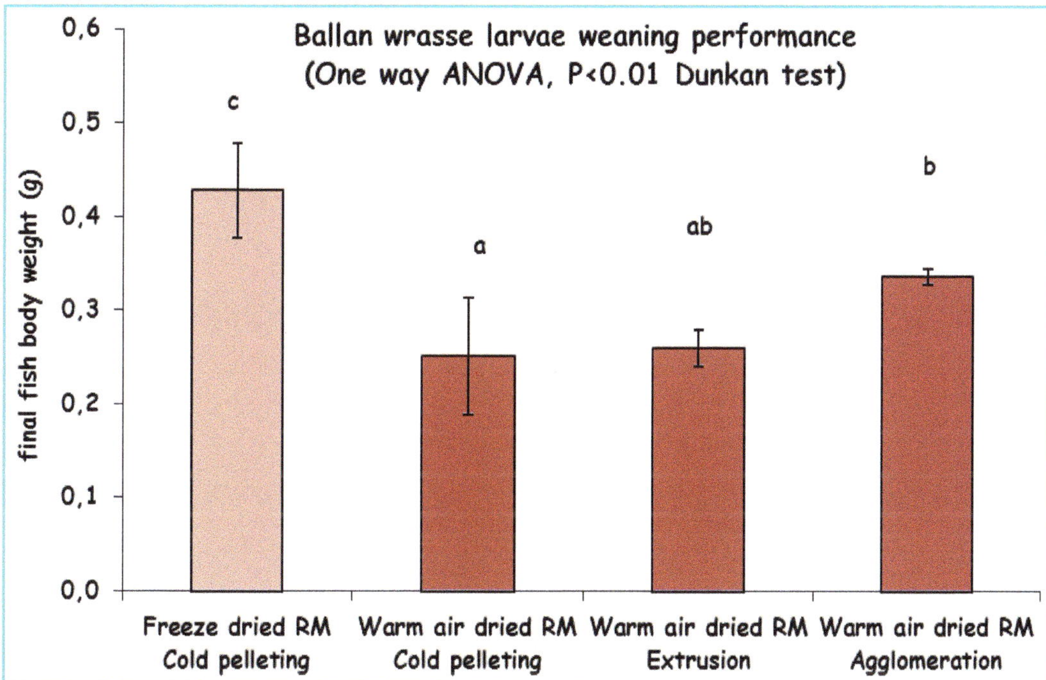

Fig. 10.3 Ballan wrasse larvae weaning performance on diets produced with freeze- or warm air-dried raw materials (RM) and different pelleting methods. Different letters above columns indicate significant statistical differences between diets.

10.5.4 First feeding lumpfish larvae

Lumpfish larvae accept common fishmeal-based commercial artificial diets used for other marine fish species (e.g. Gemma microstarter feed by Skretting, Stavanger, Norway for first feeding). From two days after hatching there is no need for live feeds, rendering their culture much easier than for Ballan wrasse. Though first feeding of lumpfish with live feed may seem unnecessary, in practice it can have significant advantages. Several farmers report that use of artificial diets at first feeding results in deterioration of water quality and induces high larval mortalities due to bacterial infections (Schaer, 2017). For instance, lumpfish weaned and grown on cryopreserved copepods containing more than 70% lipids in the form of marine phospholipids had significantly higher body weight than when reared on a dry feed (Tokle, 2017). The author recommended feeding lumpfish larvae with 115 copepods/larvae/day from day 1, increasing to 215 copepods/larvae/day until day 20, with a mean daily need of 160 copepods/larvae/day for this period. Lumpfish exhibit high growth rates on currently available commercial aquafeeds. Thus, the challenges in rearing them in captivity are more related to maintaining good fish health and welfare. During first feeding, high water quality is required to maintain low bacterial growth. Moreover, better knowledge of the nutritional requirements of the fish and its production biology are needed in order to achieve good fish health and decrease disease and mortalities during first feeding as well as at later production stages. Moreover, there is a need for frequent size sorting of the lumpfish populations due to large individual growth variation and cannibalistic behaviour (Ulvan, 2016).

10.6 Juvenile cleaner fish nutrition and feeding practices

The nutritional requirements of cleaner fish grown in captivity are not yet established. Nevertheless, there is a great number of ongoing research projects, both in academic and production environments, which are expected to advance our knowledge on this field in the years to come. In this section, we present findings of published and yet unpublished studies on Ballan wrasse and lumpfish nutrition.

Ballan wrasse, as agastric fish, have basic pH along their short digestive tract, ranging from 7.7 to 8.2. The species is thus lacking the ability of acid digestion as well as that of efficient prehydration of formulated feed pellets. In rearing wrasse of all stages, it is common practice to offer moist formulated feeds with the aim to improve feed acceptability and digestion efficiency. However, a recent study showed no differences in either feed hydration speed in the intestine or nutrient digestibility from adding moisture to the diet prior to feeding or not (Sæle, 2017). In a screening study, Hamre *et al.* (2013) found that on-grown Ballan wrasse juveniles grow best when fed high protein diets (65%), low fat (12%) and relatively high levels of carbohydrates (16%). Moreover, when dietary lipid levels are higher (15–20%), in order to ensure optimal lipid utilisation and thus higher growth rates, 50% should be in the form of phospholipids (Sæle *et al.*, 2014). Ballan wrasse juveniles have higher intestinal trypsin and lower amylase activity than the larvae of the same species (Hansen *et al.*, 2014; Krogdahl *et al.*, 2014), whereas significant activity levels of leucine aminopeptidase and maltase were also measured.

During the whole nursery period, Ballan wrasse juveniles are fed krill- or shrimp-based feeds. As in earlier stages, the Ballan wrasse juveniles are rather fussy in their feeding preferences and do not accept well feeds that do not contain very high levels of full crustacean meals exhibiting very low feeding and high mortality rates (Kousoulaki *et al.*, 2014c). However, the growth rates of farmed Ballan wrasse juveniles are generally low, while at the same time feed cost is very high, threatening the economic viability of Ballan wrasse production. More economical Ballan wrasse grow-out feeds need to be developed. Several efforts have been made to address this problem with no success so far. Feeds containing other marine raw materials as attractants, such as shrimp

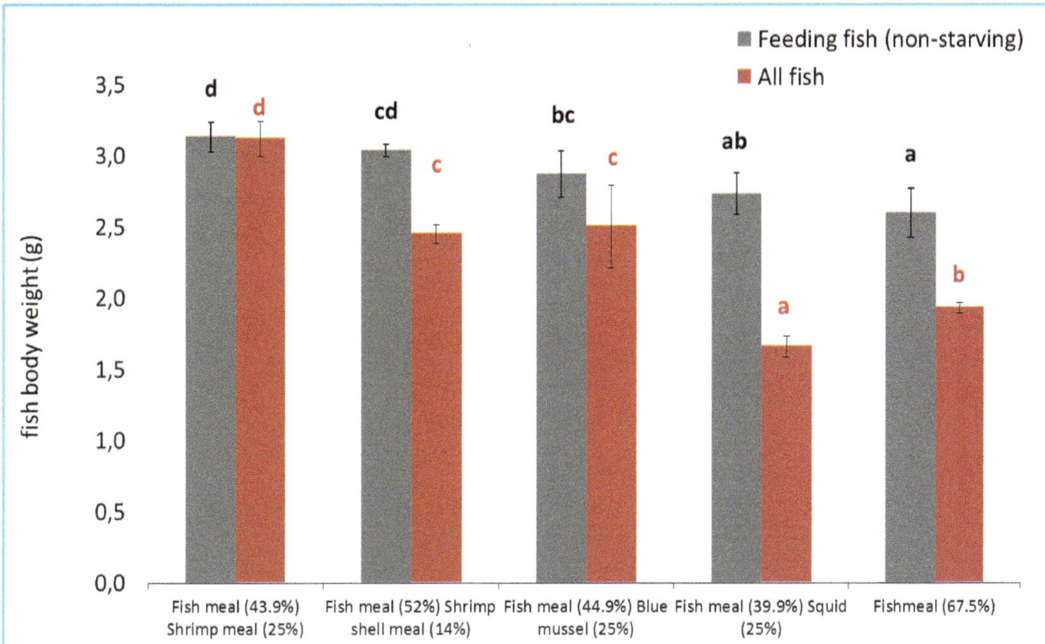

Fig. 10.4 Juvenile growth performance of Ballan wrasse on diets containing different attractants. Different letters are indicative of statistical differences (among groups of same colour).

shell meal, blue mussel meal and squid meal, are clearly inferior in performance, leading to significant mortality rates compared to a shrimp meal-based diet (Figures 10.4, 10.5; Nordgreen *et al.*, 2013). Wild adult Ballan wrasse stomach content in the Azores was composed of mostly Echinodermata (45.1%), which are not yet used in commercial or experimental wrasse feeds. Nevertheless, both decapods (26.7%) and molluscs (11.1%) also represented considerable percentages of their diet (Figueiredo *et al.*, 2005).

Farmed lumpfish juveniles are commonly fed to satiation on crushed salmon pellets or feeds especially developed for lumpfish (Imsland *et al.*, 2015). The focus towards improving production efficiency at this stage is less on the feed and more related to the tank type, high water quality, low light intensity, and the elimination of great size differences through grading and adjustment of feeding in critical phases, which helps reduce aggression between individuals in the tank. Lumpfish need more tank space compared with other commonly farmed fish species and have low disease tolerance, mainly against bacterial diseases. However, intense bacterial growth can be experienced in the production tanks during the weaning period and it was suspected that it may be explained partly by dust pollution originating from the small particles of diets used for first feeding of lumpfish (150 μm) (Ulvan, 2016). The alternative will be to return to offering first feeding lumpfish live feeds during the first life stages of production when water exchange rates are at the lowest. Relatively small fish with bad nutritional status are observed to develop a characteristic sucker deformity.

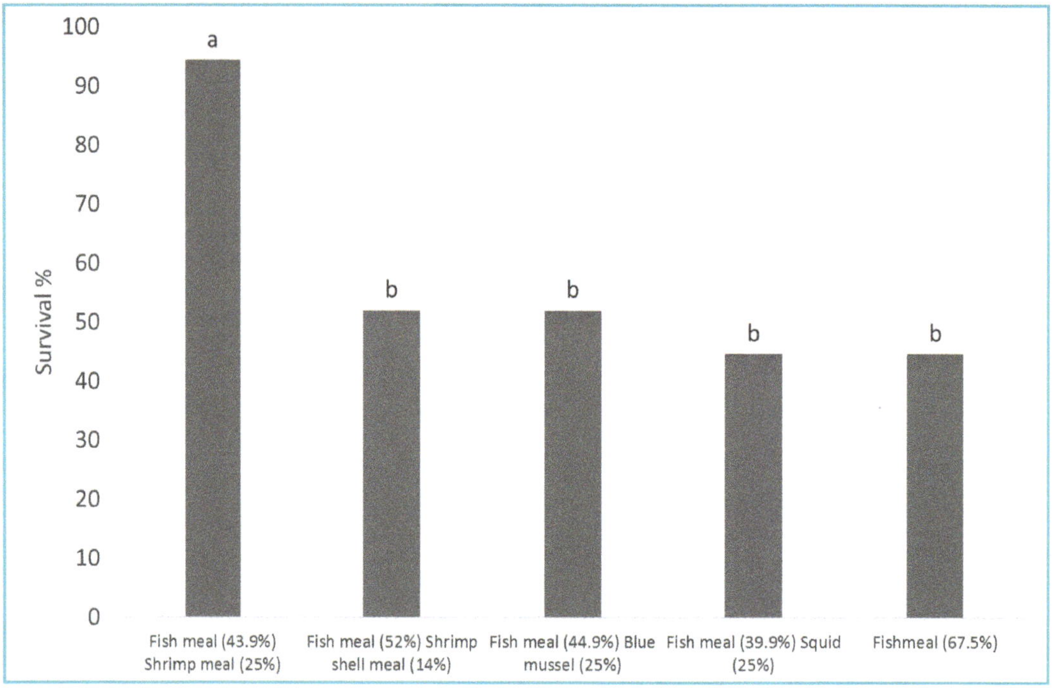

Fig. 10.5 Survival of juvenile Ballan wrasse on diets containing different attractants. Different letters are indicative of statistical differences.

10.7 Supplementary feeding of cleaner fish in salmon net pens

Cleaner fish are deployed preventively within commercial salmon net pens where lice numbers can vary and in long periods may be too low to provide adequate nutrition for the fish. Net biofouling acts as an alternative feed source for the cleaner fish but is also typically kept minimal to maintain biological delousing and the performance of the rearing system (Deady *et al.*, 1995c). Under such conditions, the condition factor and body mass of Ballan wrasse already decrease within six weeks of deployment despite relatively high initial lice numbers (nine lice/salmon) and documented delousing activity (Skiftesvik *et al.*, 2013). It is thus ethically and practically important to provide supplementary feeding of cleaner fish deployed within commercial salmon pens during the grow-out period (18–22 months) to maintain high cleaner fish nutritional status, welfare, delousing efficacy and high survival rates for as long as possible. Supplementary feeding using fresh crushed blue mussels did not reduce the delousing efficiency of farmed Ballan wrasse stocked at five wrasse/100 salmon infected with 12 motile lice per fish, and they consumed more than 75% of the available sea lice (~180 lice per wrasse) during the first 24 hours (Leclercq *et al.*, 2014).

Fresh seafood (e.g. locally collected crushed blue mussels, ready cooked shrimps, crabs, raw mackerel and Pollock, and cooked cod) and manufactured extruded pellets formulated according to natural wrasse preferences are typically offered in salmon pens with stocked cleaner fish, using submerged nets. However, this practice is laborious, entailing difficult logistics and raising biosecurity concerns. On the other hand, specially formulated wrasse feeds have been developed, for instance the commercially available Wrasse grower and Symbio Maintenance by BioMar UK[Ltd] (Grangemouth, UK) and Amber Neptun by Skretting. Those are offered in common bait bags, open pots/traps, or in trays in feeding hides (Leclercq *et al.*, 2015). Manufactured dry pellets (extruded) delivered in fine mesh bags were found to disintegrate within hours in water leading to significant wastage, further compromising validation and quantification of feed intake. A study by Leclercq *et al.* (2015) proposed a water-stable jelly feed prepared by mixing a manufactured grinded ingredient mix with a 20 g l^{-1} agar solution at a 1/1.6 (w/v) ratio and offered in blocks, which cleaner fish actively grazed on within the salmon pens during a week.

The supplementary cage feeding practices currently applied are impractical, costly and inefficient. Alternative practices to increase cleaner fish robustness and survival in the net pens need to be found, such as more efficient feeds and feeding regimes, or alternatively optimal fish condition needs to be ensured before the onset of very low water temperatures, which may allow the fish to survive the winter with less or no feeding. A practical feed for cleaner fish within salmon net pens should combine a manufactured base providing a complete and standardised nutrient profile, biosecurity and ease of procurement with high water stability for distribution as grazing substrate. In addition, it is considered very important to provide feeding stations due to the clear hierarchies developing in the wrasse populations of the same or different species, which can lead to very aggressive behaviour (Austefjord, 2012).

Lumpfish is opportunistic and will feed on salmon pellets if provided with the opportunity, potentially affecting growth and FCR of the salmon population when lumpfish are large and at high relative numbers (4–10%) (Imsland *et al.*, 2014a, b). Nevertheless, lumpfish kept in salmon net enclosures consume sea lice as well as organisms associated with net biofouling such as hydrozoans, mussels (e.g. *Mytilus edulis*), crustaceans, diptera (Imsland *et al.*, 2015). The commercial feed producers offer lumpfish diets (EWOS Lumpus by Cargill, Classic Marine by Biomar and Neptun by Skretting) for the on-growing stages including cage feeding and are recommending feeding regimes of 5% to 2% from 1 g to 20 g fish in the nursery, 2–3% of the cleaner fish biomass at later stages and at higher temperatures, and down to 1.5% at water temperatures below 6°C. The recommended cage feeding frequency for the first 60 days after sea transfer is two to three times daily, later on during the summer period two to three times per day, two to three days a week, and in the winter once or twice a week (Sandstad and Davidsen, 2016). Both wild and farmed (Morken, 2016) lumpfish store excessive energy as liver lipids and tolerate low water temperatures better than wrasse.

The production and commercial use of lumpfish as cleaner fish in salmon cages has increased tremendously during the last few years (2015–2017), with 33 production sites in Norway alone, delivering 17.5 million fish in 2016 (Waatevik, 2017). Production data from salmon farms show that lumpfish, though still considered inferior to wrasse, are effective in eating lice in salmon cages, with increasing performance over an adaptation period of approximately one month (Imsland *et al.*, 2015). Nevertheless, only about a third of the stocked lumpfish eventually perform delousing behaviour (Imsland *et al.*, 2015) and up to a third of the stocked lumpfish may die of starvation in salmon cages within a few weeks after deployment (Breck, 2015). The cause of lower delousing efficiency of some lumpfish or lumpfish batches is still unknown, though the significant cataract incidences in farmed lumpfish (Jonassen, 2016) can provide an explanation.

Powell *et al.* (2017) propose that it might be possible to increase delousing efficiency by selecting individuals that have a greater affinity for consuming sea lice, and perhaps also by conditioning them prior to deployment.

Due to high lumpfish mortality rates, restocking with lumpfish may currently be necessary almost monthly or every other month (Vold, 2017). High lumpfish mortality rates are related to disease or the inevitable chemical or mechanical delousing treatments practised in salmon stocks in net pens when lice numbers increase despite the presence of cleaner fish. As no viruses were isolated from moribund lumpfish in the pens, and most bacteria identified were opportunistic (Gulla, 2017), it is evident that the fish set out in the pens are not robust enough. It is thus imperative that in order to ethically and efficiently use lumpfish as cleaner fish for salmon we urgently need to enhance our knowledge of their nutritional and welfare requirements.

References

Almli, M. (2012) Effects of different live feed on larval growth and development in ballan wrasse (*Labrus bergylta* Ascanius, 1767). MSc-thesis, NTNU, Institutt for Biologi, 59 pp. (www.rensefisk.no/dokumenter/fagrapporter-tema.html)

Austefjord, U. (2012) Hvordan kan man best få leppefisken til å overvintre i laksemerdene? Rensefisk nytt, Nr. 2, 30 November, 2012 (*in Norwegian*).

Ben, S.S., Menif, D. and Ben Hassine, O.K. (2007) Diet of the brown wrasse, *Labrus merula* (Labridae) of the northern coasts of Tunisia. CYBIUM 31, 175–180.

Bogevik, A.S., Kousoulaki, K., Skiftesvik, A.B. and Opstad, I. (2016) Low quality fish meal and ethoxyquin have negative effect on weaning performance of cleaner fish ballan wrasse (*Labrus bergylta*). *Aquaculture Nutrition* 22, 46–50.

Boughamou, N., Derbal, F. and Kara, M.H. (2016) Feeding habits of the peacock wrasse *Symphodus tinca* (Linnaeus, 1758) (Actinopterygii: Perciformes: Labridae) from Eastern Algeria. *Cahiers de Biologiie Marine* 67, 25–33.

Breck, O. (2015) An update from Norway. Proceedings of the Cleanerfish Workshop. Scottish Salmon Producers Organisation, Inverness, 1 December 2015.

Bron, J.E. and Treasurer, J.W. (1992) Sea lice (Caligidae) on wrasse (Labridae) from selected British wild and salmon farm sources. *Journal of the Marine Biological Association* UK 2, 645–650.

Cadwallader, H.F., Turner, J.R. and Oliver, S.P. (2015) Cleaner wrasse forage on ectoparasitic digeneans (phylum Platyhelminthes) that infect pelagic thresher sharks (*Alopias pelagicus*). *Marine Biodiversity* 45, 613–614.

Carr, W.E.S., Netherton, III J.C., Glesson, R.A. and Derby, D.C. (1996) Stimulants of feeding behavior in fish: analysis of tissues of diverse marine organisms *Biological Bulletin* 190 (1996), 149–160.

Clark, W.M., Leclercq, E., Migaud, H., Nairn, J. and Davie, A. (2016) Isolation, identification and characterisation of ballan wrasse *Labrus bergylta* plasma pigment. *Journal of Fish Biology* 89, 2070–2084.

Davenport, J. (1985) Synopsis of biological data on the lumpsucker *Cyclopterus lumpus* (Linnaeus, 1958). FAO Fisheries Synopsis no. 147, 31.

Davenport, J. and Rees, E.I.S. (1993) Observations on neuston and floating weed patches in the Irish Sea. *Estuarine, Coastal and Shelf Science* 36, 395–411.

Deady, S. and Fives, J.M. (1995a) Diet of ballan wrasse, *Labrus bergylta*, and some comparisons with the diet of corkwing wrasse, *Crenilabrus melops*. *Journal of the Marine Biological Association UK* 75, 651–665.

Deady, S. and Fives, J.M. (1995b) The diet of corkwing wrasse, *Crenilabrus melops*, in Galway bay, Ireland, and in Dinard, France. *Journal of the Marine Biological Association UK* 75, 635–649.

Deady, S., Varian, S.J.A. and Fives, J.M. (1995c) The use of cleaner-fish to control sea-lice on two Irish salmon (*Salmo salar*) farms with particular reference to wrasse behaviour in salmon cages. *Aquaculture* 131, 73–90.

Denny, C.M. and Schiel, D.R. (2001) Feeding ecology of the banded wrasse *Notolabrus fucicola* (Labridae) in southern New Zealand: prey items, seasonal differences, and ontogenetic variation. *NZ Journal of Marine and Freshwater* 35, 925–933.

Dipper, F.A., Bridges, C.R. and Menz, A. (1977) Age, growth and feeding in the ballan wrasse *Labrus bergylta* Ascanius 1767. *Journal of Fish Biology* 11, 105–120.

Figueiredo, M., Morato, T., Barreiros, J.P., Afonso, P. and Santos, R.S. (2005) Feeding ecology of the white seabream, *Diplodus sargus*, and the ballan wrasse, *Labrus bergylta*, in the Azores. *Fisheries Research* 75, 107–119.

Fraser, A.J., Gamble, J.C. and Sargent, J.R. (1988) Changes in lipid content, lipid class composition and fatty acid composition of developing eggs and unfed larvae of cod (*Gadus morhua*). *Marine Biology* 99, 307–313.

Froese, R. and Pauly, D. (2012) FishBase. Version 08/12, www.fishbase.org

Gingins, S., Werminghausen, J., Johnstone, R.A., Grutter, A.S. and Bshary, R. (2013) Power and temptation cause shifts between exploitation and cooperation in a cleaner wrasse mutualism. *Proceedings of the Royal Society B-Biological Sciences* 280, Issue 1761, Article Number: 20130553, DOI: 10.1098/rspb.2013.0553.

Grant, B., Picchi, N., Davie, A., Leclercq, E. and Migaud, H. (2016) Removal of the adhesive gum layer surrounding naturally fertilised ballan wrasse (*Labrus bergylta*) eggs. *Aquaculture* 456, 44–49.

Grøntvedt, R.N. (2010) Oppdrett av berggylt – muligheter og utfordringer. Norsk fiskeoppdrett 5, 66–57.

Gulla, S. (2017) Health status of cleaner fish in 2016. FHF Cleaner fish conference 2017, 8–9 February, Trondheim, Norway. (*In Norwegian*) www.fhf.no/media/157605/07_helsestatus_for_rensefisk_sg2.pdf

Hagerman, A. (2014) Production manual for *Acartia tonsa* Dana. The copepod *Acartia tonsa* Dana as live prey for ballan wrasse larvae. In: Helland *et al.* (eds), *Production of ballan wrasse. Science and Practice* p. 36–37. www.rensefisk.no/fileadmin/Opplaeringskontoret/cropped_image/publication-Production-of-ballan-wrasse.pdf

Hamre, K. (2016) FHF Rensefiskkonferansen 2016, 8–9 February, Oslo Gardemoen, Norway. www.fhf.no/nyheter/arrangementer/rensefiskkonferansen-2016

Hamre, K., Nordgreen, N., Grøtan, E. and Breck, O. (2013) A holistic approach to development of diets for Ballan wrasse (*Labrus bergylta*) – a new species in aquaculture. *PeerJ* 1:e99; DOI 10.7717/peerj.99

Hamre, K. and Sæle, Ø. (2011) Oppdrett av leppefisk til lakselusbekjempelse: Hva star på menyen. Norsk Fiskeoppdrett 9/2011: 70–71.

Hansen, T.W., Folkvord, A., Grøtan, E., Sæle, Ø. (2014) The effect of live prey versus a formulated diet on dietary enzymes. In: Helland *et al.* (eds), Production of ballan wrasse. Science and Practice p. 72–73. www.rensefisk.no/fileadmin/Opplaeringskontoret/cropped_image/publication-Production-of-ballan-wrasse.pdf

Helland, S., Bæverfjord, G. and Lein, I. (2012) Produksjon av berggylt (*Ballan wrasse*) – Statusrapport fra Sunndalsøra, In: Leppeprod Dialogmøte. Nofima, Stord.

Henriques, M., Almada, V.C. (1997) Relative importance of cleaning behaviour in *Centrolabrus exoletus* and other wrasse at Arrabida, Portugal. *Journal of the Marine Biological Association UK* 77, 891–898.

Hilldén, N-O. (1978) On the feeding of the goldsinny, *Ctenolabrus rupestris* L. (Pisces, Laridae). *Ophelia* 17, 195–198.

Hilldén, N-O. (1983) Cleaning behaviour of the goldsinny (Pisces, Labridae) in Swedish waters. *Behavioural Processes* 8, 87–90.

Hovland, H. (2016) Utfordringer i oppdrett av Berggylt. FHF Rensefiskkonferansen 2016, 8–9 February, Oslo Gardemoen, Norway. www.fhf.no/media/129943/8_hovland_havlandet_marin_yngel_utfordringer_-_berggylt.pdf

Imsland, A.K., Foss, A., Koedijk, R., Folkvord, A., Stefansson, S.O. and Jonassen, T.M. (2006) Short- and long-term differences in growth, feed conversion efficiency and deformities in juvenile Atlantic cod (*Gadus morhua*) startfed on rotifers or zooplankton. *Aquaculture Research* 37, 1015–1027.

Imsland, A.K., Reynolds, P., Eliassen, G., Hangstad, T.A., Vigdisdatter, A., Nytrø, A.V., Foss A., Vikingstad, E. and Elvegård, T.A. (2014a) Assessment of growth and sea lice infection levels in Atlantic salmon stocked in small-scale cages with lumpfish. *Aquaculture* 433, 137–142.

Imsland, A.K., Reynolds, P., Eliassen, G., Hangstad, T.A., Nytrø, A.V., Foss, A., Vikingstad, E. Elvegård, T.A. (2014b) Notes on behaviour of lumpfish in sea pens with and without Atlantic salmon. *Journal of Ethology* 32, 117–122.

Imsland, A.K., Reynolds, P., Eliassen, G., Hangstad, T.A., Nytrø, A.V., Foss, A., Vikingstad, E. and Elvegård, T.A. (2015) Feeding preferences of lumpfish (*Cyclopterus lumpus* L.) maintained in open net-pens with Atlantic salmon (*Salmo salar* L.). *Aquaculture* 436, 47–51.

Ingólfsson, A. and Kristjánsson, B.K. (2002) Diet of juvenile lumpsucker *Cyclopterus lumpus* (Cyclopteridae) in floating seaweed: Effects of ontogeny and prey availability *Copeia* 2, 472–476.

Izquierdo, M. and Koven, W. (2011) Lipids. In: Holt, G.J. (ed.) Larval Fish Nutrition. Wiley-Blackwell, Oxford, UK. doi: 10.1002/9780470959862.ch2

Jonassen, T. (2016) Utbredelse av katarakt på rognkjeks. 8–9 February, Oslo Gardemoen, Norway. www.fhf.no/media/130051/35_jonassen_akvaplan_niva_kataraktscreening_rognkjeks.pdf

Koumoundouros, G., Divanach, P. and Kentouri, M. (1999) Osteological development of the vertebral column and of the caudal complex in *Dentex dentex*. *Journal of Fish Biology* 54, 424–436.

Kousoulaki, K, Bogevik, A.S., Skiftesvik, A.B., Jensen, P.A. and Opstad, I. (2014a) Marine raw material choice, quality and weaning performance of ballan wrasse (*Labrus bergylta*) larvae. *Aquaculture Nutrition* 21, 644–654.

Kousoulaki, K., Grøtan, E., Kvalheim, K., Høstmark, Ø., Klinge, M. and Bogevik, A.S. (2014b) Protein quality, commodity options and ballan wrasse weaning performance. In: Helland *et al.* eds *Production of ballan wrasse. Science and Practice* p. 96–100. www.rensefisk.no/fileadmin/Opplaeringskontoret/cropped_image/publication-Production-of-ballan-wrasse.pdf

Kousoulaki, K., Grøtan, E., Bjelland, R., Kvalheim, K., Bogevik, A.S. and Skiftesvik, A.B. (2014c) Protein quality and feed technical quality effects on ballan wrasse on-growing performance. In: Helland *et al.* (eds.) *Production of ballan wrasse. Science and Practice*, pp. 101–103. www.rensefisk.no/fileadmin/Opplaeringskontoret/cropped_image/publication-Production-of-ballan-wrasse.pdf

Kousoulaki, K. and Opstad, I. (2012) WP 2: Optimize technical and nutritional quality of weaning diets. 'Optimised production, nutrition and use of the cleanerfish Ballan wrasse (*Labrus bergylta*)' NFR project 200523/S40 Report. 26 pp.

Krogdahl, Å., Sæle, Ø., Lie, K.K., Kousoulaki, K., Hamre, K., Helland, S. and Lein, I. (2014) Characteristics of the digestive functions in ballan wrasse fed dry and moist diets. In: Helland et al. (eds.) *Production of ballan wrasse.* Science and Practice, pp. 74–79. www.rensefisk.no/fileadmin/Opplaeringskontoret/cropped_image/publication-Production-of-ballan-wrasse.pdf

Leclercq, E., Davie, A. and Migaud, H. (2014) Delousing efficiency of farmed ballan wrasse (*Labrus bergylta*) against *Lepeophtheirus salmonis* infecting Atlantic salmon (*Salmo salar*) post-smolts. *Pest Management Science* 70, 1274–1282.

Leclercq, E., Graham, P. and Migaud, H. (2015) Development of a water-stable agar-based diet for the supplementary feeding of cleaner fish ballan wrasse (*Labrus bergylta*) deployed within commercial Atlantic salmon (*Salmo salar*) net-pens. *Animal Feed Science and Technology* 208, 98–106.

Morken, T. (2016) Fôrutvikling til rognkjeks. FHF Rensefiskkonferansen 2016, 8–9 February, Oslo Gardemoen, Norway. www.fhf.no/media/129955/12_morken_biomar_f_rutvikling_til_rognkjeks.pdf

Navarez, P., Furtado, M., Neto, A., Moniz, I., Azevedo, J.M.N. and Soares, M.C. (2015) Temperate facultative cleaner wrasses selectively remove ectoparasites from their client-fish in the Azores. *Marine ecology progress series* 540, 217–226.

Nordgreen, A., Kvalheim, K., Grøtan, E., Hamre, K. and Kousoulaki, K. (2013) WP4: Optimization of rearing temperature and evaluation of the effect of different protein quality in the feed for juvenile fish. 'Optimised production, nutrition and use of the cleanerfish Ballan wrasse (*Labrus bergylta*)' NFR project 200523/S40 Report. 16 pp.

Øie, G. (2014) Rotifers in aquaculture – Production Manual. In: Helland *et al.*, (eds.) *Production of ballan wrasse. Science and Practice* pp. 38. www.rensefisk.no/fileadmin/Opplaeringskontoret/cropped_image/publication-Production-of-ballan-wrasse.pdf

Powell, A., Treasurer, J.W., Pooley, C.L., Keay, A.J., Lloyd, R., Imsland, A.K. and de Leaniz, C.G. (2017) Use of lumpfish for sea-lice control in salmon farming: challenges and opportunities. *Reviews in Aquaculture* (doi: 10.1111/raq.12194).

Quignard, J.P. (1966) Recherches sur les Labridae (Poissons, Teléostéens, Perciformes) des côtes européennes: systématique et biologie. Naturalia Monspeliensia (*Zoologie*) 5, 1–248.

Sæle, Ø., Solbakken, J.S., Watanabe, K., Hamre, K., Power, D. and Pittman, K. (2004) Staging of Atlantic halibut (*Hippoglossus hippoglossus* L.) from first feeding through metamorphosis, including cranial ossification independent of eye migration. *Aquaculture* 239, 445–465.

Sæle, Ø., Hamre, K., Nordgreen, A., Skiftesvik, A-B. and Opstad, I. (2014) Need for phospholipids. Description of the digestion, turnover and metabolism of lipids. Helland *et al.* (eds.), *Production of ballan wrasse. Science and Practice* pp. 36–37. www.rensefisk.no/fileadmin/Opplaeringskontoret/cropped_image/publication-Production-of-ballan-wrasse.pdf

Sæle, Ø. (2017) Hva vet vi om ernæring hos rensefisk? Rensefiskkonferansen og medikamentfri lakseluskontroll 2017 February 8-9, 2017, Trondheim, Norway. www.fhf.no/media/157728/17_erfaringer_med_bruk_av_levende_for_marco_schaer_salmar.pdf

Sandstad, M. and Davidsen, M. (2016) Fôring av rensefisk i sjøen. FHF Rensefiskkonferansen 2016, February 8-9, Oslo Gardemoen, Norway. http://www.fhf.no/prosjektdetaljer/?projectNumber=901161

Sargent, J.R., Tocher, D.R. and Bell, J.G. (2002) The lipids. In: Halver, J.E., Hardy, R.W. (eds.) *Fish Nutrition*, 3rd edition. Academic Press, San Diego, pp. 181–257.

Sayer, M.D.J., Gibson, R.N. and Atkinson, R.J.A. (1995) Growth, diet and condition of goldsinny on the west coast of Scotland. *Journal of Fish Biology* 46, 317–340.

Sayer, M.D.J., Gibson, R.N. and Atkinson, R.J.A. (1996) Growth, diet and condition of corkwing wrasse and rock cook on the west coast of Scotland. *Journal of Fish Biology* 49, 76 94.

Shepherd, S.A. and Clarkson, P.S. (2001) Diet, feeding behaviour, activity and predation of the temperate blue-throated wrasse, *Notolabrus tetricus*. *Marine and Freshwater Research* 52, 311–322.

Schaer, M. (2017) Bruk av KryoPlankton på rognkjeks. Rensefiskkonferansen og medikamentfri lakseluskontroll 2017, 8–9 February, 2017, Trondheim, Norway. www.fhf.no/media/157728/17_erfaringer_med_bruk_av_levende_for_marco_schaer_salmar.pdf

Skiftesvik, A.B. (2009) Oppdrett av berggylte. In: Norsk Fiskeoppdrett, 36–37.

Skiftesvik, A.B. and Bjelland, R. (2003) Farming of ballan wrasse (*Labrus bergylta*). In: Norsk Fiskeoppdrett, 41–44.

Skiftesvik, A.B., Opstad, I., Kvenseth, P.G. and Bjelland, R. (2011) Oppdrett av berggylte. In: Norsk Fiskeoppdrett, 42–43.

Skiftesvik, A.B., Bjelland, R.M., Durif, C.M.F., Johansen, I.S. and Browman, H.I. (2013) Delousing of Atlantic salmon (*Salmo salar*) by cultured vs. wild ballan wrasse (*Labrus bergylta*). *Aquaculture* 402–403, 113–118.

Sørøy, M.O. (2012) Dietary effects of different prey on growth and functional development in ballan wrasse (*Labrus bergylta*) larvae and juveniles. MSc-thesis, NTNU, Institutt for Biologi, 90 pp. (www.rensefisk.no/dokumenter/fagrapporter-tema.html)

Tocher, D.R., Fraser, A., Sargent, J.R. and Gamble, J.C. (1985) Fatty acid composition of phospholipids and neutral lipids during embryonic and early larval development in Atlantic herring (*Clupea harengus* L.). *Lipids* 20, 69–74.

Tokle, N. (2017) Optimal yngelkvalitet med Planktonic's levendefôr – helt enkelt! Rensefiskkonferansen og medikamentfri lakseluskontroll 2017, 8–9 February, Trondheim, Norway. www.fhf.no/media/157728/17_erfaringer_med_bruk_av_levende_for_marco_schaer_salmar.pdf

Ulvan, L.J. (2016) Landbasert produkjson av rognkjeks, det er let det? FHF Rensefiskkonferansen 2016, 8–9 February, Oslo Gardemoen, Norway. www.fhf.no/nyheter/arrangementer/rensefisk konferansen-2016

Van Hasselt, M.J.F.M. (1979) Morphology and movements of jaw apparatus of some Labrinae (Pisces: Perciformes). *Netherlands Journal of Zoology* 29, 52–102.

Vandendriessche, S, Messiaen, M., O'Flynn, S., Vincx, M. and Degraer, S. (2007) Hiding and feeding in floating seaweed: Floating seaweed clumps as possible refuges or feeding grounds for fishes. *Estuarine, Coastal and Shelf Science* 71, 691–703.

Vold, V. (2017) Rensefisk – Motivasjon og forankring. Rensefiskkonferansen og medikamentfri lakseluskontroll 2017, 8–9 February 2017, Trondheim, Norway. www.fhf.no/media/157593/02_rensefisk-konferansen_fhf_-_vidar_vold_ler_y_seafood_group.pdf

Waatevik, E. (2017) Rensefisk 2016, produksjon og utvikling. Rensefiskkonferansen og medikamentfri lakseluskontroll 2017 February 8-9, 2017, Trondheim, Norway. www.fhf.no/media/157680/01_fhf_rensefiskkonferansen_2017_produksjon_av_rognkjeks__enelig_versjon__autosaved_.pdf

Wu, T.H. and Bechtel, P.J. (2012) Screening for low molecular weight compounds in fish meal solubles by hydrophilic interaction liquid chromatography coupled to mass spectrometry. *Food Chemistry* 130, 739–745.

CHAPTER 11

Population genetics of wild wrasse species of interest to the aquaculture sector

Joana Robalo[1] *and Luca Mirimin*[2]

[1]MARE – Marine and Environmental Sciences Centre, ISPA – Instituto Universitário de Ciências Psicológicas, Sociais e da Vida, Lisbon, Portugal

[2]Marine and Freshwater Research Centre, Galway-Mayo Institute of Technology, Galway, Ireland

Variability and heritability of DNA molecules are among the key characteristics that make genetic data extremely useful in studying living organisms. In fact, whether we compare a partial sequence of a gene among individuals or the whole genome between species, genetic data have aided a wide range of research fields and applications, including taxonomy, evolutionary biology, forensics, food production, resource management and conservation.

The taxonomy of many aquatic organisms is still unresolved and population connectivity and gene flow among groups of organisms can vary greatly, especially in the marine environment (Hauser and Carvalho, 2008). The analysis of genetic data can play a significant role in aiding effective management and sustainable use of wild fish stocks (Hauser and Seeb, 2008), particularly when combined with biological and ecological data (Begg and Waldman, 1999). Such a holistic approach is essential to understand distribution of stocks, abundance and adaptive potential, and hence formulate realistic predictions of the potential effect of human activities as well as changing climatic conditions on living natural resources. Understanding inter- and intraspecific genetic variation of wild stocks has become a fundamental component of successful and long-lasting aquaculture practices, especially in early-phase broodstock selection and management. In fact, different from the most commonly domesticated terrestrial animals (e.g. pigs, cattle, sheep and chicken), a great variety of wild-type specimens of aquatic organisms can still be acquired directly from the wild and subjected to domestication and evaluation of traits of interest. Exploiting such genetic variation is a key element of broodstock development in modern selective breeding programmes. However, carefully thought through practices must be put into place to allow a profitable and yet sustainable progress of the aquaculture sector (Teletchea, 2016).

Many molecular techniques and approaches have been developed to provide the tools used to obtain genetic data and hence survey genetic variation among individuals (Vignal *et al.*, 2002). Such tools are often referred to as genetic markers. While only a few decades ago the discovery of highly informative and reliable genetic markers was considered laborious (e.g. Zane *et al.*, 2002), recent technological and computational advances have made the development and collection of genetic data increasingly accessible and cost- and time-effective (Davey *et al.*, 2011). Amidst the ongoing technological revolution, the advent of high-throughput Next Generation Sequencing (NGS) technology in particular has now opened the doors to the prompt collection of powerful data sets capable of unravelling levels of variation at the genomic level (Goodwin *et al.*, 2016). Nonetheless, one aspect to take into account is that the analysis, storage and interpretation of such large and complex data sets can be challenging and requires appropriate computational and analytical power (e.g. Alkan *et al.*, 2011).

In the case of wild cleaner fish, especially wrasses (labrids), genetic studies have been of increasing scientific relevance and have contributed to the characterisation of diversity and population structure among extant stocks. Such studies can be of service to fisheries management and broodstock development, as well as provide baseline information to mitigate the potential effect of fish harvest and translocation practices in the wild, as it has been shown in other more intensively produced species (e.g. Skaala *et al.*, 2006; Glover *et al.*, 2012). The following sections provide an overview of genetic studies on wrasse species at the inter- and intraspecific levels, respectively, with particular relevance to fisheries management and the aquaculture industry sector.

11.1 Genetic studies at the interspecific level

Members of the family Labridae exhibit high phenotypic plasticity as well as strong adaptive potential. Of the approximately 519 species distributed across 71 genera (Nelson *et al.*, 2016) of the family Labridae, 27 species are currently recognised in the north-eastern Atlantic Ocean (Froese and Pauly, 2017) and only some of these have been used in conjunction with salmonid farming as well as direct fisheries (Skiftesvik *et al.*, 2014); these are goldsinny wrasse (*Ctenolabrus rupestris* Linnaeus, 1758), corkwing wrasse (*Symphodus melops* Linnaeus, 1758), cuckoo wrasse (*Labrus mixtus* Linnaeus, 1758), rock cook (*Centrolabrus exoletus* Linnaeus, 1758) and Ballan wrasse (*Labrus bergylta* Ascanius, 1767). Nonetheless, the high variation observed in morphological and coloration patterns poses questions on the integrity and level of reproductive segregation within some of the currently recognised species. For instance, coloration patterns can be diagnostic of species identity, but it may not always be true, as revealed by genetic analyses in four out of five monotypic reef species in the south-western Pacific (Drew *et al.*, 2008). In contrast, a revision of the taxonomic status of members of the genera *Centrolabrus* and *Symphodus* was invoked by genetic data, which were in this case in agreement with behavioural and meristic data (Almada *et al.*, 2002). Two morphotypes (spotty and plain) of Ballan wrasse (*L. bergylta*) are recognised in the eastern North Atlantic, however, differences in life history and nuclear genetic markers have recently argued for the taxonomic revision of this species (Quintela *et al.*, 2016), which is also discussed in more detail in section 11.2. The latter study is yet another example of the increasing usefulness of using multidisciplinary approaches and the incorporation of genetic data to resolve taxonomical uncertainties in the marine environment.

Thanks to their species-richness and high diversity, labrids can be considered as the marine antagonist to cichlids (Suborder Labroidei), a remarkably diverse *taxon* that has been the focus of evolutionary genetic studies on speciation and adaptation (Henning and Meyer, 2014). Analyses of genetic data have been used in conjunction with fossil records and behavioural as well as morphological traits in order to understand speciation events and colonisation of coastal areas of wrasse species around the world. Following early genetic studies using allozymes and restriction fragment length polymorphisms (RFLPs) (e.g. Shulman and Birmingham, 1995), more recent genetic studies have used primarily DNA sequence data (i.e. portions of gene sequences compared among individuals). Depending on the research question and the resolution power required, most studies have focused on nuclear and mitochondrial genes, ranging from comparing variation of several target DNA regions (e.g. Almada *et al.*, 2016) to sequencing whole mitochondrial genomes (Qi *et al.*, 2013). Levels of diversity among extant wrasse species have been explained with a wide range of plausible scenarios. For instance, the biogeography patterns observed in genera such as *Symphodus* and *Labrus* seem to be affected by speciation events driven by opening and closing of the Mediterranean Sea and periodic glacial fluctuations in the past million years *(Hanel et al.*, 2002). On the other hand, Barber and Bellwood (2005) used genetic data to support the hypothesis that biodiversity of the genus *Halichoeres* (a widely distributed coral reef *taxon*) predated the closure of the Isthmus of Panama, suggesting that the role of Pleistocene sea level changes in the origins of marine biodiversity within the Indo– Australian Archipelago may be less important than previously thought. An analysis of skull functionality and biomechanics among 84 species representing all major labrid clades (as revealed by a phylogenetic reconstruction) revealed patterns of evolutionary convergence on a global scale as well as divergence on local geographic scales driven by adaptation to different trophic levels and target prey types (Westneat *et al.*, 2005). Alfaro *et al.* (2009) compared a phylogeny of 131 labrid species to morphological characteristics to come to the conclusion that labrid diversification rates are likely to be driven by extreme dichromatism and behavioural traits associated with sexual selection rather than pharyngeal jaw apparatus modifications. The latter finding is not unprecedented as, in fact, genetic studies suggested that mechanisms such as sexual selection play a significant role in species diversification of sympatric cichlid species in the African Great Lakes (Meyer, 1993).

Thus, as highlighted by these few examples, it is important to take into account that wrasse species (as it is the case for any wild species) are not taxonomic entities that are fixed through time, but rather temporal dynamic units that result from the response to environmental and sometimes anthropogenic pressure. Furthermore, the north-east Atlantic represents the northernmost limit of the distribution range of several species, such as Ballan and goldsinny wrasse. This is an important consideration because, at such distributional extremes, environmental conditions and potential isolation can affect adaptive plasticity and hence lead to important speciation events. It is therefore imperative to conduct further multidisciplinary studies to consolidate wrasse taxonomy for management of wild stocks as well as a more educated selection of wild broodstock fish in farming practices.

11.2 Genetic studies at the intraspecific level

Molecular genetic studies have revolutionised our understanding of seemingly boundless and resilient marine ecosystems (Hauser and Seeb, 2008). Specifically, genetic data analyses have revealed that heterogeneous population structure, extensive adaptive differentiation and small effective (number of breeders) *vs* large census population size are more recurrent characteristics in marine fish species than previously thought (Hauser and Carvalho, 2008). Resolution of population structure at the intra-specific level requires highly variable genetic markers, which are capable of disentangling levels of diversity from the population to the individual level. Interestingly, although widely used in other taxa, highly variable classes of markers such as microsatellite loci and single nucleotide polymorphisms (SNPs) have only been applied to a minor extent in wrasse species. While SNPs consist of many point mutations scattered throughout the genome, microsatellites loci are non-coding regions of DNA that vary in number of tandemly repeated motifs (units) that consist of up to six nucleotides (e.g. '… GAGAGAGAGA …' is an example of a di-nucleotide unit (GA) repeated five times). A single microsatellite locus can be characterised by a range of number of variants (alleles), which can range from one (monomorphic) to tens (polymorphic) alleles. The difference in frequencies of such variants is what is generally analysed in order to investigate population structure under a range of mutational models (Excoffier and Heckel, 2006). Because the location of microsatellite markers is often unknown in the genome of non-model organisms, microsatellite markers need to be discovered *de novo*. Despite the high number of species found in the family Labridae, microsatellite markers have been developed from several wrasse species. Table 11.1 summarises most microsatellite markers developed to date from wrasse species (Family Labridae). As it can be seen in Table 11.1, a total of 237 microsatellites has been isolated to date using mostly conventional marker discovery approaches (e.g. screening of enriched genomic libraries), with the exception of the three recent studies that used NGS approaches (van der Meer *et al.*, 2013; Jansson *et al.*, 2016; Priest *et al.*, 2015). The number of markers discovered per effort ranged between six and 29, although the use of NGS technology is expected to generate hundreds if not thousands of markers discovered per sequencing effort (e.g. see Jansson *et al.*, 2016). The number of markers available to study wrasse species is, however, bound to increase dramatically as ongoing NGS approaches are going to expose whole species' genomes. Additionally, microsatellite markers isolated from one species can be used in closely related taxa with varying degrees of success, indicating some level of transferability (e.g. Arigoni and Largiader, 2000). One important consideration to be made is that while microsatellite marker discovery is becoming faster and more economically affordable, newly discovered markers still need to be tested and validated prior to being reliably utilised, which can be costly and time-consuming. Nonetheless, new approaches are being established for validation and scoring of large numbers of markers at relatively low costs and with high reliability (e.g. Vartia *et al.*, 2014).

Patterns of larval dispersal, life history and post-glacial colonisation are generally invoked as potential mechanisms for shaping population structure of wrasse species. However, the hypothesis that population differentiation is greater in species characterised by shorter larval dispersal is not always supported in wrasse species (Froukh and Kochzius, 2007; Robalo *et al.*, 2012; Almada *et al.*, 2016; Aurelle *et al.*, 2003; Chen *et al.*, 2004; D'Arcy *et al.*, 2013; Drew *et al.*, 2008). Furthermore, some questions on population structure can only be answered by analysing individuals from their whole natural distribution range. A number of genetic studies have been carried out on species of interest to aquaculture in the wild, such as *S. melops* (Knutsen *et al.*, 2013; Robalo *et al.*, 2012) and *L. bergylta* (Almada *et al.*, 2016, 2017; D'Arcy *et al.*, 2013; Quintela *et al.*, 2016). *S. melops* is a rocky shore species that inhabits temperate Atlantic waters from Norway to Morocco and also the Portuguese archipelago of the Azores (e.g. Froese and Pauly, 2017). The adults display a reduced dispersal ability, thus dispersal is achieved mostly or exclusively by the planktonic larvae. Eggs are demersal and males guard them until hatching

Table 11.1 Details of microsatellite markers discovered in wrasse species to date. 'Screening of genomic libraries' refers to methods that use a range of genomic library enrichment steps followed by the incorporation of target fragments into competent bacteria and Sanger sequencing of positive clones; 'Next Generation Sequencing' refers to approaches that use high throughput sequencing platforms. N.a.: not available.

Reference	Target taxa	Number of markers	Methodology of discovery[#]	Accession numbers
(Wooninck *et al.*, 1998)	*Thalassoma bifasciatum*	6	Screening of genomic libraries	AF061789 – AF061794
(Arigoni and Largiader, 2000)	*Symphodus ocellatus*	7	Screening of genomic libraries	AJ278566 – AJ278572
(Guillemaud *et al.*, 2000)*	*Coris julis*	8	Screening of genomic libraries	AF190802 – AF190809
(Williams *et al.*, 2004)	*Thalassoma bifasciatum*	7	Screening of genomic libraries	AY253189 (...) AF061789
(Galarza *et al.*, 2006)	*Symphodus tinca*	8	Screening of genomic libraries	DQ324536 – DQ324543
(Carlon and Lippé, 2007)	*Scarus rubrioviolaceus*	17	Screening of genomic libraries	EF066612 – EF066628
(Knutsen and Sannæs, 2009)	*Symphodus melops*	12	Screening of genomic libraries	GQ922106 – GQ922117
(Poortvliet *et al.*, 2009)	*Semicossyphus pulcher*	11	Screening of genomic libraries	EU781576 – EU781586
(Seyoum *et al.*, 2012)	*Lachnolaimus maximus*	29	Screening of genomic libraries	FJ844445 – FJ844474
(Gardner *et al.*, 2011)	*Choerodon rubescens*	22	Screening of genomic libraries	HM754266 – HM754287
(Hu *et al.*, 2013)	*Cheilinus undulatus*	15	Screening of genomic libraries	HQ995574 (...) HQ995654
(van der Meer *et al.*, 2013)	*Coris bulbifrons*	17	Next Generation Sequencing	JX952210 – JX952226
(Quintela *et al.*, 2014)	*Labrus bergylta*	20	Screening of genomic libraries	KF971335 – KF971354
(Priest *et al.*, 2015)	*Bolbometopon muricatum*	29	Next Generation Sequencing	KJ489028 – KJ489056
(Jansson *et al.*, 2016)	*Ctenolabrus rupestris*	17	Next Generation Sequencing	N.a.
(Cox *et al.*, n.d.)	*Sparisoma viride*	10	N.a.	KR053431 – KR053440

*: two additional markers are available on Genbank (AF190810–AF190811), but not included in the original study.

(Potts, 1984). *L. bergylta* is a protogynous hermaphrodite species common in rocky shores and kelp beds in the north-eastern Atlantic from Norway to Morocco, including the Macaronesian archipelagos (Canaries, Madeira and Azores) (e.g. Froese and Pauly, 2017). Muncaster *et al.* (2010) in their study on the reproductive cycle of this species found evidence for it to be a group-synchronous multiple spawner with spawning in spring, and Ottesen *et al.* (2012) described its larval development. It is a commercially important species to local fisheries in south-western

Europe (Treasurer, 1994; Villegas Ríos *et al.*, 2013a). It has been increasingly targeted as a cleaner fish due its adult size, being the largest wrasse of this region (e.g. Bañon *et al.*, 2010), achieving 60 cm TL and therefore able to be kept in cages with salmon of 3–5 kg (Muncaster *et al.*, 2010).

The studies on corkwing wrasse revealed a marked geographic difference between Atlantic (from south Portugal to England and Ireland) and Scandinavian locations (Knutsen *et al.*, 2013; Robalo *et al.*, 2012). Genetic analysis using one mitochondrial (control region) and one nuclear (first intron of the S7 ribosomal protein) gene revealed genetic variability decreasing drastically to the north. In fact, all fish from Scandinavian locations displayed exactly the same haplotype for the mitochondrial control region. Nevertheless, this haplotype was not present anywhere else, suggesting the existence of glacial refugia somewhere in the North Sea. The high variability present in the southern populations (Iberian Peninsula) and the somehow intermediate position of Roskoff, Belfast and Plymouth samples suggests a retreat of the species to the south during the Last Glacial Maximum. Even if a species contraction happened, it retained a lot more variation than the hypothesised northern refugia (Robalo *et al.*, 2012). Knutsen *et al.* (2013) evaluated the connectivity among the same sample locations with microsatellites. Their analysis revealed small-scale differences between sample localities both in Norway and in the UK. As mentioned above, *L. bergylta* is being increasingly used due to its large size and therefore ability to predate on larger parasites. This species presents two different morphotypes (with no relation whatsoever with sexual dimorphism): plain and spotted (Villegas-Ríos *et al.*, 2013a, b). Several hypotheses can explain the existence of two morphotypes: evolutionary divergence by assortative mating; allopatric divergence followed by recolonisation by both forms; epigenetic regulation, and environmental effects (Almada *et al.*, 2016). The two morphotypes occur in sympatry along the entire distribution area and no microhabitat segregation or behavioural differences have been reported (Villegas-Ríos, 2013). Nevertheless, differences in several parameters of life history were reported (Villegas-Ríos *et al.*, 2013a, 2014). Genetic data have shown contrasting results in terms of differentiation between the two morphotypes (Almada *et al.*, 2016; Quintela *et al.*, 2016). A study using mitochondrial (control region, COI gene) and nuclear (18S, S7) genes and including fish from almost the entire distribution range found no differences between the plain and spotted morphotypes recognised in the literature (Almada *et al.*, 2016). Sequences from both morphotypes shared haplotypes for all markers (15% CR, 100% COI; 100% 18S and 27% S7) along the entire sampling area. On the contrary, a study using both the barcoding approach (mitochondrial COI gene) and 20 microsatellites in fish only from Galicia, revealed that some differences exist but they are not enough to unambiguously discriminate the two morphs (Quintela *et al.*, 2016). These results suggest that the segregation between the two morphotypes is not strong enough to justify the presence of two distinct species. Nonetheless, there is enough evidence that some level of segregation is present, which at present may reflect plasticity at the intraspecific level and potential adaptation to distinct ecological niches, but it also indicates potential speciation processes leading to separate species if environmental or other conditions will favour one morphotype over the other.

The phylogeographic structure of *L. bergylta* was investigated by D'Arcy *et al.* (2013) for some locations in Ireland, Scotland and Norway (using mitochondrial control region) and more recently by Almada *et al.* (2017) including fishes from 14 locations along almost the entire distribution range of the species and adding the nuclear S7 gene (first intron). D'Arcy *et al.* (2013) found high haplotypic diversity in Irish and Scottish sampled locations, as opposed to low diversity encountered in the southern Norwegian locations. It is interesting to note that a similar pattern was described by Robalo *et al.* (2012) for *S. melops*. For *L. bergylta* the haplotype network obtained by D'Arcy *et al.* (2013) revealed the presence of two divergent clades, with each clade including fish from all the locations sampled, which lead the authors to suggest that both clades should have been already present when the sampled areas were colonised. Clade 1 consisted of one more frequent haplotype and several unique or low-frequency haplotypes, being the more frequent one abundant in southern Norway and not in Ireland or Scotland. Clade 2 presented several medium frequency haplotypes frequent in Irish and Scottish sampling locations but rare in Norway. The same authors found that Clade 1 probably had a more recent origin in time than Clade 2, Norway being colonised recently and by fewer individuals than the more southern locations. Nevertheless, as a whole, Ballan wrasse in these northern areas seems to have undergone population expansion before the Last Glacial Maximum.

Almada *et al.* (2017) found the same two clades across the entire distribution area, but also revealed a new very distinct group, including the specimens from the Azores. This marked genetic structure could reflect both recolonisation from three distinct glacial *refugia* (North Sea, western and southern Iberian coasts and Azores) and current and past oceanographic circulation patterns. These authors emphasized the isolation and singularity of the Azorean population, which is likely to be maintained and/or accentuated by current circulation patterns in the North Atlantic, possibly leading to an incipient speciation event in this archipelago. The same authors consider future monitoring of *L. bergylta* crucial to evaluate how this species is coping with current environmental changes. These studies did not control for morphotype sampling, therefore it was not possible to infer eventual genetic differences in population structure.

11.3 Conservation and management of wild wrasse populations

As is the case for all exploitable resources, fisheries require appropriate and effective management measures in order to guarantee the long-lasting availability of the target stocks. Such measures depend on the availability of accurate information on environmental characteristics and biological aspects of the target species, which determine the optimal amount, timing and location of harvest. In this regard, genetic data and state-of-the-art technologies have been increasingly useful in providing resources to assist fishing industries (Ovenden *et al.*, 2015). Understanding levels of diversity prior to exploitation and the identification of natural boundaries to gene flow between populations can, in fact, inform which stocks are being targeted and to what extent. Fish aggregations must be seen as dynamic entities through time as they change in composition (often seasonally) during spawning, migrating and breeding. Similarly, the establishment of profitable and sustainable aquaculture practices depends on accurate information on genetic structure and diversity of the wild source populations upon which base broodstock populations and strains are selected and established (Liu and Cordes, 2004). Adaptation to natural environmental and ecological conditions are at the base of the levels of diversity that form the starting blank canvas from which traits of interest are selected in captive breeding programs. Furthermore, wild and (escaped) farmed organisms can interact in many ways by exchange of parasites, diseases and genes (interbreeding) or by competing for resources. This is particularly relevant to species that are subject to translocation outside of their native habitat or that have undergone several generations of breeding in captivity (e.g. Glover *et al.*, 2012; Glover *et al.*, 2013). Thus, prior knowledge of levels of diversity and natural barriers to gene flow can be critical in mitigating potential detrimental effects of fishing, translocating and aquaculture practices. One of the most relevant examples of aquaculture practices that have boomed in the past decades is that of the Atlantic salmon farming industry, with yearly production of farmed fish exceeding in the order of hundreds of times the amount of wild-caught fish in the North Atlantic. At the forefront of improving such industry is the need to reduce losses due to diseases and to increase growth by suitable diets, as well as reducing escapes of farmed fish to limit potential negative effects to wild populations. In fact, numbers of farmed salmon that inadvertently 'leak' into the wild every year are rather large compared to their wild counterparts and evidences of direct interactions, such as interbreeding between wild and farmed fish, have started to emerge (e.g. Glover *et al.*, 2012, 2013; McGinnity *et al.*, 1997; Roberge *et al.*, 2008).

As can be seen in sections 11.1 and 11.2, wrasses are characterised by high phenotypic plasticity as well as adaptive potential, which have enabled them to occupy a variety of ecological niches over a wide geographical range. Thus, the effect of harvesting and translocating large amounts of individuals right at the northernmost edge of their distribution range (e.g. southern Norway) may result in significant impacts on wild local populations. In fact, wrasses have been used to control parasites in salmonid farms since the late 1980s and such practice involves not only the harvesting of large numbers of wild fish, but also the relocation to farms of interest. As shown in this chapter, population structuring can occur at different geographical scales, hence assumptions that fish from the same hydrographic region may belong to the same stock can only be confirmed following a genetic investigation. A study performed on the Hardangerfjord in Norway reported between 86,000 and 251,000 wild wrasse individuals employed in salmon and rainbow trout farms from 2002 to 2006, increasing to 1.1 million in 2009 and 2010 (Skiftesvik *et al.*, 2014). According to the same authors, around 1.54 million labrids were reported landed in this fjord in the years between 2000 and 2010; the estimation for Norway is that 10 million wild fish were caught. However, measuring the extent of wild-escaped interactions is a difficult task, thus a preventative approach is the focus of current research, including improved infrastructures to prevent escapes and sterile strains to avoid unintentional interbreeding. Therefore, advancing knowledge and understanding local wrasse stocks in terms of genetic connectivity, variation and effective population size (i.e. number of effective breeders) is crucial in order to identify tipping points at which the number of farmed escapees may become significant. In particular, when stocks are not large enough to satisfy local demand additional specimens must be sourced from more distant areas, as is the case between northern and southern Norway (Skiftesvik *et al.*, 2014). When taking into account the life history of Ballan wrasse and reproductive mode (i.e. only a portion of large individuals develop male testes) caution should be exercised to avoid potential fisheries-induced selective pressure, a phenomenon that seems to have been overlooked in fisheries worldwide (Heino *et al.*, 2015). Furthermore, although the genetic distinctiveness of the two Ballan wrasse morphotypes (spotted *vs* plain) remains to be further investigated (Almada *et al.*, 2016; Villegas-Ríos *et al.*, 2013b), spotted individuals tend to be larger and they are specially targeted by local fisheries as they are sold at a higher price and even commercialised under a different name (Villegas-Ríos *et al.*, 2013b).

In order to cope with the increasing demand for cleaner fish in salmonid farming, a number of wrasse breeding programmes have been initiated. Culture of some cleaner fish species is already a reality (Ottesen *et al.,* 2012; Skiftesvik *et al.*, 2014), though such practices provide the usual challenges of rearing new species in captivity, as is the case of Ballan wrasse for which the control of the annual reproductive cycle at high latitudes remains a challenge (Muncaster *et al.*, 2010). Despite the emergence of such breeding initiatives, wild wrasse specimens are still being harvested in more abundant populations, normally to the south of their distribution range (Skiftesvik *et al.*, 2014). Overall, the following recommendations can be formulated for wrasse species that are of interest to aquaculture practices: 1) the taxonomic status of closely related species should be resolved by a multidisciplinary approach involving molecular markers (both nuclear and mitochondrial) as well as morphological, behavioural and ecological parameters; 2) levels of genetic diversity and population genetic structure should be elucidated in areas affected by related fisheries and/or farming industries; 3) potential impacts of escapees should be evaluated, taking into account levels of diversity and the extent of production against local effective populations sizes; 4) appropriate measures of limiting potential impacts of escapees should be adopted, including the preferential use of local stocks (as opposed to translocated individuals), the estimation of population size and sustainable quotas for harvest of wild stocks, the establishment of sterile strains (to avoid interbreeding between farmed and wild fish), and the adoption of rigorous quarantine procedures (to avoid the spread of diseases and parasites), among others.

References

Alfaro, M.E, Brock, C.D., Banbury, B.L. and Wainwright, P.C. (2009) Does evolutionary innovation in pharyngeal jaws lead to rapid lineage diversification in labrid fishes? *BMC Evolutionary Biology* 9, 255. Available at: www.ncbi.nlm.nih.gov/pmc/articles/PMC2779191

Alkan, C., Sajjadian, S. and Eichler, E.E. (2011) Limitations of next-generation genome sequence assembly. *Nature Methods*, 8(1), 61–65. Available at: http://dx.doi.org/10.1038/nmeth.1527.

Almada, F. *et al.* (2017). Historical gene flow constraints in a northeastern Atlantic fish: phylogeography of the ballan wrasse *Labrus bergylta* across its distribution range. *R. Soc. open sci.* 4: 160773. Available at: http://dx.doi.org/10.1098/rsos.160773.

Almada, F. *et al.* (2016) On the absence of genetic differentiation between morphotypes of the ballan wrasse *Labrus bergylta* (Labridae). *Marine Biology* 163(4), 1–6. Available at: http://dx.doi.org/10.1007/s00227-016-2860-8.

Almada, V.C. *et al.* (2002) On the phylogenetic affinities of *Centrolabrus trutta* and *Centrolabrus caeruleus* (Perciforms: Labridae). Molecular and meristic evidence. *Arquipelago Life and Marine Sciences*, 19, 85–92.

Arigoni, S. and Largiader, C.R. (2000) Isolation and characterization of microsatellite loci from the ocellated wrasse *Symphodus ocellatus* (Perciformes: Labridae) and their applicability to related taxa. *Molecular Ecology* 9(12), pp. 2166–2169. Available at: http://doi.wiley.com/10.1046/j.1365-294X.2000.10537.x [Accessed 19 April 2016].

Aurelle, D. *et al.* (2003) Genetic study of *Coris julis* (Osteichtyes, Perciformes, Labridae) evolutionary history and dispersal abilities. *Comptes Rendus Biologies* 326(8), 771–785. Available at: www.sciencedirect.com/science/article/pii/S1631069103001720 [Accessed 19 April 2016].

Bañon, R., Villegas-Ríos, D., Serrano, A., Mucientes, G. and Arronte, J.C. (2010) Marine fishes from Galicia (NW Spain): an updated checklist. *Zootaxa* 2667, 1–27

Barber, P.H. and Bellwood, D.R. (2005) Biodiversity hotspots: evolutionary origins of biodiversity in wrasses (Halichoeres: Labridae) in the Indo-Pacific and new world tropics. *Molecular Phylogenetics and Evolution* 35(1), 235–253. Available at: www.sciencedirect.com/science/article/pii/S1055790304002969.

Begg, G.A. and Waldman, J.R. (1999) An holistic approach to fish stock identification. *Fisheries Research* 43(1–3), 35–44. Available at: www.sciencedirect.com/science/article/pii/S016578369900065X.

Carlon, D.B., and Lippé, C. (2007) Isolation and characterization of 17 new microsatellite markers for the ember parrotfish *Scarus rubroviolaceus*, and cross-amplification in four other parrotfish species. *Molecular Ecology Notes* 7(4), 613–616. Blackwell Publishing Ltd. doi:10.1111/j.1471-8286.2006.01650.x.

Chen, C.A. *et al.* (2004) Population Structure and Genetic Variability of Six Bar Wrasse (*Thallasoma hardwicki*) in Northern South China Sea Revealed by Mitochondrial Control Region Sequences. *Marine Biotechnology* 6(4), pp. 312–326. Available at: http://dx.doi.org/10.1007/s10126-003-0028-2.

D'Arcy, J., Mirimin, L. and Fitzgerald, R. (2013) Phylogeographic structure of a protogynous hermaphrodite species, the ballan wrasse *Labrus bergylta*, in Ireland, Scotland, and Norway, using mitochondrial DNA sequence data. *ICES Journal of Marine Science* 70(3), 685–693.

Davey, J.W. *et al.* (2011) Genome-wide genetic marker discovery and genotyping using next-generation sequencing. *Nature Reviews Genetics*, 12(7), 499–510. Available at: http://dx.doi.org/10.1038/nrg3012.

Drew, J. *et al.* (2008) Endemism and Regional Color and Genetic Differences in Five Putatively Cosmopolitan Reef Fishes. *Conservation Biology*, 22(4), 965–975. Available at: http://dx.doi.org/10.1111/j.1523-1739.2008.01011.x.

Excoffier, L. and Heckel, G. (2006) Computer programs for population genetics data analysis: a survival guide. *Nature Reviews Genetics* 7(10), 745–758. Available at: http://dx.doi.org/10.1038/nrg1904.

Froese, R. and Pauly, D. (2017) FishBase. www.fishbase.org

Froukh, T. and Kochzius, M. (2007) Genetic population structure of the endemic fourline wrasse (*Larabicus quadrilineatus*) suggests limited larval dispersal distances in the Red Sea. *Molecular Ecology* 16(7), 1359–1367. Available at: http://doi.wiley.com/10.1111/j.1365-294X.2007.03236.x [Accessed 19 April 2016].

Galarza, J.A., Carreras-Carbonell, J., Macpherson, E., Turner, G.F., and Rico, C. (2006) Isolation and characterization of polymorphic microsatellite markers for peacock wrasse (*Symphodus tinca*). *Molecular Ecology Notes* 6(3), 747–749. Available at: http://dx.doi.org/10.1111/j.1471-8286.2006.01328.x.

Gardner, M.J., Chaplin, J.A., and Shaw, K.M. (2011) Development and characterisation of novel microsatellite loci for the baldchin groper (*Choerodon rubescens*) and cross-amplification in seven other labrid species. *Conservation Genetic Resources* 3(3), 461–466. Available at: http://dx.doi.org/10.1007/s12686-010-9379-y.

Glover, K.A. *et al.* (2013) Atlantic salmon populations invaded by farmed escapees: quantifying genetic introgression with a Bayesian approach and SNPs. *BMC Genetics* 14(1), 1–19. Available at: http://dx.doi.org/10.1186/1471-2156-14-74.

Glover, K.A. *et al.* (2012) Three decades of farmed escapees in the wild: a spatio-temporal analysis of population genetic structure throughout Norway. *PLoS One*, 7. Available at: http://dx.doi.org/10.1371/journal.pone.0043129.

Goodwin, S., McPherson, J.D. and McCombie, W.R. (2016) Coming of age: ten years of next-generation sequencing technologies. *Nature Reviews Genetics*, 17(6), 333–351. Available at: http://dx.doi.org/10.1038/nrg.2016.49.

Guillemaud, T., Streiff, R., Santos, R.S., Afonso, P., Morato, T. and Cancela, M.L. (2000) Microsatellite characterization in the rainbow wrasse *Coris julis* (Pisces: Labridae). *Molecular Ecology* 9(5), 631–632. Available at: http://dx.doi.org/10.1046/j.1365-294x.2000.00882-1.x.

Hanel, R., Westneat, W.M. and Sturmbauer, C. (2002) Phylogenetic Relationships, Evolution of Broodcare Behavior, and Geographic Speciation in the Wrasse Tribe Labrini. *Journal of Molecular Evolution* 55(6), 776–789. Available at: http://dx.doi.org/10.1007/s00239-002-2373-6.

Hauser, L. and Carvalho, G.R. (2008) Paradigm shifts in marine fisheries genetics: ugly hypotheses slain by beautiful facts. *Fish and Fisheries* 9(4), 333–362. Available at: http://dx.doi.org/10.1111/j.1467-2979.2008.00299.x.

Hauser, L. and Seeb, J.E. (2008) Advances in molecular technology and their impact on fisheries genetics. *Fish and Fisheries* 9(4), 473–486. Available at: http://dx.doi.org/10.1111/j.1467-2979.2008.00306.x.

Heino, M., Díaz Pauli, B. and Dieckmann, U. (2015) Fisheries-Induced Evolution. *Annual Review of Ecology, Evolution, and Systematics* 46(1), 461–480. Available at: http://dx.doi.org/10.1146/annurev-ecolsys-112414-054339.

Henning, F. and Meyer, A. (2014) The Evolutionary Genomics of Cichlid Fishes: Explosive Speciation and Adaptation in the Postgenomic Era. *Annual Review of Genomics and Human Genetics* 15(1), 417–441. Available at: http://dx.doi.org/10.1146/annurev-genom-090413-025412.

Hu, J., Zhu, X.P., Luo, J., Yin, S.W., Peng, Y.H., Hu, Y.L., and Zhu, F. (2013) Development and characterization of microsatellite loci in a threatened marine fish, *Cheilinus undulatus* (humphead wrasse). *Genetic Molecular Resources* 12(3), 2633–2636.

Jansson, E. *et al.* (2016) Development of SNP and microsatellite markers for goldsinny wrasse (*Ctenolabrus rupestris*) from ddRAD sequencing data. *Conservation Genetics Resources* 8(3), 201–206. Available at: http://dx.doi.org/10.1007/s12686-016-0532-0.

Knutsen H, Jorde PE, Gonzalez EB, Robalo J, Albretsen J, Almada V (2013) Climate change and genetic structure of leading edge and rear end populations in a northwards shifting marine fish species, the corkwing wrasse (*Symphodus melops*). *PLoS ONE*, 8(6): e67492. https://dx.doi.org/10.1371/journal.pone.0067492

Knutsen, H., and Sannæs, H. (2009) Development of twelve microsatellite loci in the corkwing wrasse (*Symphodus melops*). *Conservation Genetic Resources* 1(1), 433. doi:10.1007/s12686-009-9100-1.

Liu, Z.J. and Cordes, J.F. (2004) DNA marker technologies and their applications in aquaculture genetics. *Aquaculture* 238(1), 1–37.

McGinnity, P. *et al.* (1997) Genetic impact of escaped farmed Atlantic salmon (*Salmo salar L.*) on native populations: use of DNA profiling to assess freshwater performance of wild, farmed, and hybrid progeny in a natural river environment. *ICES Journal of Marine Science*, 54, 998–1008.

van der Meer, M.H. et al. (2013) Identification of seventeen microsatellite loci for conservation genetic studies of the endemic wrasse *Coris bulbifrons*. *Conservation Genetics Resources* 5(2), 363–366. Available at: http://dx.doi.org/10.1007/s12686-012-9804-5.

Meyer, A. (1993) Phylogenetic relationships and evolutionary processes in east African cichlid fishes. *Trends in Ecology and Evolution* 8, 279–284.

Muncaster, S., Andersson, E., Kjesbu, O. S., Taranger, G. L., Skiftesvik, A. B. and Norberg, B. (2010) The reproductive cycle of female Ballan wrasse *Labrus bergylta* in high latitude, temperate waters. *Journal of Fish Biology* 77, 494–511.

Nelson, J.S., Grande, T.S and Wilson, M.V.H. (2016) *Fishes of the World*. 5th edition. Wiley, Upper Saddle River, NJ.

Ottesen, O.H., Dunaevskaya, E., and D'Arcy, J. (2012) Development of *Labrus bergylta* (Ascanius 1767) larvae from hatching to metamorphosis. *Journal of Aquaculture Research and Development* 3, 1–4.

Ovenden, J.R. *et al.* (2015) Ocean's eleven: a critical evaluation of the role of population, evolutionary and molecular genetics in the management of wild fisheries. *Fish and Fisheries* 16(1), 125–159. Available at: http://dx.doi.org/10.1111/faf.12052.

Poortvliet, M., Olsen, J.L., Selkoe, K.A., Coyer, J.A., and Bernardi, G. (2009) Isolation and characterization of 11 microsatellite primers for a temperate reef fish, the California sheephead (*Semicossyphus pulcher*). *Molecular Ecology Resources*, 9(1), 429–430. Available at: http://dx.doi.org/10.1111/j.1755-0998.2008.02467.x.

Potts, GW. (1984) Parental behaviour in temperate marine teleosts with special reference to the development of nest structures. In: Potts, G.W. and Wootton R.J., (eds) *Fish Reproduction: Strategies and Tactics*. London: Academic Press, pp. 223-244.

Priest, M.A. *et al.* (2015) Isolation and characterization of 29 microsatellite markers for the bumphead parrotfish, *Bolbometopon muricatum*, and cross amplification in 12 related species. *Marine Biodiversity* 45(4), 861–866. Available at: http://dx.doi.org/10.1007/s12526-014-0278-4.

Qi, X.Z. *et al.* (2013) Complete mitochondrial genome sequence of the humphead wrasse, *Cheilinus undulatus*. *Genetics and Molecular Research* 12(2), 1095–1105.

Quintela, M., Danielsen, E.A., Svåsand, T., Knutsen, H., Skiftesvik, A.B., and Glover, K.A. (2014) Isolation and characterization of twenty microsatellite loci for the ballan wrasse, *Labrus bergylta*. *Conservation Genetic Resources*, 6(2), 425–428. Available at: http://dx.doi.org/10.1007/s12686-013-0114-3.

Quintela, M. *et al.* (2016) Is the ballan wrasse (*Labrus bergylta*) two species? Genetic analysis reveals within-species divergence associated with plain and spotted morphotype frequencies. *Integrative Zoology* 11(2), 162–172. Available at: http://dx.doi.org/10.1111/1749-4877.12186.

Robalo, J.I. *et al.* (2012) Northern refugia and recent expansion in the North Sea: the case of the wrasse *Symphodus melops* (Linnaeus, 1758). *Ecology and Evolution* 2(1), 153–164. Available at: http://dx.doi.org/10.1002/ece3.77.

Roberge, C. *et al.* (2008). Genetic consequences of interbreeding between farmed and wild Atlantic salmon: insights from the transcriptome. *Molecular ecology* 17(1), 314–24. Available at: http://www.ncbi.nlm.nih.gov/pubmed/18173503 [Accessed 27 July 2016].

Seyoum, S., Tringali, M.D., Puchulutegui, C., Davis, M.C., Collins, A.B., and McBride, R.S. (2012) Isolation and characterization of 30 microsatellite loci for genetic analyses of the hogfish *Lachnolaimus maximus* (Permanent Genetic Resources added to Molecular Ecology Resources Database 1 February 2012 – 31 March 2012). *Molecular Ecology Resources* 12, 779–781.

Shulman, M.J. and Birmingham, E. (1995) Early life histories, ocean currents, and the population genetics of Caribbean reef fishes. *Evolution* 49, 897–910.

Skaala, O., Wennevik, V. and Glover, K.A. (2006) Evidence of temporal genetic change in wild Atlantic salmon, *Salmo salar* L., populations affected by farm escapees. *ICES Journal of Marine Science*, 63. Available at: http://dx.doi.org/10.1016/j.icesjms.2006.04.005.

Skiftesvik, A.B., Blom, G., Agnalt, A-L.., Durif, C.M.F., Browman, H.I., Bjelland, R.M., Harkestad, L.S., Farestveit, E., Paulsen, O.I., Fauske, M., Havelin, T., Johnsen K. and Mortensen, S. (2014) Wrasse (Labridae) as cleaner fish in salmonid aquaculture – The Hardangerfjord as a case study. *Marine Biology Research* 10(3), 289-300. Available at: http://dx.doi.org/10.1080/17451000.2013.810760

Teletchea, F. (2016) Is Fish Domestication Going Too Fast? *Natural Resource*, 7, 399–404.

Treasurer, J. W. (1994) The distribution, age and growth of wrasse (Labridae) in inshore waters of west Scotland. *Journal of Fish Biology* 44, 905–918.

Vartia, S. *et al.* (2014) Multiplexing with three-primer PCR for rapid and economical microsatellite validation. *Hereditas* 151(2–3), 43–54.

Vignal, A., Milan, D., SanCristobal, M. and Eggen, A. (2002) A review on SNP and other types of molecular markers and their use in animal genetics. *Genetics Selection Evolution* 34(3), 275. Available at: http://dx.doi.org/10.1186/1297-9686-34-3-275.

Villegas-Ríos, D. (2014) Life-history and behaviour of *Labrus bergylta* in Galicia. *Cybium, International Journal of Ichthyology* 38, no. 1, 2014, p. 60. Academic OneFile, Accessed 17 Nov. 2017

Villegas Ríos, D., Alonso-Fernández, A., Fabeiro, M., Bañón, R, Saborido-Rey F (2013a) Demographic variation between colour patterns in a temperate protogynous hermaphrodite, the ballan wrasse *Labrus bergylta*. *PLoS ONE*, 8(8), e71591. https://doi.org/10.1371/journal.pone.0071591.

Villegas-Ríos, D., Alonso-Fernández, A., Domínguez-Petit, A. and Fran Saborido-Rey, A. (2013b) Intraspecific variability in reproductive patterns in the temperate hermaphrodite fish, *Labrus bergylta*. *Marine and Freshwater Research*, 64(12), 1156-1168. Available at: https://doi.org/10.1071/MF12362.

Westneat, M.W. *et al.* (2005) Local phylogenetic divergence and global evolutionary convergence of skull function in reef fishes of the family Labridae. *Proceedings of the Royal Society B: Biological Sciences* 272(1567), 993–1000. Available at: www.ncbi.nlm.nih.gov/pmc/articles/PMC1599880.

Williams, D.A., Purcell, J., Cowen, R.K., and Hughes, C.R. (2004) Characterization of microsatellite multiplexes for population genetic studies of bluehead wrasse (*Thalassoma bifasciatum*, Pisces: Labridae). *Molecular Ecology Notes*, 4(3), 525–527. Available at: http://dx.doi.org/10.1111/j.1471-8286.2004.00713.x.

Wooninck, L., Strassmann, J.E., Fleischer, R.C., and Warner, R.R. (1998) Characterization of microsatellite loci in a pelagic spawner: the bluehead wrasse, *Thalassoma bifasciatum* L. *Molecular Ecology*, 7(11), 1613–1621. Available at: http://dx.doi.org/10.1046/j.1365-294X.1998.00474.x.

Zane, L., Bargelloni, L. and Patarnello, T. (2002) Strategies for microsatellite isolation: a review. *Molecular Ecology* 11(1), 1–16. Available at: http://dx.doi.org/10.1046/j.0962-1083.2001.01418.x.

CHAPTER 12

Lumpfish biology and population genetics

Ólöf Dóra Bartels Jónsdóttir,[1] Albert K. D. Imsland,[1,2] and James Kennedy[3]

[1]Akvaplan-niva Iceland Office, Kópavogur, Iceland
[2]Department of Biology, University of Bergen, High Technology Centre, Bergen, Norway
[3]Marine Research Institute, Reykjavik, Iceland

12.1 General biology of lumpfish

Lumpfish, *Cyclopterus lumpus* (Linnaeus, 1758) is a non-shoaling, cold water marine fish in the family Cyclopteridae (lumpsuckers) and it is the only member of the genus Cyclopterus, living mainly in high latitudes. It is widely distributed across the North Atlantic (Fig. 12.1) from Cape Cod (USA) to Canada on the western side, to Greenland and Iceland, and all the way to Spitsbergen on the eastern side (Stein, 1986) as well as in the North Sea, the Baltic Sea and Barents Sea (Lindberg, 1973; Wheeler, 1985; Kudryavtseva and Karamushko, 2002). The southward range extends as far as the southern North Sea (Heessen *et al.*, 2015) but a few specimens have been caught along the coast of Portugal (Vasconcelos *et al.*, 2004), in Galicia, Spain (Bañón *et al.*, 2008) and in the Mediterranean Sea (Dulčić and Golani, 2006).

There has been much confusion over whether the lumpfish is a pelagic or demersal species. They have several characteristics that indicate a demersal lifestyle, namely pelvic fins that have evolved into a suction disc and the lack of a swimbladder. However, they have features, such as an almost uncalcified cartilaginous skeleton, which brings them close to neutral buoyancy, indicating the potential for a pelagic lifestyle (Davenport and Kjørsvik, 1986). There are many reports that describe the catches of lumpfish in pelagic trawls (Holst, 1993; ICES, 2012; Eriksen *et al.,* 2014); however, lumpfish are also frequently caught in bottom trawls around Iceland, Newfoundland and in the North Sea and the Barents Sea (Knijn *et al.*, 1993; Casey and Myers, 1998; Wienerroither *et al.*, 2013). This confusion was partly resolved by the use of data storage tags attached to lumpfish that showed that, during their spawning migration, female lumpfish spend time both within the water column and also associated with the seabed, and gradually become more pelagic through time from March to May (Kennedy *et al.*, 2016).

Lumpfish feed on a variety of organisms, mainly large planktonic organisms living in the surface/mid-waters such as gelatinous zooplankton, fish eggs and small crustaceans. They will

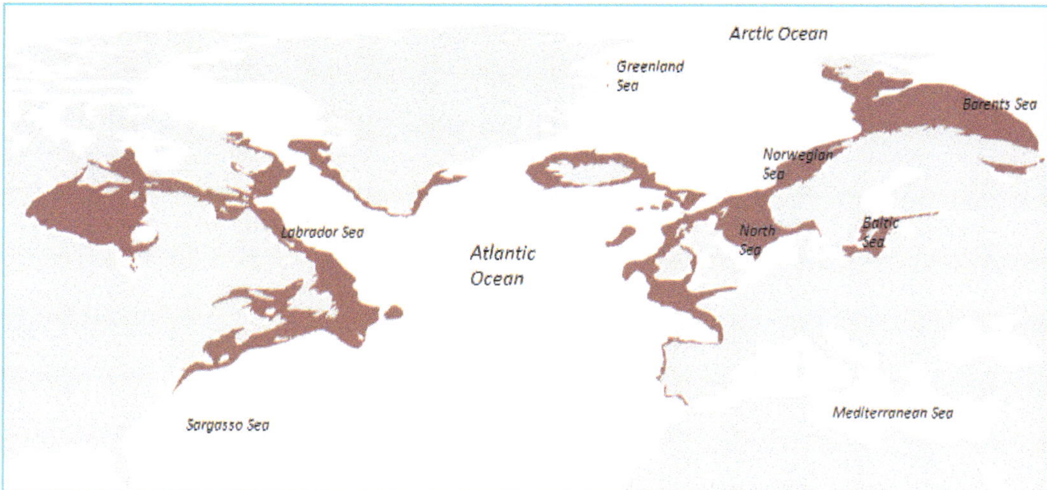

Fig. 12.1 Distribution map of lumpfish in the north Atlantic. From http://www.fao.org/fishery/species/2537/en (modified by Jonsdottir, accessed 5 July 2017).

also occasionally browse upon benthic organisms (Davenport, 1985). Their extensive vertical movements, often exceeding 100 m within one day, have been attributed to their search for prey (Kennedy *et al.,* 2016).

The age at which lumpfish breed for the first time remains unclear. Age readings of otoliths indicate that the majority of the spawning lumpfish in Greenland are three to four years old (Hedeholm *et al.*, 2014). This is similar to the estimates for Norway and Iceland (Albert *et al.*, 2002). However, previous work indicated that the majority of the fish in Iceland started to spawn at five years, and fish up to 13 years of age were present within the population (Thorsteinsson, 1981). Tagging data show that lumpfish exhibit rapid growth until maturity, and it then slows to approximately 1 cm per year (Kasper *et al.*, 2014). This slow growth after maturity may make accurate interpretation of the otoliths difficult, and it must be noted that age readings from lumpfish otoliths are currently unverified.

Lumpfish may undertake long annual distance migrations from spawning areas to winter grounds and vice versa (Schopka, 1974). In fact, tagging data revealed extensive movements during this time with the majority of tagged lumpfish exhibiting a displacement distance (shortest distance between tagging and recapture location) up to 350 km; however one individual was recaptured 587 km from its tagging location. Tagging data has also revealed repeated homing behaviour of lumpfish, either natal homing or homing to where they had spawned previously (Kennedy *et al.*, 2015).

As the spawning season approaches (March–August) the mature fish migrate to coastal areas on both sides of the North Atlantic. Males are the first to arrive at the coast, during January–February in Iceland, and the females will arrive in late February–March. The earlier arrival of males is presumed to allow them to find suitable spawning sites and prepare a nesting area before the females arrive. After arrival to coastal areas, the females will migrate around the coast. Spawning occurs in the sublittoral zone on rocky substrate, rich in seaweed (Lindberg,

1973; Muus and Nielsen, 1998; Zhitenev, 1970). Before the eggs are laid, lumpfish will perform complex mating rituals with nest cleaning, fin brushing, and body quivering (Goulet *et al.*, 1986). Females release their eggs in two or three egg masses, at intervals of eight to 14 days (Ehrenbaum, 1904; Andriyashev, 1954), then leave the spawning areas and play no further part in the rearing of the eggs. The territorial males guard the eggs; cleaning and fanning them throughout (Fulton, 1907), until the young hatch six to eight weeks later. Each male may guard egg masses from several females (Davenport, 1985). After hatching, a portion of the larvae and juveniles are known to spend the first few months in tidal pools before migrating out into open water (Mochek, 1973; Moring, 2011) but yolk sac larvae and juveniles, <30 mm are also found associated with floating seaweed clumps between 10 and 30 km from land (Ingólfsson, 2000).

12.2 Larval distribution

Despite the economic importance of lumpfish (for their roe), parts of their life cycle are not well known. The spawning migration of the species is still not fully understood, and the knowledge about larval behaviour is still limited. There has been some dispute about the larval phase immediately after hatching. For well over a century it has been suggested that newly hatched lumpfish fry attach themselves to the male fish to benefit from its protection, an idea that has been supported by scuba-divers in Norway. It has also been suggested that newly hatched lumpfish remain in shallow water, often attached by their suckers to substratum in the intertidal zone (Davenport, 1985). A third possibility is based on studies in near natural conditions, which suggested that the larvae rapidly disperse with water currents (Fulton, 1907; Mochek, 1973; Davenport, 1985).

Growing larvae and juveniles are certainly found in intertidal pools throughout the summer after spring spawning, but larvae (<one year old) have also been found in colonised clumps of floating seaweed in the open sea of bays (Daborn and Gregory, 1983; Ingólfsson, 2000). It appears that older juveniles (>one year old) migrate to open sea for feeding, thus adopting the habitat of the adult fish (Davenport, 1985).

12.3 Genetic structure of lumpfish

Knowledge about the genetic structure of lumpfish along the species distribution area is currently scarce. Only recently the first genetic markers, microsatellites, were described for the species. Skírnisdóttir and co-workers (2013) developed and characterized 22 microsatellite loci, of which all proved to be polymorphic with numbers of alleles per locus ranging from 3 to 27, and observed and expected heterozygosity ranging between 0.0625–0.979 and 0.0618–0.946, respectively (Skírnisdóttir et.al., 2013). Up to 14 of these microsatellites have been used for genetic structure analysis of lumpfish in the North Atlantic (Pampoulie *et al.*, 2014; Garcia-Mayoral *et al.*, 2016; Jonsdottir *et al.*, 2018). In the study of Pampoulie and co-workers (2014) the sampling area ranged from Canadian waters in the west to Baltic Sea in the east, including several sampling sites in Icelandic waters. The study revealed significant genetic divergence on large geographic scale, but not at a relatively small geographical scale. A total of three genetically distinct populations were found: Maine-Canada-Greenland (Northwest), Iceland-Norway (Northeast) and the Baltic Sea, but no genetic distinct populations were found in eleven sampling areas around Iceland (Pampoulie *et al.*, 2014). Different environmental conditions might contribute to the divergence of the three populations. The southward flowing cold Polar Water west of Greenland may act as a barrier to gene flow between the Northwest samples and the Northeast samples, whereas sharp salinity gradient across the Skagerrak, might act as a barrier to gene flow between the Baltic Sea population and the Northeast sample. The differentiation of the Baltic Sea population from their north Atlantic counterparts is common among marine species (Case *et al.*, 2005; Pampoulie *et al.*, 2008; Gaggiotti *et al.*, 2009). Congruent with the study by Pampoulie and co-workers (2014), a more recent study by Garcia-Mayoral and co-workers (2016) revealed closer relationship between Greenland and Canada, than Greenland and Iceland. Further, the results showed significant structuring with isolation by distance in the West Greenland samples, thus two major subpopulations, north and south, were suggested. In the most recent study by Jónsdóttir and co-workers (2018) the stock structure along the coastline of Norway was investigated. This was done by using 14 microsatellite loci and 297 specimens collected at five fishing grounds, covering most of the Norwegian coastline from south to north, with additional 18 samples of first generation reared fish from a fish farm outside Tromsø (North Norway). In accordance with the study of Pampoulie and co-workers (2014), no significant genetic divergence was found on a relatively small geographical scale (i.e. Icelandic and Norwegian waters. Pampoulie *et al.*, 2014 and Jónsdóttir *et al.*, 2018, respectively) but, in contrast to the study of Garcia-Mayoral and co-workers (2016), there was no indication of significant spatial genetic structuring (Fig. 12.2) or of positive correlation between geographic and genetic distance among the wild lumpfish samples.

Lumpfish has shown to be effective 'cleaner fish' in Atlantic salmon aquaculture in Norway, (Imsland *et al.*, 2014a, b, c; Imsland *et al.*, 2015a) and therefore of great interest for the aquaculture industry. As a result, the use of lumpfish, for delousing of salmon, has increased dramatically and hence, the production of juvenile lumpfish. Currently lumpfish is translocated along the Norwegian coast to serve as cleaning fish in aquaculture and, as a first step to assess if this practice may potentially pose a conservation problem, e.g., by promoting increased admixture between genetic clusters, it was urgent to determine the stock structure of lumpfish in Norwegian waters. Based on the results of no genetic structure in the wild Norwegian lumpfish population (Jónsdóttir *et al.*, 2018), current farming practices, including transport of juveniles, will probably not have a negative impact on the wild stock of lumpfish. Thus translocated individuals that escape from aquaculture in Norway will apparently have a similar allele distribution as the local fish.

12.4 DNA microsatellites

Microsatellites are simple sequence tandem repeats (SSTRs) of DNA units, generally ranging in length from two to five base pairs, and repeated in bead-like fashion a variable number of times, ranging typically from a few to a few hundred. They tend to be evenly distributed in the genome on all chromosomes and all regions of the chromosome, and have been found inside gene coding regions as well as in the non-gene sequences (Liu *et al.*, 2001).

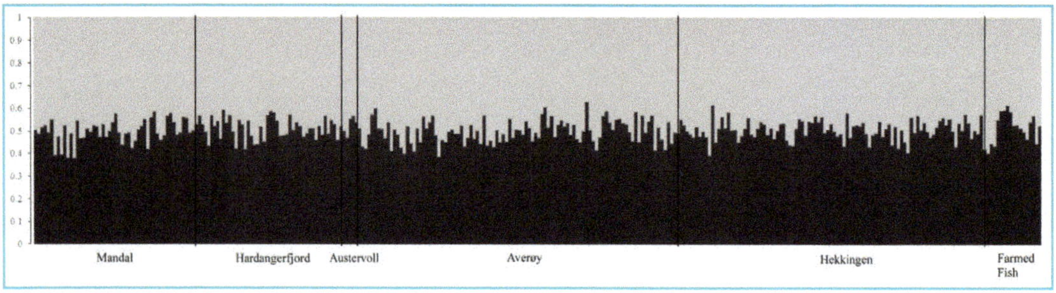

Fig. 12.2 Bayesian cluster analyses performed in STRUCTURE (Pritchard *et al.*, 2000) on lumpfish samples from five sample localities along the Norwegian coast and from F1 farmed fish. Each vertical bar represents an individual, while colours denote inferred clusters. The y-axis indicates the probability of an individual belonging to a particular cluster. As all samples are located around the 50% line there is no apparent sub-population structure.

Microsatellites, which are often referred to as short tandem repeats (STRs) or as simple sequence repeats (SSRs) are useful genetic markers because they tend to be highly polymorphic as well as having high diversity in populations of species, the polymorphism being based on size differences due to varying numbers of repeat units (Fig. 12.3) contained by alleles at a given locus (Liu and Cordes, 2004).

Microsatellite mutation rates have been reported as high as 10^{-2} per generation (Weber and Wong, 1993; Crawford and Cuthbertson, 1996), and are believed to be caused by polymerase slippage during DNA replication, resulting in differences in the number of repeat units (Levinson and Gutman, 1987; Tautz, 1989; Liu and Cordes, 2004). Generally speaking, microsatellites containing a larger number of repeats are more polymorphic, though polymorphism has been observed in microsatellites with only five repeats (Karsi *et al.*, 2002; Liu and Cordes, 2004). They are abundant in all species studied to date, and have been estimated to occur as often as once every 10 kb in fishes (Wright, 1993).

Microsatellite markers have been used extensively in fisheries research, including studies of genome mapping, parentage, kinship, and stock structure (see O'Connell and Wright, 1997 for review), and have been shown to be excellent tools due to their abundance, even genomic distribution, small locus size, high polymorphism and Mendelian inherited fashion as co-dominant markers (Liui and Cordes, 2004). The large number of microsatellites needed for many of these applications can often generate high costs due to materials and personnel time, a cost that can be reduced by multiplexing (Fig. 12.4). Microsatellite multiplexing is the combination of polymerase chain reaction (PCR) amplification product from multiple microsatellites into a single lane of an electrophoretic gel (Olsen *et al.*, 1996; Neff *et al.*, 2000; Renshaw *et al.*, 2006).

12.5 Single nucleotide polymorphism (SNP)

SNP describes polymorphisms caused by point mutations, resulting in a single nucleotide variation occurring at a specific position in the genome (Fig. 12.5).

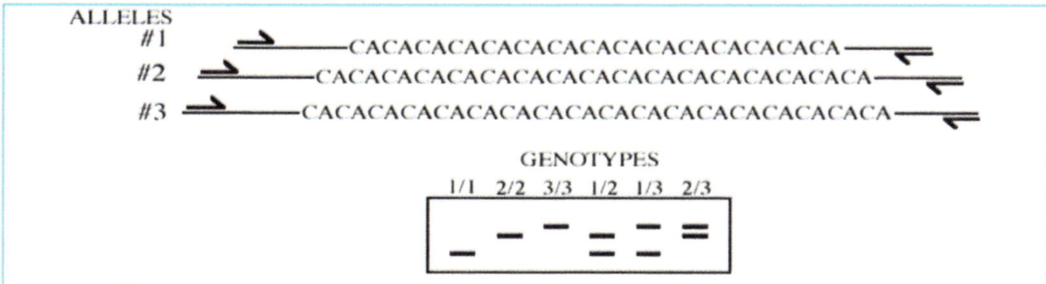

Fig. 12.3 Example of DNA microsatellite.

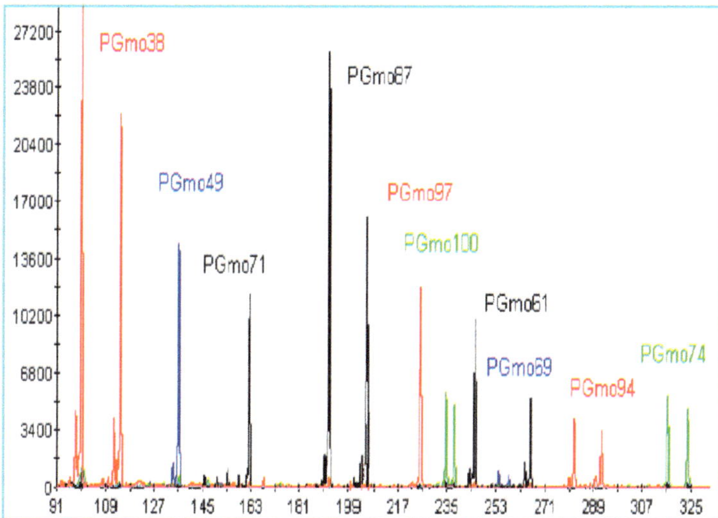

Fig. 12.4 Electropherogram from a single multiplex reaction of Atlantic cod, *Gadus morhua*. The name of each loci is shown at each respective peak. The *y* axis displays fluorescence intensity in arbitrary units and the *x* axis displays allele size range (modified from Imsland *et al.*, 2011).

Such mutation gives rise to different alleles, containing alternative bases at a given nucleotide position within a locus. Most SNPs so far discovered are bi-allelic, where either the two pyrimidines C/T or the two purines A/G reveal two different alleles, and are inherited as co-dominant markers. The shortcoming of SNPs scarce allele numbers is balanced by their great abundance in the genome and, in fact, SNPs are becoming a focal point in molecular marker development since they are the most abundant polymorphism in any organism, adaptable to automation. They reveal hidden polymorphism not detected with other markers and methods (Liu and Cordes, 2004).

Microsatellites and SNPs are powerful tools for a variety of aquaculture applications, genetic mapping in particular. SNP is also promising for identification of candidate genes for QTL (Quantitative trade loci). Microsatellites, in particular, have proven to be very useful when it comes to strain and hybrid identification, genetic diversity analysis, and mapping of quantitative trait loci, and it is particularly powerful when it comes to parental assignment (Lui and Cordes, 2004).

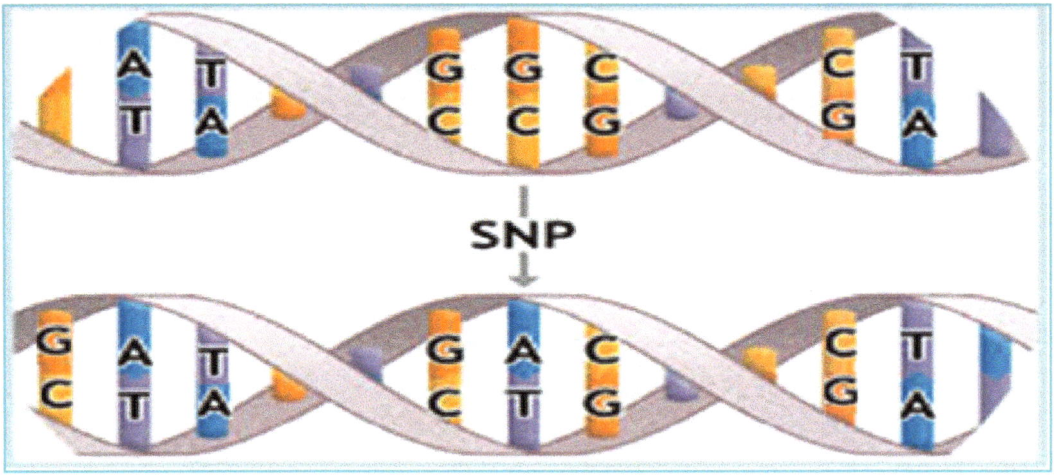

Fig. 12.5 Example of a single nucleotide polymorphism SNP (modified by Jónsdóttir).

12.6 Search for QTLs for lice grazing

Quantitative traits are observable phenotypes that depend on the cumulative actions of multiple genes and the environment. Quantitative trait locus (QTL) is thus a section of DNA that correlates with these variable phenotypes, being linked to, or containing genes that control, these phenotypes. Since most, if not all, performance and production traits are controlled by multiple genes and therefore inherited as quantitative traits, analysis of their associated quantitative trait loci (QTL) is emerging as a very important part of aquaculture genetics/genomics. Performance traits such as growth rate, disease resistance and feeding preferences are important to breeders, all being controlled by multiple genes and therefore good candidates for QTL analysis. Relative chromosomal positions of QTL in a species genome needs to be identified for this purpose, and microsatellites and SNPs have been shown to be important tools (Liu and Cordes, 2004).

Feeding preferences of juvenile lumpfish reared together with Atlantic salmon in sea pens were assessed by Imsland *et al.* (2014a, b, c, 2015a, b). A high percentage (61–96%) of lumpfish was found to have ingested salmon feed pellets on each sampling day. In addition, large individual differences in feeding preferences between lumpfish and their sea lice grazing activity have been indicated (Imsland *et al.*, 2014a). In a semi-commercial trial, it was found that only 10% of the lumpfish grazed on lice from salmon at the beginning of the trial, increasing to 36% toward the end. Data from the trial showed that there are individuals grazing more actively on sea lice compared with their conspecifics. This indicates that there is a large variation in foraging behaviour, hence delousing efficiency, of lumpfish and it is still unknown what the cause may be.

Focus has been set on numerous different quantitative traits using QTL, one of which is feeding preference in association with sea lice grazing of lumpfish on Atlantic salmon. Imsland and co-workers (2016) investigated a possible family influence on sea lice grazing of lumpfish, and found significant differences between families; moreover, both maternal and paternal effect was found for sea lice grazing, indicating a genetically inherited trait (Imsland *et al.*, 2016). Furthermore, significant differences in behaviour and mortality between genetically distinct families of lumpfish were observed. The fish with lowest exploitation of natural food sources had significantly higher behavioural interaction with the salmon (swimming along and competing for pellets with salmon) and less resting behaviour compared to the other families. Observed differences in mortality rates between the nine families was connected to *Pasteurella* spp. outbreak, suggesting that some families were more resistant than others. If these observed differences have a genetic basis then selective breeding programmes would allow for these favourable traits to be maintained in future lumpfish populations. Currently there are at least two initiatives under way in Norway for breeding programmes for lumpfish focusing on improving sea lice grazing both through the application of classical individual- and family-based selection as well as through use of marker-associated selection to search for QTLs.

References

Albert, O.T., Torstensen, E., Bertelsen, B., Jonsson, S.T., Pettersen, I.H. and Holst, J. C. (2002) Age-reading of lumpsucker (*Cyclopterus lumpus*) otoliths: dissection, interpretation and comparison with length frequencies. *Fisheries Research* 55, 239–252.

Andriyashev, A.P. (1954) Ryby severykh morey S.S.S.R. (Fishes of the northern seas of the U.S.S.R.) Moscow, AN S.S.S.R.

Bañón, R., Garazo, A. and Fernández, A. (2008) Note about the presence of the lumpsucker *Cyclopterus lumpus* (Teleostei, Cyclopteridae) in Galician waters (NW Spain). *Journal of Applied Ichthyology* 24, 108–109.

Case, R.A.J., Hutchinson, W.F., Hauser, L., Van oosterhout, C. and Carvalho, G.R. (2005) Macro- and micro-geographic variation in pantophysin (Pan I) allele frequencies in NE Atlantic cod *Gadus morhua*. *Marine Ecology Progress Series* 301, 267–278.

Casey, J.M. and Myers, R.A. (1998) Diel variation in trawl catchability: is it as clear as day and night? *Canadian Journal of Fisheries and Aquatic Sciences* 55, 2329–2340.

Crawford, A.M. and Cuthbertson, R.P. (1996) Mutations in sheep microsatellites. *Genome Research* 6, 876– 879.

Daborn, G.R. and Gregory, R.S. (1983) Occurrence, distribution, and feeding habits of juvenile lumpfish, *Cyclopterus lumpus* L. in the Bay of Fundy. *Canadian Journal of Zoology* 61, 797–801.

Davenport, J. (1985) Synopsis of biological data on the lumpsucker *Cyclopterus lumpus* (Linnaeus, 1758). Food and Agriculture Organization of the United Nations, Rome, Italy. *FAO Fisheries Synopsis* 147, 31.

Davenport, J. and Kjørsvik, E. (1986) Buoyancy in the lumpsucker *Cyclopterus lumpus*. *Journal of the Marine Biological Association of the United Kingdom* 66, 159–174.

Dulčić, J. and Golani, D. (2006) First record of *Cyclopterus lumpus* L., 1758 (Osteichthyes: Cyclopteridae) in the Mediterranean Sea. *Journal of Fish Biology* 69, 300–303.

Ehrenbaum, E. (1904) Eier und larven von fischen der deutschen bucht. 3. Fische mit festsitzenden eiern. *Wiss.Meeresunters.* 6, 127–200.

Eriksen, E., Durif, C.M.F. and Prozorkevich, D. (2014) Lumpfish (*Cyclopterus lumpus*) in the Barents Sea: development of biomass and abundance indices, and spatial distribution. *ICES Journal of Marine Science* 71, 2398–2402.

Fulton, T.W. (1907) On the spawning of the lumpsucker (*Cyclopterus lumpus*) and the paternal guardianship of the eggs. *Report of the Fishery Board of Scotland* 24, 169–178.

Gaggiotti, O.E., Bekkevold, D., Jørgensen, H.B.H., Foll, M., Carvalho, G.R., Andre, C. and Ruzzante, D.E. (2009) Disentangling the effect of evolutionary, demographic, and environmental factors influencing genetic structure of natural populations: Atlantic herring as a case study. *Evolution* 63, 2939–2951.

Garcia-Mayoral, E., Olsen, M., Hedeholm, R., Post, S., Nielsen, E.E. and Bekkevold, D. (2016) Genetic structure of West Greenland populations of lumpfish *Cyclopterus lumpus*. *Journal of Fish Biology* 89, 2625–2642.

Goulet, D., Green, J.M. and Shears, T.H. (1986) Courtship, spawning, and parental care behavior of the lumpfish, *Cyclopterus lumpus* L., in Newfoundland. *Canadian Journal of Zoology* 64, 1320–1325.

Heessen, H.J.L., Daan, N., and Ellis, J.R. (2015). Fish atlas of the Celtic Sea, North Sea, and Baltic Sea. Wageningen Academic Publishers, The Netherlands.

Hedeholm, R., Blicher, M.E. and Grønkjær, P. (2014) First estimates of age and production of lumpsucker (*Cyclopterus lumpus*) in Greenland. *Fisheries Research*, 149, 1–4.

Holst, J.C. (1993) Observations on the distribution of lumpsucker (*Cyclopterus lumpus* L.) in the Norwegian Sea. *Fishery Research* 17, 369–372.

ICES. (2012) Report of the Working Group of International Pelagic Surveys (WGIPS), 3–7 December 2012, ICES Headquarters, Copenhagen, Denmark. ICES CM2012/SSGESST: 22, 317 pp.

Imsland, A.K., Koedijk, R., Stefansson, S.O., Foss, A., Hjörleifsdóttir, S., Hreggviðsson, G.Ó., Otterlei, E. and Folkvord, A. (2011) A retrospective approach to fractionize variation in body mass of Atlantic cod *Gadus morhua*. *Journal of Fish Biology* 78, 251–264.

Imsland, A.K., Reynolds, P., Eliassen, G., Hangstad, T.A., Foss, A., Vikingstad, E. and Elvegård, T.A. (2014a) The use of lumpfish (*Cyclopterus lumpus* L.) to control sea lice (*Lepeophtheirus salmonis* Krøyer) infestations in intensively farmed Atlantic salmon (*Salmo salar* L.). *Aquaculture* 425-426, 18–23.

Imsland, A.K., Reynolds, P., Eliassen, G., Hangstad, T.A., Nytrø, A.V., Foss, A., Vikingstad, E. and Elvegård, T.A. (2014b) Notes on behaviour of lumpfish in sea pens with and without Atlantic salmon. *Journal of Ethology* 32, 117–122.

Imsland, A.K., Reynolds, P., Eliassen, G., Hangstad, T.A., Nytrø, A.V., Foss, A., Vikingstad, E. and Elvegård, T.A. (2014c) Assessment of growth and sea lice infection levels in Atlantic salmon stocked in small-scale cages with lumpfish. *Aquaculture* 433, 137–142.

Imsland, A.K., Reynolds, P., Eliassen, G., Hangstad, T.A., Nytrø, A.V., Foss, A., Vikingstad, E. and Elvegård, T.A. (2015a) Feeding preferences of lumpfish (*Cyclopterus lumpus* L.) maintained in open net-pens with Atlantic salmon (*Salmo salar* L.). *Aquaculture* 436, 47–51.

Imsland, A.K., Reynolds, P., Eliassen, G., Hangstad, T.A., Nytrø, A.V., Foss, A., Vikingstad, E. and Elvegård, T.A. (2015b) Assessment of suitable substrates for lumpfish in sea pens. *Aquaculture International* 23, 639–645.

Imsland, A.K., Reynold, P., Eliassen, G., Mortensen, A., Hansen, Ø.J., Puvanendran, V. Hangstad, T.A., Jonsdottir, O.D.B., Emaus, P.A., Elvegård, T.A., Lemmens, S.C.A., Rydland, R., Nytrø, A.V. and Jonassen, T.M. (2016) Is cleaning behaviour in lumpfish (*Cyclopterus lumpus*) parentally controlled? *Aquaculture* 459, 156–165.

Ingólfsson, A. (2000) Colonization of floating seaweed by pelagic and subtidal benthic animals in southwestern Iceland. *Hydrobiologia* 440, 181–189.

Jónsdóttir, Ó.D.B., Schregel, J., Hagen, S., Tobiassen, C., Aarnes, S.G., Imsland, A.K.D. (2018) Population structure of lumpfish along the Norwegian coast: aquaculture implications. *Aquaculture International* 26, 49–60, DOI 10.1007/s10499-017-0194-2).

Karsi, A., Patterson, A., Feng, J. and Liu, Z.J. (2002). Translational machinery of channel catfish: I. A transcriptomic approach to the analysis of 32 40S ribosomal protein genes and their expression. *Gene* 291, 177–186.

Kasper, J.M., Bertelsen, B., Ólafsson, H.G., Holst, J.C., Sturlaugsson, J. and Jónsson, S.P. (2014) Observations of growth and postspawning survival of lumpfish *Cyclopterus lumpus* from mark-recapture studies. *Journal of Fish Biology* 84, 1958–1963.

Kennedy, J., Jonsson, S.T., Kasper, J.M. and Olafsson, H.G. (2015) Movements of female lumpfish (*Cyclopterus lumpus*) around Iceland. *ICES Journal of Marine Science* 72, 880–889.

Kennedy, J., Jónsson, S.Þ., Ólafsson, H.G. and Kasper, J.M. (2016) Observations of vertical movements and depth distribution of migrating female lumpfish (*Cyclopterus lumpus*) in Iceland from data storage tags and trawl surveys. *ICES Journal of Marine Science* 73, 1160–1169.

Knijn, R.J., Boon, T.W., Heessen, H.J.L. and Hislop, J.R.G. (1993) Atlas of North Sea Fishes. *ICES Cooperative Research Report*, 194.

Kudryavtseva. O.Y. and Karamushko, O.V. (2002) Biology of the lumpfish *Cyclopterus lumpus* during the reproduction in the Murman near shore zone. *Journal of Ichthyology* 42, 309-313.

Levinson, G. and Gutman, G.A. (1987) High frequency of short frameshifts in poly-CA/GT tandem repeats borne by bacteriophage M13 in Escherichia coli K-12. *Nucleic Acids Research* 15, 5323–5338.

Lindberg, G.U. (1973) Cyclopteridae In.: Hureau, J.-C. and Monod, T. (eds). *Check-list of the fishes of the north-eastern Atlantic and of the Mediterranean* (CLOFNAM). Vol. I. UNESCO, Paris, France. pp. 607–608.

Liu, Z.J., Li, P., Kocabas, A., Ju, Z., Karsi, A., Cao, D. and Patterson, A. (2001) Microsatellite-containing genes from the channel catfish brain: evidence of trinucleotide repeat expansion in the coding region of nucleotide excision repair gene RAD23B. *Biochemical and Biophysical Research Communications* 289, 317–324.

Liu, Z.J. and Cordes, J.F. (2004) DNA marker technologies and their applications in aquaculture genetics. *Aquaculture* 238, 1–37.

Mochek, A.D. (1973) Spawning behaviour of the lumpsucker *Cyclopterus lumpus L. Journal of Ichthyology* 13, 615–9.

Moring, J. R. (2011) Intertidal growth of larval and juvenile lumpfish in Maine: A 20-year assessment. *Northeastern Naturalist* 8, 347–354.

Muus, B.J. and Nielsen, J.G. (1998) Peces de Mar del Atlántico y del Mediterráneo. Biología, Pesca, Importancia Económica. Ediciones Omega, S.A., Barcelona, Spain.

Neff, B.D., Fu, P. and Gross, M.R. (2000) Microsatellite multiplexing in fish. *Transactions of the American Fisheries Society* 129, 584–593.

O'Connell, M. and Wright, J.M. (1997) Microsatellite DNA in fishes. *Reviews in Fish Biology and Fisheries* 7, 331–363.

Olsen, J.B., Wenburg, J.K. and Bentzen, P. (1996) Semiautomated multilocus genotyping of Pacific salmon (Oncorhynchus spp.) using microsatellites. *Molecular Marine Biology and Biotehnology* 5, 259–272.

Pampoulie, C., Stefánsson, M.Ö., Jörundsdóttir, T.D., Danilowicz, B.S. and Daníelsdóttir, A.K. (2008) Recolonization history and large scale dispersal in the open sea: the case study of the North Atlantic cod, *Gadus morhua* L. *Biological Journal of the Linnean Society* 94, 315–329.

Pampoulie, C., Skirnisdottir, S., Olafsdottir, G., Helyar, S.J., Thorsteinsson, V., Jónsson, S.Þ., Fréchet, A., Durif, C.M.F., Sherman, S., Lampart-Kałużniacka, M., Hedeholm, R., Ólafsson, H., Daníelsdóttir, A.K. and Kasper, J.M. (2014) Genetic structure of the lumpfish *Cyclopterus lumpus* across the North Atlantic. *ICES Journal of Marine Science* 71(9), 2390–2397.

Pritchard, J.K., Stephens, M., and Donnelly, P. (2000) Inference of population structure using multilocus genotype data. *Genetics* 155, 945-959.

Renshaw, M.A., Saillant, E., Bradfield, S.C. and Gold, J.R. (2006) Microsatellite multiplex panels for genetic studies of three species of marine fishes: red drum (*Sciaenops ocellatus*), red snapper (*Lutjanus campechanus*), and cobia (*Rachycentron canadum*). *Aquaculture* 253, 1–4, 31, 731–735.

Schopka, S.A. (1974) Preliminary results from tagging of lumpsucker (*Cyclopterus lumpus*), in Icelandic waters 1971–1974. *ICES CM.*1974/F:18. Demersal Fish Northern Committee. 6 pp.

Skírnisdóttir, S., Olafsdottir, G., Olafsson, K., Jendrossek, T., Lloyd, H.a.D., Helyar, S., Pampoulie, C., Danielsdottir, A.K. and Kasper, J.M. (2013) Twenty-two novel microsatellite loci for lumpfish (*Cyclopterus lumpus*). *Conservation Genetics Resources* 5, 177–179.

Stein, D.L. (1986) Cyclopteridae. In: Whitehead, P.J.P., Bauchot, M.L., Hureau, J.C., Nielsen, J.G. and Tortenese, E., (eds.) F*ishes of the North-eastern Atlantic and the Mediterranean, III*, UNESCO, Paris. pp. 1269–1274.

Tautz, D. (1989) Hypervariability of simple sequences as a general source for polymorphic DNA markers. *Nucleic Acids Research* 17, 6463–6471.

Thorsteinsson, V. (1981) The ageing validation of the lumpsucker (*Cyclopterus lumpus*) and the age composition of the lumpsucker in Icelandic lumpsucker fisheries. *ICES C.M.*, G58, 12 pp.

Vasconcelos, P., Monteiro, C.C., Santos, M.N. and Gaspar, M.B. (2004) First record of the lumpfish (*Cyclopterus lumpus* Linnaeus, 1758) off the Algarve coast (southern Portugal): southward extension of the species distributional range. *Journal of Applied Ichthyology* 20, 159–160.

Weber, J.L. and Wong, C. (1993) Mutation of human short tandem repeats. *Human Molecular Genetics* 2, 1123–1128.

Wheeler, A. (1985) *The world encyclopaedia of fishes.* Macdonald & Co., London, UK.

Wienerroither, R., Johannesen, E., Dolgov, A., Byrkjedal, I., Aglen, A., Bjelland, O. and Drevetnyak, K. (2013) Atlas of the Barents Sea Fishes based on the winter survey. *IMR-PINRO Joint Report Series* 2-2013. ISSN 1502–8828. 220 pp.

Wright, J.M. (1993) DNA fingerprinting in fishes. In: Hochachka, P.W. and Mommsen, T. (Eds.), *Biochemistry and Molecular Biology of Fishes*. Elsevier, Amsterdam, pp. 58–91.

Zhitenev, A.N. (1970) Ecological and morphological features of the reproduction of the lumpsucker (*Cyclopterus lumpus* (L.)). *Journal of Ichthyology* 10, 77–84.

CHAPTER 13

Cleaner fish health

Felix Scholz,[1,3] Henriette Glosvik,[2] and Mar Marcos-López[1,3]

[1]FishVet Group Ireland, Oranmore Business Park, Oranmore, Co. Galway, Ireland
[2]Marine Harvest Norway AS, Bergen, Norway
[3]Marine and Freshwater Research Centre, Galway-Mayo Institute of Technology, Galway, Co. Galway, Ireland

13.1 Introduction

The large-scale use and aquaculture of cleaner fish (wrasse species and lumpfish) is comparatively recent and is of associated commercial and scientific interest. Not enough is known about the diseases of these species and the situation is complicated by the cohabitation of different species at sea, all with specific requirements and health challenges. This ultimately leads to a compromise in the ideal conditions for cleaner fish, challenge with pathogens of the other species and stress resulting in heightened disease susceptibility and clinical disease for previously subclinical infections. Infections occur in wild fish stocks that could potentially cause disease under farm conditions and be transmitted to farmed fish. New diseases, bacterial, parasitic, fungal and viral, are rapidly being identified in cleaner fish species under these conditions and infectious diseases are currently one of the most important restraints for the sustainability and efficacy of cleaner fish use (Norwegian Veterinary Institute NVI, 2016). The development of best health management practices and research on optimal rearing conditions is necessary. Possible disease transmission to Atlantic salmon (*Salmo salar*), though not recorded under farm conditions to date, is a concern. This chapter describes some of the most important diseases encountered in captive cleaner fish to date, as well as briefly addressing health management factors. Unless otherwise specified, salmon will refer to Atlantic salmon.

13.2 Bacteria

Bacterial diseases are currently the most significant cause of mortality in cleaner fish, both in land-based facilities and in salmon pens at sea, which resembles the first years of salmon farming prior to the development of effective vaccines. Many fish-infecting bacteria are considered opportunistic pathogens. These species can be widespread in the marine environment and even be present on healthy fish, becoming pathogenic only when their hosts are rendered susceptible through predisposing factors. Other bacteria, such as *Aeromonas salmonicida,* are considered primary pathogens and largely rely on a fish host for multiplication. But, even with these, disease outbreaks are usually linked to predisposing factors. A third group of bacteria are saprophytic species that play a part in natural degradative pathways. These species are often isolated from dead or moribund fish in high numbers but are not responsible for mortality. Interpreting the significance of some bacterial species can be challenging, especially if isolated from dead or moribund fish. Different subtypes of the same species can show host specificity and differences in immunogenic factors, which makes their identification highly relevant in evaluating the threat they pose to cleaner fish and salmon as well as for the development of effective vaccines. The taxonomy of fish-infecting bacteria is updated regularly as research progresses.

Aeromonas salmonicida

Aetiological agent and host: *Aeromonas salmonicida* is a gram-negative, rod-shaped bacterium that infects a wide range of fish species worldwide with the potential to cause severe disease. Several subspecies exist that have been divided into typical (ssp. *salmonicida*) and atypical (ssp. *achromogenes, mascoucida, smithia* and *pectinolytica)* subspecies. Typical *A. salmonicida* (ssp. *salmonicida*) is the aetiological agent of the disease furunculosis in salmonids, which caused high economic losses in salmon aquaculture before effective vaccines were developed. This subtype is rare in cleaner fish species. Typical *A. salmonicida* infections in wrasse, with mortalities up to 19% (Treasurer and Cox, 1991) and 30% (Treasurer and Laidler, 1994), are believed to have occurred due to cohabitation with infected salmon, through water transmission and/or via wrasse feeding on infected salmon mortalities (Treasurer and Cox, 1991; Treasurer and Laidler, 1994; Hjeltnes *et al.*, 1995). Typical furunculosis has caused mortality in lumpfish in northern Norway (Johansen *et al.*, 2016; NVI, 2016), but no signs of disease were found in the salmon on the same site. The other *A. salmonicida* subtypes cause the disease 'atypical furunculosis' and are the biggest single disease problem for wrasse and lumpfish in many areas. The collective term 'atypical *Aeromonas salmonicida*' however, is now outdated. Research on sequence variations in the virulence array protein gene (vapA), the gene encoding the 'A-layer', showed 14 distinct A-layer groups; four of which were made up of isolates from the known subspecies, while ten others showed the existence of new subspecies (Gulla *et al.*, 2016). In the same study, single isolates were placed outside of these 14 groupings, indicating the potential existence of more subspecies. The presence of the A-layer protein is highly relevant to pathogenicity. A-layer positive strains have caused up to 100% mortalities in cleaner fish, while strains lacking the A-layer only lead to mortalities when other stress factors occurred simultaneously (Rønneseth *et al.*, 2015). Many of the new 'atypical' subspecies appear to be more host-specific than previously assumed (Gulla *et al.*, 2016). Cleaner fish isolates are almost exclusively subtypes 5 and 6; subtype 6 being prominent in lumpfish, and both subtypes 5 and 6 infect wrasse species. These bacteria are present in wild wrasse populations, but the majority of wrasse become infected after stocking in salmon pens (Gulla *et al.*, 2015a). In lumpfish, outbreaks have occurred both in hatcheries and at sea (Johansen *et al.*, 2016). The principal mode of transmission is horizontal, most commonly oral.

Clinical signs and pathology: In cleaner fish species, atypical furunculosis can occur as an acute or chronic infection. Acute outbreaks are often triggered by stress events and are characterised by high mortality. Clinical signs and gross lesions can include mortality, lethargy, loss of appetite, skin ulcers (Fig. 13.1A), haemorrhaging/hyperaemia at the fin bases, internal granulomas and haemorrhaging in visceral organs; although in some cases external symptoms can be inconspicuous. In lumpfish, darkening of the skin has been described. On histopathology, bacterial microcolonies are seen in gills, skin, muscle and/or internal organs. Bacterial colonies can be associated with tissue necrosis, haemorrhaging and/or granulomatous inflammation (Fig. 13.1B), but can also induce no host response.

Diagnosis and health management: The observation of prominent bacterial colonies in organs on histopathology may lead to a presumptive diagnosis of furunculosis but a final diagnosis requires bacteriology and molecular analysis, since other bacterial species (i.e. *Pasteurella* sp. in lumpfish) can cause similar pathology. *A. salmonicida* can be isolated on tryptone soya agar (TSA). The ssp. *salmonicida* produces a characteristic diffusible brown pigment on TSA plates. Other ssp. do not always produce this pigment. Classical bacteriology is still useful to distinguish the ssp. *salmonicida*, an important legal distinction in some countries, but molecular sequencing is necessary to identify other subspecies. Early attempts to vaccinate cleaner fish species against atypical furunculosis were not successful, probably because the vaccine isolates belonged to different A. *salmonicida* subspecies, and it seems likely that subtype/A-layer specific vaccines are essential (Gulla *et al.*, 2016). However, recent findings strongly indicate that vaccination of lumpfish will be successful (Rønneseth *et al.*, 2017) and commercial vaccines made from the relevant A-layer types are being developed. Autogenous vaccines are being used in some sites and first results are promising. Mutations can lead to a dysfunctional expression of the A-layer (subtype 6 appears to be especially prone), possibly influencing vaccine protection. Further research is needed to find out if mutated strains retain their virulence (Gulla *et al.*, 2016). Resistances to different antibiotics have been documented, and an antibiotic sensitivity test before treatment is recommended. Antibiotic treatments have been successful in treating outbreaks, but the granulomatous nature of chronic infections makes this condition difficult to eradicate and recurrent disease can arise after treatment. In addition, the presence of these bacteria in the environment means recurring outbreaks can be expected.

Impacts: The apparent species specificity of atypical subtypes makes their transmission from cleaner fish to salmon unlikely. Attempts to infect salmon with atypical isolates from wild corkwing (*Symphodus melops*) and goldsinny wrasse (*Ctenolabrus rupestris*) were unsuccessful (Frerichs *et al.*, 1992; Costello *et al.*, 1996). Transmission between wrasse and lumpfish however, is likely as they are susceptible to the same subtypes. Atypical *A. salmonicida* spp. are present in wild wrasse populations (Costello *et al.*, 1996; Gulla *et al.*, 2015a) and their cohabitation with naïve farmed cleaner fish is considered a serious threat for the latter, particularly as they are more susceptible following transport. *A. salmonicida* ssp. *salmonicida* infects both salmon and cleaner fish species, which could act as a reservoir. The chances of salmon infecting cleaner fish is considered higher than vice versa (Bricknell *et al.*, 1996). This bacterial species is rare in cleaner fish and salmon aquaculture nowadays, but sporadic cases still occur and the effect of prolonged exposure on salmon vaccine efficacy, as could occur through infected cleaner fish, is unknown (Johansen *et al.*, 2016).

Vibrio anguillarum

Aetiological agent and host: *Vibrio anguillarum* is a gram-negative, curved rod-shaped bacterium and the aetiological agent of classical vibriosis in salmon, wrasse, lumpfish and other marine fish species. The disease can be severe and mortalities high in unvaccinated fish. Based on the variability of the O antigen, 23 different O serotypes (O1 to O23) are distinguished. Of these, only serotypes O1 and O2 (O2a, O2a-biotype II and O2b), and to a lesser extent serotype O3, are associated with disease in fish (Pedersen *et al.*, 1999). Different serotypes vary in host specificity and pathogenicity (Egidius and Andersen, 1978). Serotype O1, which also affects salmonids, is the most frequently isolated from wrasse and lumpfish (Biering *et al.*, 2016), but serotypes O2a, O2a-biotype II and O2b have also been detected in cleaner fish (NVI, 2013; NVI, 2014; NVI, 2015; NVI, 2016). The transmission of *V. anguillarum* is horizontal and infection can occur via ingestion or through the skin and gills, hence open wounds can facilitate and exacerbate the infection. Water temperature is considered a risk factor, with higher water temperatures (>15°C) increasing disease occurrence (Austin and Austin, 2007). However, outbreaks of vibriosis in lumpfish have been reported at temperatures as low as 6°C (Marcos-López *et al.*, 2013).
Clinical signs and pathology: Clinical signs of disease include lethargy, loss of appetite and mortalities. Obvious gross lesions can be absent in acute cases, but when present these can include congestion and haemorrhaging around the mouth, operculum, fins and vent, exophthalmia, ascites, fluid-filled intestine and stomach, and/or internal haemorrhages (Marcos-López *et al.*, 2013). Skin discoloration and erosion around the caudal/peduncle area and tail rot are also frequently described in cases of vibriosis in Norwegian lumpfish hatcheries (Fig. 13.1C). On histopathology, high numbers of the bacteria can be observed in the parenchyma and vessels of all organs, in association with tissue necrosis, haemorrhage and/or oedema (Fig. 13.1D) (Marcos-López *et al.*, 2013).
Diagnosis and health management: Complete diagnosis includes histopathology and bacteriology with biochemical characterisation and serotyping. *V. anguillarum* can be cultured in different rich media. Vaccination has proven effective in preventing outbreaks, and vibriosis is seldom observed in vaccinated lumpfish at sea. Different vaccination regimes are used. Dip vaccines are used in the younger stages, while intra-peritoneal vaccination is used in older fish and this is considered crucial in protecting fish from vibriosis after deployment. As serotypes vary in host specificity and pathogenicity, the serotype is a key consideration for an effective vaccine. Cleaner fish vibriosis vaccines should include the serotype O1. Differences in pathogenicity between strains of the same serotype have been reported and therefore strains should be screened for virulence prior to selection of a vaccine strain (Biering *et al.*, 2016). Vibriosis in cleaner fish can be treated with antibiotics with good results if initiated early.
Impacts: In Norway and Scotland several lumpfish hatcheries have suffered from recurrent *V. anguillarum* O1 infections, leading to significant stock losses. The transmission of serotypes O1 and O2a between cleaner fish and salmon is considered a significant risk (Johansen *et al.*, 2016). However, clinical vibriosis in salmon is rare and most salmon stocks are vaccinated.

Other Vibrio species

Many *Vibrio* species are part of the normal flora in the marine environment and they are frequently isolated from cleaner fish. The most commonly isolated species include *V. tapetis*, *V.*

logei, *V. wodanis* and *V. splendidus*. Their pathogenicity is still largely unclear and their role as disease agents needs to be further investigated. *V. splendidus*-related strains are among the most commonly identified bacteria from cleaner fish and are frequently detected in cases of fin rot or in dead and moribund cleaner fish (authors' observation). A study of *V. splendidus*-related strains found that most isolates were closely related to the *V. splendidus* type strain, but displayed significant genetic and antigenic (O-antigen) diversity, possibly explaining the variable results regarding their capacity to induce clinical disease in infection trials (Gulla *et al.*, 2015b). The variability between isolates complicates the selection of representative candidate strains for infection trials or vaccines (Gulla *et al.*, 2015b). There is also significant intra-outbreak genetic and antigenic diversity, suggesting direct infection from seawater, rather than fish-to-fish transmission, as the main route of infection (Gulla *et al.*, 2015b). Mortality in corkwing wrasse has been attributed to *V. splendidus* and *V. tapetis* (Jensen *et al.*, 2003). Challenge trials have indicated that both are pathogenic to corkwing wrasse but not to goldsinny wrasse or salmon, though mortalities in the control group complicated analysis (Bergh and Samuelsen, 2007). *V. tapetis* is generally recognised as an opportunistic pathogen in cleaner fish, causing disease only in weakened individuals. Pilot studies indicate that *V. tapetis* is not a primary pathogen of ballan wrasse (Gulla *et al.*, 2017). Wrasse isolates have been found to belong to the same subspecies (*V. tapetis* subsp. *tapetis*) and to four O-antigen serotypes (Gulla *et al.*, 2017). *V. splendidus*, *V. ichthyoenteri* and *V. pacinii* have been isolated from the digestive tract of larval ballan and goldsinny wrasse, though their significance is uncertain (Birkbeck and Treasurer, 2014). *V. ordalii* (formerly classified as *V. anguillarum* biotype 2) causes atypical vibriosis in salmonids and has recently emerged as a significant pathogen in lumpfish in Norway, causing significant losses (NVI 2013; Breiland *et al.*, 2015). *V. ordalii* has been associated with white lesions around the eyes (Bornø *et al.*, 2016). Unlike *V. anguillarum*, *V. ordalii* shows antigenic homogeneity (Toranzo et al., 2005).

Pasteurella *sp.*

Aetiological agent and host: A gram-negative, short rod-shaped bacterium of the genus *Pasteurella,* not yet characterised to species level, is responsible for disease outbreaks in farmed lumpfish in Norway and Scotland. Phylogenetic analysis shows a close relation to *P. skyensis,* a species affecting salmon in Scotland, and *P. varracalbmi* described in salmon in Norway (Alarcón *et al.*, 2016). The first case of pasteurellosis in Norwegian lumpfish was described in 2012, and several cases have been reported yearly since (NVI, 2012; NVI, 2013; NVI, 2014). The disease can affect lumpfish juveniles and broodstock, and has been described in both hatcheries and sea pens. Reported cases mostly occurred during late summer and autumn, and were associated with prior stressful events (Alarcón *et al.*, 2016).

Clinical signs and pathology: Clinical and pathological descriptions are limited to what is reported by Alarcón *et al.* (2016). Mortalities up to 95–100% have occurred (Alarcón *et al.*, 2016; authors' observation). Externally, affected fish can show tail rot, skin lesions (ulcers appearing as white patches on head and/or along the side of the fish), bleeding gills and congestion/haemorrhage of fins. White nodules (granulomas) and haemorrhages can be seen on the internal organs. The histopathology assessment reveals multifocal bacterial microcolonies in skin, gills and internal organs. Reaction in surrounding tissue varies from none to focal necrosis with inflammatory cell infiltration and granulomatous responses (Alarcón *et al.*, 2016). Histopathology can resemble cases of atypical furunculosis. Further bacteriology and molecular analysis are therefore necessary to distinguish between both diseases.

Diagnosis and health management: Complete diagnosis includes histopathology and bacteriology followed by phenotypic and/or genotypic characterisation. The *Pasteurella* sp. from lumpfish exhibits slow growth and requires media containing both blood and salt (2%) (Alarcón *et al.*, 2016). Antibiotic treatments have been tried with mixed success. Florfenicol appears effective at reducing mortalities, but disease can reoccur in the same stock.

Impacts: The potential for mortalities up to 100% both in hatcheries and in sea pens makes *Pasteurella* sp. a serious threat to the production and use of lumpfish. This is exacerbated by the current lack of vaccines and effective treatments. Further research and in-vivo challenges are needed. The potential of transmission to salmon is unknown but considered low (Johansen *et al.*, 2016).

Moritella viscosa

Aetiological agent and host: *Moritella viscosa (*previously *Vibrio viscosus)* is a short rod-shaped, gram-negative bacterium and the primary aetiological agent of winter ulcer disease, a commercially highly significant and difficult to control disease in farmed salmon in Europe and Canada, typically associated with water under 10°C; Gudmundsdóttir and Björnsdóttir, 2007; Løvoll *et al.*, 2009). Non-salmonid species susceptible to the disease include Atlantic cod (*Gadus morhua*) and turbot (*Scophthalmus maximus)* (Colquhoun *et al.*, 2004; Björnsdóttir *et al.*, 2004*). M. viscosa* has been isolated from a variety of marine species with unknown significance, including a low to non-virulent strain from healthy wild lumpfish in Iceland and Norway (Benediktsdóttir *et al.*, 2000) and from wrasse (Johansen *et al.*, 2016). *M. viscosa* can be phenotypically and genetically split into two largely homogenous strains, a 'variant strain' and a 'type strain' (Grove *et al.*, 2010). The type strain is made up almost exclusively of salmon isolates and appears to show a degree of host specificity, while the variant strain infects a wide variety of fish, including lumpfish and salmon (Grove *et al.*, 2010). *M. viscosa* has caused high mortalities in lumpfish at sea in Norway.

Clinical signs and pathology: In salmon, *M. viscosa* generally causes low-level mortalities and large skin ulcers that can lead to welfare issues and loss of economic value (Løvoll *et al.*, 2009). Other clinical symptoms in susceptible species include darkening, fin rot, gill pallor, disseminated skin haemorrhages, lethargy and loss of appetite. Internally, liver ecchymosis resulting in dark spots is typical. Ecchymosis on pyloric caeca and perivisceral fat or generalised haemorrhaging are also described, as are hyperaemia, ascites, congestion and necrosis in spleen and kidneys (Björnsdóttir *et al.*, 2004; Gudmundsdóttir *et al.*, 2006; Bruno *et al.*, 1998). On histopathology, skin lesions/ulcers can show rod-shaped bacteria, inflammation, oedema, necrosis and disseminated haemorrhages in the skeletal muscle and dermis. Necrosis, inflammation and haemorrhaging can be present in internal organs, usually in kidney and spleen. Congestion and dilation of blood-filled sinusoids can occur in the liver (Bruno *et al.*, 1998; Gudmundsdóttir *et al.*, 2006; Grove *et al.*, 2010). Infection can induce granulomatous inflammation in cod (Gudmundsdóttir *et al.*, 2006). Clinical disease with systemic infection and superficial lesions occurs in lumpfish. Lumpfish have shown varying grades of kidney necrosis and superficial lesions can be colonised by secondary bacterial growth, for example with *Tenacibaculum sp.* (G. Jónsson, Iceland, 2015, personal communication).

Diagnosis and health management: *Moritella viscosa* can be cultured both from inoculates of superficial lesions and the kidney, but culture is complicated by a long incubation period and specific requirements to growth media (Benediktsdóttir *et al.*, 2000). Colonies show a typical viscosity. The antigenic variability within *M. viscosa* poses a potential problem for vaccine efficacy (Benediktsdóttir *et al.*, 2000; Heidarsdóttir *et al.*, 2008) and disease outbreaks occur in vaccinated salmon stocks (Gudmundsdóttir and Björnsdóttir, 2007). No data exists on vaccine efficacy in cleaner fish, but a salmon vaccine was not effective in protecting turbot or Atlantic cod (Björnsdóttir *et al.*, 2004, Gudmundsdóttir and Björnsdóttir, 2007). Antimicrobial therapy has not always been effective in salmon aquaculture (Løvoll *et al.*, 2009).

Impacts: It is likely that *M. viscosa* is going to pose a significant disease challenge for lumpfish and there is potential for interspecies transmission to salmon. To date only the 'variant strain' has been described from lumpfish. The variant strain is capable of causing disease in salmon, though this occurs only in isolated cases in Norway where the typical strain is a significant problem in salmon (Grove *et al.*, 2010). There are indications of strain conservation in lumpfish isolates and differences to salmonid isolates in Iceland (G. Jónsson, unpublished; Einarsdóttir *et al.*, 2015).

Pseudomonas anguilliseptica

Aetiological agent and host: *Pseudomonas anguilliseptica* is a long rod-shaped, gram-negative bacterium that was first economically relevant as the aetiological agent of red spot disease in Japanese eel (*Anguilla japonica*) aquaculture in Japan, but has since emerged as a pathogen for a wide variety of fish species in Europe, including salmon and lumpfish. Mortalities of 50% in salmon are documented in the Baltic Sea in Finland (Wiklund and Bylund, 1990). Two genetic groups can be distinguished in this species, one linked to the majority of eel isolates and one to isolates from other species (López-Romalde *et al.*, 2003). In lumpfish, *P. anguilliseptica* is not as widespread as some other bacterial diseases, but outbreaks are linked to high mortality (authors' observation). *P. anguilliseptica* has not been described in wrasse.

Clinical signs and pathology: *P. anguilliseptica* causes haemorrhagic septicaemia in different susceptible fish species. The most consistent symptoms are keratitis, petechial haemorrhaging on the peritoneum, liver and adipose tissue, and external haemorrhaging ventrally, around the mouth, fin bases and vent (Berthe *et al.*, 1995; Ferguson *et al.*, 2004; Wiklund and Bylund, 1990). Lethargy and darkening are reported (Ferguson *et al.*, 2004). In lumpfish, symptoms include external and internal haemorrhaging, congestion, swollen spleen, emaciation, fin erosion and skin irritations (Fig. 13.1E). On histopathology, the bacteria can be seen in the blood vessels (Fig. 13.1F) and internal organs, partly associated with necrosis and inflammation. Other findings include leucocytosis and haemorrhaging in the exocrine pancreas and the lamina propria of the gut (authors' observation). The typical eye infection reported in other species has not been confirmed for lumpfish. Outbreaks seem to occur early in the year, though there are not yet enough data to strongly support this observation (authors' observation).

Diagnosis and health management: *P. anguilliseptica* can be cultured on standard media, but colonies grow comparatively slowly and are easily overgrown by faster-growing species (Vennerström, 2015). Antibiotic treatments have been successful (Vennerström, 2015), but resistances have been documented making a sensitivity test necessary (Wiklund and Bylund, 1990; López-Romalde *et al.*, 2003). Vaccination trials with other species have been successful, but currently there is no commercial vaccine available (Pridgeon and Klesius, 2013; Vennerström, 2015).

Impacts: *P. anguilliseptica* is an important emerging disease and a significant cause of mortality in lumpfish at sea. It has been considered a hazard to salmon farming in the Baltic Sea (Wiklund and Bylund, 1990), but there is no documented case of disease transmission to salmon despite severe disease outbreaks in cohabited lumpfish populations.

Tenacibaculum maritimum (*and* T. dicentrarchi)

Aetiological agent and host: *Tenacibaculum maritimum* (formerly *Flexibacter maritimus)* is a gram-negative, long rod-shaped to filamentous bacterium that causes the disease tenacibaculosis in a wide variety of marine fish species and is a major cause of economic loss to aquaculture worldwide. Juvenile fish are especially susceptible (Toranzo *et al.,* 2005). Genetic variability within this species exists (Avendaño-Herrera *et al.*, 2004; Habib *et al.,* 2014) and different serotypes have been identified, which may pose the need for categorisation of *T. maritimum* into subspecies (Avendaño-Herrera *et al.*, 2006). There is evidence suggesting that *T. maritimum* strains isolated from lumpfish in Norway are unique (Småge *et al.*, 2016). *T. maritimum* has caused high mortality in farm-reared juvenile lumpfish in Norway (Småge *et al.*, 2016) and also infects lumpfish at sea (Johansen *et al.*, 2016), often following transfer. Primary infection occurs on the skin and gills. In salmon aquaculture, tenacibaculosis is commonly seen as a secondary infection following mechanical injury, jellyfish-induced damage or other primary bacterial ulceration. The natural reservoir of *T. maritimum* is unknown. It has been isolated from sediment, water samples and tank walls following disease outbreaks on fish farms (Avendaño-Herrera *et al.*, 2006), as well as from different jellyfish species, including the mauve stinger *(Pelagia noctiluca)* (Delannoy *et al.*, 2011), and from sea lice *Lepeophtheirus salmonis* (Barker *et al.*, 2009). It has been suggested that *T. maritimum* is part of the natural fish skin microbiome and only opportunistically pathogenic (Avendaño-Herrera *et al.,* 2006). Disease occurs more frequently at water temperatures above 15°C and in low water quality conditions (Toranzo *et al.*, 2005). *T. finnmarkense* and *T. dicentrarchi* are other *Tenacibaculum* species that have been isolated from lumpfish in Scotland, Norway, the Faroe Islands and Ireland *T. dicentrarchi* and *T. soleae* are described in wrasse (Olsen et al. 2017). *T. dicentrarchi* infects a variety of marine fish and is an emerging problem in salmon aquaculture in Chile (Avendaño-Herrera *et al.*, 2016). *Tenacibaculum* isolates closely related to *T. dicentrarchi* have caused disease in salmon in Norway (Habib *et al.*, 2014).

Clinical signs and pathology: *T. maritimum* generally causes ulcerative or erosive lesions in the skin and mouth, fin rot and frayed fins, though infection can become systemic (Toranzo *et al.*, 2005). Macroscopic symptoms are white necrotic lesions, especially around the head, on the eyes and bony tubercles, and excess mucus production. Behavioural symptoms include lethargy and loss of appetite (Småge *et al.*, 2016). Brownish-yellow growth can occur. On histology, *Tenacibaculum* bacteria are typically visible as bacterial mats associated with epithelial ulcers. The skin over the bony tubercles is often affected in lumpfish (Småge *et al.*, 2016). Internal pathology is not described for lumpfish.

Diagnosis and health management: Different pathogens, including *Vibrio ordalii*, *Pasteurella* sp. and scuticociliates, can cause similar lesions to *T. maritimum* and secondary or mixed infections with other bacteria or opportunistic parasites can complicate the diagnosis (Avendaño-Herrera *et al.*, 2006). Presumptive diagnosis is currently based on clinical pathology and the presence of filamentous bacteria in wet mounts of skin and gill lesions that display a gliding motility on wet surfaces (Avendaño-Herrera *et al.*, 2006), but diagnosis should always include culture or molecular analysis (Toranzo *et al.*, 2005). Isolation can be difficult, slow growing *T. maritimum* colonies are often overgrown or inhibited by faster-growing species (Pazos *et al.*, 1996) and mixed infections are common. Suitable media includes marine agar (MA) and *Flexibacter maritimus* medium (FMM) (Pazos *et al.*, 1996). Vaccine development for *T. maritimum* is complicated by antigenic variation between isolates (Romalde *et al.*, 2005). PCR screening for *T. maritimum* and the routine use of suitable media for bacteriology when investigating lumpfish mortalities or skin disorders is advisable (Småge *et al.*, 2016). Eliminating predisposing factors is considered vital in preventing disease.

Impacts: *T. maritimum* shows a lack of strict host specificity (Avendaño-Herrera *et al.,* 2006) and the wide host range includes salmon and lumpfish (Småge *et al.*, 2016). There are indications for a possible interspecies transmission and an associated risk to multispecies aquaculture (Habib *et al.*, 2014), but this is not proven. *T. dicentrarchi* is a potential biosecurity risk but there is not yet enough data available to properly evaluate this.

Piscirickettsia salmonis

Aetiological agent and host: *Piscirickettsia salmonis* is a facultative intracellular, coccoid, gram-negative bacterium and the aetiological agent of piscirickettsiosis or salmonid rickettsial septicaemia (SRS) in several salmonid species. *P. salmonis* was first detected in a lumpfish hatchery in Ireland in 2015 (Marcos-López *et al.*, 2017). Other non-salmonid farmed fish species known to be susceptible to *P. salmonis* are the European sea bass (*Dicentrarchus labrax*) (McCarthy *et al.*, 2005) and the white seabass (*Atractoscion nobilis*) (Arkush *et al.*, 2005). *P. salmonis* has also been detected in a number of wild marine fish from the coast of Chile (Contreras-Lynch *et al.*, 2015).

Clinical signs and pathology: Affected lumpfish display marked abdominal distension due to severe ascites. Diseased fish also show very pale gills, pale and swollen kidneys, green discoloration of the liver with multifocal white-creamy patches (Fig. 13.1G), and green discoloration of the heart. Fish can also show multifocal petechial haemorrhages in the adipose tissue around pyloric caeca and hind gut. The histopathology analysis reveals multifocal necrosis and inflammation of the affected organs, mainly liver, pancreas and visceral adipose tissue, with numerous *Piscirickettsia*-like organisms present in affected tissues (Fig. 13.1H). The bacteria can also be seen in gills, kidney and spleen. In Giemsa-stained sections, the coccoid bacteria stain dark blue. The bacteria are intracellular but can be observed extracellularly in necrotic areas (Marcos-López *et. al.*, 2017).

Diagnosis and health management: The diagnosis is based on internal gross and microscopic lesions, combined with the demonstration of the presence of *P. salmonis* using serological and/or molecular methods. *P. salmonis* was previously considered to be cultivable only in cell lines, but more recent reports show that it can also be cultured using specific bacteriology media (Rozas and Enriquez, 2014). In the case that occurred in Ireland, mortalities were very low and no treatment was undertaken. In salmonids, piscirickettsiosis is difficult to control due to lack of efficacy of antibiotic treatments and variable long-term efficacy of currently available vaccines (Rozas and Enriquez, 2014).

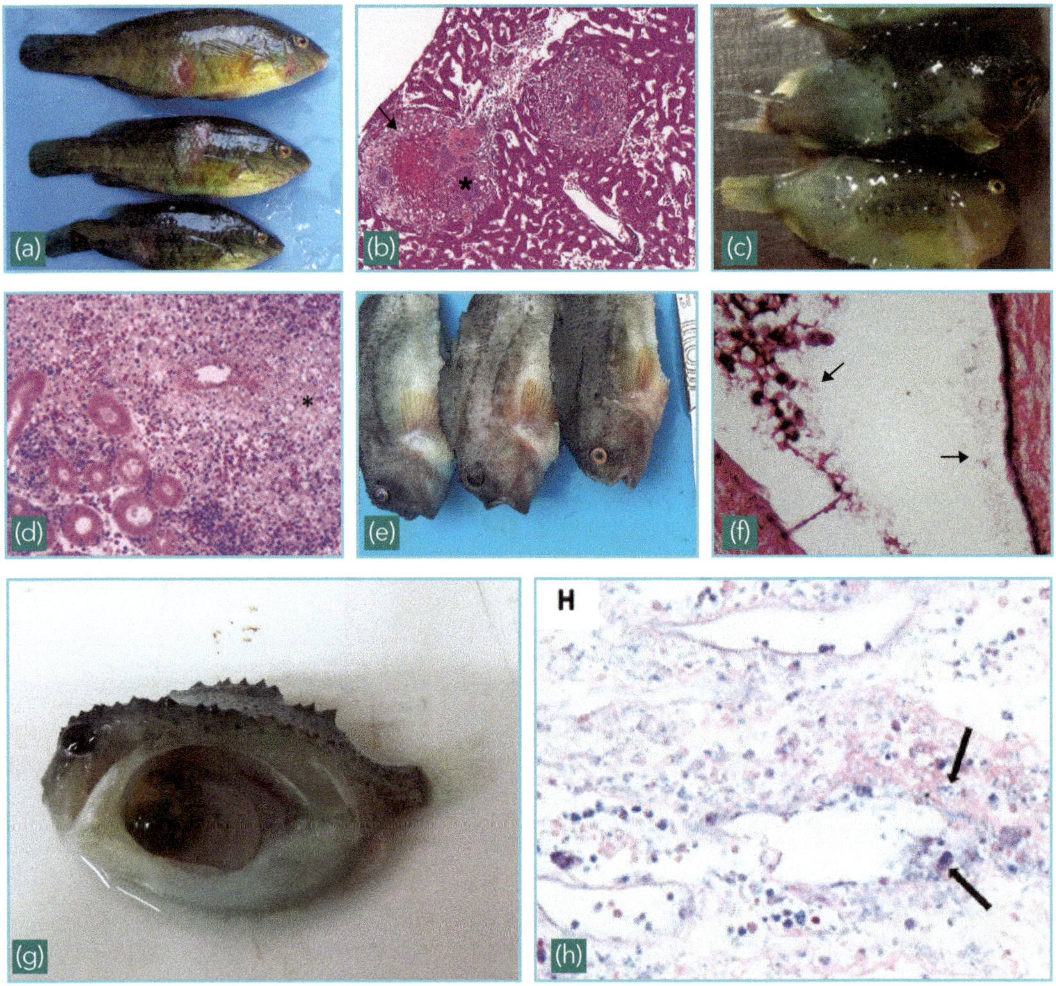

Fig. 13.1 (A) Corkwing wrasse with skin lesions on the flanks due to atypical *Aeromonas salmonicida* vapA layer type 6. (B) Atypical furunculosis in ballan wrasse. *Aeromonas salmonicida* colonies can be seen in liver parenchyma (*) surrounded by granulomatous inflammatory reaction (arrow) and infiltration of eosinophilic granular cells (EGCs) (×40 magnification). (C) Lumpfish juveniles suffering from classical vibriosis. Note skin discoloration and erosion mainly around caudal/ peduncle area and tail rot. (D) *Vibrio anguillarum* infection in lumpfish. Note focal necrosis of the renal interstitium (*) with presence of rod shaped bacteria (×100 magnification). (E) Lumpfish with clinical *P. anguilliseptica* infection. Note congestion and reddening of opercula and fin bases, as well as haemorrhaging in eye on right fish. (F) Bacterial septicaemia (*P. anguilliseptica*) with numerous bacteria (arrows) in branchial blood vessel of lumpfish (×1000 magnification). (G) *Piscirickettsia salmonis* infection in lumpfish. Note pale gills and green colour liver with white-creamy patches. (H) *Piscirickettsia salmonis* infection in lumpfish. Note focal necrosis of the liver with presence of blue stained coccoid organisms (arrow) (Giemsa stain, ×400 magnification). (Credit: M. Marcos-López (B, D, F, G, H); F. Scholz (A, E); L. Stranden, Marine Harvest Norway (C))

Impacts: Piscirickettsiosis is one of the main health issues for the salmon aquaculture industry in Chile, but it also occurs sporadically in Ireland, Scotland and Norway. To the authors' knowledge, *P. salmonis* has only been detected from lumpfish in Ireland, but new cases could potentially occur in other locations. Mortalities in the affected Irish stock were very low, but it is possible infected lumpfish could increase the infectious pressure in salmon pens.

13.3 Parasites

Wild fish naturally harbour parasites and it is normal to find a diverse range of parasites during clinical examination of wild-caught cleaner fish. In most cases these findings are not clinically significant, but under stressful conditions parasitic infections may develop into a serious health challenge. Several articles have reviewed the parasitic flora present in cleaner fish species in different geographical locations including Ireland (Costello *et al.*, 1996), Scotland (Treasurer, 1997), Norway (Karlsbakk *et al.*, 1996) and Poland (Rolbiecki and Rokicki, 2008). In general, most parasites described in wrasse are either host-specific, require an intermediate host (usually an invertebrate) or require the cleaner fish to be eaten to potentially infect salmon. Even if this was to occur in single cases, there is little danger from wrasse parasites to salmon stocks in the pen environment (Treasurer, 1997). A notable exception is the protozoan *Neoparamoeba perurans*. Other parasites of concern are *Anisakis simplex, Pseudoterranova* spp. and *Hysterothylacium aduncum,* which occur in cleaner fish species, can infect salmon and are potentially zoonotic diseases. However, they occur in the marine environment and have not been problematic in salmon farming to date (Johansen *et al.*, 2016). *Gyrodactylus cyclopteri* and *Caligus elongatus* are examples of wild lumpfish parasites that are potentially problematic under aquaculture conditions.

Protozoa

Neoparamoeba perurans

Aetiological agent and host: The marine protozoan *Neoparamoeba perurans* is the causative agent of amoebic gill disease (AGD) in salmon and other farmed marine fish species, including wrasse (Karlsbakk *et al.*, 2013) and lumpfish (Oldham *et al.*, 2016; Haugland *et al.*, 2016). AGD is a problem both at sea and in land-based facilities.

Clinical signs and pathology: Upon colonisation of the fish gills, *N. perurans* induces proliferation of the lamellar epithelium and increased mucus production, resulting in macroscopically visible white mucus patches. The salmon industry uses a standardised gill scoring system to evaluate the presence and severity of these lesions. Both wrasse and lumpfish can develop typical gill pathology (Fig. 13.2A) but, unlike in salmon, they are not a reliable indicator in these species and the classical gill scoring system is unsuitable. In wrasse, typical lesions can be visible only in advanced stages of infection (authors' observation). Lumpfish can show no or only mild macroscopic lesions (Haugland *et al.*, 2016), or a non-raised and diffuse disease presentation with gill paleness being the only symptom at times (authors' observation). In combination with the small operculum opening, this complicates the clinical examination and diagnosis. AGD is fatal for both wrasse and lumpfish if they are not treated. The histopathology assessment shows multifocal hyperplasia and fusion of the lamellar epithelium with formation of inter-lamellar vesicles and varying numbers of amoebae (Fig. 13.2B). Histopathology is also useful to detect subclinical infections not visible macroscopically and should be part of routine examinations. Amoeba numbers can be underestimated on histopathology as they can be washed off in the fixing process.

Diagnosis and health management: In clinical cases, a complete diagnosis includes gross observation, histopathology, fresh smears of AGD lesions or gill surface (Fig. 13.2C) and molecular identification of *N. perurans*. All stocks, both at sea and in land-based facilities, should be monitored for AGD and species-specific protocols are needed. PCR screening of gill swabs is the recommended method for the detection of subclinical or early infections, with the advantage of being sensitive and non-lethal. In breeding facilities, monitoring the intake water is an option. Treatment plans should be prepared in advance. Freshwater (<3 ppt for three hours) and hydrogen peroxide (800–1200 ppm for 20 minutes) baths are the only treatments available to treat AGD in salmon. Both treatments have been tried in cleaner fish with the following results. Lumpfish can be treated with pure freshwater or with brackish water. During treatment it is important to keep the water well aerated and monitor the fish for signs of distress. Timely treatment is advisable as mortalities can occur when treating AGD in an advanced state (Fig. 13.2D). Wrasse do not tolerate freshwater treatment, but both adults and juveniles can be treated with brackish water for extended periods of time. Salinities of 15 ppt for six days (with slow down and up regulation of salinity, approximately two days on each side), have given a 100% clearance of *N. perurans* (authors' observation). A treatment trial with wrasse using 15 ppt brackish water showed no effect after three hours, mild effect after 24 hours and good results after three days or more (Dahle, 2015). Using 20 ppt for five days has given unsatisfactory results (authors' observation). Wrasse tolerate H_2O_2 treatments (up to 1200–1400ppt), but it has been reported that these do not give 100% clearance and fish soon become re-infected. It should be noted that while fish survive the treatments as described, no study on the effects or potential damage caused to the fish have been done to date.

Impacts: The impact of AGD has been high on both wrasse and lumpfish and the ubiquitous nature of the pathogen in many areas means AGD will continue to be a problem. Effective treatment protocols are necessary, which is especially challenging at sea. Lumpfish are very susceptible and clinical pathology in net pens usually occurs around the same time as it is seen in salmon. To date, *N. perurans* is the only disease that has been experimentally transmitted from lumpfish to salmon (Haugland *et al.*, 2016) and cleaner fish can act as a reservoir, re-infecting the salmon if treatments are not coordinated.

Scuticociliate species

Aetiological agent and host: Scuticociliates (phylum Ciliophora) are widespread in marine and estuarine environments and can be histophagous parasites under certain circumstances, resulting in high mortalities in aquaculture worldwide (Piazzon *et al.*, 2013). No species identification has been made on scuticociliates infecting lumpfish but macroscopically they resemble *Uronema marinum*. *U. marinum* has been recorded as causing superficial as well as severe systemic infections in gills, skeletal muscle and viscera of a wide variety of marine fish (Lom and Dyková, 1992). The condition is quickly lethal once systemic, but death has also been attributed to high parasite loads interfering with normal gill function.

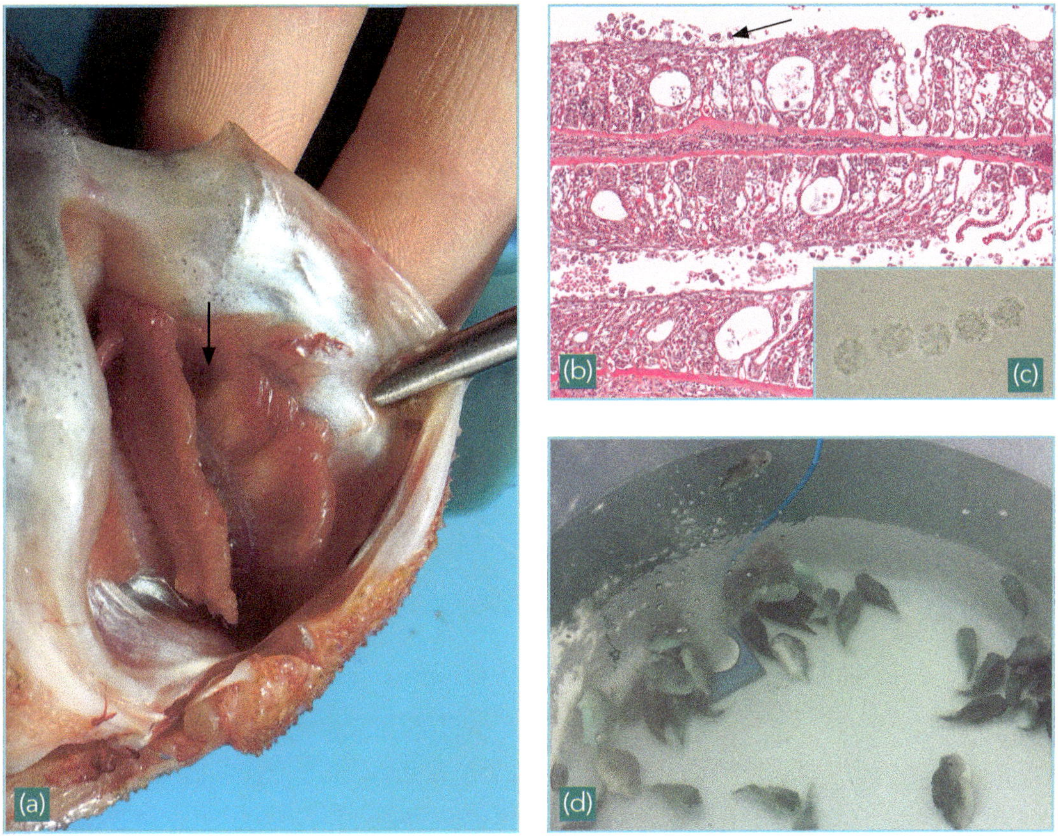

Fig. 13.2 (A) Raised mucus AGD lesions (arrow) in lumpfish, gill opening enlarged surgically. (B) Histopathology of AGD in ballan wrasse. Amoeba (arrow) can be seen in association with hyperplasia and fusion of the lamellar epithelium (×40 magnification). (C) *N. perurans* in a gill scrape from a lumpfish with clinical AGD. Note pseudopodia and granular appearance (×400 magnification). (D) Adverse effects of freshwater treatment on lumpfish broodstock with severe AGD. Note fish lying on their sides (Credit: F. Scholz (A, C, D); M. Marcos-López (B)).

Clinical signs and pathology: Scuticociliate infection usually presents externally as fin and skin lesions, exophthalmia and abdominal distension, but mortalities without external symptoms can occur. Internally, muscle lesions, haemorrhages and liquefaction of the brain are described (Piazzon *et al.*, 2013). In lumpfish, infection has been linked to white ulcerative lesions, mostly located on the head and around the eyes (authors' observation). Scuticociliates have been isolated from the gills of lumpfish with clinical AGD and from skin lesions of primarily fungal origin (authors' observation). Protozoan ciliates have been described as a secondary infection on the gills of lumpfish in Norway (Alarcón *et al.*, 2015).

Diagnosis and health management: Scuticociliates are easily visible as small, fast-moving parasites in fresh scrapes of skin lesions, gills or wet mounts of organs (Fig. 13.3A), and can be seen on histology. The species can be identified by special staining followed by morphometric analysis, or by PCR analysis (Piazzon *et al.*, 2013). Early stage infections have been treated with

freshwater or formalin baths. The disease is usually associated with predisposing factors (Lom and Dyková, 1992), such as other primary diseases or stress.
Impacts: Scuticociliates are significant secondary pathogens in lumpfish but they have not been associated with disease in healthy salmon.

Trichodina *species*

Aetiological agent and host: *Trichodinidae* (phylum Ciliophora) are mobile ciliate organisms, commonly found on a wide variety of aquatic vertebrates and invertebrates, either as parasites or as commensals (Basson and Van As, 2006). Low numbers of *Trichodina* are common on wild fish species but in immunologically impaired hosts they can proliferate, damage epithelial cells and reduce the functional surface area of skin and gills (Basson and Vas As, 2006). *T. cyclopteri* and *T. galyae* are host-specific and significant parasites of lumpfish (Johansen *et al.*, 2016) and *Trichodina* sp. have been a problem in lumpfish hatcheries. Wild-caught wrasse in some areas have a very high prevalence of infection, though numbers are usually low and not associated to any pathology.
Clinical signs and pathology: Symptoms can include excessive mucus production with associated greyish appearance, peeling epithelium and frayed fins as well as behavioural symptoms such as lethargy and anorexia (Basson and Van As, 2006). Infections in lumpfish hatcheries have been associated with lesions on the skin. On histology, *Trichodina* are most commonly seen between gill lamellae (Fig. 13.3B). They can detach from the tissue during fixing and processing, which can lead to an underestimation of numbers.
Diagnosis and health management: *Trichodina* are easily identified in skin and gill scrapes and numbers can be very high in clinical cases. Species identification, however, is complicated and involves special staining to highlight morphological features (Lom and Dyková, 1992). Elevated numbers of *Trichodina* are often associated with poor water quality or other predisposing factors. Bath treatments with formalin and freshwater have been effective.
Impacts: In wrasse, *Trichodina* have not been significant as primary pathogens. They are a potential problem in lumpfish hatcheries and fish should be treated if numbers are high.

Ichthyobodo necator *complex*

Ichthyobodo necator (syn. *Costia necatrix)* is a flagellate that alternates between a free swimming stage and an attached feeding stage. Different species exist (Todal *et al.*, 2004; Callahan *et al.*, 2005), leading some authors to use the term *I. necator* complex. *I. necator* is an obligate pathogen most problematic in juvenile fish, while in older fish the disease is usually associated with predisposing factors (Lom and Dyková, 1992). Commercially the disease is most significant in freshwater hatcheries, but a wide variety of marine species are also susceptible. The attached form fixes on to epithelial cells of skin and gills where they can cause widespread pathology, leading to death in severe cases. Clinical symptoms include 'flashing' (scraping on the bottom) and a greyish cast caused by hyperplasia of the body epithelium. *Ichthyobodo* spp. are regularly seen on wrasse, though not associated to significant pathology. The parasites are easily identified in wet mounts of skin and gill scrapes or on histology (Fig. 13.3C).

Myxozoa (Myxosporea)

Myxozoa are metazoan parasites with an extremely reduced morphology belonging to the phylum Cnidaria (Takeuchi *et al.*, 2015) that infect either tissues (histozoic Myxozoa) or cavities (coelozoic myxozoa) of a wide variety of marine hosts. The life cycle typically includes two phases. The sexual actinosporean phase normally takes place in the definitive host, an invertebrate, and results in the production of an actinospore. The actinospore infects the intermediate host, usually a fish, orally or through the skin or gill epithelium. Spore formation takes place in plasmodia or pseudoplasmodia (Gruhl, 2015), which can be visible macroscopically. Host specificity in the Myxozoa ranges from strictly host-specific to low host specificity (Feist and Longshaw, 2006; Kristmundsson and Freeman, 2014). Most myxozoan infections remain chronic and subclinical (Cavin *et al.*, 2012), but several species are problematic in aquaculture, such as *Myxobolus cerebralis* (whirling disease), *Tetracapsuloides bryosalmonae* (proliferative kidney disease) or *Kudoa thrystis* (soft flesh disease). The infection usually occurs through an invertebrate host, but direct transmission between fish is documented (Cavin *et al.*, 2012). Diagnosis is still largely reliant on spore morphology, but molecular analysis is preferable (Feist and Longshaw, 2006). To date, there are no consistently effective treatments against myxozoan infections. Infective environmental stages can be killed by disinfecting water with UV, light ozone, hypochlorite and hydrogen peroxide (Cavin *et al.*, 2012; Feist and Longshaw, 2006).

Kudoa islandica

Aetiological agent and host: The genus *Kudoa* comprises 95 species, mostly host-specific parasites in the skeletal muscle of fish, though other organs can be infected. *Kudoa* are commercially significant as some species cause post-mortem myoliquefaction, known as 'soft flesh disease' as well as white nodules in the muscle. *Kudoa islandica* was first described in wild lumpfish populations in Iceland, where the phenomenon of spoiled lumpfish meat with white nodules has been known for decades (Kristmundsson and Freeman, 2014). Recently, this parasite was also diagnosed in farm-reared lumpfish in sea pens in Norway (Alarcón *et al.*, 2015). *K. islandica* has been reported in lumpfish, spotted wolffish (*Anarhichas minor*) and Atlantic wolffish (*Anarhichas lupus*) (Kristmundsson and Freeman, 2014); the life cycle and mode of transmission are not known. *Kudoa*-like myxozoa are regular incidental findings in the skeletal muscle of wrasse (Fig. 13.3D), but so far have not been linked to any clinical symptoms and the species has not been identified.
Clinical signs and pathology: *K. islandica* is histozoic, infecting the skeletal muscle. Fresh plasmodia are white, tubular structures measuring up to 11 × 1.4 mm and in severe cases they can make up a large proportion of the trunk muscle (Kristmundsson and Freeman, 2014). In general, *Kudoa* spp. do not have an obvious impact on fish health (Moran *et al.*, 1999), but severe infection can be expected to impact the swimming capability and welfare of fish. Impaired swimming capability and loss of weight in a case of co-infection with *Nucleospora cyclopteri* is described (Alarcón *et al.*, 2015). Post-mortem myoliquefaction is indicative, though other pathogens, such as *Tetramicra brevifilum*, can cause a similar condition and mature female lumpfish can have a naturally soft flesh, not to be confused with the pathological condition. On histology, plasmodia are seen intracellularly in white muscle fibres. *K. islandica* differs from most other kudoids in that often multiple plasmodia are located in one muscle cell (Kristmundsson and Freeman, 2014).

Typically, there is no host response, but focal necrosis can occur in combination with ruptured plasmodia (Kristmundsson and Freeman, 2014). Pseudocysts develop when large numbers of plasmodia virtually substitute the sarcoplasm of grossly enlarged infected muscle cells and are encased by muscle fibre remnants (Moran *et al.*, 1999).

Diagnosis and health management: Grossly visible plasmodia in the skeletal muscle can give a strong indication of the aetiology. The genus *Kudoa* is easily identified through mature spore morphology, as seen on histology or fresh imprints/squashes under a light microscope. Spores are made up of four, equal-sized, pyriform polar capsules. Special stains such as gram, silver nitrate, PAS and Giemsa help in the identification. PCR protocols are available (Kristmundsson and Freeman, 2014). No known treatments exist.

Impacts: The presence of *K. islandica* in wild fish populations make it likely that this myxozoan will present a recurring problem in sea cages. *Kudoa* will not likely cause high mortality, but may impact fish welfare and decrease delousing efficiency as well as presenting a problem in regards to possible further use of the fish. The possibility of transmission to salmon exists, as *K. islandica* is one of the few *Kudoa* species known to infect fish from different taxonomic orders (Kristmundsson and Freeman, 2014).

Myxobolus aeglefini (*syn.* M. albi)

Myxobolus aeglefini is a cartilage-infecting myxozoan that has been described in a range of gadoids and pleuronectids as well as in lumpfish. The species was found to be synonymous with *M. albi* (Karlsbakk *et al.*, 2017). *M. albi* was reported from wild-caught lumpfish in Maine, USA (Cavin *et al.*, 2012). In lumpfish, it infects cartilage throughout the host body, including gill arches, skull, vertebra, tongue, fin insertions and pectoral girdle. It also infects cartilage in the sclera of the eye, where it is macroscopically visible as white or tan nodules (1–2 mm diameter) covering up to 70–90% of the sclera. This is associated with uni- or bilateral exophthalmos. Hyphaema in the eye, spiral swimming and increased respiratory rate can occur. An inflammatory response is only associated with ruptured cysts and free spores (Cavin *et al.*, 2012). *M. aeglefini* is possibly underdiagnosed as subclinical infections may go unnoticed in routine clinical examination and histology. Infection of salmon is considered unlikely.

13.4 Fungi

Microsporidia

Microsporidia are unicellular, spore-forming, obligate intracellular organisms (Dyková, 1995), now widely accepted as belonging to the kingdom fungi (Capella-Gutiérrez *et al.*, 2012), that can infect a wide range of fish species (Lom and Nilsen, 2003). Microsporidia can cause focal or systemic infections, which can occur as subclinical or severe disease depending on parasite, host and/or environmental factors. Fish microsporidia are found in the cytoplasm of the host cells, except the genera *Nucleospora*, which is found in the nucleoplasm (Lom and Nilsen, 2003). Some cytoplasmic species induce hypertrophic growth of the host cell, ending in the formation of a xenoparasitic complex known as xenoma (Lom and Dyková, 2005). Fish-infecting microsporidia

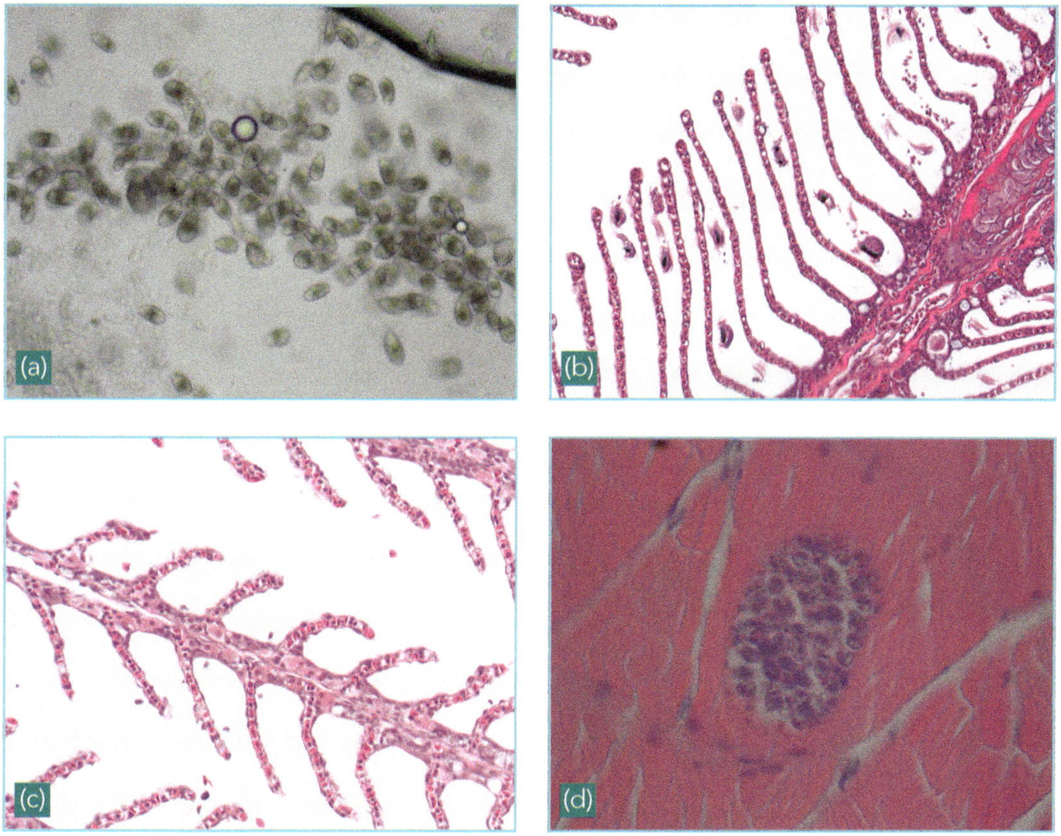

Fig. 13.3 (A) *Uronema*-like parasites in a skin scrape of lumpfish with skin ulcers (x600 magnification). (B) Histology of *Trichodina* sp. between the gill lamellae of ballan wrasse (x100 magnification). (C) *Ichthyobodo* sp. in association with the gill epithelium in wrasse (x200 magnification). (D) Plasmodium with *Kudoa*-like spores in muscle of corkwing wrasse (x600 magnification) (Credit: F. Scholz (A, D); M. Marcos-López (B, C)).

are usually transmitted orally and microsporidian spores can retain their viability in water for at least a year at 4°C (Lom and Dyková, 1992). The role of crustaceans as reservoirs or vectors for microsporidian species has been demonstrated (Freeman and Sommerville, 2011; Lom and Dyková, 2005) and the involvement of benthic species such as oligochaetes is likely. Genera and species determination are based on spore morphology, characteristics of developmental stages (Dyková, 1995) and molecular analysis.

Nucleospora cyclopteri

Aetiological agent and host: *Nucleospora cyclopteri* is an intranuclear microsporidian, most closely resembling the salmon parasite *Nucleospora salmonis* (96% identity). Lumpfish are the only known host for this species. *N. cyclopteri* is common in wild lumpfish populations around Iceland (Freeman *et al.*, 2013), it has been diagnosed in wild (Karlsbakk *et al.*, 2014) and farmed

(Alarcón *et al.,* 2015) lumpfish in Norway, wild-caught broodstock from England (authors' observation) and most likely in a recirculation facility in Canada (Mullins *et al.,* 1994; Freeman and Kristmundsson, 2013). The life cycle and mode of transmission are not yet known, but most fish-infecting microsporidia show horizontal transmission and have a simple, direct life cycle (Lom and Nilsen, 2003). Freeman *et al.* (2013) indicated some evidence for direct transmission of *N. cyclopteri*, but also suggested that the parasitic copepod *Caligus elongatus* could play a role in the infection of lumpfish.

Clinical signs and pathology: Clinical signs of disease in wild fish are very difficult to observe and there are no data from the wild infections. In the case reported in farmed lumpfish from Norway (Alarcón *et al.,* 2015), affected fish (which also harboured other parasites) showed erratic swimming near the water surface, emaciation and mortalities up to 65%. External lesions reported include bilateral exophthalmus, branchial pallor and small skin lesions (Freeman *et al.,* 2013; Mullins *et al.,* 1994). Internally, the kidney is the most affected organ, with presentation ranging from normal-sized kidneys with a patchy appearance or with white nodules, to extensive enlargement and pallor of the whole kidney (Fig. 13.4A) (Alarcón *et al.,* 2015; Freeman *et al.,* 2013). Cardiac pallor and serosanguinous peritoneal fluid have also been documented (Mullins *et al.,* 1994). By histopathology, microsporidia are seen in the nuclei of lymphocytes or lymphocyte precursor cells in all organs (Fig. 13.4B), although the kidney and the spleen are the most severely affected. In severe cases, extensive necrosis and infiltration of the affected organs occur in association with the presence of high numbers of infected cells.

Diagnosis and health management: Disease diagnosis is achieved through gross pathology, histopathology (×1000 magnification) and PCR. Kidney enlargement and pallor in severe cases is indicative, but the definitive diagnosis requires histopathology and PCR. Spores are uniformly oval along two lines of symmetry and measure approx. 3.12 × 1.3 μm in a fresh state (Freeman *et al.*, 2013). Nested-PCR analysis is significantly more sensitive and suited for population screening and detection of subclinical infections, with kidney and heart being the most suited tissues followed by skin, gills and spleen (Freeman *et al.*, 2013). There is no known treatment for *N. cyclopteri* and effective treatment under farm conditions is unlikely. Screening of wild-caught broodstock and populations at sea is advisable considering the prevalence in wild fish and common subclinical infections (Freeman *et al.*, 2013).

Impacts: *N. cyclopteri* is a proven pathogen of lumpfish and capable of causing severe clinical disease and mortality (Alarcón *et al.*, 2015; Freeman *et al.*, 2013). Impairment of the immune system is likely considering the infection of lymphocytes. The high prevalence in some wild fish populations makes it likely that lumpfish at sea will be challenged with this pathogen regularly. Vertical transmission is common in the phylum Microspora (Terry *et al.*, 2004) and the regular detection of *N. cyclopteri* in milt, eggs, and egg fluid of lumpfish broodstock indicates that there may be a risk of vertical transmission (authors' observation). Salmon held with infected lumpfish were not affected (Alarcón *et al.*, 2015) and the transmission to salmon is considered unlikely.

Tetramicra brevifilum

Aetiological agent and host: *Tetramicra brevifilum* is a xenoma-forming microsporidian (Matthews and Matthews, 1980) that has caused significant mortality in farmed turbot in Spain (Figueras *et al.*, 1992) and has been found in wild turbot and black anglerfish (*Lophius budegassa*)

(Maíllo *et al.*, 1998; Matthews and Matthews, 1980). The first case in lumpfish was diagnosed in Ireland, where *T. brevifilum* caused significant disease in wild-caught lumpfish broodstock (Scholz *et al.*, 2016). Nothing is known about this species outside of the fish host or the route of infection, though intramuscular spore injection has experimentally infected turbot and there is evidence that immune cells are involved in spreading the infection within the host (Matthews and Matthews, 1980).

Clinical signs and pathology: External clinical symptoms in lumpfish include bloating, exophthalmos and the formation of blister-like lesions with a recorded diameter of up to 7 mm on the body surface. White patches can cover parts of the corneal surface. Behavioural symptoms such as lethargy and loss of appetite occur only in late stages of the disease. Cysts and white nodules are documented on all internal organs, skin, muscle, the peritoneum and around the optical nerve (Fig. 13.4C) (Scholz *et al.*, 2016). At necropsy, affected fish can show large quantities of white nodules in the skeletal muscle, partly associated with myoliquefaction, and ascites. Ascetic fluid can contain high numbers of aggregated spores macroscopically visible as white material. The liver is usually the most heavily affected organ with cysts covering up to half of the organ surface in severe cases. On histopathology the infection is associated with the formation of xenomas in all internal organs, skin, skeletal muscle and the retina. Xenomas are often branched in form and found inside cysts measuring up to 220 µm in diameter (Fig. 13.4D). These cysts often form space occupying complexes (Scholz *et al.*, 2016). Diagnostic features of xenomas are the reticulated host cell nucleus and the microvillous xenoma surface (Matthews and Matthews, 1980). An inflammatory response occurs only in combination with ruptured xenomas and free microsporidian stages in the host tissue (Scholz *et al.*, 2016). In the muscle tissue, infection is partly associated with necrosis, fibrosis, hyaline degeneration and the mechanical separation of myofibrils. Free microsporidian stages have been seen in blood vessels and the spleen (Scholz *et al.*, 2016).

Diagnosis and health management: The gross pathology and the histopathology are strongly indicative, though the species should be confirmed through molecular analysis. In fresh smears of ascetic fluid or cyst content, ovoid microsporidian spores with a prominent posterior vacuole containing an inclusion body can be seen under a light microscope. There is no known treatment for *T. brevifilum* infection. Careful management and removal of severely infected fish to lower infective pressure was the only way of controlling the disease demonstrated in turbot (Figueras *et al.*, 1992).

Impacts: *T. brevifilum* is a significant pathogen of lumpfish and is seen commonly in lumpfish at sea around the Irish coast (authors' observation). Vertical transmission is described for microsporidia (Dunn and Smith, 2001) and the presence of *T. brevifilum* in the gonads implies this possibility. It is likely that the infection can remain subclinical until triggered, so screening broodstock is advised. The danger of transmission to salmon is considered remote, though it has not yet been investigated.

Exophiala *species*

Aetiological agent and host: *Exophiala* species are melanised fungi, also known as black yeasts. They are widespread, opportunistic pathogens causing systemic or superficial disease in both warm- and cold-blooded animals. *Exophiala* describes an anamorphic element, or the asexual life cycle stage, of some fungi in the order Chaetothyreales, while the teleomorphic relations,

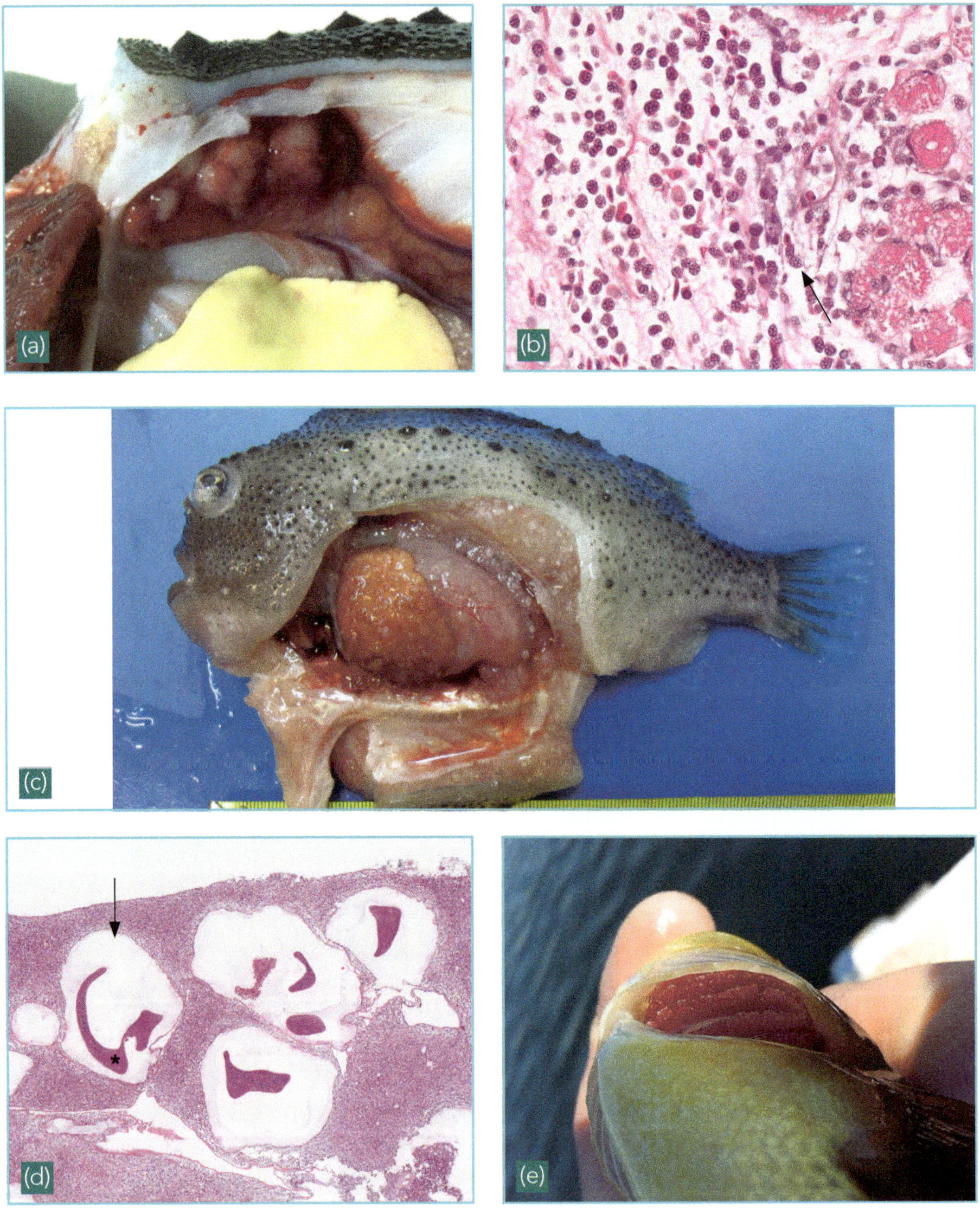

Fig. 13.4 (A) Female lumpfish broodstock with kidney pathology due to *N. cyclopteri*. (B) *Nucleospora cyclopteri* infection in dermis of lumpfish. Microsporidia can be seen within the nuclei of lymphocyte-like cells (arrow) (×1000 magnification). (C) Severe *T. brevifilum* infection. Note ascites, microsporidian xenomas in internal organs and skeletal muscle, white discoloration in sclera and vesicles on skin. (D) Xenomas in liver (*) with cyst formation (arrow) (×40 magnification). (E) Severe *Loma*-like infection visible as white nodules on the gills of a ballan wrasse. (Credit: F. Scholz (C, D, E); M. Marcos-López (B); L. Stranden, Marine Harvest Norway (A)).

meaning the equivalent sexual life cycle stages, lie in the genus *Capronia* (De Hoog *et al.*, 2011). The asexual stage, *Exophiala*, is the only stage pathogenic to fish. Two species have been diagnosed from lumpfish to date, *E. angulospora* and *E. psychrophila* (authors' observation), usually associated with chronic mortality. *E. angulospora* has been found worldwide in cold fresh- and saltwater environments, but also in polluted soil, human skin and nails (De Hoog *et al.*, 2011). Besides their capability of causing systemic infection in lumpfish, *E. angulospora* can affect Atlantic cod (Gjessing *et al.*, 2011) and halibut (*Hippoglossus hippoglossus*) (Overy *et al.*, 2014), and *E. psychrophila* can infect salmon.

Clinical signs and pathology: Fungal infection is externally visible as multifocal dark skin discoloration or dark, ulcerative lesions that can reach several centimetres in diameter (Fig. 13.5A). These often first occur on exposed parts of the body such as bony tubercles, possibly following mechanical damage (authors' observation). Ulcers can occur together with secondary bacterial growth that can alter the appearance of the lesions. Fungal growth can also affect the gills (Fig. 13.5A), appearing as brownish or grey patches. Internally, pathology ranges from large dark spots to raised, dark grey lumps in different visceral organs (Fig. 13.5C) and on internal surfaces. The infection is usually first seen in skin, spleen and kidney, but eventually all organs get infected. Internal pathology is usually severe by the time external lesions are noticed (authors' observation). In histology, septate, branching, brown-pigmented fungal hyphae are seen in the affected organs (Fig. 13.5B) accompanied by multifocal to confluent necrosis, and high numbers of inflammatory cells. A granulomatous inflammatory reaction has been described (Gjessing *et al.*, 2011).

Diagnosis and health management: A presumptive diagnosis can be reached after gross examination, wet mount microscopy and histopathology, but molecular characterisation is required for species identification. *E. angulospora* and *E. psychrophila* grow on malt extract agar, potato dextrose agar and *E. angulospora* also grows on blood agar (Fig. 13.5D) (authors' observation). There is no known treatment for *Exophiala* infections in fish. Biosecurity measures should be put in place to prevent infection and spreading. Screening of broodstock populations is advised.

Impacts: The diagnosis of *Exophiala* has severe consequences and stocks have been culled following the diagnosis of both *E. angulospora* and *E. psychrophila*. The significance of *E. angulospora* isolates from human skin and nails is not known as there is no proof of disease available (De Hoog *et al.*, 2011). The significance of *E. angulospora* to salmon is unknown. *E. psychrophila* has caused chronic infection with high mortality in salmon smolts in Norway (Pedersen and Langvad, 1989), but reports are rare.

Other fungi

Loma are xenoma-forming microsporidia that infect a wide range of teleost hosts and are a common incidental finding on the gills of wrasse. The species is unknown. Xenomas are approximately 1 mm in diameter and can be visible as white nodules (Fig. 13.4E). Even heavy infections have not been linked to clinical symptoms and the host response seen on histology is mild to non-existent. An *Ichthyophonus hoferi*-like infection has been reported from lumpfish in Norway (NVI, 2014). This fungus has been reported from more than 80 marine species and has been responsible for high mortalities in feral fish stocks, but has not been a serious problem in aquaculture (McVicar, 1999). The species can be distinguished by the presence of big fungal spores with a thick double wall and multinucleated cytoplasm on histology.

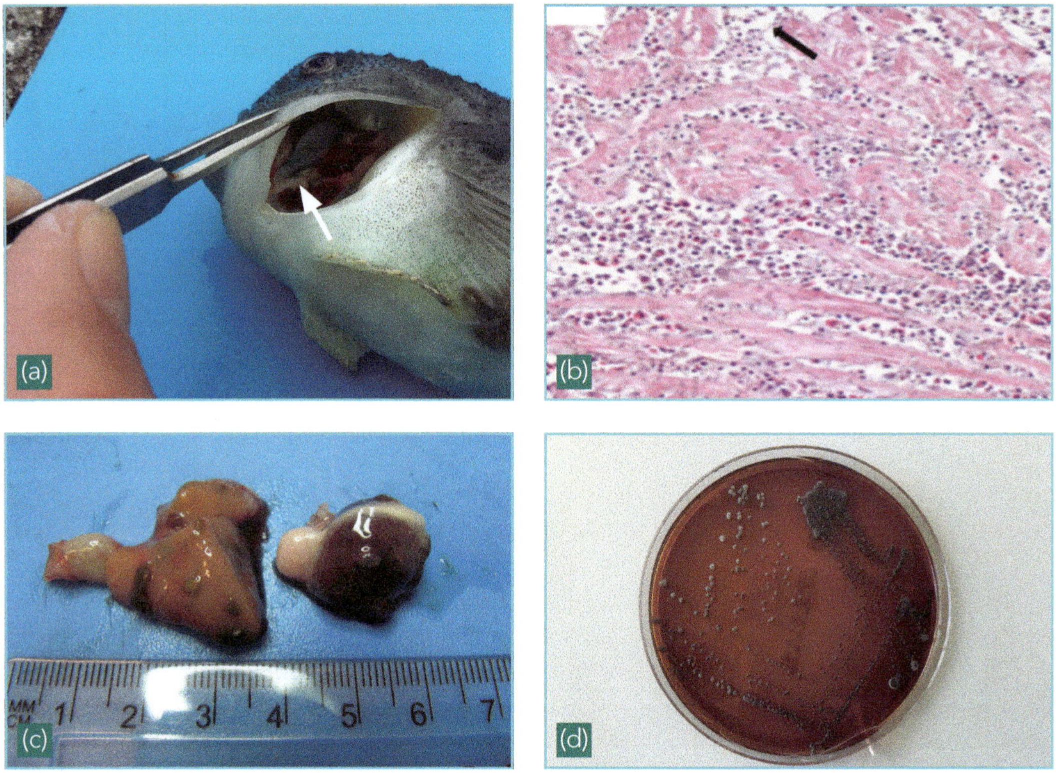

Fig. 13.5 (A) *Exophiala angulospora* infection in lumpfish. Fungal lesions are visible on the ventral flank and the gills (arrow). (B) *Exophiala* sp. infection in heart of lumpfish. Fungal hyphae are within the ventricle's lumen (arrow) and in association with cardiomyocytes (×200 magnification). (C) Fungal nodules on heart and spleen of lumpfish with severe *Exophiala angulospora* infection. (D) *Exophiala angulospora* growing on Columbia blood agar plus 2% salt (Credit: F. Scholz (A, C, D); M. Marcos-López (B).

13.5 Viruses

Viral diseases are a major problem in aquaculture as vaccines are often less effective than in bacterial diseases and antiviral therapies are not commercially viable (Noga, 2010; Robertsen, 2011). Though some viral vaccines exist, management strategies are often the only practical way of reducing impact. Not many viruses are known from cleaner fish species, but more are likely to be identified as research progresses. Uncertainty regarding the potential for cleaner fish to transmit salmonid viruses has also been a concern when using wild-caught fish (Treasurer, 2002). The risk to salmon posed by novel viruses of cleaner fish and by the capacity of viruses to adapt to new hosts is unknown but cannot be neglected.

Viral haemorrhagic septicaemia virus

Aetiological agent and host: The viral haemorrhagic septicaemia virus (VHSv) is a negative-stranded RNA virus (*Rhabdoviridae*), and agent of viral haemorrhagic septicaemia (VHS), a highly significant disease in aquaculture. VHSv has been detected in at least 80 marine and freshwater species and can be divided into four genotypes (Einer-Jensen *et al.*, 2005; Snow *et al.*, 1999). Subtypes are described for genotypes I and IV, and are associated to geographic area rather than fish species (Elsayed *et al.*, 2006; Gagné *et al.*, 2007). Genotype III (European marine strain) has been isolated from all wrasse species commonly used as cleaner fish in Shetland and Scotland (Hall *et al.*, 2013; Munro *et al.*, 2015). This genotype is endemic in wild fish stocks in the North Sea and the North Atlantic (Gagné *et al.*, 2007; Dale *et al.*, 2009), and has caused clinical disease in farmed turbot (Ross *et al.*, 1995) and farmed rainbow trout (*Oncorhynchus mykiss*) (Dale *et al.*, 2009). Investigations following the detection in wrasse indicated that these were most likely infected by wild fish locally (Hall *et al.*, 2013). Genotype IV (North American strain) has been isolated from lumpfish in Iceland, though clinical VHS was not confirmed (Guðmundsdóttir, 2016). In a challenge trial, high mortality occurred in injected lumpfish and in naïve cohabited lumpfish (Cuenca *et al.*, 2017). Other genotype IV subtypes have caused high mortality in different species, such as Pacific herring (*Clupea pallasii*) and sardine (*Sardinops sagax*) (IVa) (Garver *et al.*, 2013), wild fish in the North American great lakes (IVb) (Elsayed *et al.*, 2006) and wild fish on the North American Atlantic coast (IVc) (Pierce and Stepien, 2012). The isolate from lumpfish in Iceland may present a new subtype based on molecular sequencing (Guðmundsdóttir, 2016). The pathogenicity factors of VHSv are not well known (Gagné, 2007). Infection usually occurs from fish to fish and infected fish shed the virus with urine and ovarian fluids (Skall *et al.*, 2005). Species not susceptible to the disease can also carry and shed the virus, and viral particles can survive for two weeks in the marine environment. RNA viruses are prone to mutation and VHSv demonstrates a high ability of host adaptation and the development of new, virulent strains.

Clinical signs and pathology: Gross pathology of VHS in susceptible fish species includes haemorrhaging on the skin, internal organs, gills, muscle and eyes, as well as pale gills, exophthalmia, ascites and darkening of the body (Skall *et al.*, 2005). High mortalities can occur with no external symptoms in acute cases (Emmenegger *et al.*, 2013). Chronic or subclinical infections are common, and most wild fish isolates came from clinically healthy fish. External lesions and internal haemorrhaging were described in VHSv-positive moribund wrasse (Munro *et al.*, 2015), but subsequent experimentally infected goldsinny wrasse did not develop the typical haemorrhagic lesions (Matejusova *et al.*, 2016). Heart pathology (multifocal myocardial necrosis, degeneration, infiltration and endocardial proliferation) was the most consistent finding in naturally and experimentally infected wrasse (Matejusova *et al.*, 2016; Munro *et al.*, 2015), and the atrium was found to be the target tissue for VHSv in goldsinny wrasse (Matejusova *et al.*, 2016).

Diagnosis and health management: It is impossible to protect against infection in net pens considering the presence of the virus in the marine environment, but the risk can be minimised by not feeding unprocessed fish to cleaner fish. Screening of wrasse and lumpfish populations by virological and molecular biological methods is advisable.

Impacts: VHS is on the OIE list of notifiable diseases, and for the OIE the presence of the virus is synonymous with the disease. The diagnosis has severe consequences in certain regions, including culling of the affected fish and potentially the loss of VHS-free status in an area. VHSv genotype III was evaluated as posing a risk to multispecies aquaculture (Matejusova *et al.*, 2016). While not pathogenic to salmon in bath challenges, this genotype has produced high mortalities in salmon following i.p injection (Dale *et al.*, 2009). However, following the outbreak in Shetland no salmon were found infected and to date VHSv has not been found in salmon in European waters where this strain is endemic. Genotype IVa isolates have repeatedly demonstrated a low virulence for salmonids (Emmenegger *et al.*, 2013), but have been isolated from farmed salmon on the North American east coast (Garver *et al.*, 2013). The lumpfish strain did induce clinical signs and morbidity in injected salmon, but was less pathogenic than reference strains Ia and IVa. Salmon cohabited with shedding lumpfish were not affected (Cuenca *et al.*, 2017).

Other viruses of cleaner fish

A new ranavirus (*Iridoviridae*), closely related to the notifiable epizooitic hematopoietic necrosis virus (EHNv), has been isolated from lumpfish in Ireland, Scotland and Iceland (Stagg *et al.*, 2016; authors' observation). Ranaviruses occur globally, cause severe systemic infection in finfish and amphibians, and typically lack host specificity. No clinical symptoms were reported in infected lumpfish stocks in Scotland or Iceland. In Ireland, the virus was isolated from a mortality event in lumpfish fry. Moribund fry were lying on their side and displayed laboured breathing, but were able to recover temporarily when disturbed. On histology, findings included inflammation and necrosis, most notable in skeletal muscle, liver, kidney and gastro-intestinal tract. In a challenge trial, the virus did not induce significant mortality or consistent clinical disease in 0.2g or 15g lumpfish following a bath challenge, or in salmon smolts following a bath challenge or intraperitoneal injection (*authors*' observation). This indicates that the virus is unlikely to be a primary pathogen, though further studies are considered necessary.

A virus belonging to the *Flaviviridae* has been detected in all stages of lumpfish production along the Norwegian coast and has been detected in Scotland. Significant mortality has been attributed to the virus in Norway, though it has also been detected in clinically healthy fish. Viral RNA levels are highest in livers and kidneys, but can be detected in most organs in acute cases. Livers of affected fish have been described as pale and firm. The predominant presentation on histopathology is liver necrosis and inflammation in acute cases and liver fibrosis in chronic cases. Screening broodstock is recommended, though vertical transmission has not been proven (Skoge et al., 2017; www.pharmaq-analytiq.com).

Lymphocystis virus (*Iridoviridae*) is a common disease agent in both fresh and saltwater fish characterised by severely enlarged fibroblasts, visible on histology and at times macroscopically as masses of varying size. The disease is superficial and rarely causes economic loss (Whittington *et al.*, 2010). Lymphocystis-like lesions have been seen on skin, fins and gills of wild-caught ballan wrasse.

Nervous necrosis virus (NNV) (Betanodaviridae) is known to affect a high number of marine species and is the aetiological agent of viral nervous necrosis (VNN) (syn. viral encephalopathy and retinopathy (VER)). Vertical transmission can occur. A screening of wild corkwing, ballan and goldsinny wrasse in Norwegian and Swedish waters found an overall prevalence of 6.7% (Korsnes *et al.*, 2017). No clinical symptoms are described and the significance is unknown.

Salmonid viruses

Infectious pancreatic necrosis virus (IPNv) (*Birnaviridae*) causes the disease infectious pancreatic necrosis (IPN), a highly virulent and commercially significant disease, especially of young salmonids. IPNv has been isolated from a wide variety of fish species, but also from some molluscs and crustaceans (Gibson *et al.*, 1995). Transmission is both horizontal and vertical, and non-clinical carrier fish can shed the virus. Bath-challenged goldsinny wrasse appeared to be susceptible to the disease, with infected fish showing darkening of the body surface and pancreas necrosis (Gibson *et al.,* 1995), but they quickly recovered and did not show further changes associated with the disease in salmon. Infected wrasse shed the virus in the faeces for up to five weeks post-infection, making their role as reservoir and vector conceivable (Gibson *et al.*, 1995). In further observations, however, naïve salmon in co-habitation with exposed wrasse failed to become infected (Treasurer, 2012). The virus has been detected in gills of wrasse in salmon pens but with no associated clinical symptoms; it is assumed they were infected by salmon (Johansen *et al.,* 2016). Experiments have shown that lumpfish can potentially act as reservoirs for IPNv (Johansen *et al.*, 2016; NVI, 2014). Although the overall risk is considered low, it is advisable to screen stocks of concern.

Salmonid alphavirus (SAV) (*Togaviridae*) consists of six subtypes (Fringuelli *et al.*, 2008) and is notifiable in parts of Norway. SAV is one of the most commercially significant pathogens of farmed salmonids in Europe, causing pancreas disease (PD) in sea-reared salmon and rainbow trout, and sleeping disease in freshwater-reared rainbow trout (McLoughlin and Graham, 2007). Challenge studies with ballan wrasse and goldsinny wrasse did not induce pathology in these species and the authors came to the conclusion that they do not develop PD and the risk of acting as reservoirs for salmon is low (Gibson and Sommerville, 1996; Røsæg *et al.*, 2017). Goldsinny wrasse were challenged with kidney homogenate from infected salmon and naïve salmon did not show signs of infection after cohabitation or after injection with challenged goldsinny wrasse kidney homogenate (Gibson and Sommerville, 1996). The ballan wrasse were challenged with SAV subtypes 2 and 3 (Røsæg *et al.*, 2016). SAV, however, has been isolated from wrasse stocked with salmon during a PD outbreak (NVI, 2014).

Infectious salmon anaemia virus (ISAv) (*Orthomyxoviridae*) is the aetiological agent of the notifiable disease infectious salmon anaemia (ISA), a threat to salmon farming in Europe, North America and Chile. The virus has been isolated from wild salmonids (Raynard *et al.*, 2001), and herring (*Clupea harengus*) are potential asymptomatic carriers (Nylund *et al.*, 2002). Concerns regarding the potential role of wrasse species in the ISA outbreak in Scotland in 1998 led to a sudden cessation in their use at that time, though no indication of a connection existed (Treasurer, 2002). Goldsinny wrasse could not be infected through virus injection or cohabitation with diseased salmon, and naïve salmon could not be infected with kidney homogenate from challenged goldsinny wrasse (Kvenseth, 1998). It was concluded that the virus does not establish an infection in goldsinny wrasse and their role as reservoirs or vectors is highly unlikely.

Piscine myocarditis virus (PMCV) (*Totiviridae*) is the aetiological agent of cardiomyopathy syndrome (CMS), which has a very high economic impact on infected salmon stocks (Wiik Nielsen *et al.,* 2016) and has been diagnosed in several north European countries. The virus has been detected by PCR in corkwing and ballan wrasse with unknown clinical significance. Phylogenetic and amino acid analysis did not indicate differences between wrasse and salmon PMCV isolates and it was considered likely that wrasse were infected horizontally by cohabited salmon with clinical CMS. Wrasse have to be considered potential reservoirs and vectors for PMCV (Scholz *et al.*, 2017).

Piscine orthoreovirus (PRV) *(Reoviridae)* is associated with the disease 'heart and skeletal muscle inflammation' (HSMI) in salmon. The virus has been detected in the gills and heart of wrasse held together with PRV positive salmon. The significance of this is unknown and contamination could not be excluded (Johansen *et al.*, 2016).

13.6 Non-infectious diseases

Non-infectious diseases can include nutritional, hereditary, environmental, physical and management-related aetiologies. This is an extensive topic and only a few common issues will be listed in the following section.

Nutrition has been a challenge in the culture of ballan wrasse, with the live feed period and the weaning on to dry feed being bottle necks in production. Problems of nutritional aetiology can include mortality, low growth rate, abnormal organ development, deformities, oxidative stress, and skeletal myopathies (McLoughlin, UK, 2016, personal communication; Kousoulaki *et al.,* 2014). Lumpfish are comparatively easily weaned on to a dry feed. Swimbladder-related problems occur in wrasse and swimbladder development should be monitored in hatcheries (McLoughlin, UK, 2016, personal communication). Wrasse use a gas gland to regulate swimbladder inflation, but rely on gulping air for initial inflation. This occurs in a finite window around seven days post-hatch and can be complicated by abiotic factors such as photoperiod and oil film or scum/particles on the water surface, leading to mortality, buoyancy problems, higher energy needs and deformities (Woolley and Qin, 2010). Swimbladder over-inflation can also occur in hatcheries and is a common problem where wild wrasse are brought up in pots from deep water.

Deformities are not a widespread issue in lumpfish, though their anatomy makes spinal deformities hard to recognise and these are possibly underdiagnosed. A notable exception is a deformity of the ventral suction disc (Fig. 13.6A, B) that results in the inability to attach to surfaces and rest, leading to stress, exhaustion and ultimately high mortality in affected fish. There are indications that this condition may be hereditary (Ashby, UK and Sturluson, Iceland, 2016, personal communication). Gas supersaturation has led to clinical disease in lumpfish (authors' observation). Cataracts are a common finding in lumpfish and are highly relevant as they are sight feeders. It is likely that disturbed nutrient metabolism or malnutrition are the widespread aetiology, but mechanical injury and infectious diseases can contribute (Jonassen *et al.*, 2017).

Behavioural problems seen in both ballan wrasse and lumpfish culture include fin nipping and tail biting (Fig. 13.6C). In lumpfish the starvation period before transport, insufficient grading and fin discoloration through bacterial colonisation have been suggested as predisposing factors, and stress has been associated with the behaviour in ballan wrasse. Ballan wrasse can also develop hierarchical structures and sufficient feeders are important to reduce the impact of this behaviour. Nephrocalcinosis is a relatively common finding of unknown aetiology in both wrasse and lumpfish. A variety of factors such as photoperiod, light intensity, stocking density, feeding regime, water current and temperature are important factors in the culture of all fish species and are not yet understood sufficiently for cleaner fish species.

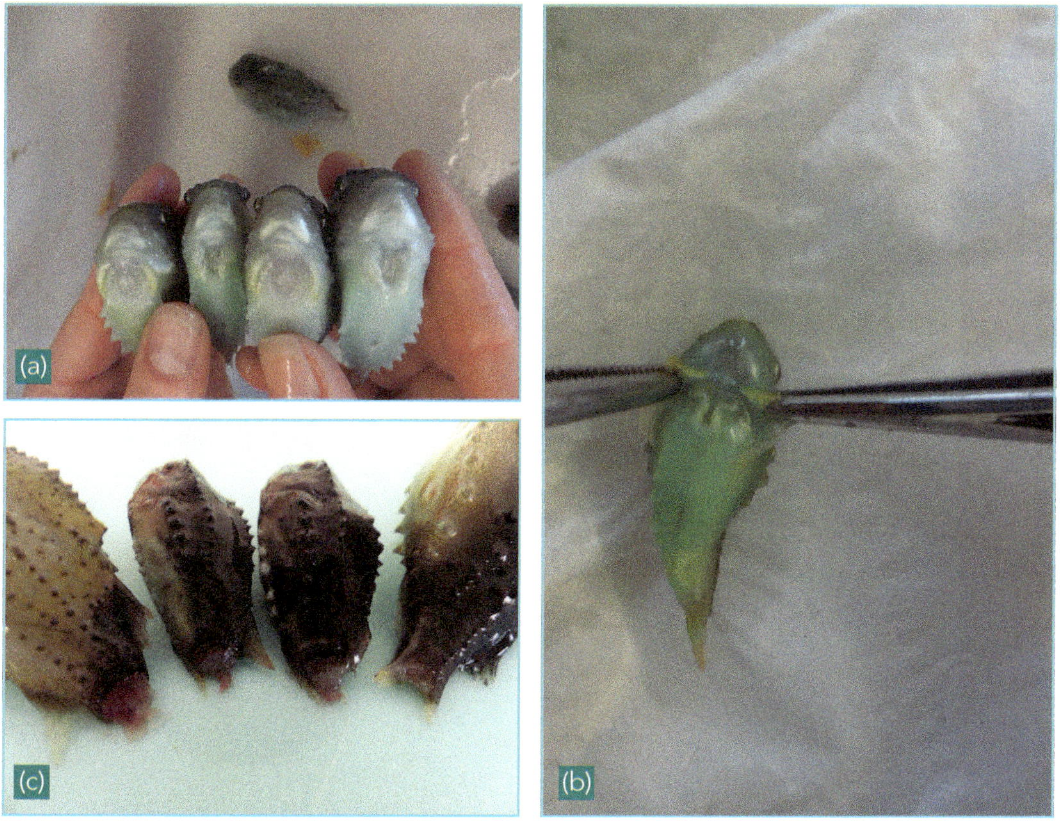

Fig. 13.6 (A) Normal and abnormal ventral suction discs in juvenile lumpfish. (B) Severe deformity of ventral suction disk. (C) Lesions due to tail biting in juvenile lumpfish. (Credit: A. Ashby (A, B); F. Scholz (C))

13.7 General health management

Preventing diseases is preferable to treatment and fish welfare is not only an ethical priority, but decreases disease occurrence. Stress and damage through events such as capture, handling, transport or grading increase disease susceptibility and high mortality can follow these events. Lumpfish, despite their robust appearance, are very susceptible to skin damage and incidence is increased by their habit of adhering to transport tanks, pipe walls, vaccination tables, from which they are at times forcefully removed. A wide range of factors influence the welfare of cleaner fish and it is often the sites that put most effort into welfare that have the least disease problems (authors' observation).

Health management of cleaner fish in salmon pens can be challenging. The stress and handling during transfer and stocking, together with the inevitable disease challenges in the pen environment, make the first few weeks at sea a critical time and acute mortalities often occur (Bornø *et al.*, 2016). Mortalities should be assessed very regularly as small cleaner fish do not remain in the net pen mortality collection sock for long, leading to an underestimation of mortalities. A complete disease diagnosis is recommended as early as possible, as diagnostic procedures take time, mortality can rise exponentially, and an early intervention is key to effective treatment. Continuously documenting observations can be very helpful.

Cleaner fish can act as carriers and reservoirs for salmon pathogens. To date, AGD is the only disease for which transmission from cleaner fish to salmon is proven, but there are several diseases with this potential and new diseases remain a risk. Salmon may be infected by pathogens that they are not normally sensitive to if they are challenged at high doses, which can occur if they eat infected cleaner fish or if high morbidity in cleaner fish leads to increased shedding (Murray, 2016). Rapid diagnosis of diseases in cleaner fish is thus also relevant in protecting salmon. The biosecurity risk cannot be accurately judged with our current knowledge, but a modelling approach indicates that the benefit of lice control outweighs the risk to salmon (Murray, 2017) and the overall risk increase posed by their use is not likely substantially higher if high biosecurity standards are maintained (Murray, 2016).

Effective treatment protocols for cleaner fish are urgently needed. Treatments at sea are challenging, as separating cleaner fish from salmon is often required. Wrasse can be caught with baited traps, but only if they are feeding. Different approaches have been tried for lumpfish, such as casting nets around hides after attracting lumpfish with feed or netting them by hand after reducing the pen volume, as they tend to come to the sides as opposed to schooling with the salmon. These methods are labour intensive, stressful and damaging to the fish, and they will not catch all, leaving a reservoir for reinfection. In-feed treatment of cleaner fish in salmon pens is complicated as feeding selectively is not always possible. Protocols that allow salmon and cleaner fish to be treated together for certain diseases (for example AGD) would minimise stress and avoid mutual reinfection.

In hatcheries, environmental conditions and pathogen introduction can be controlled and regular fish examination is practicable. Sources of pathogen introduction into land-based facilities can include people, equipment, introduced fish and eggs, wildlife, intake water and feed (especially live feed). Both wrasse and lumpfish eggs are difficult to disinfect. Internal spread of pathogens can occur through people, pipework, water, feed produced on site or lack of separation. An all in, all out strategy with thorough cleaning and disinfection between batches is advisable and any crossover of age groups should be avoided.

It is advisable for all sites to conduct a risk assessment and establish a health management plan, which can include implementation of zones and separation measures, disinfection protocols, site/zone specific gear, general hygiene measures, disease screening, quarantine protocols and water treatment. Preparing treatment plans for diseases of concern is recommended. Traceability of fish stock movements and origin is necessary for establishing infection sources and identifying other possibly infected stocks, so detailed records should be kept. Wild-caught fish, both broodstock and stock for use in sea pens, and any fish destined for transfer should undergo a thorough screening, though screening all stocks regularly is advisable. This can include a clinical and post-mortem examination, parasitology, bacteriology, virology, histology and PCR. The reuse of cleaner fish, especially on other sites, significantly increases the biosecurity risk (Murray, 2017) and any reuse on the same site should be preceded by extensive screening (Johansen *et al.*, 2016). If wild-caught wrasse are used, these should be sourced locally to avoid the risk of spreading pathogens to new hydrographic areas.

Vaccination is essential for the success and sustainability of cleaner fish use. Commercial and autogenous vaccines are in use and preliminary results indicate that vaccination is effective in protecting against several bacterial pathogens (Johansen *et al.*, 2016). New vaccines for novel agents and more specific vaccines for cleaner fish isolates, however, need to be developed and research on optimal fish size, route of administration and immunisation period is needed.

The use of only farmed and vaccinated fish with known disease status is the long-term aim in cleaner fish use, both to minimise disease risks and to take pressure off wild stocks.

Acknowledgements

This work was funded by FishVet Group Ireland and the Irish Research Council. The authors would like to thank Dr. Hamish Rodger, Dr. Marian McLoughlin, Dr. Susie Mitchell, Angela Ashby and Bernhard Laxdal (FishVet Group), Leon Stranden (Marine Harvest Norway), Sölvi Sturluson (Stofnfiskur), Gísli Jónsson (Icelandic Food and Veterinary Authority) and Dr. Neil Ruane (Marine Institute Ireland) for their important contributions, expertise and help on this chapter.

References

Alarcón, M., Thoen, E., Poppe, T.T., Bornø, G., Mohammad, S.N. and Hansen, H. (2015) Co infection of *Nucleospora cyclopteri* (Microsporidia) and *Kudoa islandica* (Myxozoa) in farmed lumpfish, *Cyclopterus lumpus* L., in Norway: a case report. *Journal of Fish Diseases,* 39(4), 411–418.

Alarcón, M., Gulla, S., Røsæg, M.V., Rønneseth, A., Wergeland, H., Poppe, T.T., Nilsen, H. and Colquhoun, D.J. (2016) Pasteurellosis in lumpsucker *Cyclopterus lumpus*, farmed in Norway. *Journal of Fish Diseases*, 39(4), 489–495.

Arkush, K.D., McBride, A.M., Mendonca, H.L., Okihiro, M.S., Andree, K.B., Marshall, S., Henriquez, V. and Hedrick, R.P. (2005) Genetic characterization and experimental pathogenesis of *Piscirickettsia salmonis* isolated from white seabass *Atractoscion nobilis. Diseases of Aquatic Organisms*, 63, 139–149.

Austin, B. and Austin, D.A. (2007) *Bacterial Fish Pathogens: Diseases of Farmed and Wild fish, 4th edn.* Springer-Praxis Publishing, New York-Chichester.

Avendaño-Herrera, R., Rodríguez, J., Magariños, B., Romalde, J.L. and Toranzo, A.E. (2004) Intraspecific diversity of the marine fish pathogen *Tenacibaculum maritimum* as determined by randomly amplified polymorphic DNA-PCR. *Journal of Applied Microbiology*, 96(4), 871–877.

Avendaño-Herrera, R., Toranzo, A.E. and Magariños, B. (2006) Tenacibaculosis infection in marine fish caused by *Tenacibaculum maritimum*: a review. *Diseases of Aquatic Organisms*, 71(3), 255–266.

Avendaño-Herrera, R., Irgang, R., Sandoval, C., Moreno-Lira, P., Houel, A., Duchaud, E., Poblete-Morales, M., Nicolas, P. and Ilardi, P. (2016) Isolation, characterization and virulence potential of *Tenacibaculum dicentrarchi* in salmonid cultures in Chile. *Transboundary and Emerging Diseases*, 63(2), 121–126.

Barker, D.E., Bradensantos, L.M., Coombs, M.P. and Boyce, B. (2009) Preliminary studies on the isolation of bacteria from sea lice, *Lepeophtheirus salmonis,* infecting farmed salmon in British Columbia, Canada. *Parasitology Research*, 105(4), 1173–1177.

Basson, L. and Van As, J.G. (2006) *Trichodinidae* and other ciliophorans (phylum ciliophora) In: Woo (ed) *Fish diseases and disorders. Volume 1: protozoan and metazoan infections. 2nd ed.* CABI Publishing, Wallingford, pp. 154–182.

Benediktsdóttir, E., Verdonck, L., Spröer, C., Helgason, S. and Swings, J. (2000) Characterization of *Vibrio viscosus* and *Vibrio wodanis* isolated at different geographical locations: a proposal for reclassification of *Vibrio viscosus* as *Moritella viscosa* comb. nov. *International Journal of Systematic and Evolutionary Microbiology,* 50(2), 479–488.

Bergh, Ø. and Samuelsen, O.B. (2007) Susceptibility of corkwing wrasse *Symphodus melops*, goldsinny wrasse *Ctenolabrus rupestis*, and Atlantic salmon *Salmo salar* smolt, to experimental challenge with *Vibrio tapetis* and *Vibrio splendidus* isolated from corkwing wrasse. *Aquaculture International*, 15(1), 11–18.

Berthe, F.C., Michel, C. and Bernardet, J.F. (1995) Identification of *Pseudomonas anguilliseptica* isolated from several fish species in France. *Diseases of Aquatic Organisms*, 21(2), 151–155.

Biering, E., Vaagnes, Ø., Krossøy, B., Gulla, S. and Colquhoun, D.J. (2016) Challenge models for atypical *Aeromonas salmonicida* and *Vibrio anguillarum* in farmed ballan wrasse (*Labrus bergylta*) and preliminary testing of a trial vaccine against atypical *Aeromonas salmonicida*. *Journal of Fish Diseases*, 39(10), 1257–1261.

Birkbeck, T.H. and Treasurer, J.W. (2014) *Vibrio splendidus, Vibrio ichthyoenteri* and *Vibrio pacinii* isolated from the digestive tract microflora of larval ballan wrasse, *Labrus bergylta* Ascanius, and goldsinny wrasse, *Ctenolabrus rupestris* (L.). *Journal of Fish Diseases*, 37(1), 69–74.

Björnsdóttir, B., Gudmundsdóttir, S., Bambir, S.H., Magnadóttir, B. and Gudmundsdóttir, B.K. (2004). Experimental infection of turbot, *Scophthalmus maximus* (L.), by *Moritella viscosa*, vaccination effort and vaccine-induced side-effects. *Journal of Fish Diseases* 27(11), 645–655.

Bornø, G., Alarcón, M., Linaker, M., Colquhoun, D., Nilsen, H., Gu, J., Gjerset, B., Haakon, H., Thoen, E., Gulla, S. and Jensen, B. (2016) Akutt dødelighet hos rognkjeks (*Cycloterus lumpus*) I 2015. Veterinærinstituttets rapportserie. Veterinærinstituttet (The Norwegian Veterinary Institute), Oslo. www.vetinst.no

Breiland, M.S.W., Johansen, L.H., Mikalsen, H.E., Hansen, Ø.J., Drangsholt, T.M.K. and Mortensen, A. (2015) Susceptibility of different lumpfish (*Cyclopterus lumpus*) families to *Vibrio ordalii* infection. In: Book of Abstracts, 17th International Conference on Diseases of Fish and Shellfish 2015, p. 111.

Bricknell, I.R., Bruno, D.W. and Stone J. (1996) *Aeromonas salmonicida* infectivity studies in goldsinny wrasse, *Ctenolabrus rupestris* (L.). *Journal of Fish Diseases* 19(6), 469–474.

Bruno, D.W., Griffiths, J., Petrie, J. and Hastings, T.S. (1998) *Vibrio viscosus* in farmed Atlantic salmon *Salmo salar* in Scotland: field and experimental observations. *Diseases of Aquatic Organisms* 34, 161–166.

Bullock, G.L., Rucker, R.R., Amend, D., Wolf, K. and Stuckey, H.M. (1976) Infectious pancreatic necrosis: transmission with iodine-treated and nontreated eggs of brook trout (*Salvelinus fontinalis*). *Journal of the Fisheries Board of Canada* 33(5), 1197–1198.

Cali, A. and Takvorian, P.M. (1999) Developmental Morphology and Life Cycles of the Microsporidia. In: Wittner, M. and Weiss, L.M. (eds.) *The Microsporidia and Microsporidiosis* ASM Press, Washington, pp 85–128.

Callahan, H.A., Litaker, R.W. and Noga, E.J. (2005) Genetic relationships among members of the *Ichthyobodo necator* complex: implications for the management of aquaculture stocks. *Journal of Fish Diseases*, 28(2), 111–118.

Capella-Gutiérrez, S., Marcet-Houben, M. and Gabaldón, T. (2012) Phylogenomics supports microsporidia as the earliest diverging clade of sequenced fungi. *BMC biology*, 10, 47.

Cavin, J.M., Donahoe, S.L, Frasca, S.Jr., Innis, C.J, Kinsel, M.J, Kurobe, T., Naples, L.M., Nyaoke, A., Poll, C.P. and Weber, E.P. (2012) *Myxobolus albi* infection in cartilage of captive lumpfish (*Cyclopterus lumpus*). *Journal of Veterinary Diagnostic Investigation*, 24(3), 516–524.

Colquhoun, D.J., Hovland, H., Hellberg, H., Haug, T. and Nilsen, H. (2004) *Moritella viscosa* isolated from farmed Atlantic cod (*Gadus morhua*). *Bulletin of the European Association of Fish Pathologists,* 24, 109–114.

Contreras-Lynch, S., Olmos, P., Vargas, A., Figueroa, J., González-Stegmaier, R., Enríquez, R., and Romero, A. (2015) Identification and genetic characterization of *Piscirickettsia salmonis* in native fish from southern Chile. *Diseases of Aquatic Organisms,* 155, 233–244.

Costello, M.J., Deady, S., Pike, A. and Fives, J.M. (1996) Parasites and diseases of wrasse being used as cleaner-fish on salmon farms in Ireland and Scotland In: Sayer, M.D.J., Treasurer, J.W. and Costello, M.J. (eds.) *Wrasse: Biology and use in aquaculture.* Fishing News Books, Oxford, pp. 240–246.

Cuenca, A., Vendramin, N., Sigurdardöttir, H., Moesgaard Iburg, T. Gudmundsdöttir, S. and Olesen, N. (2017) Emergence of a novel VHSV genotype IV in lumpfish (*Cyclopterus lumpus*): Risk assessment for European aquaculture. In: *Book of Abstracts, 18th International Conference on Diseases of Fish and Shellfish,* p. 140.

De Hoog, G.S., Vicente, V.A., Najafzadeh, M.J., Harrak, M.J., Badali, H. and Seyedmousavi, S. (2011) Waterborne Exophiala species causing disease in cold-blooded animals. *Persoonia-Molecular Phylogeny and Evolution of Fungi,* 27(1), 46–72.

Dale, O.B., Ørpetveit, I., Lyngstad, T.M., Kahns, S., Skall, H.F., Olesen, N.J. and Dannevig, B.H. (2009) Outbreak of viral haemorrhagic septicaemia (VHS) in seawater-farmed rainbow trout in Norway caused by VHS virus genotype III. *Diseases of Aquatic Organisms,* 85(2), 93–103.

Dahle, O.M.V. (2015) Experimental infections with *Paramoeba perurans* and AGD development in Atlantic salmon (*Salmo salar*) and ballan wrasse (*Labrus bergylta*). Are there host and *P. perurans* strain-related differences in infectivity and virulence? Thesis for the degree MSc, University of Bergen.

Delannoy, C.M., Houghton, J.D., Fleming, N.E. and Ferguson H.W. (2011) Mauve Stingers (*Pelagia noctiluca*) as carriers of the bacterial fish pathogen *Tenacibaculum maritimum. Aquaculture* 311, 255–257.

Dunn, A.M. and Smith, J.E. (2001) Microsporidian life cycles and diversity: The relationship between virulence and transmission. *Microbes and Infection,* 3, 381–388.

Dyková, I.V. (1995) In: Woo (ed) *Fish diseases and disorders. Volume 1: protozoan and metazoan infections.* CABI Publishing, Wallingford, pp. 149–179.

Egidius, E. and Andersen, K. (1978) Host-specific pathogenicity of strains of *Vibrio anguillarum* isolated from rainbow trout *Salmo gairdneri* Richardson and saithe *Pollachius virens* (L.). *Journal of Fish Diseases* 1, 45–50.

Einarsdóttir, T., Benediktsdóttir, E., Rau, J. and Hjartardóttir, S. (2015) *Moritella viscosa* in Icelandic lumpfish. In: Book of Abstracts, 17th International Conference on Diseases of Fish and Shellfish 2015, p. 69.

Einer-Jensen, K., Winton, J. and Lorenzen N. (2005) Genotyping of the fish rhabdovirus, viral haemorrhagic septicaemia virus, by restriction fragment length polymorphisms. *Veterinary Microbiology,* 106(3), 167–178.

Elsayed, E., Faisal, M., Thomas, M., Whelan, G., Batts, W. and Winton, J. (2006) Isolation of viral haemorrhagic septicaemia virus from muskellunge, *Esox masquinongy* (Mitchill), in Lake St Clair, Michigan, USA reveals a new sublineage of the North American genotype. *Journal of Fish Diseases,* 29(10), 611–619.

Emmenegger, E.J., Moon, C.H., Hershberger, P.K. and Kurath, G. (2013) Virulence of viral hemorrhagic septicemia virus (VHSV) genotypes Ia, IVa, IVb, and IVc in five fish species. *Diseases of Aquatic Organisms,* 107(2), 99–111.

Feist, S.W. and Longshaw, M. (2006) Phylum Myxozoa. In: Woo (ed) *Fish diseases and disorders. Volume 1: protozoan and metazoan infections.* 2nd ed. CABI Publishing, Wallingford, pp 230–296.

Ferguson, H.W., Collins, R.O., Moore, M., Coles, M. and MacPhee, D.D. (2004) *Pseudomonas anguilli-septica* infection in farmed cod, *Gadus morhua* L. *Journal of Fish Diseases*, 27(4), 249–253.

Figueras, A., Novoa, B., Santarém, M.M., Martínez, E., Álvarez, J.M., Toranzo, A.E. and Dyková I. (1992) *Tetramicra brevifilum*, a potential threat to farmed turbot *Scophthalmus Maximus*. *Diseases of Aquatic Organisms* 14, 127–135.

Freeman, M.A. and Sommerville, C. (2011) Original observations of *Desmozoon lepeophtherii*, a microsporidian hyperparasite infecting the salmon louse *Lepeophtheirus salmonis*, and its subsequent detection by other researchers. *Parasites and Vectors*, 4(1), 1.

Freeman, M.A, Jacob, M.K. and Kristmundsson, Á. (2013) *Nucleospora cyclopteri* n. sp., an intranuclear microsporidian infecting wild lumpfish, *Cyclopterus lumpus* L., in Icelandic waters *Parasites and Vectors*, 6, 49.

Freeman, M.A., and Kristmundsson, Á. (2013) Ultrastructure of *Nucleospora cyclopteri*, an intranuclear microsporidian infecting the Atlantic lumpfish (*Cyclopterus lumpus L.*) *Bulletin of the European Association of Fish Pathologists,* 33(6), 194–198.

Frerichs, G.N., Millar, S.D. and McManus, C. (1992) Atypical *Aeromonas salmonicida* isolated from healthy wrasse (*Ctenolabrus rupestris*). *Bulletin of the European Association of Fish Pathologists* 12(2), 48–49.

Fringuelli, E., Rowley, H.M., Wilson, J.C., Hunter, R., Rodger, H. and Graham, D.A. (2008) Phylogenetic analyses and molecular epidemiology of European salmonid alphaviruses (SAV) based on partial E2 and nsP3 gene nucleotide sequences. *Journal of Fish Diseases*, 31(11), 811–823.

Gagné, N., MacKinnon, A.M., Boston, L., Souter B., Cook-Versloot, M., Griffiths, S. and Olivier G. (2007) Isolation of viral haemorrhagic septicaemia virus from mummichog, stickleback, striped bass and brown trout in eastern Canada. *Journal of Fish Diseases*, 30(4), 213–223.

Garver, K.A., Traxler, G.S., Hawley, L.M., Richard, J., Ross, J.P. and Lovy J. (2013) Molecular epidemiology of viral haemorrhagic septicaemia virus (VHSV) in British Columbia, Canada, reveals transmission from wild to farmed fish. *Diseases of Aquatic Organisms,* 104(2), 93–104.

Gibson, D.R., Smail, D. and Sommerville, C. (1995) Infectious pancreatic necrosis virus: experimental infection of goldsinny wrasse *Ctenolabrus rupestris* L (Labridae). *Journal of Fish Diseases* 21, 399–406.

Gibson, D.R. and Sommerville, C. (1996) The potential for viral problems related to the use of wrasse in the farming of Atlantic salmon. In: Sayer, M.D.J., Treasurer, J.W. and Costello, M.J. (eds.) *Wrasse: Biology and use in aquaculture.* Fishing News Books, Oxford, pp. 240–246.

Gjessing, M.C., Davey, M., Kvellestad, A. and Vrålstad, T. (2011) *Exophiala angulospora* causes systemic inflammation in Atlantic cod *Gadus morhua. Diseases of Aquatic Organisms*, 96 (3), 209–219.

Grove, S., Wiik-Nielsen, C.R., Lunder, T., Tunsjø, H.S., Tandstad, N.M., Reitan, L.J., Marthinussen, A., Sørgaard, M., Olsen, A.B. and Colquhoun, D.J. (2010) Previously unrecognised division within *Moritella viscosa* isolated from fish farmed in the North Atlantic. *Diseases of Aquatic Organisms,* 93(1), 51–61.

Gruhl, A. (2015). Myxozoa. In: Wanninger, A. (ed) *Evolutionary Developmental Biology of Invertebrates 1: Introduction, Non-Bilateria, Acoelomorpha, Xenoturbellida, Chaetognatha.* Springer-Verlag, Wien, pp 165–177.

Guðmundsdóttir, S. (2016) Detection of Ranavirus and VHS genotype IV in lumpfish in Iceland. *Report: 20th Annual Workshop of the National Reference Laboratories for Fish Diseases,* Copenhagen, Denmark.

Gudmundsdóttir, B.K. and Björnsdóttir, B. (2007) Vaccination against atypical furunculosis and winter ulcer disease of fish. *Vaccine,* 25(30), 5512–5523.

Gudmundsdóttir, B.K., Björnsdóttir, B., Gudmundsdóttir, S. and Bambir, S.H. (2006) A comparative study of susceptibility and induced pathology of cod, *Gadus morhua* (L.), and halibut, *Hippoglossus hippoglossus* (L.), following experimental infection with *Moritella viscosa*. *Journal of Fish Diseases*, 29(8), 481–487.

Gulla, S., Duodu, S., Nilsen, A., Fossen, I, and Colquhoun, D.J. (2015a) *Aeromonas salmonicida* infection levels in pre- and post-stocked cleaner fish assessed by culture and an amended qPCR assay. *Journal of Fish Diseases*, 39 (7), 867–877.

Gulla, S., Sørum, H., Vågnes, Ø, and Colquhoun, D.J. (2015b) Phylogenetic analysis and serotyping of *Vibrio splendidus*-related bacteria isolated from salmon farm cleaner fish. *Diseases of Aquatic Organisms*, 117(2), 121–131.

Gulla, S., Lund, V., Kristoffersen, A.B., Sørum, H. and Colquhoun D.J. (2016) vapA (A-layer) typing differentiates *Aeromonas salmonicida* subspecies and identifies a number of previously undescribed subtypes. *Journal of Fish Diseases*, 39 (3), 329–342.

Gulla, S., Rønneseth, A., Sørum, H., Vågnes, Ø., Balboa, S., Romalde, J. L. and Colquhoun, D. J. (2017) *Vibrio tapetis* from wrasse used for ectoparasite bio-control in salmon farming: phylogenetic analysis and serotyping. *Diseases of Aquatic Organisms* 125, 189-197.

Habib, C., Houel, A., Lunazzi, A., Bernardet, J.F., Olsen, A.B., Nilsen, H., Toranzo, A.E., Castro, N. Nicolas, P. and Duchaud, E. (2014) Multilocus sequence analysis of the marine bacterial genus *Tenacibaculum* suggests parallel evolution of fish pathogenicity and endemic colonization of aquaculture systems. *Applied and Environmental Microbiology*, 80(17), 5503–5514.

Hall, L.M., Smith, R.J., Munro, E.S., Matejusova, I., Allan, C.E.T., Murray, A.G., Duguid, S.J., Salama, N.K.G., McBeath, A.J.A., Wallace, I.S., Bain, N., Marcos-Lopez, M. and Raynard, R.S. (2013) Epidemiology and control of an outbreak of viral haemorrhagic septicaemia in wrasse around Shetland commencing 2012. *Scottish Marine and Freshwater Science*, 4(3), 1–45.

Haugland, G.T., Olsen, A.B., Rønneseth, A. and Andersen, L. (2016) Lumpfish (*Cyclopterus lumpus* L.) develop amoebic gill disease (AGD) after experimental challenge with *Paramoeba perurans* and can transfer amoebae to Atlantic salmon (*Salmo salar* L.). *Aquaculture*, in press. doi: 10.1016/j.aquaculture.2016.04.001

Heidarsdóttir, K.J., Gravningen, K. and Benediktsdottir, E. (2008) Antigen profiles of the fish pathogen *Moritella viscosa* and protection in fish. *Journal of Applied Microbiology* 104, 944–951.

Hjeltnes, B. K., Bergh, Ø., Wergeland, H. and Holm, J.C. (1995) Susceptibility of Atlantic cod *Gadus morhua*, halibut *Hippoglossus hippoglossus* and wrasse (*Labridae*) to *Aeromonas salmonicida* subsp. *salmonicida* and the possibility of transmission of furunculosis from farmed salmon *Salmo salar* to marine fish. *Diseases of Aquatic Organisms*, 23 (1), 25–31.

Jensen, S., Samuelsen, O.B., Andersen, K., Torkildsen, L., Lambert, C., Choquet G., Paillard, C. and Bergh, O. (2003) Characterization of strains of *Vibrio splendidus* and *V. tapetis* isolated from corkwing wrasse *Symphodus melops* suffering vibriosis. *Diseases of Aquatic Organisms*, 53(1), 25–31.

Johansen, L. H., Colquhoun, D., Hansen, H., Hildre, S., Wergeland, H. I. and Mikalsen, H.E. (2016) Analyse av sykdomsrelatert risiko forbundet med bruk av villfanget og oppdrettet rensefisk for kontroll av lakselus (in Norwegian). Rapport 9/2016, Nofima.

Jonassen, T., Hamadi, M., Remø, S. C., and Waagbø, R. (2017) An epidemiological study of cataracts in wild and farmed lumpfish (*Cyclopterus lumpus* L.) and the relation to nutrition. *Journal of Fish Diseases* 40, 1903–1914. Available at: http://dx.doi.org/ 10.1111/jfd.12664.

Karlsbakk, E., Hodneland, K. and Nylund, A. (1996) Health status of goldsinny wrasse, including a detailed examination of the parasite community at Floedevigen, southern Norway. In: Sayer, M.D.J., Treasurer, J.W. and Costello, M.J. (eds.) *Wrasse: biology and use in aquaculture*. Fishing News Books, Oxford, pp. 240–246.

Karlsbakk, E., Olsen, A.B., Einen, A.C.B., Mo, T.A., Fiksdal, I.U., Aase H., Kalgraff, C., Skar, S.A. and Hansen, H. (2013) Amoebic gill disease due to *Paramoeba perurans* in ballan wrasse (*Labrus bergylta*). *Aquaculture*, 412, 41–44.

Karlsbakk, E., Alarcón, M., Hansen, H. and Nylund, A., (2014) *Sykdom og parasitter i vill og oppdrettet rognkjeks* (in Norwegian). Havforskningsrapporten, pp. 37–39.

Karlsbakk, E., Kristmundsson, Á., Albano, M., Brown, P. and Freeman, M.A. (2017) Redescription and phylogenetic position of *Myxobolus 'eglefini'* and *Myxobolus platessae* n. comb. (Myxosporea), parasites in the cartilage of some North Atlantic marine fishes, with notes on the phylogeny and classification of the Platysporina. *Parasitology International*, 66(1), 952–959.

Korsnes, K., Karlsbakk, E., Skår, C.K., Sælemyr, L., Nylund, A., Kvamme, B.O and Mortensen, S. (2017) High nervous necrosis virus (NNV) diversity in wild wrasse (Labridae) in Norway and Sweden. *Diseases of Aquatic Organisms* 126, 43–50.

Kousoulaki, K., Grøtan, E., Kvalheim, K, Høstmark, Ø, Klinge, M. and Bogevik, A. (2014) Protein quality, commodity options and ballan wrasse weaning performance. In: Synnøve, H., Wiborg, D.S.V, Courtney, H., and Jørgen, B. (eds) *Production of ballan wrasse (Labrus bergylta). Science and Practice*, SINTEF research report, Trondheim, pp. 96–100.

Kristmundsson, Á. and Freeman, M.A. (2014) Negative effects of *Kudoa islandica* n. sp. (Myxosporea: *Kudoidae*) on aquaculture and wild fisheries in Iceland. *International Journal for Parasitology: Parasites and Wildlife*, 3(2), 135–146.

Kvenseth, A.M. (1998) Wrasse – do they transfer diseases to salmon? Caligus 5, EUFAIR programme newsletter, pp. 20.

Lom, J. and Dyková, I. (1992) In: *Protozoan parasites of fishes*, Developments in Aquaculture and Fisheries Science, 26. Elsevier, Amsterdam.

Lom, J. and Dyková, I. (2005) Microsporidian xenomas in fish seen in a wider perspective. *Folia Parasitogica*, 52, 69–81.

Lom J., and Nilsen F. (2003) Fish microsporidia: fine structural diversity and phylogeny. *International Journal for Parasitology*, 33(2), 107–127.

López-Romalde, S., Magariños, B., Núñez, S., Toranzo, A.E. and Romalde, J.L. (2003) Phenotypic and genetic characterization of *Pseudomonas anguilliseptica* strains isolated from fish. *Journal of Aquatic Animal Health*, 15(1), 39–47.

Løvoll, M., Wiik-Nielsen, C.R., Tunsjø, H.S., Colquhoun, D., Lunder, T., Sørum, H. and Grove S. (2009) Atlantic salmon bath challenged with *Moritella viscosa*–pathogen invasion and host response. *Fish & Shellfish Immunology*, 26(6), 877–884.

Maíllo, P.A., Amigó, J.A., Baena, R., Salvadó, H. and Gracia, M.P. (1998) *Tetramicra brevifilum* (Matthews & Matthews, 1980) (Microsporida: *Tetramicriidae*) in a new fish host, *Lophius budegassa* (Spinola, 1807) in Spain. *Parasitology Research*, 84, 208–212.

Marcos-López, M., Donald, K., Stagg, H. and McCarthy, U. (2013) Clinical *Vibrio anguillarum* infection in lumpsucker *Cyclopterus lumpus* in Scotland. *The Veterinary Record*, 173(13), 319–319.

Marcos-López, M., Ruane, N.M., Scholz, F., Bolton-Warberg, M., Mitchell, S.O., Murphy O'Sullivan, S., Irwin Moore, A. and Rodger, H.D. (2017) *Piscirickettsia salmonis* infection in cultured lumpfish (*Cyclopterus lumpus* L.). *Journal of Fish Diseases*, Available at: http://dx.doi.org/10.1111/jfd.12630.

Matejusova, I., Noguera, P.A., Hall, M., McBeath, A.J.A., Urquhart, K., Simons, J., Fordyce, M.J. and Bruno, D.W. (2016) Susceptibility of goldsinny wrasse, *Ctenolabrus rupestris* L. (*Labridae*), to viral haemorrhagic septicaemia virus (VHSV) genotype III: Experimental challenge and pathology. *Veterinary Microbiology*, 186, 164–173.

Matthews, R.A. and Matthews, B.F. (1980) Cell and tissue reactions of turbot *Scophthalmus maximus* (L.) to *Tetramicra brevifilum* gen. n., sp. n. (Microspora). *Journal of Fish Diseases*, 3, 495–515.

McCarthy, U., Steiropoulos, N.A., Thompson, K.D., Adams, A., Ellis, A.E. and Ferguson, H.W. (2005) Confirmation of *Piscirickettsia salmonis* as a pathogen in European sea bass *Dicentrarchus labrax* and phylogenetic comparison with salmonid strains. *Diseases of Aquatic Organisms* 64, 107–119.

McLoughlin, M.F. and Graham, D.A. (2007) Alphavirus infections in salmonids–a review. *Journal of Fish Diseases*, 30(9), 511–531.

McVicar, A.H. (1999) *Ichthyophonus* and related organisms. In: Woo, T.K. and Bruno, D.W. (eds) *Fish Diseases and Disorders, Vol. 3 Viral, Bacterial and Fungal Infections*, CAB International, Wallingford. pp. 661—687.

Moran, J.D.W., Whitaker, D.J. and Kent, M.L. (1999) A review of the myxosporean genus *Kudoa* Meglitsch, 1947, and its impact on the international aquaculture industry and commercial fisheries. *Aquaculture*, 172(1), 163–196.

Mullins, J.E., Powell, M., Speare, D.J. and Cawthorn, R. (1994) An intranuclear microsporidian in lumpfish *Cyclopterus lumpus*. *Diseases of Aquatic Organisms*, 20(1), 7–13.

Munro, E.S., McIntosh, R.E., Weir, S.J., Noguera, P.A., Sandilands, J.M., Matejusova, I., Mayes, A.S. and Smith, R. (2015) A mortality event in wrasse species *(Labridae)* associated with the presence of viral haemorrhagic septicaemia virus. *Journal of Fish Diseases*, 38(4), 335–341.

Murray, A. G. (2017). A model of the process of spillover and adaption leading to potential emergence of disease in salmon held with cleaner fish used to control lice. *Aquaculture* 473, 283-290.

Murray, A. G. (2016). A modelling framework for assessing the risk of emerging diseases associated with the use of cleaner fish to control parasitic sea lice on salmon farms. *Transboundary and Emerging Diseases*, 63(2).

Noga, E. J. (2011) *Fish disease: diagnosis and treatment*, John Wiley & Sons.

Norwegian Veterinary Institute (NVI) Fish Health Report 2012 (2013) Available at: www.vetinst.no

Norwegian Veterinary Institute (NVI) Fish Health Report 2013 (2014) Available at: www.vetinst.no

Norwegian Veterinary Institute (NVI) Fish Health Report 2014 (2015) Available at: www.vetinst.no

Norwegian Veterinary Institute (NVI) Fish Health Report 2015 (2016) Available at: www.vetinst.no

Nylund, A., Devold, M., Mullins, J. and Plarre, H. (2002) Herring (*Clupea harengus*): a host for infectious salmon anemia virus (ISAV). *Bulletin of the European Association of Fish Pathololologists*, 22(5), 311–318.

Oldham, T., Rodger, H. and Nowak, B. F. (2016) Incidence and distribution of amoebic gill disease (AGD) – An epidemiological review. *Aquaculture* 457, 35–42.

Olsen, A. B., Gulla, S., Steinum, T., Colquhoun, D. J., Nilsen, H. K. and Duchaud, E. (2017) Multilocus sequence analysis reveals extensive genetic variety within *Tenacibaculum* spp. associated with ulcers in sea-farmed fish in Norway. *Veterinary Microbiology* 205, 39–45.

Overy, D.P., Groman, D., Giles, J., Duffy, S., Rommens, M. and Johnson G. (2014) *Exophiala angulospora* causes systemic mycosis in Atlantic Halibut: a case report. *Journal of Aquatic Animal Health*, 27(1), 12–19.

Pazos, F., Santos, Y., Macías, A.R., Núñez, S. and Toranzo, A.E. (1996) Evaluation of media for the successful culture of *Flexibacter maritimus*. *Journal of Fish Diseases*, 19(2), 193–197.

Pedersen, K., Grisez, L., van Houdt, R., Tiainen, T., Ollevier, F. and Larsen, J.L. (1999) Extended serotyping scheme for *Vibrio anguillarum* with the definition and characterization of seven provisional O-serogroups. *Current Microbiology*, 38, 183–189.

Pedersen, O.A. and Langvad, F. (1989) *Exophiala psychrophila* sp. nov., a pathogenic species of the black yeasts isolated from farmed Atlantic salmon. *Mycological Research*, 92(2), 153–156.

Piazzon, M.C., Leiro, J. and Lamas, J. (2013) Fish immunity to scuticociliate parasites. *Developmental & Comparative Immunology*, 41(2), 248–256.

Pierce, L.R. and Stepien, C.A. (2012) Evolution and biogeography of an emerging quasispecies:

diversity patterns of the fish Viral Hemorrhagic Septicemia virus (VHSv). *Molecular Phylogenetics and Evolution*, 63(2), 327–341.

Pridgeon, J.W. and Klesius, P.H. (2013) Major bacterial diseases in aquaculture and their vaccine development. *Animal Science Reviews* 2012, 141.

Raynard, R.S., Murray, A.G. and Gregory, A. (2001) Infectious salmon anaemia virus in wild fish from Scotland. *Diseases of Aquatic Organisms*, 46(2), 93–100.

Robertsen, B. (2011) Can we get the upper hand on viral diseases in aquaculture of Atlantic salmon? *Aquaculture Research*, 42(1), 125–131.

Rolbiecki, L. and Rokicki, J. (2008) Helminths of the lumpsucker (*Cyclopterus lumpus*) from the Gulf of Gdańsk and Vistula Lagoon (Poland). *Oceanological and Hydrobiological Studies*. Available at: https://doi.org/10.2478/v10009-008-0012-2.

Romalde, J.L., Ravelo C., López-Romalde, S., Avendaño-Herrera, R., Magariños, B. and Toranzo, A.E. (2005) Vaccination strategies to prevent emerging diseases for Spanish aquaculture. *Developments in biologicals* (*Basel*) 121, 85-95.

Rønneseth, A., Haugland, G.T., Brudal, E. and Wergeland, H.I. (2015) Bacterial diseases and vaccination of farmed lumpfish (*Cyclopterus lumpus l.*). In: Book of Abstracts, 17th International Conference on Diseases of Fish and Shellfish 2015, p. 97.

Rønneseth, A., Haugland, G. T., Colquhoun, D. J., Brudal, E. and Wergeland, H. I. (2017) Protection and antibody reactivity following vaccination of lumpfish (*Cyclopterus lumpus* L.) against atypical *Aeromonas salmonicida. Fish and shellfish Immunology* 64, 383–391.

Røsæg, M.V., Sindre, H., Persson, D., Breck, O., Knappskog, D., Olsen, A.B. and Taksdal, T. (2017) Ballan wrasse (*Labrus bergylta* Ascanius) is not susceptible to pancreas disease caused by salmonid alphavirus subtype 2 and 3. *Journal of Fish Diseases* 40, 975–978. 10.1111/jfd.12572

Ross, K., McCarthy, U., Huntly, P.J., Wood, B.P., Stuart, D., Rough, E.I., Smail, D.A. and Bruno, D.W. (1995) A outbreak of viral haemorrhagic septicaemia (VHS) in turbot (*Scophthalmus maximus*) in Scotland. *Bulletin of the European Association of Fish Pathologists,* 14(6), 213–214.

Rozas, M. and Enríquez, R. (2014) Piscirickettsiosis and *Piscirickettsia salmonis* in fish: a review. *Journal of Fish Diseases*, 37(3), 163–188.

Scholz, F., Fringuelli, E., Bolton-Warberg, M., Marcos-Lopez, M., Mitchell, S., Prodhol, P., Moffet, D., Savage, P., Murphy O'Sullivan, S., O'Connor, I., McCarthy, E. and Rodger, H.D. (2016) First record of *Tetramicra brevifilum* in lumpfish (*Cyclopterus lumpus*, L.) *Journal of Fish Diseases*, Available at: http://dx.doi.org/10.1111/jfd.12554

Scholz, F., Ruane, N. M., Morrissey, T., Marcos-López, M., Mitchell, S., O'Connor, I., Mirimin, L., MacCarthy, E and Rodger, H. D. (2017) Piscine myocarditis virus detected in corkwing wrasse (*Symphodus melops*) and ballan wrasse (*Labrus bergylta*). *Journal of Fish Diseases* 41, 147–152. doi: 10.1111/jfd.12661

Shaw, R.W. and Kent, M.L. (1999) Fish microsporidia. In: Wittner, M. and Weiss, L.M . (eds.) *The Microsporidia and Microsporidiosis* ASM Press, Washington, pp 418–446.

Skall, H.F., Olesen, N.J. and Mellergaard, S. (2005) Viral haemorrhagic septicaemia virus in marine fish and its implications for fish farming – a review. *Journal of Fish Diseases,* 28(9), 509–529.

Skoge, R. H., Brattespe, J., Økland, A. L., Plarre, H. and Nylund, A. (2017) New virus of the family laviviridae detected in lumpfish (*Cyclopterus lumpus*). *Archives of Virology,* 1–7. https://doi.org/10.1007/s00705-017-3643-3

Småge, S.B., Frisch, K., Brevik, Ø.J., Watanabe, K. and Nylund, A. (2016) First isolation, identification and characterisation of *Tenacibaculum maritimum* in Norway, isolated from diseased farmed sea lice cleaner fish *Cyclopterus lumpus* L. *Aquaculture,* 464, 178–184.

Snow, M., Cunningham, C.O., Melvin, W.T. and Kurath, G. (1999) Analysis of the nucleoprotein gene identifies distinct lineages of viral haemorrhagic septicaemia virus within the European marine environment. *Virus Research*, 63(1), 35–44.

Stagg, H.E.B., Black, J., Murray, W., Savage, P., Pflaum, S. and Munro, E.S. (2016) *Ranavirus* isolated from lumpfish, *Cyclopterus lumpus*. In: Book of Abstracts, EAFP UK and Ireland Branches Second Meeting 2016, p.40.

Takeuchi, F., Sekizuka, T., Ogasawara, Y., Yokoyama, H., Kamikawa, R., Inagaki Y., Nozaki T., Sugita-Konishi, Y., Ohnishi T. and Kuroda, M. (2015) The mitochondrial genomes of a myxozoan genus *Kudoa* are extremely divergent in metazoa. *PloS ONE*, 10(7), e0132030.

Terry, R.S., Smith, J.E., Sharpe, R.G., Rigaud, T., Littlewood, D.T.J., Ironside, J.E., Rollinson, D., Bouchon, D., MacNeil, C., Dick, J.T.A. and Dunn, A.M. (2004) Widespread vertical transmission and associated host sex–ratio distortion within the eukaryotic phylum Microspora. *Proceedings of the Royal Society of London B: Biological Sciences,* 271(1550), 1783–1789.

Todal, J.A., Karlsbakk, E., Isaksen, T.E., Plarre, H., Urawa, S., Mouton, A., Hoel, E., Koren, C.W R. and Nylund, A. (2004) *Ichthyobodo necator* (Kinetoplastida) a complex of sibling species. *Diseases of Aquatic Organisms,* 58(1), 9–16.

Toranzo, A.E., Magariños, B. and Romalde, J.L. (2005) A review of the main bacterial fish diseases in mariculture systems. *Aquaculture,* 246(1), 37–61.

Treasurer, J. and Cox, D. (1991) The occurrence of *Aeromonas salmonicida* in wrasse (Labridae) and the implication for Atlantic salmon farming. *Bulletin of the European Association of Fish Pathologists*, 11, 208–210.

Treasurer, J.W. and Laidler, L.A. (1994) *Aeromonas salmonicida* infection in wrasse (Labridae), used as cleaner fish, on an Atlantic salmon, *Salmo salar* L., farm. *Journal of Fish Diseases*, 17(2), 155–161.

Treasurer, J.W. (1997) Parasites of wrasse (Labridae) in inshore waters of west Scotland stocked as cleaner fish of sea lice (*Caligidae)* on farmed Atlantic salmon. *Journal of Fish Biology*, 50(4), 895–899.

Treasurer, J.W. (2002) A review of potential pathogens of sea lice and the application of cleaner fish in biological control. *Pest Management Science*, 58(6), 546–558.

Treasurer, J.W. (2012) Diseases of north European wrasse (Labridae) and possible interactions with cohabited farmed salmon, *Salmo salar* L. *Journal of Fish Diseases*, 35, 555–562.

Van As, J. and Basson, L. (1990) An articulated internal skeleton resembling a spinal column in a ciliated protozoan. *Naturwissenschaften*, 77(5), 229–231.

Vennerström, P. (2015) ICES identification leaflets for diseases and parasites of fish and shellfish. Leaflet No. 63 Pseudomoniasis (*P. anguilliseptica*) in farmed fish. Available at: www.ices.dk

Whittington, R.J., Becker, J.A. and Dennis, M.M. (2010) Iridovirus infections in finfish – critical review with emphasis on ranaviruses. *Journal of Fish Diseases* 33(2), 95–122.

Wiik-Nielsen, J., Alarcón, M., Jensen, B. B., Haugland, Ø. and Mikalsen, A. B. (2016) Viral co-infections in farmed Atlantic salmon, *Salmo salar* L., displaying myocarditis. *Journal of Fish Diseases*, 39, 1495–1507.

Wiklund, T. and Bylund G. (1990) *Pseudomonas anguilliseptica* as a pathogen of salmonid fish in Finland. *Diseases of Aquatic Organisms,* 8(1), 13–19.

Woolley, L.D. and Qin, J.G. (2010) Swimbladder inflation and its implication to the culture of marine finfish larvae. *Reviews in Aquaculture*, 2(4), 181–190.

CHAPTER 14

Immunology and vaccinology of lumpfish and wrasse

Gyri Teien Haugland, Anita Rønneseth and Heidrun I. Wergeland

Department of Biology, University of Bergen, Norway

14.1 Immunology

The immune system consists of immune cells and molecules that protect the body from pathogens, harmful substances and from altered self-components. To obtain protection, the immune system fulfils four steps: 1) immunological recognition of pathogens, 2) effector functions in order to restrict and possibly eliminate the pathogens, 3) immunological regulation to ensure response to the pathogens without self-damage and finally 4) immunological memory to protect the organism against recurring disease due to the same pathogen. As teleost fish are the earliest phylogenetic group with an adaptive immune system, these tasks are valid for fish as well as mammals. Step 1 and 2 are performed by the innate immune system, step 3 initiates an adaptive immune response and step 4 is performed by the adaptive immune system by producing memory cells and pathogen-specific antibodies.

14.1.1 Immune organs

Head kidney, thymus, spleen

Teleost fish do not have bone marrow or lymph nodes as mammals. The major immune organs in fish are head kidney, thymus, spleen and mucosa-associated tissues (MALTs) (Fig. 14.1A) (Salinas *et al.*, 2011; Tacchi *et al.*, 2014; Castro and Tafalla, 2015). In teleost fish, the two first mentioned are primary lymphoid organs where immune cells are generated, while spleen, kidney and the MALTs are secondary lymphoid organs where antigen-presentation takes place. The head kidney in both lumpfish (*Cyclopterus lumpus* L.) (Fig. 14.1B) and Ballan wrasse (*Labrus bergylta* Ascanius) is bi-lobed.

Mucosa-associated lymphoid tissues (MALTs)

The MALTs are immunological active sites important for protecting the fish against microbes. In addition to recognising and trapping microbes, MALTs also contain a variety of antimicrobial

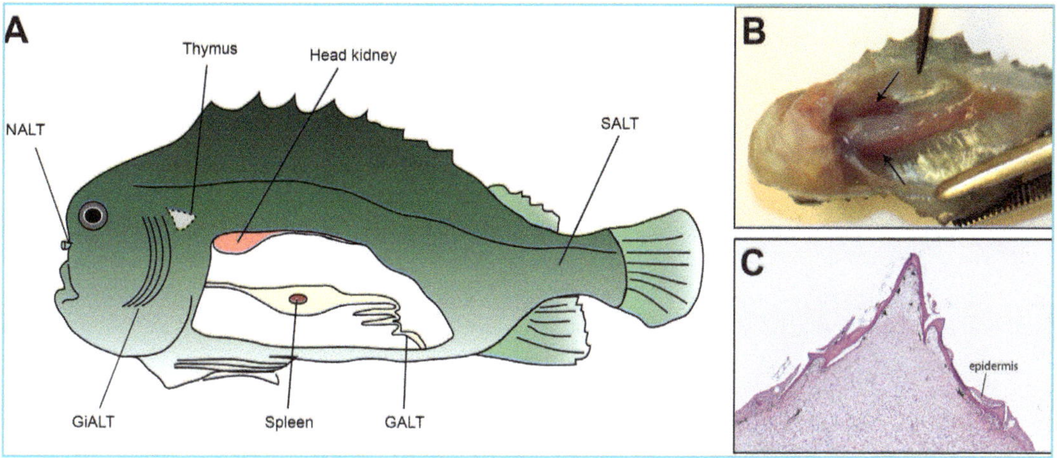

Fig. 14.1 Schematic overview of the lymphoid organs in lumpfish. A) Location of lymphoid organs and MALTs in lumpfish. There is currently no information about the MALTs in lumpfish, but it is likely that lumpfish has NALT, GiALT, GALT and SALT as in other fish species. B) Localisation of the bi-lobed kidney in lumpfish. C) Histological picture of a spike in lumpfish. Note that the spike penetrates through the epidermis and if not covered by mucosa this site can be vulnerable for entrance of pathogens. Credit: Renate Johansen, PHARMAQ Analytiq AS.

factors such as lectins, pentraxins, complement proteins, anti-bacterial peptides and immunoglobulins (Igs). The MALTs are also shown to be important sites for adaptive immunity in fish (Rombout *et al.*, 2014; Xu *et al.*, 2016). The MALTs includes skin-associated tissue (SALT), gut-associated tissue (GALT), gill associated tissue (GiALT) and the recently discovered nasopharynx-associated lymphoid tissue (NALT) (Salinas *et al.*, 2011; Xu *et al.*, 2013; Tacchi *et al.*, 2014; Salinas, 2015; Parra *et al.*, 2016; Sepahi and Salinas, 2016; Xu *et al.*, 2016). The common features of the teleost MALTs are: 1) presence of diffuse lymphoid cells with the absence of organised tissue, 2) predominance of IgT (termed IgZ in zebrafish) and IgT^+ B cells, 3) presence of a diverse microbial community and coating of commensals by mucosal Igs (Salinas *et al.*, 2011; Xu *et al.*, 2013). A review of the different MALTs and functions in teleost fish is also covered in the book *Mucosal Health in Aquaculture* (Beck and Peatman, 2015). There are currently few studies on mucosa immunology in lumpfish or wrasse (Patel and Brinchmann, 2017). It is however likely that the mucosa also plays a major role for immune protection in these species. In particular for lumpfish as the spikes can penetrate the epidermis making these sites vulnerable to the entry of pathogens if not covered by mucus (Renate Johansen, PHARMAQ Analytiq AS, March 2017, personal communication). The mucosa is not visible on the image, but covers the epidermis and the spike (Fig. 14.1C).

14.1.2 Immune cells

If the physical and chemical barriers, such as the skin or antimicrobial proteins secreted at the mucosal surfaces, are being overcome or evaded, other components of the immune system come

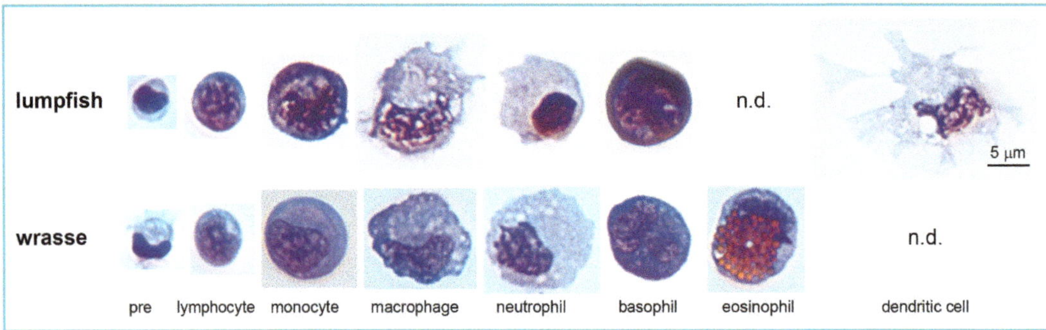

Fig. 14.2 Different types of isolated immune cells in lumpfish and Ballan wrasse. Representative cells of different subtypes of leukocytes isolated from lumpfish and Ballan wrasse using discontinues Percoll gradients; small agranular cells with high phagocytic activity that are possible precursor cells (pre) or immature cells, lymphocytes, monocytes, macrophages, granulocytes, dendritic-like cells in lumpfish and eosinophils in wrasse. In lumpfish, eosinophils have been observed in tissue (Haugland *et al.*, 2016), but not among isolated leukocytes. Dendritic-like cells with characteristic morphology have not yet been described from wrasse. n.d. = non-detected.

into play. The complement system can be activated within minutes to recognise and destroy pathogens and phagocytic cells of the innate immune system, such as macrophages and neutrophils, can ingest and kill pathogens by producing reactive oxygen and nitrogen species and degrading enzymes. In fish, B cells are also professional phagocytic cells (Li *et al.*, 2006; Øverland *et al.*, 2010; Rønneseth *et al.*, 2015) and dendritic-like cells have also been described (Ganassin and Bols, 1996; Lovy *et al.*, 2009; Aghaallaei *et al.*, 2010; Lugo-Villarino *et al.*, 2010; Wittamer *et al.*, 2011; Bassity and Clark, 2012; Haugland *et al.*, 2012b; Granja *et al.*, 2015). The immune cells important for immune protection are also known as white blood cells or leukocytes. Similar to other fish species, lumpfish and Ballan wrasse have all the common subsets of leukocytes such as lymphocytes, monocytes, macrophages, and granulocytes (Fig. 14.2). One striking observation is the eosinophils in Ballan wrasse (Fig. 14.2), which are highly abundant among isolated leukocytes (Haugland *et al.*, 2014). Cells with dendritic-like morphology have been observed among lumpfish leukocytes, but not among leukocytes isolated from wrasse.

Lymphocytes (B cells, T cells and non-specific cytotoxic cells)

B and T cells are the major players of adaptive immunity. They are activated when encountered with their cognate antigen. B cells can recognise antigens alone, while T cells require that the antigens are presented for them via a major histocompatibility complex (MHC) molecule. When activated, B and T cells differentiate into effector cells and memory cells essential for long-lasting protection. Activated B cells become plasma cells that produce antibodies (also known as immunoglobulins (Ig)) and execute the humoral arm of the adaptive immune system, while T cells are the players in the cellular part of adaptive immunity.

B cells in teleost fish are divided into three subtypes based on surface Ig expression; 1) $IgM^+/IgD^+/IgT^-$ B cells, which are most abundant in blood, 2) $IgM^-/IgD^+/IgT^-$ B cells, which thus far are only described in channel catfish and European rainbow trout and 3) $IgM^-/IgD^-/IgT^+$ (termed IgZ in zebrafish) B cells, the predominant subtype in mucosa (Fillatreau *et al.*, 2013).

For lumpfish, B cells have been isolated from peripheral blood and characterised functionally (Rønneseth *et al.*, 2015), but the genotype/phenotype has not yet been described. It is known that the B cells in lumpfish have phagocytic activity similar to B cells in other teleost fish, suggesting that they play a role in innate immunity (Rønneseth *et al.*, 2015), but further characterisation of the B cells is required. Characterisation of B cells from any species of wrasse has not yet been performed.

T cells differentiate into either T helper (Th) cells or T cytotoxic (Tc) cells. The subtype of effector T cell that is activated depends on the antigen-presenting molecules. If the antigen is presented via MHC class II at the surface of professional antigen-presenting cells, Th cells are activated, while if the antigen is presented via MHC class I, Tc cells are activated. All cells have the ability to present antigens via MHC class I and will do so if the cell is infected with virus. The Tc cells are able to destroy infected cells. Th cells are important for activation of other cells. In mammals, several T subtypes are known; Th1, Th2, Th17, Th22, Tc and also invariant T cells (iT cells). Due to a lack of reagents to isolate live T cells in teleost fish, there is less knowledge about fish T cell subtypes and functions compared with the B cells. However, based on transcriptomic and genomic analyses, genes for T helper cells including Th1/Th2/Th17/Th22, T regulatory and T cytotoxic cells have been reported from teleosts (Castro and Tafalla, 2015; Somamoto *et al.*, 2014; Nakanishi *et al.*, 2015). IgM^- lymphocytes in blood of lumpfish have been described by Rønneseth and co-workers (Rønneseth *et al.*, 2015), but it is currently not known if these are T cells, common lymphocyte precursors/blast cells or non-specific cytotoxic cells (NCC). In the transcriptome of lumpfish (our unpublished data), markers for T cells are present, but further analysis is required to characterise the T cell markers and T cell subtypes in lumpfish. The genome of Ballan wrasse, which was recently sequenced in Norway (funded by the Norwegian Seafood Research Fund (FHF)), will provide further information about T cell markers.

Another lymphoid cell in fish, NCC, is functional equivalent to a mammalian natural killer cell and plays an important role in innate immunity where it can recognise and spontaneously kill altered cells such as stressed or virus infected cells without antigen-presentation via MHC class I. NCC have been described from several fish species such as catfish, rainbow trout, common carp, damsel fish and tilapia (Uribe *et al.*, 2011). Hitherto, there is no characterisation or functional description of NCC in lumpfish or any species of wrasse.

Granulocytes (neutrophils, eosinophils, basophils, mast cells)

Granulocytes are characterised by the presence of granules in their cytoplasm. There are three principal types of granulocytes: neutrophils, eosinophils and basophils. Neutrophils are the most common granulocytes and are present in all teleost fish described to date, while eosinophils and basophils are rarer and not present in all species. In lumpfish, neutrophils and basophils have been described (Rønneseth *et al.*, 2015). The neutrophils have a bi-lobed or kidney-shaped nucleus (Fig. 14.2), similar to neutrophils in cod and wrasse, but different from the polymorphonucleated neutrophils present in salmonids. Eosinophils have not been observed among Colorrapid-stained leukocytes isolated from head kidney, spleen or peripheral blood in lumpfish. However, in hematoxylin and eosin (HE) stained histology sections of gills from lumpfish infected with the amoeba *Paramoeba perurans*, numerous eosinophils were present (Haugland *et al.*, 2016). In contrast, in Ballan wrasse, eosinophils with intense red granules after Colorrapid-staining were abundant among isolated leukocytes from head kidney, spleen and peripheral blood (Haugland *et al.*, 2014). The significance of this has not been elucidated. Neutrophils are the most abundant granulocyte in Ballan wrasse, followed by eosinophils and basophils.

The neutrophils are the first cells that are attracted to a site of injury/infection, following chemical signals such as interleukin 8 (IL-8) and chemokines secreted from macrophages. The neutrophils are professional phagocytic cells and play an important role in innate immunity by trapping and degrading microbes. Their granules contain a variety of components with antimicrobial properties, such as enzymes. They can also release reactive oxygen species (ROS). ROS are produced from oxygen taken up by the cells during phagocytosis in a process called the respiratory burst. By use of flow cytometry, studies have shown that both lumpfish and Ballan wrasse have efficient innate immune mechanisms including phagocytosis and a respiratory burst (Haugland *et al.*, 2012a; Haugland *et al.*, 2014). The neutrophils in lumpfish and wrasse, like other fish species, contain myeloperoxidase, which is an enzyme involved in the respiratory burst process.

Mast cells/eosinophilic granule cells within tissues have been described in several fish species including Ballan wrasse, and cuckoo wrasse (*Labrus mixtus* L.) (Reite and Evensen, 2006).

Macrophages/monocytes

Macrophages are professional phagocytic cells that engulf and digest microbes and cellular debris in a process called phagocytosis. Macrophages are found in organs and tissues, and in mammals they are named after where they are located. They have different kinds of sensors (receptors) on their surface that initiate phagocytosis and signalling to other cells. Upon microbial infection, macrophages in the tissue sense invading microbes and send signals to neutrophils to come to the infection site. Monocytes, present in blood, differentiate into macrophages or dendritic cells after migrating into the tissue. Macrophages are also, similar to B cells and dendritic cells, professional antigen-presenting cells that present antigens to T cells and initiate an adaptive immune response. Macrophages and monocytes are present in teleost fish, including lumpfish and Ballan wrasse (Fig. 14.2 and (Haugland *et al.*, 2012a; Haugland *et al.*, 2014; Rønneseth *et al.*, 2015)).

Dendritic cells

Dendritic cells in mammals are the main orchestrators of the immune system involved in both innate and adaptive immunity. They play an important role as phagocytes in innate immunity and initiate and regulate the adaptive immune responses to ensure the responses are tailored to the microbial infection. Similar to macrophages, they are located in different tissues and they have receptors to recognise a range of danger signals. Some of these receptors are unique for different subsets of dendritic cells. Dendritic-like cells have also been described from teleost fish (Ganassin and Bols, 1996; Lovy *et al.*, 2009; Aghaallaei *et al.*, 2010; Lugo-Villarino *et al.*, 2010; Wittamer *et al.*, 2011; Bassity and Clark, 2012; Haugland *et al.*, 2012b; Granja *et al.*, 2015). Similar to DCs in mammals, functional analyses of fish DCs have reported that they have the ability to stimulate T cell proliferation and they have phagocytic activity (Bassity and Clark, 2012, Lugo-Villarino *et al.*, 2010, Granja *et al.*, 2015), but it is currently not known if different subtypes of dendritic cells are present in fish, or if fish have a primordial dendritic type. In mammals, DCs are the most potent of the antigen presenting cells, with a unique ability to present antigens to naïve T cells (immature T cells that have not yet encountered their cognate antigen). It has been reported that DC-like cells in fish are MHC class II$^+$ and have the ability to activate T cells similar to their mammalian counterparts, but further functional characterisation is required to know if these cells are true homologues of DCs in mammals. Cells with DC-like morphology have been observed among isolated leukocytes in lumpfish (Fig. 14.2), but not yet in wrasse.

14.1.3 Innate and adaptive immunity

Immune cells involved in innate immunity are the first line of defence against infection. They are located in the tissues and bloodstream, where they can detect, trap and degrade foreign microbes (Fig. 14.3). The cell types in fish that are professional phagocytes are monocytes, macrophages, neutrophils, dendritic cells and also B cells. The discovery that B cells are professional phagocytic cells in fish (Li *et al.*, 2006; Øverland *et al.*, 2010), including lumpfish (Rønneseth *et al.*, 2015), has shed light on evolution of the phagocytic cells and has led to new discoveries in the mammalian immune system. The two main types of lymphocytes, B cells and T cells, are the main players of adaptive immunity. As mentioned in section 14.1.2, B cells have the ability to

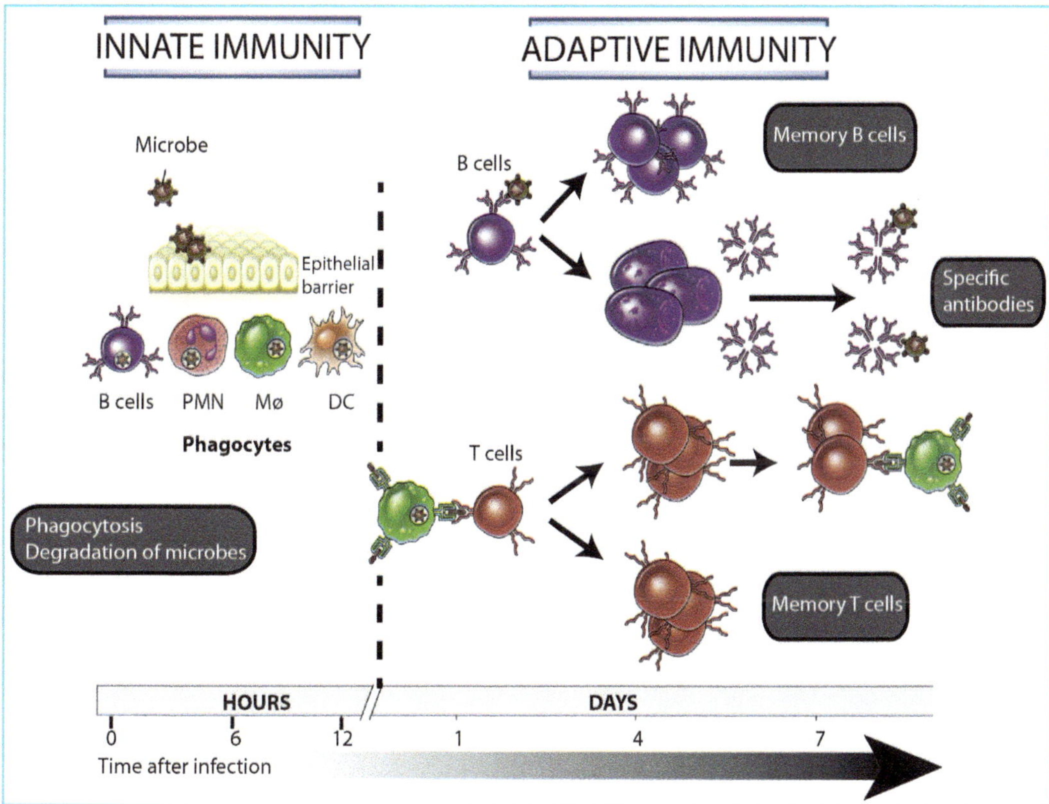

Fig. 14.3 Schematic illustration of innate and adaptive immune systems in teleosts and the interaction between them. Microbes are recognised, ingested and degraded by phagocytic cells of the innate immune system. B and T lymphocytes, the major players of the adaptive immunity are activated by antigens. B cells can recognise antigens alone (which can be either whole microbes or parts of pathogens), while T cells require presentation of the antigens by either professional antigen-presenting cells or infected cells. The results of antigen recognition are effector cells and formation of memory cells.

recognise soluble antigens, while T cells can only recognise antigens presented by antigen-presenting cells via MHC class I or class II molecules. Upon antigen encounter, B cells differentiate into antibody producing plasma cells and memory cells, while T cells differentiate either into T helper cells or T cytotoxic cells and memory cells. The memory cells are essential for long-lasting immunity, obtained after natural infection or vaccination. The antibodies produced by plasma cells are identical to the immunoglobulin (Ig) on the surface of the particular B cell. The Ig on the cell surface is termed B cell receptor (BcR). While BcR is a monomer, the secreted IgM by plasma cells in fish is a tetramer (Sunyer, 2013).

Functional assay measuring innate immune responses

Phagocytosis

Mechanisms of the innate immunity aim to recognise and eliminate a broad range of microbes. As opposed to lymphocytes that recognise a particular antigen among a billion others in a very specific manner, the immune cells of innate immunity recognise molecular patterns present on large groups of invading pathogens, such as flagellin, sugar structures or nucleic acids. These patterns are called pattern-associated molecular patterns (PAMPs). In addition, the immune cells recognise cell debris from dead or dying cells, termed damage-associated molecular patterns (DAMPs). Professional phagocytic cells have structures, so-called pattern recognition receptors (PRRs), which recognise the PAMPs and DAMPs and initiate uptake (phagocytosis) and degradation (respiratory burst) of microbes and debris.

Phagocytosis is a mechanism for ingestion of particles larger than 1 μm. The process was first described by the Russian microbiologist Ilya Ilyich Mechnikov, who discovered phagocytosis in starfish larvae. For his work on phagocytosis, he received the Nobel Prize in 1908. For measuring the phagocytic activity of immune cells in the laboratory, it is common to use fluorescent plastic beads and measure the uptake by flow cytometry. Using this approach, it is possible to measure phagocytic activity of leukocytes (isolated from any species) without the requirement of specific reagents such as antibodies or available genome data. For cleaner fish, these methods have been used for lumpfish (Haugland *et al.*, 2012a) and Ballan wrasse (Haugland *et al.*, 2014). We found that the phagocytic activity of lumpfish leukocytes was highly efficient, and that the leukocytes had already ingested a large proportion of the beads after one hour (Haugland *et al.*, 2012a). All cells can ingest particles to some degree, but the professional phagocytic cells are distinguished from other phagocytic cells by the capacity, speed and size of the particles they can ingest. Also, the professional phagocytic cells use the PRRs to recognise microbes that are ingested. In lumpfish and wrasse several cell types were able to ingest beads, but those with the highest capacity were macrophages/monocytes and also some small agranular cells. The identity of the latter is currently not known, but similar cells have been described from Atlantic salmon, lumpfish and Ballan wrasse (Haugland *et al.*, 2012b; Haugland *et al.*, 2014; Rønneseth *et al.*, 2015). It has been suggested that they may possibly be precursor cells or immature cells.

Respiratory burst

Following ingestion of microbes and debris, the phagocytic cells have different degradation mechanisms taking place in the phagolysosome, either dependent on oxygen (in the process called respiratory burst where oxygen is converted to reactive oxygen species) or independent of oxygen. The oxygen-independent mechanisms are different types of enzymes such as acid phosphatase. Similar to phagocytosis, it is possible to measure respiratory burst activity without any species cell specific reagents. One common approach that has proven suitable for different teleost species including Atlantic salmon and Atlantic cod (Kalgraff *et al.*, 2011), is to measure oxidation of the probe dihydrorhodamine, which is oxidised to the green fluorescent rhodamine 123 in the presence of the receptor independent activator phorbol 12-myristate 13-acetate (PMA). Positive cells can thus be observed in a fluorescent microscope or quantified by flow cytometry. The respiratory burst assay has been used for lumpfish and Ballan wrasse. The cell type with highest respiratory activity in both lumpfish and Ballan wrasse is the neutropils (Haugland *et al.*, 2012a; Haugland *et al.*, 2014). Different studies on teleost fish leukocytes have shown that some of the commercially available cytochemical kits for human leukocytes can be used to stain leukocytes from fish. These are used to identify cell components related to cell structure or function. Although there are differences as to which kits are most suitable to stain leukocytes from fish, and the utility varies for different species, staining of myeloperoxidase and acid phosphatase that are both involved in microbial killing are suitable for all reported fish species. Myeloperoxidase catalyses a process in the respiratory burst whereby hydrogen peroxidase (H_2O_2) is converted to hypochlorous acid (HOCl) and chloride anion (Cl^-). Hypochlorous acid is toxic for the microbes and is used by the immune cells to degrade bacteria and other pathogens. Both lumpfish and wrasse leukocytes stained positive for these enzymes (Haugland *et al.*, 2012a; Haugland *et al.*, 2014; Rønneseth *et al.*, 2015).

Immune components

In order to protect the organisms from invading microbes, a range of soluble components are present in fish, including the components of the complement system, immunoglobulins, cytokines, antimicrobial peptides, lectins and enzymes in the mucosa. An overview of these factors in different fish species is given by Castro and Tafalla (2015). As there is currently no knowledge of the cytokines, antimicrobial peptides, lectins and mucosa in cleaner fish, only the complement system and immunoglobulins will be described here.

The complement system

The complement system represents an evolutionary ancient part of the host defence. Many of the mammalian complement components are found in teleost fish (Nakao *et al.*, 2011) and complement analogues have also been discovered in ancient phyla, including *Cephalochordata* (lancelets), *Urochordata* (tunicates) and *Echinodermata* (sea urchins) (Dunkelberger and Song, 2010). The complement system includes 30 different proteins, including serum proteins and cell membrane-associated proteins (Dunkelberger and Song, 2010). It plays a major role in the innate immune system by 'tagging' pathogens and altered self-components for clearance, but it also influences T and B cell biology and adaptive responses (Dunkelberger and Song, 2010).

The complement system, which can be fully active within seconds, is activated through three major pathways: classical, lectin and alternative (Fig. 14.4A, B). The classical pathway is initiated when the complement component C1q, together with C1r and C1s, recognise antibodies bound to pathogens (Fig. 14.4B). The lectin pathway is analogous to the classical pathway, but is immunoglobulin-independent. It is initiated when germ-line-encoded pattern recognition receptors

(PRRs) such as mannose-binding lectin (MBL) and ficolins (F) recognise carbohydrates at the surface of microbes and form complexes with MBL-associated serine proteases (MASPs). The alternative pathway is different from the two aforementioned pathways and is initiated when C3(H_2O), formed after spontaneous hydrolysis of the complement factor C3, interacts with factor B and D. Properdin (P) is a positive regulator for the alternative pathway. Although different, all three pathways lead to formation of so-called C3 convertases and thereafter C5 convertases, resulting in production of a range of effector molecules that induce different immune responses (Fig. 14.4A, B).

The effector molecules are: 1) anaphylatoxins (C3a, C4a and C5a), which are small protein fragments generated during activation of complement factors that trigger a range of inflammatory and immune-stimulating responses by binding to their cognate receptors (C3aR and C5aR) on various effector cells, 2) opsonins (C4b, C3b, iC3b, C1q and MBL), which attach to the microbes. The opsonin-microbe complexes are recognised by receptors (CR1-4, CR3, CR4, CRIg and C1qR) facilitating clearance of microbes by phagocytes and promote secretion of proinflammatory molecules, 3) the membrane attack complex (MAC) formed at the surface of pathogens and which causes direct lysis of targeted cells (Fig. 14.4A, B). In addition, the complement system function as a disposal system and it is important for removal of immune complexes and foreign bodies, potentially dangerous and toxic substances including apoptotic cells, necrotic cells and cell debris.

Recently, it became apparent that the complement system is not confined to innate immunity. The complement factor C3 also plays an important role in activation and maturation of B cells, including generation of effector and memory B cells (Ricklin *et al.*, 2010). Depletion of C3 results in impaired production of antibodies, which is important for neutralising and opsonisation, and affects the immunological memory necessary for protection against reinfection (Dunkelberger and Song, 2010). Also, crosstalk between Toll-like receptors (TLR), complement receptors and regulators modulate IL-12 production in antigen-presenting cells (APC) and thereby induce the activation and differentiation of T cells (Kemper and Atkinson, 2007; Ricklin *et al.*, 2010).

The complement system is tightly regulated by several regulatory proteins (Zipfel and Skerka, 2009). These proteins prevent complement from being activated on the surfaces of healthy host cells. The regulatory factors are termed factor H, I, D and B, membrane cofactor protein (MCP) and decay-accelerating factor (DAF). For lumpfish, *de novo* RNA sequencing and transcriptome assembly has been performed (our unpublished data) and has given valuable information about components of the immune system and immune responses. Many of the known complement factors and receptors have been identified in lumpfish (Fig. 14.4B), such as C1q, C3, C1r, C1s, the MBL-associated serine proteases MASP-1 and MASP-2, factor B, factor F and properdin. Further, the complement components C2, C3, C4A, C4B, C5, C6, C7, C8A, C8B and C9 are present in the lumpfish transcriptome, in addition to the regulatory factors H, I and D, membrane cofactor protein (MCP) and decay-accelerating factor (DAF). Of the complement receptors, homologues to CR1, C5AR1 and C3AR have also been identified. Further characterisation of these components in lumpfish is needed however, to verify their function.

Immunoglobulins/antibodies

In mammals there are five immunoglobulin classes; IgG, IgM, IgD, IgA, IgE, while in teleost fish there are only three classes: IgM, IgD and IgT (also called IgZ) (Sunyer, 2013). The

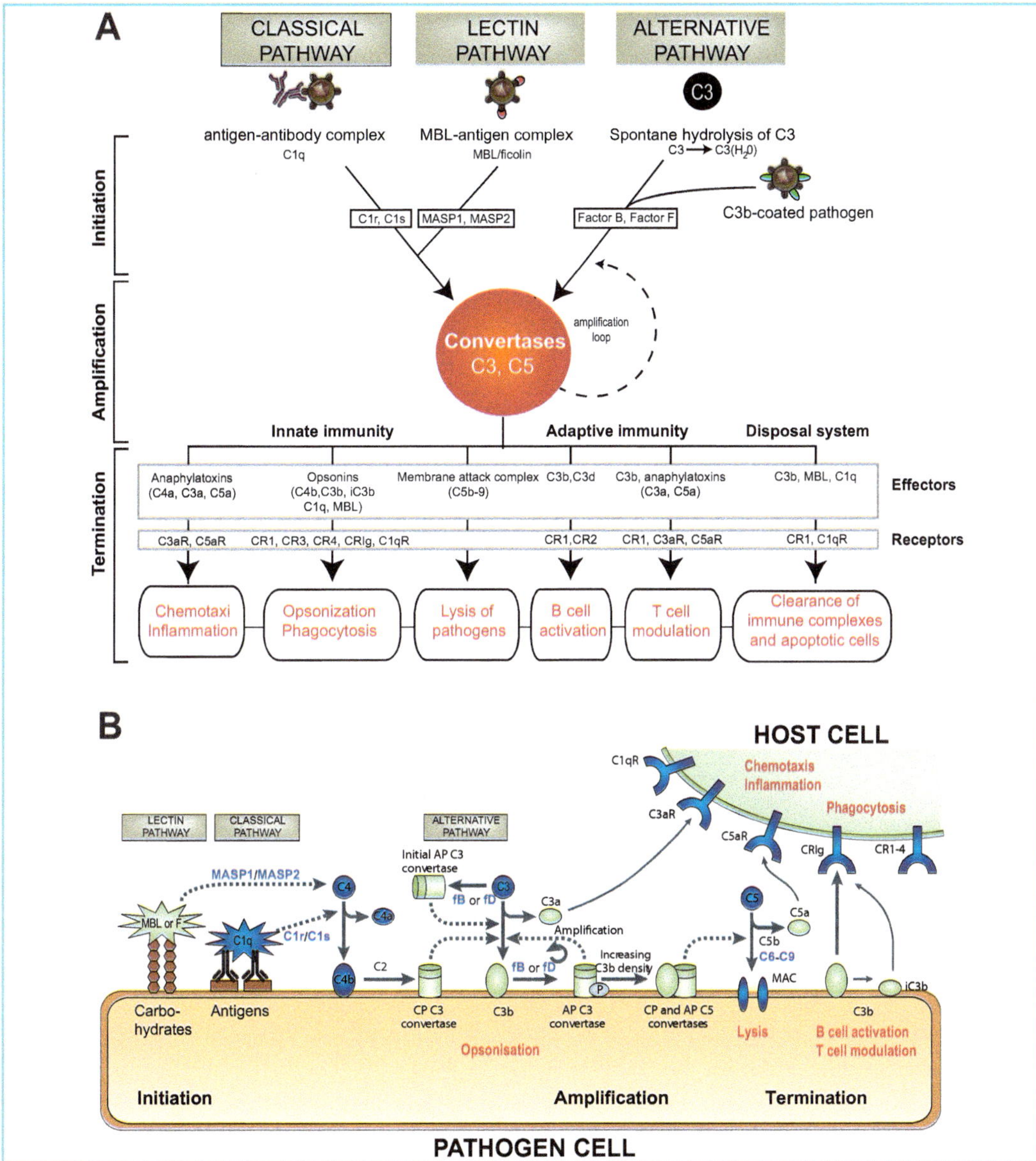

Fig. 14.4 Overview of the complement system in mammals and the complement components present in lumpfish. A) The immune response of the complement system can be divided into 3 phases: 1) Initiation, 2) Amplification and 3) Termination. In the initiation process, the three pathways of complement activation are activated. The classical pathway is initiated when the receptor complex C1q recognises antigen–antibody complexes, the lectin pathway is initiated when mannose-binding lectin (MBL) or Ficolins (F) recognise carbohydrate at the surface of the pathogen, and the alternative pathway is initiated by $C3(H_2O)$ or C3b-coated pathogens. All three pathways result in formation of C3 and C5 convertases and further activation is initiated. Depending on the pathway that is activated, different immune responses are initiated in the termination phase. B) Overview of the components of the complement system. The complement factors coloured blue have been identified in the lumpfish transcriptome.

immunoglobulin family is a protein family to which antibodies and B cell receptors belong. In plasma of teleost fish, soluble IgM (sIgM) is the most prevalent immunoglobulin (Sunyer, 2013), although the concentration varies between species (0.6–20 mg mL^{-1}). In lumpfish, the concentration varies between 0.14 +/- 0.099 mg mL^{-1} in small fish (13.2 +/- 5 g) (Rønneseth *et al.*, 2015) to 1 – 2.6 mg mL^{-1} in large fish (0.6 – 3.6 kg) (Bilal *et al.*, 2016). Serum IgM concentration in adult ballan wrasse is 13.4 mg mL^{-1} (Bilal *et al.*, 2016) IgT, on the other hand, is a monomer in plasma and is present in low concentration. The function of IgD is still an enigma, but it has been suggested that it may function as a PRR or possibly be involved in the adaptive recognition of antigens. In mucosa of gut and gills, the dominant Ig is IgT (IgZ in zebrafish) and IgM is present at low concentrations. IgT in mucosa is a tetramer and IgT^+ B cells are found to be the main B cell subset in the GALT, GiALT, SALT and NALT (Zhang *et al.*, 2010; Xu *et al.*, 2013; Tacchi *et al.*, 2014; Xu *et al.*, 2016).

In lumpfish, sIgM has been purified from serum and membrane-bound forms of IgM and IgD has been identified from transcriptomic data (our unpublished data). The sIgM in serum from lumpfish is a tetramer (Rønneseth *et al.*, 2015), similar to other fish species (reviewed in Salinas *et al.*, 2011), but different from humans where the secreted IgM is a pentamer. IgM has been identified and characterized in different species of wrasse (Bilal *et al.*, 2016) is ongoing and the recent genome sequencing of Ballan wrasse will provide valuable information about the other immunoglobulins and immune components in this species.

The immunoglobulins are involved in several immunological processes involving recognition of pathogens and altered-self components, which are thus targeted for destruction. Professional phagocytic cells have Fc-receptors on their surface where antibodies in immune complexes are bound and results in phagocytosis and degradation of microbes and particles by the cell. As mentioned above, immunoglobulin–pathogen/altered cell complexes also initiate the complement system.

Activation of B cells leads to differentiation into plasma cells that produce antibodies and also memory cells. The antibodies produced are specific for the presented antigen and this is the reason why selection of the correct antigen included in the vaccines is of major importance. After booster immunisation, an increase in specific IgM titer has been reported in teleost fish, but the process is slower than for mammals (weeks instead for days). The production of specific antibodies upon immune-stimulation has been measured in lumpfish (Rønneseth *et al.*, 2015). It was found that the level of specific antibodies was significantly higher in fish injected with atypical *A. salmonicida* (two strains), *Pasturella* isolated from lumpfish and *V. anguillarum* (Fig. 14.5). The only exception was fish injected with *V. ordalii*, where the levels of specific antibodies were not significantly higher than the controls.

14.2 Vaccinology

14.2.1 History, background and status

Wild-caught wrasse has been used as cleaner fish for many years in Norway and Scotland (Treasurer, 1996; Treasurer, 2002). In Norway, the estimated use of wild-caught wrasse in 2016 was 20 million (www.fiskeridir.no/Akvakultur/Statistikk-akvakultur/Akvakulturstatistikk-tidsserier/Rensefisk). These have not been vaccinated with only few exceptions. In Norway,

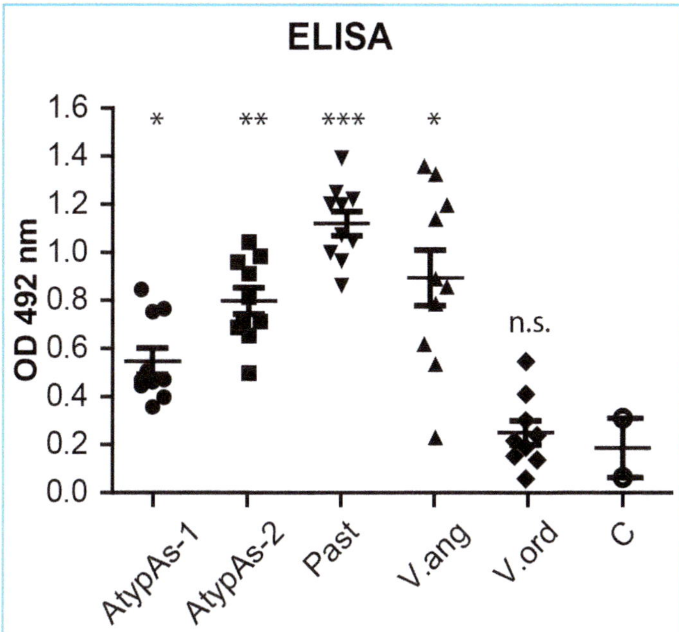

Fig. 14.5 In lumpfish, there is a significant upregulation of specific antibodies after immunisation. The level of specific antibodies after immunisation with bacteria was measured by ELISA. The average with standard deviation is shown in the diagram (n = 10 fish for each bacteria, except for the non-immunised controls where n = 2 fish). Statistical significance: P 0.05 = n.s., *P ≤0.05, **P ≤0.01, ***P ≤0.001. Source: Rønneseth *et al.*, 2015 (reprinted with permission from DCI).

vaccination of farmed wrasse was tested in 2015. Another species that has shown to efficiently eat sea lice is lumpfish (Imsland *et al.*, 2014). In contrast to wrasse, lumpfish is easy to farm and the number of farmed lumpfish in Norway has increased rapidly in the last few years from 0.4 million fish in 2012 to 15.5 million in 2016. (www.fiskeridir.no/Akvakultur/Statistikk-akvakultur/Akvakulturstatistikk-tidsserier/Rensefisk). Unfortunately, in 2012–2013 large-scale mortality was reported among lumpfish due to bacterial infections (www.vetinst.no/eng/Publications/Fish-Health-Report-2012,-2013). Thus, it became clear that vaccination was required to improve the fish welfare and the efficiency of the cleaner fish strategy for sea lice control. Vaccination of lumpfish was initiated in 2012 using a vaccine developed for marine species. Although the vaccines gave some protection, more efficient vaccines tailored to lumpfish had to be developed.

14.2.2 Development of vaccines for cleaner fish

Efficient vaccines protecting against pathogenic bacteria have to contain antigens that stimulate an immune response in order to initiate production of antibodies and the immune reactions necessary to eliminate pathogens. Determination of the antigens to be included in the vaccines, either whole bacteria or bacteria components, is the first step in development for novel vaccines (Fig. 14.6).

The currently available vaccines provide good protection against vibriosis. *V. anguillarum* isolated from lumpfish and wrasse belong to serotype O1 and O2a. These serotypes are included in the currently available lumpfish vaccine and seem to provide adequate protection in lumpfish. The main challenge is to develop a vaccine that provides protection against atypical *Aeromonas salmonicida*, the causative agent of atypical furunculosis, and efforts are now ongoing in Norway and Scotland to achieve this. It is well known that the atypical *A. salmonicida* is a heterogeneous

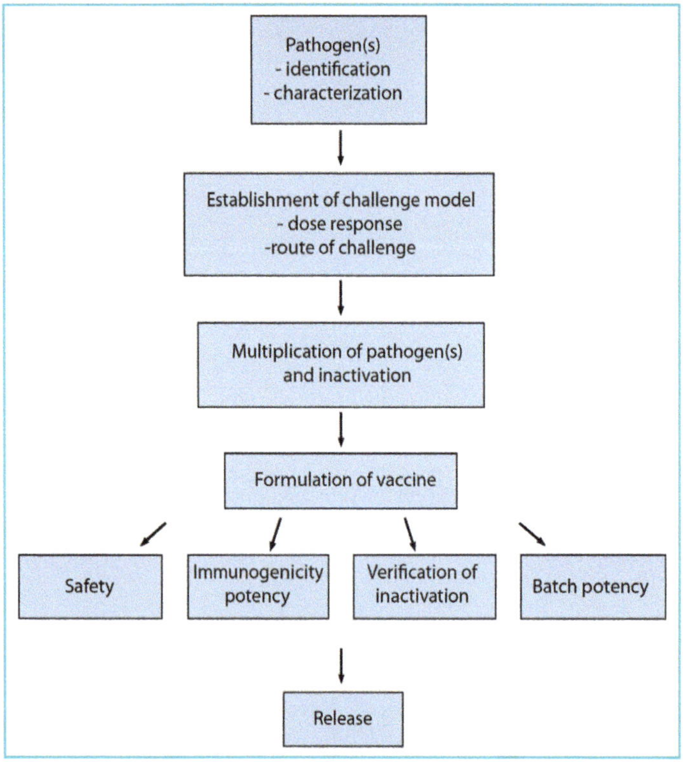

Fig. 14.6 General steps in vaccine production from identification and characterisation of the pathogen(s) until release of a vaccine. Lumpfish fall out of the requirements within the specific monographs of the European Directorate for the Quality of Medicines (EDQM), but still general monographs must be compiled.

group of bacteria and the different strains have a different ability to induce specific or cross-reactive antibodies when used as vaccine antigens (Gudmundsdottir and Bjornsdottir, 2007). Difficulties in obtaining efficient vaccines protecting against atypical furunculosis are also known for other species. Recently, Gulla and co-workers (Gulla *et al.*, 2016) grouped the atypical *A. salmonicida* into 14 groups based on the gene encoding the A-layer (*VapA*). So far, isolates from disease outbreaks among lumpfish have grouped into group VI and isolates from diseased wrasse to group V. The consequences of these differences in A-layer upon vaccine development and protection are not known. Basic research to characterise different strains of bacteria and bacterial components has been initiated to identify which strain or component(s) to include in the vaccine. Optimally, the chosen bacterial strain(s) or component should initiate production of antibodies that recognise and protect against the actual disease causing strains.

At present there are no vaccines developed that provide protection against pasteurellosis, caused by *Pasteurella* sp. isolated from lumpfish, in spite of frequent isolation of this pathogen from diseased fish. This is likely due to technical challenges in growing the bacteria.

14.2.3 Vaccines and vaccination strategies

Vaccination is regarded as the most important single factor for prevention of infectious diseases and is thus essential for improving fish health (Gudding, 2014). Cleaner fish can be vaccinated by injection and/or immersion methods (Fig. 14.7). The currently available vaccines for cleaner fish are shown in Table 14.1.

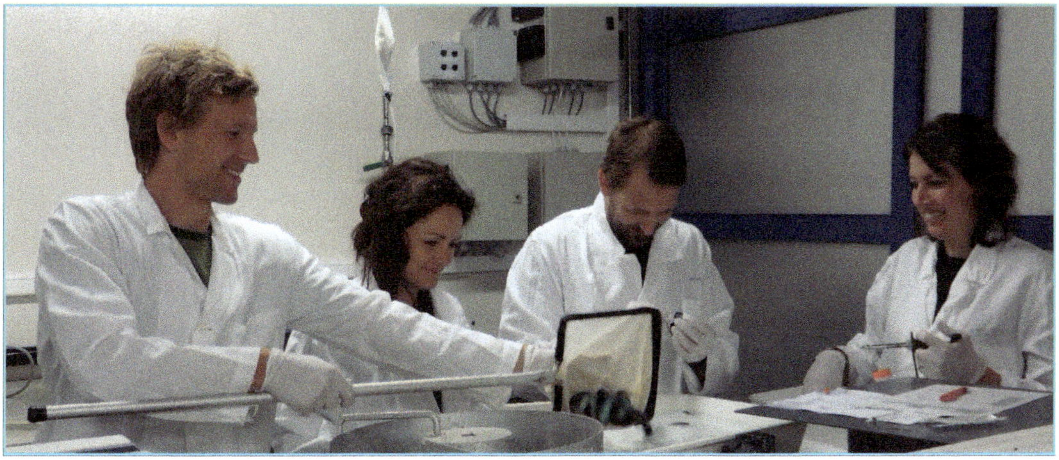

Fig. 14.7 Vaccination of lumpfish. The lumpfish is vaccinated by injection manually when they are about 10 g. After 500 degree days, the lumpfish is transferred to the challenge unit to test the effect of the vaccine.

Bacteria that are not considered to be primary pathogens, such as *Vibrio splendidus* and *Vibrio logei,* may not pose a problem under optimal rearing conditions. Among factors that may affect the health of the fish is water quality and management procedures concerning cleaning, feeding; there may be other factors. Improved feeding and optimised handling of cleaner fish may thus be enough to avoid problems with such bacteria. Bacteria that are not proven to be primary pathogens should therefore not be recommended as vaccine components.

A prerequisite for obtaining good protection after vaccination is that the fish is healthy and immunocompetent. As there is currently little knowledge of immunocompetence in cleaner fish, it is not known at which developmental stage the fish is able to produce an immune response upon exposure to an antigen. Ontogenetic studies of lumpfish larvae are ongoing and results from these studies will be valuable for vaccination strategies. Immersion vaccination can be used for very small fish that cannot be injected but the effect of bath exposure and bath followed by later intraperitoneal injection is currently not documented. At the moment, the suggested fish size for vaccination of cleaner fish, given in Table 14.2, is based on present knowledge and experience.

The injection vaccines currently available for cleaner fish are oil-based emulsion vaccines in which an aqua-phase containing bacteria or bacterial components is mixed with or dispersed within an oil phase. Use of oil will increase the effect of the vaccine, but it may also cause side-effects as for other fish species. Since vaccination of lumpfish is in an initial phase, knowledge of side-effects is limited. Some minor degree of adhesions between the body wall and internal organs near the injection site has been reported. The adhesions are scored using the Spielberg scale (Midtlyng, 1996). The experience so far is that the side-effects after vaccination decrease when the fish size increases, and thus the recommended size for vaccination is 10 g (min 8 g). Although generally of low severity, the significance of the adhesions with regards to growth rate and fish welfare is hitherto not known.

Vaccination of wrasse is in an experimental phase and there is thus no knowledge regarding

Table 14.1 Overview of currently available vaccines to cleaner-fish.

Species	Type of vaccine	Administration	Country	Content
Wrasse	Ordinary (VAXX ON MARINE 3 vet.*)	injection	Norway	2 strains of atypical *Aeromonas salmonicida* *Vibrio anguillarum* O1
	Ordinary (AMarine micro 3-1*)	injection	Scotland	*Vibrio anguillarum* O1, O2a 2 strains of atypical *Aeromonas salmonicida*
Lumpfish	Autogen, different combinations	dip	Norway, Iceland, Faroe Islands	*Vibrio splendidus, Vibrio logei, Vibrio ordalii, Vibrio anguillarum* O1, *Vibrio wodanis, Pseudomonas anguilliseptica*, 2 strains of atypical *Aeromonas salmonicida*
	Autogen	Injection	Iceland, Faroe Islands	1 strain of Atypical *Aeromonas salmonicida, Vibrio anguillarum* O1, *Pseudomonas anguilliseptica*
	Ordinary (VAXX ON MARINE 3 vet.*)	injection	Norway	2 strains of atypical *Aeromonas salmonicida* *Vibrio anguillarum* O1
	Ordinary	dip	Norway	*Vibrio anguillarum* O1, O2a, O2b
	Ordinary (AMarine micro 3-1*)	injection	Norway, Iceland, Faroe Islands, Scotland	*Vibrio anguillarum* O1, O2a 1 strain of atypical *Aeromonas salmonicida*

*AMarine micro 3-1 and VAXX ON MARINE 3 vet. are vaccines which are sold under a special license scheme (application for special exemptions from requirement for marketing authorization).

Table 14.2 Overview of fish size when vaccinated.

Species	Administration	Fish size
Wrasse	Injection	>15 g
Lumpfish	Dip, autogen	>1 g
	Dip, ordinary	>1 g*
	Injection, autogen	>10 g
	Injection, ordinary	>10 g

*an additional round of dip should be performed at least 4 weeks after 1st dip vaccination.

side-effects in Ballan wrasse. The injection point of vaccines in Ballan wrasse and lumpfish is shown in Fig. 14.8. For Ballan wrasse, the site of injection is between the base of the pelvic fin and the anus for small fish (20–40 g) and at the end of the pelvic fin for larger fish (>40 g). During vaccination of the small fish, the needle should be slanting forward, while for larger fish, the angle for the needle should be orthogonal (Henriette Glosvik, Marine Harvest Labrus,

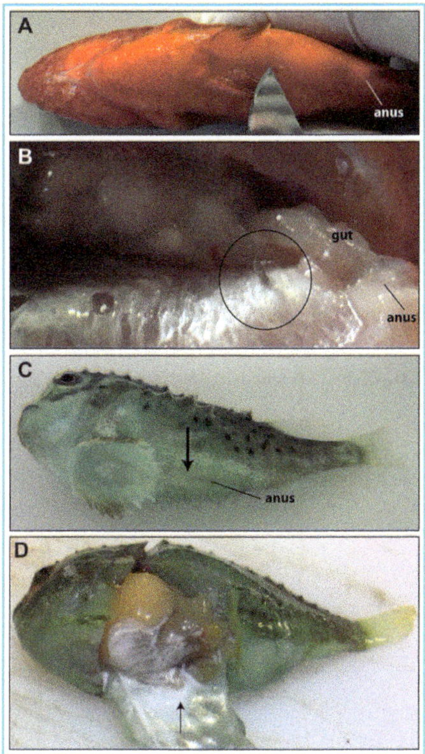

Fig. 14.8 Injection points during vaccination in Ballan wrasse and lumpfish. A) The injection point (shown by the razor blade) in Ballan wrasse is at the base of the pelvic fins for fish >40 g. Credit: Henriette Glosvik, Marine Harvest Labrus. B) The needle should be inserted in front of the anus. Credit: Henriette Glosvik, Marine Harvest Labrus. C) In lumpfish, the vaccine should be injected between the back of the suction disc and anus as indicated by the arrow. The needle should be slanting forward for injection. D) Deposition of the vaccine in the cavity in lumpfish (Credit: Nils Steine, PHARMAQ). The arrow shows the injection point.

personal communication). For lumpfish, the common size when vaccinated by injection is 10 g (min 8 g). The injection point is in the middle between the back edge of the suction dish and anus. The needle should be slanting forward during injection.

Guidelines for vaccination of lumpfish are given by PHARMAQ (www.pharmaq.no/forms-manuals/forms). The measurements included in Table 14.3 should be considered to ensure that the fish is of the correct size and is in good health prior to vaccination.

14.2.4 Vaccine testing

Challenge models

A prerequisite to evaluate the effect of vaccines in fish is to first establish reliable and reproducible challenge models. Different challenge doses and models are tested to find a suitable model for each pathogen (Fig.14.9). Bath and cohabitant models mimic a more natural infection route compared to challenge by injection, thus these models are preferred.

Ideally infection should be established and mortalities start after one to two weeks and result in approximately 80% mortality. Challenge models for *Vibrio anguillarum* (injection, cohabitant and bath) and atypical *A. salmonicida* (injection, cohabitant and bath) have been established for lumpfish (Rønneseth *et al.*, 2017) and Ballan wrasse (Biering *et al.*, 2016). For the lumpfish pathogens Koch's postulates have been fulfilled, meaning that the causative relationship between the microbe and the disease has been verified. The isolate used for challenge was obtained from

Table 14.3 Vaccination control.*

<table>
<tr><th>Measurement</th><th colspan="2">Definition</th><th colspan="2">Score</th></tr>
<tr><td>High</td><td colspan="2">The highest point on the fish</td><td colspan="2">cm</td></tr>
<tr><td>Length</td><td colspan="2">From tip of nose to base of tail</td><td colspan="2">cm</td></tr>
<tr><td>Weight</td><td colspan="2">Weight</td><td colspan="2">g</td></tr>
<tr><td></td><td colspan="2"></td><td>Value</td><td>Definition</td></tr>
<tr><td rowspan="4">Caudal fin</td><td rowspan="4" colspan="2">Erosion of the fin</td><td>0</td><td>Intact</td></tr>
<tr><td>1</td><td>Erosion</td></tr>
<tr><td>2</td><td>Severe erosion</td></tr>
<tr><td>3</td><td>No tail left</td></tr>
<tr><td rowspan="2">Other fins</td><td rowspan="2" colspan="2">Erosion on dorsal fin/pectoral fin</td><td>0</td><td>Intact</td></tr>
<tr><td>1</td><td>Erosion</td></tr>
<tr><td rowspan="2">Deformity</td><td rowspan="2" colspan="2">Suction disc, spine, other</td><td>0</td><td>Intact</td></tr>
<tr><td>1</td><td>Erosion</td></tr>
<tr><td rowspan="4">Skin</td><td rowspan="4" colspan="2">The condition of the skin regarding damage, wounds, diseases</td><td>0</td><td>Intact</td></tr>
<tr><td>1</td><td>Erosion on skin/spikes, nose</td></tr>
<tr><td>2</td><td>More severe erosion</td></tr>
<tr><td>3</td><td>Open wounds, severe damage</td></tr>
<tr><td rowspan="5">Cataract</td><td rowspan="5" colspan="2">The condition of the eyes</td><td>0</td><td>Normal</td></tr>
<tr><td>1</td><td>Pinpoint–10%</td></tr>
<tr><td>2</td><td>10–50%</td></tr>
<tr><td>3</td><td>50–75%</td></tr>
<tr><td>4</td><td>>75%</td></tr>
<tr><td rowspan="6">Injections point (length/side)</td><td colspan="2">Longitudinal direction</td><td colspan="2">Lateral direction</td></tr>
<tr><td>2</td><td>Faulty injection, too far forward</td><td>-2</td><td>Faulty injection, too far to the left</td></tr>
<tr><td>1</td><td>Far forward, but acceptable</td><td>-1</td><td>Acceptable, but to the left of the centre line</td></tr>
<tr><td>0</td><td>Optimal</td><td>0</td><td>Optimal</td></tr>
<tr><td>-1</td><td>Far behind, but acceptable</td><td>1</td><td>Acceptable, but to the right of the centre line</td></tr>
<tr><td>-2</td><td>Faulty injection, too far behind</td><td>2</td><td>Faulty injection, too far to the right</td></tr>
<tr><td rowspan="2">Feed in intestine</td><td rowspan="2" colspan="2">Feed in stomach/ intestine</td><td>0</td><td>No feed</td></tr>
<tr><td>1</td><td>Feed</td></tr>
<tr><td rowspan="4">Rest of vaccine</td><td rowspan="4" colspan="2">Remaining vaccine in the abdominal cavity</td><td>0</td><td>Non-visible</td></tr>
<tr><td>1</td><td>Little remaining, small vesicles</td></tr>
<tr><td>2</td><td>More, larger vehicles</td></tr>
<tr><td>3</td><td>Just vaccinated</td></tr>
<tr><td rowspan="3">Vaccine deposition
</td><td>Region 1</td><td colspan="3">Oesophagus/stomach, liver, foremost part of peritoneal cavity</td></tr>
<tr><td>Region 2</td><td colspan="3">Hindmost part of peritoneal cavity, dorsal parts of the hindgut</td></tr>
<tr><td>Region 3</td><td colspan="3">Ventrally in the abdominal cavity, near the injection site, areas not included in region 1 or 2.</td></tr>
</table>

*The table is copied and modified with permission from PHARMAQ

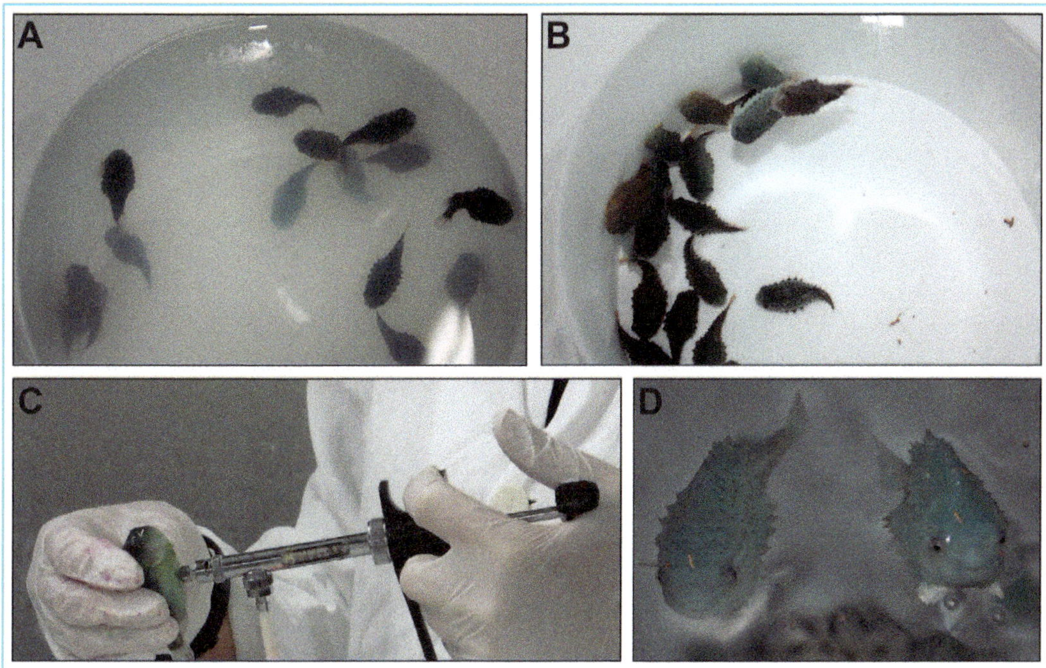

Fig. 14.9 Challenge of lumpfish with different strategies. A) Bath challenge with bacteria, B) Control fish (non-infected) are treated similarly to the fish in the challenged group, C) vaccination by injection, D) VIE tag is suitable to label lumpfish. These two fish have an orange tag, but other colours (green, purple, red and blue) are available.

diseased lumpfish in the field, grown in pure culture, caused disease after challenge and was re-isolated from the diseased fish after experimental challenge.

After vaccination of lumpfish, it is recommended to wait 500 degree days (for example six weeks if the water temperature is 12°C) before the fish is challenged with bacteria. This recommendation is based on measurement of antibody levels at different time points after vaccination (our unpublished measurements). For wrasse, 500–600 degree days have been recommended.

Testing of experimental vaccines against atypical *A. salmonicida* has been performed in both lumpfish (Rønneseth *et al.*, 2017) and Ballan wrasse (Biering *et al.*, 2016) and the results indicate that protection is provided when challenge was performed with the bacteria used as vaccine antigen. Further studies are currently ongoing to test cross-protection provided by the strain(s) selected for use in the vaccine. This is particularly for atypical *A. salmonicida* that is a heterogeneous group of bacteria. In addition, experimental vaccines targeting *Pasteurella* sp. in lumpfish have been tested after experimental challenge (our unpublished results). Challenge models for *V. anguillarum* isolated from lumpfish have also been established (Rønneseth *et al.*, 2013). The level of protection is given in Relative Percent Survival (RPS). For example, biological activity of a dose is given as RPS60 for each antigen using the current formula: [1-(% dead vaccinated fish/60% dead control fish)] × 100. The RPS for the commercially available vaccine currently available for cleaner fish is not yet determined. More vaccine trials and information from the field are needed to verify the level of protection of these vaccines.

Measurement of antibody levels after vaccination

The ability to mount specific antibodies has been verified for lumpfish (Rønneseth *et al.*, 2017), but not yet for wrasse. Due to the heterogeneity of bacterial strains belonging to atypical *A. salmonicida* one aim is to induce production of cross-reactive antibodies after vaccination to all actual variants by selecting, ideally, one or two strains as vaccine antigens. The method suitable for analysis of antibody specificity and the presence of cross-reactive antibodies to various bacterial strains is the enzyme-linked immunosorbent assay (ELISA). To further identify the immuno-reactive antigens, immunoblotting of SDS-PAGE-separated antigens can be performed, and the immuno-reactive band(s) then identified by liquid chromatography-tandem mass spectrometry (LC-MS/MS). Total amount of IgM can be measured by, for example, single radial immune diffusion (SRID), also called the Mancini method (Mancini *et al.*, 1965). SRID has been used for several fish species including lumpfish and wrasse (Rønneseth *et al.*, 2015, Bilal et al. 2016).

14.2.5 Vaccination procedures

Lumpfish and Ballan wrasse are vaccinated manually with injection vaccines. The use of an anaesthetic during vaccination varies. Some lumpfish producers anaesthetise the lumpfish prior to vaccination, while others do not. Our experience is that it is not necessary to anaesthetise lumpfish for vaccination. Ballan wrasse is, however, anaesthetised during vaccination as this species seems to be more easily stressed. Established protocols for full anaesthesia and euthanasia of lumpfish have been developed (Skår *et al.*, 2017). A concentration of 200 mg L^{-1} metacaine can be used independently of fish size and sea water temperature. For vaccination of lumpfish, it is recommended that the fish is starved 24 hours prior to vaccination (source: PHARMAQ). In our experience lumpfish eat as normal within a few hours after vaccination.

14.2.6 Concluding remarks

Cleaner fish have been shown to be an efficient biological alternative to medicines to reduce the problems with sea lice. Wild-caught wrasse have been mainly used until recently, but it is recommended to use farmed fish, both wrasse and lumpfish, to avoid further reduction of the stock of wrasse, and better control of possible disease transfer and also fish welfare (Johansen *et al.*, 2016). The number of farmed lumpfish has increased tremendously during the last few years, but it is clear that efficient vaccines are required to protect against the most pathogenic bacteria. Similarly, vaccines must be provided for farmed wrasse. Better characterisation of the pathogenic bacteria, the immune system and responses in these species will give the basis for development of more efficient vaccines than those currently available. One must keep in mind that vaccine development in cleaner fish commenced very recently and further progress in vaccine efficacy is expected. The improved welfare and health of the cleaner fish is a pre-requisite for further and successful use of cleaner fish.

Acknowledgements

Nils Steine (PHARMAQ/Zoetis), Nils Vestvik (PHARMAQ/Zoetis), Espen Brudal (PHARMAQ/Zoetis), Rune Wiulsrød (PHARMAQ/Zoetis), Henriette Glosvik (Marine Harvest Labrus), Renate Johansen (PHARMAQ Analytic AS/Zoetis), Eivind Isdal (Vaxxinova).

Partial funding: The RCN project 244148 (GTH), FHF projects (HIW).

References

Aghaallaei, N., Bajoghli, B., Schwarz, H., Schorpp, M. and Boehm, T. (2010) Characterisation of mononuclear phagocytic cells in medaka fish transgenic for a cxcr3a:gfp reporter. *Proceedings of the National Academy of Science of the United States of America* 107, 18079–18084.

Bassity, E. and Clark, T.G. (2012) Functional identification of dendritic cells in the teleost model, rainbow trout (*Oncorhynchus mykiss*). *PLoS One* 7, e33196.

Beck, B.H. and Peatman, E. (2015). *Mucosal health in aquaculture.* Academic press, Elsevier, Oxford, U.K.

Biering, E., Vaagnes, O., Krossøy, B., Gulla, S. and Colquhoun, D.J. (2016) Challenge models for atypical *Aeromonas salmonicida* and *Vibrio anguillarum* in farmed Ballan wrasse (*Labrus bergylta*) and preliminary testing of a trial vaccine against atypical *Aeromonas salmonicida. Journal of Fish Diseases* 39, 1257–1261.

Bilal, S., Lie, K.K., Karlsen, O.A. and Hordvik, I. (2016) Characterization of IgM in Norwegian cleaner fish (lumpfish and wrasse). *Fish and Shellfish Immunology* 59, 9–17.

Castro, R. and Tafalla, C. (2015) Organs, tissues, and general structures. In: Beck, B.H. and Peatman, E. (eds.) *Mucosal health in aquaculture.* Academic press, Elsevier, Oxford, UK, pp. 3–54.

Dunkelberger, J.R. and Song, W.C. (2010) Complement and its role in innate and adaptive immune responses. *Cell Research* 20, 34–50.

Fillatreau, S., Six, A., Magadan, S., Castro, R., Sunyer, J.O. and Boudinot, P. (2013) The astonishing diversity of Ig classes and B cell repertoires in teleost fish. *Frontiers Immunology* 4, 28.

Ganassin, R. and Bols, N. (1996) Development of long-term rainbow trout spleen cultures that are haemopoietic and produce dendritic cells. *Fish and Shellfish Immunology* 6, 17–34.

Granja, A.G., Leal, E., Pignatelli, J., Castro, R., Abos, B., Kato, G., Fischer, U. and Tafalla, C. (2015) Identification of teleost skin CD8alpha+ dendritic-like cells, representing a potential common ancestor for mammalian cross-presenting dendritic cells. *Journal of Immunology* 195, 1825–1837.

Gudding, R. (2014) Vaccination as a preventive measure. In: Gudding, R., Lillehaug, A. Evensen, Ø (eds.) *Fish Vaccination.* Wiley Blackwell, Oxford.

Gudmundsdottir, B.K. and Bjornsdottir, B. (2007) Vaccination against atypical furunculosis and winter ulcer disease of fish. *Vaccine* 25, 5512–5523.

Gulla, S., Lund, V., Kristoffersen, A.B., Sørum, H. and Colquhoun, D.J. (2016) vapA (A-layer) typing differentiates *Aeromonas salmonicida* subspecies and identifies a number of previously undescribed subtypes. *Journal of Fish Diseases* 39, 329–342.

Haugland, G.T., Jakobsen, R.A., Vestvik, N., Ulven, K., Stokka, L. and Wergeland, H.I. (2012a) Phagocytosis and respiratory burst activity in lumpsucker (*Cyclopterus lumpus* L.) leucocytes analysed by flow cytometry. *PLoS One* 7, e47909.

Haugland, G.T., Jordal, A.E. and Wergeland, H.I. (2012b) Characterisation of small, mononuclear blood cells from salmon having high phagocytic capacity and ability to differentiate into dendritic like cells. *PLoS One* 7, e49260.

Haugland, G.T., Olsen, A.-B., Rønneseth, A. and Andersen, L., (2016) Lumpfish (*Cyclopterus lumpus* L.) develop amoebic gill disease (AGD) after experimental challenge with *Paramoeba perurans* and can transfer amoebae to Atlantic salmon (*Salmo salar* L.). *Aquaculture*. 478, 48–55.

Haugland, G.T., Rønneseth, A. and Wergeland, H.I. (2014) Flow cytometry analyses of phagocytic and respiratory burst activities and cytochemical characterisation of leucocytes isolated from wrasse (*Labrus bergylta* A.). *Fish and Shellfish Immunology* 39, 51–60.

Iliev, D.B., Thim, H., Lagos, L., Olsen, R. and Jorgensen, J.B. (2013) Homing of antigen-presenting cells in head kidney and spleen – salmon head kidney hosts diverse APC types. *Frontiers Immunology* 4.

Imsland, A.K., Reynolds, P., Eliassen, G., Hangstad, T.A., Foss, A., Vikingstad, E. and Elvegård, T.A. (2014) The use of lumpfish (*Cyclopeterus lumpus* L.) to control sea lice (*Lepeophtheirus salmonis* Krøyer) infestation is intensively farmed Atlantic salmon (*Salmon salar* L.). *Aquaculture* 424–425, 18–23.

Johansen, L.-H., Colquhoun, D., Hansen, H., Hildre, S., Wergeland, H.I. and Mikalsen, H.E. (2016) Analyse av sykdomsrelatert risiko forbundet med bruk av villfanget og oppdrettet rensefisk for kontroll av lakselus. *In norwegian.*

Kalgraff, C.A., Wergeland, H.I. and Pettersen, E.F. (2011) Flow cytometry assays of respiratory burst in Atlantic salmon (*Salmo salar* L.) and in Atlantic cod (*Gadus morhua* L.) leucocytes. *Fish and Shellfish Immunology* 31, 381–388.

Kemper, C. and Atkinson, J.P. (2007) T-cell regulation: with complements from innate immunity. *Nature Review Immunology* 7, 9–18.

Li, J., Barreda, D.R., Zhang, Y.A., Boshra, H., Gelman, A.E., Lapatra, S., Tort, L. and Sunyer, J.O. (2006) B lymphocytes from early vertebrates have potent phagocytic and microbicidal abilities. *Nature Immunology* 7, 1116–1124.

Lovy, J., Savidant, G.P., Speare, D.J. and Wright, G.M. (2009) Langerin/CD207 positive dendritic-like cells in the haemopoietic tissues of salmonids. *Fish and Shellfish Immunology* 27, 365–368.

Lugo-Villarino, G., Balla, K.M., Stachura, D.L., Banuelos, K., Werneck, M.B. and Traver, D. (2010) Identification of dendritic antigen-presenting cells in the zebrafish. *Proceedings of the National Academy of Science of the United States of America* 107, 15850–15855.

Mancini, G., Carbonara, A.O. and Heremans, J.F. (1965) Immunochemical quantitation of antigens by single radial immunodiffusion. *Immunochemistry* 2, 235–254.

Midtlyng, P.J. (1996) A field study on intraperitoneal vaccination of Atlantic salmon (*Salmo salar* L.) against furunculosis. *Fish and Shellfish Immunology* 6, 553–565.

Nakanishi, T., Shibasaki, Y. and Matsuura, Y. (2015) T Cells in Fish. *Biology (Basel)* 4, 640–63.

Nakao, M., Tsujikura, M., Ichiki, S., Vo, T.K. and Somamoto, T. (2011) The complement system in teleost fish: progress of post-homolog-hunting researches. *Development and Comparative Immunology* 35, 1296–1308.

Øverland, H.S., Pettersen, E.F., Rønneseth, A. and Wergeland, H.I. (2010) Phagocytosis by B-cells and neutrophils in Atlantic salmon (*Salmo salar* L.) and Atlantic cod (*Gadus morhua* L.). *Fish Shellfish Immunology* 28, 193–204.

Parra, D., Korytar, T., Takizawa, F. and Sunyer, J.O. (2016) B cells and their role in the teleost gut. *Development and Comparative Immunology* doi:10.1016/j.dci.2016.03.013.

Patel, D.M. and Brinchmann, M.F. (2017) Skin mucus proteins of lumpsucker (Cyclopterus lumpus). *Biochemistry and Biophysics Reports* 9, 217-225

Reite, O.B. and Evensen, O. (2006) Inflammatory cells of teleostean fish: A review focusing on mast cells/ eosinophilic granule cells and rodlet cells. *Fish and Shellfish Immunology* 20, 192–208.

Ricklin, D., Hajishengallis, G., Yang, K. and Lambris, J.D. (2010) Complement: a key system for immune surveillance and homeostasis. *Nature Immunology* 11, 785–97.

Rombout, J.H.W.M., Yang, G.W. and Kiron, V. (2014) Adaptive immune responses at mucosal surfaces of teleost fish. *Fish and Shellfish Immunology* 40, 634–643.

Rønneseth, A., Ghebretnsae, D.B., Wergeland, H.I. and Haugland, G.T. (2015) Functional characterisation of IgM⁺ B cells and adaptive immunity in lumpfish (*Cyclopterus lumpus* L.). *Development and Comparative Immunology* 52, 132–143.

Rønneseth, A., Haugland, G.T., Colquhoun, D., Brudal, E. and Wergeland, H.I. (2017) Protection and antibody reactivity following vaccination of lumpfish (*Cyclopterus lumpus* L.) against atypical *Aeromonas salmonicida. Fish and Shellfish Immunology* 64, 383–391.

Rønneseth, A., Nilsen, H., Haugland, G.T., Colquhoun, D. and Wergeland, H.I. (2013) Tapsårsaker og smittemodeller (Ap 3). *In Norwegian.*

Salinas, I. (2015) The Mucosal Immune System of Teleost Fish. *Biology (Basel)* 4, 525–539.

Salinas, I., Zhang, Y.A. and Sunyer, J.O. (2011) Mucosal immunoglobulins and B cells of teleost fish. *Developmental and Comparative Immunology* 35, 1346–1365.

Sepahi, A. and Salinas, I. (2016) The evolution of nasal immune systems in vertebrates. *Molecular Immunology* 69, 131–138.

Skår, M.W., Haugland, G.T., Powell, M.D., Wergeland, H.I. and Samuelsen, O.B. (2017) Development of anaesthetic protocols for lumpfish (*Cyclopterus lumpus* L.): Effect of anaesthetic concentrations, sea water temperature and body weight. *PLoS One* 12(7): e0179344-.

Somamoto, T., Koppang, E.O. and Fischer, U. (2014) Antiviral functions of CD8(+) cytotoxic T cells in teleost fish. *Development and Comparative Immunology* 43, 197–204.

Sunyer, J.O. (2013) Fishing for mammalian paradigms in the teleost immune system. *Nature Immunology* 14, 320-326.

Tacchi, L., Musharrafieh, R., Larragoite, E.T., Crossey, K., Erhardt, E.B., Martin, S.A., LaPatra, S.E. and Salinas, I. (2014) Nasal immunity is an ancient arm of the mucosal immune system of vertebrates. *Nature Communication* 5, 5205.

Tacchi, L., Lowrey, L., Musharrafieh, R., Crossey, K., Larragoite, E.T. and Salinas, R. (2015) Effects of transportation stress and addition of salt to transport water on the skin mucosal homeostasis of rainbow trout (*Oncorhynchus mykiss*). *Aquaculture* 439, 66–67.

Treasurer, J.W. (1996) Wrasse (*Labridae*) as cleaner fish of sea lice on farmed Atlantic salmon in west Scotland. In: Sayer, M.D.J., Treasurer, J.W. and Costello, M.J (eds.) *Wrasse: biology and use in aquaculture*, Fishing News Books, Oxford, UK, pp. 74–90.

Treasurer, J.W. (2002) A review of potential pathogens of sea lice and the application of cleaner fish in biological control. *Pest Management Science* 58, 546–558.

Uribe, C., Folch, H., Enriquez, R. and Moran, G. (2011) Innate and adaptive immunity in teleost fish: a review. *Veterinarni Medicina* 56, 486–503.

Wittamer, V., Bertrand, J.Y., Gutschow, P.W. and Traver, D. (2011) Characterisation of the mononuclear phagocyte system in zebrafish. *Blood* 117, 7126–7135.

Xu, Z., Parra, D., Gomez, D., Salinas, I., Zhang, Y.A., von Gersdorff Jorgensen, L., Heinecke, R.D., Buchmann, K., LaPatra, S. and Sunyer, J.O. (2013) Teleost skin, an ancient mucosal surface that elicits gut-like immune responses. *Proceedings of the National Academy of Science of the United States of America* 110, 13097–13102.

Xu, Z., Takizawa, F., Parra, D., Gomez, D., von Gersdorff Jorgensen, L., LaPatra, S.E. and Sunyer, J.O. (2016) Mucosal immunoglobulins at respiratory surfaces mark an ancient association that predates the emergence of tetrapods. *Nature Communication* 7, 10728.

Zhang, Y.A., Salinas, I., Li, J., Parra, D., Bjork, S., Xu, Z., LaPatra, S.E., Bartholomew, J. and Sunyer, J.O.

(2010) IgT, a primitive immunoglobulin class specialized in mucosal immunity. *Nature Immunology* 11, 827–835.

Zipfel, P.F. and Skerka, C. (2009) Complement regulators and inhibitory proteins. *Nature Reviews Immunology* 9, 729–740.

Zwollo, P., Cole, S., Bromage, E. and Kattaari, S. (2005) B cell heterogeneity in the teleost kidney: evidence of a maturation gradient from anterior to posterior kidney. *Journal of Immunology* 174, 6608-6616.

CHAPTER 15
Cleaner fish welfare

Jim Treasurer,[1] *Chris Noble,*[2] *Velmurugu Puvanendran*[2]*, Sonia Rey Planellas*[3] *and Martin Haugmo Iversen*[4]

[1]FAI Aquaculture, Ardtoe, UK
[2]Nofima, Tromsø, Norway
[3]Institute of Aquaculture, University of Stirling, UK
[4]Faculty of Biosciences and Aquaculture, University of Nordland, Bodø, Norway

15.1 Fish welfare in aquaculture

The welfare of fish is a critical aspect of fish rearing (Scott and Ellis, 2007) and has received an increased amount of attention in recent years. This attention has stemmed from the aquaculture industry itself, in terms of optimising the husbandry and stewardship practices in relation to the species-specific requirements of the fish (Dykes, 2012) and also due to concerns from the wider stakeholder community (Martins *et al.*, 2012). Numerous aquaculture acts, policy and certification schemes now explicitly state that fish welfare must be incorporated into general husbandry and aquaculture practices. For example, at a policy level the Council Directive 98/58/EC (http://eur-lex.europa.eu/legal-content/EN/TXT/?uri=CELEX:31998L0058) exists and the World Organisation for Animal Health (OIE) has a series of guiding principles for farmed fish welfare in its 'Aquatic Animal Health Code' that was updated in 2016 (www.oie.int/en/international-standard-setting/aquatic-code/access-online). Further, national aquaculture acts and legislatory frameworks are also available that account for farmed fish welfare, see for example Voas (2008) for a review of farmed fish legislation in the UK and the Norwegian Animal Welfare Act (LOV 2009-06-19 nr 97: Lov om dyrevelferd https://lovdata.no/dokument/NL/lov/2009-06-19-97). Niche aquaculture sectors including organic aquaculture also address and regulate fish welfare at the national and EU level, for example. EU Regulation No. 834/2007 (http://eur-lex.europa.eu/legal-content/EN/TXT/PDF/?uri=CELEX:32007R0834&rid=6) and there are now certification schemes that are specifically devoted to fish welfare, such as the RSPCA Assured welfare standards for Atlantic salmon *Salmo salar* (RSPCA, 2015).

This increased focus on fish welfare has also been seen in the wider scientific community. There are numerous research and review papers covering fish welfare (Ellis *et al.*, 2002; Huntingford *et al.*, 2006; Ashley, 2007; Martins *et al.*, 2012; Noble *et al.*, 2012a; Segner *et al.*, 2012 and Stien *et al.*, 2013) and also books (e.g. Branson, 2008). This means farmers and other interested stakeholders have a wealth of information at hand regarding fish welfare, but sometimes little in the way of a framework for assessing and optimising this on the farm.

15.2 Measuring welfare out on the farm

How to measure fish welfare

To incorporate fish welfare considerations into farming practices, they must first be capable of being evaluated and appraised. This can be done using Welfare Indicators (WIs), that can be based upon the animal or the resources available to it, which are termed either direct (animal-based) or indirect (resource-based) WIs. Both categories of WI can also be broken down into either short- or long-term indicators of welfare and, in the case of animal-based WIs, it can be centred upon the individual or the group. Direct animal-based WIs can focus upon the physiology, health, morphology and behaviour of the fish (Huntingford *et al.*, 2006). Indirect resource-based WIs can be based upon parameters of the rearing environment such as water quality (Huntingford and Kadri, 2009). Key Performance Indicators (KPIs) such as short- or long-term changes in growth, feeding efficiency, appetite and mortality (Dykes, 2012) are also suitable as WIs in animal production systems.

However, from the suite of WIs available for assessing welfare, not all are appropriate for measuring welfare on the farm as they may be too cumbersome or difficult to measure robustly in a practical setting. For example, stress can be measured directly with cortisol from blood samples, mucus or faeces or non-invasively by sampling the water in which the fish are present (Scott and Ellis, 2007), but these, especially the latter, can be time-consuming and impractical for routine monitoring. A number of non-intrusive techniques that can be applied readily on farms as practical measures on farmed fish have been suggested and these vary with fish species and circumstances (Huntingford *et al.*, 2006). These are generally termed Operational Welfare Indicators, or OWIs (North *et al.*, 2008), and must be robust indicators of fish welfare, be repeatable and suitable for use out on the farm (Noble *et al.*, 2012b), and be suited for different life stages and rearing systems/husbandry routines.

A range of OWI tools can be used to assess welfare on the farm and can be selected from the different WI categories outlined above, for example growth (a performance WI), stress (animal-based physiological WI), and skin or fin damage (animal-based morphological WI). Huntingford *et al.* (2006) stated OWIs can include, but are not limited to, changes in ventilation rate, behaviour, reduced food intake, condition, poor growth, abnormalities, injuries including fin damage and disease. Stien *et al.* (2013) have also provided a comprehensive list of the potential WIs and OWIs that are available for measuring fish welfare. These OWIs can then be combined (e.g. Turnbull *et al.*, 2005) to help the farmer or other interested stakeholders get a better overview of the welfare of the fish from a multitude of angles (integrated welfare assessment in fish). They can also be utilised in welfare assessment or assurance protocols and schemes (Stien *et al.*, 2013; Main *et al.*, 2014; Pettersen *et al.*, 2014; RSPCA, 2015).

15.3 WIs and OWIs for cleaner fish

As cleaner fish are a group of less established aquaculture species, especially lumpfish *Cyclopterus lumpus*, there is limited information regarding welfare, WIs and OWIs compared to other farmed species such as Atlantic salmon *Salmo salar* or rainbow trout *Onchorhynchus mykiss*. However, we will attempt to break down current known WIs and OWIs into i) direct indicators (morphological, behavioural, physiological) and ii) indirect indicators (based upon the rearing environment) according to the species and, where appropriate, the life stage as well.

15.3.1 Morphological WIs and OWIs for wrasse and lumpfish

Morphological welfare indicators can be widely applicable across a range of life stages and rearing systems. The range of morphological WIs and OWIs that are suitable for cleaner fish in tanks and cages and from juveniles to broodstock include i) fin splitting, ii) fin erosion (both active and healed), iii) jaw and snout damage, iv) eye damage, v) epidermal damage such as scale loss and vi) opercular damage. Other potential morphological OWIs include malformations, skin colouration and, for lumpfish, deformities to the ventral suction disc.

Wrasse

OWIs based upon fish morphology and injuries have been previously reported for five wrasse species (Treasurer and Feledi, 2014). Fin damage is a well-established OWI in numerous aquaculture species and a modified fin damage index based on Hoyle *et al.* (2007, rainbow trout) and Noble *et al.* (2008, Atlantic salmon) could be a suitable tool for assessing fin damage in wrasse. However, fin damage classification scales for salmon and trout are currently being updated to account for either active or healed fin damage, simplified to a 0–3 damage scale (J. Turnbull, Stirling, March 2017, personal communication) and these classifications may be considered in future fin damage assessments on both wrasse and lumpfish. Treasurer and Feledi (2014) suggested that the erosion and splitting of the fins were the most commonly identified signs of physical damage in wrasse and these indices can be used by farmers to routinely assess the condition of wrasse on receipt after wild capture, during the hatchery rearing phase or while stocked in salmon cages. The indices were low for fin erosion but fin splitting was more prevalent and was more pronounced on the caudal fin. Rock cook *Centrolabrus exoletus* and corkwing *Symphodus melops* were more prone to fin splitting compared with the other wrasse species (Figs 15.1, 15.2). The occurrence of eye damage was low, recorded in only one rock cook, and none of the fish examined exhibited any jaw damage. Further, there was evidence of filamentous bacterial involvement in the more advanced levels of fin erosion in goldsinny wrasse *Ctenolabrus rupestris* during the breeding season. Bacterial infection was secondary to minor lesions arising from physical abrasion or fin nipping. However, these bacteria were of mixed origin and could not be identified. Measures of welfare varied between these wrasse species held together in the same tanks. The levels of fin damage differed between wrasse species and rock cook and corkwing were more susceptible to damage than others. The condition factor was also been found to be a critical OWI for wrasse (Fulton's Index, Treasurer and Feledi, 2014) that is quantitative, easily measured and gives a good indication of the physical condition of wrasse. The condition factor has also been used successfully in both farmed and wild Ballan wrasse *Labrus bergylta* and can be an indicator of acclimation to, for example, the cage rearing environment as it was found to decline by 2.5–3.6% in the six weeks after fish were stocked in salmon cages (Skiftesvik *et al.*, 2013).

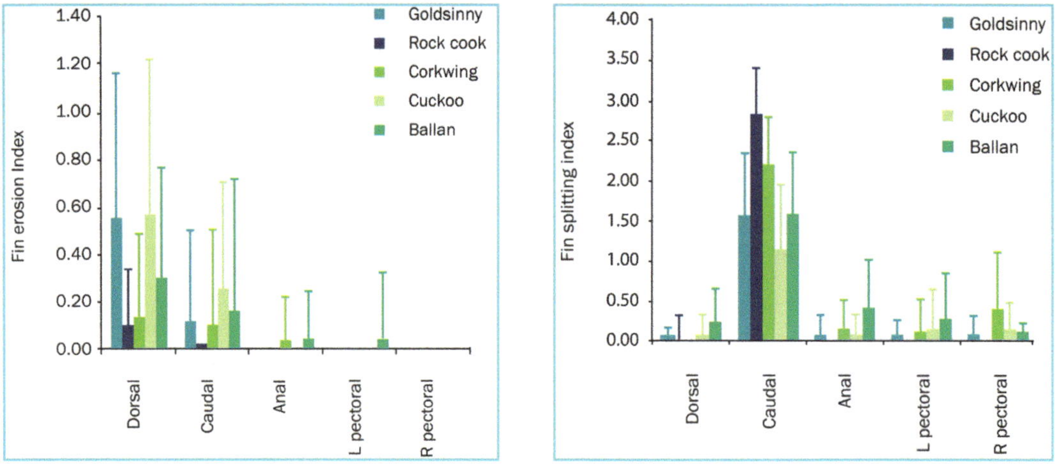

Fig. 15.1 Fin erosion and splitting in five species of wrasse. SD = vertical bars. The index of splitting of the caudal fin was significantly higher ($\underline{P}$ <0.05) than the other fin-splitting indices. Sample size = 50 fish for all species with the exception of cuckoo wrasse = 25. Modified after Treasurer and Feledi, 2014.

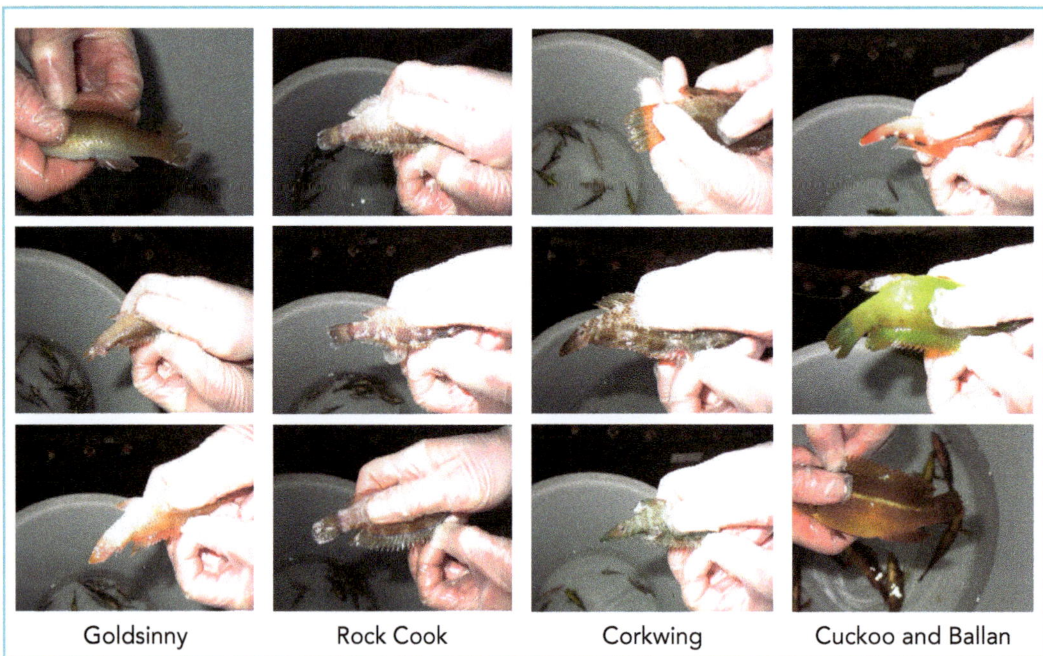

Fig. 15.2 Examples of the range in fin erosion and splitting indices in five wrasse species. From the left top, goldsinny wrasse with grade 4 fin splitting, below two goldsinny shown with high and grade 3 FEI; top, rock cook with grade 1 both erosion and fin splitting, middle rock cook with grade 4 tail splitting, bottom rock cook with anal fin erosion grade 1; top, corkwing with FSI grade 1, middle, corkwing dorsal FEI grade 2, bottom, corkwing with dorsal FEI grade 1; top, cuckoo wrasse with very minor dorsal fin split grade 1; middle Ballan wrasse with one split in tail, FSI grade 1; bottom, Ballan no fin damage. Based on Treasurer and Feledi, 2014.

Lumpfish

As far as the authors are aware, little in terms of morphological OWIs have been published regarding lumpfish. In this respect, we will present unpublished data on the suitability of different morphological WIs and OWIs for different life stages of lumpfish that can be used in the hatchery, cage deployment and also for lumpfish broodstock. In this current study an assessment was carried out on whether fin erosion (FEI) and fin splitting (FSI) indices that have been developed for wrasse (Treasurer and Feledi, 2014) could be applied in lumpfish. This 0-4 classification scheme is currently being reviewed with a view to a 0-3 grading as indicated in 15.3.1 (Rey *et al.*, 2017) Examination of 150 lumpfish in a preliminary study, and 340 in a detailed study, showed that similar FEI and FSI of wrasse can be applied to juvenile lumpfish (Figs 15.3, 15.4) and it is likely that these could also be used when the fish are deployed in sea cages. Unlike wrasse, fin erosion was a major issue, especially to the caudal fin, with a high incidence over several months. Caudal fin erosion also appears to be life stage-specific, occurring in both the juvenile and maturation phase. For example, observations during the early phase of rearing (first two to three months) indicated that lumpfish juveniles are aggressive to each other and the prevalence of caudal fin nipping is very high (V. Puvanendran, Tromsø, March 2017, personal observations). Beyond this stage, this behaviour ceased and once the lumpfish juveniles reached about 10 g in weight, the caudal fin damage disappeared and juveniles maintained healthy and complete fins under optimal rearing conditions. However, from another study at normal fish farming rearing densities of 10 up to 20 kg m^{-3} with environmental enrichment to increase surface area (S. Rey, Stirling, April 2017, personal communication) the occurrence of fin damage was observed in juvenile fish up to 40 g. However, by grading the fish, aggression decreased and, as a consequence, overall fin condition improved for all fish sizes. Other welfare indices for juveniles were physical lesions in the jaw area, above the head and on the head, although these were not seen in other populations (S. Rey, personal observation). A further example has been observed during gonadal maturation (Puvanendran V., Tromsø, March 2017, personal observations) and spawning of one year old or older lumpfish where higher stocking densities affected the welfare status. Severe fin damage and lesions appeared, and subsequently higher mortalities occurred. From these observations, Puvanendran and colleagues recommend lower stocking densities of less than 20 kg m^{-3} during gonadal development for a tank without stacked resting places and higher stocking densities if more resting places are available. The dorsal and anal fins had occasional low-level erosion but were largely unaffected.

Morphological indices were also used to assess welfare in a small study on broodstock lumpfish at Ardtoe (J. Treasurer, Ardtoe, March 2017, personal communication). Unlike juvenile lumpfish, fin erosion and splitting were negligible and the tails were undamaged. However, jaw erosion and physical damage (Fig. 15.5), seen as lesions and in the appearance of cataracts, were common. The abrasions were present on both sides of the lower jaw and beside the eye (Fig. 15.5). There was no physical damage to eyes or exophthalmia, but cataracts were often seen. Some fish also showed the signs of fungal infection with hyphae. The welfare issues of lumpfish broodstock therefore appear different from juveniles. There was no physical or behavioural evidence of aggression between adults, but the jaw damage and lesions indicated physical damage. While juveniles might benefit from continuous feeding and a strict grading plan to reduce aggression, damage in adult fish may be reduced by shade screening of the tank and a reduction in stocking density. Both juveniles and adults also require enough surface area on which to attach. A minimum individual surface area index required at different developmental stages would be of great value.

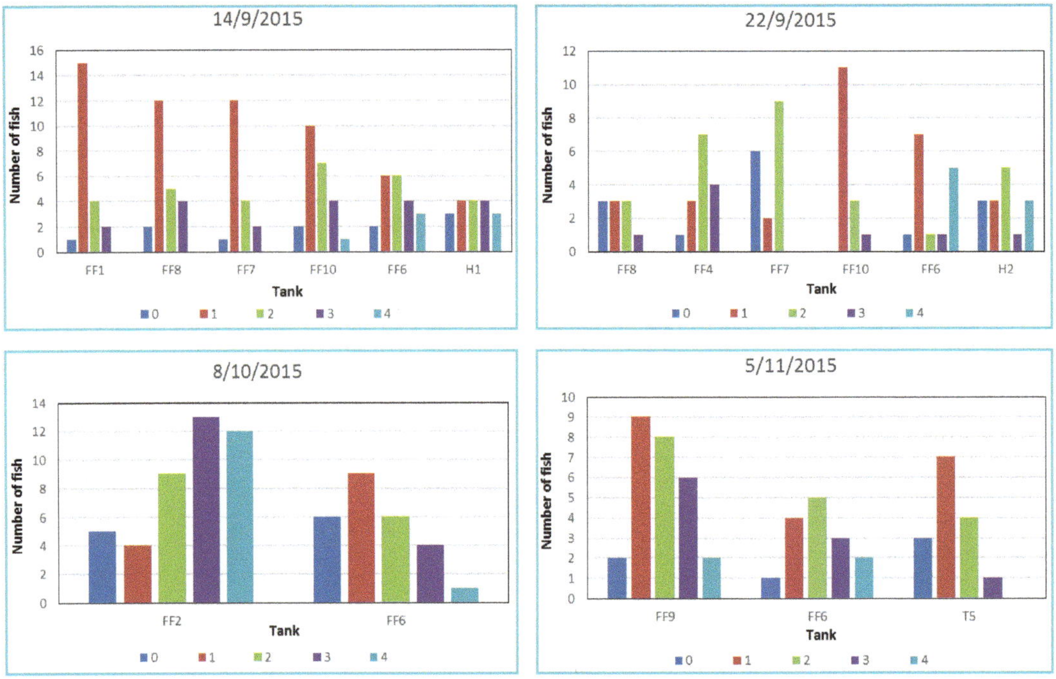

Fig. 15.3 Fin erosion indices of lumpfish juveniles from 30 to 40 g weight from several rearing tanks on four dates, n = 127 on 14/9/15, n = 87 on 22/9/15, n = 69 on 8/10/15, n = 57 on 5/11/15. Scale of 0, 1, 2, 3, and 4, from 1 = ¼ fin to 4 =severe erosion and most of caudal fin missing, indices identified by colour coding 0=blue, 1=red, 2=green, 3=purple, 4=light blue. J. Treasurer, unpublished data.

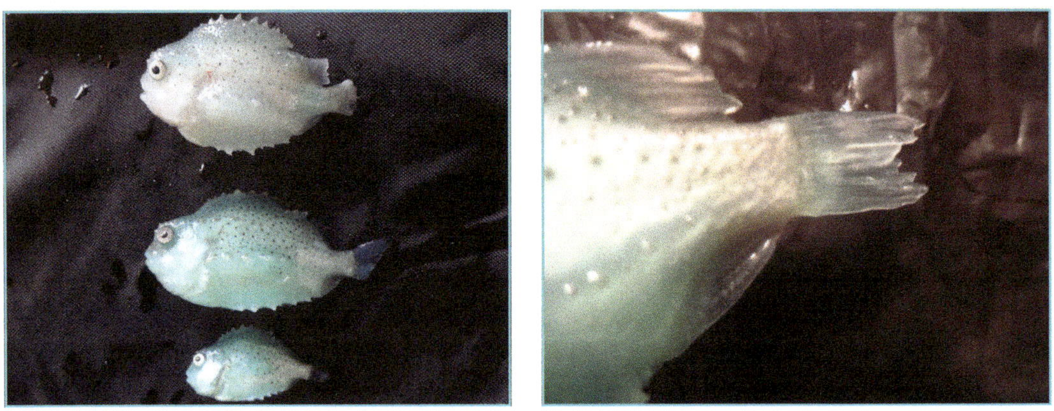

Fig. 15.4 Examples of fin erosion in lumpfish juveniles. Photos: J. Treasurer.

A further species-specific OWI for lumpfish can be for recording potential deformities to the ventral suction disc. It can be diagnosed or quantified out of the water or during handling (a normal ventral sucker disc would be a complete circle while the deformed disc would show a discontinuity) (Fig. 15.6), or it can easily be observed on living fish in the rearing systems as

Fig. 15.5 Physical damage in lumpfish broodstock: (a) fin lesion, (b) gill cover, (c) head, (d) cataract. Credit: J. Treasurer

they cannot attach to the substrate. Observations by V. Puvanendran (personal communication) reported that during the 2016 production year at Nofima's marine rearing facility, 65% of the juvenile lumpfish had a deformed ventral sucker disc. The sucker deformity was not evident at the early stages but the problem became apparent when the fish were about two to three months old. At this stage, the juvenile lumpfish were unable to attach to the substrate and they were observed laying on their sides on the tank bottom or were constantly swimming. The reason for this deformity is not clear.

Current evidence therefore suggests the morphological WIs and OWIs that are suitable for wrasse are also suitable for lumpfish across a range of life stages and rearing systems, although fin erosion (particularly to the caudal fin) is a particular concern, especially in the juvenile and maturation phases. Condition factor can be a difficult OWI to apply to lumpfish due to their shape. However, a recent study on developing a specific lumpfish condition factor for juvenile lumpfish (Rey *et al.*, 2018, in preparation) gave a different exponential factor of 2.7 in contrast with the exponential factor 3 based upon the commonly used Fulton's condition index. More data should be gathered from wild and farmed lumpfish to obtain a species-specific condition index based on relative weight.

The colour of the lumpfish may be a potential indicator of distress and the onset of morbidity. 'Healthy' active fish are vivid green and dark blue in colour, and the loss of colour and transition to grey and pink hues was consistent with less active fish and fish that then displayed morbidity (personal observations by J. Treasurer). More work needs to be done to quantify the suitability of fish colour as an OWI in this species across a range of conditions and life stages.

Fig. 15.6 Sucker disc abnormalities in lumpfish. Credit: J. Treasurer

A liver colour grading scale can also be used to detect possible unhealthy lumpfish. Liver colour in farmed lumpfish ranges from white to intense orange and then to dark red. The wild healthy lumpfish seem to have the intense orange colour liver that can also be seen on some deployed lumpfish in sea farms (S. Rey, Stirling, March 2017, personal observation). Different liver tones may be a sign of poor (or maybe not entirely adequate) nutrition, poor welfare or disease. More research is currently being carried out on the nutritional needs of lumpfish in hatcheries and in salmon sea cages. For example, an hepatosomatic index that can be correlated with a potential liver colour index is also being used as condition index for lumpfish in order to assess the nutritional state of the animals (Rey *et al.*, in prep).

15.3.2 Behavioural WIs and OWIs for wrasse and lumpfish

With regard to behavioural OWIs, a comprehensive review carried out by Martins *et al.* (2012) has stated that behavioural indicators of welfare are robust candidates for OWIs as they give a rapid indicator of welfare state and are suitable for use on the farm. Indeed, many farmers qualitatively use short-term changes in fish behaviour as an early warning system for potential health and welfare threats (e.g. water quality has deteriorated or a predator is around the cages). Deviations from normal behaviour may be an output of an underlying existing problem and may indicate physiological imbalance within the fish that may be frequently irreversible. However, quantitatively assessing behavioural OWIs on the farm can be labour intensive due to the protracted amount of data gathering and processing that is required. Large-scale rearing systems may also prevent the farmers from obtaining a good behavioural overview of the majority of the fish and different rearing or environmental conditions, such as high turbidity or algal blooms, can limit the ability of the farmers to visually monitor the fish. This can be remedied by the increasing use of cameras for feeding and husbandry monitoring and also the emerging development of automated hardware and software for the monitoring and processing of behavioural data from aquaculture (see Pinkiewicz *et al.*, 2011; Zion, 2012). Potential behavioural indicators of welfare include aggression, changes in swimming activity and tank distribution, ventilation frequency (Huntingford *et al.*, 2006; Canon Jones *et al.*, 2010; Martins *et al.*, 2012) and feeding responses. Monitoring latencies on feeding responses could be a good operational behavioural test (OBT) to be implemented in farms to detect changes in fish behaviour due to poor welfare conditions (Ibarra-Zatarain, 2016).

Wrasse

i) Swimming activity can be a potential behavioural OWI for wrasse and studies are currently being carried out in relation to the vertical and horizontal distributions of wrasse during sea cage deployment (A. Davie, Stirling, March 2017, personal communication).

ii) Aggression is not an acute problem in larval and juvenile wrasse, but has been observed in broodstock fish and, in extreme cases, leading to scale picking and mortality during the spawning season (Treasurer, personal observation), and particularly in males during the spawning period when the scales of rival males may be picked and, in some cases, result in the death of males.

iii) Balling/clumping is an abnormal tank behaviour that seems to be related to poor welfare and high stress levels. Shoaling in fish is related to social interactions, which provides foraging and reproductive advantages, hydrodynamic efficiency and anti-predator functions. Wrasse are not a shoaling species in nature and the artificial tank environment could favour this potential sheltering behaviour to decrease their stress levels. It could also be a WI of poor welfare conditions if this happens frequently. Experiments show that shoaling has, in other fish species such as zebrafish *Zebra danio*, anxiolytic and prosocial effects (Braida *et al.*, 2012). Unpublished data (Å Espmark, March 2017, personal communication) suggest this can be ameliorated by better water flow, natural photoperiods, shading and improved feeding strategies, but the phenomenon still warrants further investigation.

iv) Ventilation rate is commonly used as an indicator of a response to stress in fish, especially during hypoxic conditions (Kramer, 1987; van Rooij and Videler, 1996) but may also be a response to increased aerobic activity. It is a robust OWI candidate (Martin *et al.*, 2012) as it is relatively easy to monitor in high-visibility conditions but it has the requirement that the fish should be slow moving or static.

Lumpfish

i) In general, swimming activity is limited in all life stages of lumpfish. From personal observations (Sonia Rey, personal observations), swimming in newly hatched larvae is very limited and they attach to the tank walls most of the time. Feeding starts within a few days of hatching and swimming activity increases when food is offered, but larvae still spend most of the time resting by attaching to the tank walls. Swimming activity increases during the juvenile stage and is maintained at this level until sexual maturation starts. During maturation the swimming activity of both males and females is reduced and sometimes fish display chasing activity.

ii) Aggression. Early juvenile stages are known to be cannibalistic in nature (Ingolfsson and Kristjansson, 2002) with smaller larval lumpfish an important proportion of their diet. Juvenile lumpfish can be very aggressive during the first three months of life and they nip the caudal fins regularly. This aggression occurs even if they are fed continually and thus it seems unrelated to lack of food but rather to their inherent behavioural repertoire (V. Puvanendran, personal observations). This fin-nipping behaviour seems to become more prominent when the size disparity within the cohort increases with age. Regular size grading would help in minimising aggression at this stage. Environmental enrichment should also be considered, not only for increasing substrate surface but also to provide different environments to hide or explore within the tanks. This decreases aggression in larvae (Turner, 2016, MSc thesis) and juvenile lumpfish prefer to hide inside or under pipes or hides than to be in open spaces if they can choose these objects (S. Rey, personal observation). However, if enrichment is provided, it is important that all the individuals have the opportunity to have some space to avoid competition for preferred places. Some aggression has also been seen during the late sexual maturation phase, especially among males chasing each other, but it has been relatively minor.

iii) Ventilation rate is a robust OWI for lumpfish as they are either static and attached to the substrate or can be slow-moving swimmers allowing for easy observation of opercular beat frequency. A recent study of lumpfish showed that ventilation rates increased with decreasing oxygen levels in the water (Jørgensen *et al.*, 2017). Lumpfish juveniles kept at 55% oxygen saturation significantly increased ventilation rate compared to fish kept at 81% and 96% oxygen saturation. Although hyperventilation increases the oxygen perfusion over the gills, this increased ventilation comes with a cost, and the extra cost of the hyperventilation restricts the energy available for voluntary activities such as feeding and thereby growth (Kramer, 1987; Perry *et al.*, 2009). Under stressful conditions, oxygen consumption tends to increase and a similar trend was also observed in lumpfish kept at 55% oxygen saturation (Barton and Schreck, 1987). Jørgensen *et al.* (2017) also suggested that to compensate the energy expenditure in hypoxic conditions, lumpfish reduce their feed intake, which subsequently affects the growth and welfare of the fish. Severe hypoxia not only suppresses growth, but may also reduce oxygen uptake rate, as well as other fundamental needs such as osmoregulation and cardiac function, and thereby cause poor welfare (Stien *et al.*, 2013). Thus, rapid and irregular respiration rate can be a good indicator of respiratory issues, with opercular rate and 'gulping' movements more rapid in fish in tanks affected by AGD, for example (J. Treasurer, personal observation).

15.3.3 Physiological WIs and OWIs in wrasse and lumpfish

Physiological WIs and OWIs are interlinked with the concept of allostasis and stress. Stress is defined as a condition in which the dynamic equilibrium of an organism, called homeostasis, is threatened or disturbed as a result of the actions of internal or external stimuli, commonly defined as stressors (Selye, 1950, 1973; Varsamos *et al.*, 2006; Wendelaar Bonga, 1997, 2011). The physiological response to stressors has received great attention over the last decade and a great amount of data have attempted to complete, or even replace, Selye's non-specific stress concept, also known as the General Adaptation Syndrome (GAS), which describes an alarm, a resistance and an exhaustion stage (Varsamos *et al.*, 2006). Over the last few years the concept of allostasis has been introduced to complement the concept of stress, and more precisely to try and describe the role of primary mediators (for example glucocorticosteroids) in response to a stressor (Goymann and Wingfield, 2004; McEwen, 1998, 2005; McEwen and Wingfield 2003; Wingfield, 2005). Allostasis is achieving stability through change. This means that the set points and other boundaries of control must also change. McEwen (2003) felt that allostasis clarifies an inherent ambiguity in the term homeostasis and distinguishes between the systems that are essential for life (homeostasis) and those that maintain these systems in balance (allostasis), as environment and life history stage change. The concept is mostly applied to mammalian physiology but several authors have also adapted this concept to fish (Schreck, 2010; Iversen and Eliassen, 2014).

Stress plays a key role in the ability of vertebrates to survive and perform necessary life functions during unfavourable situations (Schreck *et al.*, 1993). Physiological responses of fish to environmental stressors can be grouped broadly as primary, secondary and tertiary responses (Fig. 15.7). Primary responses involve the initial neuroendocrine responses, including the release of catecholamines (CA) from chromaffin tissue, and the stimulation of the hypothalamic–pituitary–interrenal (HPI) axis culminating in the release of corticosteroid hormones. Secondary responses include, amongst others, changes in plasma and tissue ions (such as sodium, chloride, magnesium) and metabolite levels (such as lactate, glucose), and changes in haematological features. All of these relate to physiological adjustments related to metabolism, respiration, acid-base status, hydro-mineral balance, immune function and cellular responses (Barton, 2002; Mommsen *et al.*, 1999; Selye, 1950, 1973; Wendelaar Bonga, 1997, 2011). Tertiary responses are aspects of whole animal performance such as changes in growth, disease resistance, metabolic scope for activity, behaviour, and survival (Fig. 15.7). This is a simplified classification as it is also dependent upon the magnitude and duration of the response, which may affect fish at all levels of organisation, from molecular and biochemical to population and community (Wedemeyer, 1996). Cortisol is often associated with the detrimental effects of stress, including decreased growth rates and reproductive dysfunction (Mommsen *et al.*, 1999; Morgan *et al.*, 1999; Schreck *et al.*, 2001), increased incidence of disease and parasite resistance (Barton, 2002; Davis *et al.*, 2002, 2003; Einarsdottir *et al.*, 2000a, 2000b; Finstad *et al.*, 2007; 2012; Schreck et al., 1993), reduced seawater tolerance (Iversen *et al.*, 1998; Liebert and Schreck, 2006) and survival (Finstad *et al.*, 2003; Hasan and Bart 2007; Iversen *et al.*, 1998, 2005). Knowledge of cumulative stress is important for fish farmers because, while sub-lethal disturbances may not appear to be harmful,

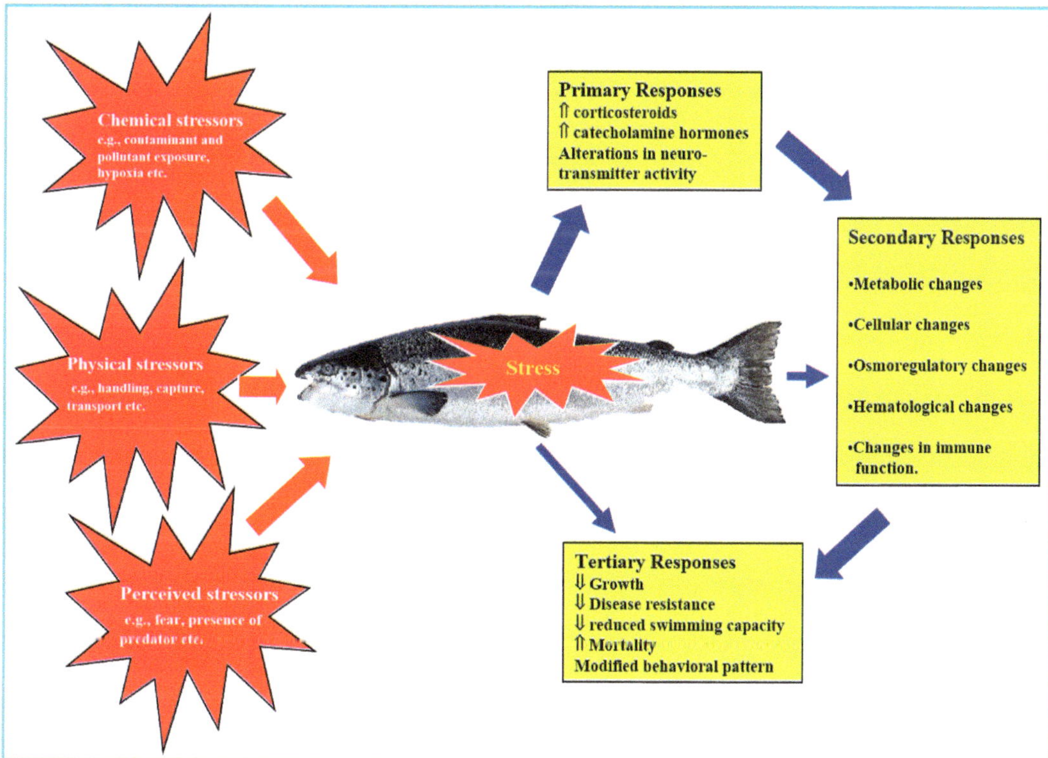

Fig. 15.7 Physical, chemical and other perceived stressors act on fish to evoke physiological and related effects, which commonly are grouped as primary, secondary and tertiary responses. The arrows indicate the various response routes. Iversen redrawn from Barton (2002).

the accumulation of responses to such disturbances without a proper recovery period could be lethal for the fish (Barton *et al.*, 1986; Barton and Iwama, 1991; Iversen *et al.*, 2005; 2009; Iversen and Eliassen, 2014). As stated earlier, most physiological OWIs are interlinked with cumulative stress (distress, negative stress, maladaptive stress), which has been well documented in salmonids (Barton, 2002; Wendelaar Bonga, 1997, 2011).

However, in relation to Ballan wrasse and lumpfish little or no data exist describing physiological OWIs in the context of distress. In this section we present experimental data on different physiological WIs and OWIs from wrasse, lumpfish and Atlantic salmon (for comparative purposes) from a series of experiments carried out at Nord University, Faculty of Aquaculture and Bioscience, in Bodø, Norway. Differing groups of fish (salmon, wrasse or lumpfish) were acclimated at least three weeks prior to the start of the experiment. Fish were subjected to a crowding stress (lowering water level abruptly to 10 cm (density 262 kg m^{-3}) and kept there for 30 minutes), and afterwards divided into ten isolated tanks. The blood was sampled before exposure to the stressor (pre-stress) and then 1, 2, 3, 4, 6, 12, 24, 48, and 72 hours post-exposure. Blood plasma was then analysed for plasma cortisol (primary stress response), glucose, lactate, osmolality, chloride and magnesium (secondary stress responses). Data from the physiological WIs plasma cortisol and lactate are shown here. Figures 15.8 and 15.9 show Ballan wrasse and salmon have a similar primary and secondary stress response as indicated via plasma cortisol and lactate. Lumpfish, however, only show a moderate increase in their primary stress response, and only a minor increase in lactate (Figs 15.8 and 15.9). The stress profile of lumpfish is very similar to the flatfish Atlantic halibut (*H. hippoglossus*) (M. Iversen, personal observation).

Does this mean that the lumpfish is less susceptible to stress-related diseases and mortality? The answer to this is not straightforward. Early reports from production of cleaner fish in Norway suggest that lumpfish seems to be more 'robust' compared to different wrasse species. Lumpfish also seems to be bolder in general and more resilient to stress, and this could explain the lower basal levels of cortisol before and after stressful events. Bolder fish personalities in general have lower levels of cortisol after stress and also in their basal levels (S. Rey, personal

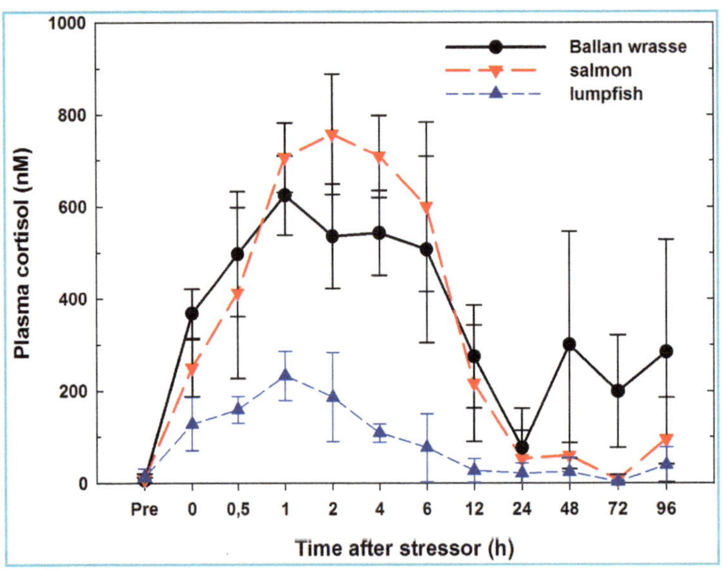

Fig. 15.8 The average plasma cortisol response (nmol l^{-1}) after 30 minutes of crowding stress in Ballan wrasse, Atlantic salmon and lumpfish (Data: Iversen).

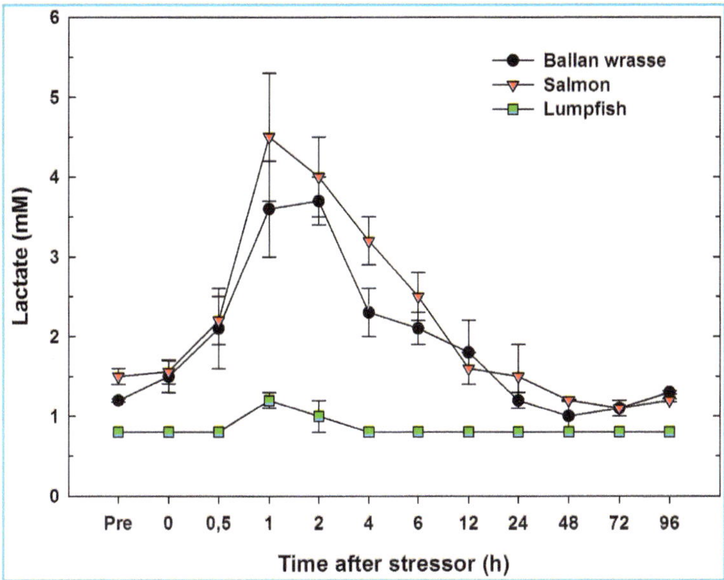

Fig. 15.9 The average lactate (mmol l^{-1}) after 30 minutes of crowding stress in Ballan wrasse, Atlantic salmon and lumpfish (Data: Iversen).

observation). However, lumpfish do experience compromised welfare in commercial production in relation to diseases, fin erosion, physical damage and increased mortality (Hanssen *et al.*, in prep.). A plausible explanation for the differing physiological response of wrasse and lumpfish subjected to an identical stressor could be related to their ecology, behaviour and morphology. In an escape response most teleost fishes have an escape reflex or C-start response, which is triggered by a pair of large neurons located in the hindbrain of fish, known as the flee Mauthner cells (Webb, 1976; Bierman *et al.*, 2009). The escape reflex is divided in two stages where the fish i) orient the head away from the threat, and ii) initiate a tail beat to the opposite side of the body and instigate a propelling trajectory away from the threat (Eaton *et al.*, 1988). Lumpfish seem to lack these Mauthner cells (Hale, 2000) and the primary response to a threat can be to attach to a solid object with its suction disc and hide, rather than initiating a flight response. The lumpfish therefore rely on their hardened skeleton for protection, morphological spines and suction disc as a first defence against predators (Davenport, 1985).

This evolutionary difference in reaction to perceived threats between wrasse and lumpfish will most likely manifest itself in some of the differences in physiological WIs and OWIs. The elevation of lactate concentrations observed in salmon and wrasse immediately after stress is likely due to muscle glycolysis (Moon and Foster, 1995), and will be triggered by high muscle activity elicited by the escape reflex. Since lumpfish do not swim away from danger but instead hide, they do not elicit high levels of glucose or lactate in the blood during stress. This will also potentially have an effect on the primary stress response, such as plasma cortisol. This hiding response is perhaps an evolutionary adaptation to threat, and this kind of physiological response is also seen in halibut and wolfish *Anarhichas lupus*. However, even if plasma cortisol is three to five times higher in wrasse compared to lumpfish during exposure to a stressor, this should be interpreted carefully as the wrasse is more susceptible to stress-related side effects. In wrasse this could just be an evolutionary adaption to different ways of life. Hansen *et al.* (in prep.) showed that lumpfish subjected to daily stressors exhibit signs of chronic stress with a diminished functioning of the hypothalamus–pituitary–interrenal axis. The authors suggested this was the driving mechanism for reduced growth and increased fin erosion. More work needs to be done to understand the subtle differences in the stress response between the different species of cleaner fish to identify the most appropriate and fit-for-purpose WIs and OWIs for each species, in addition to both the optimal and detrimental ranges of the different WIs and OWIs. This is also further evidence that species-specific WIs and OWIs are needed to make sure welfare is assessed and quantified in an appropriate way and to ensure that cleaner fish thrive in aquaculture environments.

15.3.4 OWIs related to production

The feeding response is perhaps one of the most widely used OWIs in aquaculture which is usually qualitative (Huntingford *et al.*, 2006). The farmer monitors the fish for any sudden dip in appetite and will also monitor both the short- and long-term feeding response as potential OWIs. However, poor appetite may not just be due to potentially poor welfare. For example, it may be because the fish has just fed and is already satiated or is exhibiting a reduced feeding response due to longer-term factors such as decreasing water temperature in relation to changes in the season (e.g. Noble *et al.*, 2007 for Atlantic salmon). Further, quantitative analysis of appetite as an OWI is more problematic as appetite is known to vary within and between days for a wide range of farmed and wild fish (e.g. Jobling *et al.*, 2012) so quantitatively assessing potential decline in appetite can be challenging. Appetite (and changes in it) can therefore be used as a robust early warning system for potential welfare threats, but it must be considered with other potential OWIs to identify the source and severity of the threat.

Growth is another parameter that can be used as a potential welfare indicator (Huntingford *et al.*, 2006). Once again, an abrupt or longer-term reduction in growth rate can be indicative of poor welfare, but longer-term changes may also be related to life stage and not due to welfare risks. It can also be highly variable within and between individuals, for example in relation to appetite or feed access (Johansen and Jobling, 1998) meaning the coefficient of variation in growth rate can also be a potential OWI for cleaner fish.

Short- or long-term changes in mortality can also be used as an instantaneous or retrospective OWI in animal production systems (e.g. Soares *et al.*, 2011; Stien *et al.*, 2016). However, it is a crude welfare indicator (Nilsson, Stien, Iversen *et al.*, 2017) and we would like to identify potential welfare threats and implement mitigation measures before it gets to this stage. An increasing number of authors are benchmarking mortalities (e.g. Soares *et al.*, 2011; 2013; Stien *et al.*, 2016) at various levels from either the rearing unit or production site to the regional or national level. This approach can help the user identify periods where mortality is higher than usual and can then be used to implement actions to prevent further increases or, if used retrospectively, can be used to investigate potential trends or causes. However, the use of mortality as an OWI in both lumpfish and wrasse is not without its problems. It has been suggested anecdotally that both wrasse and lumpfish mortalities may break down rapidly in the cage, meaning accurate monitoring of mortalities may be a problem for the farmer. Also, mortality trends in all animal production systems may not help the farmer clearly identify welfare threats, for example if the welfare threat is cumulative or is only reflected in delayed mortalities.

15.3.5 Indirect – environmental OWIs

With regard to environmental OWIs, a large number of rearing system parameters can influence the welfare of cleaner fish. Indirect, environmental OWIs largely include water quality parameters, for example temperature, oxygen levels, levels of CO_2 and water pH, nitrate, nitrite, and TAN, but can also cover other factors such as lighting, water velocity and rearing density. If thresholds or limits for both optimal holding conditions and also conditions where welfare is threatened have been identified, environmental OWIs can be valuable indirect tools for identifying rearing situations that can be detrimental for welfare. Where the factors can be monitored in real-time and also remotely using sensors it allows for the rapid identification of welfare risks and the early deployment of mitigation methods. However, very little information, especially peer-reviewed information, exists on what is acceptable or optimal in terms of environment OWIs and cleaner fish welfare, irrespective of species.

Preliminary data regarding water quality requirements for cleaner fish

As stated earlier, water quality requirements are still difficult to pinpoint on a species basis as wrasse and lumpfish rearing are still in their relative infancy. Different life history stages, such as broodfish, eggs, larvae, juveniles and growers, should also be considered. Known/suggested water requirements for specific life and rearing stages are outlined in this section. The majority of these data are based upon practical experience as peer-reviewed publications on water quality and the potential impacts upon welfare are limited. In this absence, we can also look at the welfare of fish under other procedures, such as during transport (see Chapter 16) to obtain pointers regarding their welfare needs. However, much research still remains to be done on this field.

Juvenile rearing

With regard to juvenile rearing the most appropriate representative data may be found in RAS systems used to rear both wrasse and lumpfish. However, the water is not recirculated in most wrasse hatcheries (e.g. flow-through rearing at Øygarden, Stord and Havelandet, partial water recirculation at Machrihanish, J. Treasurer, personal communication). The water quality ranges for lumpfish and wrasse in RAS and flow through systems are shown in Table 15.1. Particular features for lumpfish are maintaining temperatures below 14°C as the fish are sensitive to higher temperatures. Stress from elevated temperatures has also resulted in increasing mortalities due to bacterial disease, for example Vibriosis, and this requires treatment with antibiotics (J. Treasurer, personal communication). Wrasse are more tolerant of temperatures up to 20°C, but the lower thermal limit is higher than for lumpfish, with fish being routinely reared at >10°C to encourage growth. Temperatures below 4°C may impact on welfare and especially in more sensitive species such as corkwing. Oxygen levels of 80–90% saturation have been recommended for both lumpfish and wrasse (Tables 15.1, 15.2). Other parameters such as CO_2, and total gas are comparable with other marine species and the suggestions recommended by the manual NIVA report 6782-2015 may be applied, https://brage.bibsys.no/xmlui/handle/11250/285802 (Hess-Erga *et al.*, 2015). The thermal preferenda of older stages was higher. Specific measurements for wrasse production were quoted at MHN Øygarden wrasse hatchery over several weeks. This included pH from 7.9–8.0, CO_2 from 0–12 mg l^{-1}, NH_4 from negligible to 2.8, NO_2^- up to 2.8, and alkalinity as mmol l^{-1} of 2400. Values of Al, Cu, Fe and Zn were also quoted, and also a turbidity range of 0.3–0.5 FNu, total suspended solids TSS of 0.3 to 1.8, and bacteria from 100 to 10,000 as CFU ml^{-1}. Water flow is an important factor for wrasse, as they have problems with high current speeds or any changes, and areas with low current speed should be made available. This will result in a reduction in self cleaning and therefore a dirty environment is more likely compared with other species.

General recommendations for total ammonia of <0.5 mg l^{-1}, nitrite <200 mg l^{-1}, nitrate <50 mg l^{-1} (in seawater) are critical in the range routinely applied in aquaculture (Richard Prickett, Weymouth, March 2017, personal communication). If pH declines below 7.3, then sodium hydroxide should be added to the system. Total Ammonia (NH_3/NH_4^+) and unionised ammonia (NH_3) depend on pH and temperature. Unionised ammonia in RAS systems is normally about 0.8% of total ammonia at a pH of 7.8 and a temperature of 14°C. In the case of Nitrite (NO_2^-), the chlorinity of seawater reduces toxicity by 20 times. Nitrate (NO_3^-) is not usually a problem. Also, redox can be increased by using ozone if required (Richard Prickett, personal communication). As with other RAS systems, most of these criteria can be improved by stopping feeding.

15.4 Key welfare challenges for cleaner fish

15.4.1 Wild fish capture and handling

Capture of wild fish is the area where farmers have the least direct control as a large proportion are purchased from fishermen. While experienced fishermen who fish almost exclusively for cleaner fish have well-established routines, newcomers to this area require the welfare guidelines for capture, holding and transport set out in the Royal Society for the Prevention of Cruelty to Animals Freedom Foods. These stipulate environmental conditions during transfer, conditions for holding, soft nets for handling cleaner fish, and advice on pre-transport fasting and notes on transport. These conditions are specified by the salmon farmers within contractual arrangements. While mortality associated with capture and conveyance can be as low as 5% (Callaghan and Pering, 2016) this is attributed to care provided by the fishermen and use of specialised companies to transport fish. Lorries are fitted with oxygen control equipment, air blowing to minimize ammonia accumulation and also buffers can be added to the transport tanks to maintain pH (Chapter 16). Additional shelving can also be added to provide attachment areas for lumpfish, although this does not seem to be commonly adopted.

Wrasse

While many fish are captured within easy reach of farms, there has been transport of wrasse over large distances from southern Norway to farms around the coast and in the UK, of wrasse shipped from the English Channel to Scotland, involving a long road journey. There have also been reports of relatively high mortalities of wrasse during the storage of wrasse prior to shipping and fish being kept in small containers at high density (Callaghan and Pering, 2016; Riley *et al.*, 2017). Callaghan and Pering also reported the outbreak of bacterial disease in some of the stored cleaner fish batches, which may be due to poor storage conditions and handling, and fish

Table 15.1 Parameters for cleaner fish cultivation in flow-through and RAS systems in relation to welfare requirements. Recommended water quality from experience with RAS systems and lumpfish rearing of larvae and juveniles, LS = lumpfish, WR = wrasse; Richard Prickett, Weymouth, March 2017, personal communication; Jim Treasurer, Ardtoe = lumpfish flow through, OFS = flow-through lumpfish, Antonios Chalaris, personal communication; Machrihanish = Paul Featherstone, personal communication; MHLabrus = flow-through Ballan wrasse, Anette Lekva, personal communication; Fiskaaling flow through, Ása Johannesen, personal communication. +/-= standard deviation.

Species: Parameter: Water supply: Location:	LS Recommended RAS	LS Actual Flow-through Ardtoe	LS Actual RAS Fiskaaling Faroes	WR Actual Flow-through MHLabrus	WR Actual RAS Machrihanish
Oxygen %	80%+	80–100%	91.7+/-1.9		>7mg l^{-1}
Temperature °C	<14°C	8–16	11.6+/-0.3		4–16°C
Salinity ppt		31–33.5	33.4+/-0.3		35*
pH	7.3–7.8	7.8–7.9	8.1+/-0.1	7.5–7.8	7.2–8.0
Total ammonia	<0.5 mg l^{-1}	0.21–0.44		0.4–0.6	<0.5 mg l^{-1}
Nitrate NO_3-N mg^{-1}	<50 mg l^{-1} (in seawater)			0.1–0.2	<50 ppm
Nitrite NO_2-N mg l^{-1}	<200 mg l^{-1}		0.2+/-0.1	0.1–1.1	<0.1mg l^{-1}
Ammonia NH_3-N mg^{-1}		0.25–0.45	0.004+/-0.001		
REDOX	200–350 MV				>150 mv
CO_2					<15 ppm
Supersaturation					<102%
NH_4+		0.27–0.45			
Alkalinity mmol l^{-1}				2400	~3meq/l-1
TSS mg l^{-1}				0.36-1.8	
TOC mg l^{-1}				1.1-1.3	
Bacteria CFU ml^{-1}				100-10000	

*but can take down to 12 ppt for a few days

Table 15.2 General environmental conditions in marine hatcheries. Data from manual, NIVA 6782-2015

Life history stage:	Broodfish	Eggs	Larvae	Juveniles	Growers
Oxygen	80–90%	80–90%	80–90%	80–90%	80–90%
Temperature °C	8.0–11.5	11	start at 12, increase to 16	16	14–16
Light (L)	Natural light regimes: 7–19 h L, depending on date	12L:12D	dark first 3 days, 24L from 4 dph	24L	24L
Water flow (l/min)	60 and 300	15	40 and 100 (increasing from 20 dph)	300	
Tank size (m^3)	9 and 70	0.9	9 and 30	27	
Water exchange in tank (min)	150 and 230	60	225 and 270 (increasing from 20 dph)	90	

may also have been stored for too long without an adequate food source. Of further concern is the use of harsh fishing methods in the English Channel using trammel nets and trawl (Riley *et al.*, 2017). Fish may be taken from deeper water and, if not taken to the surface gradually, may suffer decompression issues (Chapter 17). The by-catch of other species and shore crab may also cause damage to fish if the traps are not emptied regularly, and a trap that was set off the bottom in Norway has been developed to reduce this type of damage.

Lumpfish

Currently the industry relies on wild-caught lumpfish broodstock, although the intention is to move to hatchery-reared broodstock to enhance biosecurity and to guarantee egg supply. Soon after capture, broodstock are transferred to the land facility and eggs and sperm are stripped and the broodstock are released back to the sea, or sacrificed for disease screening. In some cases, the availability of males is limited and the females are temporarily stocked in sea pens until more males become available.

15.4.2 Nutrition

There is a lack of knowledge of broodstock nutritional requirements of Ballan wrasse, thus suitable commercial diets are not available (Chapter 10). Current commercial diets used during captive on-growing and broodstock maintenance should be further improved in terms of their nutrient composition. The nutritional quality of rotifers may be poor for first feeding Ballan wrasse larvae as they frequently develop skeletal anomalies and have poor growth and low survival rates (Helland *et al.*, 2012). Rotifers can be substituted by live copepods. A fishmeal-free extruded weaning diet offered to Ballan wrasse larvae has led to almost complete head and jaw deformities (Kousoulaki *et al.*, 2014). More work is required to identify nutritional requirements.

15.4.3 Supplementary feeding of cleaner fish in salmon cages

Sea lice numbers can vary and may be too low over prolonged periods to give adequate nutrition for the cleaner fish. Therefore the condition factor and body mass of cleaner fish may decline within six weeks of deployment despite relatively high initial lice numbers (nine lice/salmon) and documented delousing activity (Skiftesvik *et al.*, 2013 on ballan wrasse). Supplementary feeding of cleaner fish should therefore be provided during the grow-out period to maintain high cleaner fish nutritional status, welfare, delousing efficacy and high survival rates for as long as possible. Special formulated wrasse feeds have been developed for wrasse. Those are offered in bait bags, open pots/traps, or in trays in feeding hides (Leclercq *et al.*, 2015). Manufactured dry pellets (extruded) delivered in fine mesh bags were found to disintegrate within hours in water, leading to significant wastage, further compromising validation and quantification of feed intake. Agar blocks made with marine premix have been found to be physically water stable for one week with no nutrient leaching and are well accepted by the wrasse (Leclercq *et al.*, 2015). Such a water-stable, agar-based diet is presented to wrasse within fine mesh feed bags at different points in the net pen and close to the hides (Chapter 4). Lumpfish are also usually hand fed and the diet is a general marine fish diet. A more specialised type of diet according to their requirements (for example levels of carotenoids) is needed.

15.4.4 Health

Events such as capture, handling, transport or grading of fish increase susceptibility to disease and elevated mortality can follow these events. Lumpfish, despite their robust appearance, are susceptible to skin damage and this incidence is increased by their habit of adhering to transport tanks (Chapter 13). The stress and handling during transfer and stocking, together with the inevitable disease challenges in the pen environment, make the first few weeks at sea a critical time for lumpfish and acute mortalities often occur (Bornø *et al.*, 2016). High lumpfish mortality rates can be related to disease or potential chemical or mechanical delousing treatments (Chapter 9). Vaccination is essential for the success and sustainability of cleaner fish use (Chapters 13, 14). Systemic bacterial infections (sometimes, secondary to physical damage caused by aggression or during routine farming operations) and Amoebic Gill Disease (AGD) are the two main causes of mortality in wrasse populations (Chapter 4). Vaccines for bacterial infection have been improved but further progress is required in meeting a range of pathogens from Vibrio species, atypical furunculosis and Pasteurella. Site operations should include regular health and condition checks and designation of a member of staff to ensure farm operations take due regard of the welfare of stocked cleaner fish.

15.4.5 Potential predation by salmon

This may be an issue when cleaner fish are first stocked in pens and may be disorientated. Cleaner fish should be introduced within or next to hides if possible. The salmon should have been well fed prior to stocking the cleaner fish to ensure they are not searching for food.

15.4.6 Deployment and cohabitation

Successful deployment of cleaner fish relies on good preparation from the sites receiving them, such as adequate provision of hides, clean netting, and adequate supplementary feeding when fish are introduced. Farmed wrasse in particular require some adaptation to the new environment of cages by stocking in nursery cages within the main fish cage to adapt to cage and environmental conditions. Hides give some protection from currents, and salinity and temperature changes with tidal regimes. There are many farm operations such as grading, sea lice treatments and net changing that can be stressful and these can be accommodated by removing cleaner fish from pens or being careful when fish are transferred to clean nets. Vaccination will give some protection from bacterial disease but further improvement in vaccines is required.

15.4.7 Optimal stocking time

Wrasse are essentially acclimated to warm summer temperatures (Sayer *et al.*, 1996) so it is recommended that they are stocked in sea cages with Atlantic salmon smolts, from early May to late September/early November, when the sea water temperatures are above 10°C. Lumpfish are usually stocked in commercial sea cages during the winter, between November and April, when the wrasse are not sufficiently active to maintain sea lice numbers. Thus, lumpfish may have the potential to survive winters in cages with salmon (Imsland *et al.,* 2016b) and contribute to the cleaner fish delousing ability.

15.4.8 Mortalities post-transfer

Special concerns have been raised related to large losses of cleaner fish and especially lumpfish in the weeks post-transfer. During the past few years, mortalities have been seen all along the coast, in some cases with >50% losses within a few months (Sveier and Breck, Chapter 18). A recent review (Bornø *et al.*, 2016) has pointed at infectious diseases, especially atypical Furunculosis, causing some of the losses, emphasising the need for better vaccines. A precondition for further large-scale use of cleaner fish is required to solve the challenges related to fish health and welfare of the wrasse species and the lumpfish. Lowering the mortalities and losses is crucial to continuing and defining the use of cleaner fish as a sustainable and environmentally friendly method (Sveier and Breck, Chapter 18). The greatest challenges in the production of lumpfish have been related to outbreaks of bacterial diseases, which have caused high mortalities, especially in warmer periods. It has not been possible to single out a genus of bacteria as the causative agent in each case of mortality. Rather, a diverse mix of different bacteria has been found. It is therefore possible that the underlying cause may be multifactorial and related to stress, suboptimal environmental and rearing conditions, which in turn may lead to the growth of pathogens and secondary infections. The most common pathogens found in lumpfish on the Faroes are atypical furunculosis (*Aeromonas salmonicida*) and vibriosis (*V. wodanis, V. logeii* and *V. splendidus*), though Pasteurella sp and Tenacibaculum sp have also been found in connection with infections in lumpfish (Johannesen, Chapter 22).

15.4.9 Basic husbandry

It is important to follow good husbandry measures to ensure a good delousing efficacy by cleaner fish eating sea lice. Farmers must insist that fishermen adhere to the aim of good fish welfare during capture in the wild and transfer to sea cages to ensure the cleaner fish work efficiently in the salmon cages (Chapter 4). A strict schedule in maintaining clean nets and collecting fish mortalities on a daily basis is crucial for the cleaner fish to be effective and to avoid possible distractions. Also, when hides are removed to be washed, they should be lifted very slowly. Cage nets and hides should be lifted at a speed that does not compromise the swimbladder in wrasse, and any compromised fish should be placed in a creel or open fish trap and lowered slowly to 5 m. During crowding operations, site staff must be made available to transfer any cleaner fish from the crowded to the non-crowded section of the pen. Alternative and efficient methods of removal of cleaner fish from pens need to be formulated. An example of a recovery of fish during a husbandry procedure when wrasse are collected before salmon are pumped aboard a wellboat for a freshwater bath treatment.

15.4.10 Overwintering strategies

Wrasse are susceptible to low temperatures over winter and are less active. Adequate hides to afford protection from temperature and salinity fluctuations and strong currents should be provided. Removal of wrasse from pens and storing them onshore in tanks has resulted in survival of 70 to 90% (J. Treasurer, personal communication). Another approach was to store the fish in cages on the seabed.

15.4.11 Inter-cycle storage

Sites retrieve fish at harvest and cleaner fish can frequently be stored until the next salmon production cycle. Cleaner fish are removed and can be held in small-scale nets with hides and cover, and fed regularly. The issue with this is that wrasse may be susceptible to strong tides and therefore suitable cover should be provided.

15.4.12 Inter-species considerations

A size differential between wrasse and lumpfish has been reported (Imsland *et al.* 2016) and larger lumpfish can be aggressive toward smaller wrasse, so the sizes of each species have to be assessed when both lumpfish and wrasse are kept together in same cage. Also, the relative size of salmon at deployment has to be considered. Sufficient experimental information is available on this to ensure size grades of wrasse and lumpfish are in harmony. The size of wrasse relative to salmon is considered in stocking plans by producers, with smaller wrasse species such as goldsinny and rock cook being used in the first year of the salmon production cycle and lumpfish over the first winter.

15.5 Current cleaner fish welfare requirements of certification schemes

Several certification schemes are used to accredit the welfare of farmed salmon, for example, the RSPCA Assured welfare scheme (formerly Freedom Foods, Anon, 2012), and also the Code of Good Practice for Scottish Finfish Aquaculture (SSPO). In the UK, the RSPCA Assured welfare standard for salmon was a guideline but, from November 2017, became a scheme complaint condition (E. Rodgers, RSPCA, May 2017). However, the RSPCA feels that all of the welfare requirements have not been adequately studied. The current RSPCA Assured salmon welfare standards suggest the welfare standards for cleaner fish must address numerous factors including, but not limited to i) nutritional and potential supplementary feeding requirements (also covered here in Chapter 10), ii) environment/enrichment needs, iii) hatchery and on-growing requirements, iv) transport, v) handling, vi) slaughter, vii) stocking requirements, densities and strategies (RSPCA Assured, 2015). Further, the use of wrasse or lumpfish is not allowed without prior permission from the RSPCA (RSPCA Assured, 2015).

The Code of Good Practice for Scottish Finfish Aquaculture (CoGP) (revised 2015) has been adopted by the Scottish industry and, under an agreement with Marine Scotland and the SSPO, Scottish farmers agreed to comply with the code. The code refers to salmon but farmers are expected to follow the spirit of the guidelines for cleaner fish. An amendment of February 2015 relating to welfare stipulates that hatchery-reared cleaner fish should be used where available. If hatchery-reared fish are not available and wild fish are sourced, i) appropriate baits must be used in their capture, ii) predator control measures such as otter exclusion devices must be deployed on all fyke nets, iii) there are maximum and minimum sizes for all fish that are retained. The code further stipulates that prior to stocking i) a health check should be carried out to ensure the cleaner fish are free from pathogens that can affect both the salmon and themselves, ii) appropriate hides/environment enrichment should be deployed, iii) there should be provisions for supplementary feeding, and iv) fish should be humanely slaughtered at the end of the production cycle or reused according to the requirements of the CoGP regarding, for example, health checks and the health status of the fish (CoGP, 2015).

15.6 Optimising welfare in relation to the species-specific needs of the cleaner fish

15.6.1 Acclimation

The acclimatisation to sea pen conditions of farmed Ballan wrasse when transferred into seawater cages is one of the key features to achieve success (C. Gutierrez, Chapter 4). In the case of farmed wrasse, fish should be acclimatised in the hatcheries to ambient water temperature, and also hides and feed, around five to six weeks before they are deployed to commercial sea cages. Conditioned farmed Ballan wrasse were more active in trials and covered a large area around the perimeter of a pen compared with non-acclimated farmed wrasse (Chapter 4). Therefore, companies such as MHS use acclimation pens for one to two weeks when stocking farmed wrasse into seawater cages. These are from 6–12 m^3 in volume and are provided with hides made of artificial kelp in each corner together with feed bags that offer agar blocks.

15.6.2 Environmental enrichment – Species-specific refuges and shelter for cages and tanks

A challenge for the cleaner fish producer and also the salmon farmer is bringing the rearing environment closer to the species-specific welfare needs of the fish. During rearing of juvenile and broodstock cleaner fish in tanks, the farmer only has a single species to deal with. When cleaner fish are deployed in open sea cages with salmon, the farmer is faced with the challenge of duoculture and matching an established rearing environment for the salmon to the introduction of various species, sizes and ratios of cleaner fish.

A way to address this challenge is to use environmental enrichment (EE), which has previously been described as 'a deliberate increase in environmental complexity with the aim to reduce maladaptive and aberrant traits in fish reared in otherwise stimuli-deprived environments' (Näslund and Johnsson, 2016). The use of physical enrichment materials to increase habitat complexity has been shown to decrease aggression in fish raised in the lab (Batzina and Karakatsouli, 2012), can improve foraging behaviour and efficiency (Brown *et al.*, 2003) or decrease cortisol levels (Näslund *et al.*, 2013) and injuries (Persson and Alanärä, 2014), thus improving welfare. However, the benefits of EE are not universal, e.g. if fish are provided with inappropriate EE materials it can increase aggression (Nijman and Heuts, 2000), meaning farmers have to be careful and provide the cleaner fish with enrichment materials that are appropriate for the species and life stage. Enrichment could also generate aggression between individuals that compete for preferred areas in the tank or cages. Irrespective of the species-specific requirements of the fish, all enrichment materials must be easy to deploy and maintain, easy to clean, not pose a biosecurity risk to the fish and must bring the cleaner fish in close proximity with their potential salmon clients in a way that neither interferes with the farming of the salmon but also provides the cleaner fish with a secure and protected environment.

While the culture of cleaner fish is still in its relative infancy, and especially where wild-caught fish are still widely used, it is appropriate to look to the ecology of wild fish to get an insight into the types of materials that may be suitable for improving their welfare (this topic is also covered by Imsland *et al.*, (2016a, Chapter 9). For example, lumpfish are semi-pelagic with juvenile lumpfish being found in both shallow waters and offshore (Eriksen *et al.*, 2014) but usually among kelp in shallow coastal environments (Imsland *et al.*, 2014). Adult lumpfish are generally pelagic but can be observed around floating seaweeds (see Imsland *et al.*, 2015 and references therein) before returning to shallower coastal areas to spawn (Eriksen *et al.*, 2014). This suggests floating or hanging substrates would be a suitable type of environmental enrichment for both tank- and cage-held lumpfish, as they provide the fish with an appropriate substrate on which to adhere. Imsland *et al.* (2015) confirmed that smooth plastic floating vertical enrichment materials, in addition to natural kelp (Imsland *et al.*, 2014), were suitable and preferred types of EE when juvenile lumpfish were held in cages. Indeed, floating plastic strips and artificial kelp are some of the common types of enrichment materials one generally finds in lumpfish aquaculture. Further, in exposed environments with high wave action or height, solid 'blade-like' enrichment materials deployed either vertically (Hiddenfjord, May 2017, personal communication) or horizontally can provide the fish with a stable substrate on which to adhere and can also shelter them from high water flow or wave action if the fish attach to the lee side of the material. In more sheltered environments, flexible kelp-like enrichment materials or natural kelp themselves (Hiddenfjord, May 2017, personal communication), deployed in, for example, parallel curtains are suitable.

Ballan wrasse generally inhabit areas <30 m deep and are benthic, with a preference for rocky areas and kelp (Villegas-Ríos *et al.*, 2013). Goldsinny also have a preference for rocky areas in the wild (Skiftesvik *et al.*, 2015 and references therein). This suggests both floating (cage-rearing of wrasse) and floating and benthic enrichment materials (tank-rearing of wrasse) would be suitable for these species, and this has been found to be the case in both scientific studies (e.g. Skiftesvik *et al.*, 2013; Leclercq *et al.*, 2015) and also in commercial aquaculture. As with the lumpfish, recent advice suggests the deployment of floating 'kelp curtains' is proving to be an effective enrichment strategy for wrasse. In addition, the hanging 'artificial kelp' shelters have also been adapted into species-specific feeding stations for Ballan wrasse (see Leclercq *et al.*, 2015).

The use of substrates/structures to provide shelter or an attachment substrate can therefore be beneficial to cleaner fish welfare, as long as it is both life stage and species appropriate. In the case of lumpfish, it provides a resting substrate for attachment during the nocturnal period (Imsland *et al.*, 2015) and in wrasse it provides shelter and may increase survival, especially over-winter. The enrichment materials that are currently available (floating artificial kelps or plastic strips) appear to be suitable candidates for enhancing the rearing environment for cleaner fish species. They also provide the added benefits of being operational, easy to install/remove, easy to clean and relatively cheap. Additional developments could refine enrichment materials further to increase their effective use as a refuge and to provide rest areas.

15.7 Reducing stress in wrasse and lumpfish via anaesthesia

Chemical anaesthetics are often used in aquaculture research to ease handling and reduce stress. In addition to preventing physical injury, certain anaesthetics may reduce or block activation of the hypothalamo–pituitary–interrenal (HPI) axis associated with handling stress (Bosworth *et al.*, 2007; Davis and Griffin, 2004; Iversen *et al.*, 2003; Sandodden *et al.*, 2001). Failure to suppress activation of the HPI axis during stress can result in a release of cortisol, which in turn causes various secondary and tertiary stress responses (Ackerman *et al.*, 2000; Ellis *et al.*, 2012; Wendelaar Bonga, 1997, 2011). The importance of controlling the stress response may differ depending on the acute or chronic nature of the stressor, and on the ultimate use of the fish (Davis, 2006). However, during critical commercial handling routines such as vaccination and transport, it could be beneficial to keep allostatic load under control. A possible method to keep this type of 'stress' under control is to use a low dosage of anaesthetics during potentially stressful procedures such as handling and induce a sedatory effect, but the fish is still swimming and is kept in an upright position. The sedation reduces the consciousness of the fish, which then exhibit only minor responses to disturbance and handling (Iversen *et al.*, 2003; Ross and Ross, 2008). Iversen *et al.* (2015) examined long-term sedation of Aqui-S on Ballan wrasse and lumpfish. This showed that lumpfish had a higher tolerance for Aqui-S compared to Ballan wrasse. The 24-hour LD50 dose was between 8 to 13 mg l^{-1} and 5 to 10 mg l^{-1} Aqui-S for lumpfish and

Ballan wrasse, respectively. Ballan wrasse sedated with Aqui-S also showed a significant dose-related reduction in plasma cortisol. No such dose-related changes to primary stress responses were observed in lumpfish. In fact, Aqui-S seemed to be a mild stressor in lumpfish, but sedation may still have beneficial effects as a sedated fish is easier to handle, thus reducing the occurrence of handling-related injuries (Iversen *et al.*, 2003, 2009, Ross and Ross, 2008). The authors are aware that some producers and transporters have already established routines where 10 mg l^{-1} Aqui-S is added a few minutes before emptying the tanks with lumpfish. This causes the fish to drop from the vessel wall and they are pumped out of the tanks without the need to flush or scrape the fish away from the walls of tanks and pipes (Iversen *et al.*, 2015). However, further studies are required to establish best practice protocols adapted to the individual behaviour of each of the cleaner fish species.

15.8 Conclusions

Welfare is the key ethical issue in the use of cleaner fish in salmon production as the public and retailers generally back the use of cleaner fish because of environmental and efficacy benefits, but only if the welfare of the cleaner fish is maintained. It should also be reflected that quality assurance schemes increasingly highlight and spotlight the capture, welfare and rearing conditions of cleaner fish and provide guidance, and cleaner fish are also mentioned in the CoGP in Scotland. It is acknowledged that these accreditation schemes often suggest recommendations but agree that further quantitative work is required to verify these and to enable implementation in a practical sense. Welfare is therefore an important area that requires quantitative assessment to determine the progressive threats to the well-being and condition of the fish, and therefore the efficacy of sea lice cleaning ability. Progressive changes in the welfare indices of the fish may follow stocking in sea cages when their condition may deteriorate due to onset of winter (in the case of wrasse) or due to inadequate supplementary feeding. Losses can be an area of concern, especially after deployment in the sea cages. This current review has indicated the early nature of cleaner fish cultivation and, in particular, deployment and the need to develop robust species-specific OWIs that quantitatively measure any change in fish condition and welfare. This will enable farmers to identify critical points and practices in cleaner fish use and will permit preventive and remedial action to be taken as required. These OWIs should be measured and reported on a regular basis throughout the rearing and deployment cycle. The current review and presentation of new data has shown that stress is different in wrasse and lumpfish. This is not comparable and different supplementary feeding regimes and environmental enrichment are required. This review also indicates that different OWIs may be required for different production phases, such as capture, rearing, transport and deployment of cleaner fish. Also, a range of different OWIs and welfare measures may be required for different developmental stages and many of these OWIs need to be defined. This chapter has therefore endeavoured to highlight the initial findings with welfare and measures of stress in cleaner fish.

References

Ackerman, P.A., Forsyth, R.B., Mazur, C.F. and Iwama, G.K. (2000) Stress hormones and the cellular stress response in salmonids. *Fish Physiology and Biochemistry* 23 (4), 327–336.

Adams, C.E., Turnbull, J.F., Bron, J.E. and Huntingford, F.A. (2007) Multiple determinants of welfare in farmed fish: stocking density, disturbance, and aggression in Atlantic salmon (*Salmo salar*). *Canadian Journal of Fisheries and Aquaculture Science* 64, 336–344.

Ashley, P.J. (2007) Fish welfare: current issues in aquaculture. *Applied Animal Behaviour Science* 104(3), 199–235.

Balm, P.H.M. (1997) Immune-endocrine interactions. In: G.K. Iwama, A.D. Pickering, and C. B. Schreck, (Eds.) *'Fish Stress and Health in Aquaculture'*, Cambridge Univ. Press, Cambridge, pp. 195–221.

Barton, B.A. (2002) Stress in fishes: A diversity of responses with particular reference to changes in circulating corticosteroids. *Integrative and Comparative Biology* 42 (3), 517–525.

Barton, B.A., Schreck, C.B., Sigismondi, L.A. (1986) Multiple Acute Disturbances Evoke Cumulative Physiological Stress Responses in Juvenile Chinook Salmon. *Transactions of the American Fisheries Society* 115 (2), 245–251.

Barton, B.A. and Iwama, G.K. (1991) Physiological changes in fish from stress in aquaculture with emphasis on the response and effects of corticosteriods. *Annual Review of Fish Diseases*, 3–26.

Barton, B. and Schreck, C. (1987) Metabolic cost of acute physical stress in juvenile steelhead. *Transactions of the American Fisheries Society* 116, 257–263.

Batzina, A. and Karakatsouli, N. (2012). The presence of substrate as a means of environmental enrichment in intensively reared gilthead seabream Sparus aurata: growth and behavioral effects. *Aquaculture* 370, 54–60.

Bierman, H.S., Zottou, S.J. and Hale, M.E. (2009) Evolution of the Mauthner axon cap. *Brain Behaviour Evolution* 73, 174–187.

Bjordal, A. (1990) Sea lice infestation on farmed salmon: Possible use of cleaner fish as an alternative method for de-lousing. *Canadian Technical Report Fisheries Aquatic Science*, Department of Fisheries and Oceans, St. Andrews.

Bornø, G. *et al.* (2016) Akutt dødelighet hos rognkjeks (*Cycloterus lumpus*) i 2015. *Norwegian Veterinary Institute Report Series.* 2016–2. 45 pp.

Bosakowski, T. and E.J. Wagner, E.J. (1994) Assessment of fin erosion by comparison of relative fin length in hatchery and wild trout in Utah. *Canadian Journal of Fisheries and Aquaculture Science* 51, 636–641.

Bosworth, B.G., Small, B.C., Gregory, D., Kim, J., Black, S. and Jerrett, A. (2007) Effects of rested-harvest using the anesthetic AQUI-S (TM) on channel catfish, *Ictalurus punctatus*, physiology and fillet quality. *Aquaculture* 262 (2–4), 302–318.

Braida, D., Donzelli, A., Martucci, R., Capurro, V., Busnelli, M., Chini, B. and Sala, M. (2012) Neurohypophyseal hormones manipulation modulate social and anxiety-related behavior in zebrafish. Psychopharmacology (Berl). 220, 319–330. doi:10.1007/s00213-011-2482-2

Branson, E.J. (Ed.). (2008) *Fish welfare.* John Wiley & Sons.

Brown, C., Davidson, T. and Laland, K. (2003) Environmental enrichment and prior experience of live prey improve foraging behaviour in hatchery-reared Atlantic salmon. *Journal of Fish Biology* 63(s1), 187–196.

Callaghan, A. and Pering, D. (2016) NMC Wrasse Cleanerfish project. Unpublished report, NMC.

Cañon Jones, H.A.C., Hansen, L.A., Noble, C., Damsgård, B., Broom, D.M. and Pearce, G.P.

(2010) Social network analysis of behavioural interactions influencing fin damage development in Atlantic salmon (*Salmo salar*) during feed-restriction. *Applied animal behaviour science* 127(3), 139–151.

Code of Good Practice (2015) Code of Good Practice for Scottish Finfish Aquaculture. thecodeof good-practice.co.uk

Davenport, J. (1985) *Synopsis of Biological data on the Lumpsucker Cyclopterus lumpus* (*Linnaeus*, 1758). FAO Fisheries Synopsis N 147 (Vol. 9).

Davis, K.B. (2006) Management of physiological stress in finfish aquaculture. *North American Journal of Aquaculture* 68 (2), 116–121.

Davis, K.B., Griffin, B.R. and Gray, W.L. (2002) Effect of handling stress on susceptibility of channel catfish *Ictalurus punctatus* to *Icthyophthirius multifiliis* and channel catfish virus infection. *Aquaculture* 214 (1–4), 55–66.

Davis, K.B., Griffin, B.R. and Gray, W.L. (2003) Effect of dietary cortisol on resistance of channel catfish to infection by *Ichthyopthirius multifiliis* and channel catfish virus disease. *Aquaculture* 218 (1–4), 121–130.

Davis, K.B. and Griffin, B.R. (2004) Physiological responses of hybrid striped bass under sedation by several anesthetics. *Aquaculture* 233 (1–4), 531–548.

Domenici, P. and Blake, R. (1997) The kinematics and performance of fish fast-start swimming. *Journal of Experimental Biology* 200, 1165–1178.

Dykes, A. (2012) Cost and benefits of fish welfare–A producer's perspective. *Aquaculture Economics and Management* 16(4), 429–432.

Eaton, R.C., Didomenico, R. and Nissanov, J. (1988) Flexible body dynamics of the goldfish C-start: implications for reticulospinal command mechanisms. *Journal of Neuroscience* 8, 2758–2768.

Einarsdottir, I.E., Nilssen, K.J., Iversen, M. (2000a) Effects of rearing stress on Atlantic salmon (*Salmo salar* L.) antibody response to a non-pathogenic antigen. *Aquaculture Research* 31 (12), 923–930.

Einarsdottir, I.E., Nilssen, K.J., Oren, S.O. and Iversen, M. (2000b) Temperature influence on Arctic charr (*Salvelinus alpinus* L.) antibody response to a cellular antigen. *Polar Biology* 23 (4), 231–235.

Ellis, T., North, B., Scott, A.P., Bromage, N.R., Porter, M. and Gadd, D. (2002) The relationships between stocking density and welfare in farmed rainbow trout. *Journal of Fish Biology* 61(3), 493–531.

Ellis, T., Yildiz, H.Y., Lopez-Olmeda, J., Spedicato, M.T., Tort, L., Overli, O. *et al.* (2012) Cortisol and finfish welfare. *Fish Physiology and Biochemistry* 38 (1), 163–188. doi: 10.1007/ s10695-011-9568-y

Eriksen, E., Durif, C.M. and Prozorkevich, D. (2014) Lumpfish (*Cyclopterus lumpus*) in the Barents Sea: development of biomass and abundance indices, and spatial distribution. *ICES Journal of Marine Science: Journal du Conseil* 71, 2398–2402.

European Animal Welfare Platform (2011) Salmon Production Strategic Approach Documents, EAWP. www.animalwelfareplatform.eu/documents/EAWPStrategicApproachDocuments-Salmon Production.pdf [Accessed 13 February 2013].

Finstad, B., Iversen, M. and Sandodden, R. (2003) Stress-reducing methods for releases of Atlantic salmon (*Salmo salar*) smolts in Norway. *Aquaculture* 222 (1–4), 203–214.

Finstad, B., Kroglund, F., Strand, R., Stefansson, S.O., Bjørn, P.A., Rosseland, B.O. *et al.* (2007) Salmon lice or suboptimal water quality – Reasons for reduced postsmolt survival? *Aquaculture* 273, 374–383.

Finstad, B., Kroglund, F., Bjørn, P.A., Nilsen, R., Pettersen, K., Rosseland, B.O. *et al.* (2012) Salmon lice-induced mortality of Atlantic salmon postsmolts experiencing episodic acidification and recovery in freshwater. *Aquaculture* 362–363, 193–199.

Goymann, W. and Wingfield, J.C. (2004) Allostatic load, social status and stress hormones: the costs of social status matter. *Animal Behaviour* 67, 591–602. doi:10.1016/j.anbehav.2003.08.007

Hale, M.E. (2000) Startle responses of fish without Mauthner neurons: escape behavior of the lumpfish (*Cyclopterus lumpus*). *Biological Bulletin* 199, 180–182.

Hanssen, A.B., Patel, D., Brinchmann, M.F. and Iversen, M.H. (2017) The effect of long-term stress on basal levels of plasma cortisol and hypothalamic–pituitary–Interrenal (HPI) Axis in lumpsucker (*Cyclopterus lumpus*). *Aquaculture Research* submitted.

Hasan, M. and Bart, A.N. (2007) Improved survival of rohu, *Labeo rohita* (Hamilton-Buchanan) and silver carp, *Hypophthalmichthys molitrix* (Valenciennes) fingerlings using low-dose quinaldine and benzocaine during transport. *Aquaculture Research* 38 (1), 50–58.

Helland, S, Bæverfjord, G. and Lein, I (2012) Produksjon av berggylt (*Ballan wrasse*) Statusrapport fra Sunndalsøra, In: Leppeprod Dialogmøte. Nofima, Stord.

Hess-Erga, O.J., Johannessen, T.V. and Åtland, Å. (2015) Monitoring and training program related to water quality and fish health in marine juvenile facilities. Norwegian Institute for Water Research, ISBN No.: ISBN 978-82-577-6517-0

Hoyle, I., Oidtmann, B., Ellis, T., Turnbull, J., North, B., Nikolaidis, J. and Knowles, T.J. (2007) A validated macroscopic key to assess fin damage in farmed rainbow trout *Onchorhynchus mykiss*. *Aquaculture* 270, 142–148.

Huntingford, A., Adams, C.V.A., Braithwaite, V., Kadri, S., Pottinger, T.G., Sandøe, P. and Turnbull, J.F. (2006) Current issues in fish welfare. *Journal of Fish Biology* 68, 32-372.

Huntingford, F. A. and Kadri, S. (2009) Taking account of fish welfare: lessons from aquaculture. *Journal of Fish Biology* 75(10), 2862–2867.

Ibarra-Zatarain, Z., Fatsini, E., Rey, S., Chereguini, O., Martin, I., Rasines, I. and Duncan, N. (2016) Characterization of stress coping style in Senegalese sole (*Solea senegalensis*) juveniles and breeders for aquaculture. *Royal Society Open Science* 3(11), 160495. doi:10.1098/rsos.160495

Imsland, A.K., Reynolds, P., Eliassen, G., Hangstad, T.A., Nytrø, A.V., Foss, A., Vikingstad, E. and Elvegård, T.A. (2014) Notes on the behaviour of lumpfish in sea pens with and without Atlantic salmon present. *Journal of ethology* 32(2), 117 122.

Imsland, A.K., Reynolds, P., Eliassen, G., Hangstad, T.A., Nytrø, A.V., Foss, A., Vikingstad, E. and Elvegård, T.A. (2015) Assessment of suitable substrates for lumpfish in sea pens. *Aquaculture International* 23(2), 639–645.

Imsland, A.K., Reynolds, P., Eliassen, G., Mortensen, A., Hansen, Ø.J., Puvanendran, V., Hangstad, T.A., Jónsdóttir, Ó.D.B., Emaus, P.A., Elvegård, T.A., Lemmens, S.C.A., Rydland, R., Nytrø, A.V. and Jonassen, T.M. (2016b) Is cleaning behaviour in lumpfish (*Cycloptherus lumpus*) parentally controlled? Aquaculture 459, 156-165.

Imsland, A.K., Reynolds, P., Eliassen, G., Hangstad, T.A., Jónsdóttir, Ó.D.B., Elvegård, T.A., Lemmens, S.C.A., Rydland, R. and Nytrø, A.V. (2016a) Investigation and quantification of behavioural interactions between lumpfish (*Cyclopterus lumpus*) and Goldsinny (*Ctenolabrus rupestris*) under controlled conditions. *Aquaculture International* 24, 1509–1521.

Ingólfsson, A. and Kristjánsson, B.K. (2002) Diet of juvenile lumpsucker Cyclopterus lumpus (Cyclopteridae) in floating seaweed : effects of ontogeny and prey availability. *Copeia* 2002, 472–476. doi:10.1643/0045-8511(2002)002[0472:DOJLCL]2.0.CO;2

Iversen, M., Finstad B. and Nilssen K.J. (1998) Recovery from loading and transport stress in Atlantic salmon (*Salmo salar* L.) smolts. *Aquaculture* 168 (1–4), 387–394.

Iversen, M., Finstad, B., McKinley, R. S. and Eliassen, R. A. (2003) The efficacy of metomidate, clove oil, Aqui-S^{TM} and Benzoak® as anaesthetics in Atlantic salmon (*Salmo sala*r L.) smolts, and their potential stress-reducing capacity. *Aquaculture* 221, 549–566.

Iversen, M., Finstad, B., McKinley, R.S., Eliassen, R.A., Carlsen, K.T. and Evjen, T. (2005) Stress responses in Atlantic salmon (*Salmo salar* L.) smolts during commercial well boat transports, and effects on survival after transfer to sea. *Aquaculture* 243 (1–4), 373–382.

Iversen, M. and Eliassen, R.A. (2009) The Effect of AQUI-S R Sedation on Primary, Secondary, and Tertiary Stress Responses during Salmon Smolt, *Salmo salar* L., Transport and Transfer to Sea. *Journal of the World Aquaculture Society* 40 (2), 216–225. doi:10.1111/j.1749-7345.2009.00244.x

Iversen, M., Eliassen, R.A. and Finstad, B. (2009) Potential benefit of clove oil sedation on animal welfare during salmon smolt, *Salmo salar* L. transport and transfer to sea. *Aquaculture Research* 40, 233–241.

Iversen, M.H. and Eliassen, R.A. (2014). The effect of allostatic load on hypothalamic-pituitary-interrenal (HPI) axis before and after secondary vaccination in Atlantic salmon postsmolts (*Salmo salar* L.). *Fish Physiology and Biochemistry* 40, 527–538.

Iversen, M.H., Jakobsen, R., Eliassen, R. and Ottesen, O. (2015) Sedasjon av berggylt og rognkjeks for å redusere stress og dødelighet. *Norsk Fiskeoppdrett 2015. NFexpert*, 42–46.

Jobling, M., Alanärä, A., Noble, C., Sánches-Vázquez, J., Kadri, S. And Huntingford, F. (2012) Appetite and Feed Intake. In: Huntingford, F. A., Jobling, M., Kadri, S. (eds.). *Aquaculture and Behavior*. Wiley-Blackwell, Oxford. ISBN: 978-1-4051-3089-9, 183-219.

Johansen, S. J. S. and Jobling, M. (1998) The influence of feeding regime on growth and slaughter traits of cage-reared Atlantic salmon. *Aquaculture International* 6, 1–17.

Jørgensen, E.H., Haatuft, A., Puvanendran, V. and Mortensen, A. (2017) Effects of reduced water exchange rate and oxygen saturation on growth and stress indicators of juvenile lumpfish (*Cyclopterus lumpus* L.) in aquaculture. *Aquaculture* 474, 26–33.

Kousoulaki, K., Grøtan, E., Kvalheim, K., Høstmark, Ø., Klinge, M., Bogevik, A.S. (2014) Protein quality, commodity options and ballan wrasse weaning performance. In: Helland et al., (eds) *Production of ballan wrasse. Science and Practice* p. 96–100. www.rensefisk.no/fileadmin/Opplaeringskontoret/cropped_image/publication-production-of-ballan-wrasse.pdf

Kramer, D. (1987) Dissolved oxygen and fish behavior. *Environmental Biology of Fishes* 18, 81–92.

Leclercq, E., Graham, P. and Migaud, H. (2015) Development of a water-stable agar-based diet for the supplementary feeding of cleaner fish ballan wrasse (*Labrus bergylta*) deployed within commercial Atlantic salmon (*Salmon salar*) net-pens. *Animal Feed Science and Technology* 208, 98–106.

Liebert, A.M. and Schreck, C.B. (2006) Effects of acute stress on osmoregulation, feed intake, IGF-1, and cortisol in yearling steelhead trout (*Oncorhynchus mykiss*) during seawater adaptation. *General and Comparative Endocrinology* 148 (2), 195–202.

Main, D.C.J., Mullan, S., Atkinson, C., Cooper, M., Wrathall, J.H.M. and Blokhuis, H.J. (2014) Best practice framework for animal welfare certification schemes. *Trends in Food Science and Technology* 37(2), 127–136.

Martins, C.I., Galhardo, L., Noble, C., Damsgård, B., Spedicato, M.T., Zupa, W., Beauchaud, M., Kulczykowska, E., Massabuau, J.C., Carter, T. and Rey-Planellas, S. (2012) Behavioural indicators of welfare in farmed fish. *Fish Physiology and Biochemistry* 38(1), 17–41.

McEwen, B.S. (1998) Stress, adaptation, and disease – Allostasis and allostatic load. Neuroimmunomodulation, *Annals of the New York Academy of Sciences* 840, 33–44.

McEwen, B.S. (2003) Mood disorders and allostatic load. *Biological Psychiatry* 54, 200–207.

McEwen, B.S. (2005) Stressed or stressed out: What is the difference? *Journal of Psychiatry and Neuroscience* 30 (5), 315–318.

McEwen, B.S. and Wingfield, J.C. (2003) The concept of allostasis in biology and biomedicine. *Hormones and Behavior* 43 (1), 2–15. doi:10.1016/s0018-506x(02)00024-7

Mesa, M.G. (1994) Effects of Multiple Acute Stressors on the Predator Avoidance Ability and Physiology of Juvenile Chinook Salmon. *Transactions of the American Fisheries Society* 123 (5), 786–793.

Mommsen, T.P., Vijayan, M.M. and Moon, T.W. (1999) Cortisol in teleosts: dynamics, mechanisms of action, and metabolic regulation. *Reviews in Fish Biology and Fisheries* 9 (3), 211–268.

Moon, T.W. and Foster, J.D. (1995) Tissue carbohydrate metabolism, gluconeogenesis and hormonal and environmental influences. *Biochemistry and Molecular Biology of Fishes* 4, 65–100.

Morgan, J.D. and Iwama, G.K. (1996) Cortisol-induced changes in oxygen consumption and ionic regulation in coastal cutthroat trout (*Oncorhynchus clarki clarki*) parr. *Fish Physiology and Biochemistry* 15, 385–394.

Morgan, M.J., Wilson, C.E. and Crim, L.W. (1999) The effect of stress on reproduction in Atlantic cod. *Journal of Fish Biology* 54, 477–488.

Näslund, J., Rosengren, M., Del Villar, D., Gansel, L., Norrgård, J.R., Persson, L., Winkowski, J.J. and Kvingedal, E. (2013) Hatchery tank enrichment affects cortisol levels and shelter-seeking in Atlantic salmon (*Salmo salar*). *Canadian Journal of Fisheries and Aquatic Sciences* 70(4), 585–590.

Nijman, V. and Heuts, B. A. (2000) Effect of environmental enrichment upon resource holding power in fish in prior residence situations. *Behavioural Processes* 49(2), 77-83.

Noble, C., Kadri, S., Mitchell, D.F. and Huntingford, F.A. (2008) Growth, production and fin damage in cage-held 0+ Atlantic salmon pre-smolts (*Salmo salar* L.) fed either (a) on demand, or (b) to a fixed satiation-restriction regime: data from a commercial farm. *Aquaculture* 275, 163–168.

Noble, C., Jones, H.A.C., Damsgård, B., Flood, M.J., Midling, K.Ø., Roque, A., Sæther, B.S. and Cottee, S.Y. (2012a) Injuries and deformities in fish: their potential impacts upon aquacultural production and welfare. *Fish Physiology and Biochemistry* 38(1), 61–83.

Noble, C., Berrill, I.K., Kankainen, M., Setälä, J., Waller, B., Honkanen, P., Mejdell, C.M., Turnbull J.F., Damsgård, B., Schneider. O., Toften, H., Kole, A. and Kadri, S. (2012b). A multi-disciplinary framework for bio-economic modeling in aquaculture: A welfare case study. *Aquaculture Economics and Management* 16(4), 297–314.

North, B.P., Turnbull, J.F.T., Ellis, T., Porter, M.J., Migaud, H., Bron, J. and Bromage, N.R. (2006) The impact of stocking density on the welfare of rainbow trout (*Oncorhynchus mykiss*). *Aquaculture* 255, 466–479.

North, B. P., Ellis, T., Bron, J., Knowles, T. G., and Turnbull, J. F. (2008) The use of stakeholder focus groups to identify indicators for the on-farm assessment of trout welfare. In: Branson, E.J. (ed.) *Fish welfare*. Blackwell, Oxford, pp. 243-267.

Pelis, R.M. and McCormick, S.D. (2003) Fin development in stream- and hatchery-reared Atlantic salmon. *Aquaculture* 220, 525–536.

Perry, S., Jonz, M. and Gilmour, K. (2009) Oxygen sensing and the hypoxic ventilatory response. In: Richards, J., Farrell, J. and Brauner C. (eds) *Fish Physiology* 27, Hypoxia, 193–253.

Persson, L. and Alanärä, A. (2014) The effect of shelter on welfare of juvenile Atlantic salmon Salmo salar reared under a feed restriction regimen. *Journal of Fish Biology* 85(3), 645–656.

Pettersen, J.M., Bracke, M., Midtlyng, P.J., Folkedal, O., Stien, L.H., Steffenak, H., and Kristiansen, T.S. (2014) Salmon welfare index model 2.0: an extended model for overall welfare assessment of caged Atlantic salmon, based on a review of selected welfare indicators and intended for fish health professionals. *Reviews in Aquaculture* 6(3), 162–179.

Pinkiewicz, T.H., Purser, G.J., and Williams, R.N. (2011) A computer vision system to analyse the swimming behaviour of farmed fish in commercial aquaculture facilities: a case study using cage-held Atlantic salmon. *Aquacultural Engineering* 45(1), 20–27.

Rey, S., McAdam, B., Pattillo, C., Mackenzie, S., Appleyard, B. and Treasurer, J. (2017) Health and welfare of the lumpfish (*Cyclopterus lumpus*) in hatchery production and deployed in Scottish salmon cages for use as cleaner fish. EAS poster abstract 244, October, Dubrovnik.

Riley, A., Jeffery, K., Cochrane-Dyet, T., White, P. and Ellis, J. (2017) Northern European Wrasse

– Summary of commercial use, fisheries and implications for management. Retrieved from https://secure.toolkitfiles.co.uk/clients/15340/sitedata/Wrasse/CEFAS-Wrasse-report-v8.pdf

Rodgers, E. (2017). The development of welfare standards for cleaner fish. SAIC/FHF International cleaner fish summit, 10 May 2017, Glasgow.

Ross, L.G. and Ross, B. (2008) *Anaesthetic and sedative techniques for aquatic animals.* Third edition, Blackwell Science, London, UK. 222 pp.

RSPCA. (2015). RSPCA Welfare Standards for Farmed Atlantic Salmon. RSPCA, Horsham, UK www.rspca.org.uk/sciencegroup/farmanimals/standards. Accessed 13 February 2016.

Sandodden, R., Finstad, B and Iversen, M. (2001) Transport stress in Atlantic salmon (*Salmo salar* L.): anaesthesia and recovery. *Aquaculture Research* 32, 87–90.

Sayer, M. J., Reader, J.P. and Davenport, J. (1996) Survival, osmoregulation and oxygen consumption of wrasse at low salinity and/or low temperature. In: Sayer, M.D.J., Treasurer, J.W. and Costello, M.J. (eds) *Wrasse: biology and use in aquaculture*, Fishing News Book, Oxford, 283 pp.

Scott, A. and Ellis, T. (2007) Measurement of fish steroid in water – a review. *General Comparative Endocrinology* 153, 392–400.

Schreck, C.B. (2010) Stress and fish reproduction: The roles of allostasis and hormesis. *General and Comparative Endocrinology* 165 (3), 549–556.

Schreck, C.B., Contreras-Sanchez W. and Fitzpatrick M.S. (2001) Effects of stress on fish reproduction, gamete quality, and progeny. *Aquaculture* 197 (1–4), 3–24.

Schreck, C.B., Maule, A.G. and Kaattari, S.L. (1993) Stress and disease resistance. In: Roberts R.J. and Muir J.F. (eds) *Recent Advances in Aquaculture*, IV. Blackwell Scientific Publications, Oxford, pp. 170–175.

Segner, H., Sundh, H., Buchmann, K., Douxfils, J., Sundell, K.S., Mathieu, C., Ruane, N., Jutfelt, F., Toften, H. and Vaughan, L. (2012) Health of farmed fish: its relation to fish welfare and its utility as welfare indicator. *Fish physiology and biochemistry* 38(1), 85-105.

Selye, H. (1950) Stress and the general adaption syndrome. *British Medical Journal* 1 (4667), 1383–1392.

Selye, H. (1973) Homeostasis and heterostasis. *Perspectives of Biological Medicine* 16, 441–445.

Skiftesvik, A.V., Bjelland, R.M., Durif, C.M.F., Johansen, I.S. and Browman, H.I. (2013) Delousing of Atlantic salmon (*Salmo salar*) by cultured vs. wild ballan wrasse (*Labrus bergylta*). *Aquaculture* 402–403, 113–118.

Skiftesvik, A.B., Durif, C.M., Bjelland, R.M. and Browman, H.I. (2015) Distribution and habitat preferences of five species of wrasse (Family Labridae) in a Norwegian fjord. *ICES Journal of Marine Science: Journal du Conseil* 72(3), 890–899.

Soares, S., Green, D. M., Turnbull, J. F., Crumlish, M. and Murray, A. G. (2011) A baseline method for benchmarking mortality losses in Atlantic salmon (*Salmo salar*) production. *Aquaculture* 314, 7–12.

Soares S., Murray A. G., Crumlish M., Turnbull J. F. and Green D. M. (2013) Factors affecting variation in mortality of marine Atlantic salmon *Salmo salar* in Scotland. *Diseases of Aquatic Organisms* 103, 101–109.

Stien, L.H., Bracke, M., Folkedal, O., Nilsson, J., Oppedal, F., Torgersen, T., Kittilsen, S., Midtlyng, P.J., Vindas, M.A., Øverli, Ø. and Kristiansen, T.S. (2013) Salmon Welfare Index Model (SWIM 1.0): a semantic model for overall welfare assessment of caged Atlantic salmon: review of the selected welfare indicators and model presentation. *Reviews in Aquaculture* 5(1), 33–57.

Stien, L. H., Oppedal, F. and Kristiansen, T. S. (2016) Dødelighetsstatistikk for lakseproduksjon (eng: Mortality statistics for salmon production). In: Svåsand, T., Karlsen, Ø., Kvamme, B. O., Stien, L. H., Taranger, G. L. and Boxaspen, K.K. (eds.) Risikovurdering norsk fiskeoppdrett 2016. Havforskningsinstituttet, Fisken og Havet, særnr; 2-2016, 129–134.

Treasurer, J.W. (2011) *Wrasse farming developments in Scotland.* Presentation at the Scottish Sea Lice Symposium, Edinburgh. October 2011. CD publication, SSPO, Perth.

Treasurer, J. and Feledi, T. (2014) The Physical Condition and Welfare of Five Species of Wild-caught Wrasse Stocked under Aquaculture Conditions and when Stocked in Atlantic Salmon, *Salmo salar*, Production Cages. *Journal of the World Aquaculture Society* 45(2), 213-219.

Turnbull, J., Bell, A., Adams, C. Bron, J. and Huntingford, F. (2005) Stocking density and welfare of cage farmed Atlantic salmon: application of multivariate analysis. *Aquaculture* 243, 121–132.

Turner, J. (2016) Video tracking of behaviour and tank distribution of lumpfish (*Cyclopterus lumpus*) related to fin damage to develop operational welfare indicators. Unpublished MSc Thesis, University of Stirling.

van Raaij, M., Pit, D., Balm, P., Steffens, A. and van den Thillart, G. (1996) Behavioral Strategy and the Physiological Stress Response in Rainbow Trout Exposed to Severe Hypoxia. *Hormones and Behavior* 30, 85–92.

van Rooji, J.M. and Videler, J.J. (1996) Estimating oxygen uptake rate from ventilation frequency in the reef fish *Sparisoma viride. Marine Ecology Progress Series* 132, 31–41.

Varsamos, S., Flik, G., Pepin, J.F., Bonga, S.E.W. and Breuil, G. (2006) Husbandry stress during early life stages affects the stress response and health status of juvenile sea bass, *Dicentrarchus labrax. Fish and Shellfish Immunology* 20 (1), 83–96.

Villegas-Ríos, D., Alós, J., March, D., Palmer, M., Mucientes, G. and Saborido-Rey, F. (2013) Home range and diel behavior of the ballan wrasse, *Labrus bergylta*, determined by acoustic telemetry. *Journal of Sea Research* 80, 61–71.

Voas, A. (2008) Welfare legislation applying to farmed fish in the UK. In: Branson, E.J. (ed.) *Fish Welfare*, Blackwell Publishing.

Webb, P.W. (1976) The effect of size on the fast-start performance of rainbow trout *Salmo gairdneri*, and a consideration of piscivorous predator-prey interactions. *Journal of Experimental Biology* 65, 157–177

Wedemeyer, G.A. (1996) *Physiology of fish in intensive culture systems.* Chapman & Hall, New York, USA. 232 pp.

Wendelaar Bonga, S.E.W. (1997) The stress response in fish. *Physiological Reviews* 77 (3), 591–625.

Wendelaar Bonga, S.E.W. (2011) Hormonal responses to stress. In: Anthony, P.F. (ed) *Encyclopaedia of Fish Physiology*. Academic Press, San Diego, USA. Pp. 1515–1523.

Wingfield, J.C. (2005) The concept of allostasis: Coping with a capricious environment. *Journal of Mammalogy* 86 (2), 248–254.

Zion, B. (2012) The use of computer vision technologies in aquaculture – a review. *Computers and electronics in agriculture* 88, 125–132.

CHAPTER 16

Transport of lumpfish and wrasse

Thor Magne Jonassen,[1] Mette Remen,[1] Anette Lekva,[2] Tómas Árnason[3] and Agnar Steinarsson[3]

[1]Akvaplan-niva, Tromsø, Norway
[2]Marine Harvest Norway, Bergen, Norway
[3]Marine Research Institute, Reykjavik, Iceland

16.1 Introduction

Transport is probably one of the most complicated and stressful operations in fish farming. The most obvious challenges for the fish, particularly in closed systems typical for road transport, are the transfer between environments that may differ in water quality, the effect of repeated handling during transport, and challenges with accumulation of metabolites in transport tanks with no water exchange (Southgate, 2008; Harmon, 2009; King, 2009). An investigation of mortality in lumpfish (Bornø *et al.*, 2016) showed that among 23 salmon farms where lumpfish *Cyclopterus lumpus* was used, 80% reported mortality rates of 25% or more, and that some of the acute mortalities were associated to the fish transport. It is conceivable that such initial mortalities, at least in part, may be related to transport conditions and handling (Remen and Jonassen, 2017). Research on the effect of handling and transport conditions for lumpfish on stress physiology has only been available recently, and this will help to define optimal transport protocols (Remen and Jonassen, 2017). No such research has until now been conducted for wrasse. There is a lack of knowledge for both species, particularly on water quality, where much has to rely on experience from transports of other cultured fish species, particular salmon *Salmo salar*. It is therefore particularly important to examine the current transport practice and to build a knowledge base for further optimisation of transport protocols for lumpfish and wrasse.

This chapter focuses mainly on current protocols and important considerations for road transport of lumpfish. More elaboration on general principles for transport of fish and other publications are available, and the following publications are recommended:

1) Rosten and Kristensen (2011). Best practice in live fish transport. NIVA-report SNO 6102-2011. 22 pp.
2) Anon (2008). Transportation of fish within a closed system. Norwegian Scientific Committee for Food Safety, 63 pp. ISBN: 978-82-8082-242-0.
3) Harmon, T.S. (2009). Methods for reducing stressors and maintaining water quality

associated with live fish transport in tanks: a review of the basics. Reviews in Aquaculture (2009) 1, 58–66.

16.2 The transport vessel

Lumpfish have been successfully transported in closed tanks with oxygenation and aeration by truck for long distances (up to 28 hours), in traditional wellboats and, for shorter distance (one to two hours), by boat in closed fish tanks on deck with oxygenation (Remen and Jonassen, 2017). Good practice has also been developed for air freight of juvenile fish and eggs in plastic bags with oxygen placed in Styrofoam boxes. The most common transport method for lumpfish, which will be described here, is truck transport in tanks (of approx. size 2 m^3), normally stretching from the hatchery to the shore close to the cage farming site, and from there by boat to the salmon on-growing sites (Fig. 16.1).

Fin and skin erosion is commonly observed in lumpfish after transport (S. Ag, Aqua Kompetanse, Norway, 2016, personal communication), and the frequency may be high. It is unclear what causes this, but the physical impact from transport should be minimised by ensuring that all surfaces in the tanks are smooth. This also applies for equipment used for handling the fish. Tanks should be equipped with an oxygenation system, preferably with automatic regulation of oxygen injection and with aeration for CO_2 degassing and pH stabilisation. This is probably also important for shorter transport in closed tanks on boat. Sensors and automatic logging systems for oxygen, temperature and pH are required. Unless the sensors are protected, i.e. with a rigid plastic net cover, the lumpfish may attach to the sensors and the sensors may be prevented from logging appropriate water quality.

Truck transport normally requires a secondary transport to reach the salmon farming sites, hence the fish are reloaded over to fish tanks placed on the boat deck for the last transport haul. This extra handling shortly followed by a new handling when the fish are transferred to sea cages may cause additional stress and fin and skin erosions. The use of trucks with flexible tank systems where the tanks can be detached from the truck and moved on to a boat would be a considerable advantage and would reduce labour and fish stress related to this operation.

Fig. 16.1 Unloading lumpfish from boat to salmon cage. Credit: Erika Kunickiene.

16.3 Planning and preparation of transport

A specific delivery plan should be part of the long-term planning, and specific criteria for each delivery should be defined, i.e. size and size distribution, vaccination status, delivery time and place. Some recommended checkpoints for a delivery plan are listed in Table 16.1.

When numbers, size distribution and time of delivery for the fish are decided, the group(s) have to be graded and counted, normally from one or more groups (tanks) of fish. Special concern on traceability of the fish has to be noticed and recorded for documentation of the fish background. Grading, counting and placement in delivery tanks should be done preferably no less than two weeks prior to transport. The closer this operation is to delivery, the higher is the risk of compromising the tolerance of lumpfish towards transport and transfer to the new environment in salmon sea cages.

Only healthy and vaccinated fish should be delivered. A withdrawal period of 500 degree days after vaccination is recommended to ensure good effect of the vaccine before transfer to sea cages. To prevent escapes after transfer, fish have to be graded. Individual weight samples should be taken shortly before transport to verify that the smallest fish delivered cannot pass through the cage net meshes.

Temperature affects the metabolism of the fish, and tolerance to handling and transport stress is normally higher at the lower end of the temperature tolerance range. In lumpfish, optimal temperature for growth is in the range 9–16°C, depending on the fish size (Nytrø *et al.*, 2014). In commercial transports of lumpfish temperature ranged from 3.3 to 15.9°C (Remen and Jonassen, 2017). There was less transport stress at the higher temperatures, and fewer problems with transferring fish to higher temperatures (from approx. 8 to 15°C) than from medium temperatures to

Table 16.1 Checklist for planning and preparation of fish transport.

Planning and preparation	Delivery	Documentation
• Early agreements between producer, customer and carrier • Veterinary inspection and health certificate • Vaccination plan • Grading and size distribution • Feed withdrawal prior to transport • Temperature acclimation • Checklist for equipment needed for transport	• Disinfection of transport tanks • Calibration of water quality sensors in transport tanks • Crowding, pumping/netting and counting fish on to transport tanks • Distribution according to max. biomass/number of fish per tank • Regulation and monitoring of oxygen in fish tanks and transport tanks • Waybill with loading time, fish data and distribution on tanks • Receiving inspection and control of size according to cage mesh size before transfer to cages (prevent escape)	• Commercial contracts • Health certificate and data on production history (CV) • Documentation of transport tank disinfection • Water quality logs from transport • Freight letter (waybill) • Receiving control letter • Mortality data 30 days after transport

low (from approx. 8 to 3°C). Farms with the possibility of cooling water should aim at a transport temperature of around 8°C and to acclimatise the fish to this temperature one to two weeks prior to transport. If the temperature is lower than 8°C at the farming site, acclimation to even lower temperatures is preferable, as a large and abrupt drop may induce a 'cold shock' response that involves stress, and mortalities at worst (Donaldson *et al.*, 2008).

Feed withdrawal before transport is important to reduce the loading of faeces and metabolites in the water during transport, which may give deterioration in water quality. Hence, starved fish will tolerate a longer transport time. Fasted fish will normally have higher tolerance to stress and handling. The normal procedure for lumpfish is two to three days feed withdrawal in darkness before transport at normal rearing temperatures of approx. 8°C at the juvenile stage (20–30 g). At lower temperatures, fasting time should be increased since the rate of digestion is reduced with temperature.

Fasted lumpfish tend to become more aggressive, and agonistic behaviour may result in stress and physical damage, particularly on the tail fin. This negative effect of fasting may be prevented by keeping the lumpfish in total darkness during fasting in the last days before transport. Analyses of the stress hormone cortisol from blood plasma indicates no increase in stress in lumpfish after two to three days fasting in darkness (Remen and Jonassen, 2017).

16.4 Loading the truck

The most critical factors during loading of fish are time (duration of handling stress), oxygen monitoring and control, physical impact from equipment/procedures (loss of mucus and skin erosion) and biomass control in transport tanks (fish counting and distribution).

The truck should be prepared for loading by filling the transport tanks with fresh seawater, normally from the same source where the fish have been reared to ensure similar water quality and temperature. The water source should be known free of pathogens. Before transferring the fish to the truck, either by pumping or gentle hand netting, the water level should be reduced slowly in the fish tanks to crowd the fish and make them easier to collect without chasing. The best way to control this is to replace the standard standpipe in the tank with a short one, which allows for water flow-through in the tanks at a low water level. The oxygen level and fish behaviour should be monitored closely and additional oxygen added if necessary. Oxygen should be kept around 90–100% and the same oxygen level should be maintained in the transport tanks. It is important to be aware that oxygen may drop abruptly in the transport tanks during loading due to the handling stress, unless oxygen is closely monitored and regulated.

Care should be taken to ensure that all equipment used is gentle for the fish, without rough surfaces and sharp edges that may result in skin damage and loss of skin mucus. A closed net that brings water together with the fish when hand netting is recommended, or a fine-meshed soft net without knots.

It is critical for minimising stress that the whole operation with crowding and transfer of fish is carried out as quickly as possible. Hence it is advantageous to distribute the fish in several smaller holding tanks before transport rather than a few large ones to minimise the time for emptying a tank. The longer the operation the more stress is accumulated in the last fish transferred. Ideally one holding tank ('delivery tank') of graded and counted fish should be used for

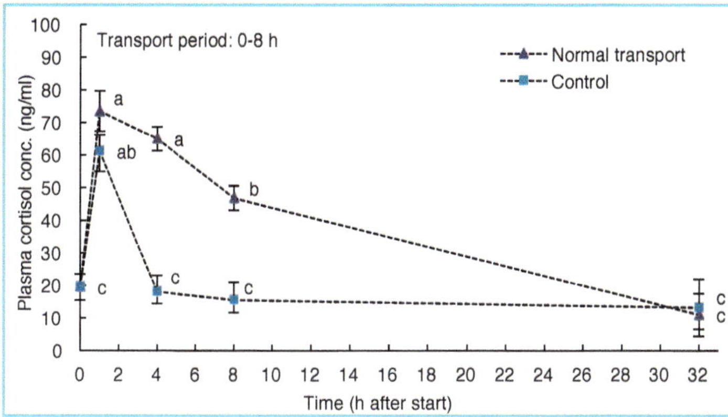

Fig. 16.2 Change in plasma cortisol concentrations (ng ml^{-1}, mean ± SE) over time in groups of lumpfish subjected to 'normal transport' (blue triangles) or subjected to netting only (at O h, termed 'control', light blue squares). The transport conditions in the transport group were 8°C, ~100% O_2, eight hours duration, 30 kg m^{-3} stocking density, aeration and darkness.

each transport tank, which would give better control of fish size, numbers and biomass in the transport tanks.

Unless oxygen is monitored and controlled in all transport tanks, which is normally not the case, it is important to keep a similar fish biomass in all tanks to ensure appropriate water quality. Oxygen consumption will increase if biomass is too high and will give an increased accumulation of total ammonia (TAN) and CO_2 (monitored via pH). Control of fish weight and numbers in each tank is therefore crucial. Fish may be pumped via a fish counter before entering the transport tank or, if water is drained from the fish via a grid before entering the transport tank, biomass may be estimated volumetrically by measuring the change in water level in the tank.

A series of controlled transport simulations with lumpfish showed maximum stress response shortly after loading (Remen and Jonassen, 2017), where plasma cortisol increased in average from 27 to 83 ng ml^{-1} during the first hour after transfer to the transport tanks (Fig. 16.2). Thereafter the average cortisol levels decreased to 53 ng ml^{-1} during the eight-hour transport simulation, despite a gradual deterioration of water quality. This points to the loading procedure as a major stressor during transport, and therefore this critical point should be optimised, particularly with respect to handling time.

16.5 Road transport of lumpfish

There are different risks and challenges associated with water quality changes during transport, such as transport time and acclimatisation of fish to the new environment, and these may be avoided by early and appropriate planning. The main priority during transport is to monitor and ensure a stable and suitable water quality for the fish. Since truck transport in closed tanks lacks the possibility of water exchange or water purification by filtration (for particles or TAN), maximum transport time depends on how long suitable and stable water quality can be maintained. This again depends on the fish sensitivity to different water quality parameters and the rate at which CO_2 and TAN are accumulating. This will also be related to temperature during transport, fish species and size, and how well the fish has been fasted (emptied gut and reduced metabolic rate). Attempting to change water during transport when water quality is deteriorating is associated with an increased risk of stress as a result of increasing ammonia toxicity (increased pH

Table 16.2 Water quality parameters at the start and end of transport simulations of lumpfish from experiment I-V (average values ±SE). Characteristics of normal transport: eight hours' transport time, 30 g fish, 30 kg m^{-3} fish density, 100% oxygen, water temperature 8°C and 48 hours' fasting before transport. For experimental groups only one parameter was changed each time.

Experiment	Group	pH	TAN mg N/l	NH_3-N µg/l	CO_2 mg/l	TOC mg C/l
I-V	Before transport	7.8±0.1	0.0±0.0	0.0±0.0	2.0±0.1	1.3±0.2
I-V	Normal transport	7.5±0.1	0.5±0.0	1.3±0.1	6.4±0.5	3.1±0.3
I	20 h transport	7.4±0.0	1.6±0.1	4.4±0.1	8.6±2.8	4.9±0.1
II	60 kg/m^3	7.1±0.0	1.3±0.0	2.3±0.1	7.5±0.3	8.2±0.3
III	12°C	7.8±0.0	0.8±0.0	3.3±0.1	4.9±0.3	3.4±0.0
IV	60 g fish	7.6±0.0	0.3±0.0	0.7±0.0	5.8±0.6	2.5±0.2
V	150% O_2	7.5±0.0	0.5±0.0	1.1±0.0	6.4±0.8	3.7±0.2

in new water shifts NH_4^+ to the more toxic form NH_3), increased metabolic rate and metabolite build-up (as a result of the disturbance) and extended transport time.

The outcomes of trials examining increased transport time (increasing from eight to 20 hours), fish density (increasing from 30 kg m^{-3} to 60 kg m^{-3}), temperature (increasing from 8°C to 12°C), fish size (increasing from 30 g to 60 g) and oxygen saturation (increasing from 100% to 150%) are given in Table 16.2 (from Remen and Jonassen, 2017). Only one of the criteria was changed each time and standard conditions were eight hours transport, 30 g fish, 30 kg/m^{-3} fish density, 100% oxygen, 8°C and 48 hours fasting. The pH in the transport water declined with increased fish density. In particular, increased transport time, fish density and temperature affected TAN and NH_3-N measured at the end of the transport. Fish density had the strongest effect on TOC. There was not a clear change in CO_2 with changing transport conditions, which was probably more influenced by the regulation of water aeration.

As previously mentioned, loading stress is a main stressor during transport, while the actual transport gives resting time where the fish slowly recover despite gradually deteriorating water quality. In a simulated eight-hour transport plasma cortisol at the end of the transport was reduced to approx. 50% of the initial levels (measured one hour after loading), and 24 hours after finished transport, cortisol levels were comparable to levels in undisturbed fish (Fig. 16.3). Transport simulations with increased oxygen saturation (from 100% to 150%), increased temperature (from 8°C to 12°C) or increased fish size (from 30 g to 60 g), did not give any change in the lumpfish stress response. However, increased stocking density from 30 to 60 kg m^{-3} gave higher plasma cortisol and pH after four hours' transport, but there were no differences at the end of the transport (eight hours). The stress response was lower after 20 hours' transport compared to eight hours, probably as a result of longer acclimation time at 20 hours. These observations suggest that, as long as fish are provided with enough oxygen (~100% O_2), aeration is sufficient and stocking density is acceptable (<60 kg m^{-3}), the fish are able to tolerate the transport conditions mentioned above. However, it is possible that the short-term period of stress is rendering the fish more vulnerable to other challenges related to sea cage transfer. Further studies should examine the combined effect of transport and other stressors in order to elucidate possible accumulative effects.

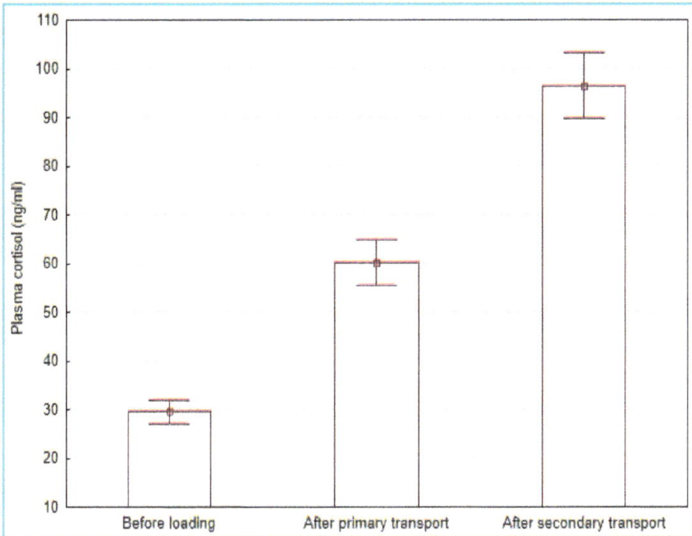

Fig. 16.3 Average plasma cortisol concentration (ng ml^{-1}) of 15 groups of lumpfish (n = 6 per group and sampling time) through three stages of the transport. Vertical lines indicate standard error of mean (±SEM).

16.6 Lumpfish transport in Iceland

As south-west Iceland and the Faroes are separated by about 800 km of open ocean, the safe transport of live fish between the two countries is by no means a straightforward operation. The cargo has to be moved either by air or sea but the choice of transport method may depend on a number of factors such as logistics, weather conditions or scale of production.

The air transport method involves the packing of lumpfish into plastic bags inside Styrofoam boxes. Feed is withheld for the last three days before packing and the mean weight of the juveniles is measured on the day before transport. The packing line is prepared approximately six to seven hours before flight departure and the juveniles are collected and brought to the packing area in a mesh-lined fish tub filled with seawater. The first step in the packing line involves adding 20 l of oxygenated (>400% saturation) and buffered (crystalline Trizma base buffer, 0.6 g per litre) seawater (4°C) into a double plastic bag. The open bags are then placed in a bucket on a scale and 3–5 kg of lumpfish juveniles added before the inner bag is filled up with pure oxygen and both bags sealed tightly with plastic strips. Higher loading densities are possible but are risky in terms of post-transport mortality. The double bag is then placed into a Styrofoam box (80 × 40 × 26 cm) and the box sealed with tape (Fig. 16.4). Finally, the boxes are stacked on pallets (15 boxes per pallet) and wrapped with cellophane wrapping. The total weight of each packed box is about 25 kg, depending on the fish biomass. The aeroplane can usually take up to 3 tonnes of boxes per shipment, so up to 120 boxes can be shipped per flight. Each box can safely hold at least 120–200 fish (18–30 g mean weight) which gives a safe total of 14,000–24,000 fish per shipment.

Results from box deliveries

The transit time from Reykjavík or Keflavík, Iceland, to Vágar, Faroes, is about 75 minutes and the total transport time from tank to cage is only six to nine hours. The unloading of boxes was

Fig. 16.4 An open airfreight transport box containing a bag filled with 20 l of seawater and approximately 160 live lumpfish juveniles ready for transport to the Faroe Islands.

usually finished within only one hour after arrival in the Faroes and, for most shipments, the immediate survival was close to 100%. In some of the early shipments, with loading densities of up to 6 kg per box, there was a low immediate mortality but a high post-shipment mortality, perhaps caused by the high packing density. Air transport may be more expensive per shipped juvenile than sea transport but, on the other hand, it has several advantages. The transport time is much shorter and the boxes can be unloaded directly into the sea cages.

Results from air transport simulations

In transport simulations the oxygen level in the bags started to fall rapidly after the bags were sealed. If a box with 4 kg of lumpfish juveniles was left undisturbed the oxygen saturation fell from >400% to critical levels (<30%) in approximately two to three hours. However, if the boxes were gently shaken for one to two minutes at regular intervals the oxygen quickly increased (Fig. 16.5).

There was a steep initial reduction in oxygen saturation before the bag was sealed, probably due to the initial stress associated with the handling of the fish. However, the fish soon recovered and the oxygen level stabilised, probably as a result of better mixing of oxygen and water when the boxes were moved and shaken.

Sea container transport

Live transport container units for sea transport similar to those for salmon juveniles have also been used successfully in Iceland, with some adaptations for lumpfish juveniles. Each container is equipped with air pumps and oxygenation system and can safely hold 30,000–40,000 lumpfish juveniles during the two-day journey by cargo ship from Iceland to the Faroes. After some initial difficulties, especially during bad weather conditions, this transport method has been optimised and is now being used routinely by Stofnfiskur for the large-scale transport of lumpfish juveniles.

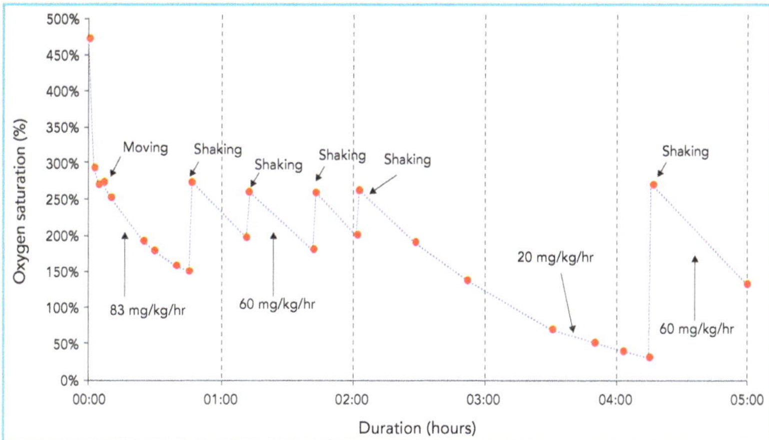

Fig. 16.5 Oxygen saturation during a transport simulation with 4 kg of lumpfish juveniles (25 g mean weight) inside a double plastic bag enclosed in a Styrofoam box. The bag contained 20 litres of oxygenated (>400%) seawater (4°C) and a similar volume of pure oxygen. Shaking or moving the boxes is indicated on the figure and the numbers show the calculated oxygen consumption of the juveniles based on the fall in oxygen saturation.

Air freight in closed containers

The FishPacTM live transportation system, which is a purpose-designed container for live aquatic transport, was used by the MRI during the latest production season in year 2016 (Fig. 16.6). The system consists of an 800 l insulated plastic container, and integrated oxygenation. The containers were stacked with plastic shelves to increase the available dwelling area. An even distribution of the fish reduced crowding-induced abrasion caused by the sharp dorsal spikes of the lumpfish. The lumpfish were fasted for two days before transport (at 7°C). The transport bins were filled with 5–7°C seawater. A specialised transport buffer (900 g per container of Trizma base buffer from Sigma) was added to the water to regulate the pH level to 7.2 during transport. If the pH level became too high before stocking the level was lowered by adding some diluted hydrochloric acid. An option is to reduce the salinity of the transport water, but any effects on fish tolerance and welfare would have to be assessed.

Fig. 16.6 A FishPac™ transport tank fully stocked with live lumpfish juveniles, showing the top shelf. Oxygen levels are being monitored and adjusted before transport.

The lumpfish were stocked on the bottom and successive shelves based on a specific fish density of 1080 kg m^{-3}. Around 140–180 kg fish were stocked per container without a problem. Oxygen level was monitored and adjusted before the container was closed, sealed, labelled and tilt-tested before being delivered by transport truck to the airport.

The lumpfish were sluggish on arrival in some cases, and this may be due to high CO_2 levels that induce narcosis and stun the fish. However, they recovered immediately when transferred to the sea cages. This transport method was considered less work-demanding than the box transport method and more convenient for maintaining fish welfare during transport. There is still some room for improvement in the transport of lumpfish, i.e. by stocking smaller fish (<25 g), lowering temperature and salinity, and maintaining desirable pH and oxygen levels during transport.

16.7 Transport of eggs and milt

The production of lumpfish today is to a large degree still dependent on the eggs and milt from wild fish. Many farms choose not to keep wild-caught broodfish for biosecurity reasons. Hence there is an extensive transport of ova and milt, and air freight using plastic bags placed in chilled Styrofoam boxes is the most convenient and common method. Fertilised and unfertilised eggs can be shipped. Unfertilised eggs in egg fluid from one female (normally 400–800 ml) are packed in plastic bags filled with oxygen. Fertilised eggs (300–900 ml) are packed in plastic bags with 7 l of seawater and filled with about the same amount of oxygen. It is very important not to contaminate the egg fluid of unfertilised eggs with water or urine, which makes the eggs stick together and then they cannot be fertilised. Crushed ice is placed at the bottom of the box for chilling, and separated from the ova by a soft pad or a Styrofoam plate. This will maintain the temperature at around 2–4°C in properly sealed boxes. Movement of the box helps oxygen exchange to the eggs. Fresh cooled milt is packed the same way as the eggs in boxes with ice and sent separately in small bottles. Eggs have been successfully transported for up to 24 hours. The normal fertilisation rate of transported eggs is 90% and hatching rate is around 90%. Fresh milt may be stored cooled for at least two weeks provided it is regularly stirred/moved to exchange oxygen to the milt. This time window gives the opportunity to screen the milt for diseases prior to use.

16.8 Transport of wrasse

Wrasse are mostly transported by truck in 2 m^3 holding tanks similar to those used for salmon. The fish are loaded with fish pumps to the truck or boat and via a fish counter. A water separator is used when loading to a truck, and then the wrasse number may also be measured by a volumetric calculation when the average weight is known. Temperature differences between the hatchery/nursery and the boat must be taken into account when loading to the temperature in the wellboat before transportation. Long transportation in an open circulation wellboat increases the risk of acquiring infections from the external water and bringing these to a new location. Hence, closed wellboat transportation or truck is preferred. The tank volume is reduced during pumping, and oxygen level and fish behaviour must be monitored regularly.

Table 16.3 Examples of successful transportations of Ballan wrasse by truck. Duration of transports from 3 to 10 hours.

Journey time 3 hours with truck up to 10 hours									
Fish weight (g average)	1	3	4	6	9	12	17	7	18
Biomass (kg)	40	45	52	90	90	90	85	70	90
Number fish in tank	40,000	15,000	13,000	15,000	10,000	7,500	5,000	10,000	5,000
Density (kg pr m^3)	20	23	26	26	45	45	45	35	45
Tank size, m^3	2	2	2	2	2	2	2	2	2

The stress level may be reduced with sedation in closed transportation tanks, to ensure improved fish welfare and water quality, as stressed fish will produce more CO_2 and other waste products such as faeces. Aqui-S (Isoeugenol 540 mg ml^{-1}) can be used to calm fish, and wrasse will show less escape behaviour with the correct doses, which are 2.5–3 ml m^{-3} for Ballan wrasse (H. Glosvik, Marine Harvest, Norway, 2016, personal communication). Correct doses are important to ensure light sedation and prevent deep anaesthesia, and the oxygen level must be monitored continuously. Aqui-S may be considered for shorter transportations (less than six hours) for use in tanks before pumping commences. Aqui-S takes effect after ten to 15 minutes, and is more beneficial when added early in the process before the fish are exposed to stress. The maximum dosing time is 24 hours, and sedation has been shown to work well in transportations for at least around ten hours by truck for wrasse of 5 g.

Wrasse have been transported at different stages; eggs, larvae, juvenile (1–50 g). Eggs and larvae have only been transported a few times in relation to production trials at other facilities without broodstock. The transport of juvenile wrasse from hatchery to other on-growing facilities has been carried out when fish were 0.5 to 20 g. The policy of MHN has been to transfer fish of 20 g to sea cages when stocked with smolts and around 50 g when stocked with larger salmon. Transportation to other on growing sites is conducted at a size from 5 to 20 g. Trucks with around 13 tanks of 2 m^3 each are normally used. Small wellboats are often used to transport Ballan wrasse to sea sites. Maximum stocking densities are about 50 kg m^{-3} for fish of 5 to 50 g, but the density is reduced for smaller fish. Successful transports of Ballan wrasse by truck are illustrated (Table 16.3).

16.9 Water quality under transport

The tolerance of lumpfish to different water quality parameters is still largely unknown, and transport guidelines must so far be based on knowledge of other species and practical experience. In closed containment fish transport the water quality deterioration due to the fish metabolism sets limits to transport time. Mucus loss and faeces may also affect water quality. The metabolism increases with temperature, stress level, handling, swimming speed, digestion rate and with reduced fish size.

In particular, four water quality parameters are important under fish transport; oxygen (O_2), carbon dioxide (CO_2), total ammonia (TAN = the sum of NH_4^+ and NH_3) and total organic carbon (TOC). In addition, temperature acclimation of the fish and transport temperature has a large impact on the result of the transport. Temperature is important in the way it affects the metabolism of the fish and hence water quality, but also in the way it affects the solubility of O_2 and CO_2 (solubility decreases with increased temperature). The toxicity of ammonia (NH_3) increases with reduced temperature and the amount of NH_3 increases with increasing temperature. All fish species have their optimal temperature, and small juveniles have a higher temperature optimum than adults. Although optimal temperature for growth in young lumpfish of 20–40 g is approx. 16.1°C (Nytrø *et al.*, 2014), stress is lowered by lowering temperature. Lumpfish are normally transported at 8–10°C, and increase of temperature from 8°C to 12°C in eight-hour simulated transports at 30 kg m^{-3} did not result in any changes in plasma cortisol or several other secondary stress responses (Remen and Jonassen, 2017).

Oxygen is the primary parameter to control and regulate, and should be kept around 100% saturation. High O_2 saturation may reduce the toxicity of ammonia and increase tolerance for CO_2, but may also damage the fish, i.e. for salmon it may damage the gill chloride cells. A classic example is that oxygen is kept high for salmon to increase the ammonia tolerance under long transport, which often gives good survival under transport, but high mortality 24–48 hours after sea transfer. For lumpfish it has been speculated that the increased mortality sometimes seen after sea transfer may be related to the reduction in oxygen level when transferred to sea cages. Oxygen super-saturation may occur in transport tanks, while oxygen saturation in sea cages may typically vary from 85–95 %. Increase of oxygen saturation from approx. 100% to 150% in eight-hour simulated transports at 30 kg m^{-3} did not result in any changes in plasma cortisol or several other secondary stress responses (Remen and Jonassen, 2017).

Oxygen use is 1:1 in relation to CO_2 excretion and 1:10 in relation to the excretion of TAN. The practical consequence is that CO_2 accumulates ten times quicker than TAN. When CO_2 accumulates in water it releases H^+ and pH drops. Seawater has a relatively high buffering capacity for CO_2 and when the transport water is aerated (degassed) pH and CO_2 are normally easy to stabilise under transport of lumpfish.

A maximum level of 60 mg l^{-1} CO_2 is set for transport of salmon smolts in freshwater (Rosten, 2000). The highest CO_2 level reported in a lumpfish transport was 29 mg l^{-1}, which was in a 14-hour transport of 18 g fish at a fish density of 32 kg m^{-3} and temperature of approx. 10°C (Remen and Jonassen, 2017). The pH at the end of the transport was 7.1.

In general, long before CO_2 reaches harmful levels the NH_3 will pose a serious risk for fish welfare. The best way to keep low NH_3 is to keep the pH low, which happens through the accumulation of CO_2. NH_3 also increases with reduced temperature (i.e. NH_3 doubles when temperature falls from 20 to 10°C as a result of change of the NH_4^+–NH_3 equilibrium), which may be of concern under closed transport at low temperatures. Increased salinity gives a small decrease in NH_3. It is also worth noting that tolerance for NH_3 is reduced with fish size, stressed fish are more sensitive to NH_3 than unstressed fish and swimming fish are more sensitive than resting fish. Most importantly, the tolerance for NH_3 varies largely among fish species, and there is no information on the tolerance in lumpfish. The average of the mean acute toxicity values for 17 seawater species is 1.86 mg $NH_3 l^{-1}$ (Randall and Tsui, 2002). While turbot tolerate 100–400 µg NH_3 l^{-1} salmon only tolerate 25–75 µg NH_3 l^{-1}. An investigation of commercial truck transports of lumpfish (Remen and Jonassen, 2017) showed that TAN accumulate to 4,700 µg N l^{-1}

after 28 hours (at approx. 5°C, 27 g fish, 29 kg m^{-3}, two days' starvation), while an eight-hour transport gave a TAN concentration of 530 µg N l^{-1} (at approx. 8°C, 43 g fish, 23 kg m^{-3}, two days' starvation). At the end of the transport pH was 7.2 and 7.5 respectively, giving an NH_3 concentration of 4.79 and 1.74 µg N l^{-1} respectively. These were examples of successful transports. However, the survival shortly after transfer to the sea cages was not recorded. The relatively low TAN accumulation in lumpfish transports may be a result of low activity of the fish. It is observed that they normally stay attached to the tank surface during transport.

Particularly in seawater, there is a risk of acute NH_3 poisoning to the fish when transport water (with low pH) is mixed with fresh seawater (at high pH, normally around pH = 8.2). In the mixing zone between fresh and 'old' seawater, ammonium (NH_4^+) with low toxicity is converted to highly toxic ammonia (NH_3). For example, when transferring lumpfish from primary transport (truck) to new tanks in secondary transport (boat) 'old' transport water should be drained from the fish before it enters new tanks with fresh seawater, or the fish should be kept in the 'old' transport water from the start until the fish enters the sea cages.

Another issue in closed containment transport is accumulation of waste products such as mucus and faeces observed as an increase in total organic carbon (TOC). There is a risk that such substances may attach to the fish gills, hence disturbing the respiration and stress the fish. There is little known about this in lumpfish, but care should be taken to reduce the risk, primarily by fasting the fish for two to three days depending on temperature prior to transport to empty the alimentary tract.

16.10 Discharging fish to sea pens

When best practice is applied it is expected to find the fish in a moderately stressed condition, and the fish should be observed closely to consider their welfare and health status. A fish veterinarian should be consulted if health issues are observed. Pipes have to be carefully secured with at least a doubled security attachment to prevent escapes when unloading fish transport. Fish size should be measured and compared with the mesh size of the cage net to ensure there is no risk of even the smallest fish passing through the net. Fish should also be checked for damage and normal behaviour after transfer to the cages. Normally the fish hides are moved close to the point where the fish are released so it is a place to hide and rest that can be easily accessed.

Unloading fish after a truck transport is normally a quick process, but there are some critical points. First, the process should be gentle for the fish, but the faster the process the less stress. The truck does not usually get direct access to the sea cages and a secondary transport with boat to transfer the fish from the truck to the cages is required. The ideal process would be to unload the whole transport tank from the truck and move over to a secondary boat transport. However, such solutions have not yet been developed. The process of transferring the fish to new transport tanks for secondary transport exposes the fish to extra handling, which accumulates significant additional stress on the fish. If the procedure is not optimised, this may potentially be exhausting for the fish since transport time after this handling (secondary transport) is often too short to reduce stress before a new handling of the fish during transfer to the net cages. The accumulation of stress through the different stages of 15 commercial transports is given in Fig. 16.3 (from

Fig. 16.7 Process of secondary transport from truck to the cages (from left to right): 1: Transfer from truck to transport tanks for on-deck boat transport to sea cages. Note the end piece on transport hose with grids for dewatering. Tanks are lined with a fine-meshed net that prevents the lumpfish attaching to the tank surface. 2: Lumpfish after transfer to the transport tank. 3: Transport tanks with fish on deck. Oxygenation or flow through water by pumping for maintaining good oxygen conditions. 4 and 5: Transport tank lifted and fish transferred to sea cage.

Remen and Jonassen, 2017), showing a threefold increase in stress after the secondary transport compared to the levels before the fish are loaded to the truck. A secondary transport process also adds additional technical and biological risks related to the handling operation (Fig. 16.7).

The risk of mixing transport water with high total ammonia (TAN) with fresh seawater is of particular concern, as discussed above. A dewatering device (grid) at the end of the transport pipe prevents this problem when unloading from the truck. Alternatively, the fish may be transported in the same water all the way to the cages.

Lumpfish lack a swimbladder and prefer to rest during transport by attaching to different surfaces in the tanks. When stressed the lumpfish may stick very hard to the surface and then need to be released by force, either with a hand net or by flushing gently with water. This may potentially result in loss of mucus, fin erosion or wounds to the fish, which makes it more exposed to osmotic stress and infections. A technique to avoid such problems is to lower the water in the tanks slowly when unloading, and the fish will passively be moved by the water. Another method is to sedate the fish by adding low doses of anaesthetics (Fig. 16.8, sedated fish). The equipment used should have types of material and shapes that minimise the risk of physical damage to the fish.

All transports should be planned in such a way that change in water temperature on sea transfer is minimised. Practical considerations will modify this, but the establishment of some main principles may contribute to improved fish robustness after sea transfer. One is temperature transitions, which should be minimised, and abrupt change to a lower temperature may be particularly challenging for lumpfish. The transport may be at very low temperature in winter that may give temperature fluctuations during transport or fish may be exposed to very cold air temperatures when unloaded from the truck. There are examples of such transports that resulted in increased mortality and disease problems after transfer to sea (Remen and Jonassen, 2017).

Observations of the fish behaviour in the transport tanks before loading or in the cages afterwards may give some idea of the health status of the fish. Good routines for following up fish after transport are important, both for control, management and learning. Observing the fish in

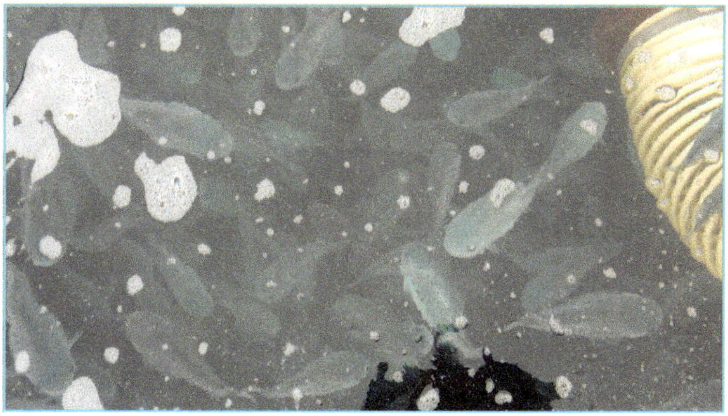

Fig. 16.8 Lumpfish sedated to prevent the fish from attaching to the tank surface when unloading the fish tank. Credit: Erika Kunickiene

the cages gives the opportunity to consider fish welfare, viability, appetite, amount of mucus loss, schooling pattern and mortality, which may be categorised and used as operational fish welfare indicators.

All equipment, fish tanks and the truck should be washed and disinfected after transport to ensure that the vessel is clean and disinfected between every new group of fish transported. Only approved chemicals and dosages must be used. This should be quality assured through documentation and records.

16.11 Quality control and documentation

Atlantic salmon farmers have been experiencing highly variable effects after releasing lumpfish into the net pens, with respect to mortality and lice grazing ability, and the reasons for this are still unclear. Water temperature, lumpfish size and parental origin may contribute to this variation (Nytrø *et al.*, 2014; Imsland *et al.*, 2016), and outbreaks of diseases are thought to be important for the observed mortalities (Bornø *et al.*, 2016). The current development and optimisation of suitable vaccines may help to reduce mortalities, but the extent to which other factors contribute, such as stress or injuries related to handling, nutrition and environmental conditions, is still largely unknown. Generally, stress is known to increase disease susceptibility and reduce appetite in fish. Hence a suitable production protocol, including transport, that ensures good fish welfare and a robust lumpfish that is able to perform well as a grazer of sea lice in the cages is necessary.

The documentation and registration of conditions and methods under different operations is valuable for improvements, learning and quality assurance. An example checklist for quality control and documentation is given in Table 16.4.

A veterinary control (transport water quality, injuries, mortalities, samples to screen for particular diseases) is recommended when the cleaner fish arrive at the farm site, both to document the health status of the fish, and to detect possible problems related to the transport, such as injuries caused by handling, mortalities or poor water quality. Once the lumpfish enter the net pen, injured or sick fish may attach to a substrate and stay there until they expire. Problems related to the transport may therefore not be detected. After this, veterinary controls should be

Table 16.4 Checklist for quality control and documentation in the transport of cleaner fish.

Fish transport log	Registrations of temperature O_2 and pH from transport
Escape prevention checklist	Security points during unloading, measures of fish size, check of cage nets, etc.
Waybill from carrier	Fish origin, numbers, weight, distribution on tanks etc.
Veterinary report on fish	Health status, treatments, production history (CV)
Delivery report	Check list of fish condition and environmental conditions (truck and sea) at delivery
HSE and deviation report	Registration of deviations from normal procedures and requirements
Cleaning and disinfection routine	After transport
Calibration report	Sensors in tanks and monitoring system
Follow-up report	Mortality, diseases, observations of behaviour (video) until 30 days post-delivery

repeated regularly (e.g. bi-weekly controls) to follow up the nutritional and health status of the cleaner fish.

Recordings of transport conditions (e.g. temperature, duration, stocking density, survival, water quality) and results from veterinary controls may, together with recordings of net pen husbandry protocols and daily registrations of environmental conditions and mortalities, build a database that will become useful for optimising procedures. Such experience-based optimisation of lumpfish culture may prove to be important, as large numbers of lumpfish will be produced and placed into net pens within the next few years. Such databases may also be useful to detect factors that are important for susceptibility of lumpfish to disease in the salmon net pens.

References

Bornø, G., Alarcón, M., Linaker, M.L., Colquhoun, D., Nilsen, H., Gu, J., Gjerset, B., Hansen, H., Thoen, E., Gulla, S. and Jensen, B.B. (2016) Akutt dødelighet hos rognkjeks (*Cyclopterus lumpus*) i 2015. Veterinærinstituttets rapportserie 2-2016, ISSN 1890–3290.

Donaldson, M.R., Cooke, S.J., Patterson, D.A, and Macdonald, J.S. (2008) Cold shock and fish. *Journal of Fish Biology* 73(7), 1491–1530.

Harmon, T.S. (2009) Methods for reducing stressors and maintaining water quality associated with live fish transport in tanks: a review of the basics. *Reviews in Aquaculture* 1, 58–66.

Imsland, A.K., Reynolds, P., Eliassen, G., Mortensen, A., Hansen, Ø.J., Puvanendran, V., Hangstad, T.A., Jónsdóttir, Ó.D.B., Emaus, P.A., Elvegård, T.A., Lemmens, S.C.A., Rydland, R., Nytrø, A.V. and Jonassen, T.M. (2016) Is cleaning behavior in lumpfish (*Cyclopterus lumpus*) parentally controlled? *Aquaculture* 459, 156–165.

King, H.R. (2009) Fish transport in the aquaculture sector: An overview of the road transport of Atlantic salmon in Tasmania. *Journal of Veterinary Behaviour: Clinical Applications and Research* 4(4), 163–168.

Nytrø, A.V., Vikingstad, E., Foss, A., Hangstad, T.A., Reynolds, P., Eliassen, G., Elvegård, T.A., Falk-Petersen, I-B. and Imsland, A.K. (2014) The effect of temperature and fish size on growth of juvenile lumpfish (*Cyclopterus lumpus* L.). *Aquaculture* 434, 296–302.

Randall, D.J. and Tsui, T.K.N. (2002) Ammonia toxicity in fish. *Marine Pollution Bulletin* 45, 17–23.

Remen, M. and Jonassen, T. (2017) Utvikling av transport – og mottaksprosedyrer for rognkjeks basert på kartlegging av miljø og stress. Akvaplan-niva rapport 7707-1, 71 pp.

Rosten, T. (2000) The evaluation of the cause of mortality during a 30 h closed transport of smolt from Western Norway to Northern Norway. KPMG Centre for Aquaculture and Fisheries, Norway.

Southgate, P.J. (2008) Welfare of fish during transport. In: Branson, E.J. (ed.), *Fish Welfare.* Blackwell Publishing Ltd., Oxford, U.K, pp. 185–194.

CHAPTER 17

Fisheries for cleaner fish species in Europe

Katerina Kousoulaki,[1] Jim Treasurer,[2] and Richard FitzGerald† [3]

[1]Nofima AS, Bergen, Norway
[2]FAI Aquaculture, Ardtoe, Acharacle, UK
[3]Carna Research Station, Ryan Institute, National University of Galway, Ireland

17.1 Introduction

Wrasse and lumpfish represented mostly random bycatch before their use as cleaner fish and not target fisheries species. Nevertheless, lumpfish were and still are fished for their roe in some countries. The method of using wrasse as cleaner fish against sea lice in farmed salmon net pens was first developed by the Norwegian Institute of Marine Biology in the end of the 1980s (Bjørdal, 1988). Wrasse fishing for direct use as cleaner fish in salmon pens has intensified since then, and also more recently for lumpfish for collection of fertilised eggs or broodstock for hatcheries. The growth of the salmon farming industry and increasing demand for non-medicinal delousing methods created and currently still supports a rather significant small-scale cleaner fishery of great importance for the income of local fishermen. This may at the same time exert pressure on the wild stocks. Thus, wrasse and lumpfish fisheries are regulated in several countries, both in terms of quotas on total allowable number (TAC) of fish caught, and also as regards maintaining responsible husbandry welfare standards.

Much of the scientific literature examining wrasse fisheries emanated in the 1990s and relates to catch per unit effort (CPUE) and the effects of the fishery. This was described in a book series covering Norway (Maroni and Andersen, 1996), Scotland (Sayer *et al.*, 1996; Treasurer, 1996) and Ireland (Varian *et al.*, 1996). An experimental fishery was worked over three years to examine the possible effects of exploitation (Sayer *et al.*, 1996), and a nearby site acted as a control with no fishing. The biology of goldsinny wrasse *Ctenolabrus rupestris* varied between the three localities and it was concluded that local management measures would be required. It was felt that the self-regulation of the fishery in retaining only fish over 100 mm would ensure that breeding populations would be resilient, although there could be alterations to the population age/size structure and the sex ratio. Most recently, the CPUE was compared in four Marine Protected Areas (MPAs) with fished areas in southern Norway (Halvorsen *et al.*, 2017), and CPUE tended to be higher for targeted wrasse species than in fished areas. The effects of fishing on life history were unclear in goldsinny but there were fewer larger and older corkwing *Symphodus melops* in the fished areas. It was recommended that MPAs would be useful as conservation measures for wrasse. However, a later review suggested that the positive benefits of MPAs may have been over-estimated as larger wrasse in control areas and especially males had been fished prior to the start of the study (Gonzalez and de Boer, 2017). In the most extensive discussion of management measures Darwall *et al.* (1992) outlined a range of conservation measures including also the establishment of marine nature reserves protected from fishing impacts, to fish in the autumn and keep fish alive and active over winter, and also farming wrasse. Since the first concerns regarding stock sustainability of cleaner fish species were raised, both farming and fisheries regulations and husbandry best practices have been developed.

This chapter describes the cleaner species fishery in Northern Europe, its effects on the wild stocks, the operational standards that have been developed and the regulatory framework.

17.2 Lumpfish fisheries

Significant commercial and artisanal fisheries for lumpfish (*Cyclopterus lumpus*) exist along the coasts of the species' natural habitats, both in the western and eastern Northern Hemisphere (Fig. 17.1). Powell *et al.* (2017, and in Chapter 6) detail lumpfish fishery statistics from Greenland and Iceland. A commercial fishery for lumpfish has existed in the northern part of Norway since the 1950s. Mature females are caught in the spawning grounds by small vessels mainly deploying gillnets. The fishery is based on the catch of pre-spawned fish and is, therefore, highly seasonal: lasting generally five to six weeks from April to mid-June (Bertelsen, 1994) and from mid-March to the end of May/early June (Torstensen, 1988). Every year, 200–800 boats participate in the lumpfish fishery, with vessel quotas of 1.5–2.5 t of roe in 2008, and increased to a maximum quota of 3500 kg unprocessed roe per vessel in 2015 (Fiskeridirektoratet nr. 2014000233). The Russian catch was *c.* 15,000 t of lumpfish in 1940 (Rusyaev, 2000). However, today, the Russian coastal fishery in the Barents Sea and the White Sea is relatively low, with the total amount usually lower than 15–50 t annually.

The stock numbers of the lumpfish declined in the 1990s and then stabilised to more than a third the levels in the 1980s, probably due to overexploitation (Sunnanå, 2007). Based on the registered volumes, commercial and artisanal fisheries contribute most to this. Nevertheless, the

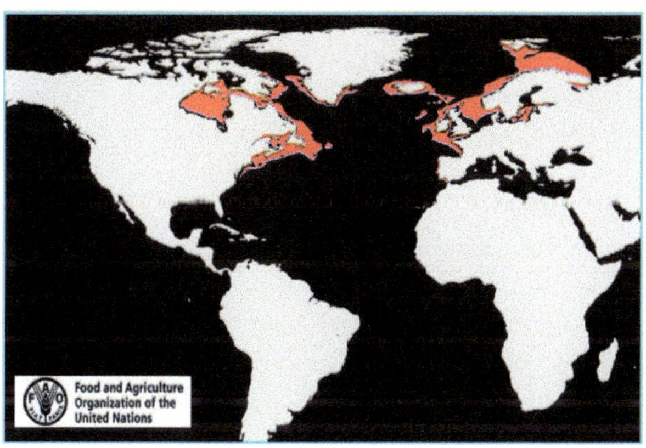

Fig. 17.1 Global lumpfish distribution. The species inhabits suitable habitats north of 45° N latitude, from Portugal and southern Bay of Biscay to Spitsbergen, including Iceland, and along the coasts in western Greenland, Hudson Bay, and the coasts of North America down to Maryland (FAO).

cleaner fish industry is removing older lumpfish for broodstock, which may have a relatively higher impact on the wild lumpfish populations.

17.2.1 Upcoming juvenile lumpfish fishery

Lumpfish have largely been fished in commercial fisheries mainly for roe, and small numbers have specifically been captured live for egg supply. These fish often do not survive first spawning, so there is a regular requirement to replace them. However, recently, there have been moves from some fishermen to harvest juvenile lumpfish from buoys, seaweeds and other floating material. Numbers initially cropped have been in the region of thousands of fish per boat but this is likely to increase. Disease screening of these fish and especially for AGD is required.

17.2.2 Wild stocks of lumpfish in Ireland

Very little is known about the status of wild stocks of lumpfish around the coast of Ireland. It is not the subject of a commercial fishery, as in many other countries, and they are rarely caught as bycatch in fisheries operations. However, targeted fishing for broodstock using gillnets proved successful in the north and west coasts of Ireland in 2016 (F. Sproll, Marine Harvest Ireland, personal communication). Anecdotal evidence has revealed that there are stocks in the south-west (Kerry), west (Galway) and north-west (Donegal) at the very least, although extensive surveys have not been conducted. Groups of small (~1 g) wild juveniles have been found sheltering among seaweed longlines in southern Ireland (e.g. in Ventry Bay; M. Edwards, personal communication). Notably, wild lumpfish have been found both inside of and in close proximity to salmon cages (F. Sproll, Marine Harvest Ireland, personal communication). In addition, wild egg masses have been found along the shoreline on spring tides during the spawning season (late March to May) (Chapter 20).

The wild-caught lumpfish broodstock in Ireland undergo disease screening to address potential biosecurity issues and this is also carried out to determine the health status of the wrasse stock in each designated fishing area. This prerequisite testing is undertaken prior to the commencement of fishing.

17.3 Wrasse fisheries

Unlike fisheries for lumpfish, wrasse fisheries were associated from the early days with the use of several wrasse species as cleaner fish of sea lice infesting farmed salmon in net pens. Cleaner fish are among the most important non-medicinal measures against salmon lice. As a result, a significant industry for fishing mainly wrasses, but also lumpfish, for the collection of eggs and broodfish, has built up in the recent years to serve the needs of salmon farming. The fishermen are well paid for delivering wrasse to the salmon farmers, that is 85–300 NOK per kg or 6–15 NOK per fish (Espeland *et al.*, 2010). A total of 24,467,000 wild wrasse, 379,000 farmed wrasse and 3,457,000 farmed lumpfish were used as cleaner fish in salmonid open cage farming in Norway in 2014. A total of 850 t of wrasse with a value of 220 million NOK, or 258 NOK/kg, were fished in 2015 and 26.6 million in 2017. The respective numbers of farmed wrasse and lumpfish had multiplied by 2017, whereas those of wild-caught wrasse have remained stable through the last few years.

17.3.1 Evidence of fishing effects

Several previous studies on the distribution and abundance of wrasse on the north and west coasts of Scotland and Ireland have reported the presence of the five common northern European species with a predominance of goldsinny and/or corkwing and/or rock cook. Densities for goldsinny have been reported in the region of 1–4 m^2 and Catch Per Unit Effort (CPUE: number of fish per pot per hour) has varied between 0.11 and 0.256 for goldsinny and 0.04 and 0.01 for corkwing (Sayer *et al.*, 1993; Sayer *et al.*, 1996a; Varian *et al.*, 1996). In one of the more recent wrasse fishery surveys conducted in Ireland, Ratcliff *et al.* (2006) investigated the wrasse populations of Clew Bay, Mayo, Ireland. It was clear that the smaller species (corkwing, goldsinny and rock cook *Centrolabrus exoletus*) were most affected. The species abundance data were generally low and heavily affected by the type/choice of the different gears (shrimp pots, creels and fyke nets), use (deployment, soakage time) and selectivity (e.g. size bias) in different areas and for the various wrasse species. A summary of distribution and relative abundance of north European wrasse, and basic population parameters such as age, size, sex ratio, are collated (Table 17.1). This table quotes CPUE figures and these are similar in studies in Ireland, Scotland and Norway. CPUE of 0.1–3.9 fish per day was indicated by Maroni and Andersen (1996) using fyke nets and purpose-designed wrasse traps compared with 0.98–1.92 wrasse per 24 hours with fyke nets. However, these catches were not directly comparable with other studies (Sayer *et al.*, 1996a; Varian *et al.*, 1996) as these quoted catch rate per hour. Clearly, comparisons between various studies are difficult to make due to the variety of fishing techniques used and differences in the methods of reporting numbers. However, differing key population biology parameters have been consistently reported – growth rate, age at maturity, natural mortality Z – of various populations of the same species within a small geographical range. This led Sayer *et al.* (1996a) to conclude that 'local variation [in biology and life history] would have to be taken into account for all fisheries', and highlighted 'the potential risks in applying site-specific data to other areas where exploitation occurs'. Hence, robust data from the specific population to be exploited is required to implement successful management of a local fishery.

17.4 The Norwegian wrasse fisheries

Fishing for goldsinny wrasse in Norway commenced in 1988 with *c.* 1000 individuals fished followed by a rapid increase to 1997, before efficient alternative delousing methods, such as emamectin benzoate (SLICE) and cypermethrin (Excis), became available. Wrasse fisheries were resumed on a large scale following lice resistance development to the applied chemical treatments (Jimenez *et al.*, 2012). Catches increased until recent years and have stabilised to just over 20 million fish due to regulations. It is noteworthy that in earlier years there was an inconsistency in the reported wrasse numbers caught and the number of fish stocked in net pens.

17.4.1 Catch composition

Several wild-caught wrasse species are used simultaneously as cleaner fish in the salmon pens. These consist mainly of five different species: goldsinny wrasse, Ballan wrasse, corkwing wrasse,

Table 17.1 Resume of some published population biology data for common, temperate European wrasse species (G = goldsinny, RC = rock cook, CW = corkwing, C = cuckoo, B = Ballan). Data are expressed as fish capture per hour (CPUE), and converted from fish in 24 hours from other studies to enable comparison. μ=mean fish length or age.

Species	Location	Abundance or CPUE h^{-1}	Gear and soak time (or equivalent)	Age (years)	Total Length (mm)	Z (natural mortality)	Sex Ratio ♂:♀	Notes	Reference
G	W Ireland	Max: 0.315 Min: 0.052 µ Yr 1: 0.256 µ Yr 2: 0.181	Baited shrimp pot, <24 hours.	♂ µ 3.3 ♀ µ 4.1	♂ µ 103 ♀ µ 93	Yr 1 = 0.425 Yr 2 = 0.672	Yr 1 = 1.68 : 1 Yr 2 = 1.54 : 1	CPUE decreased from the first to the second year.	Varian *et al.*, 1996.
G	W Scotland	0.022	Fyke net, 24 hours; creel, 3–4 days.	Max 16	65–165	0.1861	1.22	Goldsinny dominated catch in both fyke nets (11 mm mesh) and creels.	Treasurer, 1994a, b and 1996.
G	W Scotland	Max: 4 m^{-2} Min: 1m^{-2} + correlation with rugosity.	SCUBA survey: 84 hours.					High currents (2.1–3.6 ms^{-1}) observed at max density sites. Hypothesised – influence of salinity.	Sayer *et al.*, 1993
G	W Scotland	Yr 1 = 0.51 Yr 2 = 0.27 Yr 3 = 0.11	Fyke net; prawn creel; underwater anaesthetic.	♂ = 14+ ♀ = 20+	1+♂ 67–163 1+♀ 59–159		1 : 0.28 1 : 0.88	Goldsinny escape from the prawn creels.	Sayer *et al.* 1995 ; Sayer *et al.*, 1996.
G	W Norway	Low + correlation with rugosity.	Beach seine 5 mm mesh Trammel net 45 mm, 24 hours; Gill net 39 mm, 24 hours.		15–195			No association with salinity. Habitat various, no clear associations.	Thangstad, 1999.
G	Norwegian Skagerrak	0.9–12.7 fish 1000 m^{-2}	Beach seine.						Gjøsaeter, 2002
G	Norwegian Skagerrak	mean 18.5–55.5 pot^{-1} Max: 390 pot^{-1}	Baited 1cm^2 mesh, steel cylinders. 1–3 hauls per 24 hours.					Empty pots (<10%) not included in the estimation of CPUE: figures over-estimated.	Gjøsaeter, 2002

G	Norway, Nord-Trøndelag	0.1–3.9 $pot^{-1}day^{-1}$	Eel pots; wrasse pots.					This survey was undertaken at the suspected northern limits of wrasse distribution; 64.5-65°N	Maroni and Anderson, 1996.
RC	W Scotland	0.05	Fyke net, 24 hour; prawn creel, 3–4 days.	Max 9	75–155	0.4185	0.83		Treasurer, 1994a, b and 1996.
RC	W Scotland		Prawn creel; 3 m beam trawl; fyke net; UW anaesthetic.	♂: 8+ ♀: 4+	♂: 43–165 ♀: 43–138		1 : 1.09		Sayer *et al.*, 1995.
RC	W Norway	Low; + correlation with rugosity.	Beach seine 5 mm; Trammel net 45 mm, 24 hour; Gill net 39 mm, 24 hours.		10–200			No association with salinity. Habitat various: no clear associations.	Thangstad, 1999.
CW	N W Ireland	μ 5 $fyke^{-1}day^{-1}$ μ 8.7 $pot^{-1}day^{-1}$	Fyke nets, 24 hours; baited prawn creels, 24 hours.	♂: 6+ ♀: 7+ T2♂: 1-2	♂: 189 ♀: 160 T2♂: 97		May: 1:0.96 Sept: 1: 0.08	CPUE decreased from the first to the second year.	Donnelly, 1992
CW	W Scotland	0.007	Fyke net, 24 hours; prawn creel, 3–4 days.	Max 5					Treasurer 1994a, b and 1996.
CW	W Scotland			♂: 6+ ♀: 7+	♂: 212 ♀: 203		1:1.45 5% T2 ♂		Sayer *et al.*, 1995.
CW	W Norway	Low	Beach seine 5 mm; Trammel net 45 mm, 24 hours; Gill net 39 mm, 24 hours.		15–240			Habitat various, no clear associations.	Thangstad, 1999.
C	W Scotland	Low		Max 11					Treasurer, 1994a, b.
B	W Scotland	Low 0.002		Max 15					Treasurer 1994a, b and 1996.
B	Isle of Man		Rod and line; anaesthetic application.	♂: 29 ♀: 24	♂: 451 ♀: 416			Inversion from ♀ to ♂ observed between 6 and 14 years.	Dipper, 1976.

rock cook and cuckoo wrasse (*Labrus mixtus*). The first reports on use of wrasse contained information on how many fish in total were sold (Figs 17.2–3) and to which farm, but not on the species composition or the location where the wrasse were caught. More specific information on the species composition of the wrasse catches exists from 10 December 2010, whereas in most reports from 2008 and 2009 the wrasses caught were not reported by species but as 'other wrasse' (Figs 17.4–17.6). The volumes of Ballan, corkwing and goldsinny wrasse fished in 2010 and used as cleaner fish were similar. The most efficient lice cleaner, Ballan wrasse, represented only 7% of the wild wrasse numbers caught four years later in 2014, while corkwing and goldsinny wrasse prevailed in the catch. The mean fish weights were approximately five and two times higher, in Ballan and corkwing wrasse, respectively, compared with goldsinny wrasse.

17.4.2 The boats

The number of registered and non-registered boats involved in wrasse fisheries in Norway increased steadily during three years, from 391 vessels in 2014 to 493 in 2016 (Fig. 17.7). The

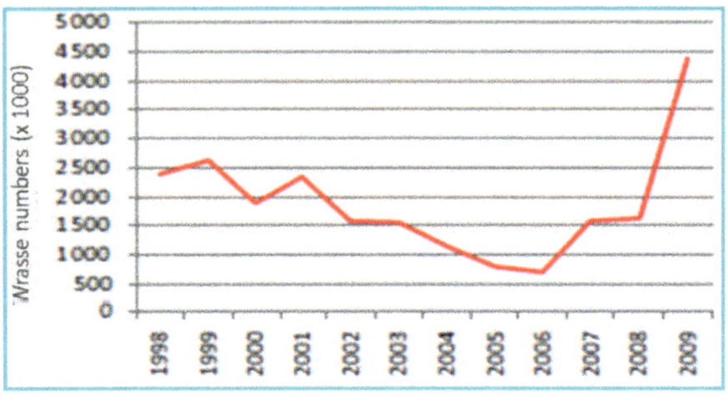

Fig. 17.2 Number of wrasse used as cleaner fish for salmon. Source: Fisheries department statistics on the use of wrasses in salmon and trout production.

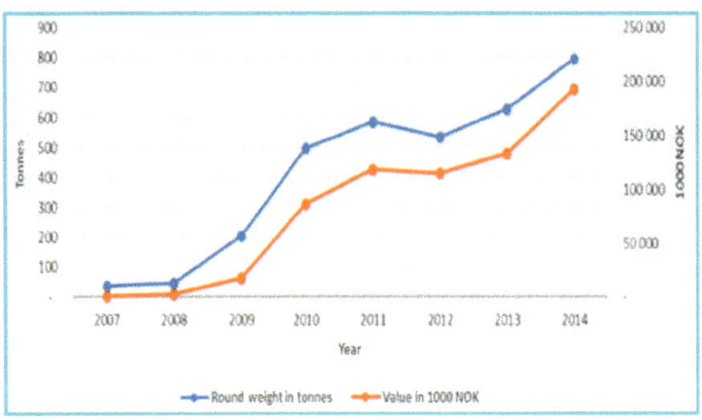

Fig. 17.3 Live wrasse turnover of the Norwegian fleet for the period 2007–2014 (Source: Landings and sales note register as of 5 February 2015). MMO data in Riley *et al.* (2017).

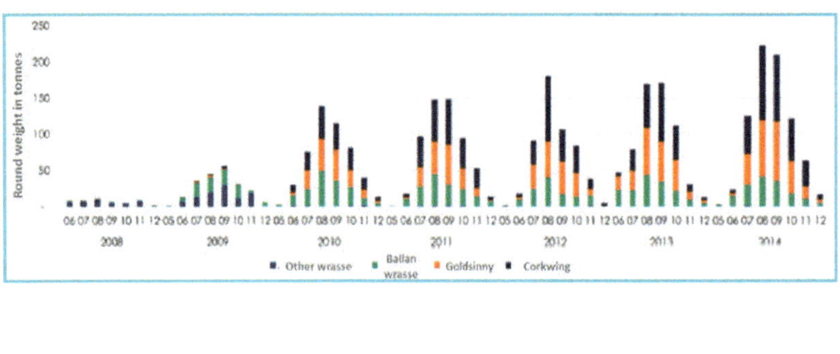

Fig. 17.4 Norwegian fleet wrasse landings (tons) by month in the period 2008–2014 (Source: Landings and sales note register as of 5 February 2015). MMO data in Riley *et al.* (2017).

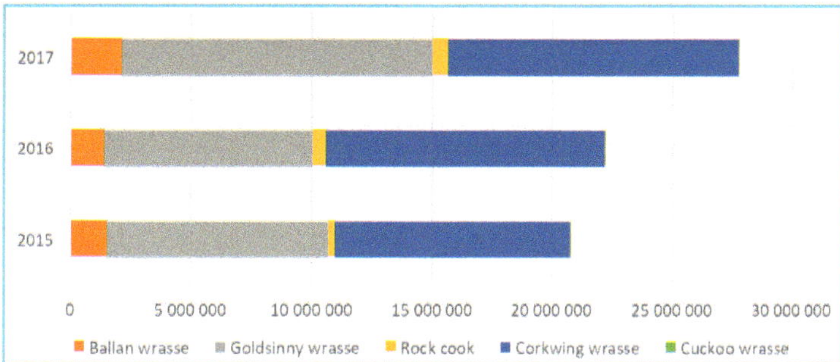

Fig. 17.5 Report data on wrasse capture (in kg) in Norway between 2015 and 2017 according to the Norwegian Fisheries Ministry's official landings register. The mean weight of the catch is 0.123 for Ballan wrasse, 0.024 for goldsinny wrasse and 0.045 for corkwing wrasse.

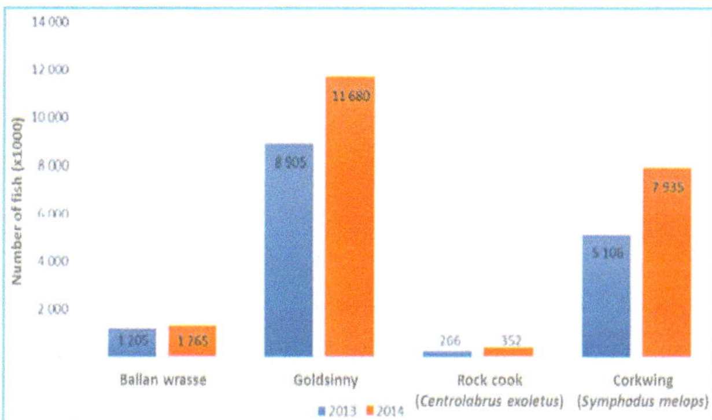

Fig. 17.6 Norwegian fleet wrasse landings (number) by species in 2013 and 2014 (Source: Landings and sales note register as of 5 February 2015).

boat sizes used range from 5 to 15 m, with the majority being under 11 m (Fig. 17.8). Most boats caught 50,000 to 100,000 fish in total per annum in the three years (2014–2016), whereas a small number of boats caught under 1000 fish per year and only one boat in 2016 fished more than 300,000 wrasse. Moreover, the increase in fleet since 2014 involves owners of either one or two boats and not consolidation to single owners of a larger fleet. There was only one owner with four and one with more than seven boats. The owners of three boats were reduced from four in 2015 to three in 2016 (Norwegian Directorate of Fisheries).

17.4.3 The fishing gear

Wrasse are caught with specially constructed pots and traps, and in a few cases also with fyke nets. Crushed crabs and shrimp shells are used as bait to attract the wrasse into the pots. Optimal net mesh size for selective catch of three wrasse species is presented in Table 17.2, and further data on mesh retention sizes for goldsinny and rock cook have been presented (Treasurer, 1996).

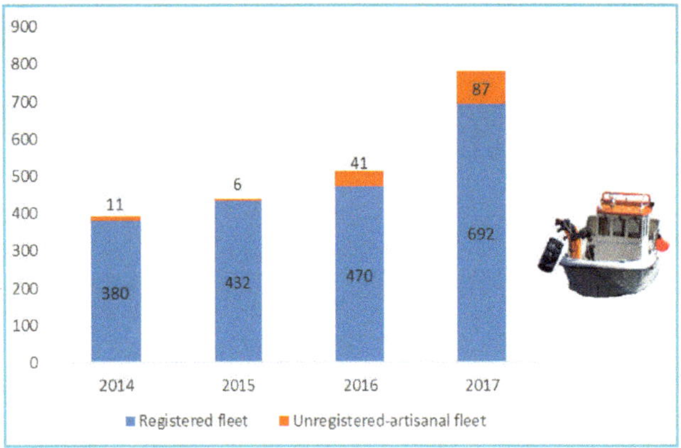

Fig. 17.7 Total Norwegian fleet involved in wrasse fisheries in Norway in 2014–2017 (Norwegian Directorate of Fisheries).

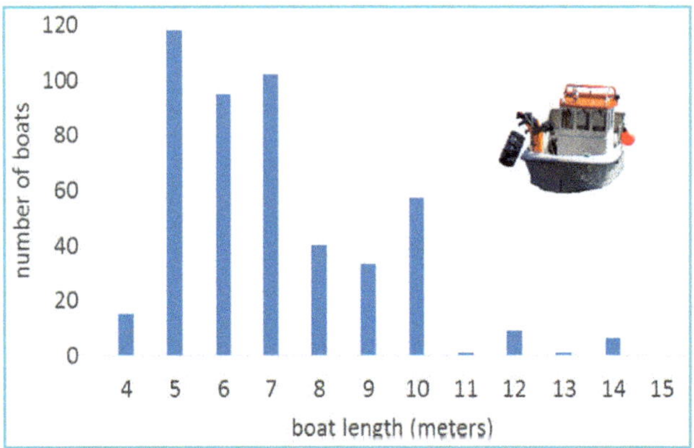

Fig. 17.8 Boat size distribution of the total Norwegian fleet involved in wrasse fisheries in 2016 (Norwegian Directorate of Fisheries).

Table 17.2 Salmon net mesh size *vs* optimal wrasse size for lower fish escapes. Source: Johan Solgaard in Espeland *et al.*, 2010, based on the study by Moe and Ildhusøy, 2006.

MESH SIZE	GOLDSINNY WRASSE	CORKWING WRASSE	BALLAN WRASSE
mm	mm	mm	mm
15.5	110+	110+	110+
16.5		110+	
18	125+	120+	130+
21		140+	
22.5			
25.5		170+	190+
29		180+	200+

17.4.4 Wrasse stock management in Norway

The increasing demand for cleaner fish has resulted in a great pressure on the wild wrasse populations and introduced considerable sustainability challenges regarding the needs for regulations to safeguard wild stocks and good fish welfare standards for captured fish. This section provides an overview of the imposed regulations on cleaner fish fisheries practices, which all aim to secure recruitment and minimise fish mortalities.

From 2015 and onwards, regulations of the wrasse fisheries came into force with the scope to protect the wild populations in Norwegian coastal waters to secure future catches and implement ethical welfare standards for the live catch. The Norwegian Directorate of Fisheries adopted changes in 2016, namely changing total quotas, fisheries opening times, exemptions, minimum sizes, handling guidelines, and rights to participate in the fisheries and equipment limitations. The fisheries directorate, moreover, retains the right to close the fishery before the total quota is fished. In this case, only registered vessels can continue fishing with up to 40 pots or traps.

Opening date for fisheries and wrasse spawning season

Wrasse fisheries are practised within the breeding period as they are only abundant and closer to the shore during this time. Nevertheless, the opening of the wrasse fisheries is intended to be as close as possible to the end of the spawning period of all the wrasse species concerned. The beginning of the fishing period for wrasses in 2016 was moved to 11 July (week 28) for south and western Norway and 26 July (week 30) for mid- and northern Norway. This represents a four-and-a-half-week delay compared to the previous year and was based on research showing that this measure was necessary to reduce the catch of spawning fish. Skiftesvik *et al.* (2016) reported no differences in the gonadal maturation status of corkwing, goldsinny and rock cook wrasse between two locations, one in southern (Flødevigen) and one in western Norway (Austevoll). Most of the goldsinny and corkwing wrasse in both sampled locations and rock cook wrasse sampled from southern Norway in the last four years had completed spawning after week 28, whereas most rock cook caught in that period from western Norway were still spawning. The temperatures at the end of the spawning season varied from 13°C to nearly 20°C in western Norway. These data show that the current regulations of the start of wrasse fishing are adequate for southern Norway and all species concerned but may apply pressure on the rock cook populations of western Norway.

Fishing quotas for wrasse species

Among the proposed conservation measures in 2016 were a decrease in total catches by 15–20% in western Norway and maintenance of the same levels compared to 2015 in southern Norway and north of Stad. The current goal of the imposed total quota regulations is ultimately to reduce the annual catch of wild individuals to 10 million by 2019. Nevertheless, the annual goal was not achieved in 2016 when the total catch was actually the highest then recorded (Table 17.3).

The minimum legal fish size remained the same from 2015 to 2016, which is 120 mm for corkwing wrasse, 140 mm for Ballan wrasse and 110 mm for other wrasse. No change in these measures was expected in 2017 (Fig. 17.9). Nevertheless, although a minimum size restriction may serve its purpose for some wrasse species, it may not do so for other. Halvorsen *et al.* (2016) showed that size selective harvesting may also be sex selective in species with sexual size

Table 17.3 Opening times, total quota and catch (in × 1000 fish) for wrasse fisheries in Norway.

Area	Year	Limits	Opening time	Total quota	Registered catch by 1 October 2016
Southern Norway	2016	Swedish borders to Varnes	11 July	x	4 287
	2015	Swedish borders to West Agder	10 June	3 700	3 741
	2014			4 300	4 351
	2013				2 764
Western Norway	2016	Varnes to 62°N	11 July	10 000	12 633
	2015	Rogaland to Song and Fjordane	1 July	12 600	12 594
	2014			11 300	11 329
	2013				9 077
Mid-North Norway	2016	North of 62°N	25 July	4 000	4 786
	2015	Møre and Romsdal to Norland	16 July	4 100	4 109
	2014			5 600	5 586
	2013				3 647

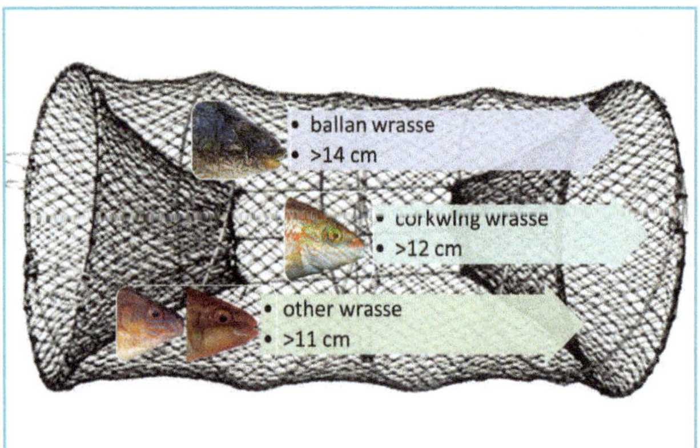

Fig. 17.9 Minimum size for fished wrasse for 2015, 2016 and 2017 (Norwegian Directorate of Fisheries).

dimorphism, such as corkwing wrasse (*Symphodus melops*), with consequences on reproduction success. In the aforementioned study, male corkwing wrasse were larger and selectively removed from the population in a size-selective fishery. According to Halvorsen *et al.* (2016), applying the current minimum legal size limit for commercial fishing in Norway (120 mm) will fail to protect mature nesting males in populations with larger fish. Thus, they recommended the implementation of spatially structured management, according to the local characteristics of the wrasse populations. Moreover, they recommended refinement of the current size regulations using either sex-specific minimum size limits or a slot limit, also protecting the larger fish.

Equipment selectivity and fish welfare standards

The capture of individuals smaller than the minimum legal size must be avoided. Moreover, according to paragraph 16 of the Law on Sea Resources, fishing must occur in the mildest possible way. Thus it is of utmost importance that the equipment used in wrasse fisheries is also adequate and selective for both fish size and species. This will in turn reduce the level of illegal catch and the need for sorting fish by the fishermen, a procedure requiring significant efforts and inducing potentially detrimental stress to the fish. However, this is challenging as different wrasse species have different behaviour, resulting in lower selectivity for some and higher for others for each type of equipment considered. For instance, Jørgensen *et al.* (2016) found that several smaller individual goldsinny wrasse (80–110 mm length) did not escape from 12 mm side or end grids, whereas few corkwing wrasse of this size were found in the same equipment. A 1 mm difference in grid distance resulted in at least 10 mm difference in minimum size caught for both species, whereas flat grids with a specific distance reduced the minimum size fish caught by nearly 5 mm compared to round grids. Most equipment is selective and can be further modified by changing the opening size and the addition of grids. There is a series of directives concerning fishing equipment selectivity, which are summarised below.

Escape opening in traps and pots

The escape opening in traps and pots used for catching wrasse should be at least 12 mm × 70 mm. It is the responsibility of the fishermen to ensure that the escape opening is correct. The pots should have at least three escape openings in the innermost chamber (catch chamber) and at least two escape openings in any of the other chambers. There must be at least three escape openings in traps into the inner chamber and at least two escape openings in the second inner chamber. The escape openings should be positioned in such a way that the wrasse can easily escape. The escape openings must have fixed frames and they can be placed in a series or individually. To prevent unwanted bycatch of particularly lobster and cod, as well as to keep otters from being trapped in and drowned, it is obligatory to install an entrance barrier (usually a simple cross is used), unless the opening is too small to fit through a cylinder with 70 mm diameter. In case one or more otters are caught, the fisheries directorate must be notified.

Participation

Regulations regarding the participation in the wrasse fisheries have existed since 2011. Leisure fishermen with a supply agreement for wrasse can also apply for a permit to fish for wrasse. From 2016 professional fishermen using traps and pots must also have a delivery agreement to fish for wrasse. Leisure fishermen, in accordance with the Marine Resources Act, can use up to 20 pots and cannot sell fish for more than NOK 50,000 each year. The existing regulations regarding the exemption for using pots by professional fishermen; the limit of maximum 100 traps and pots, in southern Norway and daily tending of traps and pots except on Sundays and holidays, were also applied in 2016.

17.4.5 Effects of fisheries on Norwegian wild wrasse populations

Initial fisheries data did not reveal any signs of overfishing of the coastal wrasse populations in the south of Norway (Fig. 17.10) (Gjøsæter, 2002), whereas catch and release studies during the same fishing period provide an estimate of approximately 10% for the fishing mortality rates of goldsinny wrasse (Julliard *et al.* unpublished data, in Espeland *et al.* 2010). The wild-caught wrasse in Norway decreased from 21,000,000 in 2014 to about 20,000,000 in 2015 (11,684,000 to 9,161,000 goldsinny wrasse, a reduction of 22%; 7957 to 9461 corkwing wrasse,

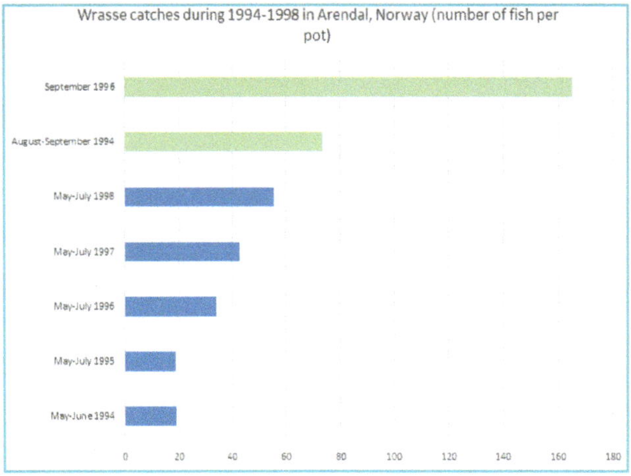

Fig. 17.10 Overview of the commercial wrasse fisheries for the years 1994 to 1998 in the area of Arendal. Source: Gjøsæter 2002, in Espeland *et al.*, 2010.

an increase of 16%; 1,268,000-1,529,000 Ballan wrasse, an increase of 20%; and 352,000 to 292,000 rock cook).

More salmon and trout farmers are using cleaner fish, still mostly wild, to fight lice infestations, with good results. For this practice to be sustainable there is need for the production of more well-vaccinated, robustly farmed cleaner fish. The production of farmed cleaner fish in 2011, was 0.4 million. In 2015, four wrasse producers (two hatcheries and two nurseries) and 30 lumpfish farmers provided the salmon cage farmers with a total of 12.5 million individual cleaner fish (mostly lumpfish, and 1 million Ballan wrasse), whereas the prognosis for 2016 was 22 million farmed fish (Waatevik, 2016). The prognosis for production of farmed lumpfish in Norway in 2017 was 30 million fish, and more than 3 million Ballan wrasse. Moreover, as other non-medicinal methods against salmon lice infestations are developed, there is need of a realistic estimation of the cost efficiency of implementing the different alternative technologies (Ervik, 2016).

17.4.6 Relevant Norwegian legislation on fisheries of species that will be kept alive

Keeping cleaner fish alive by fishermen before transferring them to the hatcheries or salmon net pens is regulated by the general regulations on the exercise of fishing in the sea (https://lovdata.no/dokument/SF/forskrift/2004-12-22-1878), paragraphs 87–97 in Chapter XVIII, titled: Capture of fish which will be kept alive, recovery and intermediate storage, as summarised below.

In general all the operations involved in fishing, keeping alive during recovery and transferring of cleaner fish are to be conducted responsibly in terms of fish welfare. All possible measures that will reduce the risk and extent of suffering, injuries and stress, must be taken. Those are in connection to fishing gear, periods of exposure to the air, crowing, handling, fishing depth (section 88), feeding and health controls. There is set a maximum period of three days from capture when the largest part of the catch should have performed a complete recovery, i.e. can swim freely. Cleaner fish fishery requires that all crew participating in the operations have received appropriate training on the current fishing method and gained knowledge about how to handle the fish properly from a fish welfare point of view (section 86).

When fishing with seines (section 87), there are requirements regarding the adjustment of the respective gear used in order to safeguard fish welfare, without reducing its selection properties. Specifically, knotless line or other suitable material, flexible side panels, and a bag filled with water are to be used when taking fish out of the water, or other equally gentle equipment is to be used. The number of fish taken up in each cast must be considered and kept at a practical minimum.

Following capture and before transfer of the fish to recovery or intermediate storage containments, the fish must be sorted (section 89). Severely injured or moribund fish that are not expected to make it through recovery and transport must be sorted out and killed by bleeding.

Besides the fishing gear, the use of pumps must also consider fish welfare, in terms of maintaining safe pressure levels, (section 90). It is encouraged to use pumps/pipes according to the siphon principle with high volume/low pressure. The use of hooks will induce injuries to the fish and thus sorting is necessary (section 89).

Hand nets used to transfer fish between containments must be designed in order to minimise fish injuries (section 91). The number of fish taken on each haul should be considered in terms of risk for reducing fish injury incidences and mortality.

Transport of fish must be as effective as possible and fish densities in transport containments must be taken into consideration (section 92). The recovery containments or cages must not be placed in the vicinity of aquaculture facilities (2.5 km at the closest) (section 93).

Prior to setting cleaner fish out in a salmon cage they must be allowed to recover in a recovery containment. The regulation provides a list of requirements regarding the specifications and the use of recovery and intermediate storage containment (sections 94 and section 95, respectively) listed in Table 17.4 below.

Table 17.4 Requirements on the specifications and the use of recovery and intermediate storage containment for fish caught and kept alive before transfer or slaughter.

Containment purpose	Recovery (§ 94)	Intermediate storage (§ 95)
Bottom	Flat and tight and of such size that the fish immediately after introduction have the opportunity to lay on it.	– Good Water/Light
Water/light	Good quality/suitable.	Good quality/suitable.
Inspections	Daily and at least one must be performed by a veterinary or fish health specialist.	Regular at least one performed by a veterinary or fish health specialist.
Design	Appropriate for the conduction of efficient inspections.	Appropriate to safeguard the behavioral and physiological needs of the fish.
Sorting	Damaged or dead fish should be picked out and killed if necessary by bleeding.	Damaged or dead fish should be picked out and killed if necessary by bleeding.
Duration	Until it is certain that fish can swim freely and respond to external stimuli.	Up to 12 weeks.
Feeding	None.	Daily max after four weeks from recovery, and after sorting of fish to equal sizes. At start feeding the Norwegian Food Safety Authority and the County Governor must by notified.

According to paragraph 96, in case of increased mortality or suspected contagious or non-contagious disease during recovery and intermediate storage, a veterinary or fish health specialist must be notified immediately and a health check must be carried out to clarify the cause. All respective incidences must be recorded.

17.5 The Irish wrasse fisheries

17.5.1 Catch composition

The five common northern European wrasse species have been reported in Irish waters with a predominance of goldsinny, corkwing and rock cook. Small-grade wrasse are predominant in sheltered locations (up to 90–95% of the catch), while larger specimens are more plentiful in more exposed locations (up to 25% of the catch). The reported Catch Per Unit Effort (CPUE: number of fish per pot per hour) varies between 0.11 and 0.256 for goldsinny and 0.04 and 0.01 for corkwing (Sayer *et al.*, 1993; Sayer *et al.* 1996; Varian *et al.* 1996). Wild wrasse are sourced from several fishing zones (Clew Bay, west; Inver Bay, Mulroy Bay and Lough Swilly, north; Bantry Bay and Kenmare Bay, south-west), chosen for their proximity to the salmon farms that are being stocked. This minimises the movement of fish between bays, which could carry disease risks. All five species of wrasse ubiquitous to Irish inshore waters are utilised with a minimum length of 12 cm set for all specimens. Fishing commences in mid-May using specialised wrasse pots.

17.5.2 Wrasse stock management in Ireland

All five species of wrasse ubiquitous to Irish inshore waters are utilised with a minimum length of 12 cm set for all specimens. Fishing commences in mid-May using specialised wrasse pots and fisherman are equipped with tanks and pumps for post-capture transport and holding pens for maintaining fish prior to collection.

17.5.3 Transport

Transport methods vary between sites and species, using smolt tanks on boats (wild wrasse caught close to the farm) or articulated lorries (longer distances, all lumpsucker) equipped with 2.3 m^3 smolt tanks. All tanks are supplied with supplemental oxygen, which is regulated and closely monitored for the duration of the trip. The maximum journey time for fish in Ireland is approximately eight hours. Stocking densities during transport are varied, depending on species, and range from 13 to 60 kg m^{-3} for lumpfish.

17.5.4 Effects of fisheries on Irish wild wrasse populations

The sustainability of wild capture for a variety of species for use as cleaner fish in Ireland, including wrasse and lumpsucker, is being investigated by the Marine Institute. At present the effects of fisheries on wild populations have not been determined. One intensive study carried out in

1991 and 1992 in Lettercallow Bay, Connemara, showed a decline in CPUE and in the size of fish captured (Varian *et al.*, 1996). Although there was agreement with Darwall *et al.* (1992) that early maturity and high fecundity of goldsinny and corkwing wrasse indicated that the stocks would be resilient to fishing, there was also a recommendation for a management strategy.

17.6 The wrasse fisheries in Scotland

A wrasse fishery in Scotland was established in the 1990s solely on the west coast when the fishery was much smaller than today, with an estimated annual catch of 150,000 based on a questionnaire to salmon farming companies regarding wrasse numbers stocked (Treasurer, 1993). Only a few fishermen were involved, principally using prawn creels and fyke nets, and both were effective techniques. There were few wrasse in Shetland and fish had to be flown to the islands (Young, 1996). At that time Ballan wrasse were not stocked in salmon pens in Scotland because it was considered that Ballan could be aggressive to salmon and removed scales and picked eyes (Bjordal, 1991). Also, cuckoo wrasse were regarded as rare and return of this species to the sea was encouraged. A range of different fishing techniques was used and a bench-marking exercise was undertaken to compare CPUE between gears (Treasurer, 1996). Today, the fishery is more widespread and fishermen will travel over a wide area during the year. However, fish in Scotland are captured as far as possible near to the fish farm in which they will be stocked, although fish are also shipped from the south coast of England by specific companies. Fish are not captured in any numbers before the end of April and wrasse only become active and move into traps at temperatures over 10°C. The fishery typically finishes around October as daylight and temperatures decline. The wrasse fishery is in shallow water <30 m, close inshore and on rocky ground, and frequently in the vicinity of kelp beds. Today around 65 boats are involved along the west coast of Scotland and the Western Isles. The gear used is almost exclusively specialist wrasse traps and prawn creels (shrimp pots). Only fish over 110 mm are retained and fish over 23 cm are also returned to the sea, although small numbers of larger Ballan wrasse may be retained as broodstock for hatcheries. The prices are generally between £1.50 and £7 per fish, although some companies pay a premium for Ballan wrasse as it is considered a more effective cleaner of lice and also robust over winter.

17.6.1 Catch composition

The species and length composition of 34,000 wrasse captured with traps and prawn creels on the west coast of Scotland was examined. The species composition varied between area, depth and the type of substrate, but the smaller wrasse species, rock cook and goldsinny, dominated catches, with rock cook more abundant in sheltered long fjord-shape sea lochs, and goldsinny more common in open water areas (Fig. 17.11). Fish less than 110 mm were returned to the sea and the bulk of the catch transferred to farms was from 110 to 150 mm.

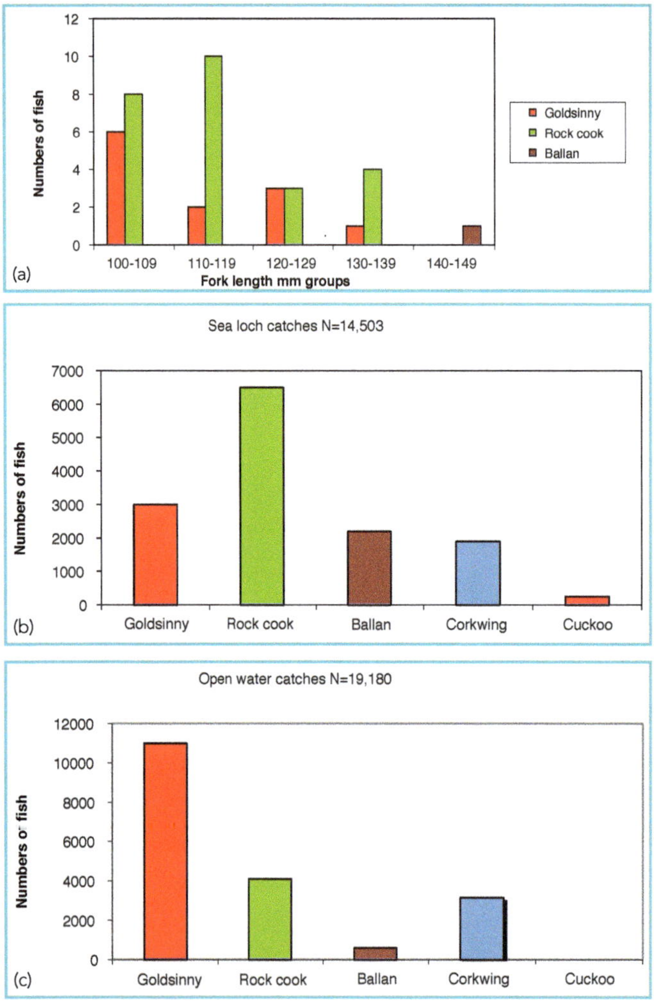

Fig. 17.11 The (a) length frequency distribution and (b, c) species composition of one wrasse fishery in west Scotland.

17.6.2 The boats

The boats used for fishing cleaner fish in Scotland are generally small work boats (Fig. 17.12) or inshore prawn creel boats. While most are owned by independent fishers, several of the salmon farming companies also operate their own vessels on a seasonal basis.

17.6.3 Wrasse stock management in Scotland

Fisheries opening times and wrasse spawning season

The wrasse fishery in Scotland was not regulated in terms of the opening of the spawning season and commenced in the spring when wrasse become more active, generally at temperatures >10°C and from April to October. The spawning of Ballan wrasse in Scotland is from late April to June, and from mid-June to early August in the smaller wrasse species such as goldsinny and rock cook wrasse. There was no closure in the fishery during the spawning season. However, following

Fig. 17.12 Inshore fishing for wrasse by Wester Ross Fisheries.

discussions with Marine Scotland and the salmon farmers it was agreed that the opening of the fishing season from 2018 would commence after the peak spawning season.

Fishing quotas for wrasse species

There are no quotas for the wrasse fishery in Scotland and no specific regulations related to the size of the fish that are retained. The boats fishing for wrasse have to be licensed with Marine Scotland and the fishermen have to submit catch returns to the local fishery office showing the weight of wrasse captured by area. Catch records were only required from 2014, and show only weight and not numbers of fish retained, with an average weight of 60 g used to calculate numbers captured. The salmon farming companies have also to maintain records of numbers of wild wrasse that have been purchased. Although there is no statutory minimum length of fish that can be retained, Marine Scotland and the industry have agreed a minimum length of 110 mm from 2018. The salmon producers also quote minimum (>110 mm) and maximum lengths (generally not >230 mm) of fish that will be purchased. The RSPCA Freedom Foods quality scheme also stipulates a minimum length of wrasse that can be stocked on farms. The fishery is being closely monitored by the quality scheme, and dialogue is open with the industry on any improvements to practice and welfare.

Equipment selectivity and fish welfare standards

The size selectivity of prawn creels and fyke nets has been examined in detail (Treasurer, 1994a, b) and goldsinny wrasse from 80–144 mm were retained in pots (mesh size 12 mm) and only fish >110 mm retained by the fyke nets (mesh size 11 mm). Rock cook of 95–149 mm length were retained using 18 mm mesh in pots. However, the fish <110 mm were returned to the sea and the % by-catch that this comprised was calculated. Marine Scotland is concerned about by-catch and fishermen have been reminded from 2018 to ensure that non target species and undersize wrasse are returned safely to the sea. Salmon producers have insisted on these requirements, and have agreed to collect data for Marine Scotland that can be used for fishery management purposes, and management measures are being constantly reviewed.

Effects of fisheries on Scottish wild wrasse populations

The total catch of wrasse of all species from 2000 to 2016 in all UK waters includes the catches from Scotland (Fig. 17.13; Riley *et al.*, 2017). However, catch records for wrasse were only instigated by Marine Scotland in 2014 and these are quoted as weight. A total of 29 tonnes of wrasse

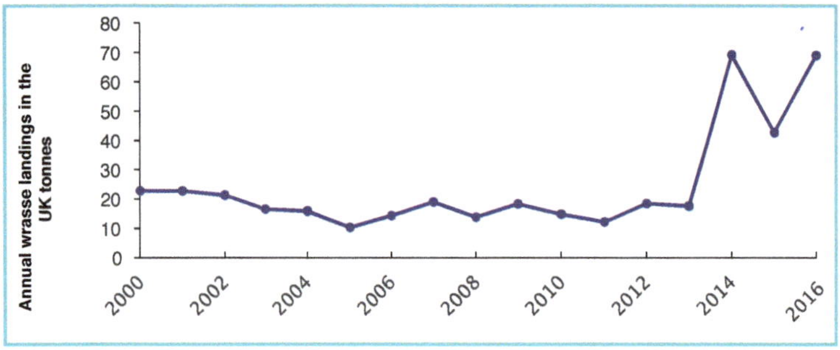

Fig. 17.13 Annual UK catches of wrasse combined for all ICES divisions 2000–2016, that is Scotland, Wales and England combined. The figures before 2014 are underestimated as no catch data were required to be reported in Scotland prior to that year.

of all species was reported in 2015 and 48 t in 2016 with the figures for 2017 similar (scottish.gov.uk) and, at an average weight of 60 g, that gives a catch of 0.5 and 0.8 million fish respectively. However, there is a discrepancy between this and the numbers of fish stocked, and the average weight of catch is probably closer to 50 g as small wrasse dominate the catch, so the annual catch in Scotland is estimated to have increased in 2016 to around 1 million wrasse. Prickett (2016) estimated that 1 to 2 million wrasse from fisheries were stocked in Scotland in 2016 and Riley et al. (2017) suggested that more than 1 million wrasse were captured on the south coast of England in that year. It is likely that the landings of the wrasse fishery in Scotland are equivalent to those in England, so the upper figure quoted by Prickett (2016) is a more realistic and conservative total. However, Marine Scotland is working on reconciling the numbers of fish reported and the purchases made by the salmon farming companies in the future to ensure improved reporting.

Sayer *et al.* (1996) carried out an experimental fishery in west Scotland on wrasse over three years. The CPUE was assessed and also a comparison made of size distribution and sex ratio between years. They found a decline in CPUE but stressed that more data were required and that fishery management arrangements would be required for each area due to biological differences in wrasse stocks. A minimum size limit of 100 mm was suggested by Darwall *et al.* (1992) for stocking in pens and this was shown to enable a proportion of both mature males and females to be returned to the sea. However, the scale of the present fishery and the numbers of boats participating is much higher than in the 1990s. It is likely that fishermen move to other areas if catches in particular bays and sea lochs are less abundant on time. There has been no recent systematic survey of wrasse populations on the west coast of Scotland. As salmon farmers wish to make the transition to using farmed fish for stocking, it is generally regarded that the fishery will remain in the current form for a limited time and is transitory. As there is no immediate prospect of farming supplying the wrasse requirements for salmon farms in Scotland, the farmers wish to retain the option of wild fish supply. Freedom Food conditions stringently define fish length retention limits and also welfare in capture, storage and transport (M. Zietz, Cleaner Fish Summit Glasgow, 9 May 2017, personal communication).

17.7 The English south coast wrasse fisheries

Wrasse species present in the south-west coast of England inhabit inshore waters among rock and algal cover. Activity is mainly limited to seawater temperatures >7°C and spawning is in spring and early summer, from May to September (Darwall *et al.*, 1992). The species captured

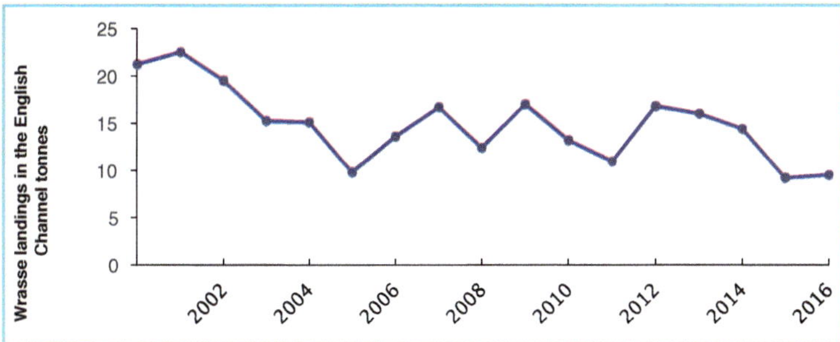

Fig.17.14 Reported wrasse landings in the English Channel, ICES divisions 7d and 7e combined, for the period 2000 to 2016.

are corkwing, goldsinny, rock cook, Ballan, cuckoo (*Labrus mixtus*), and also Baillon's wrasse (*Symphodus bailloni*) in smaller numbers (Callaghan and Pering, 2016). This Native Marine Centre (NMC) report indicated that many of the wrasse were classified as mixed species but, of the 36,900 fish that were identified, 41% were corkwing, 38% goldsinny, 13% rock cook, 7% cuckoo and only 1.3% Ballan. In contrast, Riley *et al.* (2017) suggested that Ballan wrasse was the primary species. Landings of wrasse in the English Channel have varied from 9 to 22 tonnes in the period 2000 to 2016 (Fig. 17.14), with an annual average figure of 14.8 tonnes (MMO data in Riley *et al.*, 2017 combined for ICES divisions 7d and 7e). Wrasse landed on the south coast of England were in ICES divisions 7.d–e and landed in 70 ports, with the main landings at Brixham (16% of Channel landings), Lyme Regis (~12%) and Eastbourne (~12%) (Riley *et al.*, 2017). Unlike the Scottish fishery, which is dependent on passive gears, the main gears used were gillnets (40%) and otter trawl (30%), and fish traps (pots) only accounted for 4% of recorded landings. It is estimated that more than 1 million wrasse were captured in the English Channel in 2015 (Riley *et al.*, 2017). Initial mortality rates recorded were high, but as fishermen were not required to document their mortalities, the exact figure is unknown (Callaghan and Pering, 2016). One teething problem encountered was that transport lorries were not sent to salmon farms in Scotland on regular days, but rather on a more ad hoc basis.

Fishermen were advised to catch only fish sized between 120–220 mm for the NMC, and to keep at stocking densities of no more than 200 per 210 l storage tank (Callaghan and Pering, 2016). The most abundant wrasse species landed throughout the season off Weymouth was corkwing wrasse. However, many of the goldsinny and rock cook were returned to the sea. Some were only counted by number due to lack of time and therefore a large number of fish was classified as 'assorted'. There were highly varying mortalities rates in captured wrasse between different months and batches. These were attributed to several factors, including protocol refinements, and prolonged high sea temperatures (Callaghan and Pering, 2016). Water temperatures during storage were >15°C from June to October and this could be stressful for wrasse. Advances should be made in the storage and handling of the wrasse from this fishery to ensure maximum survival and good welfare.

17.8 Status of the European wrasse fisheries and regulation of the fishery

The wrasse fisheries are unusual in that they rely almost entirely on trap methods for capture, at least in Norway, Ireland and Scotland. The fisheries in Scotland, England, Ireland and Norway are broadly similar in species captured and substrate location and timing of fisheries, although the fishery in Norway and associated nearby fishery in Swedish waters is on a much larger scale and also involves the transport of fish over large distances. The wrasse from the fishery in the English Channel are also transported over a large geographical area to northern Scotland and the fishing season is throughout most of the year. There is concern over the sustainability of the fishery and whether there is any significant effect on the abundance and population structure of stocks (Anon, 2016; Callaghan and Pering, 2016; Davies, 2016; Skiftesvik *et al.*, 2016; Halvorsen *et al.*, 2016, 2017; Riley *et al.*, 2017; Gonzalez and de Boer, 2017).

The Norwegian fishery has been in the region of 20 million fish for the last three years and the quota has been fixed at that level and split by three geographical areas. The fishery in England has been estimated to be over the 1 million fish mark for two years (Riley *et al.*, 2017). It therefore appears that the levels of fish numbers being captured remains stable, although more information on CPUE would be helpful in the context of fishing effort. It would appear that the release of larger mature Ballan and smaller species over 110 mm may enable some of the population to spawn and the wrasse stocks may be robust. However, in response to fishing there may be impacts on the wrasse stocks such as a reduced proportion of larger fish, and there may, as in many fished stocks, be an increase in fecundity and also in growth rate, and earlier maturation.

There have been calls for regulation of the cleaner fish fishery in Norway by the Ministry of Fisheries where a TAC has been imposed, with maximum catch limits being assigned for three geographical areas, and also by CEFAS and the Devon and Severn Inshore Fisheries and Conservation Authority (Anon, 2016; Riley *et al.*, 2017). However, it should be noted that in the latter case the Fisheries Inspectorate for England and Wales is separate from Scotland, and Marine Scotland has its own fisheries advisors at the Marine Scotland Laboratory in Aberdeen. Certainly, in the UK, there has been no official call from government to regulate the fishery, but precautionary management measures are agreed with industry.

The first credential in wrasse fisheries management is that the catch data have to be accurately reported. The veracity and completeness of the UK data can be questioned (Riley *et al.*, 2017). Also, no figures have been quoted for the bycatch damaged and discarded overboard. This appears strange as special buoyant traps were developed in Norway that were above the seabed and prevented predators such as crabs entering and damaging cleaner fish. The numbers of wrasse captured also do not tally with the numbers that were stocked. For example, it has been reported that in the early part of the English fishery that holding and transport mortality could be high (Callaghan and Pering, 2016), so the numbers delivered to farms could be much less than reported as catch. Also, some of the gears used in the English fishery such as trawl and gill-nets (Riley *et al.*, 2017) may entail mortality, perhaps within many days of capture. In contrast, the Norwegian, Irish and Scottish wrasse fisheries use mainly fish traps and pots (shrimp pots) and, in Norway, fyke nets are also increasingly common. Also, in Scotland, dedicated highly professional fish transporters are regularly used and carriage from capture to stocking on farms can be within two to three days and in conditions similar to those used for salmon. Consequently, mortality during capture and transport in locations such as Scotland, and in Weymouth can be <5% (Callaghan and Pering, 2016). While there is confidence in the inspectorate, the catches of wrasse in the UK may have been under-reported in previous years, as CEFAS indicated when data from 2000 to 2016 were presented (Riley *et al.*, 2017). Marine Scotland did not require a reporting system until 2014 and figures prior to that date were not reported or under-reported.

Also, Marine Scotland requires that weight (kg) of fish should be reported to fisheries offices by area, so numbers of fish caught are not available. Reconciliation is required between fisheries reporting and the numbers of wrasse purchased by farmers to come to conclusions about the fishery. It should also be acknowledged that some of the salmon farmers in Norway, Ireland and Scotland have their own dedicated wrasse fishing vessels and staff, although these too are legally obliged to report catches.

It is difficult to manage a stock until data on stock size and structure and an estimation of TAC can be made. While there are calls for fisheries management measures (Anon, 2016; Davies, 2016; Riley *et al.*, 2017), there are no data on wrasse abundance in any area of the UK nor in Norway and this is difficult in such extended coastlines. As Sayer *et al.* (1996) indicated, the growth, recruitment rate, and Z and maturation varies between wrasse in different localities, and local or area management plans are required.

17.8.1 How can the status of wrasse stocks be assessed?

The only direct method of assessing the distribution and abundance of wrasse is by diver or camera survey along marked and designated transects. This was carried out in three sea inlets in Ireland (Costello *et al.*, 1995) and multiple surveys were required over the same transects as the wrasse retreated into rock crevices when not active, and surveys therefore represented a measure of wrasse activity. However, the only documented studies in the UK have been by Sayer *et al.* (1996), when transects in two geographical areas were swam and wrasse numbers counted, that is a census. These found the highest densities of 1.1 to 3.9 wrasse m^{-2} in the summer months. This approach may be time-consuming and costly over large areas and, as an alternative, CPUE data for each species, gear and locality may have to be scrutinised. CPUE depends on season and water temperature and gear (Table 17.1), but a range of CPUEs for unfished ground or relatively unexploited shoreline have been provided for Scotland (Sayer *et al.*, 1996; Treasurer, 1996) and Ireland (Varian *et al.*, 1996). This may give the best opportunity to identify any impact on the wrasse stocks and provide a baseline for what would be expected in certain areas. CPUE needs to be standardised by gear type, species and also soak time. Some studies referred to catch per hour, whereas others have used catch per trap per day (Table 17.1). The figures have been converted therein to catch per trap per hour for comparison but catch has been shown to decline on soak time (Treasurer, 2000). The third method is to compare CPUE of gears in fished areas against MPAs (Halvorsen *et al.*, 2016). This appeared to show the effects of fishing on CPUE in corkwing, but is was less clear cut in goldsinny wrasse, and decline was not shown in all areas. By their nature MPAs may have higher natural productivity, diversity and species composition and structure. Also, the reasons for designating specific areas as MPAs need to be explored and they may not be representative of general coastal communities as they may be protected for biodiversity and special habitat status. In comparing fished areas with MPAs the topography and rock structure of the seabeds and scree and rock structure and macroalgal cover need to be described and quantified, as was done by Halvorsen *et al.* (2016).

A fourth, and often quoted method of assessing fishing effects in wrasse is in examining the size, age structure and sex ratio, and of looking at the age of sexual maturity and other responses of fishing such as changes in (density-dependent) growth rate. The proportion of larger goldsinny captured in Lettercallow Bay declined in the second year of the fishery (Varian *et al.*, 1996).

Sayer *et al.* (1996) examined this in an experimental fishery in west Scotland over three years, but found no significant reduction in fish length or difference in the male:female ratio. While, in Norway Halvorsen *et al.* (2017) showed a reduction in mean age and length of fish and in sex ratio between fished areas and MPAs, although the extent of the benefits of MPAs in that study may have been overestimated (Gonzalez and de Boer, 2017). However, the biological characteristics of wrasse populations vary between geographical areas, and the population dynamics have to be determined for each area (Sayer *et al.*, 1996).

17.8.2 Elements to incorporate in a fisheries management plan

Many details and suggestions have been made for management of wrasse fisheries (Darwall *et al.*, 1992; Sayer *et al.*, 1996; Halvorsen *et al.*, 2016). Therefore, the general approach and questions arising are considered here. Once the fishery has been measured, and the effects of fishing have been identified and the natural biomass of inshore wrasse populations have been determined, an assessment can be made of stock size and TAC. These are some of the points that can be used in a fisheries management programme.

The TAC has to be determined based on scientific evidence. Should this be 15, 20, 30, 40% of stock size based on fish within set size distributions? This will be informed by the recruitment rate of juveniles and adults coming in to the fishery, and in the case of long-lived species >30 years, such as Ballan wrasse, fish may take several years to recruit to reproduction and especially males. Exploitation can lead to increased growth rate, and reduced age at maturity (Yoneda and Wright, 2004), and fecundity can be influenced by the size of the fish with removal of the larger females having a high impact on population fecundity. Reproduction, and larval and juvenile survival in the sea need to be determined. An agreement on catch could lead to assignment of a quota for areas and/or by boats. However, a TAC on a country or a geographical area would be the obvious way to proceed. This could also translate to the number of traps that can be used per fisherman or boat.

A close season has been advised in Norway (Norwegian Directorate of Fisheries) and England (Riley *et al.*, 2017) during the spawning period. This may be a suitable break when there is a long fishing season, which may be over 12 months in England (Fig. 17.15), but the season is more restricted in Scotland, mainly from April to September. A close season that would encompass Ballan wrasse from April to June and smaller wrasse species from June to August would severely affect the wrasse fishery in Scotland, and especially when the fish are required to be stocked in cages for maximum benefit in the first year of salmon production.

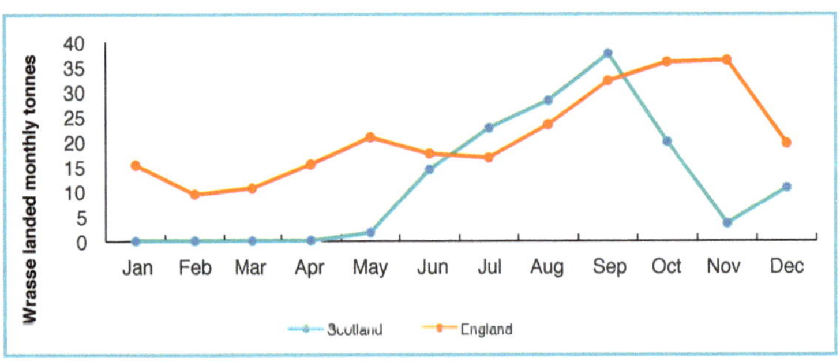

Fig. 17.15 Wrasse catches throughout the year with a comparison of the Scottish and English fisheries. Data from Riley *et al.* (2017).

The type of fishing gear that can qualify for the fishery could also be a major regulatory factor. For example, both trawl and gillnets are used in the English fishery and this can capture other species and also have stressful effects on wrasse. Traps may be a more sensitive way of capturing cleaner fish.

Once the size structure of the wrasse populations is known, a limit can be applied to the lower and upper end of the size spectrum and this is based on length rather than weight, or rather on the length of fish that can be retained by salmon net mesh of certain sizes, for example 15 to 25 mm mesh. In Norway, England and Scotland a minimum size limit of 110 mm is specified by salmon farmers and also by accreditation schemes such as Freedom Foods in the UK. This length is sufficient in the case of goldsinny to enable first spawning of a proportion of the males and a larger proportion of the slower-growing females. Many companies also specify an upper size limit of 230 mm as these fish may be regarded as too large for stocking with salmon. There is also a case for restricting the retention of Ballan wrasse >250 mm, so the males and the spawning females are not impacted by the fishery. Unlike lumpfish, there is less need to replace Ballan wrasse broodstock in hatcheries as the fish are long-lived and hatchery survival is high.

The focus has to be on minimising the stress and handling of wrasse in capture and fishing mortality, reducing discards, initiating the health screening of wild-caught fish, and reducing storage and transport mortalities. So, the overview is to record the fishery, assess the stock structure, estimate the wrasse biomass and densities, determine fishing effects and trends in CPUE, and, where necessary, when data indicate, agree and apply any fishery conservation measures. Maintenance of size retention limits and accurate catch reporting are essential to the good management of the fisheries. Nevertheless, it is concluded that the wrasse fishery in northern Europe has given additional employment and income in northern areas. It has met much of the substantial demand for wrasse from salmon farmers and helped control sea lice numbers in a high proportion of salmon farms. Improvements to wrasse storage and transport, especially over secondary journeys (Chapter 16), need to be carried out to increase survival of captured wrasse, and to improve welfare and condition of the fish and reduce the opportunity for incidence of disease.

References

Anon (2016) Summary of evidence and reasoning for the introduction of management measures to the live wrasse pot fishery. Devon and Severn Inshore Fisheries and Conservation Authority. Unpublished report.

Bertelsen, B. (1994) Bestandsutvikling og bestandsstruktur hos rognkjeks og rognkall (*Cyclopterus lumpus* L.) i Sifjorden i Troms, og maskeseleksjon og fangsteffektivitet ved garn-og rusefiske etter arten. Fiskerikandidatoppgave. The Norwegian College of Fishery Science, University of Tromsø, Tromsø. 89 pp.

Bjørdal, Å. (1998) *Cleaning symbiosis between wrasse (Labridae) and lice infested salmon (Salmo salar) in mariculture.* International Council of the Exploration of the Sea, Mariculture Committee 188, 1–8.

Bjordal, Å. (1991) Wrasse as cleaner fish for farmed salmon. *Progress in Underwater Science* 16, 17–28.

Callaghan, A. and Pering, D. (2016) NMC wrasse cleanerfish project 2016. Native Marine Centre, Weymouth, unpublished report. 23 pp.

Costello, M.J., Darwall, W.R. and Lysaght, S. (1995) Activity patterns of North European wrasse (Pisces, Labridae) species and precision of diver techniques. In: Eleftheriou, A., Ansell, A.D. and Smith,

C. (eds) *Biology and Ecology of Shallow Coastal Waters.* Proceedings of the 28th European Marine Biology Symposium, University of Crete. Olsen and Olsen.

Darwall, W.R.T., Costello, M.J., Donnelly, R. and Lysaght, S. (1992) Implications of life-history strategies for a new wrasse fishery. *Journal of Fish Biology* 41, 111–123.

Davenport, J. (1985) *Synopsis of biological data on the Lumpfish Cyclopterus lumpus (Linnaeus, 1958).* FAO Fisheries Synopsis no. 147, 31.

Davies, S. (2016) A review of wrasse ecology and fisheries interactions. Devon and Severn Inshore Fisheries and Conservation Authority. Unpublished report.

Dipper, F.A. (1976) Reproductive biology of the Manx Labridae (Picses). PhD thesis, University of Liverpool.

Donnelly, R. (1992) *The biology and use of wrasse (Labridae) as a biological control of external parasites in salmonid cultures.* MSc Thesis, University of Dublin. Unpublished.

Eriksen, E., Durif, C.M.F. and Prozorkevich, D. (2014) Lumpfish (*Cyclopterus lumpus*) in the Barents Sea: development of biomass and abundance indices, and spatial distribution. *ICES Journal of Marine Science* 71, 2398–2402.

Ervik, O-A. (2016) Motivasjon for bruk av rensefisk. Rensefiskkonferansen, Gardermoen, 8-9 Februar 2016. fhf.no/media/129925/2_ervik_salmar_motivasjon_for_bruk_av_rensefisk.pdf

Espeland, S.H., Nedreaas, K., Mortensen, S., Skiftesvik, A.B., Agnalt, A-L., Durif, C., Harkestad, L.S., Karlsbakk, E., Knutsen, H., Thangstad, T., Jørstad, K., Bjordal, Å. and Gjøsæter, J. (2010) Wrasse knowledge status. Challenges of an increasing fisheries sector. 38 pp. (in Norwegian).

Gjøsaeter, J. (2002) Fishery for Goldsinny wrasse (*Ctenolabrus rupestris*) (Labridae) with pots along the Norwegian Skagerrak coast. *Sarsia* 87(1), 83–90.

Gonzalez, E.B. and de Boer, F. (2017) The development of the Norwegian wrasse fishery and the use of wrasses as cleaner fish in the salmon aquaculture industry. *Fisheries Science* 83, 661–670.

Goulet, D, Green, J.M. and Shears, T.H. (1986) Courtship, spawning, and parental care behaviour of the lumpfish, *Cyclopterus lumpus* L., in Newfoundland. *Canadian Journal of Zoology* 64, 1320–1325.

Halvorsen, K.T., Sørdalen, T.K., Durif, C., Knutsen, H., Olsen, E.M., Skiftesvik, A.B., Rustand, T.E., Bjelland, R.E. and Vøllestad, L.A. (2016) Male-biased sexual size dimorphism in the nest building corkwing wrasse (*Symphodus melops*): implications for a size regulated fishery. *ICES Journal of Marine Science* 73 (10), 2586–2594.

Halvorsen, K.T., Larsen, T., Sørdalen, T.K., Vøllestad, L.A., Knutsen, H. and Olsen, E.M. (2017) Impact of harvesting cleaner fish for salmonid aquaculture assessed from replicated coastal marine protected areas. *Marine Biology Research* 13, 359–369.

Jimenez, D.F., Heuch, P.A., Revie, C.W. and Gettinby, G. (2012) Confidence in assessing the effectiveness of bath treatments for the control of sea lice on Norwegian salmon farms. *Aquaculture* 344–349, 58–65.

Jørgensen, T., Bjelland, R., Halvorsen, K. and Skiftesvik, A.B. (2016). Seleksjon i leppefiskredskap. Fiskeridirektoratet. Evalueringsmøte leppefisk, December 2016. (*In Norwegian*) www.fiskeridir.no/Yrkesfiske/Tema/Leppefisk/Evalueringsmoete-leppefisk-2016

Maroni, K. and Andersen, P. (1996) Distribution and abundance of wrasse in an area of northern Norway. In: Sayer, M.D.J., Treasurer, J.W. and Costello, M.J. (eds) *Wrasse: Biology and Use in Aquaculture*, Oxford: Blackwell Scientific. pp. 70–73.

Moe, M. and Ildhusøy, E. (2006) Rømming av leppefisk i forhold til maskevidde på not. Studentoppgave ved Høyskolen i Ålesund. (Wrasse escaping in relation to the net mesh size. Undergraduate study, Ålesund Academy).

Powell, A., Treasurer, J.W., Pooley, C.L., Keay, A.J., Lloyd, R., Imsland, A. K. and Garcia de Leaniz, C. (2017) Use of lumpfish for sea-lice control in salmon farming: challenges and opportunities. *Reviews in Aquaculture.* Available at: http://dx.doi.org/10.1111/raq.12194

Ratcliff, J.J., Doherty, B. and Norris, A. (2006) *Clew Bay Wrasse Populations*. Unpublished Project Report.

Riley, A., Jeffery, K., Cochrane-Dyet, T., White, P. and Ellis, J. (2017) Northern European Wrasse-Summary of commercial use, fisheries and implications for management. CEFAS, Lowestoft. 18 pp.

Rusyaev, C.M. (2000) Norwegian and Russian experience of the Lumpfish fishery. Fishery Journal, Murmansk, pp. 24–28. (in Russian).

Sayer, M.D.J., Gibson, R.N. and Atkinson, R.J.A. (1993) Distribution and density of populations of goldsinny wrasse (*Ctenolabrus rupestri*) on the west coast of Scotland. *Journal of Fish Biology* 43 (Supplement A), 157–167.

Sayer, M.D.J., Gibson, R.N. and Atkinson, R.J.A. (1995) Growth, diet and condition of goldsinny on the west coast of Scotland. *Journal of Fish Biology* 46, 317–340.

Sayer, M.D.J., Gibson, R.N. and Atkinson, R.J.A. (1996a) The biology of inshore goldsinny populations: can they sustain commercial exploitation? In: Sayer, M.D.J., Treasurer, J.W. and Costello, M.J. (eds.) *Wrasse: Biology and use in Aquaculture*. Oxford: Fishing News Books, pp. 91–99.

Sayer, M.D.J., Gibson, R.N. and Atkinson, R.J.A. (1996b) Growth, diet and condition of corkwing wrasse and rockcook on the west coast of Scotland. *Journal of Fish Biology* 49, 76–94.

Sayer, M.D.J. and Treasurer, J.W. (1996) North European wrasse: identification, distribution and habitat. In: Sayer, M.D.J., Treasurer, J.W. and Costello, M.J. (eds.) *Wrasse: Biology and Use in Aquaculture*. Fishing News Books, Oxford, pp. 3–12.

Skiftesvik, A.B., Bjelland, R., Halvorsen, K., *et al.* (2016) Gytesesong. Fiskeridirektoratet. Evalueringsmøte leppefisk, December 2016. (in Norwegian) www.fiskeridir.no/Yrkesfiske/Tema/Leppefisk/Evaluer ingsmoete-leppefisk-2016

Sunnanå, K. (2007) Rognkjeks. Kyst og Havbruk pp. 86–87. (in Norwegian)

Torstensen, E. (1988) Rognkjeks, en kommersiell ressurs på Skagerrakkysten? Fisken og Havet 15, 6.

Thangstad, T. (1999) *Spatial and temporal distribution of three wrasse species (Pisces: Labridae) in Masfjord, western Norway: habitat association and effects of environmental variables*. Thesis for the cand. Scient. Degree in fisheries biology, University of Bergen. Unpublished.

Treasurer, J.W. (1994a) The distribution, age and growth of wrasse (Labridae) in inshore waters of west Scotland. *Journal of Fish Biology* 44, 905–918.

Treasurer, J.W. (1994b) Distribution and species and length composition of wrasse (Labridae) in inshore waters of west Scotland. *Glasgow Naturalist* 22, 409–417.

Treasurer, J.W. (1993) Management of sea lice (Caligidae) with wrasse (Labridae) on Atlantic salmon (*Salmo salar* L.) farms. In: Boxshall, G.A and Defaye, D. (eds.) *Pathogens of Wild and Farmed Fish*, Ellis Horwood, Chichester, pp. 335–345.

Treasurer, J.W. (1996) Capture techniques for wrasse in inshore waters of west Scotland. In: Sayer, M.D.J., Treasurer, J.W. and Costello, M.J., (eds) *Wrasse: Biology and Use in Aquaculture*, Fishing News Books, Oxford, pp. 74–90.

Treasurer, J.W. (2000) Evaluation of the relative catching power of pots for north European wrasse (Labridae). *Journal of Ichthyology* 16, 36–38.

Varian, S.J.A., Deady, S. and Fives, J.M. (1996) The effect of intensive fishing of wild wrasse populations in Lettercallow Bay, Connemara, Ireland: implications for the future management of the fishery. In: Sayer, M.D.J., Treasurer, J.W., Costello, M.J. (eds.) *Wrasse Biology and Use in Aquaculture*. Fishing News Books, Oxford, pp. 100–118.

Waatevik, E. (2016) Tilgang på rensefisk 2016. Rensefiskkonferansen, Gardermoen, 8-9 Februar 2016. fhf.no/media/129934/5_waatevik_ewa_consulting_tilgjengelig_rensefisk_2016.pdf

Yoneda, M. and Wright, P.J. (2004) Temporal and spatial variation in reproductive investment of Atlantic cod *Gadus morhua* in the northern North Sea and Scottish west coast. *Marine Ecology Progress Series* 276, 237–248.

Young, C.M. (1996) Wrasse as cleaner-fish: the Shetland experience. In: Sayer, M.D.J., Treasurer, J.W. and Costello, M.J. (eds) Wrasse: *Biology and Use in Aquaculture*, Fishing News Books, Oxford, pp. 204–210.

PART III

Cleaner fish developments in countries

CHAPTER 18

Cleaner fish application in Norway

Harald Sveier[1] and Olav Breck[2]

[1] Lerøy Seafood Group ASA Bergen, Norway
[2] Marine Harvest Norway AS, Bergen, Norway

18.1 Sea lice control in Norway

Deployment of wrasse for sea lice control (termed 'biological delousing') has been generally known in the Norwegian salmon farming industry from back in the 1980s, with the Institute of Marine Research playing a central role in developing the method. The catch of wild-caught wrasse commenced around 1988 and increased until 1997, possibly to some extent influenced by increased prevalence of lice resistance to organophosphate therapeutants. The general use of cleaner fish by salmon farmers, however, has been rather limited until recently, as long as effective medicinal treatments have been available. Also, with the introduction of emamectin benzoate (Slice) in 1999 (Aaen *et al.,* 2015), the use of wild-caught wrasse decreased. However, wild catch has increased significantly from 2009 and onwards until today, concurrent with resistance and partly multi-resistance to lice medicines developing along the Norwegian coast. There has been a shift towards the use of hydrogen peroxide H_2O_2 when resistance to the organophosphate azamethiphos, pyrethroids and emamectin benzoate was apparent. However, resistance to hydrogen peroxide (Treasurer *et al.*, 2000) has been reported recently in Norway (Helgesen *et al.*, 2015) and has further limited the application of medicines to control lice in Norway (Aaen *et al.*, 2015).

Focus has increased since then on 'biological delousing' by the use of either wrasse species (Bjordal, 1991; Skiftesvik, 2014) or, during the last six years, lumpfish (*Cyclopterus lumpus*), which is also commonly referred to as lumpsucker in Norway (Imsland *et al.*, 2014). The wild wrasse species that are mostly used include corkwing (*Symphodus melops*), goldsinny (*Ctenolabrus rupestris*) and Ballan (*Labrus bergylta*). The number of farmed and wild-caught cleaner fish used in salmon farms rose from 4.9 million to 37.4 million fish from 2009 to 2016 (Table 18.1). In the same period, the number of salmonids in Norwegian farms increased from 335 million to 404 million (Aquaculture statistics, Directorate of Fisheries 2016). In addition, numerous non-medicinal treatment methods have been developed, with increased use in the industry. These methods are based on removing the lice from the salmon skin, for example flushing with water

(Hydro Fish Care, Norway), freshwater treatment in wellboats, and exposing the salmon and sea lice to heated water in a Thermolicer (www.steinsvik.no). These all imply having to handle the fish. Alternatively, non-handling methods are used such as either preventing lice larvae entering the cages, for example tarpaulins (www.botngaard.no, Norway) or 'electrical fences' (Seafarm Pulse Guard, Norway), or maintaining the salmon deeper in the cages to prevent infestations by larvae in the surface layers, for example by applying submerged feeding and lights (Frenzl, 2014) or 'snorkel cages' (Oppedal *et al.*, 2016). Farmers combine different non-medicinal methods in many cases, including deploying cleaner fish. The downside of many non-chemical treatment methods, as with many chemical ones, is that these operations require extensive handling of the fish, thereby introducing a risk of mortality caused by handling stress. Alternative production regimes, using large post-smolts raised up to 1 kilo in recirculating aquaculture systems (RAS) or semi-closed seawater systems (http://ctrlaqua.no), are also being developed and implemented, to shorten the period in regular production sites and thereby reducing the exposure time to lice.

Some 60% of Norwegian salmon farms use cleaner fish today as part of their strategy to control lice (lusedata.no). Cleaner fish have successfully been deployed in several farms along the Norwegian coast as the only control method for shorter or longer periods, up to the whole salmon seawater cycle. Even better, farmers in Agder county have until recently controlled lice by the use of only wrasse for several years.

The level of sea lice per fish is strictly regulated in Norway. Regulations are dynamic and frequently evaluated based on experience and scientific results. As of today (2018) regulations state that effective action needs to be taken before the average numbers of adult female sea lice reach 0.5 per fish, but for spring and early summer, average numbers should not exceed 0.2 adult females. Sea lice levels are monitored and reported weekly to the government.

Different strategies for lice control have been developed in Norway. Some industry players move towards a strategy lowering the lice treatment threshold limit below the trigger levels set by the authorities, and, in addition, monitor and control lice at a single cage level. The main aim is to keep the number of adult lice close to zero and thereby limit the reproduction and the spread of lice larvae within and between farms. Other players, however, want to move in the opposite direction, allowing higher lice treatment trigger numbers at farm level, claiming that this will reduce the problem of build-up of drug resistance. The impact of these two strategies on the cleaner fish effect is debated. Some argue that too low lice levels in the salmon cages limit the access to food, and lower the general fitness and effect of the cleaners, whereas others claim this is not a problem. They argue that the cleaner fish will maintain their delousing activity even at low lice numbers, as long as alternative feed is given to allow the fish to thrive and grow. Today it is generally recognised that cleaner fish should be given additional feed in the salmon cages, in addition to restricting the build-up of high lice numbers beyond which the cleaner fish will not be able to maintain control.

18.2 The fishery for wrasse in Norway

Wild cleaner fish used in Norwegian salmon farms are primarily caught along the Norwegian coast. Different wild-caught wrasse species are also sourced from Sweden, but these numbers represent <5% of total catch. Statistics on wild catch are yearly made available by the Directorate

Table 18.1 The use of farmed and wild-caught cleaner fish in Norway from 2012 to 2016. See www.fiskeridir.no/English for data from 1998.

	2016		2015		2014		2013		2012		2011		2010		2009	
County	Number x1000	Value NOK x1000	Number x1000	Value NOK x1000	Number x1000	Value NOK x1000	Number x1000	Value NOK x1000	Number x1000	Value NOK x1000	Number x1000	Value NOK x1000	Number x1000	Value NOK x1000	Number x1000	Value NOK x1000
Finnmark and Troms	1,200	23,585	612	10 610	123	1 493	0	0	42	495	0	0	0	0	0	0
Nordland	3,030	54,846	2 270	35 881	1 767	25 916	1 097	17 240	1 477	22 162	861	11 052	431	5 278	182	1 956
Nord-Trøndelag	1,816	34,221	1 763	30 467	2 050	24 526	1 183	13 962	1 386	18 187	819	8 880	1 294	12 514	237	1 613
Sør-Trøndelag	6,129	119,805	4 722	76 080	3 853	63 215	3 952	43 552	1 731	21 578	1 894	24 619	1 699	18 189	232	2 176
Møre and Romsdal	6,959	131,831	2 637	44 446	3 729	48 027	1 658	21 443	2 368	27 233	1 548	16 411	2 735	21 792	374	1 921
Sogn and Fjordane	3,240	61,623	1 970	31 462	4 262	10 169	921	10 382	668	6 179	847	6 816	782	5 153	175	971
Hordaland	11,736	177,836	9 629	110 572	6 236	70 637	5 167	35 560	4 696	40 542	3 209	26 884	2 462	14 501	2 482	14 689
Rogaland	2,953	43,480	2 554	27 784	2 230	27 593	2 016	17 362	1 341	11 989	1 230	12 791	1 362	11 427	1 046	6 691
Other Counties	296	5,124	252	3 401	217	2 709	212	2 342	193	2 233	230	2 875	212	1 753	154	1 267
Total	37,359	652,351	26 409	370 704	24 467	274 285	16 206	161 812	13 903	150 598	10 639	110 327	10 976	90 606	4 883	31 286

of Fisheries, responsible for the regulation of the fisheries (Table 18.1). Total catch in the previous years has steadily increased. The total catch in 2016 was 22.2 million, whereas per November 2017, 26.6 million were caught (Source: Directorate of Fisheries). The fishermen must have a valid contract of delivery with an aquaculture company to be allowed to participate in the fishery. Most of these fishermen are professionals who, except for the summer fishing for cleaner fish, engage in regular fishing for commercial species. The 477 vessels used in 2016 are small boats (typically 16–36 ft) (source: Directorate of Fisheries, Norway) with several internal wells with flow-through of seawater that can be oxygenated if necessary. The fishing gear used is pots baited with shrimp or minced crab, normally set on a maximum of 6–7 m depth. A regular cleaner fish boat is normally equipped with 200–300 pots that are controlled and emptied once each day as a minimum, but normally two times per day. The captured cleaner fish are measured to ensure that it is above a minimum size (2017) (11 cm for goldsinny wrasse, 12 cm for corkwing wrasse and 14 cm for Ballan wrasse), sorted by species, and then transported to holding cages or directly to the fish farmer.

The Norwegian Directorate of Fisheries introduced new regulations on 1 April 2016 for the catch of wrasse species based on the input from numerous consultative bodies including the Institute of Marine Research (IMR). No restrictions on catch numbers were in place until then. Based on statistics from fisheries during the last five years, IMR recommended not exceeding the 2015 catch numbers in 2016 and, in southern areas of Norway, reducing the catch by 20–25%. Following this, the Directorate set a total quota of 18 million fish for 2016; 4, 10 and 4 million for southern, western and mid-part of the Norwegian coast, respectively. This represented a 15% reduction from 2014 numbers. As actual catch in 2016 ended above the total quota, the mechanism for controlling the fisheries will have to be adjusted. The new regulation also includes postponement of the start of the fisheries to avoid capture of mature fish. The new 2016 regulations were expected to lower the availability of wild-caught fish in total and hence increase the need for farmed fish if the level of biological delousing should be maintained or rise. However, based on a set of changes in regulations prior to the 2017 catch season, actual numbers also exceeded those from 2016, reaching 26.6 million. IMR has raised concern about the increased catch and argues that 18 million should be the maximum for 2018. They also suggest catch area rotation as a possible mechanism to protect local wrasse stocks.

18.3 Hatchery production of cleaner fish in Norway: capacity and output

In 2016, 52 licenses, owned by 24 companies were registered with permission to sell cleaner fish to salmon farms (Table 18.2). Currently four licences are in use for production of Ballan wrasse, more than 90% of the farming of cleaner fish is related to lumpfish. About 50% of the companies are owned by salmon farming companies (Erlend Waatevik/EWA Consulting). The reason for the majority of new farms focusing on lumpfish is probably to a great extent related to lumpfish being an easier marine species to farm. This is partly because lumpfish can be directly start fed with commercial dry feed whereas Ballan wrasse has to be fed live feed (rotifers and artemia) before the weaning phase. Secondly, the energy demands for farming of wrasse (14–15°C water temperature) make it less expensive to build and run lumpfish farms. The successful farming of

Table 18.2 The number of licences issued for farming of cleaner fish species in Norway.

	2016		2015		2014		2013		2012	
County	Companies	Licences	Companies	Licences	Companies	Licences	Companies	Licences	Companies	Licences
Total	24	52	18	41	11	22	10	16	5	15

Table 18.3 Sale of farmed cleaner fish in Norway from 2012–2016. Numbers of fish shown, x 1000. Source: Directorate of Fisheries.

	2016		2015		2014		2013		2012	
Species	Numbers	Value	Numbers	Value	Numbers	Value	Numbers	Value	Numbers	Value
Ballan wrasse	524	15,111	1 348	30 304	379	6 740	95	1 769	270	2 000
Lumpfish	15,858	289,169	13 385	237 468	3 457	52 858	1 954	27 717	431	5 175
Other species	0	0	0	0	0	0	0	0	2	32
Total	16,382	304,280	14 733	267 771	3 836	59 598	2 049	29 486	703	7 208

lumpfish, rapidly exceeding the numbers of Ballan wrasse, has probably encouraged new players to also go for lumpfish.

In 2016, more than 40 cleaner fish farms were in operation, with permission to produce cleaner fish for scientific, educational or commercial purposes (Directorate of Fisheries). Some 30 of these are based on commercial sales.

The Norwegian Directorate of Fisheries publishes annual statistics on the number of farmed and wild-caught fish deployed in the Norwegian salmon industry (www.fiskeridir.no/English). Table 18.1 shows numbers of cleaner fish deployed from 2009–2016. The numbers include the species Ballan wrasse, being farmed from 2009, and lumpfish, with official statistics from 2012. Statistics on the sale of farmed cleaner fish (Table 18.3) show that 2016 numbers constitute some 44 % of total numbers deployed in salmon farms in 2016 (Table 18.1) compared with 16% in 2014. Some 97% percent of the farmed fish were lumpfish in 2016.

18.3.1 Ballan wrasse

Commercial farming of Ballan wrasse (Fig. 18.1) commenced in 2009 at Marine Harvest's former cod facility Cod Culture Norway in Øygarden, west of Bergen. Three more Ballan wrasse farms were established in the period 2009–2011, of which only Sagafjord, located at the Island of Stord, is still producing wrasse, based on a long-term rental agreement with Marine Harvest. Marine Harvest now also produces Ballan wrasse at a third facility at Rissa, west of Trondheim. Currently, a fourth hatchery, Havlandet Marine Yngel, west of Florø, farms wrasse, based on own broodstock as well as live feed production. In addition, Njord Salmon, situated at Tjeldbergodden, south-west of Trondheim, is on-growing Ballan wrasse from 10 g size until

Fig. 18.1 Ballan wrasse broodfish. Credit: E. Grøtan, MHN

Fig. 18.2 Lumpfish. Credit: Sunniva Wannebo Kui, PatoGen AS

transfer. One of the farms, Marine Harvest Rissa, uses a RAS unit for part of their production, whereas the others are based on flow-through systems. Heating systems are needed as Ballan wrasse are farmed at 14–15°C, implying energy costs. The farm at Tjeldbergodden uses heated seawater discharged from local industry.

The total number of Ballan wrasse transferred to salmon sites in 2016 was around 1 million fish (Olav Breck, personal communication), compared with the forecast of 1.5–2 million for 2017. A further increase in production to 3–4 million is expected in 2018.

18.3.2 Lumpsucker

From 2012 to 2016, the number of licenses for cleaner fish production has increased from 15 to 52 (Table 18.2), (Directorate of Fisheries). Around 50% of these are owned by salmon producers (Erlend Waatevik/EWA Consulting). Some 33 lumpsucker farms were operational from January 2017, including nine new farms established in 2016. Two of these farms are based on recirculation technology.

The sales of lumpfish to producers of Atlantic salmon in 2016 are reported as 15.9 million fish (Directorate of Fisheries), an increase of more than 35-fold from 2012 (see Table 18.3).

Four new farms are expected to deliver fish in 2017, and a total production of 30 million fish is foreseen. The total number, however, depends to a great extent on the impact of biological and health-related challenges. Many farms have experienced disease problems related to bacterial or viral agents in the past years, and these in some cases severely affected production and necessitated complete fallowing (H. Sveier and O. Breck, personal communication). The disease situation improved during 2017, but is still regarded as a challenge to future production numbers in spite of several disease preventative measures, such as screening of broodfish and development of new vaccines.

18.4 Future projections

A continued increase in production of lumpsucker is expected, to some extent due to more companies being established, but also due to a more effective production, with better control of the disease situation. An increase is also foreseen for Ballan wrasse, primarily based on current players having plans to increase their production. However, as more interest is shown in farming of wrasse, one or two more wrasse producers are expected to be established within the next few years.

Currently, high losses are frequently seen both in the farming period and post-transfer to salmon cages, the latter with both wild-caught and farmed cleaner fish. Especially, severe losses have been seen with lumpfish, partly within two to three months post-transfer, causing mortalities of more than 50% (Bornø *et al.*, 2016). Losses are also seen with wild-caught wrasse as well as farmed Ballan wrasse, the latter, however, seeming to be more robust (O. Breck, personal communication). Hence, there is an urgent need to improve the health and welfare of the cleaner fish. This includes further development of husbandry protocols, vaccines and breeding programmes for farmed species. The development of effective pharmaceutical as well as non-pharma lice control methods will also influence production pattern and further investments in cleaner fish farming. Many salmon producers see non-pharmaceutical-based lice control as the only viable and sustainable future control method. Effective vaccines against salmon lice are not expected within the next few years (www. Intrafish.no, January 2017).

18.5 Efficacy of the technique (supply of juveniles to the fish farms)

Ballan wrasse is produced at the moment at somewhat higher costs compared with wild-caught wrasse species. Based on figures from the Directorate of Fisheries (see Tables 18.2 and 18.3), the average price for farmed Ballan wrasse was around 22 NOK in 2014, whereas wild-caught wrasse could be estimated at 9 NOK, based on combining the two tables. The market price for wild-caught wrasse in 2016 was between 10 and 30 NOK, with Ballan wrasse the most expensive, with an average price of 32 NOK (Kyst.no, August 2016). Wild-caught wrasse species will most likely be more expensive in the future due to limitations in the fisheries, and farmed Ballan wrasse are expected to be produced at lower costs than wild caught (O. Breck, personal communication). Farmed and vaccinated cleaner fish will be available year round, representing important advantages compared with wild fish. A recent report by Johansen *et al.* (2016) on disease risk related to use of farmed and wild-caught cleaner fish concluded that, 'The most important measure for reducing the risk of disease and infection is to use only farmed, vaccinated cleaner fish. The goal must be to achieve this within a 3–5 year period.'

The average sales price for lumpfish in 2016 is estimated at 18 NOK, similar to 2015 numbers (see Table 18.3). The prices are expected to decline with increased production and competition, together with better disease control.

18.6 Farm applications

A sea lice control strategy for each site is based on historical data on lice presence during the production cycle. In some production areas in Norway (especial in the north) salmon lice is not a major challenge and therefore cleaner fish are not used, while in other areas lice are a big challenge making cleaner fish part of the daily routine. The inclusion ratio of cleaner fish normally starts at a low percentage when the salmon smolts are transferred to the production cages, typically 3–5% cleaner fish to the number of salmon in the cage. Some cleaner fish are normally lost during the autumn and winter, so replenishment is needed during late winter and spring to enable efficient control during the second year in sea. Many farmers also increase the inclusion rate up to 8% as the salmon increase in weight from 3 kg until harvesting. A total inclusion rate of 15–20% for the whole generation is regarded by many as necessary to control lice effectively (U. Austefjord, Norway, 2017, personal communication).

The cleaner fish species used for lice control varies a lot between farms, depending on the availability of wild-caught and farmed fish. Farmers in the northern part of Norway use less wild-caught wrasse due to low availability and long transport distances, in addition to unfavourable temperatures. In general, farmed Ballan wrasse and wild-caught wrasse species are deployed during the first summer and autumn, whereas farmed lumpfish are stocked during late autumn, winter and spring. Also, large farmed Ballan wrasse (>40 g) are used with salmon growers in the second spring and summer in sea cages. Farmers are becoming more dependent on farmed cleaner fish, both Ballan wrasse and lumpfish, to control lice during spring and early summer because of the new legislation with postponement of the start of the wild catch, to avoid the spawning season. Hence, the availability of farmed fish is becoming increasingly important for lice control, both for smolt and on-growers in spring and summer.

There are several key factors in succeeding with cleaner fish and, in general, they all have to be in place to ensure efficient control over time. Keeping the nets clean, providing hiding places, and additional feeding of the cleaner fish are all recognised to be crucial. Hides will give the cleaner fish protection and a place to rest. At the same time, the hide must be designed so that it is easy and tempting for salmon to approach the cleaner fish. The design must be stable in the sea, be easy to clean, and to provide the least possible risk of escape. Different hides are shown (Fig. 18.3), and the 'fake kelps' are amongst the most used. Guidelines have been developed for the use of hides (www.lusedata.no). The demand for shelter increases as the cleaner fish grows larger. For instance, in a cage holding 200,000 salmon and 8000 cleaner fish, it is recommended that there should be no more than 15 and 30 cleaner fish per metre of shelter, for large and small fish respectively.

Fig. 18.3 Cleaner fish hides of a wide range of designs.

Shelters can be organised in many ways (panels, squares and circles), but the current opinion is that the most applicable design is to arrange the shelters as a corridor in a position in the cage that forces the salmon to swim through it.

Mesh size of the nets has traditionally been adapted to the size of the salmon, increasing in size as the salmon grows. A different strategy is needed to prevent cleaner fish from escaping, entailing maintenance of smaller mesh sizes for a longer time and, in some cases, keeping the mesh size constant throughout the production period in the sea. It is crucial to keep the nets clean, as well as the shelters and all other equipment in the cage, to ensure that the cleaner fish feed on sea lice instead of organisms growing on nets or equipment. Therefore, the reduced mesh size makes good cleaning procedures even more important. Specially designed equipment is used to clean both the nets and other equipment, and this is done as often as every tenth day during the summer season.

Sea lice numbers may not be sufficient to cover the nutritional needs of the cleaner fish in the cages because of the Norwegian regulations on maximum lice levels. Therefore, successful use of cleaner fish also includes providing feed on a regular basis. Feed for cleaner fish is supplied by the feed producers, and fed to the cleaner fish by designated feeding systems. Currently, lumpfish is considered easier to feed during the period in the salmon cages, whereas the Ballan wrasse is far more 'picky' (O. Breck, personal communication). For both Ballan wrasse and lumpfish, more research is needed to ensure development of feeds of good physical quality, fulfilling the nutrition requirements of the species.

18.7 Challenges of cleaner fish use

The use of cleaner fish is a promising non-chemical sea lice control alternative with a tremendous potential. The potential is large because many sites already successfully use cleaner fish as the only control method. However, succeeding with cleaner fish relies on farm personnel really believing in the method. Cleaner fish need to be a part of the daily operations, and it requires sufficient resources and enough manpower at each site to optimise the environment for the cleaner fish. Also, it is considered important that operations are in an area with good lice control, as too high lice pressure from adjacent farms can result in higher lice levels, with legislation triggering other delousing methods. Unfortunately, many of the non-pharma treatment methods also lead to stressful handling of the cleaner fish, resulting in losses and thereby a less efficient cleaning effect post-treatment.

Special concerns have been raised related to losses of lumpfish post-transfer. During the last few years, mortalities have been seen all along the coast, in some cases with >50% losses within a few months. A recent review (Bornø *et al.*, 2016) has pointed at infectious diseases, especially atypical Furunculosis, causing some of the losses, emphasising the need for better vaccines.

18.8 Future prospects for cleaner fish use

Cleaner fish will most likely be a part of the IPM sea lice solution in the next 10–15 years, together with genetic selection of salmon for better sea lice resistance, functional feeds, and production of larger post-smolt to reduce the lice exposure time in sea water. A shift towards less stressful methods is expected as many of the other non-pharma treatment methods involve handling of the salmon. These will possibly include hi-tech solutions such as laser and ultrasound, but also passive or active methods to avoid exposing salmon to lice-infective copepodids. A precondition for further large-scale use of cleaner fish is to solve the challenges related to fish health and welfare of the wrasse species and the lumpfish. Lowering the mortalities and losses is crucial to continuing defining this as a sustainable and environmental friendly method.

References

Aaen, S.M., Helgesen, K.O., Bakke, M.J., Kaur, K. and Horsberg, T.E. (2015) Drug resistance in sea lice: a threat to salmonid aquaculture. *Trends in Parasitology* 31, 72–81.

Bjordal, Å. (1991) Wrasse as cleaner-fish for farmed salmon. *Progress in underwater science* 16, 17–28.

Bornø *et al* (2016). Akutt dødelighet hos rognkjeks (*Cycloterus lumpus*) i 2015. *The Norwegian Veterinary Institute report series.* 2016.

Frenzl, B., Stien, L.H., Cockerill, D., Oppedal, F., Richards, R.H., Shinn, A.P., Bron, J.E. and Migaud,

H. (2014) Manipulation of farmed Atlantic salmon swimming behaviour through the adjustment of lighting and feeding regimes as a tool for salmon lice control. *Aquaculture* 424–425, 183–188.

Helgesen, K.O., Romstad, H., Aaen, S.M. and Horsberg, T.E. (2015) First report of reduced sensitivity towards hydrogen peroxide found in the salmon louse *Lepeophtheirus salmonis* in Norway. *Aquaculture Reports* 1, 37–42.

Imsland, A.K., Reynolds, P., Eliassen, G., Hangstad, T.A., Foss, A., Vikingstad, E. and Elvegård, T. A. (2014) The use of lumpfish (*Cyclopterus lumpus* L.) to control sea lice (*Lepeophtheirus salmonis Krøyer*) infestations in intensively farmed Atlantic salmon (*Salmo salar L.*) *Aquaculture* 424–425, 18–23.

Johansen, L-H., Colquhoun, D., Hansen, H. and Hildre, S. (2016) *Analysis of disease related risk related to use of wild caught and farmed cleanerfish for control of salmon louse.* FHF Report 9/2016.

Oppedal, F., Dempster, T. and Stien, L.H. (2016) *Snorkelmerd: Produksjonseffektivitet, adferd og velferd.* Sluttrapport FHF prosjekt 900883. ISSN 1893-4536 (online)

Skiftesvik, A.B., Blom, G., Agnalt, A-L., Durif, C.M.F., Browman, H.I and Bjelland, R.M. (2014) Wrasse (Labridae) as cleaner fish in salmonid aquaculture – The Hardangerfjord as a case study. *Marine Biology Research* 10 (3), 289–300.

Treasurer, J.W., Wadsworth, S. and Grant, A. (2000) Resistance of sea lice, *Lepeophtheirus salmonis* (Krøyer), to hydrogen peroxide on farmed Atlantic salmon, *Salmo salar* L. *Aquaculture Research* 31, 855–860.

CHAPTER 19

Cleaner fish rearing and deployment in the UK

Jim Treasurer,[1] Richard Prickett,[2] Matthew Zietz,[3] Chris Hempleman[4] and Carlos Garcia de Leaniz[5]

[1]FAI Aquaculture, Ardtoe, UK
[2]Native Marine Centre, Portland, UK
[3]Wester Ross Salmon, Ullapool, UK
[4]Scottish Sea Farms, South Shian, Argyll UK
[5]Centre for Sustainable Aquatic Research, Swansea University, Singleton Park, Swansea, UK

19.1 Introduction

The use of wrasse in the biological control of sea lice was first developed by Åsmund Bjordal in Norway in 1988 (Bjordal, 1988, 1991; Bjordal and Costello, 1990). Wrasse of several species were stocked with salmon in Scotland two years later. The Sea Fish Industry Authority (SFIA) commenced farm trials in 1990 using wild-caught fish obtained from local fishermen and from Weymouth. The SFIA in parallel initiated a wrasse-rearing programme and focused on goldsinny wrasse *Ctenolabrus rupestris* (Sherwood, 1990; 1991). At the same time, rearing trials with both goldsinny and corkwing wrasse *Symphodus melops* were being carried out at Golden Sea Produce at Hunterston (Stone, 1996). A meeting to disseminate the results of early trials was organised by the Scottish Salmon Producers Organisation (SSPO, forerunner of SSGA) in Inverness in 1990. The SSPO reported on the use of wrasse in Norway and described many of the techniques for managing wrasse in pens (Rae, 1991, unpublished SSPO report). Wrasse were fished from the wild and introduced quickly in Scotland using four wrasse species, goldsinny, corkwing, rock cook *Centrolabrus exoletus* and the less abundant cuckoo wrasse *Labrus mixtus*. The main interest at this stage was from McConnell Salmon in Loch Sunart (Johnny Watters, Scotland, 2016, personal communication) and Marine Harvest Scotland (Ian Armstrong, Neil Crawford, Scotland, 2016, personal communication). Fishing was with prawn creels and also fyke nets (Treasurer, 1996). Various stocking densities were assessed from one wrasse to ten salmon (ratios rather than % were quoted) in the case of small numbers of broodstock salmon, 1:25, 1:50, 1:100 and 1:150, and the application of wrasse in pens was described in early leaflets and articles (Treasurer, 1991a, 1991b, 1993). Sizes of fish used were over 10 cm and percentage numbers of each wrasse species stocked varied between west coast localities, but typically ratios were goldsinny 41%, rock cook 46%, corkwing 10% and low catches of Ballan *Labrus bergylta* and cuckoo wrasse. Health screening of wild-caught wrasse was extensive with 50 to 60 fish from each sea loch or area of origin being sampled, histology taken, and samples taken for both bacterial and virus detection (Treasurer, Marine Harvest unpublished report, 1992). The results of these tests were negative for IPN, VHS, *A. salmonicida* and *V. anguillarum*. However, both typical and atypical furunculosis (causative agent *Aeromonas salmonicida*) were found in goldsinny and rock cook wrasse from salmon cages (Treasurer and Laidler, 1994; Laidler *et al.*, 1999). Wrasse of all species were prone to be affected by typical furunculosis from salmon with a suspicion that wrasse were grazing on salmon mortalities. Atypical furunculosis was common and has also been found in several species of marine finfish (Treasurer, 2012). Five species of wrasse were also screened for a range of parasites in Scotland (Treasurer, 1997). Twenty-two parasites were identified but none of these was regarded as a threat to salmon; several were host-specific, and none were transmitted to salmon. Interestingly, it was also apparent that goldsinny, Ballan and rock cook wrasse had their own species of sea louse *Caligus centrodontii* (Bron and Treasurer, 1992).

19.2 Wrasse deployment in salmon pens

Deployment trials with goldsinny, rock cook, and corkwing wrasse were carried out on several farms at various stocking densities from 1:10 to 1:150 salmon, with mixed results. A reduction in sea lice numbers was recorded in Kingairloch, Arran and Mull, as well as at the broodstock farm in Lochailort, where wrasse stocking numbers were higher (Jim Treasurer, Ardtoe, 2017, personal communication). However, efficacy was difficult to establish in large commercial pens, and the benefits of wrasse were best shown in replicated trials in 5 × 5 m square pens where it was easier to capture salmon and count sea lice numbers (Treasurer, 1994). Wrasse were effective in maintaining low sea lice numbers, and salmon had significantly fewer sea lice compared to fish in control pens with no wrasse. No medicinal treatments were required in the first year of salmon production. Wrasse selectively grazed on the largest gravid sea lice, as indicated by Ivlev's index of diet selectivity, and the occurrence of gravid lice was negligible at the end of the trial. It was also possible to calculate the rate of passage of lice through the wrasse's gut, and to quantify the daily feed intake of lice, which was found to be as high as 2% of body weight per day (Treasurer, 1994).

However, the benefits of using wrasse in Scotland during the 1990s were variable, with some sites reporting some successes, while other sites reported no difference in sea lice counts. In many cases there were unexplained declines of wrasse during this time, and atypical furunculosis is thought to have been responsible for some of the losses. At that time, the salmon industry in Scotland was not using what we now know is the most effective cleaner wrasse, Ballan wrasse, because it had been reported that this species was aggressive, removed the scales off salmon, and attacked their eyes (Bjordal, 1988). However, such level of aggression was only sporadic, probably caused by stocking Ballan wrasse of too large a size. It was shown later that Ballan wrasse were highly effective as cleaner fish in north European waters. In these early days of using cleaner fish, lumpfish were not being used, though the reasons are not clear. By the end of the 1990s, fewer wrasse were being stocked, as new medicinal treatments such as Cypermethrin (Excis) and the in-feed treatment emamectin benzoate (SLICE) (Rae, 2002) became available. However, in 1999, Infectious Salmon Anaemia (ISA) was diagnosed in salmon in Scotland. No fish were allowed to be moved from sites, all fish had to be harvested on site and many were harvested early. There was a fear that the disease might not be eradicated. It was thought that ISA had a wild fish reservoir,

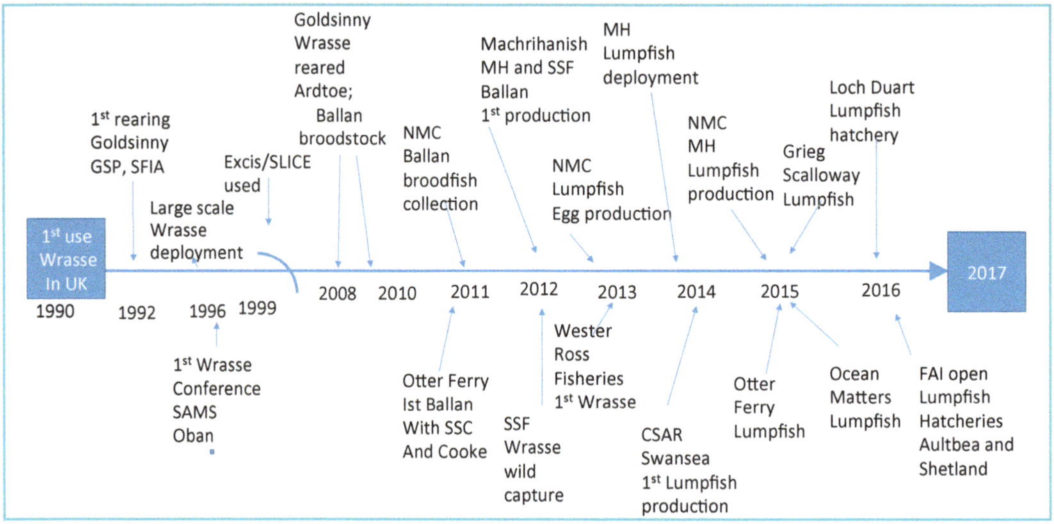

Fig. 19.1 Timeline of cleaner fish highlights and breakthroughs in the UK.

as it had been detected in wild fish such as haddock *Melanogrammus aeglefinus* and European eel *A. anguilla*. For this reason, the stocking of wild-caught fish in salmon net pens was considered too risky, and the use of wrasse ceased. From 2000 onward there was little use of wrasse on the Scottish mainland. However, several of the farms in the Western Isles stocked wrasse, including West Minch Salmon, which had organic status, and also Lewis Salmon, which was also farming organically (Treasurer SARF report, 2013).

19.3 Cleaner fish rearing

The second phase of rearing of wrasse commenced in 2008 at Ardtoe (Fig. 19.1). Initial attempts were made to breed goldsinny in captivity, and several thousand juveniles were grown (Chalaris and Treasurer, 2010). Then attention focused on Ballan wrasse as salmon producers were more interested in this species after promising results with wild-caught Ballan wrasse were reported in Norway (Chapter 18). Machrihanish Marine Farms (MMF) commenced rearing Ballan wrasse in 2010, and Otter Ferry Seafish in 2012 (Fig. 19.1). Scottish Sea Farms (SSF) started stocking wild-caught wrasse on a large and systematic scale in 2010. The idea of using lumpfish as cleaner fish came from Norway in 2014 (Fig. 19.1). In the UK, NMC Weymouth and the Cleaner Fish Company procured a few adult lumpfish, and the first lumpfish spawned in captivity at CSAR Swansea University and at NMC, which also produced the first juvenile lumpfish in the UK. Subsequently, and as part of an R&D contract between CSAR and MHS, lumpfish eggs and larvae obtained at CSAR were sent for on-growing to Ardtoe, and Otter Ferry (OFS). Meanwhile, MHS and SSF rented a marine finfish hatchery at Machrihanish and worked together with Stirling University in capturing and rearing wrasse as part of a project supported by the Technology Strategy Board in 2011. Wrasse were also reared for the SSC and Meridian at OFS from 2013. There was an outbreak of VHS in wild-caught wrasse in Shetland in December

2012 (Munro *et al.*, 2015) and no further such fish were shipped from the mainland to salmon farms on Shetland. Grieg Seafood imported lumpfish juveniles from Norway in 2015, grew them on at NAFC, and stocked them in farms at the end of the first year. The deployment of wrasse was reviewed in Scotland in a study for SARF (Treasurer, 2012) and lumpfish welfare and survival is being assessed (SAIC, 18 November 2015; scottishaquaculture.com/lumpfish-projects-could-boost-sea-lice-contol). SSF and MHS have obtained an SAIC award for another three years of development work in rearing wrasse with Stirling University (SAIC). CSAR Swansea University signed a five-year R&D contract with MHS to rear lumpfish in 2015, and was recently awarded EU funding to pursue lumpfish research with Ocean Matters (Anglesey, Wales). Today, the use of cleaner fish remains a cornerstone of Integrated Pest Management (IPM) on salmon farms in Scotland. Two new lumpfish hatcheries have opened at Aultbea and Shetland that will ensure the expansion of cleaner fish production. This will augment existing hatcheries with lumpfish at FAI Ardtoe, NMC Weymouth, CSAR-Swansea, Otter Ferry Seafish, NAFC, Ocean Matters Anglesey and a small hatchery at Loch Duart. Marine Harvest are also opening a wrasse hatchery at the former sea bass farm in Anglesey.

19.4 Cleaner fish stocking and production capacity in the UK

As the stocking ratio of cleaner fish is around 5–10% of the salmon population, there is a theoretical requirement of about 3–5 million cleaner fish in the UK (Prickett, 2016). However, some salmon farms in Norway are stocking cleaner fish at ratios of 12 to 20%, and Powell et al. (2017) suggested that up to 10 million cleaner fish would be required in Scotland. Prickett (2016) estimated that about 1–2 million wild wrasse were caught in 2015 in the UK, and that hatchery production amounted to 250,000 Ballan wrasse from three hatcheries, and 2.5 million lumpfish from eight hatcheries.

Most of the wrasse currently stocked in Scotland are wild-caught from a fishery that operates mainly by traps and prawn creels. It is difficult to obtain accurate numbers of wrasse captures, but catch has increased from about 250,000 fish in Scotland in 2012 to more than 1 million fish of all species in 2016 (Marine Scotland, 2017a). The fishery for wrasse in the south of England has also expanded in recent years, and it was estimated using government statistics that more than 1 million fish were caught (Riley *et al.*, 2017).

The numbers of Ballan wrasse and lumpfish that have been reared in the UK have grown quickly in recent years (Table 19.1). Wrasse hatchery production has increased from 102,000 in 2015 to a projected 530,000 in 2017 in three hatcheries, compared with 700,000 lumpfish in 2015 and a projected production of more than 4 million fish in eight hatcheries in 2017 (Table 19.2). These data have been reported verbally on enquiry, and it should be noted that lumpfish eggs and larvae and juveniles of 2 g have been transferred between UK hatcheries, so it has been difficult to give precise figures. Marine Scotland reported that 10 tonnes of lumpfish (c. 500,000 fish at 20 g stocking weight) were reared in Scotland in 2016 with a projected 46 tonnes in 2017, c. 2.3 million fish (Marine Scotland, 2017b). The balance of lumpfish were reared in the hatcheries in England and Wales (Table 19.2). This cleaner fish production from farms together with wild caught wrasse from England and Wales, and Scotland, gave an estimated stocking of cleaner fish on Scottish salmon farms of over 6 million fish in 2017. However, technical, health and production issues may reduce predicted farmed fish numbers.

Table 19.1 Estimated UK lumpfish production of 20 g plus juveniles transferred to sea pens 2015–2017. (*2 g fish transferred to third party sites, and other fish transferred in 2017; **plus eggs and 2 g fish shipped to third party sites; ***broodstock site).

Company	Location	2015	2016	2017
Dorset Cleaner-fish	Portland, Dorset	280 000	250 000	700 000
CSAR-Swansea University	Swansea, S. Wales	158 000	150 000**	150 000***
Ocean Matters	Anglesey, N. Wales	–	500 000	1 000 000
Otter Ferry Seafish	Loch Fyne, Scotland	230 000	600 000	750 000
FAI Aquaculture	Ardtoe, Scotland	25 000	0***	25 000
FAI Aquaculture	Aultbea, Scotland	–	0*	800 000
FAI Aquaculture	Broonies Taing, Shetland	–	0*	500 000
Hjatland	NAFC, Shetland	–	400 000	400 000
	TOTALS	693 000	1 900 000	4 325 000

Table 19.2 Production of Ballan wrasse juveniles (>50 g) transferred to sea pens in the UK in 2015–2017. *Juvenile production.

Company	Location	2015	2016	2017
Macwrasse	Machrihanish, Scotland	72 000	200 000	300 000
Otter Ferry Seafish	Loch Fyne, Scotland	25 000	50 000	200 000
FAI Aquaculture	Ardtoe, Scotland	5 000	0	(30 000)*
	TOTALS	102 000	250 000	500 000

19.5 UK cleaner fish hatcheries

Native Marine Centre Ltd (NMC)

The location of the UK cleaner fish hatcheries is shown (Fig. 19.2). The Native Marine Centre (NMC) (Fig. 19.3) based in Portland Port (Dorset) collected live fish for the aquarium industry using divers and inshore fishermen. NMC was initially contracted to catch live Ballan wrasse brood fish for projects in Scotland (e.g. Otter Ferry and Machrihanish) but this was extended to lumpfish in 2013. Marine Harvest Scotland (MHS) supported hatchery trials at NMC's Portland site in 2015, when 285,000 juvenile lumpfish of deployment size (15g) were reared, as well as 1.5 million eggs, which were sent to other partners in the MHS group. In addition, it supplied 30,000 wild Ballan wrasse of 100 g size for direct deployment in sea cages. NMC and MHS plan to form a new company to focus on lumpfish production and holding wild wrasse prior to transport to Scotland.

Fig. 19.2 Map of locations of cleaner fish hatcheries in the UK.

Lumpfish production at the Centre for Sustainable Aquatic Research (CSAR) Swansea University

The Centre for Sustainable Aquatic Research (www.swansea.ac.uk/csar) (Fig. 19.3) was commissioned in 2005 with the help of a £2 million infrastructure grant from the EU, the Welsh Government and Swansea University (Wales, UK). CSAR facilities consist of a 750 m^2 building containing two commercial grade Recirculation Aquaculture Systems (100 m^3 h^{-1}), environmentally controlled laboratories, a quarantine facility, and four dedicated fish facilities that can operate as recirculation or flow-through systems (Fig. 19.3). CSAR signed a five-year R&D contract with Marine Harvest Scotland in 2014 to optimise the production of lumpfish for the salmon farming industry. The lumpfish production hatchery consists of a 60 m^3 recirculating aquaculture system, plus a quarantine room for the reception of broodstock. The tank configuration is flexible and can be adapted to produce different stages of development, from eggs and juveniles ready for deployment, to on-growing fish through to maturation. The system can operate with 15 or 30 purposely designed 70 l upwelling incubators, eight 1,500 l (1.45 m diameter × 1 m deep) tanks, and six 4,000 l (1.85 m diameter × 1 .5 m deep) tanks, all fitted with aeration, adjustable water flow, and automatic feeders. Depending on tank configuration, the CSAR lumpfish facility has the capacity to hold 50–60 broodstock in quarantine, and to produce 1.8 million lumpfish eggs at a preferred loading density of 0.5 kg per incubator, 3.6 million weaned lumpfish larvae (0.25 g) at 25 kg m^{-3} and 112,000 × 10 g juveniles at 20 kg m^{-3}.

Fig. 19.3 UK cleaner fish hatcheries, internal and external views. a = Native Marine Centre, Weymouth; b = CSAR Swansea; c = Ocean Matters; d = FAI Aquaculture Ardtoe; e = FAI Aultbea; f = FAI Broonies Taing Shetland; g = Otter Ferry Seafish; h = Machrihanish; (i) = Grieg/NAFC Scalloway. Photos: R. Prickett, C. Garcia de Leaniz, D. Phillips, J. Treasurer, A. MacMaster, J. McEvoy, P. Featherstone, A. Chalaris, A. Rainsden.

Lumpfish production started in 2014 when eight wild females were spawned and *c.* 5,000 × 10 g juveniles were shipped to Scotland. Production increased greatly in 2015, when 120 females were spawned thanks to improved protocols for artificial reproduction and egg incubation. In total, 90.6 kg of eggs were obtained (11.3 million eggs), and 2.6 million fertilised eggs were shipped to other lumpfish establishments, along with 812,000 × four to six week old larvae (*c.* 0.2 g), and 145,682 x 10 g juveniles, making CSAR one the largest producers of lumpfish in the UK (Keay *et al.*, 2015; Powell *et al.*, 2016; 2017). Lumpfish is a novel species to aquaculture and many areas of lumpfish biology are still poorly known, and a number of research topics are being supported by the CSAR–Marine Harvest Scholarship in Sustainable Aquaculture, as well as by other funding sources.

Ocean Matters

Ocean Matters was formed in late 2015 as a dedicated lumpfish production facility (Fig. 19.3), catering for the UK and European markets. Production began in May 2016 after undertaking extensive installation of state-of-the-art RAS technologies with a volume of around 350m^3. To date more than 1 million fish have already been deployed to sea cages (Daniel Phillips, Anglesey, 31 May 2017, personal communication). Current production could reach 2 million fish per year. Only a small proportion of this large facility is currently being used, and there is thus room to expand production to higher volumes with similar systems.

Cleaner fish production at FAI Aquaculture

The Ardtoe hatchery (Fig. 19.3) was the first purpose-built marine finfish R&D hatchery in the UK and was opened in 1965 by the White Fish Authority, which then became the Sea Fish Industry Authority SFIA. An inlet from the sea was initially impounded and wild-caught plaice *Pleuronectes platessa* and turbot *Scophthalmus maximus* grown within the lagoon area. The operation then moved onshore to rear fish in tanks and the site developed over a period of 40 years. Initial work was on developing techniques for the rearing of plaice and turbot that could be transferred to industry to develop the commercial rearing of these species. Later work focused on a most valuable species of halibut, *H. hippoglossus*, with work following on shellfish such as oysters and lobsters, then rearing of cod *Gadus morhua*, and then haddock *Melanogrammus aeglefinus*. Other species farmed at the location included whiting *Merlangius merlangus* and pollock *Pollachius pollachius*. The site was managed by SAMS for two years before transfer to Viking Fish Farms, and then local management, and was finally purchased by Benchmark Ltd in October 2013. The hatchery has been used for shellfish rearing, seaweed production, trials work and integrated management IMTA in the last few years. The water supply is once use and 120 m^3 is supplied prior to drum filtration and then treated with UV light. The hatchery and nursery were rebuilt in 2015 and give a combined capacity of 1400 m^2 over two levels.

Ardtoe was approached by the SSC in 2008 to rear wrasse as cleaner fish and work focused on rearing goldsinny and latterly exclusively on Ballan wrasse. The work was supported by a joint EU project called ECOFISH and the findings were published at eco-fish.org in section leaflets. Ardtoe reared lumpfish from 2014, with broodstock selection and egg supply being the main activities. Ardtoe produced the first hatchery-reared lumpfish broodfish to spawn in 2015 at only one year old

(Towers, 2015). FAI has opened a new lumpfish hatchery at Aultbea that has a water supply capacity of 140 m^3, and potential production of 760,000 lumpfish juveniles of 20 g, and another site at Broonies Taing in Shetland with 420 m^3 tank volume with potential production of 1.5 million fish annually to supply the market in Shetland utilising mainly eggs imported from Iceland.

Otter Ferry Seafish

Otter Ferry Seafish Ltd (OFS) is one of the oldest established fish farming companies in the UK. It has always been at the forefront of aquaculture development in Scotland. Based on the shores of Loch Fyne, the company started farming trout in 1967, and since then OFS has pioneered land-based salmon culture and has specialised in smolt and egg production. OFS initiated halibut production in 1991 and currently is the only remaining Atlantic halibut hatchery in the UK. The cleaner fish production at OFS started in 2010 with the Ballan wrasse project with the support of the Scottish Salmon Company, the Meridian Group, Crown Estate, Highlands and Islands Enterprise and OFS. Parallel lumpfish production started in 2014 and the successful closure of the production cycle was achieved in 2016. Currently, OFS is one of the leading cleaner fish production hatcheries in the UK. The existing capacity is for 100 k halibut, 200 k wrasse and 600 k lumpfish, with advance stage planning to increase to 200 k halibut, 500 k wrasse and 1 million lumpfish.

Machrihanish

The Machrihanish marine hatchery (Fig. 19.3) was commissioned in 2002 as a facility to supply juvenile cod for on-growing in sea cages. Due to the demise of the cod industry the unit was closed in 2008 and mothballed. Marine Harvest took over the buildings in 2010 to investigate the feasibility of producing farmed wrasse. In the first instance all five wrasse species were considered and investigated but it became apparent Ballan wrasse was the species on which to concentrate. Substantial numbers of wild Ballan broodstock were obtained in 2011 and 2012 from around local shores and established on three photoperiods to extend the spawning time from three to nine months. Scottish Sea Farms became an additional sponsor in 2012. The first major spawning occurred in 2013. However, water quality had to be improved and recirculation systems were installed. A substantial live feed facility was developed that caters for the volume of larvae produced and, as a result, has supplied other hatcheries with larvae and live food. The maximum output at Machrihanish is about 9000 kg, or 150,000 wrasse at 60g. A larger purpose-built facility is being opened to permit a substantial increase in production. The technical challenges of the last few years have enabled Machrihanish to emerge in a strong position to supply both Marine Harvest and Scottish Sea Farms with projected requirements of Ballan wrasse by 2020.

NAFC Marine Centre and Grieg Seafood Shetland Ltd

In late 2014 Grieg Seafood Shetland began a cleaner fish project focusing on the use of lumpsuckers as a biological control agent. Since the start of the project, significant investments have been made to adapt the hatchery at the NAFC Marine Centre in Scalloway (Fig. 19.3), Shetland, to provide lumpfish for deployment – optimising the facility for the on-growing of pellet-weaned lumpfish fry. The hatchery now has the infrastructure to deliver 15 μm drum-filtered and UV-dosed marine water at 40 $m^3 h^{-1}$ to six 18.5 m^3 and 15 2.7 m^3 tanks (total 666 l min^{-1}, 151 m^3). Fry are acquired in batches of 150,000+ at 0.5 g and grown on to 10g+ before being released into farm sites, with the entire process taking three to six months from start to finish (Adam Rainsden, Shetland, 2017, personal communication). He reported that individuals from the last batch of 2014 were held within the facility to determine age at maturation in captivity. Some individuals weighed in excess of 2 kg within 12 months of their arrival and had begun to produce large quantities of eggs and milt. The male lumpfish had also developed bright colours (green, orange, red and pink) and iridescent markings along the flanks were observed. Males became very territorial and appeared to be more agitated than the female fish, and clear signs of aggression were noted. Despite successful spawning within the tanks, a lack of available space, flow and funding required for the production of fry prevented attempts to hold broodstock within the NAFC. The focus of cleaner fish work by Grieg in Shetland is to continue the on-growing of fry at the NAFC while concentrating efforts towards the study of deployed lumpfish stocks, and so developing best practice for their future use within farm sites.

19.6 UK Hatchery methods

The main rearing features and requirements of wrasse are compared with lumpfish (Table 19.3). Wrasse broodstock have been caught locally and acclimatised over a one to two year period. A spawning stock comprises from 20 to 40 individuals with one to two males being present in each tank (Fig. 19.3). The social structure is important and may change with time. Generally Ballan wrasse broodstock are long-lived, up to 30 years (Dipper and Pullin, 1979) with fish undergoing gender reversal around 35 cm fork length. The survival rates of broodstock have been high with few health issues. Fish spawn at ambient water temperature from late April over ten to 12 weeks and with no requirement for hormonal treatment (Grant *et al.*, 2016). The demersal sticky eggs (Fig. 19.4) are collected by placing mats in the tanks and attempting to maintain good tank hygiene. The eggs are collected daily and disinfected with formalin and to a lesser extent Pyceze. Egg mats can be held in separate incubation tanks until shortly before hatch, which occurs after 100 degree days (D'Arcy *et al.*, 2012). Hatching can be stimulated and synchronised by mechanical rubbing of the mat. Enriched rotifers are offered at around 5 ml^{-1} after three days and for the first 28 days post-hatch, followed by feeding with Artemia for up to eight weeks. Weaning to dry diet can be slow and extended. Fish are graded and vaccinated prior to transport. Overall survival varies greatly between egg batches from 2 to 20% and with low levels of losses normally due to aggression throughout. Fish are from 50 to 70 g mean weight prior to stocking on salmon farms, and with relatively slow growth this can take from 18 to 24 months.

Wild-caught lumpfish broodfish are collected in January–April when temperatures are <10°C, using inshore nets close to the spawning grounds. At the NMC, fish are allowed to spawn naturally in submerged barrels and the eggs incubated with a male fish present for at least two weeks. Other hatcheries acclimatise the broodfish, provide rocky nest areas, and mainly strip the eggs manually and sacrifice some of the males for milt collection (see Powell *et al.*, 2017). For the final two weeks of incubation the eggs are disinfected and placed in upwelling systems until hatch (Fig. 19.4). Larvae are collected in a floating net, counted and stocked into 2000 l conical first feeding tanks at between 50–100 larvae per litre.

Table 19.3 Hatchery characteristics of cleaner fish species in the UK.

Criteria	Ballan wrasse	Lumpfish
Spawning Behaviour	Substrate spawner	Substrate spawner
	Difficult to strip eggs	Females easy to strip; males difficult
Spawning season	May–June	January–May
Incubation period	100 deg days	280 deg days
Size of larvae at hatch	3.6mm	5mm
Rotifers	Day 4–20	N/A
Artemia	Day 15–50	Day 4–21 (optional)
Weaning	Day 30–60	Day 4+
Average Survival	1–20%	50%+
Deployable size	40–60g	15–20g

Artemia are fed for the first two weeks, after which fish are weaned to dry larval feeds. Fish are graded for the first time at about 0.5 g average weight in a 5.5 mm hand-grading box. Larger fish are counted and stocked into 7 m^3 circular tanks, where they remain until 15 g size. During this period they are graded a further one to two times and vaccinated (dip and injection). Prior to shipment, fish are fasted for at least 48 hours and are transported in conventional transport tanks at a density of about 35–50 kg m^{-3}. Lumpfish husbandry methods and technical challenges are reviewed in Powell *et al.* (2017)

19.7 Deployment of cleaner fish

Wester Ross Fisheries

Companies such as Wester Ross Fisheries (WRF) operate within the RSPCA Freedom Food welfare standard (RSPCA, 2015) that applies to both salmon and cleaner fish. Wrasse are mainly caught locally and they are thus conditioned to local water conditions and early acclimation stress is minimised. As the natural instinct of wrasse is to hide in crevices such as folds in nets, swim-through net changing procedures have been adapted to prevent damaging wrasse. A simple inversion of the net corners before lifting the net out of the water flushes wrasse into the main body of the pen, avoiding capture in net folds and damage. This extra step has had an added welfare benefit for the salmon.

Experienced wrasse are such a valuable asset it is important to retain these at the end of a production cycle, so they may be reused for a subsequent salmon intake. Retaining wrasse that have learned lice cleaning behaviour is vital so that a quick response to sea lice recruitment is available. Naïve wrasse may take up to eight weeks to become effective cleaners. As many wrasse as possible are removed from pens that are earmarked for harvest, using baited creels that are placed in the pens a few days prior to harvest. Wrasse so captured may be either redeployed into other pens

Fig. 19.4 Wrasse spawning broodstock (at FAI Ardtoe).

Fig. 19.5 Hatchery methods for lumpfish at (a, b) NMC egg incubation unit and larval rearing tanks, and (c, d) incubation and larval rearing at CSAR Swansea.

or kept in a holding pen until required. However, as some wrasse may evade capture using this method, it is important to recover these fish at harvest.

Hides

Wrasse live in a sublittoral rocky habitat, or kelp beds. For wrasse to settle into the farm environment, minimise stress and become effective lice eaters it is important to give them adequate refuge. Hides simply made from lengths of PVC pipe bonded together (Fig. 19.6), with some artificial kelp provide adequate shelter. As these units are modular, more units can be added to the pen to suit the number of wrasse present.

Fig. 19.6 Cleaner fish deployment, (a) hides, (b) refuges, and (c) feeding stations.

Wrasse Feeding

Wrasse are grazers and require small frequent meals, and so are ideally suited to the role of eating sea lice. However, when lice levels are too low to provide adequate nutrition or maintain body condition, supplementary feeding is essential. Green crab (*Carcinus maenas)* or blue mussel (*Mytilus edulis)* are both known to make up part of the natural diet of wrasse, are locally abundant, and are readily taken by wrasse. Recently, some commercially formulated wrasse diets have become available that may offer a more complete nutritional profile. However, since these diets are often adaptions of hatchery feeds, it will take time to assess their effectiveness for adult fish in the farm environment. The method of delivery for feed, both natural and formulated, needs further consideration. Weighted net bags or open top trays can be used as a method for feed delivery that can be grazed upon, as wrasse do not take sinking feed.

Stocking density

Optimising stocking of wrasse is important to regulate wrasse requirement and help to continue a sustainable fishery into the future. Through experimenting with various stocking densities and with a better understanding of wrasse behaviour and husbandry the effective stocking density has been reduced by Wester Ross Fisheries from 5% in 2013 to 2.5% in 2016 while still achieving zero numbers of adult female sea lice.

As wrasse are deployed as part of a mixed species strategy and not alone, it has proven difficult to assess the survival and cleaning efficacy of each species.

Deployment of cleaner fish by Scottish Sea Farms SSF

SSF began to work on the challenges of making wrasse an effective tool in the management of sea lice from 2010 when agreement was reached with Freedom Foods to permit stocking with salmon. It was clear that this would take time but would also have the capability of delivering an exciting and environmentally friendly method of dealing with this industry problem. There are still many questions that remain unanswered, with the most important factor being that wrasse need to be managed and cared for as a second species within the salmon pen. Therefore it is important to have trained and dedicated staff who have an understanding of the requirements of wrasse and the ability to recognise and maintain good health and welfare.

Until recently wrasse have been caught from the wild and stocked directly into salmon pens.

Farmed wrasse are just becoming available and it is expected that wrasse will only be sourced from farms within the next three to five years. This is the ideal solution as this will meet the numerical demands of farmers as the wrasse will be available as required rather than waiting on the success of fishermen. Also, farmed wrasse will be vaccinated against common wrasse diseases and, in particular, atypical furunculosis.

Some key points that should be considered in the transfer of farmed cleaner fish to farms are:

- Farmed cleaner fish should be in good health prior to transfer. A health surveillance check for AGD and atypical furunculosis should be carried out.
- Cleaner fish should be conditioned prior to transfer, with the same temperature, light and feed as in the receiving pen.
- They should be graded.
- Wrasse from farms should be 40–50 g at transfer; and lumpfish *c.* 15–30 g.
- The nets on receiving pens should be clean, so removing sources of wild feed.
- Cleaner fish should be stocked directly into hides to provide immediate shelter (Fig. 19.6b).
- Cleaner fish feed should be introduced soon after transfer to encourage them to start feeding (Fig. 19.6c).
- The condition factor K and welfare indicators should be recorded on arrival so that there is a benchmark to monitor performance. These should be measured regularly.
- Salmon should be well fed before stocking with cleaner fish and fed at a distance from the cleaner fish hides and shelters.
- A routine should be established for cleaning the hides.

19.8 Key points in rearing

Rearing of Ballan wrasse has been more difficult than lumpfish production (Chapter 3; Prickett, 2016). The main issues in Ballan wrasse production include disease control, particularly against atypical furunculosis, swimbladder over-inflation (probable symptom of stress), feed formulation (wrasse are fussy eaters and are susceptible to dietary deficiencies, e.g. selenium), slow growth and deformities. Lumpfish issues include poor milt production from male fish, sucker deformities preventing fish from settling, fin erosion and sensitivity to bacterial infections post-deployment (Chapter 13). Disease issues in both species are being addressed by the use of autogenous vaccines against atypical furunculosis, vibriosis and pasteurellosis with encouraging results. This is an important part of preparing fish for deployment as well as improving management of cleaner fish in the pens by using optimally designed hides and feeding strategies. In the hatchery there is a need to produce eggs throughout the year, which has been achieved for Ballan wrasse through the manipulation of photoperiod and temperature. Lumpfish egg supplies are still dependent on wild-sourced broodstock and out of season egg supplies have to be imported from Iceland and Norway where fish spawn for longer periods. Topics currently being researched include lumpfish genetics and traceability, factors underlying egg and larval survival, health management, welfare and growth. The development of a selection programme for improved disease resistance and better delousing efficiency are the next challenges for the incipient lumpfish industry. Genetic selection programmes are also starting in the case of lumpfish, where selection for lice eating ability and slower growth will be the required traits (reviewed in Powell *et al.*, 2017). Finally, increasing Ballan wrasse production and reducing production costs is of great importance as summer temperatures in the UK will not be optimal for lumpfish.

19.9 Cleaner fish use strategy in Scotland

Sea lice management is of high impact to the salmon farming industry and associated partners in the UK as a whole, both economically and for public image. The collaboration between industry partners, the sharing of ideas and resources, successes and failures has permitted more rapid progression on cleaner fish deployment than would have been possible otherwise. Companies such as WRF and SSF are committed to using cleaner fish as the primary sea lice management tool, and have demonstrated effectiveness on a commercial scale. Research and development will continue between industry and academic partners into areas such as optimising wrasse stocking densities, better understanding of wrasse dietary requirements, commercial scale production of farmed wrasse, and health screening of wrasse. However, despite the continuing success the industry is achieving in Scotland in utilising biological sea lice control, it is important to retain a range of sea lice management options (Rae, 2002). These should include access to efficacious medicines, as well as newer mechanical lice removal systems (Powell and Pooley, 2015). Farmers need access to an extensive toolbox of management options to effectively deal with the sea lice challenge in what is a challenging natural environment.

References

Bjordal, Å. (1988) Cleaning symbiosis between wrasse (Labridae) and lice infested salmon (*Salmo salar*) in mariculture. *ICES Mariculture Committee 1988*/F17, 8 pp.

Bjordal, Å. (1991) Wrasse as cleaner fish for farmed salmon. *Progress in Underwater Science* 16, 17–28.

Bjordal, Å. and Costello, M. (1990) How good is this natural control of sea-lice? *Fish Farmer* 13 (3), 44–46.

Bron, J. and Treasurer, J.W. (1992) Sea lice (Caligidae) on wrasse (Labridae) from selected British wild and salmon farm sources. *Journal of the Marine Biological Association of the UK* 72, 645–650.

Chalaris, A. and Treasurer, J.W. (2010) Developing the commercial rearing of goldsinny wrasse for the biological control of sea lice. *Fish Farmer* 33, 59-61.

D'Arcy, J., Dunaevskaya, E., Treasurer, J. W., Ottesen, O., Maguire, J., Zhuravleva, N., Karlsen, A., Rebours, C. and FitzGerald, R. D. (2012) Embryonic development in ballan wrasse *Labrus begylta*. *Journal of Fish Biology* 81, 1101–1110.

Dipper, F.A. and Pullin, R.S.V. (1979) Gonochorism and sex-inversion in British Labridae (Pisces). *Journal of Zoology* 187, 97–112.

Grant, B., Davie, A., Taggart, Selly, S.-L.C., J.B., Picchi, N., Bradley, C., Prodohl, P., Leclercq, E. and Migaud, H. (2016) Seasonal changes in broodstock spawning performance and egg quality in ballan wrasse. *Aquaculture* 464, 505–514.

Keay, A., Powell, A. and Garcia de Leaniz, C. (2015) Cleaner Fish Research. Passing the pest test. *Fish Farmer Magazine*, November, pp. 68–69: www.fishfarmer-magazine.com

Laidler, L.A., Treasurer, J.W., Grant, A.N. and Cox, D.I. (1999) Atypical *Aeromonas salmonicida* infection in wrasse (Labridae) used as cleaner fish of farmed Atlantic salmon, *Salmo salar* L., in Scotland. *Journal of Fish Diseases* 22, 209–213.

Marine Scotland (2017a) Landings by Scottish vessels into Scotland. www.gov.scot/Resource/0052/00524991.pdf

Marine Scotland (2017b) Aquaculture production statistics Scotland. /gov.scot/Publications/2017/09/5208/6

Munro, E.S., McIntosh. R.E., Weir, S.J., Noguera, P.A., Sandilands, J.M., Matejusova, I., Mayes, A.S. and Smith, R. (2015) A mortality event in wrasse species (Labridae) associated with the presence of viral haemorraghic septicaemia virus. *Journal of Fish Diseases* 38, 335–341.

Powell, A. and Pooley, C. (2015) *Innovative technologies for control of Salmon Lice.* Conference Poster: Biopesticides – innovative technologies and strategies for pest control; Swansea, UK.

Powell, A., Howes, P.N., Stringwell, R. and Garcia de Leaniz, C. (2016) Aquatic research at Swansea University. *Hatchery International,* May/June, 24–25.

Powell, A., Treasurer, J.W., Pooley, C.L., Keay, A.J., Lloyd, R., Imsland, A.K. and Garcia de Leaniz, C. (2017) Use of lumpfish for sea-lice control in salmon farming: challenges and opportunities. *Reviews in Aquaculture* DOI:10.1111/raq.12194

Prickett, R. (2016) A review of cleaner-fish production in the UK. *Aquaculture Association of Canada Bulletin* 2016–2, 53–58.

Rae, G.H. (1991) *The use of wrasse as cleaner fish in salmon farming.* SSGA report.

Rae, G.H. (2002) Sea louse control in Scotland, past and present. *Pest Management Science* 58, 515–520.

Riley, A., Jeffery, K., Cochrane-Dyet, K., White, P. and Ellis, J. (2017) *Northern European Wrasse – Summary of commercial use, fisheries and implications for management.* CEFAS report, Lowestoft. 18 pp.

RSPCA (2015) *RSPCA Welfare Standards for Farmed Atlantic Salmon*. RSPCA, Horsham, UK. Available at: http://www.rspca.org.uk/sciencegroup/farmanimals/standards. Accessed 13 May 2017.

Sherwood, J.P.B. (1990) *Wrasse breeding trial.* Sea Fish Industry Authority Marine Farming Unit. 11 pp.

Sherwood, J. (1991) *Wrasse breeding trial.* Sea Fish Industry Authority Marine Farming Unit. 10 pp.

Stone, J. (1996) Preliminary trials on the culture of goldsinny and corkwing wrasse. In: Sayer, M.D.J., Treasurer, J.W. and Costello, M.J. (eds). *Wrasse biology and use in aquaculture.* Fishing News Books, Oxford, pp. 142–170.

Towers, L. (2015) Ardtoe reports first spawning of captive reared lumpsuckers. www.thefishsite.com/articles/ardtoe-reports-first-spawning-of-captive-reared-lumpfish (accessed18/11/17)

Treasurer, J. (1991b) Limitations in the use of wrasse. *Fish Farmer* 14, 12–13.

Treasurer, J. (1991a) Wrasse need due care and attention. *Fish Farmer* 14, 24–26.

Treasurer, J.W. (1992). *Screening of five wild caught wrasse species from multiple sea lochs for bacterial and viral diseases.* Marine Harvest, Scotland. Unpublished report.

Treasurer, J. (1993) More facts on the role of wrasse in louse control. *Fish Farmer* 16, 37–38.

Treasurer, J.W. (1996) Capture techniques for wrasse in inshore waters of west Scotland. In: Sayer, M.D.J., Treasurer, J.W. and Costello, M.J. (eds) *Wrasse biology and use in aquaculture.* Fishing News Books, Oxford, pp. 74–90.

Treasurer, J.W. (1997) Parasites of wrasse (Labridae) in inshore waters of west Scotland stocked as cleaner fish of sea lice (Caligidae) on farmed Atlantic salmon. *Journal of Fish Biology* 50, 895–899.

Treasurer, J.W. (2012) Diseases of north European wrasse (Labridae) and possible interactions with cohabited farmed salmon, *Salmo salar* L. *Journal of Fish Diseases* 35, 555–562.

Treasurer, J.W. (2013) *Use of wrasse in sea lice control.* SARF068 Report, Birnam, www.sarf.org.uk

Treasurer, J.W. and Laidler, L.A. (1994) *Aeromonas salmonicida* infection in wrasse (Labridae) used as cleaner fish, on an Atlantic salmon, *Salmo salar* L., farm. *Journal of Fish Diseases* 17, 155–161.

CHAPTER 20

Cleaner fish use in Ireland

Majbritt Bolton-Warberg,[1] Stephen Murphy O'Sullivan,[1] Amber Irwin Moore,[1] Anne Marie Power,[2] Luke Wilson,[3] Felix Sproll,[3] Sandra Schlittenhardt,[3] Mohd Faizzi Bin Mohd Yusof,[4] Jack D'Arcy,[5] Dave Jackson[5] and Richard D. FitzGerald[†1]

[1]Carna Research Station, Ryan Institute, National University of Ireland, Galway, Ireland
[2]School of Natural Sciences, Ryan Institute, National University of Ireland, Galway, Ireland
[3]Marine Harvest Ireland, Letterkenny, Co. Donegal, Ireland
[4]Centre for Foundation Studies in Science, University of Malaya, Kuala Lumpur, Malaysia
[5]Marine Institute, Oranmore, Co. Galway, Ireland

20.1 Cleaner fish species in Ireland

This chapter reviews recent research on cleaner fish (labrid wrasse species and the lumpsucker, *Cyclopterus lumpus*) in Ireland and available literature on the biology and use of cleaner fish occurring naturally in Irish waters. An overview is also given of deployment of cleaner fish on salmon farms in Ireland, principles applied and information on production trials.

20.1.1 Wrasse

Wild stocks

The five species of commonly occurring wrasse in Ireland are goldsinny (*Ctenolabrus rupestris*), rock cook (*Centrolabrus exoletus*), corkwing wrasse (*Symphodus melops*), Ballan wrasse (*Labrus bergylta*) and cuckoo wrasse (*Labrus mixtus*). All of these are widely distributed along the eastern margins of the North Atlantic Ocean on a north–south axis from North Africa to the Arctic (mid-Norway).

It is evident that the four, shallow water inshore wrasse species, excepting cuckoo wrasse which may occur in deeper waters, appear to be very widely distributed and common on rocky shores along the southern, western, and northern coastlines of Ireland. They are often associated with macroalgal habitats. High densities of goldsinny and rock cook have been associated with complex rocky habitats containing many refuges (Thangstad, 1999; Sayer *et al.*, 1993), although rock cook have also been observed to aggregate over sandy weedy areas (Thangstad, 1999). Ballan wrasse inhabit similar areas to goldsinny and rock cook (Sayer and Treasurer, 1996), but may prefer more exposed coastline. Wrasse fishery surveys in Ireland are reported in Chapter 17, and all five species of wrasse commonly used as cleaner fish in aquaculture are found in Irish waters.

Fig. 20.1 Wild egg mass found on rocks in the lower shore on a spring tide in Moyrus, Co. Galway. Credit: M. Bolton-Warberg.

20.1.2 Lumpsucker

Wild stocks

The success of targeted fishing for lumpsucker broodfish off Irish coasts is described in Chapter 17. In addition, wild egg masses have been found along the shoreline on spring tides during the spawning season (late March to May, Figure 20.1).

Spawning (wild and farmed)

Spawning of lumpsucker in Ireland is believed to occur during the spring months (March–May) but may vary if seasonal temperatures fluctuate from normal. Mature females produce multiple batches of eggs with approximately ten days between spawning events. The number of eggs produced varies and is dependent on size, age and condition. For example, a wild-caught female (age unknown) at 1.9 kg can produce between 400 and 500 g of eggs in a single spawning (unpublished data, Carna Research Station, 2017, personal communication). A similar size of fish (2.0 kg, two+ years, first-time spawner) reared from juvenile in a laboratory environment may produce 200–300 g (unpublished data, Carna Research Station, 2017, personal communication). The general consensus is that there are roughly 100,000 eggs per kg; however, egg counts from an Irish hatchery (Carna) have shown this value can range between 97,000 and 140,000. In a hatchery environment, females as small as 500 g have been known to produce small batches of eggs (<100 g, in early June, Ireland), but, in the absence of mature males at similar sizes, it is unknown whether the eggs are viable. Males have been observed to guard the egg mass for the duration of the incubation period, protecting it from predators both in the wild (Fig. 20.2) and in captivity (Fig. 20.3). To ensure there is adequate water flow to/over the eggs, the males either blow water on the mass or fan it with their fins. The male makes indentations in the egg mass with its head shortly after fertilisation before the egg mass hardens to facilitate this ventilation (see Figs 20.2 and 20.3). Egg colour ranges from cream, to orange to purple. In laboratory tanks, males aggressively protect eggs and have been observed to attack other fish that get too close. Even in the absence of an egg mass, mature males can be aggressive to subordinate/ smaller males.

Fig. 20.2 Wild male guarding egg mass found on the shoreline on a spring tide on Muighinis, Co. Galway, Ireland. Note four distinct indentations in egg mass to facilitate 'fanning' of eggs. (Credit: M. Bolton-Warberg).

Fig. 20.3 Male lumpsucker fanning an egg mass in the hatchery.

20.2 Use of cleaner fish in Ireland

20.2.1 Historical use of cleaner fish in Ireland

In the 1990s, a significant amount of the early research on the use of wrasse species (goldsinny, rock cook and corkwing) as cleaner fish on salmon farms was undertaken in Ireland (e.g. Donnelly, 1992; Costello, 1994; Deady *et al.*, 1995; Tully *et al.*, 1996). While these early trials confirmed that these Labridae were effective at removing *Lepeophtheirus salmonis* lice, there were mixed results with *Caligus elongatus*. Several theories were put forward to explain these opposing findings, including: loss of wrasse from cages (escapees, winter mortalities, diseases); inadequate refuges; lack of true control groups and environmental factors (e.g. freshwater input reducing surface salinity).

Nonetheless, these combined studies proved the efficacy of various wrasse species at removing lice and some studies even showed improved growth rates and reduced mortality of salmon in cages stocked with wrasse compared with control groups (e.g. Darwall *et al.*, 1992). In addition, researchers highlighted:

- The need to quantify infestation pressure and to validate wrasse to salmon ratios throughout the study period (Tully *et al.*, 1996).
- The importance of not having mixed generation salmon stocked in close proximity. (Larger salmon, prior to harvest, are not treated and may act as lice reservoirs. During harvest, lice become detached and can cause 'lice blooms', which can then settle on neighbouring salmon (e.g. Deady *et al.*, 1995)).
- The importance of having a sea lice control that reduces stress.
- Fallowing of bays, to break infection cycles (it is now a condition of all marine licences in Ireland that sites are fallowed for a minimum of four weeks between generations).
- Reducing loss of wrasse by either reducing mesh size or increasing fish size (e.g. corkwing is larger than rock cook or goldsinny).

With queries over possible disease transfer from wild wrasse to farmed stocks and the development of more effective medicinal treatments for sea lice control at that time, the use of cleaner fish in Ireland was eventually discontinued. In addition, as a result of suspected atypical furunculosis in some stocks, all work on cleaner fish was suspended in the early 2000s.

20.2.2 More recent trends in cleaner fish research, production and use

Carna Research Station (CRS, National University of Ireland Galway) has been actively involved in research on Ballan wrasse over the past decade. Initially, a short pilot study in the period 2007–2009, involving Marine Harvest Ireland Ltd (MHI) and supported by BIM (Ireland's Seafood Development Agency), was conducted to establish captive broodstocks for production of juveniles for use by MHI and to explore larval development and rearing of Ballan. This preliminary project led into and partially supported a PhD work programme (D'Arcy, 2013, unpublished) and this progressed to yield a number of key papers on egg and larval development as well as significant data on the genetic structure of wild populations in Ireland and further afield (D'Arcy *et al.*, 2012, 2013; Ottesen *et al.*, 2012a). A natural continuation of this work (2010–2014), formed a further PhD study (Mohd Yusof Mohd, 2015, unpublished), looking at the general biology of wild Ballan stocks in Irish waters including reproductive maturation and spawning patterns, both natural and under induction using various hormonal analogues.

At the same time (2009–2013), the EcoFISH project proposal was successfully funded under the Northern Periphery Programme (NPP) of InterReg, with partners in Norway (Univ. i Nordland; Bioforsk), Scotland (Viking Fish Farms) and Ireland (Daithi O'Murchu Research Centre; CRS, NUI Galway), and with national, private and governmental grant aid. The project focused on developing methods for the culture and use of Ballan wrasse as a cleaner fish. This project had very ambitious objectives and, while achieving several advances (Ottesen *et al.*, 2008; Dunaevskaya *et al.*, 2012; Ottesen *et al.*, 2012b), it was only partially successful. From it a series of technical leaflets on specific aspects of the biology and rearing of Ballan wrasse were produced and these can be accessed at www.eco-fish.org.

During this latter period, Irish farms, with assistance from BIM, again commenced capturing live wild wrasse, most particularly Ballan but also other smaller species. These were stocked into commercial salmon cages to control sea lice infestations as part of pilot trials. This programme is ongoing and involves the use of inshore fishers, with appropriate gears, working in cooperation with commercial salmon farmers to provide seasonal supplies of wild-caught wrasse. Live-catch fishing in Norway and Scotland is carried out in areas of high wrasse abundance and then live stocks are transferred to farms within each country. In contrast, in Ireland, transfer of cleaner fish is subject to strict regulation and, therefore, live-catch fishing is limited to the area immediate to the farm where the wrasse are to be stocked. This regulation has two implications: firstly, seasonal fishing pressure on local wrasse stocks may be relatively intense and, secondly, because of the very localised nature of salmon production and capture fishing in limited regions of the Irish coast, the native wrasse stocks are likely not to be affected outside such areas.

20.2.3 Current national regulations (fish health and stock movements)

The Marine Institute is the designated competent authority in Ireland under EU Directive 2006/88/EC, governing the health regulations of aquaculture animals and the prevention and control of certain aquatic diseases. Every aquaculture site in Ireland must have a Fish Health Authorisation from the Marine Institute and all requirements relating to record keeping, health surveillance and biosecurity must be in place before this can be granted. Only facilities with an FHA can apply for a movement order. These applications are submitted to the Fish Health Unit of the Marine Institute in advance of moving fish offsite (five days in advance for exports, three days for internal movements) and when importing fish (five days in advance). All movement orders must be approved before transfer. A registered veterinarian is required to provide a pre-movement report on eggs or fish being moved into or around the country and in certain cases laboratory testing is also required (e.g. for wild-caught fish). For movements into Ireland from another country, health screening is also required and the stock must be certified by the authorities in the country of origin. To facilitate this, the Trade Control and Expert System (TRACES) is used. This web-based system is run by the European Commission's Directorate-General for Health and Consumer Protection. It processes all necessary documentation and sends copies to the appropriate authorities in both countries. Any site that experiences increased mortality must investigate the losses with the assistance of their veterinary practitioner. Where the presence of a listed or emerging disease is suspected, the Marine Institute must be notified.

In addition, the Marine Institute has developed a Finfish Health Handbook (see www.fishhealth.ie) to assist producers in protecting fish health and promoting welfare on Irish farms. The current version of this handbook includes a section specifically on cleaner fish and covers health screening, moving, stocking and wild capture. Risk assessments for the use of cleaner fish in Ireland (including import of ova and wild capture) have also been developed by governmental agencies and an aquatic veterinary consultancy.

Wild-caught fish

In order to address potential biosecurity issues regarding the movement of wrasse and their introduction into salmon pens, disease screening must be carried out to determine the health status of the wrasse stock in each designated fishing area. This prerequisite testing is undertaken prior to the commencement of fishing, whereby a subsample (N~60) from each bay is sampled for a variety of pathogens. Sampling for virology, including screening for the presence of VHS, IPN and IHN, and molecular testing for the presence of SAV, PRV, PMCV is carried out by the Fish

Health Unit in the Marine Institute, while histological, parasitological and bacteriological testing is performed by a veterinary consultancy.

Wild-caught broodstock lumpsucker undergo a similar screening process following the spawning season. Any fertilised egg batches obtained from these fish are maintained under biosecure conditions until the testing is completed and a healthy determination is confirmed.

Both wrasse and lumpsucker are non-quota species; however, anyone commercially fishing for cleaner fish within Ireland must hold a suitable (inshore) licence from the Department of Agriculture, Food and the Marine. Some measures put in place by Marine Harvest Ireland to reduce overexploitation of wrasse include:

- Keeping fish between 11 and 25cm only.
- Fishing after the spawning season has ended (June in Ireland).
- Bays are fished for one to two seasons (years), followed by one season of no fishing (e.g. no fishing from October 2016–May 2018).
- No fishing after September to allow pre-conditioning and due to drop in water temperatures.
- Fishermen are given a manual on cleaner fish welfare (translated from Marine Harvest Norway) and are trained in best practices during the first season.

Hatchery-reared fish

All aquaculture sites are obliged to be in compliance with such regulatory requirements as are laid down by the relevant 'aquaculture' legislation and are routinely inspected and audited by veterinary inspectors of the Department of Agriculture, Food and Marine (DAFM) acting on behalf of the Marine Institute as the relevant competent authority. In addition, a fish health management plan (FHMP) must be developed and maintained that covers all aspects of fish health and welfare, including monitoring of health (regular clinical examinations carried out by a competent veterinarian), movements, production, environmental monitoring, feeding and treatments. Within this remit, there is a requirement for a competent veterinarian to undertake regular clinical examinations of stocks (four times per annum) to ensure ongoing health of the fish. This is in addition to pre-movement inspections of stocks to be transferred.

In recognition of the enactment of S.I. No. 543 of 2012 and the designation of the Health Products Regulatory Authority (HPRA) as the competent authority for such activities, it is deemed necessary to be in compliance with the new legislation for the culture of finfish, and for any such procedures as may be undertaken on these stocks, within a nominated area/building of the existing site and facilities.

20.3 Aquaculture biology and production of wrasse in Ireland

20.3.1 Underpinning research

There is currently no active Ballan wrasse production in Ireland. However, in the past decade, two PhD studies have undertaken key investigations on the biology of *L. bergylta* that have made a significant contribution to our understanding of the biology of this species in Irish waters and underpin any future aquaculture production initiatives.

D'Arcy (2013, unpublished) focused on the biology of egg and larval development in Ballan wrasse. While the life history of adult *L. bergylta*, as protogynous hermaphrodites, is somewhat complex, it was found that the egg and embryonic development is quite conservative and clearly typical of demersal marine finfish species. The eggs are small and spherical and have an adhesive covering layer for attachment to benthic substrates. In the course of development, the eggs display standard features or landmarks and progress through a series of key developmental events, charted as a sequence of standardised stages (I to VI+). This includes, for example, meroblastic cleavage of the protoplasm, formation of an embryonic streak, rapid development of the cephalic end rather than the tail, somites forming initially around the mid-region, and emergence of a yolk sac larva. The key morphological features, following standardised nomenclature, are summarised in Table 20.1 and depicted in Figure 20.4 (see also D'Arcy *et al.*, 2012). Such features and the sequence is typical of many marine finfish with a relatively short embryonic egg life (for example, Fuiman, 2002), probably best known and documented for cod (e.g. Geffen *et al.* 2006), and a similar sequence has also been observed and recorded for an *L. bergylta* congener, the brown wrasse *L. merula* (Linnaeus), by Dulcic *et al.* (1999). This staging series provided a reference baseline for future studies on the development of Ballan eggs/embryos and for standard use in production hatcheries.

In a parallel study conducted in Norway with inputs of material and data from Ireland, age at hatching and larval ontogenesis based on external morphological features were linked to the chronological age of larvae as days post-hatching (DPH), physiological age as degree day (°C day) and standard length (SL) (Ottesen *et al.*, 2012a). The ontogeny of larvae from hatching to metamorphosis was divided into four stages: yolk sac larva (0 to 9 DPH); pre-flexion larva (10 to 25 DPH); flexion larva (26 to 33 DPH) and post-flexion larva (34 to 49 DPH). The ontogenetic development, linked to SL, chronological and physiological age, provided a baseline reference and, again, made it possible to compare developmental levels in studies of Ballan wrasse in cultured and wild populations. Other findings include:

- A significant negative (exponential) relationship between increasing temperature and time to hatch.
- Lapillar diameter (otolith) was the best predictor of age in larval Ballan wrasse.
- A single ring is deposited daily in the sagittal otoliths in larvae.
- There was an ontogenetic shift in the relative values of somatic and otolith micro-incremental growth, which corresponded to larval stage transition points.
- Ballan wrasse in southern Norway showed lower levels of genetic diversity than those from Ireland and Scotland, which appear to be the result of a population bottleneck followed by expansion (D'Arcy *et al.*, 2013). Two highly divergent clades suggest distinct recolonisation patterns in these two regions. These results have important implications for conservation and management.

Mohd Yusof Mohd's PhD thesis, (2015, unpublished) focused on different aspects of the life history of *L. bergylta,* as a precursor to developing practical hatchery production protocols, including: (i) reproductive cycles and nutritional status of stocks in the annual cycle of wild *L. bergylta*; (ii) spawning induction protocols for artificial spawning; (iii) larval weaning in co-feeding trials and (iv) age and growth determination from otolith microstructure in larval and adult otoliths.

Table 20.1 Egg development and staging – morphological characters and key features in Ballan wrasse.

Stage	Stage name	Stage Description	Circumference (degrees)	Observed morphological features
Ia	Zygote	Fertilization & zygote		
Ib	Cleavage	Early & Late Morula		
II	Blastula	Early & Late Blastula		
III	Gastrula	Pre-early gastrula		Germ-ring begins outward migration.
		Early Gastrula		Blastoderm flattens. Germ-ring clearly visible.
		Mid-gastrula		Germ ring is well defined and steadily enveloping yolk.
		Late gastrula		Blastoderm covers 3/4 yolk; Embryonic body becomes clearly visible; the enveloping layer expands uniformly over the yolk until this point.
		Early neurala	180	Oil droplets migrating toward vegetal pole; Yolk plug exposed; Rudimentary brain. Beak-like mass of cells. Kuppfer's vesicles appear.
		Late neurala	200	Rudimentary eye vesicle; Beak-like mass of cells is still visible; Kuppfer's vesicle enlarge somewhat; Blastopore at the vegetal pole
IV	Segmentation		220	2-4 Somites; Eye vesicle slightly more defined.
	Embryo extends 2200-3600 of circumference around yolk		290	Oil droplets migrate closer; Pigmentation begins; Eyes more distinct; 8-10 Somites; Flat body cavity; Kuppfer's vesicle large; Otolith vesicle; Heart appears; Beak-like mass of cells disappears from anterior of head.
			310	Oil Droplets migrate closer; Kuppfer's vesicle starts to shrink; 16-18 Somites; Heart appears; Brain distinct; Body cavity expands.
			330-340	Chevron shaped somites (24-26); Tail still attached; Kuppfer's vesicle reduces; Brain distinctly yellow; Otolith vesicle; Lens and cornea evident.
V	Segmentation		360-370	Pigmentation denser; 30-32 Somites; KV disappears; Cornea/lens distinct; Tail slightly separate from yolk; Beginning of membranes on caudal fin; Heart grows.
	Embryo extends 3600-4350 of circumference around yolk		380-410	Oil droplets coalesce; 36-38 Somites; Heartbeat 82 bpm; Membrane on tail more pronounced;
VI	Segmentation		435	40-44 Somites; Tail reached anterior of eyes; Tail membrane larger; Pectoral fin protruding; Heartbeat 87 bpm; Pigmentation denser.
	Embryo extends 4350-5100 of circumference around yolk		480	Denser pigmentation; 46-48 Somites; Heartbeat steady at 94 bpm.
VI+	Hatching (5100)		510	Oil droplet difficult to see; 46-48 Somites; Heartbeat 93 bpm; Otolith within vesicle; Cornea/lens well developed;
			510	Pigments very dark and dense; 46-48 Somites; Heartbeat 98bpm.

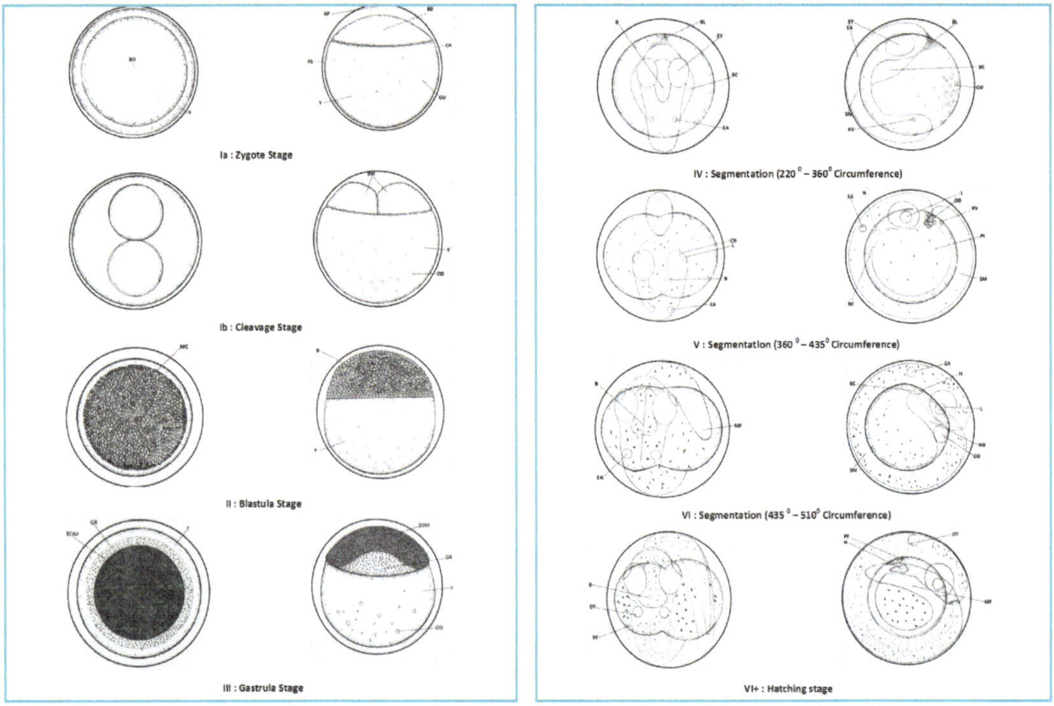

Fig. 20.4 Embryonic developmental stages in Ballan wrasse. Top (left) and lateral (right) views: Key to legend = B, brain; BC, body cavity; BD, blastodisc; BL, beak-like mass of cells; BM, blastomeres; CF, caudal fin; CH, chorion; CN, cornea; EA, otic (ear) vesicle; EY, optic (eye) vesicle; GR, germ-ring; H, heart; KV, Kuppfer's vesicle; L, lens; MC, marginal cells; MF, membranous fin; OD, oil droplet; OT, otolith; PF, pectoral fin; PS, perivitelline space; SM, somite; Y, yolk sphere.

One of the more interesting and germane findings to aquaculture production was the elucidation of gonadal maturation of *L. bergylta,* using tissue samples and histological screening of materials, and their correlation with plasma sex steroid levels and a range of parameters including gonado-somatic (GSI) and hepato-somatic (HSI) indices, over a 16-month period from April 2010 to July 2011. This study used wild-caught *L. bergylta* collected in Carna Bay, Co. Galway on the west coast of Ireland: these findings are summarised in Figure 20.5 (a), (b) and (c).

The histological data showed that ovarian recrudescence (renewal) started in November and December (Stage II ovaries) with the appearance of early maturing oocytes containing cortical alveoli accompanied by higher concentrations of oestradiol and GSI values. Increases in oestradiol, testosterone and GSI values continued towards January, along with the appearance of early vitellogenic oocytes (Stage III ovaries) in February. Oogenesis proceeded rapidly from February to April as oestradiol concentrations increased, and by March, most of the fish contained Stage IV ovaries with late vitellogenic oocytes. Prevalence of Stage IV and mature Stage V oocytes increased in April and May. Post-spawned gonads (Stage VI oocytes) were most common in May and June and resting Stage VII oocytes were seen in July. By September ovaries contained mainly Stage VIII and I oocytes (the latter being primary oocytes forming basal tissues). Some evidence of sexual transition from females to males was observed. Observation of transitional gonads was occasional, but when it occurred, this was between June and September coinciding with the post-spawning period, or the resting stages of the gonads and low plasma concentrations of oestradiol and testosterone. The *L. bergylta* in this study appeared to enter a short quiescent period between August and October, when only previtellogenic oocytes and nests of oogonia were present in the ovary (Stage I). Atresia of unspawned oocytes was only complete by September.

From the above account and as shown in Figure 20.5, it is apparent that many of the key parameters, plasma steroids concentrations, general body condition, gonad and hepatic indices and histological data, all exhibited well-defined cyclical seasonal patterns that correspond very closely to those reported for Norway by Muncaster *et al.* (2010, 2013), with peak spawning in the April–May period.

Overall, plasma sex steroid concentrations in *L. bergylta* were found to be relatively low (<5 ng ml^{-1}) and the highest concentrations of circulatory oestradiol and testosterone in the plasma coincided with the occurrence of vitellogenic follicles in the ovary and sharply increasing GSI values and, equally, falling HSI and K values (data for K not shown). This latter interplay is almost certainly due to the mobilisation of energy reserves in the liver (resulting in lowered HSI) to feed the requirement of the ovary for the development of oocytes (giving rise to increased GSI). The

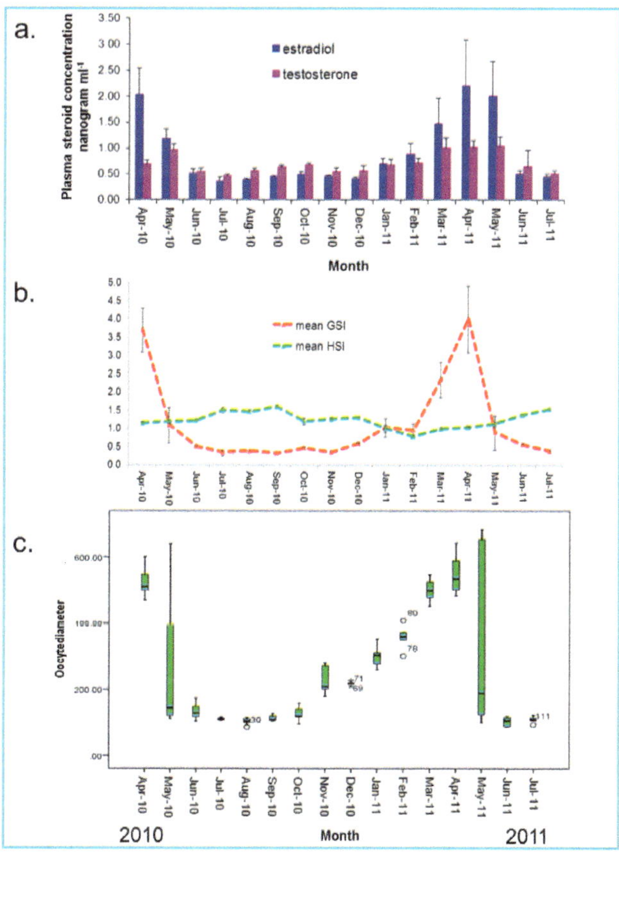

Fig. 20.5 (A) plasma steroid levels in Ballan wrasse, Mean values (+/- S.E.), (B) gonadosomatic index (GSI), and hepatosomatic index (HSI) and (C) Oocyte diameter during the period April 2010 to July 2011.

plasma concentrations of both oestradiol and testosterone started to increase in January 2011 during the pre-spawning phase, peaked in April (oestradiol) or May (testosterone) before falling sharply to their lowest levels in July of each year.

20.4 Aquaculture biology and production of lumpsucker in Ireland

20.4.1 Underpinning research

The National University of Ireland Galway has been involved in lumpsucker research and rearing since 2014 at its satellite facility, Carna Research Station. Following discussions with Marine Harvest Ireland Ltd and BIM (Ireland's Seafood Development Agency), a lumpsucker production programme (Cleaner Fish Programme) was initiated at CRS in 2015 and this continued into 2016. In 2017 CRS was a awarded a tender to develop a Lumpsucker Broodstock and Selective Breeding Programme using native Irish stocks. This project, funded under the EMFF Operational Programme 2014 under the Knowledge Gateway Scheme (KGS), is scheduled to run from 2017–2020.

In July/August 2014, researchers at Carna collected juvenile lumpsucker from seaweed longlines in the south west of the country. These juveniles (0.09–16 g) were maintained in recirculating systems for several months on a stable temperature regime (~17°C, ambient temperature at capture). Initially, a diet of ghost shrimp (unknown *spp*) was offered to the juveniles because a commercial dry diet was not accepted by the majority of the population. The fish were weaned on to a moist pellet incorporating shrimp in the beginning that was attached to cable ties suspended in the water. Temperature was gradually reduced in the recirculation systems until it coincided with the ambient temperature and the fish were subsequently moved to a system with a flow-through, ambient water supply. Juveniles were on-reared for use as future broodstock but, in the short term, feeding and behaviour were monitored and baseline growth profiles under Irish conditions were created.

The overall goal of the Cleaner Fish Programme 2015–2017 was applied research to facilitate and improve the production of juvenile lumpsucker up to 10–15 g, for transfer to sea cages on commercial salmon farms for use in an integrated management plan to control lice. In 2015 and 2016, eggs were sourced from a variety of origins (see Table 20.2) including the UK, Norway and Ireland. Hatch rate and survival were variable but considerable advances were made in 2016 with improvements to the facility (e.g. specialised cones for eggs), automatic feeders and feeding regimes.

Efforts were increased in 2016 to source native lumpsucker eggs and adults and nets were deployed in both northern and western bays with some success. Wild adult lumpsucker were caught off the Donegal and Galway coasts and transferred to holding barrels (Donegal) or tanks (Carna). Fish were housed in a ratio of 2:1 (male:female) to facilitate egg release and fertilisation. Once eggs were released and a male guarding them, the additional male and female were removed. The eggs and male were left for a further 7–10 days, then removed (or transported to Carna) and placed into incubation units in a holding system.

Of particular interest, a selection of the wild fish caught in 2014 became mature during the

Table 20.2 Eggs sourced from Ireland, Norway and Scotland in 2015 and 2016 for lumpfish production.

Year	Origin	# Eggs	Hatch Date	Hatch Rate (%)	Survival (%) 0 – 60 dph	# Transferred to Sea			Total # Produced*
						2015	2016	2017*	
2015	UK	380,000	4th April	N/A	14	31,937	51,288	0	105,631
	Ireland	104,000	20th April	N/A	21	9,315		0	
	Norway	275,000	14th May	N/A	34	13,091		0	
2016	Ireland	321,000	29th March – 1st June	45	93	0	35,844	208,004	243,848
	Norway	315,500	27-28th May	56	99				
TOTAL		**1,395,500**				**54,343**	**87,132**	**208,004**	**349,479**

*data correct as of 1st July 2017

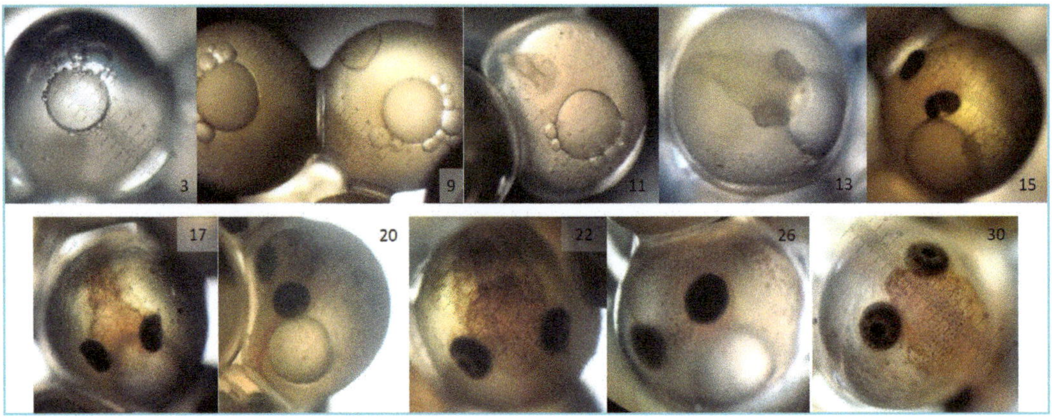

Fig. 20.6 Developmental stages of lumpsucker eggs reared at 8–9°C. Numbers are days post-fertilisation.

2016 season. As with the wild broodstock, mature females (N = 1) and males (N = 2, evident due to red colour) were housed in a single tank and allowed to spawn naturally. No conditioning diet was administered nor photoperiod manipulation employed. The eggs produced by these fish accounted for 40% of the Irish eggs incubated at Carna in 2016.

20.4.2 Lumpsucker rearing strategies employed in Ireland

20.4.2.1 Egg incubation and development

Lumpsucker eggs in Ireland are incubated at 9°C, close to ambient sea water temperature during the spawning window of native populations. All egg batches are disinfected prior to stocking into incubation units. An overview of the incubation units used is given in Table 20.3. A small portion of eggs is weighed and individual eggs counted to determine egg numbers for each batch. In addition, eggs are examined microscopically to assess fertilisation rate. Hatching of eggs is protracted, occurring between 270 and 350 degree days, and can take up to a week to complete. During the development period, the eggs are routinely examined under a compound microscope and pictures taken to create a library for comparison (Fig. 20.6).

20.4.2.2 Larval and juvenile rearing

Various larval and juvenile rearing conditions and methods have been employed at Carna since 2015 (Table 20.3). Feeding generally commences between 2–3 dph and initially small amounts of food are administered to the larval tanks to ensure fish become familiar with a food source. Food amount is gradually increased over a period of five days until it is in excess of requirements. For example, in 440 l tanks stocked with 10,000 newly hatched larvae, 10 g feed was sufficient initially.

Due to the inherent difficulties with obtaining accurate larval numbers, population sizes are estimated using batch weights starting at transfer from larval to on-growing tanks (~50 dph). More accurate estimates are obtained once grading occurs (~100 dph, approx. 1 g average). Prior to this, variations in fish size make accurate number estimates almost impossible to obtain. It is

Table 20.3 Overview of lumpsucker hatchery conditions at Carna Research Station, including flow rate (L/min), Temperature (T, °C), rearing unit volume (L) and feeding regime for 2015, 2016 and 2017.

STAGE	PARAMETER		2015			2016	2017
EGGS	Incubation cone volume (L)		70			12	12
	T (°C)		8 – 9			8 – 9	8 – 9
	Flow rate (L/min)		6			6	6+
LARVAE	Tank Volume (L)		440 – 1200			440 – 1200	440 – 1200
	T (°C)		8–9 up to ~60 dph Ambient from ~ 60 dph			8–9 up to ~40 dph Ambient from ~ 40 dph	8 – 9 up to ~40dph Ambient from ~ 40 dph
	Flow rate (L/min) [depending on tank]		10 – 30			10 – 40	10 – 40
	Feeding regime (days post hatch, dph)		UK	IR	NOR	All groups	All groups
		Artemia	Feb-21	N/A	N/A	N/A	N/A
		Weaning	14-21	N/A	N/A	N/A	N/A
		Dry Diet	From 14	From 2	From 2	2	2
	Feeding Frequency		By hand every 90 minutes			By hand every 60 minutes	Automatic feeders
JUVENILES	Tank Volume (L)		1200 – 2000			1200 – 2000	1200 – 2000
	T (°C)		Ambient			Ambient	Ambient
	Feeding Rations		1 – 10%			1 – 10%	1 – 10%

possible to estimate larval numbers at transfer from holding to rearing tanks but the methods used are problematic. In general, a container of known volume is filled and larvae added to the container. These are mixed and a subsample of the volume is sampled and all larvae counted to give a number of fish per volume, with usually three counts performed. This can then be extrapolated to account for the entire volume being transferred. This method assumes that the larvae are equally mixed within the container and that the count is accurate.

The feeding strategies employed for lumpsucker are currently evolving with each production cycle. Significant losses due to tail nipping were observed in 2015 and the feeding strategy was modified to include a period of darkness between feeds. A similar approach was undertaken in 2016 (lights out between feeds); however, feeding frequency was increased to every hour. In autumn 2016, automatic feeders were installed. As with other fish species, lumpsucker are fed a percentage of their biomass. This percentage decreases with increasing fish size but will also be dependent on temperature, stress levels, water quality, flow, tank size, system capabilities (recirculation versus flow-through) and size variation within the tank. A rule of thumb is 10% up to 0.5 g decreasing to 1% at 10 g+, but should be monitored within each system. Fish are sampled at regular intervals for comparison to previous cohorts, to evaluate the need for grading and to establish feed rations.

Size hierarchies form rapidly in lumpsucker and fish require grading every two to three weeks from approximately 100 dph. During each grade the numbers of fish are estimated, based on batch weights at each size grade, to determine feeding rates.

20.4.3 Hatchery survival and growth

20.4.3.1 Hatchery survival

Survival in lumpsucker is variable and dependent on a number of factors including quality of eggs, water quality and temperature. There can be high mortality, during egg incubation in particular, because of the adhesive nature of the eggs, making it difficult to ensure good water flow through the eggs throughout the incubation period. This can be offset by maintaining a good inflow of water, the addition of air stones to facilitate water movement and turning the egg mass periodically in the cones. In addition, it is imperative that supplemental oxygen is used, particularly in the final stages of incubation and during hatching, which utilises oxygen.

When describing survival it is necessary to define developmental stage. Survival from fertilisation to hatch in lumpsucker in Ireland has been recorded between 15 and 90% and is highly variable. These figures are calculated using egg count data instead of the presumed 100,000 eggs per kilogram used by other groups. However, hatch rates increase to almost 98% in some instances when utilising this universally employed value! Similarly, it is critical to be clear when describing survival values and the time period to which they refer. For example, survival from hatch to sea transfer was 11% from one egg batch, while the survival of the same batch could be described as 64% if referencing time from 60 dph to sea transfer.

20.4.3.2 Hatchery growth

Benchmarking growth of a new species being used in aquaculture is important to enable improvements to rearing strategies and production methodologies, including calculation of feed rations

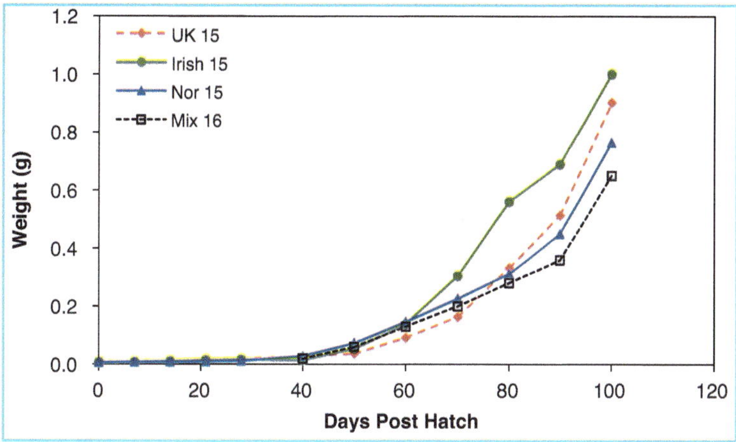

Fig. 20.7 Growth of lumpsucker in Ireland up to 100 dph for three cohorts in 2015 (UK, Irish and Norwegian) and mixed cohort in 2016 (Norwegian and Irish).

and planning. It is also important to understand how a species performs in native ambient conditions and to evaluate different populations under these same conditions. In the case of a widely dispersed species such as lumpsucker, with very little known about population differences across its range, benchmarking growth performance is a particularly useful exercise.

It is difficult to follow growth of groups after 100 dph as grading occurs regularly beyond this age and groups become mixed based on size rather than origin. A high degree of thermal control is utilised in Carna during the first 40–50 dph. Thus, it is during these early life stages that comparisons should be made. Similar growth patterns have been observed up to 50–60 dph when all groups are reared at a constant 9°C. During this time, differences in stocking density do not seem to play a large part in defining growth, although very high stocking densities are avoided during this period. After this time, when fish have been moved to ambient flow-through conditions, variations in density-dependent effects on growth become evident (Fig. 20.7). For example, the lumpsuckers of Irish origin in 2015 appear to perform better compared with other groups; however, these were the lowest stocked tanks and thus had ample space for growth. Choosing an arbitrary benchmark age of 100 dph, lumpsuckers in Irish hatchery conditions reached averages of between 0.7 g and 1 g, with individual fish size ranging from 0.05 to >10g! In addition, variations in hatch date result in differences in degree days once fish are moved into ambient conditions, attributing to variability in growth performance.

From the growth data for the 2015 cohorts produced in Carna the following general trends for growth of lumpsucker were apparent:

- For the UK stock:
    - 20% of population ready for sea (>8g) by 140dph, ~5 months
    - 5% of population still <8g (6–8g) at 250dph, ~8 months
- For the Norwegian stock:
    - 5% of population ready for sea (>8g) by 140 dph, ~5 months
    - >50% of population still <8g (3–5g) at 215 dph, ~7 months

The growth profiles for the slowest (S) and fastest (L) growing grades for both stocks are shown in Fig. 20.8 and highlight the variation in size observed in this species.

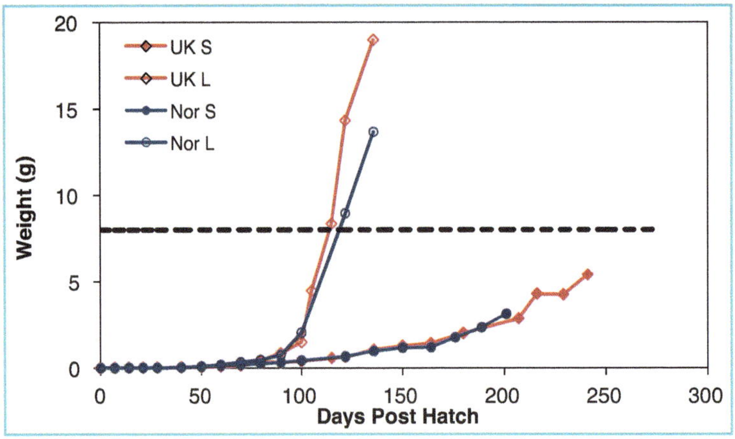

Fig. 20.8 Growth of lumpsucker in Ireland in 2015 for the largest and smallest grades of fish, dotted line denotes 8 g (sea transfer size in 2015, increased to 10 g minimum in 2016).

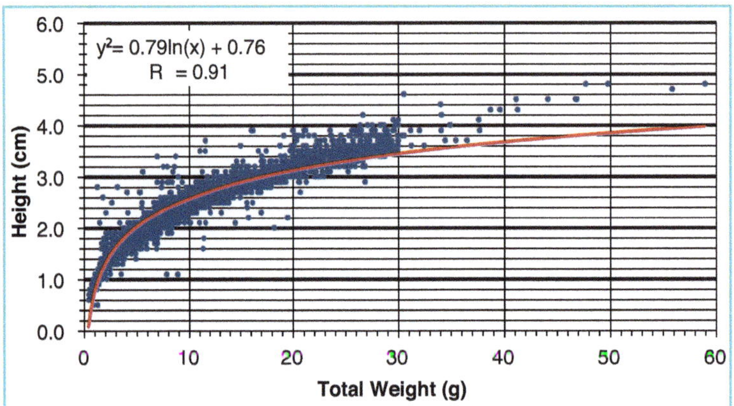

Fig 20.9 Relationship between total weight (g) and height (cm) in juvenile lumpsucker (various origins) to determine minimum weight for transfer to various salmon net sizes

20.4.4 Size at transfer

Fish are graded close to transfer because of this size variation to ensure that all are above the minimum size required. This is dependent on the mesh size of the salmon net employed. These range from 16–23 mm (side of square measurement) in Ireland, which equates to diagonal (largest opening) measurements of 22.6–32.5 mm. Carna Research Station has accumulated data on height and weight of lumpsucker juveniles from 2014 to aid in the determination of minimum size fish for transfer into the various salmon net sizes (Fig. 20.9).

20.4.5 Production issues

As previously mentioned, tail nipping caused significant problems in the first year of lumpsucker production in Carna. Smaller fish show signs of aggression, particularly when stocking densities increase. Although nipping has been mitigated by ensuring a constant supply of food to the tanks, shoals of larger fish are seen gathering in proximity to the feeders with smaller, subordinate fish adhering to the tank walls. Tail nipping frequency is observed to increase if size variation becomes too great.

Waste, both feed and faeces, accumulates on the floor of flat-bottomed tanks due to lumpsucker attachment to tanks and limited swimming capabilities, and this requires regular siphoning of tanks. Fast growth gives large size variation and the fish therefore require regular grading and handling. This potentially could prove stressful and damaging, although no mortality has been observed that is attributable to this. In addition, there are no lumpsucker-specific vaccines available in Ireland at present. Up until 2016, the practice was to use a dip vaccine for Vibrio at 1 g, followed by a second dip three to four weeks later. This potentially leaves the fish vulnerable to other diseases following transfer to sea, particularly during the warmer summer months when increased mortalities due to bacterial infections have been recorded (M. Bolton-Warberg, Carna Research Station 2017 with F. Sproll, Marine Harvest Ireland). This vaccine has been discontinued and is no longer available for use with lumpsucker. In December 2017, an autogenous vaccine (including *Pseudomonas* and *Tenacibaculum*) became available for use in Ireland.

20.5 Strategic use on salmon farms (Marine Harvest Ireland Ltd)

There are two aspects to the cleaner fish programme adopted by Marine Harvest Ireland (MHI) to control sea lice: (i) capture and introduction of wild wrasse into sea cages, predominantly during the summer months and (ii) deployment of hatchery-reared lumpsuckers from October to May when water temperatures are lower. Initial trials with lumpsucker carried out at Carna in 2014 produced promising results (see section 20.6.1). A concerted effort to use cleaner fish on Marine Harvest sites in Ireland commenced in 2015. Observations and results during this initial year increased confidence in the use of this novel technology and highlighted the integral role cleaner fish could play in MHI's sea lice control strategy.

20.5.1 Fishing and transport

Wild wrasse are sourced from several fishing zones (Clew Bay, west; Inver Bay, Mulroy Bay and Lough Swilly, north; Bantry Bay and Kenmare Bay, south-west), chosen for their proximity to the salmon farms being stocked and to minimise the movement between bays, which could carry disease risks. All five species of wrasse ubiquitous to Irish inshore waters are utilised with a minimum length of 11 cm. Fishing commences in June using specialised wrasse pots (OK Marine, Geal Force: fitted with an escape hatch for wrasse <11 cm) and fishermen are equipped with tanks and pumps for post-capture transport and holding pens for maintaining fish prior to collection. Small grade wrasse (11–15 cm) make up 90–95% of the catch in most sheltered bays, with the main species being corkwing. In contrast, in exposed locations, large wrasse (mainly ballan and corkwing, >15 cm) are more plentiful and can be up to 25–35% of the catch.

Transport methods vary between sites and species, using smolt tanks or 1000 l bins on boats when wrasse are caught close to the farm, or articulated lorries equipped with 2.3 m^3 smolt tanks for longer distances and for all lumpsucker. Tanks are set up with supplemental oxygen which is regulated and closely monitored for the duration of the trip. The maximum journey time for fish in Ireland is approximately five hours. Stocking densities during transport are varied depending on species and have ranged from 13 to 60 kg m^{-3} for lumpsucker.

Table 20.4 Overview of deployment of wrasse spp. and lumpsucker on Marine Harvest Ireland sites in 2015 and 2016, including grades, salmon stages, deployment times, stocking density and comments regarding fallowing, lice numbers at stocking and preliminary results.

Year	Cleaner Fish (Grade)	Site	Salmon Stage	Deployment	Stocking Density	Initial Lice levels	Comments
2015	Wrasse (L)	North West 1	Broodstock	Aug	100%	High	fully cleared (with lumpsucker), one cage of 1,000 salmon
	Wrasse (L)	North West 1	Grower	June – Aug	5%	High start of year	4 of 14 cages stocked, harvested out June–August due to uncontrollable lice numbers
	Wrasse (S)	North West 2	Smolt	June – Oct	5%	High	some improvement in cages with wrasse
	Wrasse (L)	North West 3	Grower	May – July	1%	High	Low stocking due to low catch rate
	Wrasse (S & L)	West 1	Smolt	May – Oct	5%	Some	next to grower site with lice
	Lumpsucker (15g+)	North West 1	Broodstock	Oct	40%	High	fully cleared (with wrasse), one cage of 1,000 salmon
	Lumpsucker (8g+)	North West 4	Smolt	Oct – Nov	6%	None	1 year fallowing, no chemical treatment
	Lumpsucker (16+g)	North West 3	Grower	Dec	8%	High	2 months fallowing, low salmon numbers (1,000) lumpsucker escapees due to strong currents. Only 3 of 10 cages stocked due to strong current
2016	Wrasse (S)	North West 2	Smolt	May – July	6%	None	3 months fallowing, 0 lice
	Wrasse (S)	North West 1	Smolt	Sep	6%	None	3 months fallowing, 0 lice
	Wrasse (L)	North West 3	Grower	May – Sept	6%	Some pre-adult	little adult females, cages not fully stocked until end of season
	Wrasse (S)	West 1	Smolt	May – Sept	5%	None	2 months fallowing
	Wrasse (L & S)	North West 4	Smolt & Grower	July	6%	None	Entire bay fallowed 1-2 months, MHI site fallowed 1 year
	Wrasse (S)	South West 1	Smolt	July	3%	Unknown	site fallowed 4 months
	Wrasse (L)	South West 2	Grower	July	3%	None	Site fallowed for 3 months
	Lumpsucker (11+g)	North West 1	Smolt	January	6%		3 months fallowing
	Lumpsucker (40+g)	West 2	Grower	March	15%	High	treated, lumpsucker cleared remaining lice
	Lumpsucker (30+g)	South West 1	Smolt	March	6%	Unknown	site fallowed 4 months

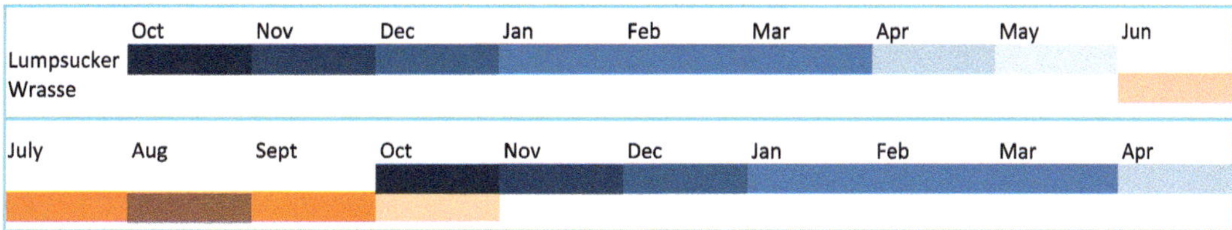

Fig. 20.10 Schematic of deployment strategy for cleaner fish used by Marine Harvest Ireland.

20.5.2 Deployment of cleaner fish

A targeted stocking density of 6% cleaner fish to salmon has been adopted, although this varies with availability, salmon size and level of lice infestation. A total of ten sites have been stocked with cleaner fish to date and an overview of deployment is provided in Table 20.4. In general, lumpsucker are deployed with (or before) S_0 smolts in October and over winter with them, with more than 50% (estimated) surviving until April. Survival declines as temperatures increase, with negligible numbers of lumpsuckers seen in summer months when wrasse fishing commences. Wrasse are stocked into cages in June, with catch numbers in cages decreasing towards autumn, when cleaner fish numbers are topped up with hatchery-reared lumpsucker to provide lice control until harvest the following spring (Fig. 20.10). Wrasse are still present in cages but, because they are less active in colder water, catch numbers fall with decreasing water temperatures.

20.5.3 On-farm husbandry

Farm staff are trained in cleaner fish husbandry through a combination of onsite courses and workshops with guest speakers from Norway. One type of artificial kelp hide with two different designs is utilised for both wrasse and lumpsucker: (i) floating frame hides and (ii) curtain hides. The selection of hide is site-specific, and dependent on the resources available and the prevailing environmental conditions (for example, tide, currents) at each site. In addition, hide depth is variable depending on the cage, ranging from 5–12 m. Floating frame hides consist of artificial strands of kelp on a plastic frame with a ring weight sinker. These hides are preferable on sites with ready access to boats with cranes and on more exposed sites. Curtain hides consist of a float line and corresponding lead line with strands of kelp running between the two. These hides are easier to handle and can be hung for cleaning without the need for machinery. A sufficient amount of cover is placed into each cage based on the number of cleaner fish stocked. In general it is assumed that 1 m of hide will accommodate 15 to 50 cleaner fish, depending on size and species.

Supplemental feeding of cleaner fish while in cages is important for ensuring healthy fish. Wrasse are fed using weighted net bags containing a block (consisting of Biomar Symbio mixed with agar) which are hung in and around the hides. Initially, green crab is crushed and mixed in with the pellets to wean wild fish on to a commercial pellet diet. Wrasse are fed approximately 0.5% of their biomass with feed bags (four per cage), which are replenished daily and moved depending on appetite and water temperature. Lumpsucker are hand fed directly into hides with Skretting Amber Neptun or Biomar at a rate of 2% every two days.

20.5.4 Problems experienced during application of cleaner fish

A variety of problems arose during 2015 and 2016, and this resulted in mixed outcomes on the farms:

- Wild wrasse fishing began slowly and fish were not captured in great numbers until mid-June. This was primarily due to low temperatures but also, in 2015, lack of experience. In addition, mature wrasse were still being caught until late May and it was decided to wait until the spawning period was over.
- Site-specific fishing challenges with one site experiencing strong tides that caused damage to pots. In addition, excessive bycatch of conger eel (*Conger conger*) was reported in this site with lower numbers of wrasse caught compared with other sites.
- The majority of wild-caught wrasse were too small to be deployed in the grower nets, and it was, therefore, a slow and arduous process to stock sites containing adult salmon. In addition, there were insufficient wrasse in these cages to counteract lice numbers and therefore medicinal treatments were required.
- Wrasse have low tolerance to freshwater and had to be removed before treatment for Amoebic Gill Disease. The onset of AGD in S_1 salmon usually occurs a few months following cage deployment, with treatments required every 1–1½ months thereafter. The preferred form of treatment is a two- to four-hour freshwater bath in a tarpaulin or wellboat and this is repeated throughout the season, April–September (F. Sproll, Marine Harvest Ireland, pers. comm). S_0 salmon sometimes require an AGD treatment one month after deployment and in month intervals until they reach 2 kg (S. Schlittenhardt, Marine Harvest Ireland, pers. comm.). The removal of wrasse from cages is time-consuming, labour intensive and often ineffective, with recovery rates ranging from 10–80%. Similarly, in some cases the replacement of wrasse was also time-consuming, sea lice numbers increased and therapeutic treatments were required. This was improved in 2016 with holding cages employed to facilitate rapid restocking. In 2016, most sites did not require any medicinal treatment for AGD.
- Lumpsucker mortalities due to *Pseudomonas anguilliseptica*. Lumpsuckers at two sites in the north and south experienced mass mortality (40%) events associated with this bacterium in summer 2016. The surviving fish were placed in holding pens and treated with an in-feed antibiotic. Lumpsucker were treated separately due to the organic status of the salmon.
- Initially there were some difficulties with removing lumpsucker from the transport tanks, but this is no longer an issue.
- During very strong tides, lumpsucker are pushed up against the net in exposed sites.

20.5.5 General farming observations

There were noticeable reductions in lice numbers when cages were stocked with sufficient levels of wrasse and nets were kept clean. Cleaner fish, in general, were found to be less effective in large cages (80 m diameter) with low salmon numbers (4000– 6000 broodstock) and high initial lice numbers. This could be due to a lower level of interaction between the two species but more likely that when lice numbers reach a certain threshold, cleaner fish alone cannot reduce the parasitic load. Thus, cleaner fish should be used as a preventative measure rather than to fight high lice numbers. In an ideal situation, there would be no adult lice on the salmon and they would approach the hide to be cleaned by either wrasse or lumpsucker.

Transported hatchery-reared lumpsuckers survived well with very little mortality reported during transport (max 1%) and post-transport (2–3% after 1 month). Lice levels in Inver Bay remained negligible through the winter and into the summer, and no medicinal treatments were required (14-month period). The Inver site had been fallowed for approximately one year previously and historically has high levels of lice infestation. Stomach content analysis revealed between 30 and 40 adult female lice in some lumpsucker, confirming their effect. In contrast, a neighbouring site (1–2 km distance) with no cleaner fish present required multiple treatments over the same 14-month period.

Lumpsuckers spend more time in the hides when using floating frame hides in comparison to curtain hides. The best method for removing them in this instance was to put a sweep net or box net around the entire hide. Lumpsucker tended to be more dispersed throughout the cage in cages with curtain hides and this made removal more difficult. As lumpsuckers are tolerant to freshwater but susceptible to AGD, they can be treated simultaneously with salmon and separation of the two species is not required. In addition, lumpsuckers (3000 initially, 50g+) were stocked with broodstock salmon at sea and transported to a facility where they were maintained at 15 ppt salinity (first 4 m of sea pens surrounded by a skirt and freshwater continually pumped in) for several weeks with no reduction in lumpsucker condition. Water temperature was between 13 and 15°C. During this time lumpsucker were observed swimming in the freshwater layer. This practice enables the holding of lice-free broodstock and in good condition, and has been used during several summers.

20.6 Studies on efficacy and optimising production

The following studies were carried out at Carna Research Station between 2014 and 2016.

20.6.1 Initial efficacy trial

A small-scale efficacy trial was conducted in November 2014, using farmed salmon produced locally (Marine Harvest Ireland) and wild lumpsuckers collected by hand from seaweed longlines in July and August (initial weights 0.09–16 g). Salmon with an average weight of 1.7 kg were stocked into two tanks (4 m^3 volume) supplied with ambient flow-through water, with supplemental oxygenation. The treatment tank had 50 salmon and ten lumpsucker with an average size of 49.7 ± 5.5 g, yielding a 20% stocking density. Initial total numbers of lice were 18 ± 7 (SD) and within one week had dropped to five in the treatment tank. During the same period the number of gravid females declined from 8 ± 4 (SD) to 0 per salmon held with lumpsucker. The control tank consisted of 50 adult salmon and no lumpsuckers. The lice numbers on these salmon increased from 18 ± 7 (SD) to 21 ± 5 (SD) total per fish. Underwater videos showed lumpsucker actively searching for lice on salmon. Despite the size difference there was no sign of stress in the lumpsucker, with one observed adhering to the head of a salmon within 30 minutes of their addition. Gastric lavage of the lumpsucker revealed lice in five (50%) of the ten lumpsucker employed in the trial, although no count data were recorded.

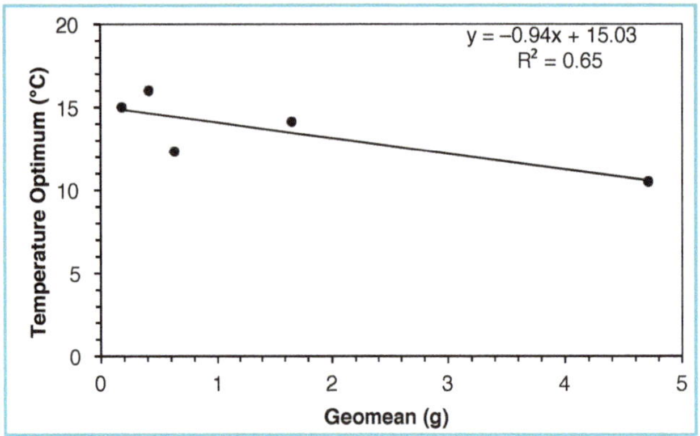

Fig. 20.11 Relationship between temperature optima (for growth) and size (geomean) of juvenile lumpsuckers. Five different size classes (0.11, 0.27, 0.44, 1.00 and 3 g initial weight) were examined at four different constant temperature regimes (nominally 9, 12, 15 and 18°C).

20.6.2 Temperature optima for lumpsucker

Variations in the optimum temperature for growth ($T_{opt.G}$) in different lumpsucker stocks are unknown. Establishing the $T_{opt.G}$ for a population is a primary objective in advance of determining the effects of other factors such as feeding and stocking density, which can facilitate improvements in commercial production of juveniles (e.g. Bolton-Warberg *et al.*, 2015). A series of experiments was therefore designed and implemented to determine the effects of temperature and size of juvenile lumpsucker on growth rate (Murphy O'Sullivan, 2016, unpublished). Five different size classes (0.11, 0.27, 0.44, 1.00 and 3 g initial weight) were examined at four different constant temperature regimes (nominally 9, 12, 15 and 18°C). Optimum temperatures for growth ranged between 10 and 16°C for the size classes studied (Figure 20.11), with a general trend of decreasing $T_{opt.G}$ with increasing size of fish from 13.40°C for 0.11 g fish to 10.79°C for 3.00 g fish. A simple rearing temperature strategy of 13°C constant for juvenile lumpsucker in this size range for optimal growth is suggested in preference to the traditional 'T-step' method adopted in other marine species because hierarchies form quickly in lumpsucker. Thus, it is preferable to employ a temperature regime that will promote growth rates that are close to maximum for all size classes.

20.6.3 Daily feed rations

Four size classes (0.54, 1.03, 2.00 and 4.89 g initial weight) of lumpsucker were examined to establish optimum feeding rations (O'Donnell, unpublished). Three different feed rations were administered (2–10% of total biomass) in triplicate tanks containing from 30–50 fish depending on size on ambient flow-through water (Table 20.5). The feed used was a commercially available crumb diet (Otohime©, C2: protein 58.3%, fat 12.9%, fibre 1.6%). Experiments were of ten and 13-day duration, with initial and final individual weights being recorded. Optimum feed ration was determined for each size class by plotting the relationship between growth rate and feed ration and solving the resulting polynomials (Figure 20.12). The average optimum feed ration for the size range examined was ~6%, with F_{opt} within size classes of 7.36% for 0.54 g fish, 5.49% for 1.03 g fish, 5.67% for 2.0 g fish and 6.54% for 4.89 g fish. The temperature for the experiment with the largest size class (4.89 g) was lower than the other groups (Table 20.5) and may have been closer to the optimum temperature

Table 20.5 Experiment, mean temperature (T, °C), feed ration (F, % biomass), number of replicates (Rep), initial and final number of fish (n1, n2), mortality (m, % of initial number), initial and final mean weights (W1, W2), geomean [GM = (W1*W2)^1/2], SGR (% per day), duration of experiment in days (d) and optimum feeding ration (Fopt).

EXPT	Start Date	T (°C)	F (%)	Rep	n1	n2	m	W1 (g)	W2 (g)	GM (g)	SGR (%)	d	Fopt (%)
1	10–Jul	15.44	2	3	160	156	2.5	0.54	0.66	0.6	1.98	10	7.36
		± 0.38	5	3	160	160	0	0.54	0.72	0.62	2.92		
			10	3	160	160	0	0.54	0.72	0.62	2.86		
2	10–Aug	16.03	2	3	100	100	0	1.03	1.25	1.13	2.21	10	5.49
		± 0.52	4	3	100	99	1	1.03	1.32	1.17	2.86		
			8	3	100	97	3	1.03	1.29	1.15	2.59		
3	28–Aug	15.37	2	3	50	50	0	2	2.69	2.32	3.36	10	5.67
		± 0.25	4	3	50	49	2	2	2.83	2.38	3.9		
			8	3	50	49	2	2	2.79	2.36	3.76		
4	30–Sep	13.85	2	3	30	28	6.7	4.89	7.51	6.06	4.88	13	6.54
		± 0.54	4	3	30	29	3.4	4.89	8.47	6.44	6.29		
			8	3	30	28	6.7	4.89	8.78	6.55	6.72		

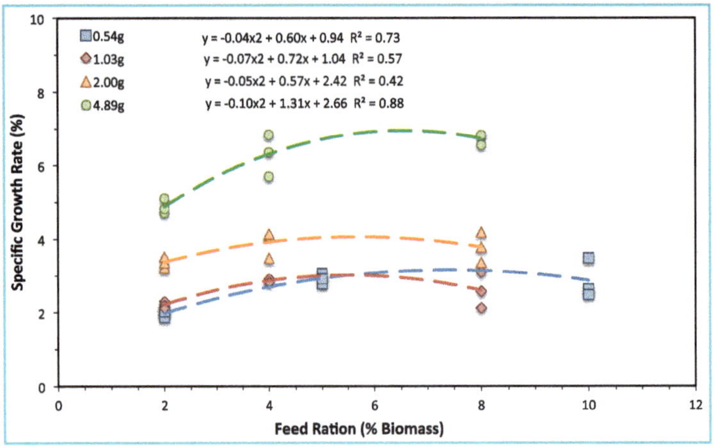

Fig. 20.12 Relationship between feed treatment (% biomass) and specific growth rate for juvenile lumpsuckers (UK origin) reared under Irish hatchery conditions.

for growth resulting in higher growth rates compared to the smaller sizes used in this series of experiments.

20.7 Future plans

In Ireland, the use of various wrasse species to control sea lice infestations on farmed salmon began in the early 1990s (e.g. Darwall *et al.*, 1992; Deady *et al.*, 1995). Despite initial successes their use was suspended due to an outbreak of atypical furunculosis in the wrasse at one site. More recently, the use of cleaner fish (wild-caught wrasse and hatchery-reared lumpsucker) has proven successful and is growing in use. Between 2015 and 2016, >600,000 wild-caught wrasse have been utilised on salmon farms with almost 250,000 lumpsucker deployed in sea cages for use as cleaner fish (see Bolton-Warberg, in press for overview of cleaner fish use in Ireland).

More than 80% of farmed salmonids are certified organic in Ireland, with the number of lice treatments restricted by quality certifying bodies (farmers must apply for special derogation to undertake them). At present, cleaner fish remain the most acceptable, effective, long-term option in an organic production cycle that is favoured by many Irish farmers and demanded by progressively more discerning consumers.

Ireland is currently dependent on catching (using a variety of gears) stocks from the wild and/or importing eggs or juveniles to fill their operational cleaner fish needs. This dependency carries with it potentially significant threats:

- An impact on the integrity of local wild wrasse stocks through the (recurring) collection of live fish for use on salmon farms with associated potential for introducing diseases.
- With the importation of lumpsucker eggs/juveniles from other jurisdictions (e.g. Norway and UK), there is potential threat to health and biosecurity through importation of a variety of disease conditions.
- The potential for the introduction to Irish waters of non-native genotypes via escapees.

These threats have been mitigated by introducing guidelines regarding wild capture (both at a national level and at producer/bay level) and strict health regulations implemented by the

Marine Institute and adhered to by aquaculture facilities (producers and research institutes alike). In addition, there are a number of strategic initiatives being undertaken to facilitate the development of cleaner fish technology as a sustainable integrated approach to mitigating the sea lice problem. These include the establishment of a reservoir of native lumpsucker stocks for use as a national base broodstock that will be used to provide a source of juvenile lumpsucker for use on Irish salmon farms in the coming years (Lumpsucker Broodstock and Selective Breeding Programme). An indigenous local stock should be better adapted to the prevailing, ambient environmental conditions and various biotic influences, thus making them the optimal choice for establishing a farmed base broodstock (Salvanes *et al.*, 2004, Bekkevold *et al.*, 2006; Nielsen *et al.*, 2009). Similarly, choosing a local population as the base population for aquaculture may reduce future negative effects on the wild population in the event of escapes occurring from the sea cages.

The Marine Institute is also carrying out studies evaluating the sustainability of wild capture for a variety of species for use as cleaner fish, including wrasse and lumpsucker. Other key areas of priority in Ireland include evaluating cleaning efficacy in sea cages (wrasse and lumpsucker) and developing a greater understanding of disease dynamics (within a species and interactions with salmon). At Carna Research Station, applied and basic research is used to advance and optimise rearing strategies to produce good quality juveniles. This research includes benchmarking growth of lumpsucker under Irish conditions, understanding the effects of stocking density on growth and conditioning captive-bred broodstock. Carna Research Station has also carried out preliminary 'proof of concept' research into the production of Ballan wrasse (*Labrus bergylta*) that demonstrated it is feasible to catch, condition and spawn wild Ballan and produce viable juveniles given the appropriate controlled conditions in Ireland (e.g. D'Arcy, 2013; Mohd Yusof Mohd, 2015). There are currently no immediate plans to develop wrasse production in Ireland; however, it is likely that this too will be re-evaluated in the coming years. Finally, research on the nutritional requirements of lumpsucker is vital to the production of robust juveniles and the development of specialised vaccines will do much to increase survival at sea. While much of this work is being undertaken in other countries, it is important that all countries working on cleaner fish maintain open lines of communication with regard to advances and ongoing research to reduce repetition of work and to facilitate development of technologies to tackle sea lice, which remains a global concern.

References

Bekkevold, D., Hansen, M.M. and Nielsen, E.E. (2006) Genetic impact of gadoid culture on wild fish populations: Predictions, lessons from salmonids, and possibilities for minimizing adverse effects. *ICES Journal of Marine Science* 63, 198–208.

Bolton-Warberg, M., O'Keeffe, D. and FitzGerald R. (2015) Exploring the temperature optima and growth rates of Atlantic cod at the south-easterly limit of its range. *Aquaculture Research* 46(3), 698–706.

Bolton-Warberg, M. in press. An overview of cleaner fish use in Ireland. *Journal of Fish Diseases.*

Costello, M. J. (1994) Development of the use of wrasse to control sea lice on salmon farms in Ireland. *Aquaculture Fisheries Management and Aquaculture Research* 25(1), 25.

D'Arcy, J., Dunaevskaya, E., Treasurer, J.W., Ottesen, O., Maguire, J., Zhuravleva, N., Karlsen, A.,

Rebours, C. and FitzGerald, R. D. (2012) Embryonic development in ballan wrasse *Labrus bergylta*. *Journal of Fish Biology* 81 (3), 1101–1110.

D'Arcy, J. (2013) *Studies on the biology of the Ballan wrasse, Labrus bergylta*. PhD Thesis, NUI Galway. Unpublished.

D'Arcy, J., Mirimin, L. and FitzGerald, R. (2013) Phylogeographic structure of a protogynous hermaphrodite species, the ballan wrasse *Labrus bergylta*, in Ireland, Scotland, and Norway, using mitochondrial DNA sequence data. *ICES Journal of Marine Science* 70 (3), 685–693.

Darwall, W., Costello, M.J. and Lysaght, S. (1992) Wrasse – how well do they work? *Aquaculture Ireland* 50, 26–29.

Deady, S., Varian, S.J.A. and Fives, J.M. (1995) The use of cleaner-fish to control sea lice on two Irish salmon (*Salmo salar*) farms with particular reference to wrasse behaviour in salmon cages. *Aquaculture* 131, 73–90.

Dipper, F.A., Bridges, C.R. and Mentz, A. (1977) Age, growth and feeding in ballan wrasse *Labrus bergylta* Ascanius 1767. *Journal of Fish Biology* 11, 105–120.

Donnelly, R. (1992) The biology and use of wrasse (Labridae) as a biological control of external parasites in salmonid cultures. MSc Thesis, University of Dublin. Unpublished.

Dulcic, J., Kozul, V., Kraljevic, M., Skaramuca, B., Glamuzina, B., Re, P. (1999) Embryonic and larval development of the brown wrasse *Labrus merula* (Pisces: Labridae) *Journal of the Marine Biological Association U.K.* 79, 327–332.

Dunaevskaya, E., Amin, A.B. and Ottesen. O.H. (2012) Organogenesis of Ballan Wrasse *Labrus bergylta* (Ascanius 1767) Larvae. *Journal of Aquaculture Research and Development* 3, 142 doi:10.4172/2155-9546.1000142.

Fuiman, L.A. (2002) Special considerations of Fish Eggs and Larvae. In: Fuiman, L.A. and Werner, R.G. (eds) *Fishery science: the unique contributions of early life stages* Oxford: Blackwell Publishing, pp. 1–32.

Geffen, A.J., Fox, C.J. and Nash, R.D.M. (2006) Temperature-dependent development rates of cod *Gadus morhua* eggs. *Journal of Fish Biology* 69, 1060–1080.

Mohd Faizzi Bin Mohd Yusof (2015) *Reproductive Biology and Growth of Ballan Wrasse Labrus bergylta in West of Ireland Waters*. PhD Thesis, NUI Galway. Unpublished.

Muncaster, S., Andersson, E., Kjesbu, O.S., Taranger, G.L., Skiftesvik, A.B. and Norberg, B. (2010) The reproductive cycle of female Ballan wrasse *Labrus bergylta* in high latitude, temperate waters. *Journal of Fish Biology* 77, 494–511.

Muncaster, S., Norberg, B. and Andersson, E. (2013) Natural sex change in the temperate protogynous Ballan wrasse *Labrus bergylta*. *Journal of Fish Biology* 82, 1858–1870.

Murphy O'Sullivan, S. (2015) *The effect of temperature on growth of juvenile lumpsucker* (*Cyclopterus lumpus*). MSc Thesis, NUI Galway. Unpublished.

Nielsen, E.E., Hemmer-Hansen, J., Poulsen, N.A., Loeschcke, V., Moen, T., Johansen, T., Mittelholzer, C., Taranger, G.-L., Ogden, R. and Carvalho, G.R. (2009) Genomic signatures of local directional selection in a high gene flow marine organism; the Atlantic cod (*Gadus morhua*). *BMC Evolutionary Biology* 9, 276.

O'Donnell, M. (2016) *Effect of variations in percentage feed rations on the growth of juvenile lumpsucker* (*Cyclopterus lumpus* Linnaeus, 1758). BSc Thesis, NUIG. Unpublished.

Ottesen, O.H., Karlsen, A., Treasurer, J., FitzGerald, R., Maguire, J., Rebours, C. and Zhuravleva, N. (2008) Ballan Wrasse Offer Efficient, Environmentally Friendly Sea Lice Control. *Global Aquaculture Advocate* Nov/Dec 2008.

Ottesen, O.H., Dunaevskaya, E. and Arcy, J.D. (2012a) Development of *Labrus bergylta* (Ascanius 1767)

Larvae from Hatching to Metamorphosis. *Journal of Aquaculture Research and Development* 3, 127 doi:10.4172/2155-9546.1000127

Ottesen, O. H., Treasurer, J., FitzGerald, R., Maguire, J. and Rebours, C. (2012b). EcoFish Project – Using Wrasse as Cleaner Fish in Salmon Farming. *Global Aquaculture Advocate* Mar/Apr 2012.

Salvanes, A.G.V., Skjæraasen, J.E. and Nielsen, T. (2004) Sub-populations of coastal cod with different behaviour and life-history strategies. *Marine Ecology Progress Series* 267, 241–251.

Sayer, M.D.J. and Treasurer, J.W. (1996) North European wrasse: identification, distribution and habitat. In: Sayer, M.D.J., Treasurer, J.W. and Costello, M.J. (eds.) Wrasse: Biology and use in Aquaculture. Fishing News Books, Oxford, pp. 3–12.

Sayer, M.D.J., Gibson, R.N. and Atkinson, R.J.A. (1993) Distribution and density of populations of goldsinny wrasse (*Ctenolabrus rupestris*) on the west coast of Scotland. *Journal of Fish Biology* 43 (Supplement A), 157–167.

Thangstad, T. (1999) *Spatial and temporal distribution of three wrasse species (Pisces: Labridae) in Masfjord, western Norway: habitat association and effects of environmental variables.* Thesis for the cand. Scient. Degree in fisheries biology, University of Bergen.

Tully, O., Daly, P., Lysaght, S., Deady, S. and Varian, S.J.A. (1996) Use of cleaner-wrasse (*Centrolabrus exoletus* (L.) and *Ctenolabrus rupestris* (L.)) to control infestations of *Caligus elongatus* Nordmann on farmed Atlantic salmon. *Aquaculture* 142, 11–24.

CHAPTER 21

Rearing of cleaner fish in Iceland

Agnar Steinarsson and Tómas Árnason

The Aquaculture Research Station of the Marine Research Institute of Iceland, Grindavík, Iceland

21.1 Introduction

Aquaculture has thus far been a relatively minor business sector in Iceland with a modest production volume of less than 10,000 tons per year (Icelandic Aquaculture Association, 2016a). Sea cage farming operations have not yet reached any significant production levels and the sector has largely been dominated by the land-based farming of Arctic char *Salvelinus alpinus*. Farming of Atlantic salmon *Salmo salar* has previously gone through two failed expansion periods and is presently in the beginning stages of a third expansion wave. Production of salmon and rainbow trout *Oncorhynchus mykiss* from sea cages is presently rising fast and was predicted to increase from 4000 tons in 2015 to 8000 tons in 2016 (IAA, 2016b).

Sea lice are considered endemic to Icelandic fjords and are commonly found as parasites on various fish species, such as salmon, trout *Salmo trutta*, char and cod *Gadus morhua*. The prevalence of lice on farmed salmon was monitored extensively during the last expansion period in 2003–2007 and increased intensities were regularly observed during late summer or autumn. Infestations were, however, almost exclusively caused by the common sea louse (*Caligus elongatus*) while the more damaging salmon louse (*Lepeophtheirus salmonis*) was rarely detected during that period (Jónsson, 2015). Present day salmon farmers do also monitor sea lice infestations and the current prevalence of sea lice is generally not considered an imminent threat to the industry (Jónsson, 2015). Two recent studies showed conflicting results regarding the effect of salmon farms on sea lice prevalence or intensities in wild salmonids within the same fjord system (Karbowski, 2015a, b), so an influence from aquaculture farms cannot be excluded at this point.

The low ambient seawater temperatures in Icelandic fjords (4–5°C mean annual temperature and 0–1°C during the winter months) are considered, by many, a natural hindrance to a possible opportunistic explosion in sea lice infestations associated with the foreseen increase in salmonid farming. At these low temperatures it takes two years to grow salmon to market size and, as a consequence, each generation is exposed to sea lice infestations for two consecutive summers. On the other hand, sea lice grow and reproduce slowly at the low winter temperatures and only have a window of five to six warmer months to do their damage. There are, however, also concerns that the predicted rise in sea water temperatures due to global warming may gradually diminish this alleged natural advantage and create a more favourable environment for sea lice in Icelandic fjords.

There is an ongoing debate among various stakeholders about the regulatory framework for sea cage farming. Fish farming regulations require a distance of at least 5 km between neighbouring farming sites and all cage sites must be rested for at least 90 days when a generation has been fully harvested. Farming is not allowed in proximity to major salmon rivers nor close to wild salmon migratory routes. Many claim that all cage sites should be fallowed for three months or even up to eight months to break the sea lice life cycle completely. The coexistence of two or more farming companies in the same fjord can, however, undermine the effectiveness of fallowing unless the companies synchronise their production. Sea lice monitoring is not mandatory for Icelandic salmon producers and no specific regulations have been issued regarding sea lice, such as tolerance thresholds for infestations. Indeed, some farmers and anglers fear that the overcrowding of farming areas under the current regulations may eventually give rise to an explosion in sea lice abundance in the fjord systems. It is a valid argument whether sea lice in Icelandic waters may be adapted to the low ambient temperatures and perhaps capable of population expansion given a greatly increased number of hosts over a long period of time. In view of the sea lice situation in other salmon farming countries it may be sensible to assume the worst but expect the best when it comes to sea lice.

21.2 Lumpfish production in Iceland

As attempts to produce lumpfish juveniles in the Faroes had not been successful, Stofnfiskur suggested a proposal to the aquaculture research station of the Marine Research Institute in Iceland (MRI) to produce and export lumpfish juveniles to the Faroes. In the spring of 2014 an agreement was made regarding the production of 250,000 lumpfish juveniles for the Faroese customer and the MRI also made an agreement with Stofnfiskur regarding the transfer of know-how and the delivery of eggs and juveniles. A commercial production of lumpfish juveniles by the MRI was, however, only planned as a temporary option until Stofnfiskur would be able to take over. Meanwhile, Stofnfiskur decided to open its own lumpfish hatchery in Iceland and started by purchasing any possible excess juveniles from MRI's first season. The MRI managed to produce and deliver the agreed number of juveniles to the Faroese customer as well as some 200,000 extra juveniles for on-growing in Stofnfiskur's new hatchery. Stofnfiskur also delivered its juveniles to its customers in the Faroes.

In spite of various initial difficulties related to transporting the juveniles, the project became a great success for the customer. Even though the lumpfish mortality was high, regular monitoring of the surviving lumpfish showed a high feeding preference for sea lice. Through the delousing effect of the lumpfish, sea lice levels were kept under regulation thresholds and, for the first time in years, the customer produced a whole generation to harvest without resorting to expensive and undesirable delousing medicines. The customer had high expectations in 2015 and also made various improvements such as adding shelters and bird nets to the cages and improving the feeding conditions. The overall result was, however, very disappointing as the lumpfish suffered high disease-related mortality and were thus rendered ineffective as delousers. Some wild-caught lumpfish breeders at the MRI tested positive for VHS (viral haemorrhagic septicaemia) in October 2015 and the facility was placed under an export ban according to EU health regulations (Towers, 2015). All remaining lumpfish were immediately discarded and the entire station was thoroughly disinfected. The virus, however, was not diagnosed in the exported lumpfish nor in the Stofnfiskur hatchery, which is located 7 km from the MRI facility. Stofnfiskur was thus able to continue with the export of lumpfish to the Faroes.

In 2016 the MRI export ban was lifted and, in view of continued interest from the customer, a new agreement was made based on a different approach. No wild breeders were now brought in to the MRI and eggs were purchased from Stofnfiskur's quarantine facility in the town of Sandgerði. No eggs were taken out from the quarantine until the parents had been tested negative for a range of viruses and bacteria. Stricter prophylactic protocols were also implemented at MRI's aquaculture station and, indeed, the production was very successful with 250,000 healthy lumpfish delivered to the Faroese buyer. During 2016, Stofnfiskur completed renovation of a former cod hatchery in the Icelandic town of Hafnir. The facility has been adapted and specialised for lumpfish production and is expected to reach a production capacity of approximately 2 million lumpfish juveniles per year in 2017. Stofnfiskur is now shipping live lumpfish to the Faroes in up to four 40 ft ocean containers per week all year round. The University of Hólar in north Iceland, in cooperation with the consultation firm Akvaplan Niva, has also been running a small-scale operation producing lumpfish juveniles for research purposes.

The future of lumpfish production in Iceland is expected to depend largely on the demand from foreign buyers, primarily the Faroe Islands. However, in view of the imminent expansion of salmon farming in Iceland combined with rising seawater temperatures, it is perhaps likely that a domestic market for lumpfish is about to open up in Iceland. Other species of cleaner fish are not native to Icelandic waters and thus the lumpfish may be the only potential biological delouser for Iceland. If sea lice abundance will indeed go hand in hand with expanding aquaculture, it will be advantageous to introduce the lumpfish as a biological delousing agent at an early stage to prevent sea lice populations from reaching epidemic proportions, with associated risks for the ecosystem.

21.3 Lumpfish production at the Marine Research Institute of Iceland

This chapter gives an overview of the methodology of lumpfish production at the Marine Research Institute of Iceland (MRI). The Aquaculture Research Station of the MRI is located in south-west Iceland in the vicinity of a large land-based Arctic char farm, 7 km west of Grindavík (Fig. 21.1). The station commenced operations in 1988 in a 560 m^2 building, but was expanded to 1350 m^2 in 2002 with the addition of a second building. The station has tanks of various sizes (up to 30 m^3) and a total indoor rearing volume of about 360 m^3. The station has an abundant supply of clean borehole seawater with a constant temperature of 7°C all year round. Slightly brackish water (4 ppt) can also be obtained from a nearby well and 60°C geothermal borehole seawater can be used to heat up the rearing water through heat exchangers. In addition, the station is equipped with a cooling machine to cool down seawater, making it ideal for studying

Fig. 21.1 The Aquaculture Research Station of the Marine and Freshwater Institute of Iceland. The station is located in Grindavík on the south-west coast of Iceland.

marine fish and invertebrates under a wide range of temperatures. Since the station was built it has been used for studies and small-scale production of many species, but in recent years the focus has been on Atlantic cod, Arctic char and lumpfish.

21.4 The Icelandic lumpfish fishery

Iceland has a well-established lumpfish fishery with highly variable landings of approximately 2000–13,000 tons per year since the 1950s. The stock declined after 1990 with average landings of 4500 tons until 2012. In recent years the stock biomass index has been on the rise and landings increased to 6400 tons in 2015. Fishing is not controlled by individual transferable quotas but instead the fishing effort is restricted by the permitted number of fishing days and fishing nets per boat. During the 2015 season, 316 small boats were registered for lumpfish fishing and a total of 32 fishing days was issued, as was the case for the 2016 fishing season (Marine Research Institute, 2015). The main emphasis of the fishery is on the female because of the roe, which is used as raw material for the production of lumpfish caviar. The roe yield is about 29% of the female live weight but, after processing, the yield decreases to about 22% (Johannesson, 2006; Þórðarson *et al.*, 2013). The roe is salted in barrels (net weight 105 kg per barrel) and sold to processing factories both domestic and abroad.

The world supply of salted lumpfish roe is highly variable but the average world production since 1992 has been 32,000 barrels per year, equivalent to about 3400 tons of roe per annum. Iceland has been the largest producer in recent years with an average of about 10,000 barrels per year (Johannesson, 2006). Iceland was also the first country to achieve a Marine Stewardship Council (MSC) certification for the sustainability of the lumpfish fishery (Marine Stewardship Council, 2015). France is by far the largest export market for Icelandic lumpfish roe and lumpfish caviar. The government issued a new regulation in 2011 forbidding the discharge of gutted lumpfish into the sea and all fish had to be brought to land (Þórðarson, 2013). No markets were known for whole lumpfish but, with a collective initiative by the industry, one was found in China. Today most of the lumpfish is landed whole and gutted in factories, creating more employment and extra income for the fishermen. The historical significance of the lumpfish

Fig. 21.2 The flipside of the Icelandic 100 króna coin with the carved image of a lumpfish.

fishery for the Icelandic community is well reflected by the lumpfish image depicted on the flipside of the Icelandic 100 króna coin (Fig. 21.2).

Broodstock procurement and husbandry

The hatchery production of lumpfish in Iceland has so far been entirely dependent on the capture of wild broodstock. The fishing grounds are divided into seven regions and the fishing is carried out with gillnets (267–292 mm mesh) during the breeding season. In north, east and south Iceland, fishing for lumpfish is allowed from 20 May until 2 June. In west Iceland the fish are caught from 1 April to 14 June, and the lumpfish are caught from 20 May to 2 August in a small location in Breiðafjörður (Kennedy et al., 2015). Moreover, mature lumpfish are a frequent incidental bycatch in Danish seines and gillnets before the start of the fishing season. Thus, mature lumpfish can be obtained relatively easily by the farming industry from February to August, allowing lumpfish hatcheries to extend the production season and increase their production capacity. The fish captured during the fishing season are already conditioned to spawn and generally require a shorter time in captivity until viable eggs and sperm can be obtained. At the MRI, wild-captured breeders have been conditioned under a continuous photoperiod (LD 24:0) and the time in captivity until the start of voluntary spawning has been approximately 14, seven and one days for breeders captured in February, April and July, respectively. The spawning of late-captured fish is much more synchronised and during the 2015 summer spawning season up to 40 females were successfully stripped in a single day.

Most of the wild lumpfish breeders caught in Iceland in 2014 and 2015 were kept at the MRI, but after Stofnfiskur took over the egg production in 2016 the captive broodfish have all been kept in a quarantine facility in the town of Sandgerði. The capture of wild lumpfish for the MRI has been carried out by two boats fishing for cod in gillnets near Grindavík, one fishing for lumpfish near Grindavík and the other in Breiðafjörður. In the three previous breeding seasons, i.e. spring 2014, spring 2015 and summer 2015, a total of 93, 125 and 176 wild breeders, respectively, were captured and brought to the MRI, about 20% of these being males. The captured broodfish were kept on-board the fishing boats in 200 l plastic barrels with running or regularly renewed seawater. The condition of the fish after capture has been variable but most fish recover quickly in the hatchery. Mortality rates during the first weeks of conditioning have generally been very low. Once the fish have been brought to the station it is common to find some fish floating on the surface of the spawning tanks. The condition arises when the fish gulp air after being brought to the surface on-board the boat. It is relatively easy to resolve this condition by inserting a plastic tube down the oesophagus and into the stomach to allow the air to escape as the fish is submerged.

At the MRI the broodfish were usually kept in grey circular tanks (7 m^3) containing two rock shelters arranged in an arcuate manner on opposite sides of each tank (Fig. 21.3). As the fish become accustomed to their surroundings during the first days or weeks after capture they begin showing typical breeding behaviour when dominant males claim their nest and defend their site from other males. Over the following days and weeks, the mature females visit the nest to mate with their chosen male and lay their eggs. The females then leave the nest and leave the parenting to the male, who meticulously tends the eggs and guards the nest from potential predators (Fig. 21.4). The dominant males were usually aggressive towards all other males who ventured close to their nests. However, an exception to this behaviour was once observed where one dominant male and two very small subordinate males teamed up in tending and defending the same nest with multiple egg batches. Interestingly, the dominant male mostly remained close to the eggs and left the guarding duties to the two smaller males, who also tended the nest. It can be speculated whether the small males had managed to sneak in and fertilise some of the eggs in the nest. Sneaking is a mating tactic sometimes used by subordinate males to achieve reproductive success in many animal species (Neff and Svensson, 2013). Alternatively, the behaviour may have been purely symbiotic with benefits for all the males involved.

The easy access to large numbers of wild spawning broodstock over a large part of the year (February–August) in Iceland has made it less crucial to speed up the build-up of hatchery-reared broodstocks. Nevertheless, it is generally agreed that all sustained aquaculture should be based on

Fig. 21.3 Two male lumpfish tending and guarding fertilised eggs in their rocky nests in a breeding tank at the MRI Aquaculture Research Station.

Fig. 21.4 A male lumpfish tending and guarding his nest of multiple egg batches in a breeding tank at the MRI Aquaculture Research Station.

such broodstock to avoid the transmittance of disease from the wild, produce eggs throughout the year and facilitate genetic enhancement through breeding programmes. The MRI started rearing broodstock in 2014, but after about 18 months from hatch the group was terminated due to disease problems (see section on diseases below). Stofnfiskur started rearing broodstock in its own hatchery from 2016 and is considering a breeding programme for lumpfish.

21.5 Spawning and incubation

The collection of fertilised lumpfish eggs at the MRI has been based on manual stripping and artificial fertilisation, rather than the collection of egg clusters from breeding tanks. The risk of transmitting pathogens from the broodfish is smaller and the fertilisation rate is generally higher in artificially fertilised eggs. The maturation status of the females was therefore checked routinely by monitoring their genital opening to avoid spontaneous spawning. Once the genital opening was protruding the females were placed on a table and the eggs stripped into a clean glass beaker by applying gentle pressure on the abdomen. A plastic tube was inserted into the mouth to prevent the gulping of air. The milt of males was stripped by firmly squeezing with one finger on each side, creating internal pressure on the gonads. The milt was either collected in a syringe or into a small glass beaker depending on volume. The amount of milt that could be stripped was highly variable but typically less than about 1 ml per male. Usually, however, only a few drops of milt are required to fertilise a large egg batch. The milt was usually not collected by stripping on busy stripping days, but rather by swiftly killing the males and placing their gonads in a garlic crusher and squeezing the milt into a glass beaker. Much more milt was collected this way so the milt from a single male could be used to fertilise multiple egg batches. Sometimes the glass beaker was placed in an airtight bag with pure oxygen, stored in a refrigerator and used the next day with good success. The eggs were fertilised by adding the available milt directly to the egg batch, stirring gently for about 30 seconds and then storing the mixture for approximately 30 minutes as the eggs were being fertilised in the ovarian fluid (dry fertilisation method).

The MRI was required by the veterinary authorities to perform a standard two-week quarantine and prophylactic protocol of all collected wild lumpfish including repeated baths with Pyceze (50% bronopol), formalin and Neomycine antibiotics. These protocols were duly carried out during the spring seasons of 2014 and 2015, but during the 2015 summer breeding season, they were interrupted by the immediate spawning of the wild collected fish. Most of these fish were stripped on the day after arriving at the MRI and had thus not been properly disinfected before they were used for breeding. In retrospect, this involuntary breach of protocol may have been the major reason behind the series of disease outbreaks in lumpfish at the MRI in 2015. The MRI was also required to produce a health certificate stating that the wild collected lumpfish were not infected by a range of diseases such as VHS, IHN, ISA, PD and BKD. All wild lumpfish breeders were killed after stripping and subsequently screened for a range of viruses and bacteria at the Institute of Experimental Pathology (IEP) of the University of Iceland.

During the first three breeding seasons various types of incubators were used for egg incubation, such as plastic hatching jars from salmon hatcheries. The majority of the eggs at the MRI were incubated in horizontal breeding troughs, each with seven incubation baskets (Fig. 21.5). The incubation protocol is: some 30 minutes after fertilisation the eggs are drained and poured

Fig. 21.5 An incubation system for lumpfish eggs at the MRI Aquaculture Research Station. The system has four horizontal breeding troughs with seven incubation baskets each.

into baskets out of water and spread in layers or mats. The baskets are subsequently weighed to determine the pre-hardened (soft) mass of each batch. The average egg count has been close to 100,000 eggs per kg soft egg mass. After an average 20% post-hardening shrinkage in weight, the egg count increases to approximately 125,000 eggs per kg hard egg mass. Immediately after weighing, the baskets are submerged very slowly into the troughs to prevent the mats from breaking up during submersion. Each basket can hold between 1–1.5 kg of hardened egg mats. Water (10 l min^{-1}, 7°C, 30 ppt salinity) is supplied into one end of the trough, flowing through successive baskets. Fluorescent light is provided according to the natural photoperiod. The eggs are disinfected with Pyceze (0.1 ml l^{-1} water, 30 minutes stagnant) on three days per week. Once the eggs reach the eyed stage and the fertilisation success becomes visibly apparent, all bad batches are removed and the remaining matching good batches are united in fewer baskets. All the baskets are removed temporarily once every two weeks to allow for proper cleaning of the troughs.

The first few larvae usually start hatching in the baskets after approximately 39–40 days' incubation at 7°C. The baskets are then moved as a routine to the larval tanks and suspended to a shallow depth to allow the larvae to swim voluntarily out from the baskets (Fig. 21.6). Hatching of the entire batch is usually completed within one to two days and the newly hatched larvae swim actively out of the baskets. Many of the larvae swarm close to the water surface at first but, before long, most of them sit comfortably on the tank walls. The eggs may sometimes take several

Fig. 21.6 A hatchery tank for larval rearing of lumpfish juveniles at the MRI Aquaculture Research Station. The tank has two submerged incubation baskets with eggs ready to hatch.

days to hatch completely if the egg mats are too thick or in the event of egg clumps. The mean incubation time at 7°C for thin egg mats has been 40 days or equivalent to ≈280 degree days.

21.6 Rearing methods and protocols

The rearing period is divided into three phases, i.e. hatchery phase (eight weeks), nursery phase (six weeks) and juvenile phase (12 weeks). The MRI uses round, black-coloured hatchery tanks equipped with a rotating cleaning arm on the bottom. The tank volume is approximately 3200 l, with a submerged wall surface area of 7 m^2 and a total surface area of 10 m^2. Water temperature is easily adjustable in each tank and lights are controlled by timers and dimmers. Photoperiods of 24:0 and 18:6 (L:D) have been used without obvious differences in survival and growth of the larvae. The central standpipe is covered with 350 µm mesh netting and constant aeration is provided from an air collar around the standpipe. Extra oxygen is added directly inside the water inlet pipe and oxygen levels are maintained at 100–120% saturation. The initial larval stocking density has been 60,000–140,000 larvae per tank or equivalent to 0.6–1.4 larvae cm^{-2} of surface area. The water level has been lowered during the first week in some cases to increase initial prey density.

The larval feeding regime involves initial start feeding with Artemia nauplii (optional) and enriched metanauplii before a gradual weaning to artificial feed is completed in three to four weeks from hatch. The rearing water is initially shaded (optional) with algal paste (Rotigreen, Reed Mariculture) during the first 14 days post-hatch. During the first three days after hatch the larvae are kept in the dark until lights are turned on and feeding is initiated. Table 21.1 shows a typical larval rearing protocol with an overview over physical and biological parameters during the hatchery phase.

During the hatchery phase, temperatures and oxygen levels are measured on a daily basis. Light intensities and flow levels are gradually adjusted and prey densities are monitored throughout the day. Artemia is fed in two equal daily rations and dry feed is fed continuously with automatic underwater feeders. During the weaning process the larvae are offered dry feed first thing in the morning when prey levels are low. When the pre-juveniles have been weaned the underwater feeders are removed and replaced with regular surface feeders. Algal paste is optionally added during the first two weeks for shading twice per day. Daily maintenance includes cleaning of surface skimmers and siphoning debris from the cleaning arm. Samples are taken to monitor the growth and development of the larvae and pre-juveniles once a week. In tanks with high survival, plastic sheets are suspended after about four to five weeks from hatch to increase the habitat surface area in the tank (Fig. 21.7). The hatchery phase is completed in about eight weeks from hatch, at which time the pre-juveniles are transferred to the nursery tanks. The free-swimming pre-juveniles are collected with fishing nets after and while lowering the water level and stocked in water buckets for immediate transfer to the nursery tanks. A forced removal of the remaining attached juveniles does not cause any noticeable damage and post-transfer mortality is very rare. The transfer buckets are lined with mesh netting to prevent the fish from attaching to the bucket walls. The fish from each hatchery tank are then graded on a 4 mm grading grid and divided into two or three nursery tanks, depending on quantity.

The MRI uses square, green-coloured nursery tanks with a tank volume of approximately 3200 l and water temperatures, lights and oxygen are adjustable as described for the hatchery

Table 21.1 A typical larval rearing protocol for lumpfish production at the MRI in Iceland based on stocking of 80,000 larvae into a 3,200 l tank volume. Numbers are indicative.

Age (days)	Temperature (°C)	Standard length (mm)	Total length (mm)	Live weight (mg)	Light (lux)	Water flow (l min^{-1})	Algal paste (ml) x2	Artemia nauplii (mill)	Artemia enriched (mill)	Dry feed (g)
0	8	4.5	5.4			3				
1	8	4.6	5.5			3				
2	8	4.7	5.6			3				
3	8	4.8	5.7		200	3	20	4		
4	8	4.9	5.8		350	4	20	6	1	
5	8	5.0	6.0		500	4	30	8	1.5	
6	8	5.1	6.1		600	4	30	6	3.5	
7	8	5.2	6.2		700	4	30	4	6	
8	9	5.3	6.3		700	5	30		8	10
9	9	5.4	6.5		700	5	30		9	10
10	9	5.5	6.6		700	5	30		10	10
11	10	5.6	6.7		700	5	30		11	10
12	10	5.7	6.9		700	6	30		12	10
13	10	5.8	7.0		700	6	30		13	10
14	10	6.0	7.1		700	6	30		14	10
15	11	6.1	7.3		700	6			16	15
16	11	6.2	7.4		700	7			18	15
17	11	6.3	7.6		700	7			20	15
18	11	6.4	7.7		700	7			22	15
19	11	6.6	7.9		700	7			24	15
20	11	6.7	8.1		700	8			24	15
21	11	6.8	8.2		700	8			22	20
22	12	7.0	8.4		700	8			20	25
23	12	7.1	8.6		700	8			18	30
24	12	7.3	8.7		700	9			16	35
25	12	7.4	8.9		700	9			14	40
26	12	7.6	9.1		700	9			12	50
27	12	7.7	9.3		700	9			10	60
28	12	7.9	9.5	25	700	10			8	70
35	12	9.1	10.9	41	700	10				150
42	12	10.4	12.5	65	700	10				200
49	12	12.0	14.4	103	700	10				250
56	12	13.8	16.6	164	700	10				300
Total							340	28	333	

Fig. 21.7 Three plastic sheets suspended in a hatchery tank to increase the habitat surface area in the tank. The sheets are densely covered with early lumpfish juveniles.

tanks. The tanks have a total rearing surface area of 10 m^2 and are usually stocked with up to 20,000 juveniles per tank. The water temperature is usually maintained at 10–12°C during the nursery phase and the feeding regime involves the automatic feeding of artificial feed at a feeding rate of 3–4% of estimated biomass during 18–24 hours of daylight. A significant part of the daily ration is hand fed and floated as the juveniles generally prefer to eat floating or sinking feed. Frequent size grading and hand feeding are essential during the nursery phase to prevent cannibalistic behaviour among the juveniles. By definition, the nursery phase ends when the size-graded juvenile groups have reached a mean live weight of 2 g (*c.* 14 weeks from hatch).

After the nursery stage is completed the juveniles are transferred to a new hatchery section with round, grey tanks (7 m^3 and 16 m^2 rearing surface) for rearing up to export size. During this final rearing phase (juvenile phase), the temperature is gradually lowered with increasing fish size down to 7–9°C at the age of five to six months post-hatch. The feeding regime involves the automatic feeding of artificial feed at a feeding rate of 2–3% of estimated biomass during 18–24 hours of daylight. At a mean live weight of about 10 g the juveniles are prepared for vaccination by withholding feed for two days. The juveniles are then manually injected with the Alpha Marine 4 vaccine according to the protocol described by the producer Pharmaq. Vaccination-related mortality has been very low but the appetite of the fish is reduced for at least a whole week after vaccination. Daily maintenance after the hatchery phase includes vortexing and flushing of debris. Routine prophylaxis includes bi-weekly treatments with formalin (1:3000 for 30 minutes) and brackish water (10‰ for six hours).

21.7 Production results

The production results can be evaluated in terms of survival, productivity, growth and quality. At the MRI the juveniles are usually counted at the end of the nursery period when the routine mortality has become very low. Counting is done by weighing and sub-sampling or, alternatively, with a Vaki micro-counter. The initial number of stocked larvae is based on drained egg weight and visual estimation of hatching percentage. An estimated 91% mean hatching percentage during the 2016 season enabled calculation of a mean survival to the end of the nursery stage of 65% based on stocked larvae and 61% based on drained egg weight. The mean productivity from the hatchery phase was 47,000 juveniles per tank, 14.7 juveniles per litre or 4,700 juveniles per m^2 rearing surface. The spawning productivity was 45,750 juveniles per female based on one batch per female and a mean drained batch weight of 600 g (75,000 eggs). The juveniles were size-graded regularly and slightly overfed to minimise aggression, and the rearing density was kept reasonably low. Water quality, tank maintenance and routine prophylaxis is especially important during the juvenile phase. If these conditions are met, the mortality is normally very low during the juvenile phase or only about 0.1% mortality per week.

The growth performance is monitored by collecting subsamples from each tank at regular intervals and measuring the length and weight of individual larvae or juveniles. At the MRI, individual length is routinely measured from hatch and throughout the nursery stage while individual live weight is first measured after the age of 30 dph. Length is optionally measured as standard length (SL (mm)) or total length (TL (mm)) and has an allometric relationship with live weight (W (mg)) in the weight range 0.1–0.5 g:

$$SL = 0.0378\ W^{3.1284}\ (R^2 = 0.98)$$
$$TL = 0.0188\ W^{3.1615}\ (R^2 = 0.98)$$

Repeated measurements have revealed a mean ratio (TL/SL) of 1.21 between TL and SL in this size range.

Growth performance during the hatchery and nursery stages was variable between tanks but Fig. 21.8 shows typical growth trajectories for both total length and live weight up until vaccination. After an initial exponential phase the length growth trajectory becomes practically linear after about two months from hatch. During the linear phase the growth increment is relatively stable at around 0.45 mm per day, or alternatively 13.5 mm per month. In this example, the daily growth rates in weight peaked at approximately 15% per day at around 20 days post-hatch (dph), after which they gradually declined to 7% per day at 70 dph and 2.7% at 150 dph. Benchmark weights were approximately 2, 5 and 10 g at 100, 125 and 150 dph, respectively (Fig. 21.8). Those numbers are only indicative as growth performance is always highly variable

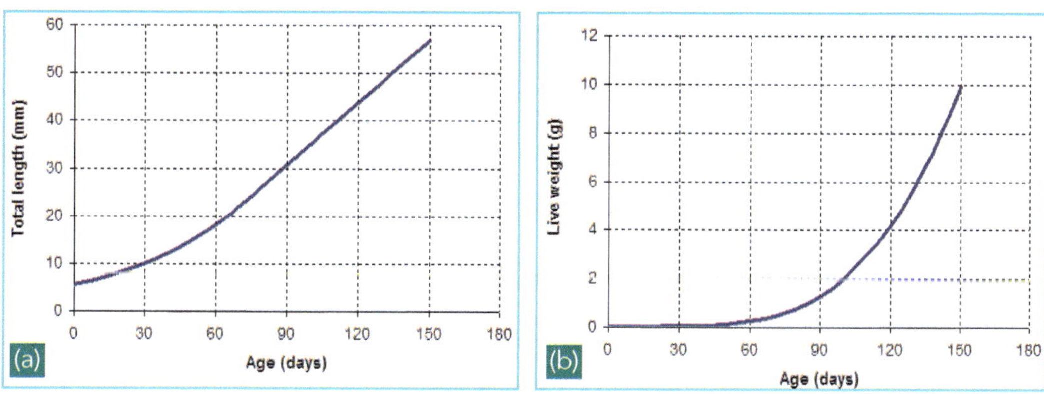

Fig. 21.8 Typical growth trajectories for hatchery-produced lumpfish at MRI Iceland. a) Mean total length during the first five months. b) Mean live weight during the first five months. Age is shown in days post-hatch (dph).

within batches. After vaccination at 10 g mean weight, the growth rates usually decrease temporarily while the fish are recovering. At the MRI the rearing density has thus far become a limiting factor soon after vaccination, sometimes resulting in suboptimal growth rates of vaccinated juveniles. Hence, the vaccinated juveniles are sometimes deliberately slightly underfed to prevent them from growing too large before they can be exported.

The quality of the juveniles can be evaluated based on features such as tails, fins or suction cups. These features are generally in very good shape at the MRI, perhaps reflecting the high water quality and frequent size grading. However, the general health status of the juveniles is a more important quality indicator and the major factor affecting the ultimate quality indicator, i.e. the delousing performance of the juveniles. The 2014 generation of MRI-lumpfish had a good overall health status and no specific diseases were reported from either the MRI-facility or from the Faroese sea cages. Unfortunately, this was not the case for the 2015 generation, which was badly ridden by various diseases.

21.8 Disease outbreaks

During the 2015 season wild breeders were brought into the farm and stripped during July–August in close proximity to the juveniles that had hatched in February–March and were being vaccinated at the time. Apparently, prophylactic measures were not sufficient to disinfect the broodstock properly as several disease outbreaks occurred among the juveniles and could be traced directly to the broodstock. Shipments of juveniles delivered in September suffered sharply increased mortality and inspection revealed that many of them had white necrotic tissue, particularly on the back and head region. These symptoms were subsequently noticed on juveniles and broodstock at the MRI and diagnosis at the Institute for Experimental Pathology (IEP) of the University of Iceland revealed bacteria belonging to the genus *Tenacibaculum* sp. In Norway, similar symptoms and high mortality rates among juvenile lumpfish have been found to be caused by *Tenacibaculum maritimum* (Småge *et al.*, 2016). Fortunately, the infection was not detected in Stofnfiskur's lumpfish farm. The disease was at first only noticed in several tanks but soon spread to other tanks as repeated treatments with antibiotics (Aquatet, feed additive), Pyceze (1:20000) and formalin (1:2000), all proved ineffective in curing the disease. Only one measure was found to be effective, i.e. long-term rearing in brackish water (10–15 ppt). This treatment was especially effective and usually after a two- to three-week long treatment the fish had regained their appetite and all external signs of the disease were gone. Access to freshwater is, however, limited at the MRI, which made it difficult to run an effective treatment protocol to eradicate the disease. In October, when many of the juveniles seemed to be recovering from the infection, the fish were struck with another health issue.

At the MRI, all wild lumpfish breeders are killed after stripping and subsequently screened for a range of viruses and bacteria (see broodstock section) at the Institute of Experimental Pathology (IEP) of the University of Iceland. In early October, several cell lines from tissue culture indicated a possible viral infection in a few samples and subsequent RT-PCR analysis indicated that the samples contained the viral haemorrhagic septicaemia virus (VHSV). The samples were sent to the European Union Reference Laboratory (EURL) for Fish Diseases in Copenhagen, Denmark, and on 19 October 2015 the IEP diagnosis was confirmed. The virus was detected subclinical, without any previous signs of symptoms or mortality. The results showed that the isolate belonged to VHSV Genotype IV: North American and Japanese/Korean isolates. This genotype has not been found before in Europe and is believed to be species-specific (Elsayed *et al.*, 2006). Immediately, the Icelandic health authorities issued an export ban on the MRI and ordered the destruction and disposal of all lumpfish on the farm. Iceland subsequently lost its VHS-free status temporarily, even though the outbreak occurred in wild-caught fish in a research laboratory. The health authorities are presently pushing for a renewal of the country's VHS-free status. Fortunately, the virus was neither detected in the lumpfish already delivered to the Faroes nor at Stofnfiskur's lumpfish farm. In transmittance tests at the EURL laboratory in Copenhagen, the virus was not transmitted to salmon or trout fingerlings via bath immersion but was found to cause some mortality after direct injection (Guðmundsdóttir, 2016).

The serious disease outbreaks that occurred at the MRI in 2015 could all be traced directly to the exposure of farmed fish to fish of wild origin. The incidents have been a valuable lesson for the lumpfish industry in Iceland and Stofnfiskur immediately aborted all plans to rear wild-caught broodstock at its lumpfish farm. Stofnfiskur made an agreement with a research facility in the town of Sandgerði to keep wild-caught lumpfish breeders in an isolated quarantine. All fish used in breeding are killed and screened for a range of viruses and bacteria. Fertilised eggs are stored in isolated incubators in the quarantine and not used until both parents have been confirmed negative by the screening process. It is hoped that this approach will greatly decrease the chances of transmitting endemic disease from wild to farmed lumpfish. The MRI hatchery now purchases fertilised and screened lumpfish eggs from Stofnfiskur and the recent success of the 2016 production season shows that this arrangement greatly improves the biosecurity of the lumpfish production. MRI also implemented stricter prophylactic protocols and a total of 250,000 healthy lumpfish was delivered to the Faroese buyer.

21.9 Future prospects

In spite of relatively high long-term lumpfish mortality and problems related to feeding and sheltering lumpfish in the Faroese salmon cages, the Icelandic lumpfish have nonetheless been very effective at delousing and have already been proven as a realistic remedy or solution of the sea lice problem. Stofnfiskur has established a key position as the biggest and perhaps soon the only producer of lumpfish for the Faroese market and is currently supplying lumpfish to all three major salmon producers in the islands. Stofnfiskur is also poised to expand its production and supply the fast-growing Icelandic salmon market with lumpfish juveniles to prevent and control future sea lice problems in Icelandic salmon farming. The MRI committed to lumpfish production for the fourth consecutive season in 2017 but no long-term plans have been made regarding continued production of lumpfish at the institute. Most likely, the institute will soon withdraw from commercial production and rather focus on lumpfish from a research point of view. MRI's Aquaculture Research Station is probably among the best equipped aquaculture research stations in Europe and ideal for conducting lumpfish research. New research projects on feeding, incubation and optimal growing temperatures of lumpfish were being conducted at the MRI in 2017–2018. The use of lumpfish as biological pest control is still in the early stages of development and will require assistance in the form of focused scientific research for years to come.

References

Elsayed, E., Faisal, M., Thomas M., Whelan, G., Batts, W. and Winton, J. (2006) Isolation of viral haemorrhagic septicaemia virus from muskellunge, *Esox masquinongy* (Mitchill), in Lake St Clair, Michigan, USA reveals a new sublineage of the North American genotype. *Journal of Fish Diseases* 29, 611–619.

Guðmundsdóttir, S. (2016) Veirur í hrognkelsum (Virus in lumpfish). Presentation at the Institute of Experimental Pathology of the University of Iceland. 4 May 2016.

IAA (2016a) Hagtölur. Available at: www.lf.is/is/hagtolur (accessed 20 February 2017).

IAA (2016b) Atlantic Salmon. Available at: www.lf.is/en/fiskeldi-a-islandi/atlantic-salmon (accessed 15 May 2016).

Johannesson, J. (2006) *Lumpfish caviar – from vessel to consumer.* FAO fisheries technical paper 485. 71 pp.

Jónsson, G. (2015) Ársskýrsla dýralæknis fisksjúkdóma (The state veterinary annual report for fish diseases). Available at: www.mast.is/library/Sk%C3%BDrslur/ Arsskyrsladyralaeknisfisksjukdoma2015.pdf (accessed 15 August 2016).

Karbowski, C.M. (2015a) *A First Assessment of Sea Lice Abundance in Arnarfjörður, Iceland. Sentinel Cage Sampling and Assessment of Hydrodynamic Modelling Feasibility.* Masters thesis. The University Centre of the Westfjords. Available at: http://hdl.handle.net/1946/22543 (accessed 20 August 2016).

Karbowski, N. (2015b) *Assessment of sea lice infection rates on wild populations of salmonids in Arnarfjörður, Iceland.* Masters thesis. The University Centre of the Westfjords. Available at: http://hdl.handle.net/1946/22539 (accessed 21 August 2016).

Kennedy, J., Jónsson, S.Þ., Kasper, J.M. and Ólafsson, H.G. (2015) Movements of female lumpfish (*Cyclopterus lumpus*) around Iceland. *ICES Journal of Marine Science* 72, 880–889.

Marine Research Institute (2015) State of marine stocks in Icelandic waters 2015/2016 and prospects for the quota year 2016/2017. Marine Research Institute in Iceland 185. 160 pp.

Marine Stewardship Council (2015) Traditional small boat Icelandic lumpfish fishery is MSC certified. Available at: www.msc.org/newsroom/news/traditional-small-boat-icelandic-lumpfish-fishery-is-msc-certified (accessed 21 February 2017).

Neff, B.D. and Svensson, E.I. (2013) Polyandry and alternative mating tactics. *Philosophical Transactions of the Royal Society B.* 368, 1–11.

Småge, S.E., Frisch, K., Brevik, Ø.J., Watanabe, K. and Nylund, A. (2016) First isolation, identification and characterization of *Tenacibaculum maritimum* in Norway, isolated from diseased farmed sea lice cleaner fish *Cyclopterus lumpus* L. *Aquaculture* 464, 178–184.

Towers, L. (2015) Viral Hemorrhagic Septicemia Reported for First Time in Icelandic Fish. 27 October 2015. Available at: www.thefishsite.com/fishnews/26621/viral-hemorrhagic-septicemia-reported-for-first-time-in-icelandic-fish/(accessed 15 August 2016).

Þórðarson, G., Pálmason, S. and Reykdal, Ó. (2013) Production development of lumpfish. Report no. 17-13. ISSN: 1670-7192. Matis publications, 24 June 2013. Available at: www.matis.is/media/matis/utgafa/17-13-Baett-nyting-grasleppuafurda.pdf (accessed 15 August 2016).

CHAPTER 22

Rearing, farm application, and behaviour of lumpfish (*Cyclopterus lumpus*) in the Faroes

Ása Johannesen,[1] Regin Arge,[2] Kirstin Eliasen[1]

[1]Fiskaaling, við Áir, Hvalvík, Faroe Islands
[2]Marine Harvest, Eysturoy, Faroe Islands

22.1 History and current situation in the Faroes

The salmon farming industry in the Faroe Islands started in the 1970s (Salmon from the Faroe Islands, 2017), and is small in comparison to many of the neighbouring countries, with a total production of 80,000 tonnes of salmon (live weight) in 2015, totalling almost 3 billion DKK in export value (Hagstova-Føroya, 2016). There are three major farming companies, two of which have partial foreign ownership (Marine Harvest and Bakkafrost) and one completely local (Hiddenfjord). Farm sites are located mostly in shallow and narrow fjords, but some are in more exposed areas with sea currents of up to 0.9 m sec^{-1} in some locations and waves reaching 6 m in others. As in the neighbouring countries, sea lice are a major health problem in the Faroes. Lice grow resistant to chemotherapeutants, so a variety of other methods are used to combat infestations. These methods include freshwater delousing, mechanical louse removal, and growing larger smolts that require less time at sea. Cleaner fish are not a new concept to the Faroese industry, as they pay attention to what their neighbours are doing. However, none of the suitable wrasse species are native to the Faroe Islands, and the authorities have denied permission to import them. Lumpfish are native to the Faroe Islands and these are presently used in salmon farms, because they are known to consume lice and are considered an important additional tool to help combat them.

Other than using lumpfish as cleaners in the salmon farming industry, no commercial exploitation of lumpfish exists on the Faroe Islands. There is no history of lumpfish being caught commercially other than as bycatch. From 2014 to 2016, lumpfish were used for delousing salmon by three different companies at six different farming locations. These have all been farmed lumpfish, mostly imported from Iceland and some were farmed at Nesvík Marine Centre, Fiskaaling.

Work on cleaner fish on the Faroes started in 2011, when Fiskaaling commenced a small lumpfish project with a series of experiments. These were mostly focused on understanding rearing requirements as well as some work on behaviour and louse consumption. During the following years, two research projects financed by governmental funds and the Faroese Fish Farmers Association were initiated. The aims of these projects were to study lumpfish behaviour and to establish best practice manuals for producing lumpfish. The work resulted in an extensive manual written in Faroese for salmon farmers. Some of the content of that manual, mostly the parts that may differ from practices elsewhere, is reproduced in English in this chapter.

The research at Fiskaaling resulted in a small local production of lumpfish in 2015/16 but, due to the small size of Nesvík Marine Centre, the production of lumpfish at that site cannot exceed 200,000 fish at a size suitable for release into salmon farms (approximately 25 g). There is a conservative estimated figure of 2 million lumpfish to cover the need of all salmon farms on the Faroe Islands. Therefore, two farming companies began importing juvenile lumpfish from Iceland in late 2014 and from 2016 onwards, all three companies have been importing lumpfish. Currently, no production of lumpfish takes place in the Faroes and the future of lumpfish production on the islands is unclear as of October 2017.

This chapter reviews current lumpfish breeding practices at Nesvík Marine Centre, which is the only site on the Faroe Islands that breeds lumpfish. It also provides some insight into the research that is carried out there and describes how lumpfish are used in industry.

22.2 Production of lumpfish

22.2.1 Livestock

Although lumpfish have been harvested in the North Atlantic for centuries (Johannesson, 2006), only a few attempts have been made in the Faroes. Thus, very little is known about the local population of lumpfish. However, the sparse data available show that the spawning season around the Faroes probably lasts from February to October – with a possible peak during spring and midsummer. Only a few mature female lumpfish from Fiskaaling's production have been used to obtain eggs in 2015 and 2016, so most eggs are obtained from individuals of wild origin (2–3 kg). Conversely, milt has mostly been obtained from captive-bred males as these start maturing at a much smaller size (200–300 g). In spite of the uncertainty concerning future production of lumpfish on the Faroes, Fiskaaling has initiated research aimed at establishing livestock of local origin for future use in Faroese production of lumpfish.

22.2.2 Stripping, fertilisation, removal of egg stickiness and cryopreservation of milt

When ready to spawn, the area around the female genital opening swells. The timing of manual egg stripping is ascertained by daily visual observation and experience. When stripping eggs, the egg mass should flow evenly and easily. Egg masses characterised as irregular thick lumps are not mature. Eggs are collected in containers and may be stored in a refrigerator for a few hours until use. Normally about 0.5–1.0 l eggs is obtained per female and, at Fiskaaling, there are usually about 90–120 eggs per ml. Lumpfish males may be stripped for milt or the milt is collected after the fish has been killed and opened. There seems to be no difference in motility of stripped milt and milt obtained from whole testes. For long-term storage, Fiskaaling has developed a protocol for freezing and preserving milt (Norðberg *et al.*, 2015). All parent fish are tested for various infectious salmon diseases, as regulated by the local fish health authorities.

Fertilisation of lumpfish eggs is done by a combination of the dry and wet method, i.e. the ovarian fluid is firstly removed and a suitable amount of milt is gently mixed with the eggs. Thereafter, one volume of seawater comparable with one egg volume is added and the mass is gently but thoroughly mixed and left for fertilisation for two minutes. Depending on egg and milt quality, fertilisation success falls within a range of about 50–90%. Normally the eggs will begin sticking together after about five minutes and normal practice at other lumpfish hatcheries is to form the egg mass into thin pancake shapes. At Fiskaaling, standard practice is to use Alcalase to degum the eggs (novozymes A/S) after fertilisation and before disinfection. Separated eggs are easier to disinfect and allow for better husbandry, such as removing dead eggs during incubation. The protocol used at Nesvík for degumming is a slightly modified version of a protocol for wrasse (Helland *et al.*, 2013), and works very well with 100% free eggs after the procedure. Fertilised lumpfish eggs are disinfected by immersion in 500 ppm glutaraldehyde in seawater for four minutes, which is slightly different to methods previously tested on marine fish (Salvesen and Vadstein, 1995). After four minutes in glutaraldehyde, the eggs are rinsed in seawater before being placed in incubators.

Eggs at Fiskaaling have been incubated in incubators of different shapes and sizes, but all incubators are designed by the same principle, i.e. the inlet is at the bottom and outlet on top. Sufficient water flow seems to be 0.2 to 0.3 incubator volume/minute and the eggs are kept at ambient temperature.

Lumpfish embryos have formed visible eyes at about 120 degree days and hatching occurs at about 300–330 degree days. Normal practice at Fiskaaling is to allow the hatched larvae to freely swim/float into a collection tank, where they are harvested daily or every other day. This practice will ensure each batch of larvae has a uniform age as well as avoiding transfer of dead larvae, egg shells, etc. into the start feeding system, which may become a source of bacterial contamination.

In the production of lumpfish, the start feeding phase is perhaps one of the most critical. Compared to other marine fish species, lumpfish larvae are well developed and are able to effectively hunt and ingest prey a week or two after hatching (Benfey and Methven, 1986; Brown, 1986). At Fiskaaling, larvae have been found with ingested prey in their stomachs before one week of age, so it is important to provide the larvae appropriate feed approximately five days after hatching.

Lumpfish larvae are about 5.5 mm long at hatching and weigh about 1 mg (DW) when start feeding commences (Fig. 22.1). The best results in growth and survival at Fiskaaling are obtained using live feed (Artemia) in the first two weeks of feeding. After 600–700 degree days the lumpfish juvenile weighs about 0.12 g (WW), is weaned to 0.5 mm dry feed pellets and is ready for transfer to larger tanks. Dry feed is introduced after one week and pellet size is increased gradually. It is important to observe appetite by hand feeding and some over-feeding is to be expected. Therefore it is important to maintain good hygiene and management routines and keep the tanks as clean as possible. During start feeding, maximum density is estimated from available surface space in the tank. As long as all larvae are able to sit comfortably (with spare surface area available) and few larvae are seen swimming, they are considered to have enough space. Oxygen levels are monitored (with a minimum acceptable 70% saturation) to ensure water exchange is high enough for the biomass.

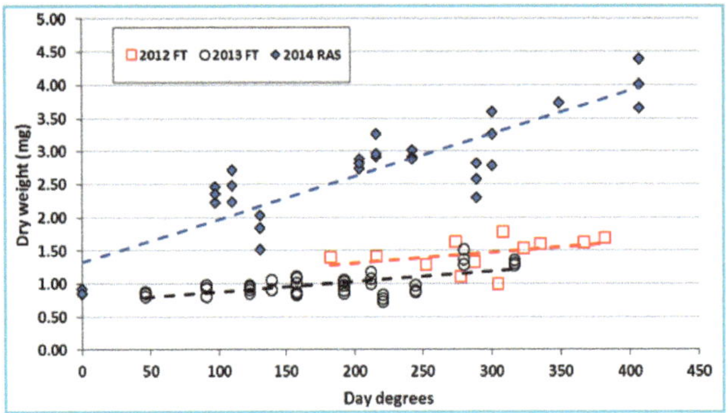

Fig. 22.1 Growth of lumpfish larvae (mg DW) in two types of start feeding systems: Flow-through (FT) and recirculation system (RAS). The data are from three separate experiments in 2012–2014. Credit: Regin Arge

22.2.3 Start feeding system

A 10 m^3 recirculation system (RAS) has been used at Fiskaaling for start feeding lumpfish larvae with remarkable results compared with the flow-through systems (Table 22.1). The benefits of using this RAS can be related to a beneficial bacterial fauna in the system as reported for other marine species (Skjermo and Vadstein, 1999). Other factors, such as improved water quality compared to flow-through water, may also be the cause. Table 22.1 shows the mean water parameters in the RAS system during start feeding. It should be mentioned that Fiskaaling has not evaluated possible optimum or toxic levels of temperature, O_2, NH_3, NH_4.

22.2.4 On-growing

After start feeding the lumpfish juveniles are transferred to larger tanks for on-growing. Fiskaaling has used 1.8 m^3 tanks for fish up to 8 g and 6 m^3 tanks for larger fish. Water exchange is one tank volume/h and the fish are kept at ambient sea temperature (6–11°C). Biomass density is kept below 8 kg m^{-3} for fish <5 g and about 12 kg m^{-3} for larger fish.

In a short experiment conducted at Fiskaaling, 1600 fish were used in a crossed design experiment with two factors: light and feed regime on a 12- or 24-hour basis. Fish were weighed and the tail fins were scored on a 1–5 scale: (1) no damage, 2) slight damage, 3) fin partially missing, 4) open sore, 5) fin completely gone (and fish severely injured) at the start and end of the experiment. The results are presented in Table 22.2 and indicate a slight improvement in weight gain at 24-hour light and poor growth in 12-hour light with 24-hour feeding. In the 12-hour light and 24-hour feed treatment, much of the feed may have gone to waste in the dark hours, which can explain the decreased growth. The small differences between the groups indicate that even with feed going to waste in the dark hours, there was enough feed available to not substantially hinder growth. However, a tendency seems to be that the fish are less aggressive, i.e. lower incidence of fin damage, when feeding has been concentrated to 12 hours, particularly in the 24-hour light regime. These values are not significantly different, though that may be explained by the short test period and the subjectivity of fin scoring.

Table 22.1 Water parameters (±SD) for start feeding lumpfish in a recirculation system (RAS). Credit: Regin Arge

Temp	°C	11.6	±	0.3
Oxygen	% saturation	91.7	±	1.9
Salinity	‰	33.4	±	0.3
pH		8.1	±	0.1
Nitrate	NO3-N mg^{-1}	0.79	±	0.12
Nitrite	NO2-N mg^{-1}	0.2	±	0.1
Ammonia	NH3-N mg^{-1}	0.004	±	0.001

22.2.5 Disease control

The fish were routinely vaccinated prior to release in salmon farms. Initially vaccinated at about 8 g (injection) but the vaccination weight has since been increased to over 10 g due to observations indicating better uptake of vaccine suspension in larger fish. Vaccine uptake is routinely monitored according to the vaccine producer guidelines. The vaccine used has been the Alpha Marine micro 4 (Pharmaq) designed to protect against furunculosis and three serotypes of *Vibrio anguillarum*. Vaccination is prescribed for all lumpfish a minimum of four weeks before transfer to sea. Fish are injected according to manufacturer's guidelines and resume feeding within 48 hours post-vaccination.

The greatest challenges in the production of lumpfish in the Faroes has probably been related to outbreaks of bacterial diseases, which have caused high mortality, especially in warmer periods. It has not been possible to single out a genus of bacteria as the causative agent in each case of mortality. Rather, a diverse mix of different bacteria is found. It is therefore possible that the underlying cause may be multifactorial and related to stress, suboptimal environment, which in turn may lead to growth of pathogens and secondary infections. Most common pathogens found in lumpfish at Nesvík are atypical furunculosis (*Aeromonas salmonicida*) and vibriosis (*V. wodanis, V. logeii* and *V. splendidus*), though Pasteurella sp. and Tenacibaculum sp. have also been found in connection with infections in lumpfish. In the most recent production cycle (2016/17), little to no mortality has been recorded at Nesvík Marine Centre. This is probably due to a combination of good hygiene practices and low fish density.

22.3 Behavioural work carried out on the Faroe Islands

Very little literature is available on lumpfish behaviour. Some work has been done on their breeding behaviour (Davenport, 1985) and their migration and diet (Brown, 1986; Pampoulie *et al.*, 2014; Williams and Brown, 1991), but important information on their welfare and utility in a farmed environment is limited in the literature (but see, for example, Imsland *et al.*, 2014, 2015). Some work has been carried out in the Faroes to ascertain some important points

Table 22.2 The mean weight, standard deviation and tail fin score of lumpfish after 60 days with either 24 or 12 hours light crossed with 24 and 12 hours continuous feeding. Credit: Ása Johannesen

Feed hours	Light hours	Weight g	SD	Tail fin score
12	12	41.40	11.11	1.85
	24	42.60	10.76	1.79
24	12	37.79	10.25	1.87
	24	42.30	10.44	1.84

regarding lumpfish behaviour as it relates to farming and cleaning. This work has focused on environmental factors such as tank colour and shelter preference, as well as internal factors such as activity levels, aggression, boldness, and cleaning behaviours. This section will summarise available knowledge and adds relevant information gathered at Fiskaaling to hopefully provide a better understanding of lumpfish behaviour.

22.3.1 General behaviour

Lumpfish is a sedentary species, and fish spend much of their time attached to smooth surfaces while resting. They are inclined to hide underneath or behind shelters, especially in small, brightly lit tanks. Anecdotal evidence suggests that seagulls and crows are lumpfish predators and will both catch juvenile lumpfish swimming near the surface as well as stranded mature males guarding their roe. It is likely that their sedentary nature and slow swimming speeds make them vulnerable to predation by birds as well as aquatic predators, so hiding is probably a good strategy.

Due to their lack of buoyancy, lumpfish preserve energy by attaching themselves to surfaces using their modified pelvic fins as a sucker disc, though after the juvenile stage, they migrate long distances in open ocean (Davenport, 1985) and are known to exhibit vertical migration of 100 m or more each day (Kennedy *et al.*, 2015). During their coastal life stage, lumpfish are sit-and-wait predators that consume any suitably-sized prey drifting/swimming by and will only switch to active foraging when prey is scarce (Brown, 1986; Killen *et al.*, 2007).

Our preliminary work on suitable lumpfish environments suggests no particular advantage or disadvantage to rearing lumpfish in black, red, green or white tanks. Fish reared in these colours for one month showed no difference in mortality, growth or stress levels as indicated by behavioural response to an external stressor. The only difference found between fish reared in these differently coloured tanks was a slight difference in skin colour, with fish in black tanks being darker than those in white tanks (personal observation). This is fairly common in a range of fish species and occurs within hours after the change in environment (Nilsson Sköld *et al.*, 2013; Rodgers *et al.*, 2013).

There are indications that lumpfish may prefer to feed in bursts rather than continually. As part of an experiment carried out at Fiskaaling, lumpfish stomachs were flushed to investigate whether they had eaten lice. These fish were provided with feed continuously and, despite this, only approximately 61% of the fish had feed in their stomachs at the time of sampling. Lumpfish also seem to tolerate one to two days of starvation well, but after this, our experience with pre-vaccination fasting indicates that lumpfish resort to tail biting and other extreme foraging behaviours.

22.3.2 Cleaning behaviour

Lumpfish that have been observed cleaning salmon at Fiskaaling have all pursued the salmon while searching for lice. However, other workers have found that salmon approach lumpfish shelters and are cleaned there. This is an interesting difference that warrants further exploration. Several different studies of cleaning have been carried out at Fiskaaling (summarised in Table 22.3) and they all provide strong indication that cleaning behaviour is not universal among lumpfish. There is some evidence that a low percentage of lumpfish consume salmon lice. Perhaps the strongest indicator of this is the 'four-day louse consumption study' (Table 22.3), that found that only 23% of the lumpfish tested consumed lice. However, acclimatisation needs to be taken into account. As part of a behavioural study mentioned in section 22.3.4, it has become clear that one can expect at least five days to pass before lumpfish resume feeding after being moved to a novel environment (Fig. 22.4). This gives the fish very little time to consume lice in a one-week study, so perhaps these numbers would be higher if acclimatisation periods were longer. The length of time taken to resume feeding probably depends on how stressful the new environment is and the coping mechanism of the fish involved. This might explain why louse numbers dropped so dramatically within the first week in the 'Emergency treatment for lice' case study in the long-term studies (Table 22.3 and Fig. 22.2). Those studies were carried out in large enriched tanks with low biomass, so were perhaps less stressful environments than the tanks these lumpfish had been transferred from.

Table 22.3 Summaries of four separate salmon cleaning studies carried out at Fiskaaling. Credit: Ása Johannesen

One-week cleaning effect study	Two lumpfish and three infected salmon cohabited for one week. Repeated nine times with different fish. Additional 9 replicates without lumpfish. Lice were counted before and after and results showed no difference.
Two-week cleaning effect study	Five lumpfish and ten infected salmon cohabited for two weeks. Repeated three times with different lumpfish. Additional three replicates without lumpfish. Lice were counted before and after and results showed lower counts with lumpfish than without.
Long-term case studies (Case study one to three)	Three sets of 21 lumpfish and 135 salmon infected with copepodid stage salmon lice weekly, cohabited in 5m diameter tanks in three case studies lasting 3 months. Lumpfish were introduced according to case study: 1) before initial infection with lice; 'Preventative use of lumpfish', 2) upon development of motile lice 'Prevention and treatment' and, 3) upon number of adult lice reaching two lice per fish on average 'Emergency treatment for lice'. See Fig. 22.2 for results.
Four-day louse consumption study	200 lumpfish in batches of 20 cohabited with 20 infected salmon for four days after a one-week habituation period. Gastric lavages carried out under anaesthetic and stomach fluid samples were collected for PCR analysis. 23% of all lumpfish had consumed lice.

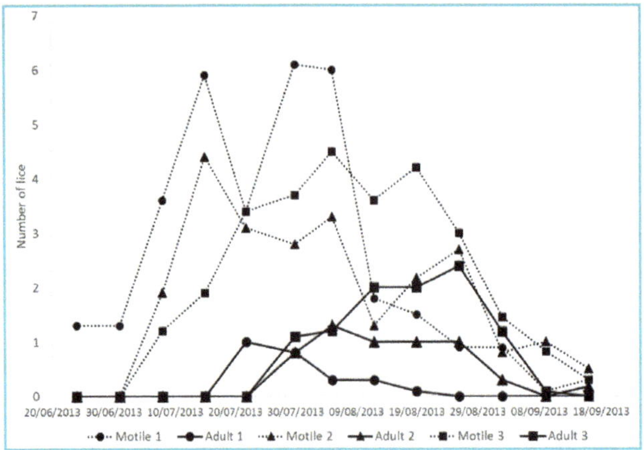

Fig. 22.2 Numbers of motile and adult salmon lice over time in three different tanks. Circles represent case study 1, triangles represent case study 2 and squares are case study 3. In case study 1) lumpfish were present from 20/06/13, 2) lumpfish were added 08/07/13, and 3) lumpfish were added 26/08/13. Credit: Ása Johannesen

22.3.3 Estimating the effect of lumpfish

Estimating the effect of lumpfish as a delousing agent might prove difficult. One way to do so is to compare the current sea lice situation where lumpfish are present with the historical data on sea lice counts from the farming site in question. However, sea lice numbers that vary unpredictably and for unknown reasons are common (unpublished data, Fiskaaling), so comparisons of current and historical sea lice data might be misleading. One factor influencing the predictability in sea lice numbers might, for instance, be the effect of local wind conditions, which are observed to affect the distribution of sea lice nauplii and copepodites in the Faroes (á Norði *et al.*, 2015).

Another way to estimate the efficacy of lumpfish is by analysing their stomach contents. However, the basis for interpretation of stomach content is to know which foraging activity the stomach content represents, i.e. the period of time the sea louse is visibly detectable in the stomach after ingestion.

To be able to make better interpretations of the lumpfish stomach content, Fiskaaling has carried out two experiments on lumpfish digestion of adult male *Lepeophtheirus salmonis* (Eysturskarð *et al.*, 2016). The first experiment investigated the digestion rate over a period of 24 hours (average lumpfish weight approximately 11 g, average water temperature 8.5°C), and at the end of the experiment, i.e. after 24 hours, the salmon louse was still visibly detectable in the lumpfish stomach (Fig. 22.3). Subsequently, a similar experiment was conducted (average lumpfish weight approximately 37 g, mean water temperature 6°C), but spanning a longer period, i.e. five days. After four days, the salmon louse was still visibly detectable in the stomach (Fig. 22.3). However, the results indicate that the digestion rate might be influenced by lumpfish size and/or temperature, i.e. the salmon louse ingested by smaller lumpfish kept at 8.5°C during digestion seemed to be considerably more digested after 24 hours compared to those ingested by larger lumpfish kept at 6°C.

22.3.4 Behavioural traits

In an ongoing study testing many behavioural parameters, a pattern is emerging that suggests lumpfish can be divided into two main types; proactive and reactive. Their differences are

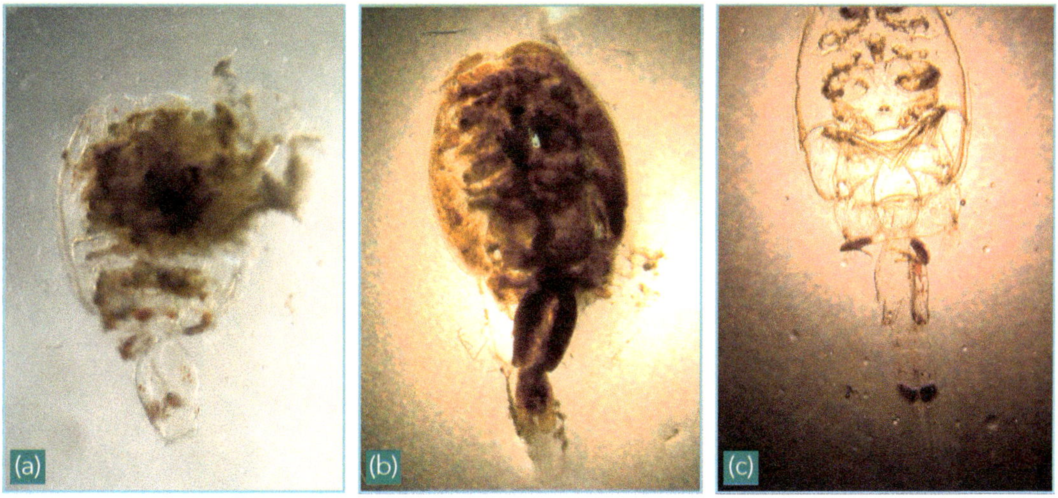

Fig. 22.3 Adult male salmon louse 24 hours after ingestion by lumpfish (~11g) incubated at 8.5°C (A), and 24 (B) and 96 (C) hours after ingestion by lumpfish (~37g) incubated at 6°C. Credit: Kirstin Eliasen

summarised in Table 22.4. The proactive fish respond to potentially harmful external stimuli with escape behaviours whereas the reactive fish respond with freezing – or potentially by not responding at all, given their inactive behaviour before the onset of an external stimulus. Generally, most of the fish fell within the 'reactive' group (~80%), though in the 'predation' test most fish moved away for a short while before resuming their previous behaviour. While the purpose of the mirror test was not to test whether fish were reactive or proactive, the general pattern of 'response through activity' *vs* 'sitting still' was also present in these tests.

Lumpfish also vary greatly in the time it takes to acclimatise to a new environment. Some fish start feeding the day after they have been moved to a new environment, and some do not resume feeding in a one-week trial (see Fig. 22.4).

22.3.5 Behavioural needs in an industrial setting

Lumpfish require smooth surfaces in a stress-free environment to rest on. Indications from fish kept in tanks and in salmon cages at sea are that lumpfish prefer to hide underneath a shelter. When provided with artificial seaweed in tanks, the fish will either sit on the tank wall under the seaweed or on the downwards facing side of the seaweed 'leaf'. At fish farms, lumpfish typically rest underneath feed tubes and on other smooth surfaces, where they can sit underneath a shelter. As mentioned earlier, birds are a likely predator of lumpfish, so perhaps there is a behavioural need to have shelter specifically overhead.

Some sort of enrichment designed to encourage exploratory and search behaviours at the juvenile stage could enhance louse cleaning behaviours once the fish enter commercial salmon farms. How to do this in an industrial setting is a difficult question to answer.

The fish may not feed for one week after release in the cage. Most fish will move very little after finding an initial resting place. Fish that find a resting place far away from the feeder may

Table 22.4 Overview of behavioural tests carried out and reactions from test fish. Credit: Ása Johannesen

Test	Proactive	Reactive
Basic behaviour	Sits still	Sits still
Novel object	Moves towards/away from object	Sits still
Predation (chasing with net)	Prolonged (>1 min) escape behaviours	Short duration of escape behaviours (<10 seconds)
Mirror	Investigates mirror	Sits still
Boldness (entering exposed area)	Moves towards exposed area	Sits still

struggle to recover after their acclimatisation period. Encouraging the fish to rest near the feeder in the first week, e.g. using an acclimatisation cage under the feeder, will improve the chances of fish learning where food is available and prevent starvation.

Proactive fish are more likely to explore the salmon cage, and thus are also more likely to encounter lice. As it is unknown at this time whether cleaning is an innate behaviour or a generalist consequence of encountering lice by chance, increasing the likelihood of encounter may be a good idea.

22.4 Implementation on salmon farms

In late 2014 cleaner fish were introduced to the Faroese salmon farming industry for the first time. Lumpfish were imported from a producer in Iceland, and transferred to all 17 cages on site. The density ranged from 8–14% of the number of salmon in each cage. Despite high mortality rates, only three of the cages, i.e. ~10% of the total biomass, had been treated with medicines for sea lice when harvest commenced, motivating further use of cleaner fish as a delousing agent in the Faroes. Approximately 25% of the Faroese salmon farming sites were using lumpfish as cleaner fish in spring 2016 and this number has increased since, with a majority of farms using lumpfish in 2017.

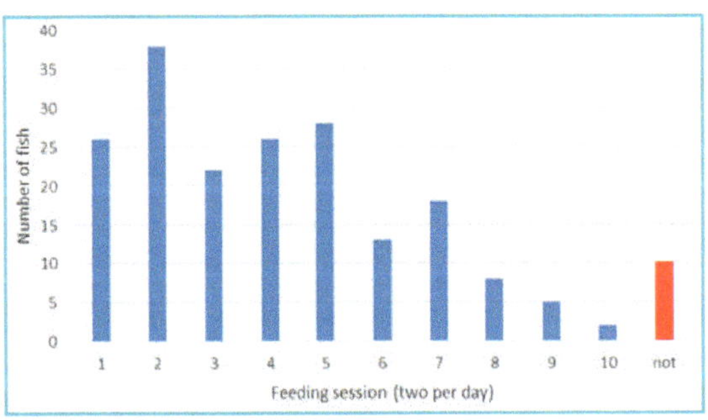

Fig. 22.4 The number of fish commencing feeding at each session. Fish were fed in two sessions each day (morning and afternoon) and feeding was considered as having commenced when the fish had taken food in three out of five consecutive sessions. The first of the three feeding sessions is registered. The red 'not' bar at the far right represents the fish that did not consume feed in at least three sessions out of five consecutive sessions. Session 1 was the morning after the fish were introduced to the new environment. Credit: Ása Johannesen

Initially, the knowledge and guidelines on lumpfish management were limited. However, the farmers on site prepared to receive lumpfish by installing artificial seaweed as an enrichment measure and automatic feed dispensers. The success of the artificial seaweed has been varied, and some farms have therefore substituted or added to the seaweed with firmer plastic to get a better substrate for attachment. Initially, none of the farms had obtained additional labour to manage the lumpfish, but currently most sites have at least one technician occupied full-time in managing the lumpfish.

The farmers' expectations of lumpfish effectiveness was initially high. However, a frequent observation is that lumpfish are often most effective immediately after introduction, but that the efficacy might diminish with time. This has led to the current acceptance of cleaner fish being a brigade in the war against sea lice, rather than the sole solution to the problem.

22.4.1 Challenges

The main obstacles for the success of lumpfish management in the Faroes are: 1) difficulty in obtaining suitable farming premises for lumpfish on the islands, 2) difficult transport from the production site in Iceland to the salmon farms in the Faroes, and 3) diseases.

Two diseases have had especially significant consequences on the survival of the lumpfish after introduction to the farming sites (as of summer 2016). The first one is characterised by visibly identifiable black dots on the exterior, as well as on the interior of the lumpfish. When introduced, this disease moderately increases daily mortality rates, but is much less aggressive compared to the latter, which is visibly identifiable by white to light brown exterior growths (Fig. 22.5, see Chapter 21 for further information). The staff at Fiskaaling who monitor lumpfish welfare in sea cages are not trained veterinarians, so the observations listed here relate to symptoms seen in connection with increased mortality. Knowledge of lumpfish pathogens is continually growing, so consulting expert veterinarians in the field is important before making any decisions in terms of treatment.

22.4.2 Biological on-site studies

Fiskaaling monitored lumpfish at three different salmon farms for approximately one year from May 2015. This work included recording weight, length, height, skin and tail erosion, as well as stomach analysis. Approximately 11% of the lumpfish on average examined had ingested sea lice. However, large spatial differences were observed in the relative number of stomachs containing sea lice, i.e. ~20% of the lumpfish stomachs sampled at site 1 (N = 195) and 2 (N = 254), while only ~5% of the lumpfish stomachs sampled at site 3 (N = 594) contained sea lice. All three sites are part of the same company, so similar husbandry practices at the three sites are to be expected. Since the lumpfish are of the same origin, the difference in stomach content is estimated to be due to the hydrographic differences among the farming sites, i.e. site 3 is positioned at a more exposed location compared to the other two farming sites. The stomach analysis has furthermore indicated, that at low sea lice loads, the effect of the lumpfish as a cleaner fish is of a more preventive, rather than of a delousing, nature, i.e. at low sea lice loads (<1 sea louse per salmon), the lumpfish prey upon the chalimus stages of sea lice as well (Fig. 22.6).

Fig. 22.5 Lumpfish with the two most challenging diseases in the Faroes. Left: black dots that can be found on the exterior, as well as on the interior of the lumpfish. Right: white or light brown exterior growth. Credit: Kirstin Eliasen

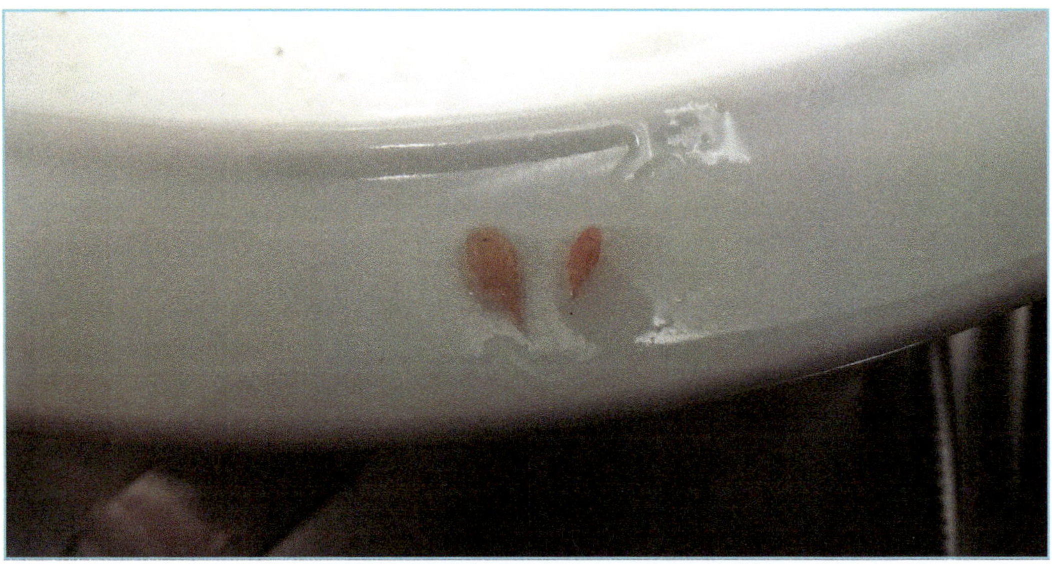

Fig. 22.6 An adult male *Caligus elongatus* (left) next to an unspecified sea lice at a late chalimus stage. Both found in the stomach of a lumpfish sampled at a Faroese salmon farming site. Credit: Kirstin Eliasen

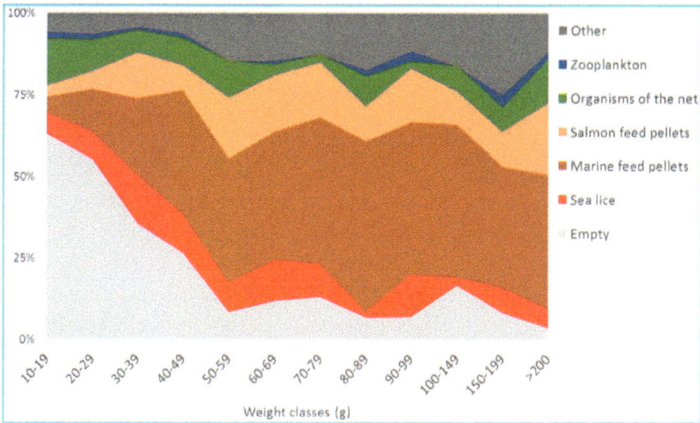

Fig. 22.7 Lumpfish stomach content as % occurrence (prevalence) in relation to size (g). Credit: Kirstin Eliasen

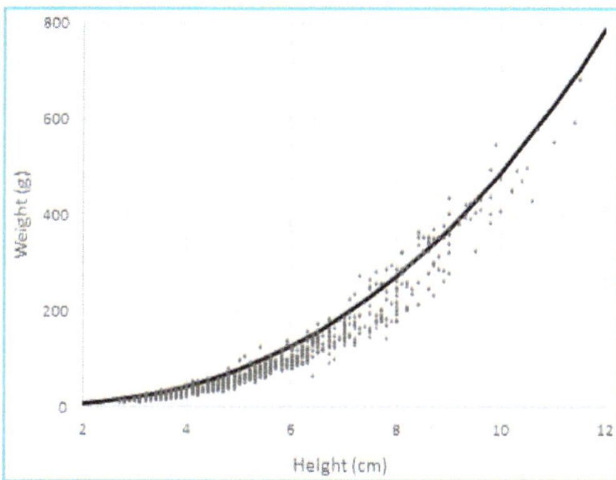

Fig. 22.8 Height–weight relationship (grey dots) of lumpfish sampled at three salmon farming sites in the Faroes. The black line indicates the reference line (Weight = exp(0.178434 + 2.61154*ln) Height) which is based on the height–weight relationship of Fiskaaling-produced lumpfish fed salmon feed. Credit: Kirstin Eliasen

In the pooled stomach analysis (N = 1043), all sizes of lumpfish seem to ingest sea lice (Fig. 22.7). The main size-related difference in the stomach content was the frequency of empty stomachs in small lumpfish (<50 g). The frequency seemed to increase with decreasing size. Based on this observation, Fiskaaling's recommendation to the industry is to put an extra effort into the routine feeding practices of newly introduced lumpfish to get the best start possible. Partly based on this, the industry has made hand feeding a larger part of its lumpfish feeding regime.

The industry mainly uses marine feed as lumpfish feed; however, despite ever increasing effort in feeding practices, it has proven difficult to maintain the lumpfish in a satisfactory physical condition (Fig. 22.8).

On the other hand, during the late summer months, lumpfish are observed to supplement their diet with alternative prey, i.e. zooplankton. These lumpfish were observed to be in a somewhat better physical state in 2015 (Fig. 22.9).

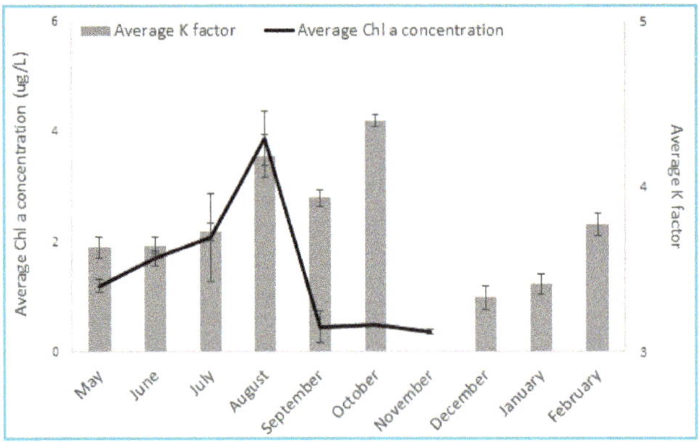

Fig. 22.9 The monthly average *K*-factor of lumpfish sampled at three different salmon farming sites in the Faroes in the period from May 2015 until February 2016, as well as the monthly average concentration of Chlorophyll *a* (µg/L) from May until November 2015 (unpublished data, Faroese Marine Research Station). The Chlorophyll *a* concentration is a measurement of the on shelf primary production, which again is an indirect estimate of the secondary production, i.e. zooplankton. Vertical bars indicate standard error. Credit: Kirstin Eliasen

References

á Norði, G., Simonsen, K., Danielsen, E., Eliasen, K., Mols-Mortensen, A., Christiansen, D., Steingrund, P., Galbraith, M. and Patursson, Ø. (2015) Abundance and distribution of planktonic *Lepeophtheirus salmonis* and *Caligus elongatus* in a fish farming region in the Faroe Islands. *Aquaculture Environment Interactions* 7 (1), 15–27. DOI:10.3354/aei00134.

Benfey, T.J. and Methven, D.A. (1986) Pilot-scale Rearing of Larval and Juvenile Lumpfish (*Cyclopterus lumpus* L.), with some Notes on Early Development. *Aquaculture* 56, 301–306.

Brown, J. A. (1986) The development of feeding behaviour in the lumpfish, *Cyclopterus lumpus*, *Journal of Fish Biology* 29 (sa), 171–178. DOI:10.1111/j.1095-8649.1986.tb05008.x.

Davenport, J. (1985) Synopsis of Biological Data on the Lumpsucker, *Cyclopterus Lumpus* (Linnaeus, 1758). *FAO Fish Synopsis*, 147, 31 pp.

Eysturskarð, J., Johannesen, Á. and Eliasen, K. (2016) Application of real-time PCR for specific detection of *Lepeophtheirus salmonis* in fluid samples from lumpfish (*Cyclopterus lumpus*) stomachs, *Aquaculture International* (7491). DOI:10.1007/s10499-016-0093-y.

Hagstova-Føroya (2016) Dátugrunnurin. Available from: www.hagstova.fo/fo/hagtalsgrunnur [Accessed 13 December 2016].

Helland, S., Dahle, S. W., Hough, C. and Borthen, J. (2013) *Production of ballan wrasse (Labrus bergylta) – Science and practice, Report.* Oslo.

Imsland, A.K., Reynolds, P., Eliassen, G., Hangstad, T., Nytrø, A.V., Foss, A., Vikingstad, E. and Elvegård, T.A. (2014) Notes on the behaviour of lumpfish in sea pens with and without Atlantic salmon present. *Journal of Ethology* 32 (2), 117–122. DOI:10.1007/s10164-014-0397-1.

Imsland, A.K., Reynolds, P., Eliassen, G., Hangstad, T.A., Nytrø, A.V., Foss, A., Vikingstad, E. and Elvegård, T.A. (2015) Assessment of suitable substrates for lumpfish in sea pens, *Aquaculture International.* 23 (2), pp. 639–645. DOI:10.1007/s10499-014-9840-0.

Johannesson, J. (2006) Lumpfish cavier – from vessel to consumer, *FAO Fisheries Technical Paper*, 485, 71 pp.

Kennedy, J., Jónsson, S.Þ., Ólafsson, H.G. and Kasper, J.M. (2015) Data Storage Tags and Trawl Surveys, *ICES Journal of Marine Science* 73 (1981), 1160–1169.

Killen, S.S., Brown, J.A. and Gamperl, A.K. (2007) The effect of prey density on foraging mode selection in juvenile lumpfish: balancing food intake with the metabolic cost of foraging., *Journal of Animal Ecology* 76 (4), 814–825. DOI:10.1111/j.1365-2656.2007.01237.x.

Nilsson Sköld, H., Aspengren, S. and Wallin, M. (2013) Rapid color change in fish and amphibians – function, regulation, and emerging applications. *Pigment Cell and Melanoma Research.* DOI:10.1111/pcmr.12040.

Norðberg, G., Johannesen, A. and Arge, R. (2015) Cryopreservation of lumpfish *Cyclopterus lumpus* (Linnaeus, 1758) milt. *PeerJ*, 3, pp. e1003. DOI:10.7717/peerj.1003.

Pampoulie, C., Skirnisdottir, S., Olafsdottir, G., Helyar, S.J., Thorsteinsson, V., Jonsson, S.T., *et al.* (2014) Genetic structure of the lumpfish *Cyclopterus lumpus* across the North Atlantic *ICES Journal of Marine Science.* DOI:10.1093/icesjms/fsu071.

Rodgers, G.M., Gladman, N.W., Corless, H.F. and Morrell, L.J. (2013) Costs of colour change in fish: food intake and behavioural decisions. *The Journal of Experimental Biology* 216, 2760–7. DOI:10.1242/jeb.080879.

Salmon from the Faroe Islands (2017) The history of aquaculture in the Faroe Islands, *salmon.fo*. Available from: http://salmon-from-the-faroe-islands.com/history.html [Accessed 18 January 2017].

Salvesen, I. and Vadstein, O. (1995) Surface disinfection of eggs from marine fish: evaluation of four chemicals. *Aquaculture International* 3 (3), 155–171. DOI:10.1007/BF00118098.

Skjermo, J. and Vadstein, O. (1999) Techniques for microbial control in the intensive rearing of marine larvae, *Aquaculture* 177 (1–4), 333–343. DOI:10.1016/S0044-8486(99)00096-4.

Williams, P. and Brown, J.A. (1991) Developmental changes in foraging-predator avoidance trade-offs in larval lumpfish *Cyclopterus lumpus. Marine Ecology Progress Series.* 76, 53–60. DOI:10.3354/meps076053.

CHAPTER 23

Cunner and lumpfish as cleaner fish species in Canada

Danny Boyce[1], *Keng Pee Ang*[2] *and Richard Prickett*[3]

[1]Department of Ocean Sciences, Memorial University of Newfoundland, St John's, NL, Canada
[2]Cooke Aquaculture Inc., Saint George, NB, Canada
[3]RSP Services Limited, Appleton, Abingdon, UK

23.1 Introduction

Unlike other countries where there is access to a broad array of approved sea lice therapeutants in salmon farming, there is limited access to these in Canada. As yet, some of those medicines are only used through emergency registration as pharmaceutical companies are working towards gaining full registration for these products. Moreover, with limited treatment options, some of the therapeutants are becoming less effective as they are more frequently used. Salmon producers in Canada consider that sea lice can be controlled without impacting the marine environment if the most effective tools can be accessed. Gaining access to an expanded array of approved products is therefore an important initiative for the country's salmon farming industry. However, there are also other initiatives aimed at developing non-therapeutant means of controlling sea lice. All of these approaches comprise the Integrated Pest Management Plan (IPMP) that is being developed and implemented by industry–government working groups, and cleaner fish are part of that strategy. This chapter reviews the latest production and research developments in hatchery rearing and deployment of cunner *Tautogolabrus adspersus* and lumpsucker *Cyclopterus lumpus* as cleaner fish in Canada.

23.1.1 Desirability of a wrasse species in salmon farming in Canada

The species of wrasse being used by the industry in Norway includes the goldsinny *Ctenolabrus rupestris*, corkwing *Symphodus melops*, cuckoo *Labrus mixtus*, rock cook *Centrolabrus exoletus* and the Ballan wrasse *Labrus bergylta*. Each of these has different merits, making them all interesting candidates to the Norwegian industry (Kvenseth *et. al.*, 2002). Unfortunately, none of the wrasse species used in Norway and Scotland is native to Atlantic Canada, and there is a need to find a different species – the cunner *Tautogolabrus adspersus*. Although this is a different species from those used in Europe, it is important to note that the cunner is from the same family, *Labridae*.

23.2 Species and geographical location

23.2.1 *Cunner* Tautogolabrus adspersus

The cunner in Atlantic Canada is a northern member of the primarily tropical *Labridae* family of wrasse. The cunner is common with a widespread range along the Atlantic coast of North America from northern Newfoundland, the southern Gulf of St Lawrence and offshore banks, south to the mouth of Chesapeake Bay (Scott and Scott, 1988a). Cunner are inhabitants of shallow inshore waters, living on or near the bottom, usually within a few kilometres of shore. They often gather near sheltered rock substrates, seaweed beds, wharves, wrecks, and reefs. Like other wrasses, the nocturnal behaviour of cunner consists of a quiescent, resting attitude accompanied by a camouflaging pattern change. Activity dwindles in the autumn until a state of dormancy is reached at about 7 to 8°C, with the larger fish becoming dormant first (Drew, 1976).

23.2.2 *Lumpfish* Cyclopterus lumpus

The lumpfish of the family *Cyclopteridae* is distributed on both sides of the Atlantic Ocean. In the eastern North Atlantic lumpfish range from east Greenland, Iceland, Faroes, White Sea, and Murman coast, southward off the British Isles to France. In the western Atlantic, lumpfish can be found from Hudson's Bay and the Labrador coast southward to New Jersey (Department of Fisheries, Forestry and Agrifoods (DFA) Emerging Species Profile Sheet). Lumpfish migrate towards shallow coastal waters during spring and early summer (April–May) to spawn and usually migrate back to deeper waters (300 m) during the late summer and early autumn. Davenport (1985) reported that males arrive in coastal spawning areas before the females and establish territories. After fertilisation, the female will begin the migration back to the deeper water while the male remains to guard and nurture the eggs. This species has a homing mechanism allowing it to return to areas where it had previously spawned (DFA Emerging Species Profile Sheet). The lumpfish is primarily a bottom-dwelling fish of cold to temperate waters, living on rocky or stony bottoms, but is frequently semi-pelagic during early life stages, found hiding under floating seaweed. It often adheres to stones, lobster pots or other solid objects by means of a pelvic adhesive disc (Scott and Scott, 1988b).

23.3 Preliminary laboratory and sea cage trials to determine if cunner are effective cleaners of sea lice

23.3.1 *Preliminary laboratory and cage trials and necessary licences*

The first preliminary work on the efficacy of the cunner in a laboratory environment was completed by Kelly Cove Salmon Ltd (KCS) at a facility owned by the Huntsman Marine Sciences Center (HMSC) in St Andrews, New Brunswick, during summer 2010. This work was carried out in tank trials with wild-caught cunner and salmon *Salmo salar* that had a sea lice infestation. Video footage of the cunner–salmon interaction was taken over multiple days. The cleaning or grazing ability of the cunner was determined by counting the number of lice on the salmon before introducing cunner and after 24, 48, and 72 hours after introduction. This preliminary trial verified that cunner are effective at removing sea lice from infected salmon.

With the positive outcome established by the proof of concept studies conducted at the HMSC, KCS proceeded to the next phase of research to determine whether cunner is a viable option as a cleaner fish in a commercial salmon cage setting. A licence to collect wild cunner was obtained from the Department of Fisheries and Ocean (DFO) by KCS to permit a contracted commercial fishing enterprise from Digby, Nova Scotia, to collect cunner from the wild. Samples of cunner were taken for health testing at different laboratories in September 2011 prior to the main collection. Test results were submitted to and approved by the New Brunswick Department of Agriculture, Aquaculture and Fisheries (NBDAAF), which subsequently gave approval to commence collection on 17 October 2011. As per DFO's requirement, a government contractor for data collection and entry was contracted to provide scientific training to the commercial fishermen with regards to data collection before they could commence fishing. Data including catch and bycatch, GPS locations, and haul times were collected and submitted to the contractor to be entered into a DFO database.

Wild cunner were successfully caught in traps by commercial fishing from the Bay of Fundy and transferred to a marine cage test site on 23 October 2011. Cunner were stocked into eight 12 m × 12 m × 7 m deep cages; stocked with 4600 salmon with natural lice infestations at different ratios i.e. two cages each with 3% cunner to salmon, 6% cunner and 9% cunner and two cages with no cunner (control cages). Sea lice counts on the salmon began prior to introduction of the cunner to establish baseline levels and counts were continued on a weekly basis through late summer and fall. A total of ten fish from each cage was randomly sampled. The trial ended on 13 September 2012 following a site inspection and review by the NBDAAF.

23.3.2 Preliminary results

The results were categorised according to the levels of three stages of sea lice recorded at each lice count, i.e. the chalimus, pre-adult and gravid stages. There was no difference in numbers of chalimus or pre-adult stages of lice between salmon in treatment and control cages, the difference was most pronounced in the levels of gravid female stages, with up to 40% fewer gravid lice on the salmon in treatment cages. No difference could be detected between the different ratios of cunner tested.

23.4 Wild broodstock collection

23.4.1 Cunner

Memorial University of Newfoundland (MUN) utilised a purpose-built cunner trap, consisting of a modified hoop net, bridle, and retrieval rope configured to be deployed vertically to the sea bottom (Fig. 23.1). The trap had a slightly funnel-shaped design, with an upper 1.2 m diameter ring and a 1.0 m lower ring, sewn with 4 cm trawl mesh hung diagonally to form a

0.8 m deep bucket-like net that collapsed flat when at rest in 'set position' on the bottom. The trap was baited with herring and lowered gently to the bottom so the bait would settle on to the mesh of the lower ring. Once a sufficient number of cunner moved in to feed on the bait, the trap was sprung by pulling up sharply on the retrieval rope attached via the bridle to the upper ring, thus surrounding the fish with netting. Fishing from a stationary structure such as a wharf or stage gave a better catch (R. O'Donnell, St John's, Newfoundland, 2013, personal communication).

23.4.2 Lumpfish

Lumpfish were collected using gillnets made with monofilament nylon. Gillnets with a minimum mesh size of 268 mm are authorised for commercial fisheries. These nets are set in depths ranging from 3–32 m, preferably on hard or rocky bottom areas that have an abundance of kelp and seaweed, and hauled every one to two days. The lumpfish become entangled in the mesh and cannot tear from the net very easily, as they are not strong swimmers. Traditional snow crab pots with a modified piece of mesh over the top have also been used to hold captured lumpfish in for a short period of time at sea until numbers warranted transport to facilities at MUN. Lumpfish (mostly males) were also collected in open water by divers. The divers searched for egg masses along the shores, generally locating male lumpfish guarding an egg mass. Both the male and egg mass were collected and transported to facilities at MUN.

23.5 Cunner

23.5.1 Broodstock management

Cunner broodstock are held in 4.4 m^3 circular deep tanks (1.68 m diameter × 2.0 m high) with one complete water exchange every two hours on a flow-through basis. Oxygen is maintained at 100–115% (10.5 mg l^{-1}) and water temperature between 10–12°C. Fish are held under an ambient photoperiod with a daytime light intensity of 60–100 lux at the water surface. All tanks are equipped with artificial hides that are designed to mimic kelp. This provides the broodstock (Fig. 23.2) with areas in which to hide and appears to reduce stress levels.

Cunner broodstock are fed to satiation using a diet of alternating baitfish of herring, capelin, mackerel, and squid supplemented with shrimp fed via feeding baskets (Fig. 23.3) until spawning time. Shrimp is included with every feeding since it is a nutrient-rich feed source high in selenium and omega 3 fatty acids, which are key components in producing good quality eggs. Just prior to spawning the fish begin to feed more aggressively, and as a result feedings are increased to five times a week. However, once spawning is completed their demand for feed decreases and as a result feeding is reduced to three times per week by mid-September, further reduced to two feeds per week by mid- to late October.

Cunner broodfish (n = 89–220 fish/tank, size range from 192–306 g; 22–25 cm length) group spawn in the tank. Cunner spawn from late spring to early summer in eastern Canada under natural conditions. Alteration of the photoperiod regime was successful in adapting a group of cunner to a two-month advanced spawning after the first season.

Fig. 23.1 A cunner trap in retrieval position.

Fig. 23.2 Cunner broodstock.

23.5.2 Egg production/spawning

Cunner spawn naturally in the holding tanks. The individual eggs float under normal saline conditions (31 ppt). Females release the eggs and are then fertilised by the males directly in the holding tanks during the spawning process. Fertilised eggs are buoyant and float on the surface of the water and, as such, can easily be collected via an external egg collector (Fig. 23.4). Eggs are incubated in the hatchery area and then transferred to first feeding post-hatch. To date we have not included manual stripping of fish. Cunner broodstock are assessed post-spawning to monitor general overall condition (Fig. 23.2). The weight (g), length (cm) and stocking density (kg m^{-3}) of spawning cunner broodstock were from 192–306 g, 22–25 cm, n = 89–220 fish/tank. Some of these fish have been domesticated for up to five years. To reduce our dependence on the natural stocks, a group of 2013 YC juveniles has been maintained. These were 25–35 g in weight at two years post-hatch. Maturation has been observed in some females. They are maintained on a dry pelleted commercial diet and on an ambient photoperiod cycle.

23.5.3 Hatchery – egg development and incubation

Cunner eggs are small (0.8 to 1.0 mm) in diameter, buoyant, transparent, and do not have an oil globule. From egg measurements there is an estimated 1.2 million eggs per litre. Egg production began in March 2015, with the three-month advanced photoperiod broodstock releasing eggs. The fish in the remaining broodstock tanks (L2–L6), which follow an ambient photoperiod began releasing eggs in mid-April. Released eggs are fertilised by the milt directly

Fig. 23.3 Feeding baskets are used to offer shrimps to broodstock cunner.

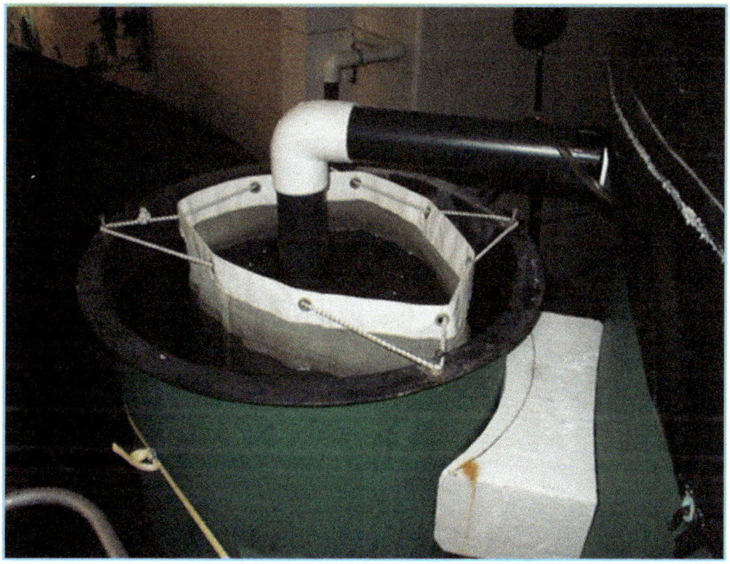

Fig. 23.4 A tank equipped with an overflow pipe and external egg collector.

in the holding tank during the spawning process. Fertilised eggs are buoyant and float on the surface of the water and, as such, can be collected relatively easy via an external egg collector (Fig. 23.4).

Cunner eggs are delicate and have a thin chorion. Therefore, egg disinfection consisted of a seawater rinse (UV-treated water) for five minutes prior to stocking the incubator with eggs. At post-gastrulation (40–50 degree days), eggs are transferred to a clean incubator and subsequently treated with a disinfectant, Chloramine T (250 ppm, 10 minutes static), to reduce any potential bacterial load within the incubator prior to hatch. This process worked well for the hatching mechanism of the eggs.

All egg batches collected are assessed for volume, fertilisation rate, symmetry and clarity. Only batches meeting the criteria of >80% fertilization and 50% symmetry rates are selected for incubation. Eggs are incubated in either 50 l or 250 l incubators depending on the volume

Fig. 23.5 Conical egg incubators of 50 l and 250 l capacity.

of the batch (Fig. 23.5). Eggs are incubated at 11–12.5°C, held under 24-hour light at a light intensity of approximately 100 lux. Each incubator is equipped with a banjo (300 μm screen) to allow for a continuous exchange of water within and gentle aeration added. Water flow rates are set at 1 l min^{-1} in the 50 l incubators and 3 l min^{-1} in the 250 l incubators. The water enters the incubator from the surface. The water and air are turned off daily in each incubator for a period of ten minutes, to separate buoyant fertilised eggs from dead or unfertilised eggs, which are heavier and sink to the bottom of the conical-shaped incubator. After the separation of live and dead eggs, a valve is opened at the bottom of the incubator for a short period and all the dead eggs are removed. The water and air are then resumed. Egg development is monitored throughout the incubation period with development progressing as expected (Fig. 23.6). Eggs hatch at 80–90 degree days.

23.5.4 Larval rearing

Water quality is monitored daily with temperatures, oxygen and nitrogen gas recorded. Temperature is maintained between 12–13°C and oxygen between 110–120% saturation. During the larval period all tanks are kept under a photoperiod of 24 hours of light and light levels were maintained at approximately 2000 lux at the water surface.

Larval tanks are equipped with a 300 μm centre drain for the delicate newly hatched 2.5–3.0 mm long larvae, then changed to a larger 500 μm drain when larvae reach a length of 8–9 mm (40 dph). The centre drain was further increased in size to 900 μm once weaning was initiated when larvae were 11–12 mm which occurred around 55–60 dph. Airstones were placed on either side of the drain to provide very gentle aeration to keep larvae and prey items circulating throughout the tank. Aeration in these larval tanks must be kept to a minimum to enable delicate larvae to maintain their position within the water column. As a result aeration is monitored and adjusted as needed daily. A protein skimmer is added to the surface of the tank to reduce water

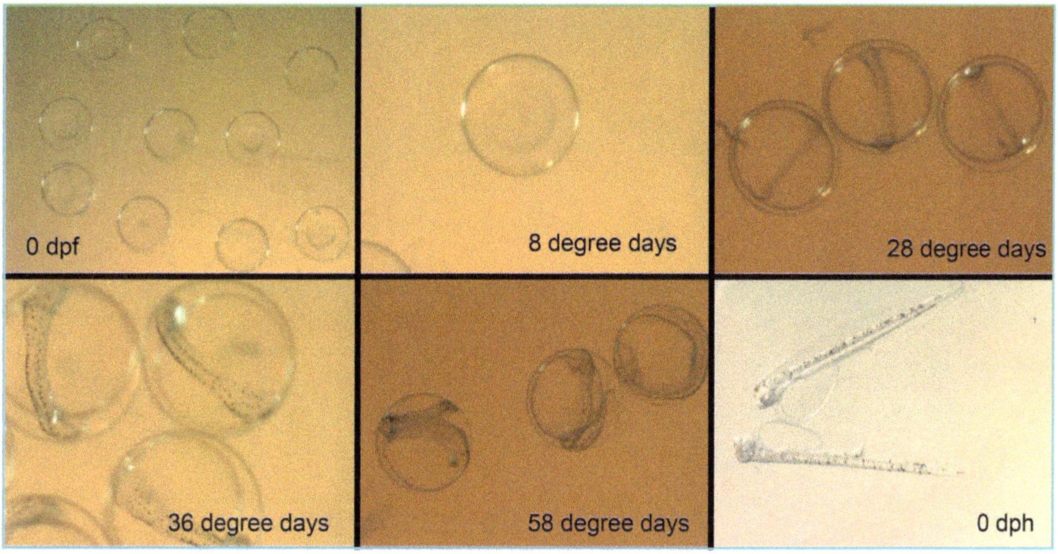

Fig. 23.6 Embryonic development of cunner (*Tautogolabrus adspersus*) from zero days post-fertilisation (dpf) to zero days post-hatch (dph).

tension and allow the cunner to break the surface and inflate their swim bladders. Cunner eggs begin hatching at 80–90 degree days. When >90% hatched is reached, larvae are transferred to 3000–6000 l larval rearing tanks at 100 larvae l^{-1}.

Larvae are small and primitive at hatch, measuring 2.5–3.0 mm in length, without a functional mouth, and have a large yolk sac (Fig. 23.7). Larvae are left undisturbed for the first four to five days post-stocking, lights are not turned on and feeding is not initiated until larvae are pigmented and the yolk sac is absorbed. Some tanks are covered and kept in complete darkness until the yolk sac is absorbed. Once the yolk sac is absorbed, feeding is initiated around five to seven days post-hatch (dph) and amounts of rotifers fed are monitored throughout the day and increased as required. Larvae are fed rotifers enriched with Ori-One (Skretting) exclusively for the first 35 days. Due to the small larval size, all rotifers are sieved through a 100 μm screen for the first 10–14 dph to ensure that there is a sufficient number of rotifers small enough for the larvae to ingest. At 36 dph larvae are introduced to AF Artemia (freshly hatched smaller Artemia nauplii), which is a specialty Artemia cyst that produces small nauplii with high HUFA levels. The AF Artemia does not require enrichment and can therefore be used as a nutritious feed source for larvae immediately after the Artemia hatch. Larvae are fed a combination of rotifers and AF Artemia up to 44 dph. At 45 dph they are switched to a larger EG Artemia (larger Artemia nauplii enriched grade), which are enriched with Ori-Green during the second instar stage. Larvae are fed three times daily with EG Artemia, with a 200 μm dry feed added twice a day one hour prior to the morning feeding and one hour prior to the afternoon feeding. This process is continued until 54 dph. At 55 dph the weaning process is initiated, the morning Artemia feeding is eliminated and a mixture of 200 and 300 μm dry food is offered to the larvae every hour until the afternoon feeding of Artemia. The afternoon Artemia feeding is discontinued at 60 dph, and night feeding is withdrawn at 65 dph and the larvae are fully

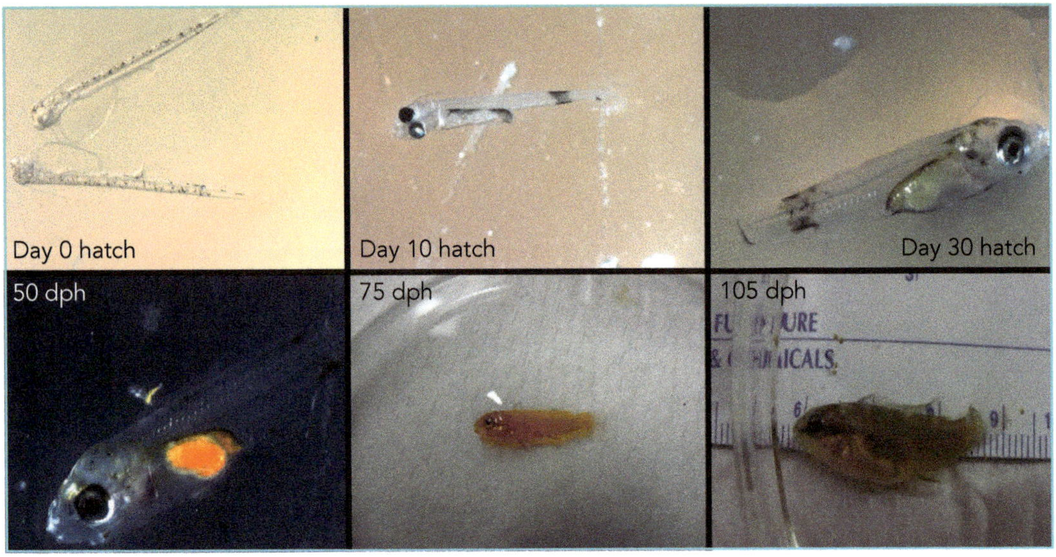

Fig. 23.7 Development of cunner (*Tautogolabrus adspersus*) larvae from 0 to 105 dph.

weaned. At this stage the larvae are offered a 300 µm dry feed every two hours via automated feeders in addition to hand feeding three times a day. Clay is added as a greening technique to help diffuse the light in the tank while providing background contrast for prey items. Since it is inert it has the added benefit of keeping bacterial levels in the tank at a manageable level. Clay was added initially twice a day and, as tank flow rates increased, a third batch was added during the night (Table 23.1).

23.5.5 Juvenile stage

Juvenile cunner are maintained in 3000 l, 6000 l and 25000 l (Fig. 23.8) tanks at 10–14°C in flow-through seawater. Submerged spray bars are added to the tanks post-weaning to increase water flow and waste removal. Centre drains extending the full depth of the tank are used throughout rearing, although hole bore size is increased periodically as the fish grow to permit better flushing of waste. Cunner instinctively seek shelter and succeed the most when supplied with tank furniture in the form of hides. Following weaning on to dry feed, grass hides are suspended throughout the tank. These consist of long, frond-like strips of heavy duty tarp weaved around a disc of PVC pipe. The cunner naturally gravitate towards the hides and form clusters of fish around each. Hides need to be removed periodically for cleaning. Upon removal, the fish will immediately swim towards another hide if available, or form a 'ball'. This consists of a tight cluster of fish rotating and 'rolling' within the water as the fish on the outside rotate back towards the inside of the cluster. This behaviour is also repeated if a group of fish are netted into a bucket. As the fish continue to grow, reliance on the grass hides is minimised, at which time they are replaced with suspended lengths of PVC pipe with large holes removed. Cunner do not tend to demonstrate cannibalism, and appear to be content living in large populations in close

Table 23.1 Rearing protocol for cunner used in this study (see text for details and descriptions).

Day (s) post hatch	Temp (°C)	Flow (l/min)	Light		Clay (batches added d^{-1})	Feed			Size (mm)	Drain Screen (μm)	Comments
			Hours	Lux		0900h	1500h	2100h			
-3 to 0	11–12.5	1.0	24 Light	100						300	
0 to 4/5	12	1.0	24 dark						2.5–3	300	100 larvae l^{-1}
5 to 14						Rotifer	Rotifer	Rotifer			Rotifers < 100μm (sieved)
15 to 35		1.5				Rotifer	Rotifer	Rotifer			Rotifer (no sieving)
36 to 38					3	AF-Art	Rotifer	Rotifer			
39 to 41		2.0		1000		AF-Art	AF-Art	Rotifer			
42 to 44			24 Light	–		AF-Art	AF-Art	AF-Art	8–9		
45 to 54				2000		EG- Art	EG- Art	EG-Art		500	
55 to 59					2	0.2mm	EG-Art	EG-Art			
60 to 64		3.0			1	0.2/0.3 mm		EG-Art			
65					0		0.3mm		11–12	900	

*AF Art: AF Artemia; EG Art: EG Artemia

Fig. 23.8 A tank with juvenile cunner (25,000 l), equipped with automated feeders.

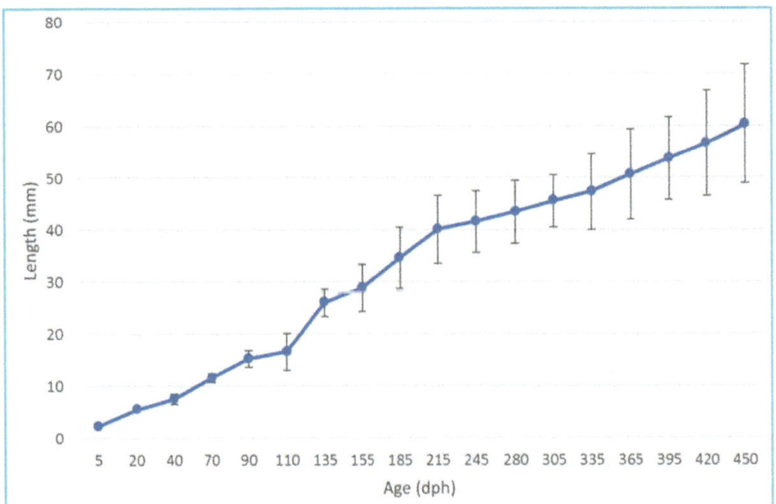

Fig. 23.9 Growth in length (mm +/-SD) of hatchery reared cunner.

proximity. As a result, grading is completed primarily to maintain size ranges within a tank to correspond with feed pellet size requirements, or optimise stocking densities. Cunner increased in standard length from 3 mm to 60 mm from 5–450 dph at 12°C (Fig. 23.9).

23.6 Lumpfish

23.6.1 Lumpfish broodstock

Lumpfish broodstock are held in large 25–45 m^3 circular tanks (Fig. 23.10) with one water exchange every two hours on flow-through. Tanks are fitted with a variety of hides and shelters. Oxygen is maintained at 100–115% (10.5 mg l^{-1}) and water temperature is maintained between

Fig. 23.10 Lumpfish broodstock in a 25,000 l holding tank.

6–10°C. Fish are held under an ambient photoperiod with a daytime light intensity of 60–100 lux at the water surface. In the event that any of the newly collected wild-captured lumpfish experience buoyancy issues (air entrapment) upon arrival, the fish is removed from the population, placed in a modified plastic barrel and then submerged in deep tanks (>3 m deep) for a few days to mimic the capture depth. This successfully eliminates the buoyancy problem and the fish can be returned to the holding tanks thereafter with no further symptoms observed. Wild lumpfish are offered a diet of moist baitfish comprising of herring, mackerel, capelin (preferred) and squid two to three times a week. Although wild fish have been held for a period of 12 months, further nutritional consideration is required to complete a formulated broodstock diet. Lumpfish spawn naturally in hides, artificial shelters and rock areas placed within the holding tanks.

23.6.2 Egg incubation in wild and domesticated lumpfish

Lumpfish eggs are relatively large (2.6 mm) in diameter, demersal, adhesive and are spawned as a single egg mass (Brown, 1992) (Fig. 23.11). Fertilised eggs are adhesive and not buoyant and lie in large spongy mass within the artificial hides or adhering to rocks in tanks until removed.

Lumpfish egg masses are typically light pink in colour when they are first released. As the eggs develop, the colour continuously darkens from pink, orange, grey, beige and finally to a chocolate brown colour just prior to hatch. Healthy egg masses tend to be firm to touch and are securely attached together. Size measurements indicate approximately 100,000 eggs kg^{-1}. Eggs

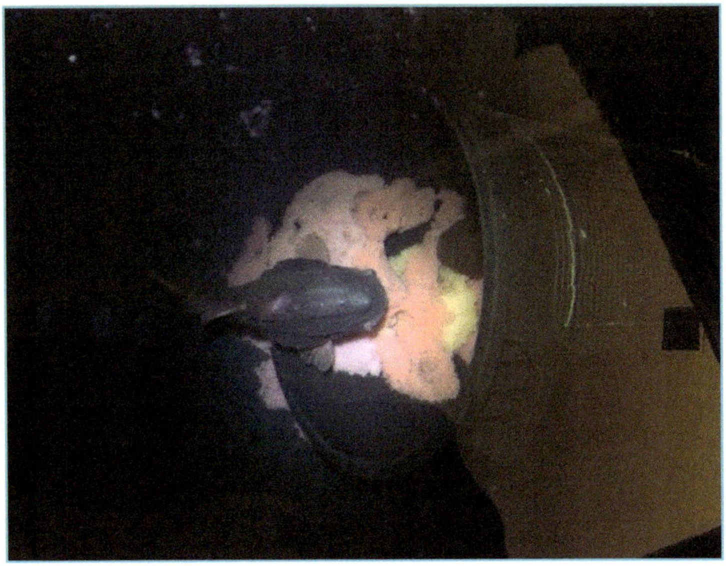

Fig. 23.11 Lumpfish egg masses inside an artificial shelter.

are disinfected with a seawater rinse (UV-treated water) for five minutes prior to stocking in the incubator tray.

Females release their eggs within the tank shelters/hides provided and males fertilise them. As indicated earlier, after fertilisation, the male remains to guard and nurture the eggs (Fig. 23.12). In contrast, a group of recently collected wild broodstock spawned successfully within an artificial hide (Fig. 23.11) only a few days after being collected. Fertilised eggs are adhesive and can be found on the bottom of the tank, as they are not buoyant. Fertilised eggs are collected from various holding tanks shortly afterwards and incubated in the hatchery trays, and then transferred to first feeding tanks post-hatch. Wild lumpfish broodstock die suddenly after spawning is complete or over a period of time in captivity. Wild fish for egg/milt during the spawning season (three to four months) continue to be captured for this reason. A number of the 2014 YC (F1) lumpfish juveniles were also retained to use as future broodstock (based entirely on size/maturity). They are held in 25–45 m^3 tanks and are fed a grower diet 6 mm pellet, 50% protein and 18% lipid, placed on an ambient photoperiod and held at a water temperature range of 6–10°C. A formulated broodstock diet is recommended for these fish.

As noted, some of these 2014 YC (F1) juveniles (16 months post-hatch) were maturing. A percentage of the larger males have exhibited nuptial colours varying from deep pink to dark red, orange. These colours fade shortly after spawning is complete. The largest females of the juveniles began releasing small batches of eggs throughout the autumn into the holding tank. However, there were no fertilised eggs from ten to 15 batches collected from the tank. This lack of fertilisation is possibly due to these fish being naïve spawners and should yield more favourable results during subsequent spawning seasons. Some of the larger females died following the naïve spawning season, which seems to be a similar trend in the wild females.

The eggs are incubated in floating baskets, which are placed in tanks with a water exchange rate of one turnover per hour, high aeration, a 12-hour light/12-hour dark photoperiod and a temperature of 9.5°C. The egg masses are periodically split into multiple pieces per tray to

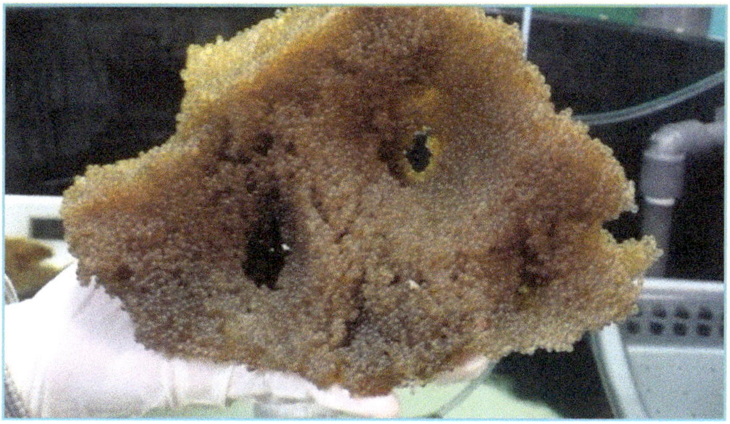

Fig. 23.12 Lumpfish egg mass showing holes inserted by male broodstock to ensure eggs are aerated.

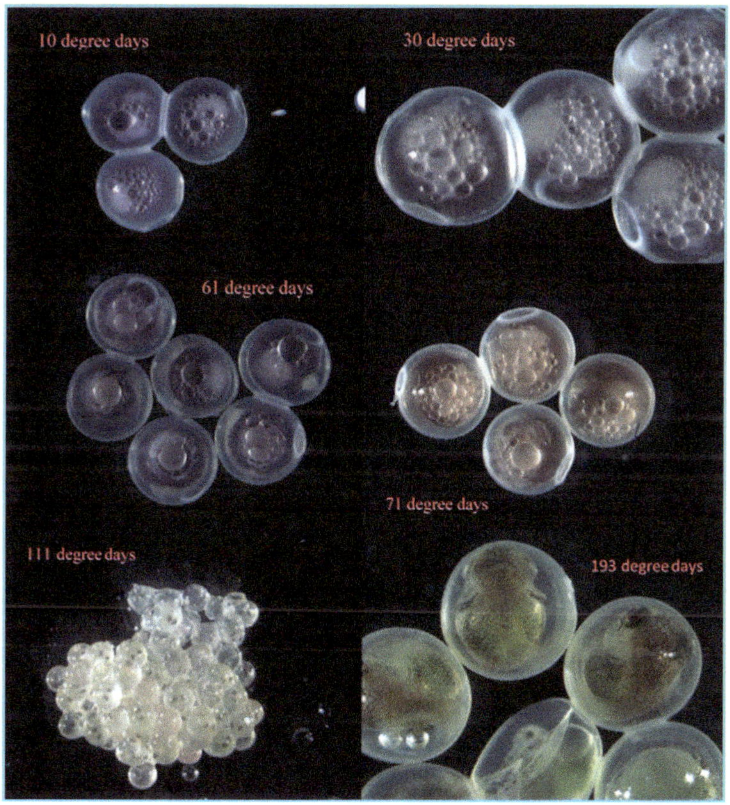

Fig. 23.13 Developmental stages of lumpfish eggs. Scale, egg diameter = 2.6 mm. Credit: Bruno Gianasi

increase aeration to eggs on the inner portions of an egg mass. Determining if eggs are fertilised after spawning can be challenging because the eggs are opaque in colour, which limits observation of internal development. However, at approximately 100 degree days fertilised eggs become 'eyed' if healthy. Lumpfish eggs hatch between 270–350 degree days, and larvae vary in colour from gold to amber to dark brown (Fig. 23.13). Immediately prior to hatch, the floating baskets holding the eggs are transferred to a clean 500 l tank to hatch. The larvae are able to swim on hatch directly into the larval tank, where they attach to the tank sides.

23.6.3 Rearing of lumpfish larvae in Canadian hatcheries

Lumpfish larvae are maintained on flow-through UV-treated seawater in 500 l to 3000 l tanks. Tank drains are equipped with a 500 µm centre drain and airstones are placed on either side of the drain to keep prey items circulating throughout the tank. Tank furniture/hides are provided at 60–65 days post-hatch and are a crucial feature in lumpfish rearing. Flat surfaces (rigid, black sheet plastic) suspended in the tank provides extra surface area for attachment but are easily removed for cleaning. These remain in place until the lumpfish reach 5 g (120–160 dph), and enter a free-swimming phase. The centre drain size can also be increased to 900 µm at this stage.

Larvae disperse within the water column, and immediately attach to any surface. Five days after the commencement of hatch, any remaining eggs (and floating basket) are removed from the tank. Larvae are fed EG Artemia from zero to 14 days post-hatch three times per day. Dry commercial microdiets (0.3 mm) are introduced at 7 dph one hour prior to feeding Artemia. Weaning begins at 15 dph, when lumpfish are 5–6 mm in length. Weaning is started by eliminating the morning Artemia feeding and replacing it with dry commercial diet. At 18 dph, a second Artemia feed is removed and replaced with dry diet. At 25 dph, the final Artemia feed is stopped, and replaced with 100% commercial dry diet. Vibrating feeders are installed to provide a constant supply of diet every 15 minutes, with additional hand feeding three times daily to ensure feed rates are sufficient. Lumpfish are subsequently moved to a 3000 l or 6000 l tank. The frequency of feeding is reduced as size increases.

Lumpfish grow rapidly. At 25 dph (end of weaning), body length is typically 6.5–8 mm. By 50 dph, length has doubled to 16–18 mm. Lumpfish enter a cannibalistic stage at approximately 2 g (110–140 dph) so an initial size grading is completed between 100–120 dph. This cannibalistic behaviour seems to recur whenever the size range is large. During grading the fish are divided into small (≤2 g), medium (3 g) and large (≥4 g). Frequent grading (every two to three weeks) is due to the rate of growth of lumpfish, and is necessary during this phase to reduce mortality. Grading is less frequent as the size differences in the tanks become less pronounced (Table 23.2).

23.6.5 Lumpfish juveniles

Juvenile lumpfish are maintained in 3000 l–25,000 l tanks on 8–10°C flow-through seawater. Tank furniture is removed at the 5 g stage and tank wall surface provides the primary area of attachment. As with cunner, centre drains extend the full depth of the tank throughout rearing. Hole bore size is increased to allow for release of waste pellets from the tank.

Due to the high lumpfish growth rates (Fig. 23.14), juvenile rearing focuses on frequent size grading and maintaining access to surface area for attachment within the tank. As previously mentioned, lumpfish enter a free-swimming state at approximately 5 g, and sufficient area for attachment along the tank surface remains important. Stocking density for these fish is 20–40 kg m^{-3} and oxygen is maintained between 110–115%. Tanks are equipped with an automated

Table 23.2 The hatchery rearing protocol for lumpfish in this Canadian study.

Day (s) post hatch	Temp (°C)	Flow (l/min)	Light		Feed			Approximate Size (mm or g)	Drain mesh (µm)
			Hours	Lux	0900h	1500h	2100h		
-5 to 0	9	8.3		100					500
0 to 14		1	24 h		EG-Art	EG-Art	EG-Art	4.5–5.0 mm	
15 to 19					0.3mm	EG-Art	EG-Art	6.0–8.0 mm	
20 to 24		1.5			0.3mm	0.3mm	EG-Art	0.01–0.05	
					0.3mm	0.3mm	0.3mm		
25+	11	2.5	24 h			0.3 mm		0.12	
						0.3mm/0.5mm		0.2–0.5	
						0.5 mm		0.5–1.0	
						0.5mm/0.8mm		1.1–2.0	
			12 h /12 h			0.8 mm		>2.0	900
120 to 160							Increase following grading		

*EG Art: EG Artemia

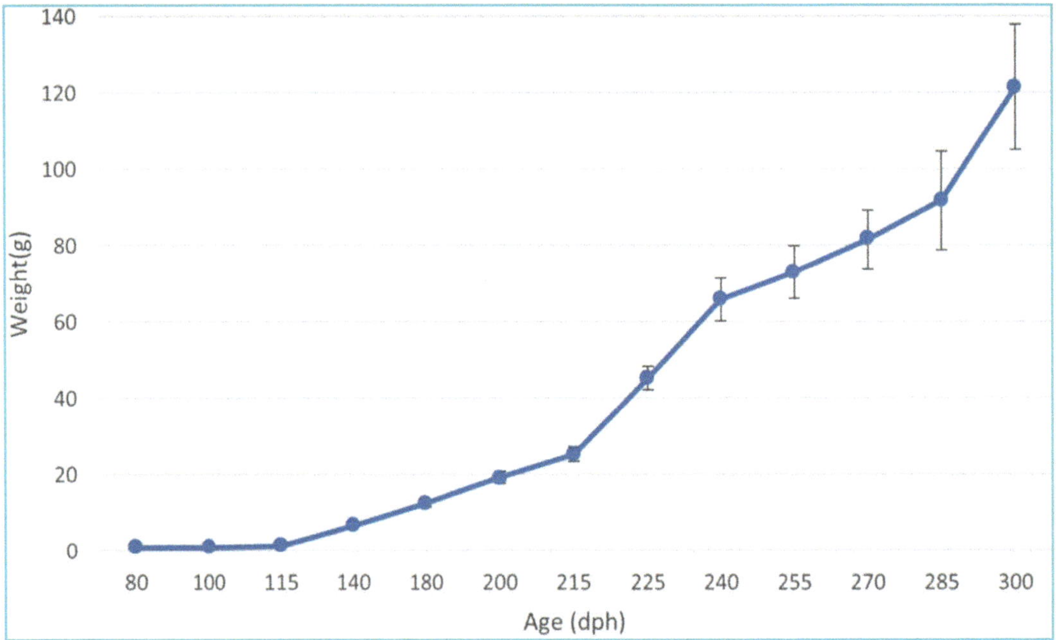

Fig. 23.14 Average weight (g) of larval lumpfish hatched from a wild egg mass.

feeder set to deliver feed to the tanks every two hours from 0800–2000 hours. All tanks are set on a 12-hour light/12-hour dark photoperiod.

23.7 Metabolic scope

Dr Kurt Gamperl, Department of Ocean Sciences, Memorial University of Newfoundland, St John's, NL. Canada, A1C 5S7 reports in this paragraph that:

The slow growth of the cunner is unsurprising based on data for wild fish, as lumpfish are *c.* 8–10 cm long at one year and 20–25 cm at two years' old, respectively (Albert *et al.*, 2002; Hedeholm *et al.*, 2014), whereas the cunner is only 4.5–6 cm and 7–8.5 cm at these two ages (Serchuk and Cole, 1974; Nitschke *et al.*, 2001). It is clear that lumpfish growth is highly temperature dependent, especially during the early life history stages. For example, for fish smaller than *c.* 10 g growth rate is 6.7% day^{-1} at 16°C compared with 4.2% day^{-1} at 10°C (and much lower at colder temperatures), whereas this difference is 0.4% day^{-1} between *c.* 10 and 20 g and the temperature–growth relationship is constant after the fish surpass 100 g (Nytro *et al.*, 2014). Given the fast growth rate of lumpfish, a size suitable for feeding on lice attached to salmon (>20–25 g) is achieved in six months, whereas cunner take up to 20–24 months to attain a suitable size.

Temperatures can range from below zero to 18–20°C in sea cages in eastern Canada, and oxygen levels in summer may fall to 4–6 mg l^{-1} (45–65% air saturation) for extended periods (Gollock *et al.*, 2006; Burt *et al.*, 2012). Given that these environmental variables are two of the most important with regards to impacts on fish physiology and mortality, a number of experiments on both these species was carried out at the Ocean Sciences Centre. As has been widely reported (e.g. Costa *et al.*, 2013; Kelly *et al.*, 2014), cunner goes into torpor (metabolic depression) when temperature decreases in the autumn, with adults initiating metabolic depression at *c.* 7°C and reaching minimal energy requirements by 3°C (Kelly *et al.*, 2014; Fig. 23.15). Clearly, this metabolic response and lack of activity will preclude them from being effective as a delouser at cooler water temperatures. The metabolic response of lumpfish to such temperatures has not been defined, but Imsland (2012) indicated that lumpfish actively feed on sea lice in the depth of winter, even at water temperatures of 4 to 7°C. Thus, they are active and feeding, not dormant like cunner, at low temperatures.

It appears that both species can tolerate summer cage site temperatures. Lumpfish have a critical thermal maximum (CT_{Max}) of 22.4°C in seawater of 100% air saturation, when acclimated to 10°C (Fig. 23.16; Rasmus *et al.*, 2016), and maxima of cunner ranging from 27.7±0.2 for 5 g fish to 26.3±0.2°C for fish >100g (Kelly *et al.*, 2014). In addition, the lumpfish has a critical oxygen tension (P_{crit}) at 10°C of 57 mm Hg O_2 (~37% air saturation; Fig. 23.16), the CT_{Max} of this species does not decrease until water oxygen levels reach 80 mm Hg (~50% air saturation), and is still 20.2°C at only 60 mm Hg (40% air saturation; 2.7 mg l^{-1}: Fig. 23.16C). The cunner is even more hypoxia tolerant than lumpfish (critical oxygen tension at 10°C of 20% air saturation; 1.75 mg l^{-1} (Fig. 23.16B)] and its CT_{Max} is unlikely to be below 20°C given its high CT_{Max} at 100% saturation and its hypoxia tolerance (P_{crit}). The cunner is both more hypoxia and temperature tolerant than the lumpfish because of basal metabolic rates and aerobic scope (Killen *et al.*, 2007; Kelly *et al.*, 2014; Rasmus *et al.*, 2016). For example, while a 25 g lumpfish has a basal metabolic rate of 80 mg kg^{-1} hr^{1} and a maximum metabolic rate of

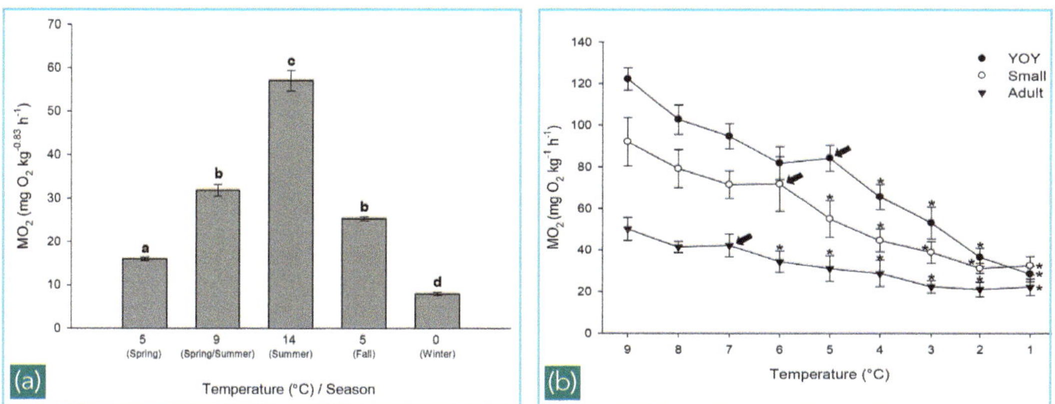

Fig. 23.15 **A**. Effect of temperature/season on the oxygen consumption (metabolic rates) in adult (>100 g) cunner. Note: these measurements were made on the same fish over one annual cycle. Dissimilar letters indicate significant differences in metabolic rate between temperatures/seasons (P < 0.05)(Costa, 2013). **B**. The effect of decreasing temperature at 2°C day^{-1} on oxygen consumption of young-of-the-year (YOY), small (3 – 5g) and adult (> 100g, large) cunner (modified from Kelly et al., 2014). Differences (P < 0.05,*) in the MO_2 between the acclimation temperature (9°C) and the other temperatures within each size class are shown. The arrows indicate the temperature at which a rapid decrease in MO_2 was initiated for each size class. Values in both panels are mean ± 1 S.E.

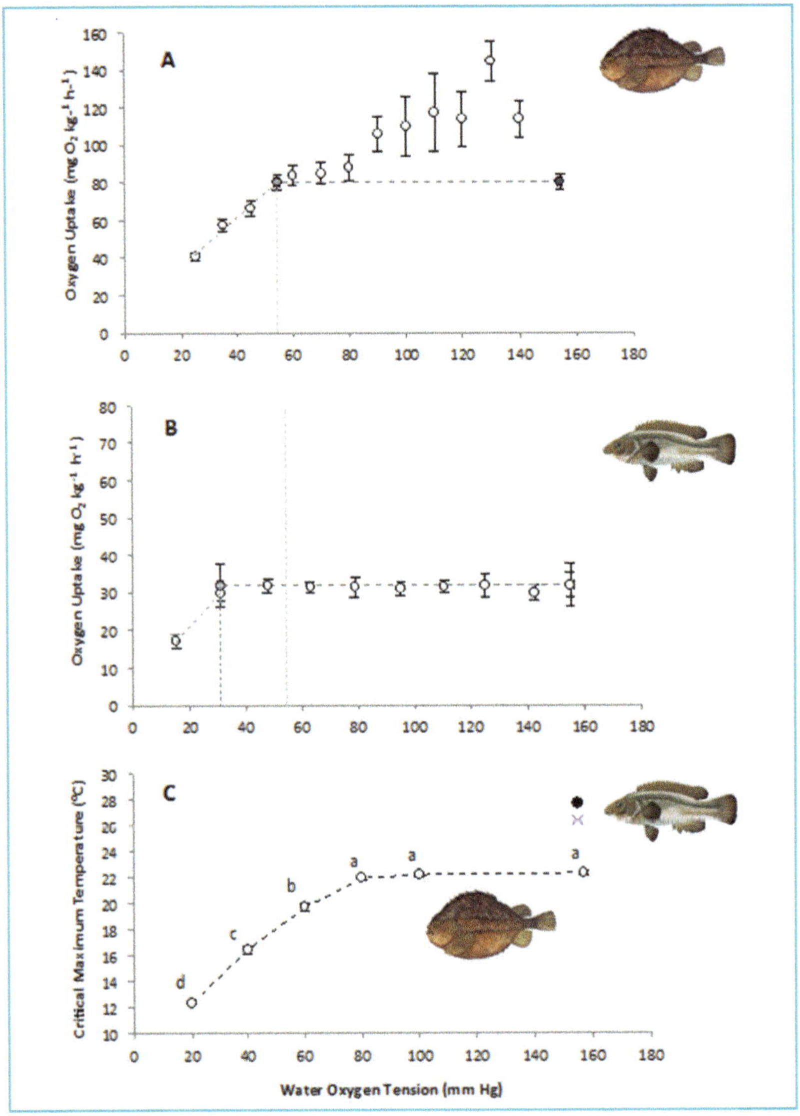

Fig. 23.16 Relationship between water oxygen levels and oxygen consumption (metabolic rate) in lumpfish (*Cyclopterus lumpus*) at 10°C (modified from Rasmus et al., 2016) (**A**) and cunner (*Tautogolabrus adspersus*) at 8°C (modified from Corkum et al., 2009). (**B**). The solid symbols (and horizontal dashed lines) indicate the basal/resting metabolic rates for each species (Note: this point at 155 mm Hg is obscured for the cunner by another data point). The vertical dashed lines represent the partial pressure of oxygen where the fish could no longer maintain metabolic rate (P_{crit}; this value was 54.4 mm Hg for the lumpfish and 31 mm Hg in cunner). In lumpfish the metabolic rate was elevated between 140 and 90 mm Hg as the fish were more active when water oxygen levels were initially lowered. (**C**) The effect of decreasing water oxygen levels on the temperature at which 10°C acclimated lumpfish lost equilibrium (i.e., reached their Critical Thermal Maximum, CT_{Max}) when warmed at 2°C h^{-1} (Rasmus et al., 2016). Data points (means ± 1 S.E) without a letter in common are significantly different at $P < 0.05$ (one-way ANOVA). Data are also shown for 3-5 g (•) and > 100 g cunner (X) at full saturation (Kelly et al., 2014).

235 mg kg^{-1} h^{-1} at 10°C [i.e. a factorial metabolic scope (i.e. MMR/BMR) of <3] (Rasmus *et al.*, 2016), cunner has a basal metabolic rate of 30–0 mg O_2 kg^{-1} h^{-1} (Figs. 23.16A and 16B; Costa *et al.*, 2013; Kelly *et al.*, 2014) and a factorial metabolic scope of 4.5 (Kelly *et al.*, 2014). Nonetheless, while the low metabolic rate/requirements of cunner make this species extremely temperature and hypoxia tolerant, cunner may be a less efficient delouser. The cunner requires fewer lice for maintenance (per g of body mass) than lumpfish, and this effect is compounded by smaller body size.

23.8 Fish health

A cleaner fish surveillance programme at the hatchery consists of diagnostic sampling every 45 days (logistics dependent). The cleaner fish will undergo a pre-transfer assessment 30–45 days prior to the transfer to the marine cage site. Pre-transfer involves histology, RT-PCR and virology sampling. Facilities have to follow guidelines of the provincial and federal authorities. There are also annual biosecurity audits. Juvenile fish are vaccinated in a static bath immersion with Vibrogen II as part of the vaccination program. This bacterin has been shown to be safe and effective when administered by immersion or intraperitoneal injection to healthy cleaner fish. Each litre of undiluted bacterin is sufficient for 100 kg of fish. One litre of vaccine is added to 9 l of clean hatchery water. Fish are netted, drained and immersed as 5 kg batches of fish (>2 g) in bacterin for 30 seconds. The bath may be reused up to 20 times before discarding. Food is withheld for 24 hours prior to vaccination. Wild cleaner fish are not permitted to be used in marine salmon sea cage sites. The Atlantic salmon smolts on the farm undergo regular surveillance by a veterinarian.

23.9 Transport by land and airfreight

Cleaner fish juveniles (cunner and lumpfish) have been transported from the hatchery to various sites in Atlantic Canada. Best welfare and husbandry practice are implemented during transport.

23.9.1 Land transport

23.9.1.1 Wild-collected broodstock transport

Transport of both species was by 5-ton vehicle equipped with a custom-constructed 2500 l fish transport tank outfitted with baffles and double access hatches. An on-board portable oxygenation system with compressed gas cylinder(s) and ceramic diffusers provided the required oxygen for the fish. Environmental conditions in the tank were monitored en route using an OxyGuard Handy Gamma™ temperature and oxygen water quality meter. The truck is equipped for both short (one- to four-hour) and long haul transportation (four-plus hours) of various fish and shellfish species, including cunner and lumpfish. Fish are transported under permit(s) and according to established introductions and transfer policies and biosecurity protocols.

Fig. 23.17 Discharge of cleaner fish from tanks on boat to individual cages at a salmon farm site.

23.9.1.2 Long-haul transport of farmed juveniles

Fish are fasted (also called conditioned) for 36 hours before transportation. The cunner (20–40 g mean) and lumpfish (184 g) were transferred to sea cage salmon smolt sites in summer 2016 (Fig. 23.17). The five transport tanks on the truck were 5000 l each, and had oxygen supplies and air blower capabilities. The transport temperature was 10–11°C and oxygen levels were maintained at 140% saturation. Transport densities were 33 kg m^{-3} for lumpfish and 16 kg m^{-3} for cunner. Transport mortality was <0.001% with a total transport time of 20 hours.

23.9.2 Air Transport

A total of 1000 juveniles was transported by air at 31 dph (0.3 g) from Newfoundland to New Brunswick. Fish were fasted for 36 hours prior to transport. They were packaged in 5 kg poly bags (double bagged) filled with 50% filtered seawater (8°C) and pure oxygen was added. Bags were secured (Styrofoam beads) and packaged in a cooler for shipment via air freight. Total transport time from packaging to being relocated in a tank of water at HMSC in New Brunswick was 12 hours. Mortality rate was minimal on arrival, oxygen was still at 150% saturation and temperature was 11.4°C. Although ice is permitted to keep the fish cool, dry ice or gel packs are preferred by most airlines. The coolers used were also well insulated and sturdy (Coleman Xtreme 5-Day Cooler with Extra ThermOZONE™).

23.10 Farm management trials and techniques

The first farm trials for cunner and lumpfish in Newfoundland were conducted in two stages to take water temperatures into consideration. The first stage started on 12 August 2015 with a first shipment by land to a KCS marine site at Sugar Loaf of 1,650 lumpfish (one cage, at 5% stocking ratio) and 11,880 cunner (four cages, at 9%). This was followed by the second shipment, also by land, on 1 September 2015 to the same site of 1,650 lumpfish (one cage, at 5%) and 14,850 cunner (five cages, at 9%).

All hides were initially constructed using tarp materials by GMG, a subsidiary of Cooke Aquacuture Inc. following designs adapted from those used at marine sites in Norway and Scotland. However, these hides proved insufficient for the lumpfish, which needed stiffer surfaces on which to attach, so plastic buckets and drums with the tops and bottoms removed were deployed in the cages with lumpfish. Supplemental marine fish feed encased in a gelatin mix was provided to both cunner and lumpfish. These were formed into tubes using 30 cm-long sections of plastic pipes that were frozen until needed. Observations made by divers showed that both cunner and lumpfish took two to three weeks to get used to consuming this feed. The size of cunner transferred to the trial cages at Sugar Loaf ranged from 25–30 g while the lumpfish at 200–250 g may have grown past the ideal transfer size because of delays in getting permits and timing of stocking of smolts at the trial site.

These trials continue but results for total lice counts (all stages) on 20 November 2015 indicated 36% fewer lice with cunner than on fish in control cages and a 20% reduction on salmon in cages stocked with lumpfish. This trend carried through to July 2016. The salmon were not treated for sea lice while salmon at two other sites holding the same year class of smolts in the same management area required medicinal treatments. Large-scale trials were also completed in 2016 on full-scale sites with 600,000 smolts in 10–150 m diameter cages. A figure of 10% lumpfish to salmon was added to all smolt cages and no treatments were required to date. Sea lice numbers were below treatable numbers.

23.11 Strategic use of cunner compared with lumpfish

Cunner and lumpfish were stocked on the same site but in different cages in 2015. This design is necessary when studying the individual efficacy of the two species but, when the respective efficacies are eventually determined, a more strategic use of both species might be to stock both species in the same cage. This will make full use of the differences in the optimal working water temperatures for each species for maximum effectiveness. This may even be necessary due to the wide range of water temperatures on the east coast of Canada and Maine where most of the salmon farms operate, such that when the temperatures fall outside the optimal range for one species, the other species will continue to perform and vice versa.

23.12 Future plans

Cleaner fish transport and deployment in sea cages in Canada is currently ongoing in multiple provinces and cage sites. The salmon farming industry has plans to increase production of cleaner fish (cunner and lumpfish) in collaboration with JBARB and HMSC. The USDA in Maine has also shown interest in setting up projects for studying the nutritional requirements and other aspects of lumpfish production. Regulations regarding the use of cleaner fish are governed by both federal and provincial authorities and, similar to other countries, fish health and animal welfare will take priority. There will be increased use of cleaner fish in Atlantic Canada in the future.

Acknowledgements

We wish to acknowledge the OSC Head Dr Garth Fletcher and dedicated and talented staff of the Dr Joe Brown Aquatic Research Building (JBARB) at Memorial University. Individuals who contributed towards the cleaner fish program are Chris Canning, John Evely, Francine Fortune, Jessica Fry, Bruno Gianasi, Kelsie Jeannot, Jennifer Monk and Denise Tucker. A special thanks to George Bishop, Robert O'Donnell and Andrew Perry of OSC Field Services, and to Sheldon George, Cold Ocean Salmon Inc. NL saltwater production manager and his dedicated team.

References

Albert, O.T., Torstensen, E., Bertelsen, B., Jonsson, S.T., Pettersen, I.H. and Holst, J.C. (2002) Age-reading of lumpsucker (*Cyclopterus lumpus*) otoliths: dissection, interpretation and comparison with length frequencies. *Fisheries Research* 55, 239–252.

Brown, J.A., Somerton, D., Methven, D.A. and Watkins, J.R. (1992) Recent Advances in Lumpfish *Cyclopterus lumpus* and Ocean Pout *Macrozoarces americanus* Larviculture. *Journal of World Aquaculture Society* 23, (4) 271–276.

Burt, K., Hamoutene, D., Mabrouk, G., Lang, C., Puestow, G., Drover, D., Losier, R. and Page, F. (2012) Environmental conditions and occurrence of hypoxia within production cages of Atlantic salmon on the South coast of Newfoundland. *Aquaculture Research* 43, 607–620.

Corkum, C.P. and Gamperl, A.K. (2009) Does the ability to metabolically downregulate alter the hypoxic tolerance of fishes?: A comparative study using cunner (*T. adspersus*) and Greenland Cod (*G. ogac*). *Journal Experimental Zoology* A. 311, 231–239

Costa, I.A.S.F, Driedzic, W.R. and Gamperl, A.K. (2013) Metabolic and cardiac responses of cunner (*Tautogolabrus adspersus*) to seasonal and acute changes in temperature. *Physiological and Biochemical Zoology* 186, 233–244.

Davenport, J. (1985) Synopsis of biological data on the lumpsucker *Cyclopterus lumpus*. *FAO Fisheries Synopsis* 147. 40 pp.

Drew, B.C. (1976) A contribution to the life history of the cunner, *Tautogolabrus adspersus*, in Fishers Island Sound, Connecticut. *Chesapeake Science* 17, 101.

Gollock, M.J., Petersen, L.H., Currie, S. and Gamperl, A.K. (2006) Cardiac function and blood oxygen carrying capacity only limit cod (*Gadhus morhua*) oxygen consumption at high temperatures. *Journal of Experimental Biology* 209, 2961–2970.

Government of Newfoundland and Labrador, Department of Fisheries and Aquaculture. (2016) Emerging Species Profile Sheet. www.fishaq.gov.nl.ca/research_development/fdp/pdf/lumpfish.pdf

Hedeholm, R., Blicher, M.E. and Grønkjaer, P. (2014) First estimates of age and production of lumpsucker (*Cyclopterus lumpus*) in Greenland. *Fisheries Research* 149, 1–4.

Imsland, A. (2012) www.forskningsradet.no/en/Newsarticle/New_promise_in_sea_liceeating_lumpfish/1253979450537

Kelly, N.I. Alzaid, A., Nash, G.W. and Gamperl, A.K. (2014) Ontogenetic effects on thermal tolerance and metabolic depression in cunner (*Tautogolabrus adspersus*). *PLoS One*. 9(12), e114765. doi: 10.1371/journal.pone.0114765

Killen, S.S., Costa, I.A.S.F. Brown, J.A. and Gamperl, A.K. (2007) Little left in the tank: metabolic scaling in marine teleosts and its implications for aerobic scope. *Proceedings of the Royal Society B: Biological Sciences* 274, 431–438.

Kvenseth, P.G., Andreassen, J. and Bergh, Ø. (2002) Leppefisk – liten rensefisk kan berge stor laks!.I: Glette, J., van der Meeren,T., Olsen, R.E. og Skilbrei, O. (red), 2002 Havbruksrapport 2002. Fisken og havet, særnr. 3-2002.

Nitschke, O., Burnett, J., and Kelly, B.C. (2001) Age and growth verification for cunner in western Cape Cod Bay, Massachusetts, using tag-recapture data. *Transactions of the American Fisheries Society* 130, 1150–1163.

Nytro, A.V., Vikingstad, E., Foss, A., Hangstad, T.A., Reynolds, P., Eliassen, G., Elvegard, T.A., Falk-Petersen, I-B. and Imsland, A.K. (2014) The effect of temperature and fish size on growth of juvenile lumpfish (*Cyclopterus lumpfish* L.). *Aquaculture* 434, 296–302.

Rasmus, E., Norin, T., Gamperl, A.K. and Esbaugh, A.J. (2016) Oxygen dependence of upper thermal limits in fishes. *Journal of Experimental Biology* 219, 3376–3383

Scott, W.B and Scott, M.G. (1988a) Atlantic Fishes of Canada. *Canadian Bulletin of Fisheries and Aquatic Sciences* 219, 398–399.

Scott, W.B. and Scott, M.G. (1988b) Atlantic Fishes of Canada. *Canadian Bulletin of Fisheries and Aquatic Sciences* 219, 518–519.

Serchuk, F.M. and Cole, C.F. (1974) Age and growth of the cunner, *Tautogolabrus adsperus* (Walbaum) (Pisces: Labridae) in the Weweantic River Estuary, Massachusetts. *Chesapeake Science* 4, 205–213.

CHAPTER 24

The Patagonian blenny (*Eleginops maclovinus*): a Chilean native fish with potential to control sea lice (*Caligus rogercresseyi*) infestations in salmonids

Juan Carlos Sánchez,[1] *Jorge Mancilla,*[2] *Martin Hevia,*[1] *Patricio J. Saez*[1]

[1]Fundación Chile, Estación Experimental Quillaipe, Puerto Montt, Chile
[2]Marine Harvest Chile, Puerto Montt, Chile

24.1 Introduction

The salmon farming industry in Chile started in 1978 and evolved over three decades to become the fastest growing industry in Chile, and the second largest producer after Norway from 1992 (Bjørndal, 2002; Niklitschek *et al.*, 2013). The main salmon production in Chile is located in the X region, around the city of Puerto Montt and Chiloé Island. However, this area extends over regions XI and XII. Due to the physical characteristics of the coast, these regions offer protected sites with ideal temperature and salinity levels. Additionally, a large number of rivers and lakes offer ideal conditions for production of salmon smolts all year around (Bjørndal, 2002). The main species farmed in Chile is Atlantic salmon (*Salmo salar*). However, rainbow trout (*Oncorhynchus mykiss*) and Pacific salmon (*Oncorhynchus kisutch*) are also produced. The total salmonid production in 2006 reached a maximum of 645,000 tonnes, with an economic return of US$2.2 billion (IFOP, 2015; SERNAPESCA, 2015). This production level was reduced in 2007 in response to the health crisis produced by the ISA virus (Infectious Salmon Anaemia). After implementation of adequate measurements, production levels recovered. In 2014, 955,000 tonnes of salmonids were harvested (SERNAPESCA, 2015).

Sea lice are copepod ectoparasites with a worldwide distribution that affect wild and farmed marine fish species (Farlora *et al.*, 2016). *Caligus rogercresseyi* is the dominant sea louse found in the salmon and trout farms in Chile (Fig. 24.1). The presence of this parasite was first described during the early stages of the Chilean salmon farming industry. *C. rogercresseyi* was transmitted by native fish species such as *Eleginops maclovinus* and *Odonthestes regia*, common wild hosts to this parasite (Gonzalez and Carvajal, 2003). Gonzalez and Carvajal (2003) determined that *C. rogercresseyi* has eight development stages, including three planktonic stages, two nauplius and a copepodid (infective stage) that settle on the host utilising a pair of antennae. Once attached to the host, the parasitic stages of *C. rogercresseyi* comprise four different chalimus stages and then adult females and males (Gonzalez and Carvajal, 2003). Although sea lice infestations do

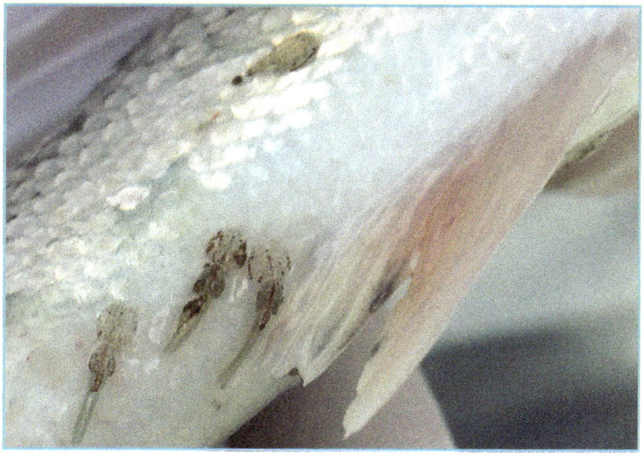

Fig. 24.1 Female *Caligus rogercresseyi* with egg strings attached to Atlantic salmon. Credit: The Copepodite Production Laboratory, Quillaipe Research Station, Fundación Chile.

not directly cause mortality, this parasite causes skin damage and induces stress response in salmonids, which generates immune depletion that makes fish more prone to further infections (Valenzuela-Miranda *el al.*, 2015). The cost of medicinal treatment alone utilised to control sea lice by the Chilean salmon industry was almost US$80 million in 2013, while the total economic disturbance related to sea lice (that is operation costs, reduction of fish quality, among others) was estimated to be about US$300 million in the same year (Agusti *et al.*, 2016) .

The Chilean National Service of Fisheries and Aquaculture (Sernapesca) classified Caligus infestations in 2007 as a High Risk Disease, due to the severity of infection, and therefore it can be included in health programmes of surveillance and control. A new vigilance and control strategy for Caligus infestations was implemented in 2012. This strategy is regulated through the Caligidosis Specific Sanitary Programme, as indicated in the regulatory resolution N° 1141. This strategy is based on the principles that Sernapesca has established for all aquaculture disease control programmes, such as early detection of cases, timely control of infested animals, and protection of healthy fish.

Currently there is no specific regulation for the use of cleaner fish in Chile, and there are no private companies using this technology. Sea louse infestations in Chile have been historically controlled with medicines applied either by bath or by oral treatments. These medicines include organophosphates, Avermectins (ivermectins and emamectin benzoate), pyrethroids (deltamethin and cypermethrin), chitin inhibitor (diflubenzuron). However, sea lice have developed resistance to most of these anti-louse medicines used in Chile (Agusti *et al.*, 2016). However, Sernapesca is encouraging the use of non-pharmacological treatments as part of the Caligus treatment strategy. Salmon farming companies, on the other hand, focus on sustainability by a reduction in use of medicines and to find a cost-effective equilibrium of non-pharmacological treatments.

24.2 The Patagonian blenny

The Patagonian blenny (*Eleginops maclovinus*) is a protandrous hermaphrodite, notothenoid fish belonging to the family Eleginopidae (Lincadeo *et al.*, 2006) (Fig. 24.2). *E. maclovinus* is distributed from Valparaiso (33° 00' S) to the Beagle Channel (54° 00' S) in the Pacific Ocean and from southern Argentina and the Falkland Islands to the coasts of Uruguay in the Atlantic Ocean (Henriquez *et al.*, 2011). This fish inhabits coastal areas, river mouths and estuaries, and exhibits characteristics of a euryhaline and eurythermic species. The Patagonian blenny during its life cycle will first sexually mature as a male and later undergo a sexual reversion process to become female individuals. This phenomenon has been described in populations from different locations. Studies have estimated size and age of sexual reversion and this varies between 35 and 53 cm length, and two and six years of age, depending on geographic locations (Ceballos, 2011). Males and females become sexually mature in August, and the main spawning season occurs during September and December in sandy estuarine areas (Brickle *et al.*, 2005). A male generally generates 9 mm^3 of seminal fluid. Spermatozoa become active when mixed with secretion from spermatic conducts. Average flagellar motility is for 1.5 hours at 17°C. Milt show a circular-shaped head (1.9 µm diameter) with an intermediate segment (1.1 µm length) and a tail of 27.4 µm average length. Patagonian blenny eggs are pelagic, spherical shaped (1.9 mm diameter), transparent and with one oily droplet.

Juvenile and adult Patagonian blenny inhabit estuarine areas and generally prey on benthic invertebrates, such as polychaetes and small crustaceans and insects. However, results from studies describing the stomach content of captured specimens have reported the presence of significant amounts of different types of vegetables. Authors agree, however, that Patagonian blenny selectively, or opportunistically, prey mainly on animal items (Sa *et al.*, 2014).

Patagonian blenny potential as cleaner fish

Trials aimed at assessing the growth potential of Patagonian blenny under sea cage conditions suggested that this species was able to graze on sea lice attached to the skin of counterparts sharing the same cage (P. Saez, Estación Experimental Quillaipe, Fundación Chile, February 2017, personal communication). Based on this finding, Fundación Chile has developed and conducted two projects aimed at assessing the sea lice grazing potential of Patagonian blenny when cohabitated with Atlantic salmon in tank conditions. Another objective was to determine the effect of

Fig. 24.2 Patagonian blenny used in cohabitation trials at the Quillaipe Research Station. Credit: Fundación Chile, Puerto Montt.

culture density and the blenny: salmon ratio on sea lice cleaning efficiency in both tanks and sea cages. The ability of Patagonian blenny to ingest salmon feed of different sizes was also assessed. The aim of this chapter is to describe the results of these trials by researchers at Fundación Chile aimed at assessing the potential of Patagonian blenny as cleaner fish against sea lice infestation.

24.3 Trials assessing the potential of Patagonian blenny as cleaner fish

24.3.1 Tank-based trials

All tank-based trials were conducted at the Estación Experimental Quillaipe, Fundación Chile, Puerto Montt, Chile. Trials were carried out by the R&D unit of the technical department of Marine Harvest Chile, and Aquadvise, during the conduct of project 12IDL1-13173 INNOVA-CORFO 'Biological treatment against sea lice (*Caligus sp.*) in farmed salmonids' and :INNOVA-CORFO project L2 'Development of a biological treatment against sea lice (*Caligus sp.*) in farmed salmonids'.

Patagonian blenny used in this trial were of wild origin, captured from the Hueihue Bay, Chiloé, X Región. Fish were kept following capture under culture conditions for a period of 12 weeks prior to commencement of the trials. During this period fish were treated with cypermethrin 1% (Betamax®) to ensure elimination of potential internal parasites.

C. rogercresseyi infestation

Fish were exposed to infestation with copepodites at the beginning of the trial. Fish were kept off feed for a period of 24 hours prior to the beginning of the infestation. Copepodites used in the trial were generated from adult females with egg strings of *C. rogercresseyi* isolated from Atlantic salmon kept at the Quillaipe Research Station. Adult females were transported to the Copepodite production laboratory and incubated in sea water (32‰, filtrated at 88 µm and UV treated). Water temperature was kept at 12–14°C with no light and constant aeration. Infestations were conducted when 90% of the individuals in the planktonic stage reached the copepodite stage. This procedure was conducted according to the protocol PO-AQ-16-LAB 'Procedures of production and infestation with C. rogercresseyi'. The experimental period for each trial was 35 days. Experimental tanks were kept in a saltwater flow-through system. Tanks were supplied with faeces collection systems provided with a mesh for parasite collection. The number of sea lice per fish was assessed at seven-day intervals.

The objective of Trial 1 was to evaluate the Patagonian blenny as a cleaner fish of Caligus infestation in Atlantic salmon. The hypothesis was: do Patagonian blenny show cleaner activity when kept in a cohabitation model with Atlantic salmon in tanks? The Patagonian blenny were of initial average body weight of 60 g (wild origin), and Atlantic salmon had an initial average body weight of 155 g. The treatments were:

- Treatment 1 (blenny:salmon ratio = 20%) = 25 Atlantic salmon + 5 P. blenny + *C. rogercresseyi*
- Treatment 2 = 30 blenny + *C. rogercresseyi*
- Treatment 3 (Control) = 30 Atlantic salmon + *C. rogercresseyi*

There was a 48.8% cleaning efficiency using Patagonian blenny (see Fig. 24.3 Treatment 1, for statistics). It was concluded that Patagonian blenny graze on *C. rogercresseyi* attached to Atlantic salmon when reared in a cohabitation model. The *C. rogercresseyi* infestation in Patagonian blenny was significantly lower than in Atlantic salmon. There were no significant differences in growth performance among Atlantic salmon reared in the cohabitation model

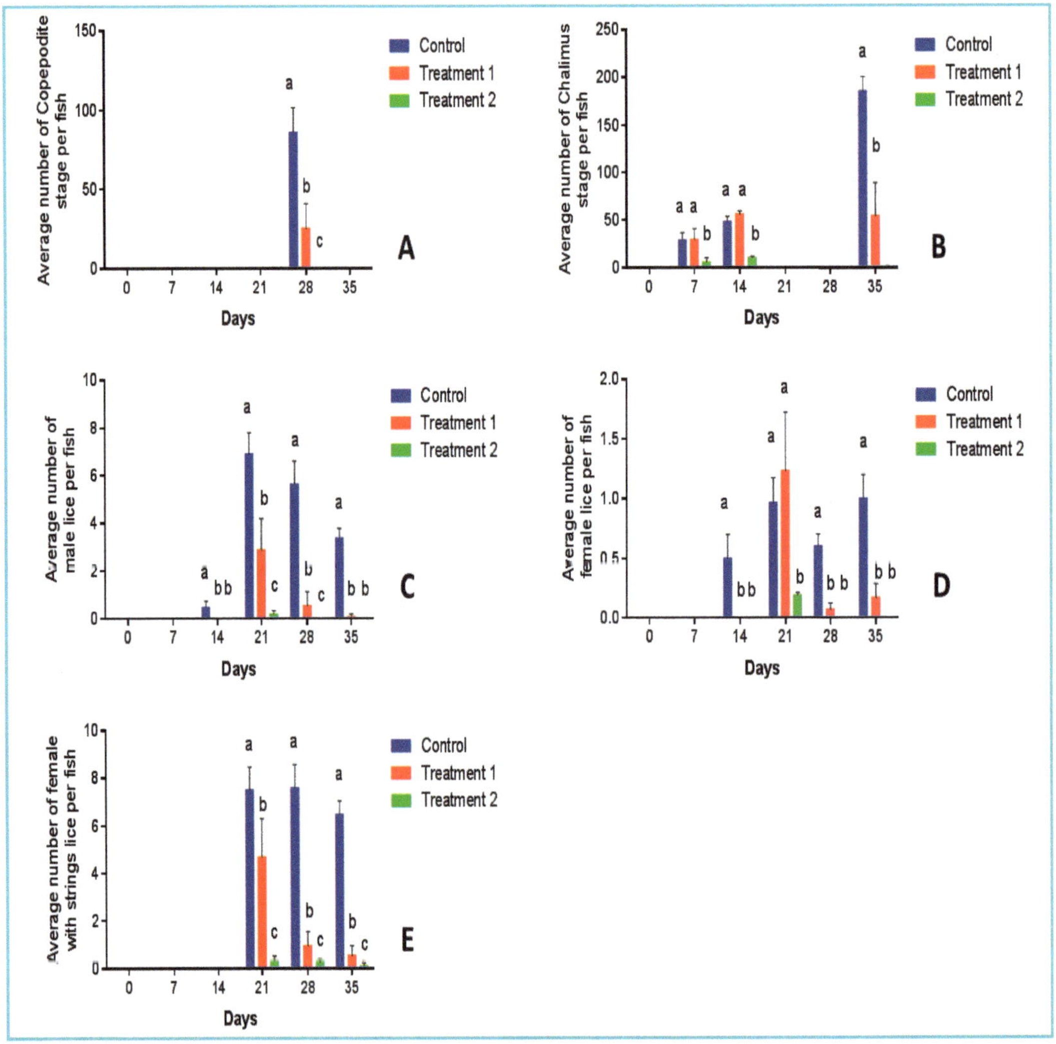

Fig. 24.3 Trial 1. Assessing sea lice cleaning activity of Patagonian blenny. Treatment 1 (P. blenny:salmon ratio = 20%) = 25 Atlantic salmon + 5 P. blenny + *C. rogercresseyi;* Treatment 2 = 30 P. blenny + *C. rogercresseyi;* Treatment 3 (Control) = 30 salmon + *C. rogercresseyi*. Average number of Copepodite (A), Chalimus (B), male (C), female (D), and female with strings (E) stages of *Caligus rogercresseyi* per fish recorded for each triplicate treatment during each sampling dates. Values are presented as means ± SD. Mean values which do not share a letter were found to be significantly different by Kruskal-Wallis and by Dunn's multiple comparisons test.

with Patagonian blenny or reared alone. Patagonian blenny grazed on *C. rogercresseyi* in the adult stage.

In Trial 2 the cleaning efficiency of Patagonian blenny on Caligus infestation in salmon reared at low culture densities was assessed. The hypothesis tested was that there are differences in the cleaning efficiency when different culture densities (based on initial and halfway culture densities observed in regular production cycles) were used. The fish used were blenny of average initial body weight = 82 g (wild origin) and Atlantic salmon were 138 g. The treatments were:

- Treatment 1 (P. blenny:salmon ratio = 10%, low culture density = 0.9 kg m^{-3}) = 50 Atlantic salmon + 5 blenny + *C. rogercresseyi*
- Treatment 2 = (P. blenny:salmon ratio = 10%, high culture density = 7.9 kg m^{-3}) = 460 Atlantic salmon + 46 blenny + *C. rogercresseyi*

The cleaning efficiency of blenny in Treatment 1 was 64.2%, compared with 93.6% in treatment 2 (Fig. 24.4). Therefore salmon stocking density had an effect on cleaning efficiency of Patagonian blenny, and on re-infestation of *C. rogercresseyi* in tanks. The greater level of sea lice reduction was observed at the higher stocking density of salmon. Not unexpectedly, salmon stocked at lower density had a higher growth performance.

In Trial 3 the cleaning efficiency of blenny in Caligus infestation was assessed at two stocking densities of blenny, 1:10 (10%) and 1:20 (20%) in tanks. The average weight of Patagonian blenny was 64 g (wild origin) and Atlantic salmon were 125 g. The treatment groups were:

- Treatment 1 = 500 Atlantic salmon only
- Treatment 2 = blenny:salmon ratio = 5% = 500 salmon + 25 blenny + *C. rogercresseyi*
- Treatment 3 = blenny:salmon ratio = 10% = 500 salmon + 50 blenny + *C. rogercresseyi*

The reduction in sea lice numbers was 60.4% in treatment 2 and 58.8% at the higher stocking ratio (Fig. 24.5). Therefore there were similar grazing levels in both treatment ratios.

The feed consumption of the blenny with different sizes of commercial salmon feed was examined in Trial 4. This work was funded by Innova Chile CORFO project L2 'Development of a biological treatment against sea lice (Caligus sp) in farmed salmonids'; with partners Marine Harvest Chile and Fundación Chile. The objective was to determine feeding rates of Patagonian blenny when fed Atlantic salmon commercial feed of different sizes during initial stages of culture cycle. The feed size 4, 6, 9, and 12 mm pellet were fed to blenny of three different body weight range. Each feed size was fed for a ten-day period. The treatments were:

- Treatment 1= blenny body weight range = 60–120 g
- Treatment 2 = blenny body weight range = 130–160 g
- Treatment 2 = blenny body weight range = 170–250 g

There was no feeding activity in any of the fish sizes when fish were fed with a pellet size of 9 and 12 mm. Fish fed pellet size 4 and 6 mm had higher feed consumption rates than those of fish fed pellet sizes 9 and 12 mm. Fish in the three weight ranges evaluated consumed the 4 and 6 mm pellet sizes.

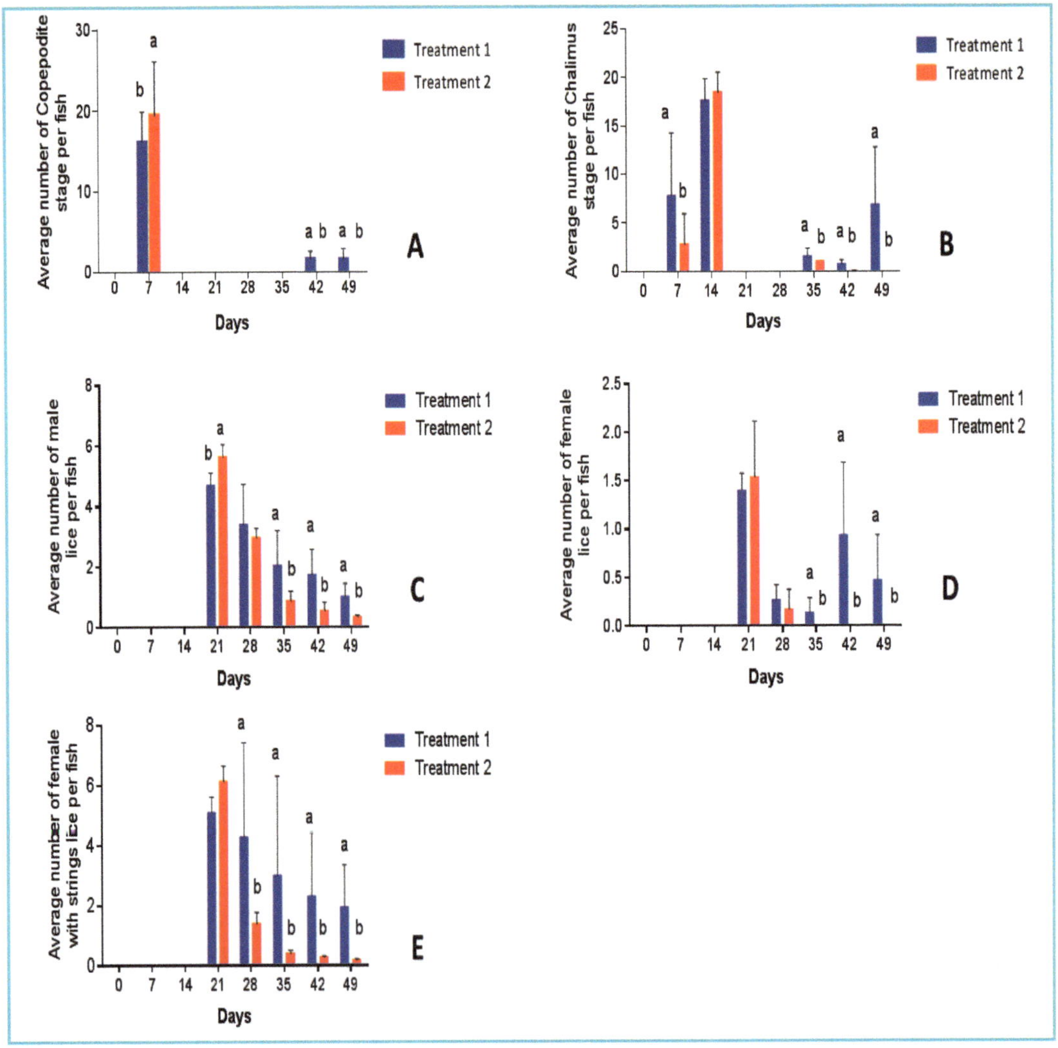

Fig. 24.4 Trial 2. Numbers of sea lice on salmon following stocking of blenny at two salmon densities. Treatment 1 (blenny:salmon ratio = 10%, low culture density = 0.9 kg m^{-3}) = 50 Atlantic salmon + 5 blenny + *C. rogercresseyi;* Treatment 2 = (blenny:salmon ratio = 10%, high culture density = 7.9 kg m^{-3}) = 460 Atlantic salmon + 46 blenny + *C. rogercresseyi*. Average number of Copepodite (A), Chalimus (B), male (C), female (D), and female with strings (E) stages of *Caligus rogercresseyi* per fish recorded for each triplicate treatment during each sampling dates in Trial 2. Values are presented as means ± SD. Mean values which do not share a letter were found to be significantly different by Kruskal-Wallis and by Dunn's multiple comparisons test.

24.3.2 Experimental trial in sea cage

All the sea cage trials (Trial 5) were conducted at the Huenquillahue Station, Marine Harvest Chile MHC, Puerto Montt by the technical department of MHC, the R&D of Quillaipe Research Station and Aquadvise, during the Innova Chile CORFO project L2 'Development of a biological

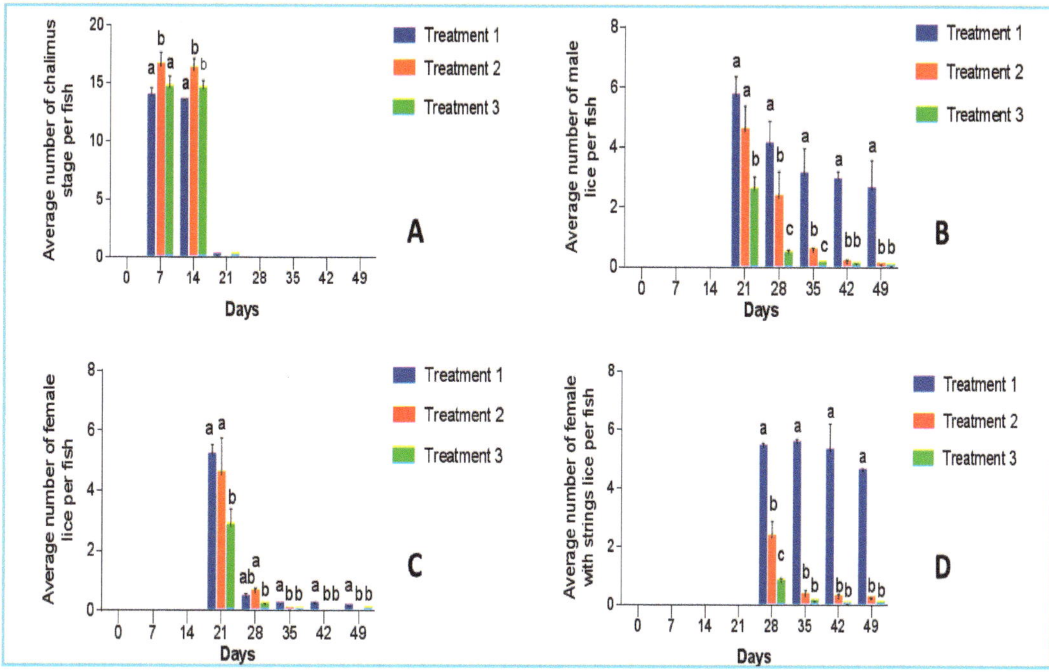

Fig. 24.5 The cleaning efficiency in Caligus infestation at two stocking densities of blenny, 1:10 (10%) and 1:20 (20%) in tanks. The average weight of Patagonian blenny was 64 g (wild origin) and Atlantic salmon were 125 g. Treatment 1 = 500 Atlantic salmon; Treatment 2 = blenny:salmon ratio = 5% = 500 salmon + 25 blenny + *C. rogercresseyi;* Treatment 3 = blenny:salmon ratio = 10% = 500 salmon + 50 blenny + *C. rogercresseyi*. Average number of Chalimus (A), male (B), female (C), and female with strings (D) stages of *Caligus rogercresseyi* per fish recorded for each duplicate treatment during each sampling dates in Trial 3. Values are presented as means ± SD. Mean values that do not share a letter were found to be significantly different by Kruskal-Wallis and by Dunn's multiple comparisons test.

treatment against sea lice (*Caligus sp.*) in farmed salmonids'. Patagonian blenny used in this trial were of wild origin captured from the Hueihue Bay, Chiloé, X Región. Fish were kept following capture under culture conditions for a period of 12 weeks prior to commencement of the trials. During this period fish were treated with brackish water (5 ppt) for 36 hours to eliminate sea lice. Fish were infected with naturally occurring sea lice at the Huenquillaue Station. The experimental period for each trial was 60 days and sea lice numbers were assessed at seven-day intervals.

In this Trial 5 the cleaning efficiency of blenny was assessed in salmon of different body weight under experimental conditions in sea cages. The treatments were:

- Treatment 1: blenny:salmon ratio = 10% = 150 Atlantic salmon (initial average body weight 500 g) + 30 blenny (initial body weight 70 g)
- Treatment 2: 150 Atlantic salmon (initial average body weight 500 g)
- Treatment 3: (blenny:salmon ratio = 20%) = 175 Atlantic salmon (initial body weight 3200 g) + 35 blenny (initial weight 70 g)
- Treatment 4: 175 Atlantic salmon (initial weight 3200 g).

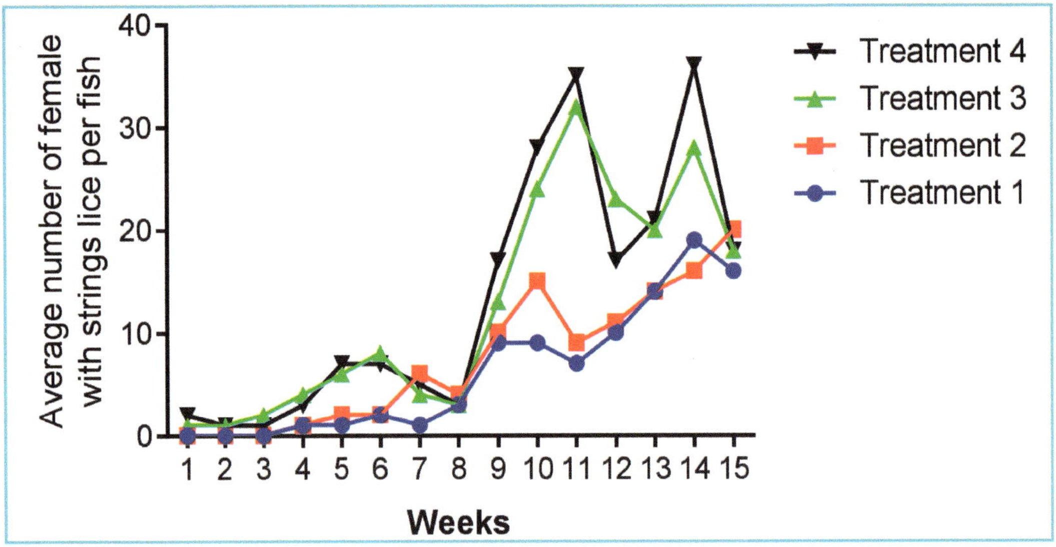

Fig. 24.6 An assessment of cleaning efficiency of the Patagonian blenny with salmon of different weights (Trial 5). Average number of female with strings of *Caligus rogercresseyi* per fish recorded for each duplicate treatment during each sampling date in Trial 5. Treatment 1: blenny:salmon ratio = 10% = 150 Atlantic salmon (initial weight 500 g) + 30 blenny (70 g); Treatment 2: 150 Atlantic salmon (initial weight 500 g); Treatment 3: (blenny:salmon ratio = 20%) = 175 Atlantic salmon (initial weight 3200 g) + 35 blenny (70 g); Treatment 4: 175 Atlantic salmon (initial weight 3200 g).

The reduction in sea lice numbers was 47.2% in treatment 1 after 35 days, compared with only 27.3% in the larger fish in treatment 3 (Fig. 24.6). The numbers of adult sea lice attached to salmon of 500 g and 3200 g/fish were reduced after 5 weeks. The numbers of sea lice removed from Atlantic salmon of 500 g was higher. There was a general reduction of the grazing activity from week 5 to week 10. This reduction in cleaning efficiency was associated with consumption of organisms fouling the nets by blenny and confirmed by gut content analysis.

In Trial 6 the cleaning efficiency of Patagonian blenny when reared with Atlantic salmon at different blenny:salmon ratios was assessed under experimental conditions in sea cages. The treatments were:

- Treatment 1: (P. blenny:salmon ratio = 10%) = 220 Atlantic salmon (initial weight 120 g) + 22 blenny (initial weight 70 g)
- Treatment 2: (blenny:salmon ratio = 30%) = 220 Atlantic salmon (initial weight 120 g) + 66 blenny (initial weight 70 g)
- Treatment 3: 220 Atlantic salmon (initial weight 120 g)

There was a reduction in sea lice numbers of 40% by day 7 at the low stocking ratio, Treatment 1, and 33% at day 49 at the high ratio (Fig. 24.7). There were no clear differences in cleaning efficiency between experimental treatments and control (Treatment 3) due to low sea lice numbers (<3 per fish).

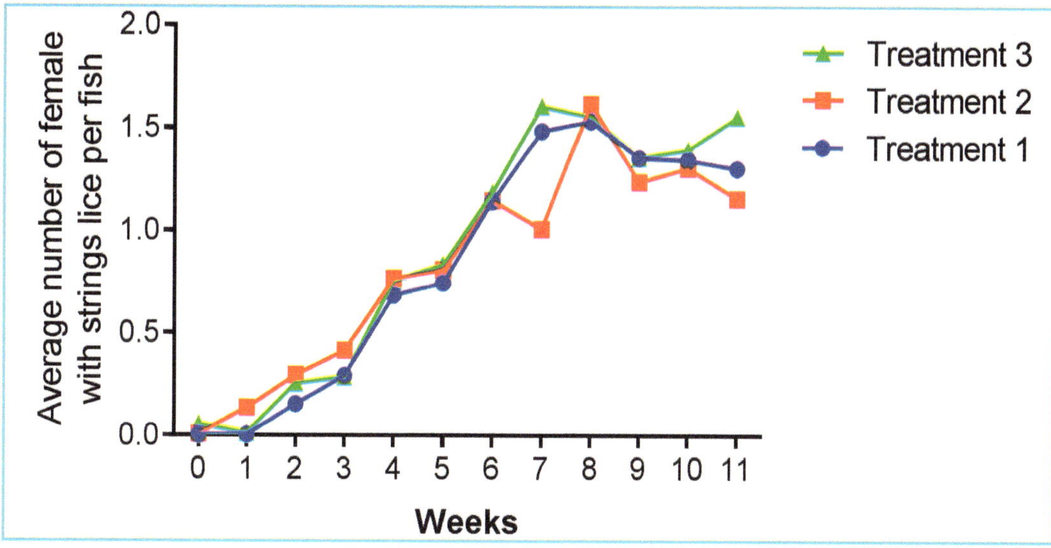

Fig. 24.7 The cleaning efficiency of Patagonian blenny when reared with Atlantic salmon at different blenny:salmon ratios under experimental conditions in sea cages. Data expressed as the average number of females with strings of *Caligus rogercresseyi* per fish recorded for each duplicate treatment during each sampling dates in the Trial 6. Treatment 1: (P. blenny:salmon ratio = 10%) with = 220 salmon (initial weight 120 g) + 22 blenny (initial weight 70 g); Treatment 2: (blenny:salmon ratio = 30%) = 220 salmon (initial weight 120 g) + 66 blenny (initial weight 70 g); Treatment 3: 220 salmon only (120 g).

From this series of six trials it is concluded that cleaning activity of the Patagonian blenny when reared with Atlantic salmon is evident under tank conditions, but is less clear in sea cages. The Patagonian blenny grazes on sea lice in the adult stage. Sea lice infection in the Patagonian blenny is significantly lower than that observed in Atlantic salmon. An increase in salmon stocking density and the P. blenny:salmon ratio showed a positive effect on cleaning efficiency under tank conditions. Blenny of weight range of 60 and <250 g consumed salmon feed sizes 4 mm and 6 mm. Higher cleaning efficiency was associated with Atlantic salmon of smaller size. Fouling of sea cage nets had a negative effect on cleaning activity of blenny.

24.4 Rearing of the Patagonian blenny

The Patagonian blenny has been successfully bred in captivity by personnel from Fundación Chile, at the Quillaipe Research Station. A brief description of this process is presented.

24.4.1 Broodstock conditioning

Wild-origin broodstock are captured during the autumn and kept in tanks maintained with a flow-through system for a period of four months. Fish are fed an extruded diet for salmonid broodstock. At the end of this period, gonad maturation is assessed.

24.4.2 Spawning, egg production and incubation

Spawning is induced using hormonal treatment through an intraperitoneal injection. After 72 hours, egg hydration takes place, and females show a pronounced belly enlargement. Eggs are then stripped after correct sedation of the fish. Immediately after spawning, eggs are fertilised by a gentle mixing with seminal fluid and water addition. Fertilised eggs are incubated in cylinder conical tanks. Egg segmentation starts one hour after fertilisation (Fig. 24.8) and extends for a

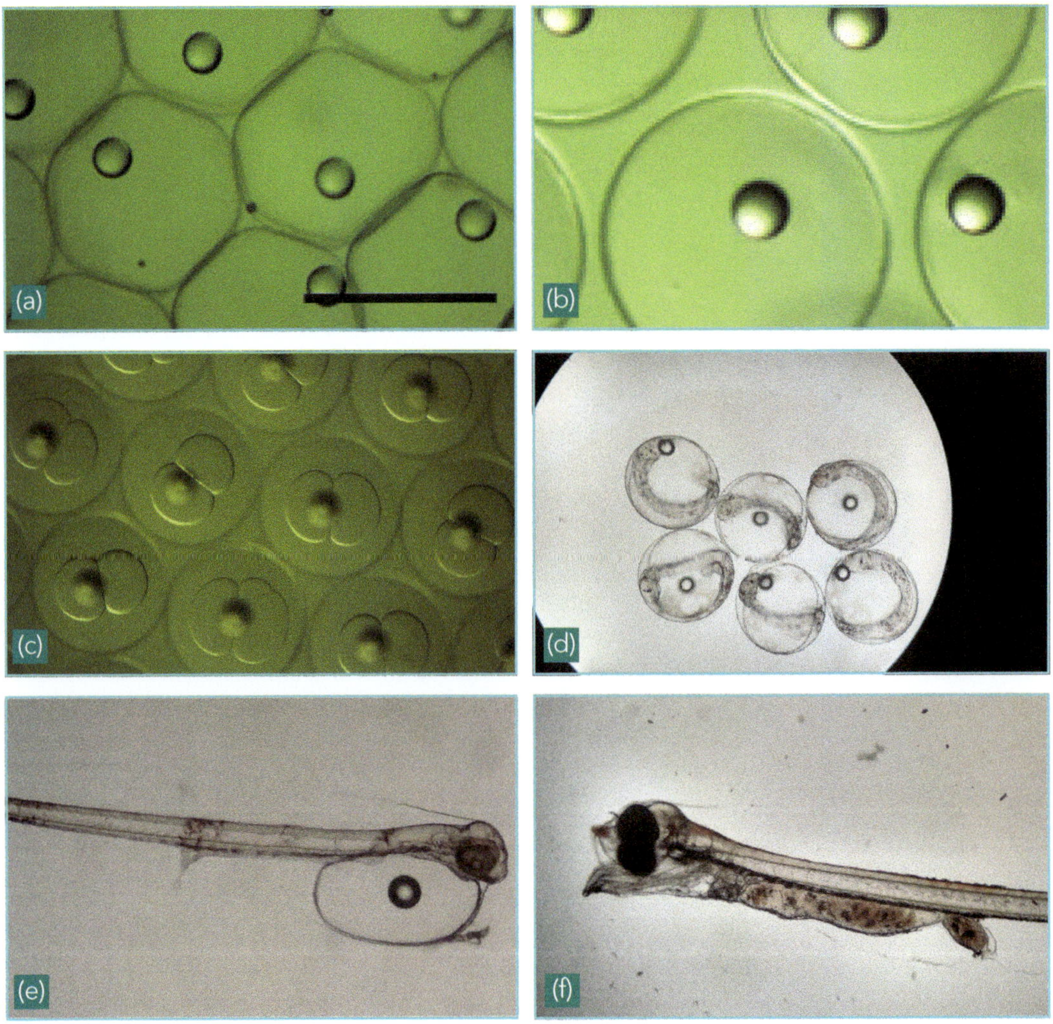

Fig. 24.8 Egg and larval developmental stages in Patagonian blenny. Clockwise from top left, Eggs (a) = unfertilised egg (scale ×56); Eggs (b) = 0.5 hours post-fertilisation (×56) 1.9 mm diameter; Eggs (c) = 3.5 hours post-fertilisation, first cleavage segmentation completed with two blastomeres (×56); Embryos (d) four days post-fertilisation (x4); Larva (e) one day post-hatching (×4), length of larvae 2.6 μm ; Larva (f) ten days post-hatch (×4). Credit: egg photos, Dr Ivan Valdebenito, Escuela de acuicultura, Universidad Católica de Temuco, Chile; embryo and larval photos, Mr Ramón Vidal, Fundación Chile, Estación Experimental Quillaipe, Puerto Montt, Chile.

period of nine hours. Hatching occurs 72 hours after fertilisation. Newly hatched larvae have an average length of 2.6 mm and, after ten days, the eyes are completely developed, as are the heart, digestive tract, mouth and pectoral fins. Enriched rotifers (*Brachionus plicatilis*) are used from first feeding to 20 days post-hatch (dph). Artemia metanauplii are incorporated from 20 dph. Live feed is gradually replaced with dry formulated feed from 40 dph.

A growth trial conducted with Patagonian blenny (initial weight = 40 g), indicated that maximum growth and nitrogen retention were obtained with 30% of dietary protein content. Optimum growth rate (expressed as specific growth rate) obtained in this trial was 0.58 (Sa *et al.*, 2014).

The hatch rate in the 2015 production season was 23% and a total of 4000 juveniles was produced. Patagonian blenny were transported by Fundación Chile in a tank with oxygen supply to the experimental cages. However, as this will be done on a commercial scale, fish should be transported by wellboat.

24.5 Future developments

Continuing work with the Patagonian blenny will include an evaluation of the effect of blenny body weight on lice cleaning efficiency, the evaluation of the effect of P. blenny origin (wild or farmed) on cleaning efficiency, the determination of blenny body size that can be stocked in sea cages, and the development of a model to determine the number of blenny to be used depending on factors such as season, farming zone and the salmon production phase, among others. Two other fish species have also been assessed as potential cleaner fish. These species include *Malapterus reticulatus* and *Mugil cephalus*. However, the results of these trials involving these species are still confidential. The industry in Chile hopes to be using cleaner fish on commercial salmon sea farms in the coming years (SalmonExpert, 2015).

References

Agusti, C., Bravo, S., Contreras, G., Bakke, M., Helgesen, K., Winkler, C., Silva, M., Mendoza, J. and Horsberg, T. (2016) Sensitivity assessment of *Caligus rogercresseyi* to anti-louse chemicals in relation to treatment efficacy in Chilean salmonid farms. *Aquaculture* 45, 195–205.

Bjørndal, T. (2002) The competitiveness of the Chilean salmon aquaculture industry. *Aquaculture Economics & Management* 6, 97–116.

Brickle, P., Laptikhovsky, V. and Arkhipkin, A. (2005) Reproductive strategy of a primitive temperate notothenioid *Eleginops maclovinus*. *Journal of Fish Biology* 66, 1044–1059.

Ceballos, S. (2011) Estructura genética poblacional, historia demográfica y variación fenotípica del róbalo, *Eleginops maclovinus* (Percifirmes). Doctoral Thesis. Facultad de Ciencias Exactas y Naturales, Universidad de Buenos Aires, Argentina. www.digital.bl.fcen.uba.ar

Farlora, R., Valenzuela-Muñoz, V., Chávez-Mardones, J. and Gallardo-Escárate, C. (2016) Aquaporin family genes exhibit developmentally-regulated and host-dependent transcription patterns in the sea louse *Caligus rogercresseyi*. *Gene* 585(1), 119–27.

Gonzalez, L. and Carvajal, J. (2003) Life cycle of *Caligus rogercresseyi*, (Copepoda: Caligidae) parasite of Chilean reared salmonids. *Aquaculture* 220, 101–117.

Henriquez, V., Gonzalez, M., Licandeo, R. and Carvajal, J. (2011) Metazoan parasite communities of rock cod *Eleginops maclovinus* along southern Chilean coast and their use as biological tags at a local spatial scale. *Journal of Fish Biology* 79, 1851–1865.

IFOP (Instituto de Fomento Pesquero). (2015) Estadísticas de exportación de productos pesqueros y acuícolas en Chile, para el periodo 2002–2014.

Lincadeo, R., Barrientos, C. and Gonzalez, M. (2006) Age, growth rates, sex change and feeding habits of notothenioid fish *Eleginops maclovinus* from the central-southern Chilean coast. *Environmental Biology of Fishes* 77, 51–61.

Niklitschek, E.J., Soto, D., Lafon, A., Molinet, C. and Toledo, P. (2013) Southward expansion of the Chilean salmon industry in the Patagonian Fjords: main environmental challenges. *Reviews in Aquaculture* 4, 1–24.

Sa, R., Gavilán, M., Rioseco, M., Llancabure, A., Vargas-Chacoff, L., Augsburger, A. and Bas, F. (2014) Dietary protein requirement of Patagonian blenny (*Eleginops maclovinus*, Cuvier 1830) juveniles. *Aquaculture* 428–429, 125–134.

SalmonExpert, (2015) Peces biocontroladores Avances en el control biológico de *Caligus rogercresseyi*. www.salmonexpert.cl/noticias/peces-biocontroladores-avances-en-el-control-biologico-de-caligus-roger-cresseyi

SERNAPESCA (Servicio Nacional de Pesca y Acuicultura). (2015) Anuarios Estadísticos de Pesca y Acuicultura. https://www.sernapesca.cl/index.php?option=com_remository&Itemid=54&func=select&id=2

Valenzuela-Miranda, D., Nuñez-Acuña, G., Valenzuela-Muñoz, V., Asgari, S. and Gallardo-Escárate, C. (2015) MicroRNA biogenesis pathway from the salmon louse (*Caligus rogercresseyi*): emerging role in delousing drug response. *Gene* 555, 231–241.

PART IV

Cleaner fish – the future

CHAPTER 25

Cleaner fish in aquaculture review and looking to the future

Jim Treasurer

FAI Aquaculture, Ardtoe, Acharacle, UK

The rapid increase in the scale of stocking of cleaner fish on salmon farms in Norway is astonishing with 37,359,000 cleaners deployed in 2016 (Fig. 25.1a). This comprised 525,000 reared wrasse, 20,658,000 wrasse from the fishery, and 15,858,000 farmed lumpfish (Fig. 25.2a; Chapter 18; Norwegian Directorate of Fisheries, 2016). The estimated stocking of cleaner fish in 2017 was 2 million reared wrasse, 30 million farmed lumpfish and 26.6 wild caught wrasse. The latter exceeded the quota limit and the fishing quota intentions for 2018 therefore had to be re-assessed. It was estimated that more than 2 million wild caught wrasse (comprising *Ctenolabrus rupestris*, *Centrolabrus exoletus*, *Symphodus melops*, *Labrus bergylta* and *Labrus mixtus*) were stocked on salmon farms in Scotland in 2017 (Fig. 25.1b; Chapter 17), and also 530,000 reared ballan wrasse (chapter 19), and potentially 4.2 million farmed lumpfish from hatcheries in Scotland, England and Wales (Fig. 25.2b). Although official records are not yet available, lumpfish numbers reared in the UK in 2017 are likely to be less due to production and quality issues. From industry sources (chapter 19) the projected total stocking of cleaner fish in Scotland in the near term is over 6 million, although it was estimated that the industry requirement would ultimately be 10 million cleaner fish (Powell *et al.*, 2017).

In both Norway (see Chapter 18) and in Scotland (fisheries Chapter 17) there is an acknowledgement that it was previously difficult to reconcile the official catch returns (Marine Scotland, 2017a) by fishermen with the numbers of wild-caught wrasse stocked on salmon farms. This is being followed up to achieve better reporting. Records of wild-caught wrasse were not mandatory in Scotland until 2014 and the weight, rather than number of fish, has been reported (Riley *et al.*, 2017). The numbers of fish would be more meaningful, and are currently calculated using an estimated figure of 60 g mean weight for wrasse captured. However, reporting the numbers of fish by geographical locality might give a better idea of the effects of the fishery and also a firmer estimate of the numbers stocked in salmon pens. This may be possible by geographical locality as, in Scotland, wrasse catches are recorded at each local fisheries office and salmon producers also request records from fishermen showing where the fish were captured. Given that there may be an upper limit in a regulated fishery, to maintain or exceed deployed numbers of cleaner fish, there is an aspiration to sustainable culture methods for both wrasse and lumpfish to conserve wild fish stocks.

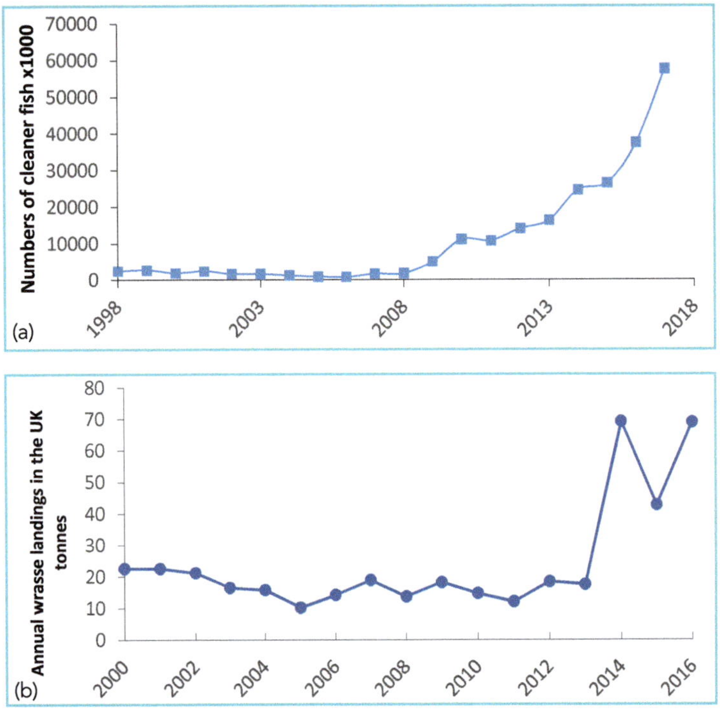

Fig. 25.1 Numbers of a) wrasse and lumpfish stocked on Norwegian salmon farms from 1998 to 2016, and an estimated figure for 2017 as official records are not yet available, and b) wrasse numbers captured in all UK fisheries for stocking on salmon farms in Scotland. Note: records for wrasse captured in Scotland prior to 2014 are underestimated as reporting to Marine Scotland was only mandatory from that year. Sources: (Norwegian Directorate of Fisheries, 2016; data for UK from Riley *et al.*, 2017)

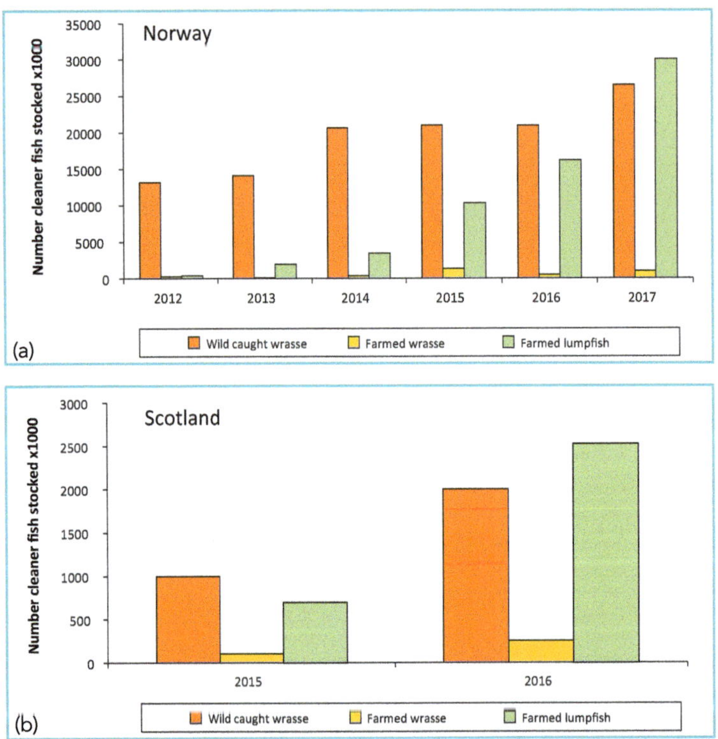

Fig. 25.2 Breakdown of cleaner fish sources and species (a) stocked on salmon farms in Norway 2012–2016 with industry estimates for 2017 in Norway, (b) Scotland from 2015–2016. Note: wrasse fisheries data for Scotland only reported from 2014 (Marine Scotland, 2017a). Sources: (Chapters 18 and 19; Source: Directorate of Fisheries, Norway; personal communications UK, Riley *et al.*, 2017, Marine Scotland, 2017b)).

25.1 Requirement to use cleaner fish: cleaner fish in sea lice control are here to stay

The trend of increased commissioning of hatcheries, from 15 in 2012 to 41 in 2016 in Norway (Chapter 18), and improved survival and growth of cleaner fish in hatcheries would suggest that the cleaner fish requirements of salmon farmers will be met from farmed production in the near future. Certainly, estimated rearing requirements of 30 million lumpfish in Norway in 2017, and almost 10 million in the UK, suggests that there will be sufficient numbers of this species for stocking to meet the needs in these salmon producing countries. Of course, it has been suggested (Boyce *et al.*, Chapter 23) that lumpfish may be produced in a marine hatchery in one country and, if health requirements are met, could be exported to other countries. This would follow the former model in the production of halibut *H. Hippoglossus* in Iceland to meet demand elsewhere, and is seen already in the shipping of more than 2 million lumpfish from Iceland to the Faroe Islands (Jónasson, 2017). The rearing of Ballan wrasse appears to be more technically challenging and the reports from Norway (Chapter 18) and Scotland (Chapter 19) of current production of 1.4 and 0.5 million fish respectively indicate a shortfall, which is met by the wild wrasse fishery of 21 million fish in Norway and >2 million in the UK. The fishing pressure that is highlighted most is the management of the wrasse fishery in the south of England (Davies, 2016; Riley *et al.*, 2017, Chapter 17). A latent gastronomic market in East Asia is also highlighted in www.cornwallgoodseafoodguide.org.uk/fish-guide/ballan-wrasse.php. However, it is likely that most of the demand for wrasse will be met for the salmon farming companies from both fishing and farming sources.

25.2 Welfare of cleaner fish

The various studies and production systems described in this book suggest that the survival and welfare of cleaner fish when stocked on salmon farms is crucial to success of the technique, and goes hand-in-hand with trained wrasse handlers on farms and with in-company education in the use of wrasse and effective oversight of the cleaner fish stocking programme. The requirements of cleaner fish welfare are just becoming understood, and the welfare review and presentation of new data (Chapter 15) indicates different responses to stress and levels of stress and parameters for measurement in wrasse and lumpfish, and that the welfare and handling requirements of the two species have to be treated differently. Chapter 15 recommends that cleaner fish welfare should be examined and defined at each stage in the capture, production and stocking chain, and there are different welfare requirements for each stage in the developmental cycle. Initial Operational Welfare Indicators (OWIs) have been proposed for both wrasse and lumpfish juveniles and broodfish, and these will need to be tested fully in the production situation and refined in the future. However, they do appear to be giving consistent results that identify aggression and stocking density effects on fin splitting and erosion. The various quality schemes and retailers will also expect the welfare of cleaner fish to have a high priority. While the public currently have a popular and positive perception of cleaner fish, this can only be maintained if the cleaner fish approach is seen as sustainable and considers animal welfare. Some engagement with NGOs may be helpful, for example with the Marine Conservation Society, who already have a policy statement on cleaner fish use (MCS, 2013).

25.3 Advances in rearing technology, protocols, handling, feeds, disease resistance, and improvements in welfare

For lumpfish, comparison of production systems in each country reveals a similarity in rearing methods (Table 25.1). Perhaps an indication and comparison of production success may be achieved in examining performance in terms of growth rates and survival figures, in OWIs and in 'quality' of juveniles; for example, measuring the extent of sucker deformations that have been reported from a number of hatcheries (Chapter 13). This impairs adhesion to surfaces, and the effect of this on survival is not yet known, although exposure to tidal conditions on high-energy sites may cause physical damage. The technology advances required for lumpfish culture have not been too excessive or arduous. Hatchery design has included special egg incubators with high flow permitting larvae to swim from the egg incubation unit into the rearing tank, and also environmental enrichment, and different stocking and environmental conditions for the broodstock.

Table 25.1 Summary and comparison of rearing methods for lumpfish in recirculation (RAS) and flow through (FLT) systems. Sources of data, this volume.

Locality	RAS or flow through FLT	Incubation temperature °C	Eggs	Hatch rate %	Disinfection of eggs	Incubation period dd	Larval stocking	Algae/Clay for turbidity
Norway	RAS	2 to 12°C or 8-13°C (heated)		>80	buffodine (2.5 ml l^{-1} water for 10 min)			
Scotland FAI	FLT	9-12°C	100,000 kg^{-1}	>50	Pyceze, 50 ppm 30 mins	300	80 l^{-1}	Algae
Scotland OFS	FLT	8-10°C	115,000 kg^{-1}		Pyceze, 50 ppm 30 mins (daily)	300	20 l^{-1}	Clay
Ireland Carna	RAS to 30/40 dph, then FLT	9	97,000-140,000 kg^{-1}	80–95	Pyceze	270-350	10-50 l^{-1}	None
Iceland IMR	FLT	7°C	100,000 kg^{-1}		Pyceze 0.1 ml l^{-1}, 30 mins 3 d/week	280		Algal paste
			120,000 kg^{-1} hardened					
Canada, HMSC		9.5°C	100,000 kg^{-1}			270-350		
Faroe Islands, Fiskaaling			90-120 eggs ml^{-1}		500 ppm glutaraldehyde for 4 mins	300-330	5.5 mm on hatch	
Lofoten Islands, Arctic Cleanerfish	FLT	from 3 to 16°C	50,000-120,000 l^{-1}	80	Pyceze 1:10000 for 30 mins	260 to 320	80,000 in 350 l^{-1} tank	None
Weymouth, Dorset Cleanerfish	RAS					280	between 50-100 larvae per litre	

Most hatcheries have achieved high survival rates in lumpfish (around 50% from egg to stocking, Table 25.1), good growth, and sizeable production figures in the first year of operation. Of course, this has relied on the know-how of marine hatcheries and has benefitted greatly from years of marine finfish development with demanding species such as halibut, and other species such as sea bass *Dicentrarchus labrax*, cod *Gadus morhua*, haddock *Melanogrammus aeglefinus*, turbot *Scophthalmus maximus* and wrasse. Live feed technology and enrichments appear to meet nutritional requirements of the pre-weaning phase. However, there have been quality issues with lumpfish with poor sucker development and also cataracts in broodfish reported frequently (Chapter 13). This may be a result of genetic issues and egg quality reflecting deficiencies in broodfish diets. Of course, several hatcheries have not used greenwater/clay techniques to increase turbidity nor enriched Artemia in larval rearing and some comparison and benchmarking of methods needs to be undertaken. However, evidence suggests that growth performance is less in both wrasse and lumpfish when Artemia are not provided and even for a short period (Chapter 7). On occasion, hatchery survival has suffered from disease issues such as the ubiquitous Amoebic Gill Disease (AGD) infection and further examination of inlet water screening systems, filters and disinfection needs to be carried out, alongside maintenance of strict biosecurity requirements with the development of biosecurity plans. Hydrogen peroxide has

Table 25.1 continued

Feeding	Live feed	Dry feed	Stocking density kg m^{-3}	Survival % from stocked larvae	25 dph Length mm	50 dph Length mm	Weight (g) at 110-140 dph	Reference
formulated feed diet, 3-5 days post hatch								Nytrø, chapter 7
within 3 d	Artemia for 2 wk	some from 10 dph	10-15 kg m^{-3}	15-45% to transfer	9.6 at 40 dph		0.5 at 70 dph	Treasurer, pers. comm.
3 dph	Artemia for 2-3 wk	30 dph		70% until after weaning; 50% to deployment				Chalaris, pers. comm.
2–3 dph	none	2–3 dph	<15	11%	7.5–8.5	9.6–13.3	0.7-1.0	Majbritt-Bolton, chapter 20; pers. comm.
	3-28 dph Artemia	1st offered at 10 ph		65%				Steinarsson and Árnason, chapter 21
on hatch	0–14 d Artemia	microdiets from 7 dph	23-43		6.5-8	16-18	2	Boyce et al., chapter 23
3-5 d after hatch	Artemia for 2 weeks	0.5 mm pellet after 600 dd					1.5-4 mg at 400 dd	Johannesen *et al.*, chapter 22
Artemia offered								
after 3 days micro feed	none	3 dph	15–30	90% to end of silo stage	8.5	15		Wittwer, chapter 8 and pers. comm.
Artemia first 2 weeks		after 2 weeks introduced						Prickett, pers. comm.

been less effective in controlling AGD with larval and juvenile lumpfish treatments, and transfer to freshwater baths has improved control of amoebae (Chapter 13; Powell *et al.*, 2017). Other disease issues include bacterial infection causing high losses in the hatchery. Vaccination and the resolution of health issues remains the largest challenge in deployment of cleaner fish, with fish being more susceptible to Vibrio, Pasteurella and atypical furunculosis infection.

Egg quality has been extensively studied in species such as Atlantic halibut, cod and haddock. Main issues have been the lack of arachidonic acid (ARA) and the balance between essential fatty acids, and this may have repercussions for larval survival and growth. Further work on cleaner fish nutrition and requirements and also the use of photoperiod to provide year-round egg supply is required. A gap analysis was carried out for future requirements for farming lumpfish (Powell *et al.*, 2017) and it suggested closing the breeding cycle, estimation of heritability of commercial traits, development of molecular markers for parentage assignment, specific lines, and control of maturation (including milt cryopreservation).

In contrast to lumpfish, the rearing of Ballan wrasse in volume remains technically challenging with slow and variable growth and low survival from hatch to stocking in salmon pens (Chapter 3). Mortalities tend to be at background levels and low and continue for a long period, with aggression and fin nipping of specific fish a long-term problem. However, many of the embryos also do not survive to hatch. Bacterial growth on eggs cause large losses, as Ballan wrasse shed eggs on a substrate and the substrate and surfaces can be covered in biofilm and faecal and other waste. Egg mats are disinfected by removing them from the spawning tanks, and suspending them in tanks containing 100 ppm formalin for up to one hour, or alternatively using repeated baths with Bronopol (Pyceze). An alternative would be to remove the gum layer surrounding the eggs and to detach the eggs from the spawning surface (Grant *et al.*, 2016). Ballan wrasse may be reared with Artemia replacement formulated diet from hatch, but offering live feeds improved growth and survival, and generally there has been success in taking fish through the larval rearing period. An issue is that weaning from live feed to dry diet can be prolonged in Ballan. This is not a large issue in some hatcheries, but it can take a long time to wean Ballan wrasse larvae to dry feed and, even at the end of this period, a proportion of fish were non-feeders. The first feed used at Ardtoe was LARVIVA and was accepted but weaning to follow up feeds took several weeks. Various trials were carried out with different feed ingredients and a shrimp-based diet was preferred. Growth of Ballan is also low in the first two years and only a proportion of fish are large enough at around 50 to 80 g at 18 months post-hatch for transfer to sea pens. Growth may be enhanced by providing additional enrichment to the tanks or by rearing with heated water, and also with selective breeding for rapid growth rate. As with lumpfish, wrasse can be susceptible to AGD and Vibrio and atypical furunculosis.

The fragmented knowledge we have about the population genetics of wrasse (Ballan and corkwing *Symphodus melops*) gives concern in the current use of fishing in one area and transporting the fish to other locations such as along the coast of Norway (Albert Imsland, personal communication, 3 April 2017). No population structuring was demonstrated in lumpfish (Imsland *et al.*, in press), but recent studies on two other cleaner fish species, Ballan and corkwing wrasse (Gonzalez *et al.*, 2016; Quintela *et al.*, 2016), commonly used in Norwegian aquaculture (Skiftesvik *et al.*, 2014), have revealed genetic divergence among populations. Corkwing wrasse in Norwegian waters show genetic differences between populations inhabiting the western and southern coast of Norway. The divergence coincides with the longest stretch of sand in the study area, suggestive of habitat fragmentation as a major driver of genetic differentiation in this species (Gonzalez *et al.*, 2016). A clear population structure has also been found between different populations of Ballan wrasse collected in Galicia. The difference is so great that consideration was given as to whether Ballan wrasse comprises two species, although other authorities question this latter conclusion (Quintela *et al.*, 2016; and Chapter 11). Approximately 21 million wrasse were used as cleaner fish in salmon sea cages in Norway in 2016, of which almost 95% were caught in the wild (Chapter 18, Anon, 2016). As mixing of genetically distinct populations should be avoided, both corkwing and Ballan wrasse should therefore be used as cleaner fish in aquaculture with care (Albert Imsland, personal communication, 4 April 2017). Particular care should be exercised when translocating wild cleaner fish from areas of origin/capture to fish farms. Two of the main potential impacts of moving wild fish out of their natal area are the inadvertent spread of pathogens and 'genetic pollution' (i.e. interbreeding of translocated fish with the wild local population). In the case of cleaner fish, knowledge gaps still exist in the population genetic structure and connectivity of wild populations along the species' distribution range, with some studies indicating that natural barriers to gene flow can exist of varying geographical scales (e.g. see Chapter 11 wrasse; Robalo and Mirimin, personal communication, 5 April 2017). Genetic structure can arise in the wild due to reproductive isolation and adaptation to local environmental, climatic fluctuations and biological pressure. It is therefore important to preserve distinctiveness and adaptive potential of local stocks. Genetic adaptation can, in fact, occur on both small and large geographical scales in the marine environment, which can be reflected in varying levels of genetic structuring (Chapter 11 and Chapter 12 lumpfish). When such information is not available, a conservative approach should be used and local stocks treated as separate units until appropriate studies are conducted. In order to evaluate potential impacts of translocations, such studies should address temporal and spatial distinctiveness at the population level in terms of both population dynamics and population size (both census and effective).

25.4 Advances in Integrated Pest Management (IPM) and in application of wrasse and lumpfish

A theme that has been suggested is that farmed cleaner fish are not as effective cleaners as wild-caught fish (Chapter 4), but quantitative work by Skiftesvik *et al.* (2013) suggests that this is not the case. However, it seems rational to propose that reared fish should be acclimatised to net cage conditions before being released into a large pen, and also offered fresh or live feed, or to share tanks with experienced cleaner fish to improve the lice cleaning ability. However, work on lumpfish (Chapter 9) has demonstrated that only a proportion of lumpfish and particular families engage in cleaning activity due to differences in behaviour ascribed to some fish being bold and others remaining near shelter areas and grazing on the nets. This may be the main trait selected in a breeding programme, along with other characteristics such as disease resistance (Chapter 12). The in-cage shelter and environmental enrichment to ensure cleaner fish health and welfare have to be improved and developed. Methods to improve grazing of sea lice should be examined, including temperature effects and further stocking density experiments. Work has been carried out on supplementary feeding of wrasse (Leclercq *et al.*, 2015) when sea lice numbers are 'low', and similar arrangements are now being developed for lumpfish. The use of feed blocks for wrasse has been useful as salmon will not feed on these, while wrasse inherently prefer to pick feed off substrates. The monitoring of survival and biomass during the deployment period should also be examined.

While there have been many hatchery-based trials to improve cultivation techniques and performance of cleaner fish, the behaviour and efficacy of cleaner fish in deployment in cages should be followed up in a more quantitative manner. For example, the effects of different types of hides, stocking density to salmon number, and management practices should be examined further to assess survival, welfare and cleaning activity. The reluctance of some farms to quote sea lice counts and the breakdown of developmental stages of lice makes it difficult to gauge the success of cleaner fish stocking on a wider scale. Behavioural studies of farmed fish in tanks are advanced, but understanding of cleaner fish–salmon interaction with other species and with salmon in pens, utilisation of cover and hides requires more study. Therefore large-scale production trials should go hand-in-hand with large-scale stocking of cleaner fish to improve techniques. Hard data are required to improve cleaner fish management and information needs to be shared. The use of cleaner fish can be micro-managed on salmon farms. The sea lice burden on salmon can vary between cages, as can cleaner fish stocking density due to losses, and some farmers remove cleaner fish from cages using baited traps and transfer these to cages with higher lice numbers. Thus, cleaner fish can be used as a roving workforce to maintain low lice numbers. This approach emphasises the importance of stockmanship to help inform overall management and targeted research priorities.

25.5 Cleaner fish survival

Hatchery survival of wrasse from hatch to weaning, *c.* 100 dph, is highly variable with figures of 2–10% at Ardtoe, 5–23% at Otter Ferry Seafish, and 10–20% in Norway (Chapter 3). Low survival has not been due to large-scale mortality from disease, but more to low long-term losses from failure to wean to dry feed and to aggression and tail nipping of individual fish. Lumpfish survival in hatcheries from egg to transfer to sea cages has been higher and with an average consistently around 50% of eggs stocked (Chapter 3, Table 25.1). Disease-related mortalities have mainly been due to Vibrios and also to AGD (Chapter 13). Conversely, survival of the two species in salmon pens may be reversed, with initial observation suggesting that lumpfish may suffer higher losses in the weeks after stocking (Imsland *et al.*, 2015).

25.6 Do cleaner fish and salmon pose a health threat to each other?

The health section (Chapter 13) summarises the range of biosecurity issues and health threats posed to cleaner fish. Receiving cleaner fish in undamaged condition from fishermen and hatcheries provides fish that are unstressed and immune competent. A number of bacterial species have been documented in wrasse and there has been no evidence of transfer to salmon, nor of typical furunculosis being transferred from salmon to wrasse (Treasurer, 2012). Viral haemorrhagic septicaemia virus (VHSV) was detected in wild-caught wrasse in Shetland in 2012 (Munro *et al.*,

2015). However, salmon were also sampled and there were no positive cases of VHSV in salmon. Goldsinny wrasse were experimentally challenged with Infectious Pancreatic Necrosis (IPN) and Infectious Salmon Anaemia (ISA) (Gibson and Sommerville, 1996; Gibson *et al.*, 1998) and, although wrasse could be carriers of IPN, there was no evidence of transfer of IPN to salmon and wrasse were also not capable of carrying ISAV.

25.7 Reuse of cleaner fish

Cleaner fish are only permitted to be used for one salmon production cycle in Norway, although Marine Scotland has a policy of permitting use over two production cycles, as Murray *et al.* (2016) demonstrated that this was statistically acceptable in reducing disease transmission between cycles. The culling of cleaner fish after one production cycle has been described as being wasteful and with welfare implications (Marine Conservation Society, 2013). Various avenues are being explored to increase sustainability, such as reusing (or reconditioning) the large fish as potential broodstock, but other ideas need to be examined. The Cornwall Good Sea Food Guide says that the commercial value of Ballan wrasse in recent years is increasing and is now being sold to Japanese restaurants, where it is prized as sashimi (www.cornwallgoodseafoodguide.org.uk/fish-guide/ballan-wrasse.php). Ballan wrasse are a culinary fish in Spain and to a lesser extent in France so there may well be an export market worth exploring. Ballan wrasse are sometimes sold in the supermarkets in the north of Spain with a price in the region of 5 Euros kg^{-1} (M. Sanchez, Ardtoe, April 2017, personal communication). Ballan are called Maragota and the sales and prices can be found at the Spanish fish market: www.pescadegalicia.gal/estadisticas. This link gives the prices in all the Galician fish markets. Lumpfish may be used for animal feeds and also for human consumption, including the roe (Stevenson and Baird, 1988). Lumpfish are being exported to Asia to be used in exotic food markets and a new market for whole and filleted frozen lumpfish has developed in China (Thordarson, 2013). Clearly greater efforts are required to recycle cleaner fish at the end of the production cycle into the food chain, not just for economic but also ecological sustainability and welfare reasons, to confirm acceptance by the wider public. Perhaps the first step is to conduct market research and a feasibility study to identify potential routes for alternative cleaner fish uses. This has been initiated by NOFIMA and Lerøy Aurora (Nøstvold *et al.*, 2016).

25.8 Expansion into other fish farming areas

The Labridae is one of the most abundant and widespread families of fishes. There was the prospect that cleaner fish could be used in other spheres of aquaculture (Sayer and Treasurer, 1995). This roll out to other farmed species has not occurred, possibly because other ectoparasites are not so pathogenic as sea lice or they perhaps are still capable of being treated with chemotherapeutants. Other examples of cleaning activity are, for example, carp *Cyprinus carpio*, which have been reported to clean *Saprolegnia* from the body and fins of pike, *Esox lucius* (Darkov and Panyushkin, 1988) in freshwater tanks. Carp have also cleaned growths, possibly *Aeromonas hydrophilla*, from other larger carp, and cleaned in a head-down vertical orientation (Soto *et al.*, 1988). Juvenile bluehead wrasse *Thalassoma bifasciatum* and the cleaning goby *Gobiosoma genie*

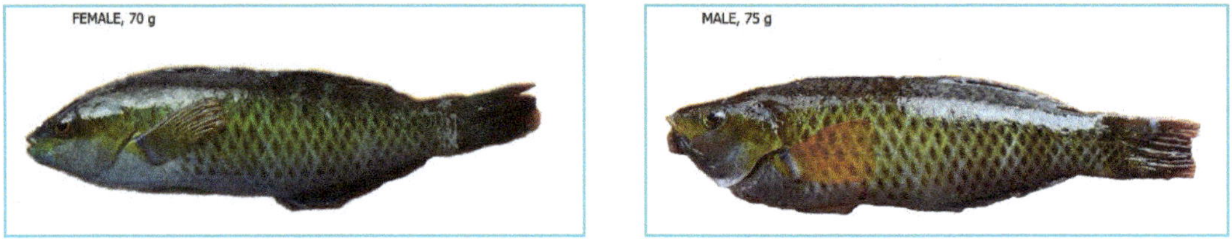

Fig. 25.3 Adult female (left) and male (right) *Malapterus reticulatus* (Credit: Daniel Nieto)

cleaned the ectoparasite flatworm *Neobenedenia melleni* from the head of Florida red tilapia, a cross between *Oreochromis urolepsis hornorum* and *O. mossambicus*. Possibly very cold winters in Atlantic Canada have deterred a more rapid introduction of cleaner fish technology and cleaner fish may be less effective against Caligus species as for *L. salmonis*. A species called the smooth lumpsucker *Aptocyclus ventricosus* is found in the Pacific and further enquiry is required to determine if this species has been assessed (Adam Powell, Sweden, 3 March 2017, personal communication). There are several studies that examined embryonic development in the smooth lumpsucker (Kyûshin, 1975). Some sea lice species have commercial impact in the Mediterranean and cleaner fish species should be explored to see if they can assist the sea bream and bass industry. While monogeneans and cymothoid parasitic isopods species are present in farmed fish these would be inaccessible to cleaner fish. Early suggestions to examine the feasibility of preliminary trials with goldsinny found in the Gulf of Lyons have not been pursued.

An issue in Chile was the availability of a native cleaner fish and there appear to be at least two distinct candidates (Chapter 24); the issue may be in rearing these species in adequate numbers. In any event, care should be taken in translocating species from different geographical areas in case of ecological impacts. Cermaq Chile is commencing a project to investigate the rearing of *Malapterus reticulatus* (Fig. 25.3) for the biological control of sea lice in the salmon farming industry as an alternative to control based on medicines (Mendoza and Nieto, Puerto Montt, personal communication, 10 April 2017). This is co-funded by the state agency CORFO-Innova Chile. *M. reticulatus* ('Vieja') is a native fish from the Juan Fernández archipelago, Chile (33.4° Lat. S, 78.5° Long. W). This species was previously (2014–2015) tested in an experimental salmon sea farm close to Puerto Montt (41.3° Lat. S, 72.6° Long. W) and gave promising results in controlling *Caligus rogercresseyi*. There is no literature regarding biological, nutritional and reproductive aspects of this species, so the project starts from a zero knowledge base. Five hundred fish have been randomly captured and transported to the experimental RAS unit comprising 16 tanks of 2 m^3 with fish acclimatised to the same temperature as the native range of the fish. Some females are already sexually mature, so reproduction should be feasible. The next step will be to develop methods for larvae culture and focusing on first feeding.

25.9 Finally

Sustainability appeared to be the most common theme in the International Cleaner Fish Summit held in May 2017 (FHF/SAIC Glasgow), and this was related to several areas, including survival of cleaner fish in sea pens, disposal of the fish after use and on harvest, and in wild fish capture. The use of cleaner fish has to be associated with good survival of wrasse and lumpfish, yet the losses, especially in lumpfish, were high and could not continue. but it could take up to 15 years to breed a fish that is disease resistant and adapted to farming (Jónasson, 2017). However, much can be done with improved vaccines and nutrition as the requirements for lumpfish have not been completely assessed. A great concern was the disposal of cleaner fish after the production cycle and some regulatory authorities were insisting that the fish had to be culled at the end of the cycle and could not be reused. A market research exercise to find consumer reaction to various lumpfish recipes and efforts has been made to find a market (Nøstvold *et al.*, 2016). The reception has been good and there is a high collagen component of lumpfish that is seen of health benefit in South Korea. This would be useful as fillet yield is low in this species at around 10%. Another concern is the sustainability of capture of wild fish. The industry is committed to using farmed fish but the salmon farmers have indicated that wild wrasse are required as there are still insufficient farmed wrasse available. There are currently few data on the natural population structure and densities of wrasse to indicate the effects of fishing on wild fish stocks and this is discussed further (Chapter 17).

The adoption of the use of cleaner fish by all major salmon farming companies in Europe and on a large and methodical scale suggests that they are making a significant contribution in the control of sea lice in an IPM approach and have become an established non-medicinal technology. The priority for the industry is to continue a reliable farmed production of cleaner fish and to deploy these successfully while maximising welfare and survival of these fish in the salmon pens. By concentrating on providing and maintaining high stocking densities of cleaner fish while maximising conditions for cleaner fish in the pens, sea lice numbers can be controlled. Further investment in rearing technology, vaccine production, feeds, and selective breeding of cleaner fish are required. There is the potential for further roll out of biological control in Canada and Chile and on to other geographical areas and farmed species. It is hoped that this discussion on cleaner fish applications will engender further improvements in the application of cleaner fish and improved collaboration between companies and researchers.

References

Anon (2016) www.fiskeridir.no/English/Aquaculture/Statistics/Cleanerfish-Lumpfish-and-Wrasse

Darkov, A.A. and Panyushkin, S.N. (1988) Cleaning symbioses in six freshwater fish species. *Journal of Ichthyology* 28, 161–167.

Davies, S. (2016) *A review of wrasse ecology and fisheries interactions.* Devon and Severn Inshore Fisheries and Conservation Authority.

Gibson, D.R. and Sommerville, C. (1996) The potential for viral problems related to the use of wrasse in the farming of Atlantic salmon. In: Sayer, M.D.J., Treasurer, J.W. and Costello, M.J. (eds.) *Wrasse: Biology and use in aquaculture.* Fishing News Books, Oxford, pp. 240–246.

Gibson, D.R., Smail, D.A. and Sommerville, C. (1998) Infectious pancreatic necrosis virus: experimental infection of goldsinny wrasse *Ctenolabrus rupestris* L. (Labridae). *Journal of Fish Diseases* 21, 399–406.

Gonzalez, E.B., Knutsen, H. and Jorde, P.E. (2016) Habitat discontinuities separate genetically divergent populations of a rocky shore marine fish. PLOS ONE 11:1–17. DOI: 10.1371/ journal. pone. 0163052.

Grant, B., Picchi, N., Davie, A., Leclercq, E. and Migaud, H. (2016) Removal of the adhesive gum layer surrounding naturally fertilised ballan wrasse (*Labrus bergylta*) eggs. *Aquaculture* 456, 44–49.

Imsland, A.K., Reynolds, P., Eliassen, G., Hangstad, T.A., Jónsdóttir, O.D.B., Elvegård, T.A., Lemmens, S.C.A., Rydland, R. and Nytrø, A.V. (2016) Investigation of behavioural interactions between lumpfish (*Cyclopterus lumpus*) and goldsinny wrasse (*Ctenolabrus rupestris*) under controlled conditions. *Aquaculture International* 24, 1509–1521.

Jónasson, J. (2017) Lumpfish production in Iceland. Presentation at International Cleaner Fish Summit, FHF/SAIC, Glasgow.

Kyûshin, K. (1975) The embryonic and larval development, growth, survival and changes in body form, and the effect of temperature on these characteristics of the smooth lumpsucker, *Aptocyclus ventricosus* (Pallas). *Bulletin of the faculty of fisheries Hokkaido University* 26, 49–72.

Leclercq, E., Graham, P. and Migaud, H. (2015) Development of a water-stable agar-based diet for the supplementary feeding of cleaner fish ballan wrasse *(Labrus bergylta)* deployed within commercial Atlantic salmon *(Salmo salar)* net-pens. *Animal Feed Science and Technology* 208, 98–106.

Marine Conservation Society (2013) Cleaner fish position paper (mcs.org). Marine Conservation Society, 7 pp. [Cited 22 March 2017.] Available from URL: www.mcsuk.org/downloads/fisheries/Cleaner_Fish_Position_Paper.pdf

Marine Scotland (2017a) Landings by Scottish vessels into Scotland. www.gov.scot/Resource/0052/00524991.pdf

Marine Scotland (2017b) Aquaculture production statistics Scotland. /gov.scot/Publications/2017/09/5208/6

Munro, E.S., McIntosh, R.E., Weir, S.J., Noguera, P.A., Sandilands, J.M., Matejusova, I., Mayes, A.S. and Smith, R. A. (2015) A mortality event in wrasse species (Labridae) associated with the presence of viral haemorrhagic septicaemia virus. *Journal of Fish Diseases* 38, 335–341.

Murray, A.G. (2016) A modelling framework for assessing the risk of emerging diseases associated with the use of cleaner fish to control parasitic sea lice on salmon farms. *Transboundary and Emerging Diseases* 63, 270–277.

Nøstvold, B.H., Kvalvik, I., Voldnes, G. and Jentoft, A.R. (2016) (ISBN 978-82-8296-428-9) Nofima rapportserie (43/2016) 45 pp. https://nofima.no/pub/1397108

Norwegian Directorate of Fisheries (2015) www.fiskeridir.no/English/Aquaculture/Statistics/Cleaner fish-lumpfish-and-Wrasse.

Powell, A., Treasurer, J., Pooley, C.L., Keay, A.J., Lloyd, R., Imsland, A.K. and Garcia de Leaniz, C. (2017) Use of lumpfish for sea-lice control in salmon farming: challenges and opportunities. *Reviews in Aquaculture* DOI: 10.1111/raq.12194

Quintela, M., Danielsen, E.A., Lopez, L., Barreiro, R., Svåsand, T., Knutsen, H., Skiftesvik, A.B. and Glover, K.A. (2016) Is the ballan wrasse (*Labrus bergylta*) two species? Genetic analysis reveals within-species divergence associated with plain and spotted morphotype frequencies. *Integrative Zoology* 11, 162–172.

Riley, A., Jeffery, K., Cochrane-Dyet, T., White, P. and Ellis, J. (2017) Northern European Wrasse – Summary of commercial use, fisheries and implications for management. CEFAS, Lowestoft.

Scottish Sea Fisheries Statistics (2015) Scottish Government. www.gov.scot/Publications/2016/09/1961/4

Skiftesvik, A.B., Bjelland, R.M., Durif, C.M.F., Johansen, I.S. and Browman, H.I. (2013) Delousing of Atlantic salmon (*Salmo salar*) by cultured vs. wild ballan wrasse (*Labrus bergylta*). *Aquaculture* 402, 113–118.

Skiftesvik, A.B., Blom, G., Agnalt, A-L., Durif, C.M.F., Browman, H.I. and Bjelland, R.M. (2014) Wrasse (Labridae) as cleaner fish in salmonid aquaculture – The Hardangerfjord as a case study. *Marine Biology Research* 10, 289–300.

Soto, C.G., Zhang, J.S. and Shi, Y.H. (1994) Intraspecific cleaning behaviour in *Cyprinus carpio* in aquaria. *Journal of Fish Biology* 44, 172–174.

Stevenson, S.C. and Baird, B.J. (1988) The fishery for lumpfish (*Cyclopterus lumpus*) in Newfoundland waters. *Canadian Technical Report of Fisheries and Aquatic Sciences*, No. C/639.805 C35/1595, p. 26.

Thordarson, G. (2013) Markets for lumpfish. Coastal Fisheries Reports. Coastal Fisheries in the North Atlantic. (Cited 22 March 2017.) Available from URL: www.coastalfisheries.net, www.coastalfisheries.net/wp-content/uploads/ 2013/06/Export-value-of-lumpfish products-from-Iceland-inc rease-around-300-million-IKR.pdf.

Treasurer, J.W. (2012) Diseases of north European wrasse (Labridae) and interactions with cohabited farmed salmon, *Salmo salar* L. *Journal of Fish Diseases* 35, 555–562.

Appendix A.1 Technical sheet – Ballan wrasse (*Labrus bergylta*)

GAMETES AND DEVELOPING EGGS	
Egg characteristics	Diameter 1 mm
Milt and sperm characteristics	Milt volume: 0.05 ml to 0.8 ml (Helland *et al.*, 2014)
Incubation time	100–110 degree days at 11.3°C
Optimal Temperature (°C)	11°C
Fertilization Success	Around 90% (Helland *et al.*, 2014)
Overall Incubation Success	Depends on sperm and egg quality
LARVAL – JUVENILE STAGE	
Larval Size at Hatching	3.7 mm at 1 dph
Optimal T °C	Start at 12°C, increase gradually to 16°C
Rearing units	9000 l, 27,000 l circular fibreglass tanks
Time until first-feeding	Rotifers 4–25 dph, Artemia 20–57 dph, formulated feed from 47 dph
Survival from hatching to weaning	10–20%, highly variable
ON-GROWING STAGE	
Commercial size	20g to >50g
Years to reach commercial size	9 to 12 months and 12 to 18 months
Rearing units	27,000 l circular fibreglass tanks

Rearing density	15–25 kg m^{-3}
Susceptibility to diseases	Low
Optimal T °C	16°C
Diet	Floating diet, 65% proteins, 12% fat and 16% carbohydrate
REPRODUCTION	
Period	Natural spawning in Norway during May to July. Spawning time can be shifted by photoperiod and temperature manipulation.
Fertilisation protocol	Males and females housed in same tanks. Collection: eggs stick to mats on the bottom of the tank, mats with eggs are incubated.
Optimal T °C	Spawning: 11°C

RESEARCH AVENUES

- Optimisation of feeding efficiency and feed composition
- Stripping or induced spawning
- Breeding programmes
- Optimisation of rearing methods
- Stress behaviour, hidings
- Fish health, development of vaccines
- Product development

Compiled by Anette Lekva, Marine Harvest Norway

Appendix A.2 Technical sheet – Goldsinny wrasse (*Ctenolabrus rupestris*)

GAMETES AND DEVELOPING EGGS	
Egg characteristics	Diameter 0.85 mm, pelagic, non-adhesive
Milt and sperm characteristics	Gonad 1 g in 60 g males. Sperm motile at stripping, but small quantity, best collected by syringe in hand-stripping
Egg yield (eggs/g female)	6300–5100 per fish, size dependent
Incubation time	45 degree days at 15°C, equivalent to 5 days at 9–10°C
Optimal Temperature (°C)	12–16°C
Fertilisation Success	40–80%
Overall Incubation Success	Variable, depends on sperm and egg quality
LARVAL – JUVENILE STAGE	
Larval Size at Hatching	1.8–2.3 mm at 0 dph
Optimal T °C	12–16°C Dissolved oxygen 110–115%
Rearing units	1000 l, 3000 l square fibreglass tanks
Time until first feeding	Rotifers start from 5–8 dph, Artemia commence from 25–30 dph, formulated feed from *c.* 50 dph onward, weaning to dry feed generally rapid
Survival from hatching to weaning	40–80%, highly variable

ON-GROWING STAGE	
Commercial size	30 to 70 g
Years to reach commercial size	18–24 months
Rearing units	1000–5000 l circular fibreglass tanks
Rearing density	10 to 20 kgm^{-3}
Susceptibility to diseases	Low
Optimal T °C	10–18°C (0–3 g)
Diet	Formulated feed shrimp-based, 55–62% proteins, 15% lipid
REPRODUCTION	
Period	Spawning in Scotland mid-June–mid-August
Fertilisation mode	External fertilisation
Fertilisation protocol	Natural spawning and egg collection of pelagic eggs; or hand strip and dry fertilise eggs in bucket
Time to reach sexual maturity	2–3 years under hatchery conditions (male matures one year earlier than females)
Optimal T °C	Spawning: 14–18°C
Control of reproduction	Spawning time can be shifted by photoperiod and temperature manipulation

RESEARCH AVENUES
– Improving first feed; use of small rotifers, shellfish larvae – Improvement of growth rate through feed, RAS and higher temperatures, and selection programmes – Development of vaccines – Improve welfare in net pens

Compiled by Jim Treasurer, FAI Aquaculture

Appendix A.3 Technical sheet – Lumpfish (*Cyclopterus lumpus*)

GAMETES AND DEVELOPING EGGS	
Egg characteristics	Diameter 2.6 mm, demersal, adhesive
Milt and sperm characteristics	Gonad 9–15 g in 500–700g males. Sperm motile at stripping. In UK could not strip males. Mature testes were larger but no data
Egg yield (eggs/kg female)	ca. 100,000 eggs kg^{-1}
Incubation time	250–275 degree days at 10°C, equivalent to 27–30 days at 9–10 °C
Optimal Temperature (°C)	8–11°C
Fertilization Success	40–80%
Overall Incubation Success	Depends on sperm and egg quality.
LARVAL – JUVENILE STAGE	
Larval Size at Hatching	4–5mm at 1dph
Optimal T °C	10–12°C for the first 150 days. 4–10°C thereafter Dissolved oxygen 110–115%. In UK on hatch started at
Rearing units	8–10°C then increase to 12oC in larval rearing/nursery 500 l, 3000 l circular fibreglass tanks
Time until first-feeding	Artemia 0–30 dph, formulated feed from *c.* 20 dph onward
Survival from hatching to weaning	40–80%, highly variable 15g commercial sales

ON-GROWING STAGE	
Commercial size	
Years to reach commercial size	5 – 8 months to 15 g
Rearing units	6000 l, 25,000 l circular fibreglass tanks
Rearing density	100–150% wall-coverage up to 100 g, thereafter 23–43 kg/m^{-3}
Susceptibility to diseases	Low (particularly in recirculation systems)
Optimal T °C	10–13°C (0–3 g); 6-10°C thereafter; in UK 8–10°C up to 1g and 10–12°C
Diet	Floating diet, 55–62% proteins, 15% fat
REPRODUCTION	
Period	Spawning in eastern Canada commence during May to July. Timing is controlled by photoperiod
Fertilisation mode	External fertilisation
Fertilisation protocol	Eggs (in ovarian fluid) and milt are incubated for 20–25 min before transfer to seawater
Time to reach sexual maturity	2–3 years under culture conditions (lower age at first maturity in males)
Optimal T °C	Spawning: 6–8°C (In Europe 8–10°C)
Control of reproduction	Spawning time can be shifted by photoperiod and temperature manipulation

RESEARCH AVENUES
– Optimisation of feeding efficiency and feed composition – Genetic characterisation and selection programs – Optimisation of rearing methods – Fish health, development of vaccines – Product development

Prepared by Danny Boyce. Memorial University of Newfoundland; Supplementary comments about European hatcheries, Richard Prickett, RSP Services

Technical sheet – Cunner (*Tautogolabrus adspersus*)

GAMETES AND DEVELOPING EGGS	
Egg characteristics	Diameter 0.8–1.0 mm, buoyant
Egg yield (eggs/kg female)	1 million
Incubation time	80–90 degree days at 12°C, equivalent to 7 days
Optimal Temperature (°C)	10.5–12°C
Fertilization Success	>80%
Overall Incubation Success	70–80%. Depends on sperm and egg quality, supported with disinfection
LARVAL – JUVENILE STAGE	
Larval Size at Hatching	2.5–3.0mm in length
Optimal T °C	12–16°C
Rearing units	3000 to 6000 l flat-bottom circular fibreglass tanks
Time until first-feeding	<100 µm rotifers 5–14 dph; >100 µm rotifers 15–35dph; AF Artemia 36–45dph, EG Artemia 46–54 dph
Survival from hatching to weaning	15–25%, highly variable
ON-GROWING STAGE	
Commercial size	30 g
Years to reach commercial size	2–3 years
Rearing units	6000 l, 25,000 l circular fibreglass tanks

Susceptibility to diseases	Low
Optimal T °C	12–16°C
Diet	Floating diet, 55–62% proteins, 15% fat
REPRODUCTION	
Period	Spawning commence during April to August. Timing is controlled by photoperiod
Fertilisation mode	External fertilisation
Fertilisation protocol	Males and females housed in same tanks. Passive collection; eggs are buoyant and are collected in external egg collection tank.
Time to reach sexual maturity	1–2 years under culture conditions
Optimal T °C	Spawning: 12°C
Control of reproduction	Spawning time can be shifted by photoperiod and temperature manipulation

Prepared by Danny Boyce, Memorial University of Newfoundland

Index